Partial Differential Equations

Mathematical Modeling and Computation

About the Series

The SIAM series on Mathematical Modeling and Computation draws attention to the wide range of important problems in the physical and life sciences and engineering that are addressed by mathematical modeling and computation; promotes the interdisciplinary culture required to meet these large-scale challenges; and encourages the education of the next generation of applied and computational mathematicians, physical and life scientists, and engineers.

The books cover analytical and computational techniques, describe significant mathematical developments, and introduce modern scientific and engineering applications. The series will publish lecture notes and texts for advanced undergraduate- or graduate-level courses in physical applied mathematics, biomathematics, and mathematical modeling and volumes of interest to a wide segment of the community of applied mathematicians, computational scientists, and engineers.

Appropriate subject areas for future books in the series include fluids, dynamical systems and chaos, mathematical biology, neuroscience, mathematical physiology, epidemiology, morphogenesis, biomedical engineering, reaction-diffusion in chemistry, nonlinear science, interfacial problems, solidification, combustion, transport theory, solid mechanics, nonlinear vibrations, electromagnetic theory, nonlinear optics, wave propagation, coherent structures, scattering theory, earth science, solid-state physics, and plasma physics.

Jianke Yang, *Nonlinear Waves in Integrable and Nonintegrable Systems*

A. J. Roberts, *Elementary Calculus of Financial Mathematics*

James D. Meiss, *Differential Dynamical Systems*

E. van Groesen and Jaap Molenaar, *Continuum Modeling in the Physical Sciences*

Gerda de Vries, Thomas Hillen, Mark Lewis, Johannes Müller, and Birgitt Schönfisch, *A Course in Mathematical Biology: Quantitative Modeling with Mathematical and Computational Methods*

Ivan Markovsky, Jan C. Willems, Sabine Van Huffel, and Bart De Moor, *Exact and Approximate Modeling of Linear Systems: A Behavioral Approach*

R. M. M. Mattheij, S. W. Rienstra, and H. H. M. ten Thije Boonkkamp, *Partial Differential Equations: Modeling, Analysis, Computation*

Johnny T. Ottesen, Mette S. Olufsen, and Jesper K. Larsen, *Applied Mathematical Models in Human Physiology*

Ingemar Kaj, *Stochastic Modeling in Broadband Communications Systems*

Peter Salamon, Paolo Sibani, and Richard Frost, *Facts, Conjectures, and Improvements for Simulated Annealing*

Lyn C. Thomas, David B. Edelman, and Jonathan N. Crook, *Credit Scoring and Its Applications*

Frank Natterer and Frank Wübbeling, *Mathematical Methods in Image Reconstruction*

Per Christian Hansen, *Rank-Deficient and Discrete Ill-Posed Problems: Numerical Aspects of Linear Inversion*

Michael Griebel, Thomas Dornseifer, and Tilman Neunhoeffer, *Numerical Simulation in Fluid Dynamics: A Practical Introduction*

Khosrow Chadan, David Colton, Lassi Päivärinta, and William Rundell, *An Introduction to Inverse Scattering and Inverse Spectral Problems*

Charles K. Chui, *Wavelets: A Mathematical Tool for Signal Analysis*

Partial Differential Equations

Modeling, Analysis, Computation

R. M. M. Mattheij
S. W. Rienstra
J. H. M. ten Thije Boonkkamp

Technische Universiteit Eindhoven
Eindhoven, The Netherlands

Society for Industrial and Applied Mathematics
Philadelphia

Universities Press

PARTIAL DIFFERENTIAL EQUATIONS: MODELING, ANALYSIS, COMPUTATION

UNIVERSITIES PRESS (INDIA) PRIVATE LIMITED

Registered Office
3-6-747/1/A & 3-6-754/1, Himayatnagar, Hyderabad 500 029, Telangana, India
info@universitiespress.com; www.universitiespress.com

Distributed by
Orient Blackswan Private Limited

Registered Office
3-6-752 Himayatnagar, Hyderabad 500 029, Telangana, India

Other Offices
Bengaluru, Bhopal, Chennai, Guwahati, Hyderabad, Jaipur, Kolkata, Lucknow, Mumbai, New Delhi, Noida, Patna, Vijayawada

Original American edition published by SIAM: Society for Industrial and Applied Mathematics, Philadelphia, Pennsylvania

First published in India by Universities Press (India) Private Limited 2018

ISBN 978 93 86235 41 1

Printed at
Rajiv Book Binding House, Delhi

Published by
Universities Press (India) Private Limited
3-6-747/1/A & 3-6-754/1, Himayatnagar
Hyderabad 500 029, Telangana, India

501378

21072017

Dedicated to Marie-Anne, Rita, and Liesbeth

Contents

List of Figures

List of Tables

Notation

Variables and Operators

t	time
x	space coordinate in $\mathbb{R}$
$\boldsymbol{x} = (x, y)^T, \boldsymbol{x} = (x, y, z)^T$	space coordinate in $\mathbb{R}^d$ $(d = 2, 3)$
$v(x, t), v(\boldsymbol{x}, t)$	scalar function
$\boldsymbol{v}(x, t), \boldsymbol{v}(\boldsymbol{x}, t)$	vector function
(u, v)	inner product of the scalar functions u and v
$\boldsymbol{u} \cdot \boldsymbol{v}$	inner product of the vectors $\boldsymbol{u}$ and $\boldsymbol{v}$
$\boldsymbol{u} \times \boldsymbol{v}$	vector product of the vectors $\boldsymbol{u}$ and $\boldsymbol{v}$
$\mathcal{L}(\boldsymbol{x}, t)[v]$	differential operator $\mathcal{L}(\boldsymbol{x}, t)$ applied to v
$\mathcal{L}^*(\boldsymbol{x}, t)$	adjoint operator of $\mathcal{L}(\boldsymbol{x}, t)$
$\frac{\mathrm{d}v}{\mathrm{d}x}, v'$	derivative of $v(x)$
$\frac{\partial v}{\partial x}, v_x$	partial derivative of $v(x, t)$
$\frac{\partial v}{\partial \boldsymbol{n}}, \boldsymbol{n} \cdot \nabla v$	directional derivative of v in the direction of $\boldsymbol{n}$
∇v	gradient of v
$\nabla \cdot \boldsymbol{v}$	divergence of $\boldsymbol{v}$
$\nabla \times \boldsymbol{v}$	curl of $\boldsymbol{v}$
$\nabla^2 v$	Laplace operator applied to v
$\int_\Omega v \, \mathrm{d}V$	generic integral over a domain $\Omega \subset \mathbb{R}^d$ $(d = 2, 3)$
$\iint_\Omega v(x, t) \, \mathrm{d}x\mathrm{d}y$	integral over a domain $\Omega \subset \mathbb{R}^2$
$\oint_{\partial\Omega} \boldsymbol{v} \cdot \boldsymbol{n} \, \mathrm{d}S$	integral over a closed surface $\partial\Omega \subset \mathbb{R}^3$
$\oint_C \boldsymbol{v} \cdot \mathrm{d}\boldsymbol{\ell}$	integral over a closed contour $\mathcal{C} \subset \mathbb{R}^d$ $(d = 2, 3)$
$[v]_-^+$	jump of v across a discontinuity
$\boldsymbol{e}_x, \boldsymbol{e}_y, \boldsymbol{e}_z$	unit vectors in the Cartesian coordinate system (x, y, z)
$\boldsymbol{e}_r, \boldsymbol{e}_\phi, \boldsymbol{e}_z$	unit vectors in the cylindrical coordinate system (r, ϕ, z)
$\boldsymbol{e}_r, \boldsymbol{e}_\theta, \boldsymbol{e}_\phi$	unit vectors in the spherical coordinate system (r, θ, ϕ)

Numerical Parameters and Variables

Δt	step size
Δx	grid size in x direction
h	generic grid size
x_j	jth grid point
$x_{j+1/2}$	location of a control volume boundary
t^n	time level $n\Delta t$
v_j^n	numerical approximation of $v(x,t)$ at grid point x_j and time level t^n
$F_{j+1/2}^n$	numerical approximation of flux $f(x,t)$ at control volume boundary $x_{j+1/2}$ and time level t^n
C, E, etc.	location names of grid points
$\boldsymbol{x}_\mathrm{C}$	generic grid point
v_C	numerical approximation of $v(\boldsymbol{x}_\mathrm{C})$
v_Δ	grid function
$\mathcal{L}_\Delta[v_j^n]$	difference operator $\mathcal{L}_\Delta$ applied to v_j^n
$\mathcal{L}_\Delta^{-1}$	inverse of $\mathcal{L}_\Delta$
$d_j^n = \mathcal{O}(\Delta x^2)$	d_j^n is of the order Δx^2 for $\Delta x \to 0$

Vectors and Matrices

$\mathbf{v} = (v_1, v_2, \ldots, v_n)^T$	column vector in $\mathbb{R}^n$
$\mathbf{v}^T = (v_1, v_2, \ldots, v_n)$	row vector in $\mathbb{R}^n$
$\mathbf{v}^k$	kth vector in a sequence
$\\|\mathbf{v}\\|_p$	p-norm of $\mathbf{v}$
$\mathbf{A} = (a_{ij})$	matrix with a_{ij} in ith row and jth column
$\mathbf{A} = (\mathbf{a}_1, \mathbf{a}_2, \ldots, \mathbf{a}_n)$	matrix with $\mathbf{a}_i$ as the ith column
$\mathbf{A}^T$	the transpose of $\mathbf{A}$
$\mathbf{A}^{-1}$	the inverse of $\mathbf{A}$
$\mathbf{I}$	identity matrix
$\mathrm{diag}(a_1, a_2, \ldots, a_n)$	diagonal matrix with a_i in ith row and column
$\det \mathbf{A}$	determinant of $\mathbf{A}$
$\\|\mathbf{A}\\|_p$	p-norm of $\mathbf{A}$
$\rho(\mathbf{A})$	spectral radius of $\mathbf{A}$

Miscellaneous

$:=$, $=:$	is defined as, defines
$\doteq$	is equal to when neglecting terms of higher order
$\sim$	is asymptotically equal to
$\mathcal{O}$	asymptotic order symbol ("big O")
o	asymptotic order symbol ("small o")
e	base of the natural logarithm (e $= 2.71828\ldots$)
i	imaginary unit
$\lvert z\rvert$	absolute value or modulus of $z \in \mathbb{C}$
$\bar{z}$	complex conjugate of $z \in \mathbb{C}$
$[v]$	dimension of variable/constant v (e.g., in SI units)
$\mathcal{C}$	characteristic/curve in $\mathbb{R}^d$ ($d = 2, 3$)
$\partial\Omega$	boundary of a domain $\Omega \subset \mathbb{R}^d$ ($d = 2, 3$)

Preface

There exist a number of good textbooks on both the analytical and the numerical aspects of PDEs. So why another book on PDEs, then? Well, most existing texts deal with either analytical or numerical aspects, but not both. There are understandable reasons for this. For one thing, it is the traditional approach. The impressive achievements in understanding a large variety of physical phenomena, long before computers came into use, have made the study of PDEs, often called applied mathematics, a well-established area in its own right, which will no doubt remain so for many years to come. At the same time, although this area has grown impressively, it has become clear that nowadays the power of this branch of mathematics does not so much lie in constructive solutions but rather in giving qualitative answers, thus having a usefulness in its own right. Notwithstanding the achievements it has brought by answering some deep questions such as those regarding existence and uniqueness, and by developing a host of instruments to grasp the solution, its machinery for concretely solving problems is still limited to simplified geometries and the use of expansions in terms of special functions and the like. Yet often the actual solution is the ultimate goal of the scientist or engineer who needs concrete answers. The advent of fast computers and the equally fast development of numerical methods filled this gap to a large extent, so that modern engineers can use a large variety of packages to find numerical approximations to the solutions of their problems. This is reflected in the literature on numerical books. In fact, the latter subject has grown to such an extent that different approaches, like finite elements and finite differences, often appear to be as far apart from each other as they are from a purely analytical text. But despite the importance of standard software, problems are more often than not standard and a thorough knowledge of at least a well-chosen subset of analytical and numerical tools and methodologies is necessary when dealing with real-life problems.

Only when dealing with PDEs in practice does it become clear that numerical treatment and analytical treatment of the subject are both needed. A numerical analyst devoted to computing a Dirichlet problem on a square by yet another method is as esoteric as an applied mathematician who is trying to see the world as built up by spheres and cylinders. This gives the main motivation for this text. We are deeply convinced that we need to treat PDEs by combining analytical and numerical aspects. The two supplement each other but have, conceptually, much in common, too. Nevertheless, concepts and theoretical problems, however sophisticated, are not enough to teach us how to deal with PDEs in practice. In a practical situation one is supposed to make a mathematical model first, obtain insights into the behaviour of the solution, and finally compute the solution (or sometimes be satisfied with a qualitative understanding of the problem). Hence, as a third component, insight into modeling real-life problems needs to complement and deepen theoretical and numerical knowledge.

This book, therefore, intends to address three aspects: the analysis of PDEs (including some basic knowledge of tools), numerical solution methods, and, last but not least, modeling. In a way, the analytical part follows some of the classical lines. As for numerics, we had to make a choice. Most numerical books focus on either finite difference methods or finite element methods (FEMs). This has a simple explanation: the setting is so large that some choices have to be made. The FEM is quite versatile with respect to the domains on which a problem can be defined. Therefore it is most appropriate for boundary value problems. It requires, however, quite a bit of preparatory work, so that a textbook with these ideals would become too big to handle. We wanted to provide a general introduction to solving PDEs numerically, and for this purpose finite differences (and finite volumes, where they come in handy as well) are sufficiently general. Their introduction is straightforward, as is their use. Moreover, given the insights gained from this text, it should not be hard to use FEMs at proper places instead. Another choice we made deals with the number of nonlinear problems one can treat. Due to lack of space (and also because of didactical constraints), we deal with nonlinearities mostly in a generic way, but certainly do not avoid them. As for the modeling part, we believe that dimensional analysis is an obligatory ingredient. Moreover, modeling can only be learned by doing, so we need a larger number of case examples. We have therefore included a separate chapter, with problems from practice, all from the personal experience of the authors.

We have not attempted to mix analytical and numerical aspects to the extreme to avoid blurring the various approaches. Instead, we have sandwiched chapters on analysis and numerics, thereby frequently cross-referencing. Apart from the aforementioned threefold setup, the book also has a second kind of partitioning. Indeed, the first five chapters are preparatory, both introducing and summarising concepts and properties from a variety of areas, in particular Fourier analysis, distributions, and difference methods. Chapters 8 to 14 deal with analytical and numerical aspects of the major types of PDEs. We have followed the traditional distinction between elliptic, parabolic, and hyperbolic equations, but we do realise that there are, of course, many mixed forms in practice. Although parabolic and hyperbolic equations often have a truly evolutionary character, hyperbolic equations may also arise in a nonevolutionary context. Elliptic equations, which give rise to boundary value problems, on the other hand, may be seen as steady states of parabolic problems. To have fewer academic problems as well, we analyse, e.g., the shallow-water equations and the Stokes equations to some extent. The modeling chapters in this book are Chapters 6, 7, 15, and 16. Chapter 6 introduces the basic equations from continuum mechanics, which are the major reasons why people study a subject like PDEs. Chapter 7 deals with the art of mathematical modeling and expands on the idea that a model should be lean and based on a systematic neglect of what is unimportant. The latter two chapters together provide some essential knowledge for dealing with practical problems. The mainly methodological analytical and numerical Chapters 8 to 14 can be followed without explicit knowledge of Chapters 6 and 7, however. Chapter 15 is a self-contained introduction to perturbation methods and complements the methodologies and tools treated in the preceding PDE chapters. It extends the modeling philosophy of "neglecting the very small" by "including the small" but utilising its smallness. By using the language of asymptotic analysis, we introduce four of the most important methods, two of which are of regular perturbation type (method of slow variation and method of strained coordinates) and two of which are of singular perturbation type (method of matched asymptotic expansions and method of multiple scales, including rays). It is shown how often remarkably sharp analytic approximations to solutions can be obtained while revealing a lot of the problem's structure. This plays an important role in actual modeling, as explicit parametric dependence clearly gives more insight than numbers often do. The last chapter, Chapter 16, contains a large number of case studies. Here a true amalgamation of all previous theories and techniques takes place. It can be used alongside other relevant chapters, where references to this chapter enable the user to work out a related practical problem. The various sections in Chapter 16 also give many suggestions for further modeling, which could typically be used in a course with room for larger projects.

The computations for this book were done using MATLAB®. There exists a host of packages and toolboxes to solve PDEs numerically. We have tried to avoid exercises for which elaborate computer usage would be required. A course taught from this book will benefit, however, from assignments dealing with some larger problems, including (less trivial) numerical computations. The exercises at the end of each chapter and in particular the case studies in Chapter 16 should give ideas for this.

The book is intended for use in courses at an advanced undergraduate or a graduate level. It has been designed so that it should be useful both for engineers, who may be more interested in the methods as such, and for more mathematically interested readers. To this end, we have made it self-contained. Teachers may judge for themselves which parts they deem more relevant, given the needs or interests of their audience. By including an extensive index and ample cross-referencing, we hope that this book is also quite suitable for self-study and for reference. The authors have had positive experiences with courses taught from the material of this book. Over the years we have benefited a great deal from the suggestions and remarks of many colleagues, and of course of our students. We owe them a debt, as we will owe to those readers who will share with us their experiences and recommendations. We would like to gratefully acknowledge, nevertheless, the advice of a few people in particular: A.E. Dahoe, J. de Graaf, R. Horvath, I. Lyulina, B. O'Malley, J. Molenaar, V. Nefedov, N.C. Ovenden, M. Patricio, N. Peake, J. Rijenga, W.H.A. Schilders, P.J. Slikkerveer, W.R. Smith, A.A.F. van de Ven, and K. Verhoeven.

We are indebted to the staff of SIAM for their help and patience with this project, which took us much longer to complete than originally anticipated.

R.M.M. Mattheij, S.W. Rienstra, and J.H.M. ten Thije Boonkkamp
Eindhoven, December 2004

Chapter 1

Differential and Difference Equations

In this chapter we give a brief introduction to PDEs. In Section 1.1 some simple problems that arise in real-life phenomena are derived. (A more detailed derivation of such problems will follow in later chapters.) We show by a number of examples how they may often be seen as continuous analogues of discrete formulations (i.e., based on difference equations). In Section 1.2 we briefly summarize the terminology used to describe various PDEs. Thus concepts like order and linearity are introduced. In Chapter 2 we shall discuss the classification of the various types of PDEs in more detail. Finally, we introduce difference equations and notions like scheme and stencil, which play a role in numerical approximation, in Section 1.3.

1.1 Introduction

Many phenomena in nature may be described mathematically by functions of a small number of independent variables and parameters. In particular, if such a phenomenon is given by a function of spatial position and time, its description gives rise to a wealth of (mathematical) models, which often result in equations, usually containing a large variety of derivatives with respect to these variables. Apart from the spatial variable(s), which are essential for the problems to be considered, the time variable will play a special role. Indeed, many events exhibit gradual or rapid changes as time proceeds. They are said to have an *evolutionary* character and an essential part of their modeling is therefore based on *causality*; i.e., the situation at any time is dependent on the past. As far as (mathematical) modeling leads to PDEs, the latter will be called evolutionary, i.e., involve the time t as a variable. The other type of problems are often referred to as *steady state*. We will give some examples to illustrate this background.

A typical PDE arises if one studies the flow of quantities like density, concentration, heat, etc. If there are no restoring forces, they usually have a tendency to spread out. In particular, one may, e.g., think of particles with higher velocities (or rather energy) colliding with particles with lower velocities. The former are initially rather clustered. The energy will gradually spread out, mainly because the high-velocity particles collide with other ones, thereby transferring some of the energy. This is called *dissipation*. A similar effect can be observed for a material dissolved in a fluid with concentrations varying in space. Brownian motion will gradually spread out the material over the entire domain. This is called *diffusion*.

Example 1.1 Consider a long tube of cross section A filled with water and a dye. Initially the dye is concentrated in the middle. Let $u(x,t)$ denote the concentration or density (mass per unit length) of the dye at position x and time t; then we see that in a small volume $A\Delta x$, positioned between $x - \frac{1}{2}\Delta x$ and $x + \frac{1}{2}\Delta x$ (Figure 1.1), the total amount of dye equals approximately $u(x,t)\Delta x$. Now consider a similar neighbouring volume $A\Delta x$ between $x + \frac{1}{2}\Delta x$ and $x + \frac{3}{2}\Delta x$, with a corresponding dye concentration $u(x + \Delta x, t)$. The mass that flows per unit time through a cross section is called the mass flux. From the physics of solutions it is known that the dye will move from the volume with higher concentration to one with lower concentration such that the mass flux f between the respective volumes is proportional to the difference in concentration between both volumes and is thus given by

$$f\left(x + \frac{1}{2}\Delta x, t\right) = \alpha\left(u\left(x + \frac{1}{2}\Delta x, t\right)\right)\frac{u(x + \Delta x, t) - u(x,t)}{\Delta x},$$

where α, the diffusion coefficient, usually depends on u. This relation is called Fick's law for mass transport by diffusion, which is the analogue of Fourier's law for heat transport by conduction.

As there is a similar flux between the centre volume and its left neighbour, we have a rate of change of total amount of mass in the centre volume equal to the difference between both fluxes given by

$$\frac{\partial}{\partial t}u(x,t)\Delta x = f\left(x + \frac{1}{2}\Delta x, t\right) - f\left(x - \frac{1}{2}\Delta x, t\right).$$

If the diffusion coefficient α is a constant, we have

$$\frac{\partial}{\partial t}u(x,t) = \alpha\frac{u(x + \Delta x, t) - 2u(x,t) + u(x - \Delta x, t)}{\Delta x^2}. \qquad (*)$$

By taking the limits for small volumes (i.e., $\Delta x \to 0$), we find

$$\frac{\partial}{\partial t}u(x,t) = \alpha\frac{\partial^2}{\partial x^2}u(x,t),$$

which is called the one-dimensional *diffusion equation*. As heat conduction satisfies the same equation, it is also called the *heat equation* if u denotes temperature. □

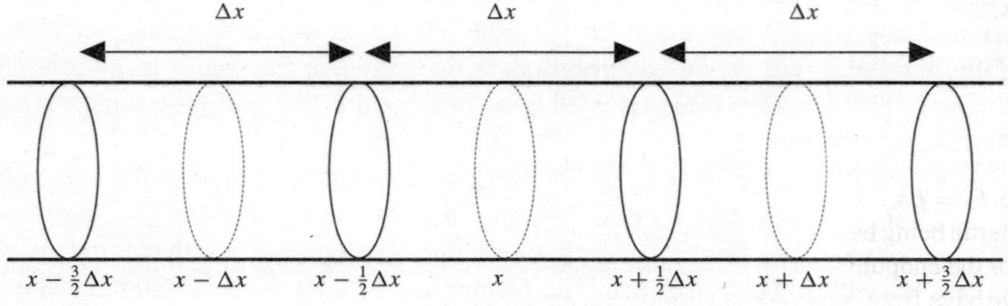

Figure 1.1. *Sketch of dye diffusion.*

Another kind of PDE occurs in the transport of particles. Here a flow typically has a dominant direction; mutual collision of particles (which is felt globally as a kind of internal friction, or viscosity) is neglected.

Example 1.2 Consider a road with heavy traffic moving in one direction, say the x direction (Figure 1.2). Let the number of cars at time t on a stretch $[x, x + \Delta x]$ be denoted by $\Delta N(x, t)$. Furthermore, let the number of cars passing a point x per time period Δt be given by $f(x, t)\Delta t$. In that period the number of cars $\Delta N(x, t + \Delta t)$ can only be changed by a difference between inflow at x and outflow at $x + \Delta x$; i.e.,

$$\Delta N(x, t + \Delta t) = \Delta N(x, t) - \big(f(x + \Delta x, t) - f(x, t)\big)\Delta t.$$

Rather than the number of cars ΔN per interval of length Δx, it is convenient to consider a *car density* $n(x, t)$, which is defined by

$$\Delta N(x, t) = n(x, t)\Delta x.$$

Hence we obtain the relation

$$\frac{n(x, t + \Delta t) - n(x, t)}{\Delta t} = -\frac{f(x + \Delta x, t) - f(x, t)}{\Delta x}.$$

Assuming sufficient smoothness (which implies that we have to allow for fractions of cars ...), this leads in the limit of $\Delta t, \Delta x \to 0$ to

$$\frac{\partial n}{\partial t} + \frac{\partial f}{\partial x} = 0,$$

which takes the form of a *conservation law*. We may recognize f again as a flux. If this flux only depends on the local car density, i.e., $f = f(n)$, and f is sufficiently smooth, we obtain

$$\frac{\partial n}{\partial t} + f'(n)\frac{\partial n}{\partial x} = 0,$$

also known as the *transport equation*. □

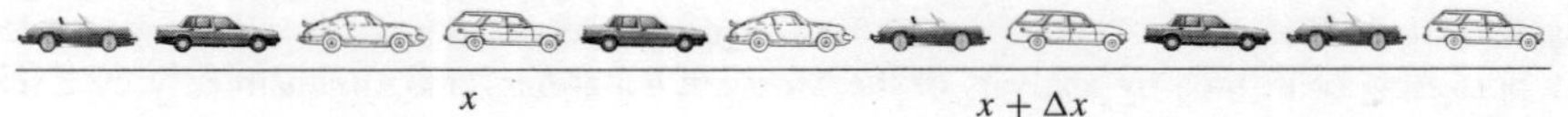

Figure 1.2. *Sketch of traffic flow.*

An important class of problems arises from classical mechanics, i.e., Newtonian systems.

Example 1.3 Consider a chain consisting of elements, each with mass m, and springs, with spring constant $\beta > 0$ and length Δx; see Figure 1.3. Denote the elements by $V_1, V_2, \ldots$ with position of the masses $x = u_1, u_2, \ldots$. Assuming linear springs, the force necessary to increase the original length Δx of the spring of element V_i by an amount $\delta_i = u_i - u_{i-1} - \Delta x$ is equal to $F_i = \beta\delta_i$. Apart from the endpoints, all masses are free to move in the x direction, their inertia being balanced by the reaction forces of the springs. Noting that each element V_i (except for the endpoints) experiences a spring force from the neighbouring ith and $(i + 1)$th springs, we have from Newton's law for the ith element that

$$m\frac{\mathrm{d}^2 u_i}{\mathrm{d}t^2} = F_{i+1} - F_i = \beta(u_{i+1} - u_i - u_i + u_{i-1}), \qquad i = 1, 2, \ldots. \qquad (*)$$

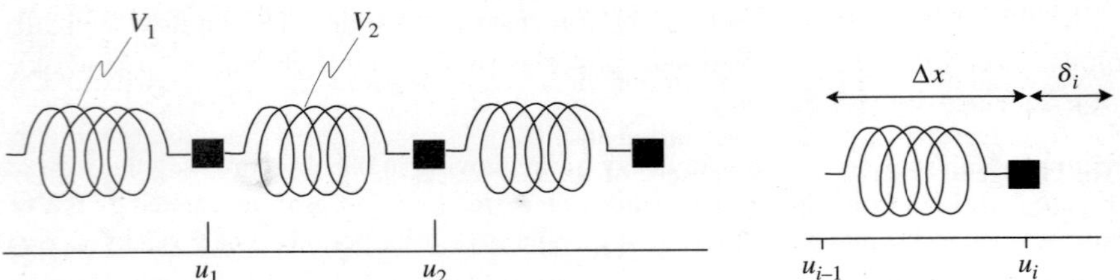

Figure 1.3. *Chain of coupled springs.*

If the chain elements increase in number, while the springs and masses decrease in size, it is natural and indeed more convenient not to distinguish the individual elements, but to blend the discrete description of $(*)$ into a continuous analogue. The small masses are conveniently described by a density ρ such that $m = \rho\Delta x$, while the large spring constants are best described by a stiffness $\sigma = \beta\Delta x$. Then we obtain from $(*)$ for the position function $u(x, t)$ the PDE

$$\frac{\partial^2 u}{\partial t^2} = \frac{\sigma}{\rho}\frac{\partial^2 u}{\partial x^2}. \qquad (\dagger)$$

As solutions of this equation are typically wave like, it is known as the *wave equation*, with a wave velocity equal to $\sqrt{\sigma/\rho}$. In our example it describes longitudinal waves along the suspended chain of masses. In the context of pressure-density perturbations of a compressible fluid like air, the equation describes one-dimensional sound waves, e.g., as they occur in organ pipes. In that case the air stiffness is equal to $\sigma = \gamma p$, where $\gamma = 1.4$ is a gas constant and p is the atmospheric pressure (see Section 6.8.2). □

In the following example we mention the analogue in electrical circuits of the motion of coupled spring-dashpot elements.

Example 1.4 The time-behaviour of electric currents in a network may be described by the variables potential V, current I, and charge Q. If the network is made of simple wires connecting isolated nodes, resistances, capacities, and coils, and the frequencies are low, it may be modeled (a posteriori confirmed by analysis of the Maxwell equations) one dimensionally by a series of elements with the material properties resistance R, capacitance C, and inductance L. Such a model is called an electrical circuit. If the frequencies are high, such that the wavelength is comparable with the length of the conductors, we have to be more precise. As the signal cannot change instantaneously at all locations, it propagates as a wave of voltage and current along the line. In such a case we cannot neglect the resistance and inductance properties of the wires. By considering the wires as being built up from a series of (infinitesimally) small elements, we can model the system by what is called a transmission line, leading to PDEs in time and space.

In or across each element we have the following relations. The current is defined as the change of charge in time, $I = \frac{\mathrm{d}}{\mathrm{d}t}Q$. The capacitance of a pair of conductors is given by $C = Q/V$, where V is the potential difference and Q is the charge difference between the conductors (Coulomb's law). The resistance between two points is given by $R = V/I$, where V is the potential difference between these points and I is the corresponding current (Ohm's law). A changing electromagnetic current in a coil with inductance L induces a counteracting potential, given by $V = -L\frac{\mathrm{d}}{\mathrm{d}t}I$ (Faraday's law). At a junction no charge can accumulate, and we have the condition $\sum I = 0$, while around a loop the summed potential vanishes. $\sum V = 0$ (Kirchhoff's laws). With these building blocks we can construct transmission line models.

A famous example is the *telegraph equation*, where an infinitesimal piece of telegraph wire is modeled (Figure 1.4) as an electrical circuit consisting of a resistance $R\Delta x$ and an inductance $L\Delta x$, while it is connected to the ground via a resistance $(G\Delta x)^{-1}$ and a capacitance $C\Delta x$.

Let $i(x,t)$ and $u(x,t)$ denote the current and voltage through the wire at position x and time t. The change of voltage across the piece of wire is now given by

$$u(x+\Delta x,t)-u(x,t)=\left[-iR\,\Delta x-\frac{\partial i}{\partial t}L\,\Delta x\right]_{x+\Delta x}.$$

The amount of current that disappears via the ground is

$$i(x+\Delta x,t)-i(x,t)=\left[-uG\,\Delta x-\frac{\partial u}{\partial t}C\,\Delta x\right]_{x}.$$

By taking the limit $\Delta x\to 0$, we get

$$\frac{\partial u}{\partial x}=-Ri-L\frac{\partial i}{\partial t},\qquad \frac{\partial i}{\partial x}=-Gu-C\frac{\partial u}{\partial t}.$$

By eliminating i, we may combine these equations into the telegraph equation for u, i.e.,

$$\frac{\partial^2 u}{\partial x^2}=LC\frac{\partial^2 u}{\partial t^2}+(LG+RC)\frac{\partial u}{\partial t}+RGu. \tag{$*$}$$

□

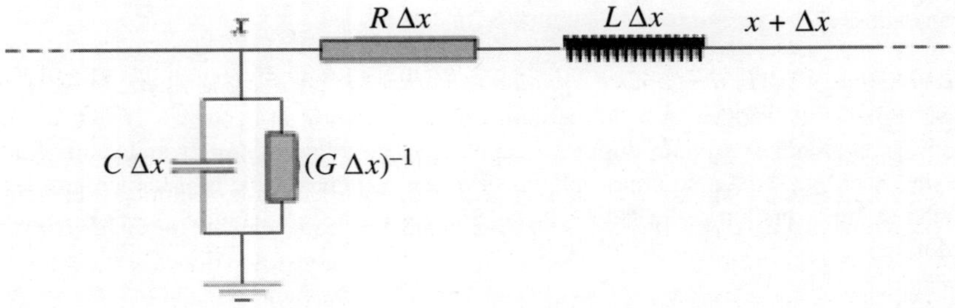

x $R\,\Delta x$ $L\,\Delta x$ $x+\Delta x$

$C\,\Delta x$ $(G\,\Delta x)^{-1}$

Figure 1.4. *A transmission line model of a telegraph wire.*

Example 1.5 Consider the following crowd of N^2 very accommodating people (Figure 1.5), for convenience ordered in a square of size $L\times L$, while each person, labelled by (i,j), is positioned at $x_i=ih$, $y_j=jh$, with $h=L/N$. Each person has an opinion given by the (scalar) number p_{ij} and can only communicate with his or her immediate neighbours. Assume that each person tries to minimize any conflict with his or her neighbours and is willing to take an opinion that is the average of their opinions. So we have

$$p_{ij}=\frac{1}{4}\left(p_{i+1,j}+p_{i-1,j}+p_{i,j+1}+p_{i,j-1}\right). \tag{$*$}$$

Only at the borders of the square are the individuals provided with information such that p is fixed.

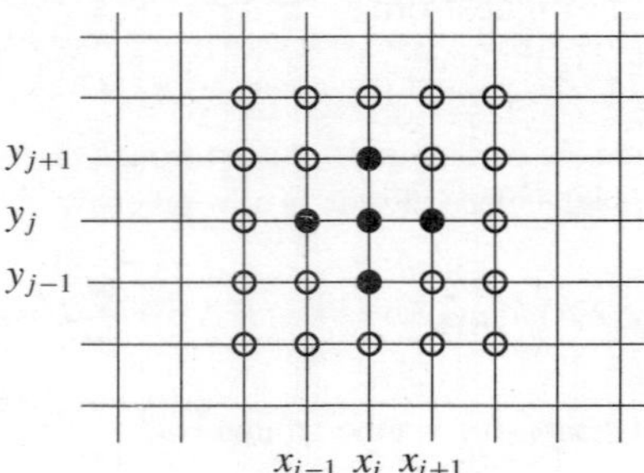

Figure 1.5. *An array of accommodating individuals.*

If the number of people becomes so large that we may take the limit $N \to \infty$ (i.e., $h \to 0$) and p becomes a continuous function of (x, y), $(*)$ becomes

$$p(x, y) = \frac{1}{4}(p(x+h, y) + p(x-h, y) + p(x, y+h) + p(x, y-h)).$$

This may be recast into

$$\big[p(x+h, y) - 2p(x, y) + p(x-h, y)\big] + \big[p(x, y+h) - 2p(x, y) + p(x, y-h)\big] = 0.$$

If this is true for any h, we may divide by h^2, and the equation becomes in the limit

$$\frac{\partial^2 p}{\partial x^2} + \frac{\partial^2 p}{\partial y^2} = 0.$$

This equation is called the *Laplace equation* and describes phenomena where, in some sense, information is exchanged in all directions until equilibrium is achieved. From the above sociological example it is not difficult to appreciate that discontinuities and sharp gradients are smoothed out, while extremes only occur at the boundary. The best-known problem described by this equation is the stationary distribution of the temperature in a heat-conducting medium. □

1.2 Nomenclature

In the previous section we met a number of equations with derivatives with respect to more than one variable. In general, such equations are called *partial differential equations*. Let x and t be two independent variables and let $u(x, t)$ denote a quantity depending on x and t. Furthermore, let

$$t \in [0, T], \quad 0 \le T \le \infty, \quad x \in [a, b] \subset \mathbb{R}. \tag{1.1}$$

For an integer n a general form for a scalar PDE (in two independent variables) reads

$$F\left(\frac{\partial^n u}{\partial t^n}, \frac{\partial^n u}{\partial t \partial x^{n-1}}, \dots, \frac{\partial^n u}{\partial x^n}, \frac{\partial^{n-1} u}{\partial t^{n-1}}, \dots, \frac{\partial^{n-1} u}{\partial x^{n-1}}, \dots, \frac{\partial u}{\partial t}, \frac{\partial u}{\partial x}, u, x, t\right) = 0. \tag{1.2}$$

The highest-order derivative is called the *order* of the PDE; not all partial derivatives (except the highest of at least one variable) need to be present. The form (1.2) is an *implicit* formulation, i.e., the highest-order derivative(s), the *principal part*, do(es) not appear explicitly. If the latter is the case, we call it an *explicit* PDE. The generalization to more than two independent variables is obvious.

Example 1.6 Some important examples of PDEs are as follows:

(i) $\dfrac{\partial u}{\partial t} + c\left(1 + \dfrac{3}{2}u\right)\dfrac{\partial u}{\partial x} + \tfrac{1}{6}ch^2\dfrac{\partial^3 u}{\partial x^3} = 0$ (*Korteweg–de Vries equation*).

This is a third order PDE.

(ii) $\dfrac{\partial u}{\partial t} + \dfrac{\partial}{\partial x}f(u) = 0$ (*nonlinear transport equation*).

If f is differentiable, we see that this is a first order PDE in u.

(iii) $\dfrac{\partial u}{\partial t} + u\dfrac{\partial u}{\partial x} = \varepsilon\dfrac{\partial^2 u}{\partial x^2}$ (*the Burgers' equation*).

If $\varepsilon = 0$, this may be referred to as the inviscid Burgers' equation, which is a special case of the transport equation.

(iv) $\dfrac{\partial^2 u}{\partial t^2} - c^2\dfrac{\partial^2 u}{\partial x^2} - \dfrac{1}{3}h^2\dfrac{\partial^4 u}{\partial x^2 \partial t^2} = 0$ (*linearized Boussinesq equation*).

(v) $EI\dfrac{\partial^4 u}{\partial x^4} - T\dfrac{\partial^2 u}{\partial x^2} + m\dfrac{\partial^2 u}{\partial t^2} = 0$ (*vibrating beam equation*).

(vi) $\dfrac{\partial u}{\partial y}\dfrac{\partial^2 u}{\partial y \partial x} - \dfrac{\partial u}{\partial x}\dfrac{\partial^2 u}{\partial y^2} = \nu\dfrac{\partial^3 u}{\partial y^3}$ (*Prandtl's boundary layer equation*). □

In quite a few cases the order can only be deduced after some (trivial) manipulation.

Example 1.7

$$\frac{\partial u}{\partial t} - \frac{\partial}{\partial x}\left(D(u)\frac{\partial u}{\partial x}\right) = f(x) \qquad \textit{(nonlinear diffusion equation)}.$$

It is clear that this PDE is second order. There is no analytical, numerical, or practical need to rework this and have $\frac{\partial^2}{\partial x^2}u$ appear explicitly. □

Usually, the variables are space and/or time. Although the variables in (1.2) are generic, we shall use the symbol t to indicate the *time* variable in general. The variable x will refer to *space*. There are major differences between problems where time does and does not play a role. If the time is not explicitly there, the problem is referred to as a *steady state problem*. If the PDE possesses solutions that evolve explicitly with t, we call it an *evolutionary problem*; i.e., there is *causality*. Most of the theory will be devoted to problems in one space variable. However, occasionally we shall encounter more than one such space variable. Fortunately, problems in more such variables often have many analogues of the one-dimensional case. We shall indicate vectors by boldface characters. So in higher-dimensional space the space variable is denoted by $\boldsymbol{x}$, or by $(x, y, z)^T$. The PDE can still be scalar. We have obvious analogues for vector-dependent variables of the foregoing.

Example 1.8 A few other examples are as follows:

(i) $\dfrac{\partial u}{\partial t} - \alpha\left(\dfrac{\partial^2 u}{\partial x^2} + \dfrac{\partial^2 u}{\partial y^2} + \dfrac{\partial^2 u}{\partial z^2}\right) = 0$ (*heat equation* in three dimensions).

We prefer to write this as $\frac{\partial}{\partial t}u - \alpha\nabla^2 u = 0$. ∇^2 is referred to as the *Laplace operator*.

(ii) $\dfrac{\partial^2 u}{\partial t^2} - c^2\nabla^2 u = 0$ (*wave equation* in three dimensions).

(iii) $\nabla^2 u + k^2 u = 0$ (*Helmholtz or reduced wave equation*).

(iv) $(1 - M^2)\dfrac{\partial^2 u}{\partial x^2} + \dfrac{\partial^2 u}{\partial y^2} + \dfrac{\partial^2 u}{\partial z^2} = 0$ (*equation for small perturbations in steady subsonic* ($M^2 < 1$) *or supersonic* ($M^2 > 1$) *flow*). □

Sometimes one also denotes a partial derivative of a certain variable by an index:

$$u_t := \frac{\partial u}{\partial t}, \qquad u_{tx} := \frac{\partial^2 u}{\partial t \partial x}. \tag{1.3}$$

If we can write (1.2) as a linear combination of u and its derivatives with respect to x and t, and with coefficients only depending on x and t, the PDE is called *linear*. Moreover, it is called *homogeneous* if it does not depend explicitly on x and/or t. If the PDE is a linear combination of derivatives but the coefficients of the highest derivative, say n, depend on $(n-1)$th order derivatives at most, then we call it *quasi-linear* [29].

For any differential equation we have to prescribe certain initial conditions and boundary conditions for the time and space variable(s), respectively. In evolutionary problems they often both appear as initial boundary conditions. We shall encounter various types and combinations in later chapters.

We finally remark that we may look for solutions that satisfy the PDE in a weak sense. In particular, the derivatives may not exist everywhere on the domain of interest. Again we refer to later chapters for further details.

1.3 Difference Equations

Initially, the actual form of the equations we derived in the examples in Section 1.1 was of a difference equation. Like a PDE, we may define a partial difference equation as any relation between values of $u(x, t)$ where $(x, t) \in \mathcal{F} \subset [a, b] \times [0, T)$, $\mathcal{F}$ being a finite set of points of the domain $[a, b] \times [0, T)$. We shall encounter difference equations when solving a PDE numerically, so they should approximate the PDE in some well-defined way. The simplest way to describe the latter is by defining a *scheme*, i.e., a discrete analogue of the (continuous) PDE. Since we shall mainly deal with finite difference approximations in this book, we perceive a scheme as the result of replacing the differentials by finite differences. To this end we have to indicate some (generic) points in the domain $[a, b] \times [0, T)$ at which the function values $u(x, t)$ are taken. The latter set of points is called a *stencil*. We shall clarify this with some examples.

Example 1.9

(i) Consider Example 1.1 again. If we replace $\frac{\partial}{\partial t}u(x, t)$ in equation (∗) by a straightforward discretisation, then we obtain the scheme

$$\frac{u(x, t + \Delta t) - u(x, t)}{\Delta t} = \alpha \frac{u(x + \Delta x, t) - 2u(x, t) + u(x - \Delta x, t)}{\Delta x^2},$$

and the stencil is the set of bullets (•) in Figure 1.6.

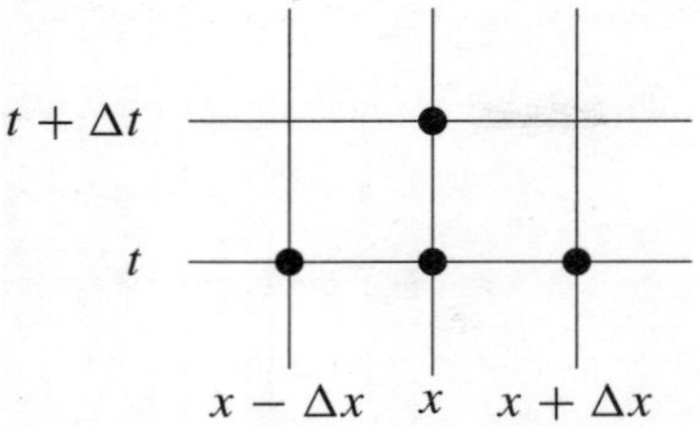

Figure 1.6. *Stencil of Example* 1.9(i).

(ii) Consider the wave equation (†) of Example 1.3. A discrete version may be found to be

$$\frac{u(x, t + \Delta t) - 2u(x, t) + u(x, t - \Delta t)}{\Delta t^2} = \frac{\sigma}{\rho} \frac{u(x + \Delta x, t) - 2u(x, t) + u(x - \Delta x, t)}{\Delta x^2}.$$

The stencil is given in Figure 1.7. □

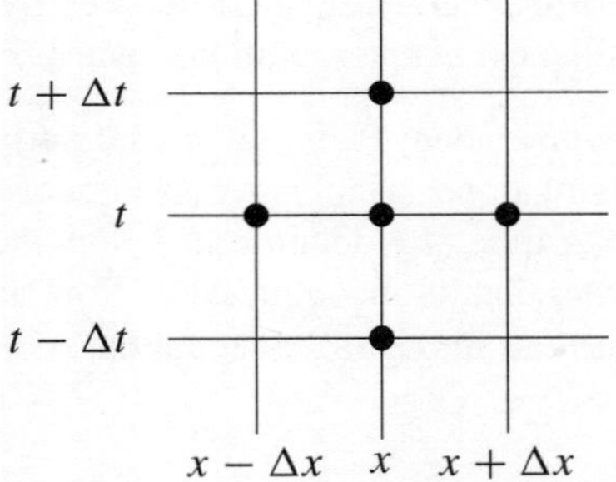

Figure 1.7. *Stencil of Example* 1.9(ii).

Given the special role of time and the implication it has for the actual computation, which should be based on the causality of the problem, we may distinguish schemes according to the number of time levels involved. If $(k + 1)$ such time levels are involved, we call the scheme a *k-step scheme*. If the scheme involves only spatial differences at earlier time levels, it is called *explicit*; otherwise it is called *implicit*.

Example 1.10

(i) The schemes in Example 1.9 are both explicit, the first being a one-step and the second a two-step scheme.

(ii) We could also approximate the u_{xx} term in the heat equation at time level $t + \Delta t$ and obtain the scheme

$$\frac{u(x, t + \Delta t) - u(x, t)}{\Delta t} = \alpha \frac{u(x + \Delta x, t + \Delta t) - 2u(x, t + \Delta t) + u(x - \Delta x, t + \Delta t)}{\Delta x^2}.$$

This scheme has the stencil given in Figure 1.8. Clearly, it is an implicit one-step scheme. □

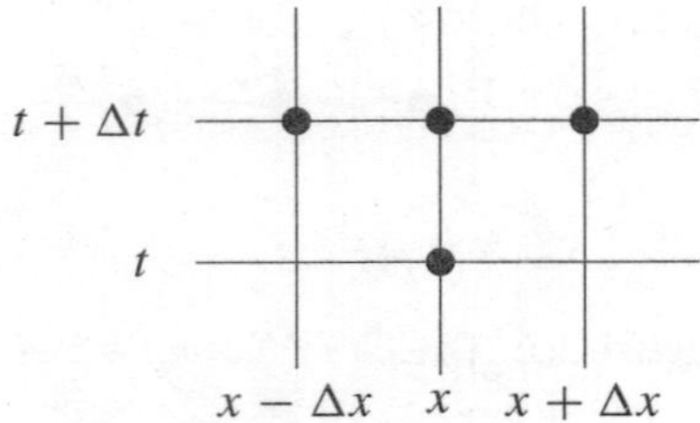

Figure 1.8. *Stencil of Example* 1.10(ii).

1.4 Discussion

- The use of the variables x and y in an equation does not mean that the PDE cannot have an evolutionary character. There are some cases where they refer to spatial coordinates, yet the corresponding equation may be hyperbolic, a type of equation we will encounter in the next chapter as an instance of evolutionary type.
- If in a system of time-dependent PDEs all spatial derivatives are replaced by suitable difference approximations, we obtain a system of ODEs in time. If one of the PDEs is independent of time, we obtain a *differential-algebraic system*. A typical example is the condition that an incompressible flow is divergence free (equivalent to conservation of mass), as in the Stokes equations. This problem will be discussed in Section 8.7.

Exercises

1.1. Show that a nonconstant diffusivity $\alpha(u)$ leads to the equation

$$\frac{\partial u}{\partial t} = \frac{\partial}{\partial x}\left(\alpha(u)\frac{\partial u}{\partial x}\right).$$

1.2. Determine the order of the *eikonal* equation

$$\left(\frac{\partial u}{\partial x}\right)^2 + \left(\frac{\partial u}{\partial y}\right)^2 + \left(\frac{\partial u}{\partial z}\right)^2 = c^2.$$

1.3. Determine the order of the PDE

$$\frac{\partial^2 u}{\partial x^2} = \frac{\partial^2 u}{\partial y^2}.$$

Derive a first order system by writing $p := \frac{\partial u}{\partial x}$, $q := \frac{\partial u}{\partial y}$.

1.4. Determine the order of the PDE (where a and b are parameters)

$$\frac{\partial u}{\partial t} = a\nabla^2 u + b\frac{\partial u}{\partial x} + c(u).$$

1.5. Verify that the solution $u = u(x,t)$ of the transport equation (cf. Example 1.2 or 1.6(ii))

$$\frac{\partial u}{\partial t} + \frac{\partial}{\partial x} f(u) = 0, \qquad u(x,0) = v(x),$$

for sufficiently smooth f is implicitly given by

$$u = v\bigl(x - f'(u)t\bigr).$$

Chapter 2

Characterisation and Classification

The study of PDEs is quite versatile. Therefore it makes sense to first characterise them according to certain properties that will provide us with guidelines for investigating them further. It is useful to start with first order equations in two independent variables. Therefore we start in Section 2.1 by describing scalar first order equations, thereby introducing the important notion of characteristics. This is generalised for first order systems in Section 2.2, leading to the definition of hyperbolicity. A well-known class of PDEs consists of second order scalar equations. In Section 2.3 we reformulate them as a first order system of equations and then discuss their classification as hyperbolic, parabolic, and elliptic equations. Quite naturally, this can be generalised to several (space) dimensions, as is shown in Section 2.4. Sometimes the underlying structure of a problem is simpler than suggested, and after a suitable transformation the PDE may be transformed into an ODE. Examples are given in Section 2.5. PDEs need further conditions to make their solutions meaningful and (we hope) unique. In Section 2.6 we briefly discuss, from a more theoretical point of view, how to properly choose the initial and boundary conditions. For this we have the Hadamard condition, which states the conditions necessary for a problem to be well posed.

2.1 First Order Scalar PDEs in Two Independent Variables

Consider the quasi-linear (explicit) first order PDE

$$a(x,t,u)\frac{\partial u}{\partial t} + b(x,t,u)\frac{\partial u}{\partial x} = c(x,t,u). \tag{2.1}$$

Usually, the independent variables x and t denote a space coordinate and time, respectively, although, strictly speaking, t might denote a space coordinate as well. Let $u = \varphi(x,t)$ be a solution of (2.1). A geometrical interpretation of this solution is as follows. The independent variables x and t and the dependent variable u constitute a two-parameter family of vectors $(x,t,u)^T$ that is lying on a surface $\mathcal{S} \subset \mathbb{R}^3$. This surface $\mathcal{S}$, given by $F(x,t,u) := \varphi(x,t) - u = 0$, is called the *integral surface* of (2.1). A normal $\boldsymbol{n}$ on $\mathcal{S}$ is given by

$$n := \nabla F = \left(\frac{\partial \varphi}{\partial x}, \frac{\partial \varphi}{\partial t}, -1\right)^T. \tag{2.2}$$

Hence, for an infinitesimal displacement $\mathrm{d}\boldsymbol{u} := (\mathrm{d}x, \mathrm{d}t, \mathrm{d}u)^T$ along the surface, we find

$$n \cdot \mathrm{d}\boldsymbol{u} = \frac{\partial \varphi}{\partial x}\,\mathrm{d}x + \frac{\partial \varphi}{\partial t}\,\mathrm{d}t - \mathrm{d}u = 0. \tag{2.3}$$

Comparing (2.1) and (2.3), we conclude that for a solution $u = \varphi(x, t)$ of (2.1) the following should hold on the integral surface $\mathcal{S}$:

$$\begin{pmatrix} a & b \\ \mathrm{d}t & \mathrm{d}x \end{pmatrix} \begin{pmatrix} \dfrac{\partial \varphi}{\partial t} \\ \dfrac{\partial \varphi}{\partial x} \end{pmatrix} = \begin{pmatrix} c \\ \mathrm{d}u \end{pmatrix}. \tag{2.4}$$

The solution of this system is unique if and only if $a\,\mathrm{d}x - b\,\mathrm{d}t \neq 0$. In the following we will simply write $u = u(x, t)$ instead of $u = \varphi(x, t)$ to indicate the solution of (2.1).

This result can be interpreted as follows (see Figure 2.1): suppose we have a smooth, one-parameter curve $\mathcal{J} = \{(x(\sigma), t(\sigma), u(\sigma) \mid \sigma \in I \subset \mathbb{R}\}$ on $\mathcal{S}$, where the condition $a\,\mathrm{d}x - b\,\mathrm{d}t \neq 0$ holds. Then the derivatives u_x and u_t are uniquely determined on $\mathcal{J}$ through (2.4). If, moreover, u is given along $\mathcal{J}$, then the solution $u = u(x, t)$ exists and

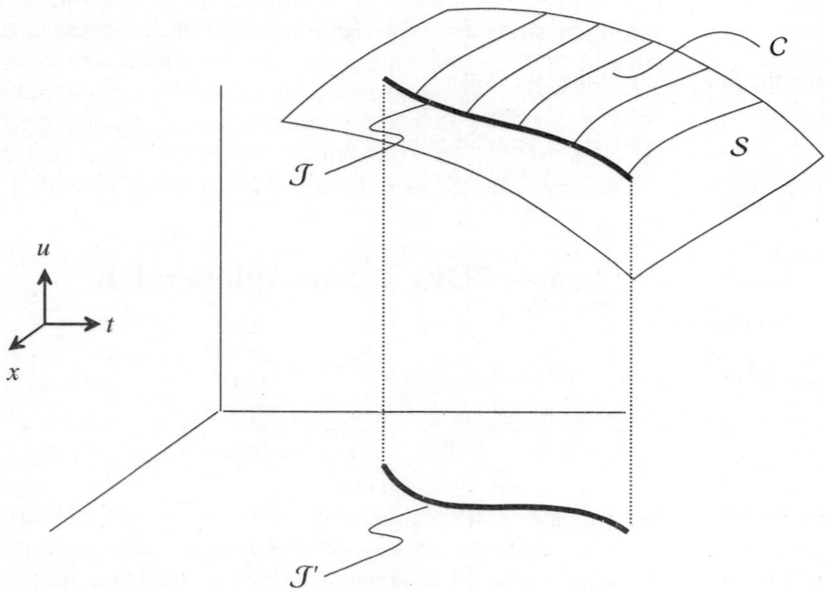

Figure 2.1. *Initial curve $\mathcal{J}$ and a characteristic $\mathcal{C}$ on the integral surface $\mathcal{S}$.*

is unique, at least in some neighbourhood of $\mathcal{J}$. The curve $\mathcal{J}$ is referred to as a curve of initial values or, briefly, an *initial curve*.

The actual construction of the solution proceeds as follows. Suppose u is given along an initial curve $\mathcal{J}$. Consider a curve $\mathcal{C}$ on $\mathcal{S}$ for which $a\,\mathrm{d}x - b\,\mathrm{d}t = 0$. Then system (2.4) has either no solution or infinitely many. In the latter case the relations

$$\frac{\mathrm{d}t}{a} = \frac{\mathrm{d}x}{b} = \frac{\mathrm{d}u}{c} \tag{2.5}$$

should hold along $\mathcal{C}$. Clearly, the vector $(a, b, c)^T$ is everywhere tangent to $\mathcal{C}$. We can now introduce a parametrization $\mathcal{C} = \{(x(s), t(s), u(s))^T \mid s \in I \subset \mathbb{R}\}$ such that $\mathrm{d}s = \mathrm{d}t/a = \mathrm{d}x/b = \mathrm{d}u/c$ and $s = 0$ on the initial curve $\mathcal{J}$. This way, we obtain the set of ODEs

$$\frac{\mathrm{d}t}{\mathrm{d}s} = a, \qquad \frac{\mathrm{d}x}{\mathrm{d}s} = b, \qquad \frac{\mathrm{d}u}{\mathrm{d}s} = c \tag{2.6a}$$

coupled with an initial condition of the form

$$u(0;\sigma) = v(x(\sigma), t(\sigma)) \qquad \text{for} \quad (x(\sigma), t(\sigma)) \in \mathcal{J}', \tag{2.6b}$$

where $\mathcal{J}'$ is the projection of $\mathcal{J}$ on the (x, t) plane. The ODEs (2.6a) are referred to as the *characteristic equations*. Consequently, the curve $\mathcal{C}$ is a solution of (2.6). $\mathcal{C}$ is called a *characteristic* and its projection on the (x, t) plane is called a *base characteristic*. Note, however, that there is no uniformity in the nomenclature in the literature; usually, no distinction is made between characteristics and base characteristics. In order to construct the integral surface, we compute for each point on the initial curve $\mathcal{J}$ the characteristic passing through that point from (2.6). We formally obtain the solution

$$t = t(s;\sigma), \qquad x = x(s;\sigma), \qquad u = u(s;\sigma).$$

Inverting the first two relations, we find $s = s(t, x)$, $\sigma = \sigma(t, x)$, and substituting these in the expression for u gives $u(x, t) := u(s(t, x); \sigma(t, x))$. This inversion is possible only if the Jacobian $t_\sigma x_s - t_s x_\sigma$ is nonzero. Thus the integral surface $\mathcal{S}$ is generated by a one-parameter family of characteristics $\mathcal{C}$ all passing through an initial curve $\mathcal{J}$. We will demonstrate this by an example.

Example 2.1 Consider the following initial value problem for the inviscid Burgers' equation:

$$\frac{\partial u}{\partial t} + u\frac{\partial u}{\partial x} = 0, \qquad x \in \mathbb{R}, \quad t > 0,$$

$$u(x, 0) = v(x) := \begin{cases} 1 & \text{if} \quad x \le 0, \\ 1 - x & \text{if} \quad 0 < x \le 1, \\ 0 & \text{if} \quad x \ge 1. \end{cases}$$

The characteristic equations and the corresponding initial conditions read

$$\frac{\mathrm{d}t}{\mathrm{d}s} = 1, \qquad \frac{\mathrm{d}x}{\mathrm{d}s} = u, \qquad \frac{\mathrm{d}u}{\mathrm{d}s} = 0,$$

$$t(0;\sigma) = 0, \quad x(0;\sigma) = \sigma, \quad u(0;\sigma) = v(\sigma).$$

The solution of this set of ODEs is given by

$$t(s;\sigma) = s, \quad x(s;\sigma) = \sigma + v(\sigma)s, \quad u(s;\sigma) = v(\sigma).$$

We can easily invert the first two relations provided that the Jacobian $t_\sigma x_s - t_s x_\sigma = -1 - v'(\sigma)s$ is nonzero. This way, we obtain $s(x,t) = t$ and $\sigma(x,t) = x - t$ for $x \le t$; $\sigma(x,t) = (x - t)/(1-t)$ for $t < x \le 1$; and $\sigma(x,t) = x$ for $x \ge 1$. Consequently, the solution is defined for $0 < t < 1$ and is given by

$$u(x,t) = \begin{cases} 1 & \text{if} \quad x \le t, \\ \dfrac{1-x}{1-t} & \text{if} \quad t < x \le 1, \\ 0 & \text{if} \quad x \ge 1. \end{cases}$$

□

Instead of using s, we can parametrize the characteristics by the variable t provided that $a \neq 0$. We thus obtain the ODEs

$$\frac{\mathrm{d}x}{\mathrm{d}t} = \frac{b}{a}, \qquad \frac{\mathrm{d}u}{\mathrm{d}t} = \frac{c}{a}. \tag{2.7}$$

We also refer to (2.7) as the characteristic equations. The first equation gives the location of the base characteristics, possibly depending on the solution u, and the second gives u along the base characteristics. We see that the existence of solutions of (2.1) can be established from studying the ODEs (2.7). It is easy to see that such a solution is composed of solutions of initial value problems defined along base characteristics. If the right-hand side of (2.7) satisfies a Lipschitz condition, then (2.7), together with an initial value for u, determines the solution u on $\mathcal{C}$.

Example 2.2 Consider the transport equation

$$\frac{\partial u}{\partial t} + b(u)\frac{\partial u}{\partial x} = 0 \tag{*}$$

subject to an initial condition of the form $u(x,0) = v(x)$. For this equation system (2.7) reduces to

$$\frac{\mathrm{d}x}{\mathrm{d}t} = b(u), \qquad \frac{\mathrm{d}u}{\mathrm{d}t} = 0,$$

implying that $u(x,t) =$ constant along the base characteristics, which, however, depend on the solution we seek. Since $u(x,t) =$ constant, $x - b(u)t =$ constant, and we obtain the following (implicit) representation of the solution:

$$u(x,t) = v(x - b(u(x,t))t).$$

In the special case of the linear advection equation, for which $b(u) = b =$ constant, this representation reduces to $u(x,t) = v(x - bt)$, i.e., the initial profile is propagated undisturbed with speed b along the base characteristics; see Figure 2.2. □

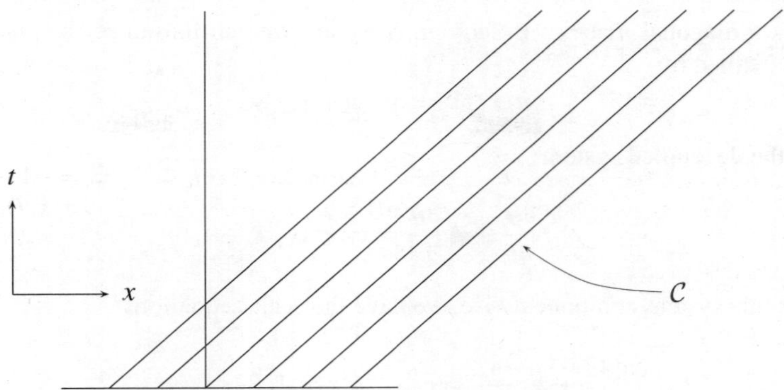

Figure 2.2. *Base characteristics of* $(*)$ *of Example* 2.2.

2.2 First Order Linear Systems in Two Independent Variables

The previous theory for scalar equations can be extended to systems. In the rest of this text we will often encounter systems of the form

$$\frac{\partial \boldsymbol{u}}{\partial t} + \frac{\partial}{\partial x} \boldsymbol{f}(\boldsymbol{u}) = \boldsymbol{c}, \tag{2.8}$$

where $\boldsymbol{u} : I_1 \times I_2 \to \mathbb{R}^m$, $I_1 \subset \mathbb{R}$, and $I_2 \subset [0, \infty)$. Introducing the Jacobi matrix

$$\boldsymbol{B} := \frac{\partial \boldsymbol{f}}{\partial \boldsymbol{u}} = \left(\frac{\partial f_i(\boldsymbol{u})}{\partial u_j} \right),$$

we find the *quasi-linear* first order system of PDEs

$$\frac{\partial \boldsymbol{u}}{\partial t} + \boldsymbol{B} \frac{\partial \boldsymbol{u}}{\partial x} = \boldsymbol{c}. \tag{2.9}$$

More generally, a quasi-linear first order system of PDEs may read

$$\boldsymbol{A} \frac{\partial \boldsymbol{u}}{\partial t} + \boldsymbol{B} \frac{\partial \boldsymbol{u}}{\partial x} = \boldsymbol{c}. \tag{2.10}$$

In this section we assume that the coefficient matrices $\boldsymbol{A}$ and $\boldsymbol{B}$ are constant and that $\boldsymbol{c} = \boldsymbol{c}(x, t, \boldsymbol{u})$. The quasi-linear case where $\boldsymbol{A} = \boldsymbol{A}(x, t, \boldsymbol{u})$ and $\boldsymbol{B} = \boldsymbol{B}(x, t, \boldsymbol{u})$ is discussed in Chapter 12.

In order to employ the theory of the previous section, we try to decouple system (2.10) into n scalar equations. For the sake of simplicity we further assume that the matrix $\boldsymbol{A}$ is nonsingular. Then we look for a nonsingular transformation matrix $\boldsymbol{S}$ such that

$$\boldsymbol{S}^{-1}(\boldsymbol{B}\boldsymbol{A}^{-1})\boldsymbol{S} = \boldsymbol{\Lambda}, \tag{2.11}$$

where $\boldsymbol{\Lambda}$ is a diagonal matrix. If such an $\boldsymbol{S}$ exists, we can introduce the *characteristic variable* $\tilde{\boldsymbol{u}}$ defined by

$$\tilde{\boldsymbol{u}}(x,t) := \boldsymbol{S}^{-1}\boldsymbol{A}\boldsymbol{u}(x,t), \tag{2.12}$$

satisfying the decoupled system

$$\frac{\partial \tilde{\boldsymbol{u}}}{\partial t} + \boldsymbol{\Lambda}\frac{\partial \tilde{\boldsymbol{u}}}{\partial x} = \tilde{\boldsymbol{c}} := \boldsymbol{S}^{-1}\boldsymbol{c}. \tag{2.13a}$$

If we write this system componentwise, we have the scalar equations

$$\frac{\partial \tilde{u}_k}{\partial t} + \lambda_k\frac{\partial \tilde{u}_k}{\partial x} = \tilde{c}_k, \qquad k = 1, 2, \ldots, m. \tag{2.13b}$$

The previous derivation gives rise to the following definition.

Definition 2.3. *The linear system* (2.10) *with nonsingular matrix* $\boldsymbol{A}$ *is called hyperbolic if* $\boldsymbol{B}\boldsymbol{A}^{-1}$ *has* m *real eigenvalues and* m *linearly independent eigenvectors.*

Consequently, when system (2.10) is hyperbolic, the matrices $\boldsymbol{\Lambda}$ and $\boldsymbol{S}$ do exist and are given by

$$\boldsymbol{\Lambda} := \mathrm{diag}(\lambda_1, \lambda_2, \ldots, \lambda_m), \qquad \boldsymbol{S} := (\boldsymbol{s}_1, \boldsymbol{s}_2, \ldots, \boldsymbol{s}_m), \tag{2.14}$$

where λ_k and $\boldsymbol{s}_k$ are the eigenvalues and corresponding eigenvectors of $\boldsymbol{B}\boldsymbol{A}^{-1}$, respectively. Thus the kth column of $\boldsymbol{S}$ is the eigenvector $\boldsymbol{s}_k$. Note that (2.10) is always hyperbolic if $\boldsymbol{B}\boldsymbol{A}^{-1}$ is symmetric; for a general matrix hyperbolicity is assured if all eigenvalues are real and distinct.

Each equation in (2.13b) induces a characteristic $\mathcal{C}_k$ corresponding to the eigenvalue λ_k and eigenvector $\boldsymbol{s}_k$. The characteristic equations (2.7) in this case read

$$\frac{\mathrm{d}x}{\mathrm{d}t} = \lambda_k, \qquad \frac{\mathrm{d}\tilde{u}_k}{\mathrm{d}t} = \tilde{c}_k. \tag{2.15}$$

The solution of (2.10) can be recovered from solutions of (2.15). This is demonstrated in the following example.

Example 2.4 Consider the initial value problem

$$\begin{pmatrix} 1 & 0 \\ 0 & 2 \end{pmatrix}\frac{\partial \boldsymbol{u}}{\partial t} + \begin{pmatrix} 1 & 4 \\ 1 & 0 \end{pmatrix}\frac{\partial \boldsymbol{u}}{\partial x} = \boldsymbol{0}, \qquad x \in \mathbb{R}, \quad t > 0,$$

$$\boldsymbol{u}(x,0) = \boldsymbol{v}(x), \qquad x \in \mathbb{R}.$$

We can easily verify that the eigenvalues and corresponding eigenvectors of $\boldsymbol{B}\boldsymbol{A}^{-1}$ are given by

$$\lambda_1 = -1, \quad \lambda_2 = 2, \quad \boldsymbol{s}_1 = \begin{pmatrix} 1 \\ -1 \end{pmatrix}, \quad \boldsymbol{s}_2 = \begin{pmatrix} 2 \\ 1 \end{pmatrix}.$$

The characteristic variable $\tilde{\boldsymbol{u}}$ defined in (2.12) is now given by

$$\tilde{u}_1 = \frac{1}{3}(u_1 - 4u_2), \qquad \tilde{u}_2 = \frac{1}{3}(u_1 + 2u_2).$$

These variables can be computed from (2.15) and we find $\tilde{u}_1(x,t) = \tilde{u}_1(x+t,0)$ and $\tilde{u}_2(x,t) = \tilde{u}_2(x-2t,0)$. Combining this result with the above relations for the characteristic variables, we obtain the final solution

$$u_1(x,t) = \frac{1}{3}(v_1(x+t) - 4v_2(x+t) + 2v_1(x-2t) + 4v_2(x-2t)),$$

$$u_2(x,t) = \frac{1}{6}(-v_1(x+t) + 4v_2(x+t) + v_1(x-2t) + 2v_2(x-2t)).$$

Clearly, the solution contains waves propagating along base characteristics $x + t =$ constant and $x - 2t =$ constant, respectively. □

In general we conclude that (2.10) should not be subject to an initial condition prescribed on a characteristic. In fact, one should prescribe $\boldsymbol{u}$ on a curve $\mathcal{J}$ that does not intersect any of these characteristics twice.

Finally, we introduce the following notions; see Figure 2.3. The domain of dependence of a point (x_0, t_0) is the region in the (x,t) plane such that $u(x_0, t_0)$ depends on all values $u(x,t)$ with (x,t) in this domain. It is bounded by the two extreme characteristics through (x_0, t_0) facing back to the initial line $t = 0$. On the other hand, the domain of influence of (x_0, t_0) is the region in the (x,t) space where the solution is influenced by $u(x_0, t_0)$.

In the next section we shall consider the special case of systems arising from scalar second order problems.

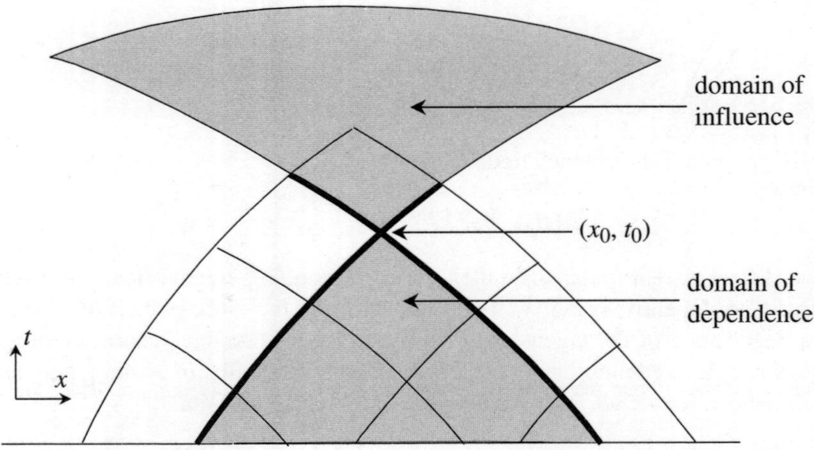

Figure 2.3. *Region of influence and region of dependence. In the case of constant coefficients these characteristics are straight lines.*

2.3 Second Order Scalar PDEs in Two Independent Variables

For PDEs with higher-order derivatives the classification may be reduced to first order systems met before, at least if they are scalar. In particular, we consider the second order linear equation

$$a\frac{\partial^2 u}{\partial t^2}+b\frac{\partial^2 u}{\partial t\partial x}+c\frac{\partial^2 u}{\partial x^2}+d\frac{\partial u}{\partial t}+e\frac{\partial u}{\partial x}=f, \tag{2.16}$$

where the coefficients $a, b, \ldots, e$ are assumed constant and where the right-hand side f possibly depends on x, t, and u. The independent variable x is a space coordinate, whereas t is either time or a space coordinate. Introducing the variables

$$p:=\frac{\partial u}{\partial t}, \qquad q:=\frac{\partial u}{\partial x}, \tag{2.17}$$

we obtain the linear system

$$\begin{pmatrix} a & 0\\ 0 & 1\end{pmatrix}\frac{\partial}{\partial t}\begin{pmatrix} p\\ q\end{pmatrix}+\begin{pmatrix} b & c\\ -1 & 0\end{pmatrix}\frac{\partial}{\partial x}\begin{pmatrix} p\\ q\end{pmatrix}=\begin{pmatrix} f-d\,p-e\,q\\ 0\end{pmatrix}. \tag{2.18}$$

Note that this form is not unique. Clearly, this system is of the form (2.10), with the coefficient matrices given by

$$\boldsymbol{A}:=\begin{pmatrix} a & 0\\ 0 & 1\end{pmatrix}, \qquad \boldsymbol{B}:=\begin{pmatrix} b & c\\ -1 & 0\end{pmatrix}. \tag{2.19}$$

We now investigate the eigenvalues and eigenvectors of this system.

In the following we assume that $a \neq 0$, so that $\boldsymbol{A}$ is nonsingular. Like in the previous section, we look for a transformation matrix $\boldsymbol{S}$ that can diagonalise the matrix

$$\boldsymbol{B}\boldsymbol{A}^{-1}=\begin{pmatrix} \dfrac{b}{a} & c\\ \dfrac{-1}{a} & 0\end{pmatrix} \tag{2.20}$$

This will succeed if the characteristic equation

$$\det(\boldsymbol{B}\boldsymbol{A}^{-1}-\lambda\boldsymbol{I})=\lambda^2-\frac{b}{a}\lambda+\frac{c}{a}=0 \tag{2.21}$$

has two different, real roots, since then the corresponding eigenvectors are linearly independent. Consequently, system (2.18) is hyperbolic if $b^2-4ac>0$. If $b^2-4ac=0$, we have a degeneracy of the eigensystem and so only one "double" characteristic exists; in fact no $\boldsymbol{S}$ and $\boldsymbol{\Lambda}$, as required in (2.11), can be found. If, finally, $b^2<4ac$, there are no real characteristic values at all. This leads to the following definition.

Definition 2.5. *The PDE* (2.16) *is called*

(i) *hyperbolic if* $b^2-4ac>0$;
(ii) *parabolic if* $b^2-4ac=0$;
(iii) *elliptic if* $b^2-4ac<0$.

The nomenclature in this definition is adopted from the theory of quadratic forms. In particular, the corresponding quadratic curve $at^2 + btx + cx^2 + dt + ex =$ constant is a hyperbola, parabola, or ellipse, depending on the value of $b^2 - 4ac$.

Next we will derive the normal, or canonical, form of (2.16) in these three different cases, which only depends on the *principal part* of the equation, i.e., the first three terms containing the second order derivatives.

In the *hyperbolic case* we find a transformation matrix

$$\boldsymbol{S} = \begin{pmatrix} \lambda_1 & \lambda_2 \\ \dfrac{-1}{a} & \dfrac{-1}{a} \end{pmatrix}, \tag{2.22}$$

where $\lambda_1 < \lambda_2$ are the eigenvalues of $\boldsymbol{BA}^{-1}$. The characteristic variables $\tilde{\boldsymbol{u}}$ defined in (2.12) are now given by

$$\tilde{\boldsymbol{u}} = \frac{a}{\lambda_2 - \lambda_1} \begin{pmatrix} -p - \lambda_2 q \\ p + \lambda_1 q \end{pmatrix}. \tag{2.23}$$

Inserting the first component $\tilde{u}_1$ into (2.13b), we obtain an equation of the form

$$\left(\frac{\partial}{\partial t} + \lambda_1 \frac{\partial}{\partial x}\right)\left(\frac{\partial}{\partial t} + \lambda_2 \frac{\partial}{\partial x}\right) u = \tilde{f}. \tag{2.24}$$

Taking into account the characteristic equations (2.15), we see that the first and second differential operators in (2.24) are just internal derivatives along characteristics of the C_1 and C_2 families, respectively. Introducing coordinates ξ and η along these characteristics, we obtain the canonical form

$$\frac{\partial^2 u}{\partial \xi \partial \eta} = \tilde{f}. \tag{2.25}$$

Note that we would obtain the same equation starting from the equation for the second characteristic variable $\tilde{u}_2$.

In the *parabolic* case there is no transformation matrix $\boldsymbol{S}$ possible that can diagonalise $\boldsymbol{BA}^{-1}$. However, we have the Jordan normal form (see Section H in the appendix)

$$\boldsymbol{S}^{-1}(\boldsymbol{BA}^{-1})\boldsymbol{S} = \boldsymbol{J} := \begin{pmatrix} \lambda & 1 \\ 0 & \lambda \end{pmatrix}, \tag{2.26}$$

with $\lambda = b/(2a)$ the eigenvalue of $\boldsymbol{BA}^{-1}$, having algebraic multiplicity two and geometric multiplicity one. A possible matrix $\boldsymbol{S}$ is given by

$$\boldsymbol{S} = \begin{pmatrix} \lambda & 1 \\ \dfrac{-1}{a} & 0 \end{pmatrix}. \tag{2.27}$$

We can now reduce system (2.18) to

$$\frac{\partial \tilde{\boldsymbol{u}}}{\partial t} + \boldsymbol{J}\frac{\partial \tilde{\boldsymbol{u}}}{\partial x} = \tilde{\boldsymbol{f}} \tag{2.28}$$

for the variable $\tilde{\boldsymbol{u}}$, given by

$$\tilde{\boldsymbol{u}} = a \begin{pmatrix} -q \\ p + \lambda q \end{pmatrix}. \tag{2.29}$$

From (2.13b) for $\tilde{u}_2$ we easily deduce

$$\left(\frac{\partial}{\partial t} + \lambda \frac{\partial}{\partial x}\right)^2 u = \tilde{f}, \tag{2.30}$$

in which we recognize an internal differentiation along a characteristic. The canonical form for (2.16) is thus given by

$$\frac{\mathrm{d}^2 u}{\mathrm{d}\xi^2} = \tilde{f}, \tag{2.31}$$

where ξ is the coordinate along the characteristic.

Finally, in the *elliptic* case, we have complex characteristics, and the transformation matrix $\boldsymbol{S}$ is also complex. Completely analogously to the hyperbolic case, we obtain the form

$$\frac{\partial^2 u}{\partial \xi \partial \eta} = \tilde{f}, \tag{2.32}$$

with ξ and η the (complex) coordinates along characteristics of the C_1 and C_2 families, respectively. One can prove that $\eta = \bar{\xi}$; see, e.g., [29]. Introducing the new coordinates

$$\mu := \frac{1}{2}(\xi + \eta) = \mathrm{Re}(\xi), \qquad \nu := \frac{1}{2\mathrm{i}}(\xi - \eta) = \mathrm{Im}(\xi), \tag{2.33}$$

we obtain the canonical form

$$\frac{\partial^2 u}{\partial \mu^2} + \frac{\partial^2 u}{\partial \nu^2} = 4\tilde{f}. \tag{2.34}$$

Example 2.6 The standard examples of hyperbolic, parabolic, and elliptic equations are, respectively,

$$\frac{\partial^2 u}{\partial t^2} = \frac{\partial^2 u}{\partial x^2} \quad \text{(wave equation)},$$

$$\frac{\partial u}{\partial t} = \frac{\partial^2 u}{\partial x^2} \quad \text{(heat equation)},$$

$$\frac{\partial^2 u}{\partial x^2} + \frac{\partial^2 u}{\partial y^2} = 0 \quad \text{(Laplace equation)}.$$

□

The classification as given in Definition 2.5 also holds for linear equations with coefficients depending on x and t and even for quasi-linear equations. The definition should then be applied pointwise, as is demonstrated in the next example.

Example 2.7 Two well-studied equations in the theory of transonic flow are the Tricomi equation and the isentropic potential flow equation. The *Tricomi equation* (see, e.g., [4]) is given by

$$y\frac{\partial^2 u}{\partial x^2} - \frac{\partial^2 u}{\partial y^2} = 0.$$

For $y > 0$ we apparently have a hyperbolic equation (related to supersonic flow), whereas for $y < 0$ the equation is elliptic (related to subsonic flow)! The *isentropic potential flow equation* for the velocity potential φ reads (see, e.g., [63, 64])

$$\left[c^2 - \left(\frac{\partial \varphi}{\partial x}\right)^2\right]\frac{\partial^2 \varphi}{\partial x^2} + \left[c^2 - \left(\frac{\partial \varphi}{\partial y}\right)^2\right]\frac{\partial^2 \varphi}{\partial y^2} - 2\frac{\partial \varphi}{\partial x}\frac{\partial \varphi}{\partial y}\frac{\partial^2 \varphi}{\partial x \partial y} = 0,$$

with the speed of sound c related to the velocity via Bernoulli's equation for compressible flow (see (7.12))

$$\left(\frac{\partial \varphi}{\partial x}\right)^2 + \left(\frac{\partial \varphi}{\partial y}\right)^2 + \frac{2c^2}{\gamma - 1} = \frac{2c_0^2}{\gamma - 1},$$

where γ is the specific heat ratio and c_0 is the sound speed for stagnant flow. Evidently, the equation is hyperbolic for supersonic flow ($\varphi_x^2 + \varphi_y^2 > c^2$) and elliptic for subsonic flow ($\varphi_x^2 + \varphi_y^2 < c^2$). □

Hyperbolic, parabolic, and elliptic equations are distinctly different. Solutions of each type of equation show an entirely different behaviour, which is also reflected in the solution methods, either analytically or numerically. In the rest of this book we will extensively address these three types of equations.

2.4 Linear Second Order Equations in Several Space Variables

The general linear second order PDE of constant coefficients is given by

$$\sum_{i=1}^{n}\sum_{j=1}^{n} a_{ij}\frac{\partial^2 u}{\partial x_i \partial x_j} + \sum_{i=1}^{n} b_i \frac{\partial u}{\partial x_i} + cu = f. \tag{2.35}$$

Here the variables $x_1, x_2, \dots, x_n$ can be time and/or any number of space variables. More precisely, for time-dependent problems we have $n = d + 1$ and $x_n = t$, whereas for stationary problems $n = d$ and all variables are space coordinates. In Section 2.3 we used characteristics to define new variables and obtained a normal form. Here we shall simply consider transformations of the variables without such a theory and look for simplified forms of (2.35) from a geometrical point of view.

To start with we may associate with (2.35) a symmetric matrix $\boldsymbol{A}$, where

$$\boldsymbol{A} := \left(\frac{1}{2}(a_{ij} + a_{ji})\right); \tag{2.36}$$

actually, we have distributed the coefficient a_{ij} symmetrically between the entries (i, j) and (j, i) of the matrix $\boldsymbol{A}$. Now also define

$$\boldsymbol{x} := (x_1, \dots, x_n)^T, \qquad \boldsymbol{b} := (b_1, \dots, b_n)^T. \tag{2.37}$$

Then (2.35) may be formulated as

$$\nabla_{\boldsymbol{x}} \cdot (\boldsymbol{A}\nabla_{\boldsymbol{x}} u) + \boldsymbol{b} \cdot \nabla_{\boldsymbol{x}} u + cu = f, \tag{2.38}$$

where $\nabla_{\boldsymbol{x}}$ denotes the gradient with respect to $\boldsymbol{x}$. This can be simplified by diagonalising $\boldsymbol{A}$. Since $\boldsymbol{A}$ is symmetric, this is possible through an orthogonal similarity transformation. So let $\boldsymbol{Q}$ be such that

$$\boldsymbol{Q}^T\boldsymbol{A}\boldsymbol{Q} = \boldsymbol{\Lambda}, \tag{2.39}$$

where $\boldsymbol{Q}^T = \boldsymbol{Q}^{-1}$ and $\boldsymbol{\Lambda}$ is the real diagonal matrix of eigenvalues of $\boldsymbol{A}$. Next we introduce a new set of variables $\boldsymbol{y} := (y_1, \ldots, y_n)^T$ by

$$\boldsymbol{y} := \boldsymbol{Q}^T\boldsymbol{x}. \tag{2.40}$$

Using the relations $\nabla_{\boldsymbol{x}} = \boldsymbol{Q}\nabla_{\boldsymbol{y}}$ and $(\boldsymbol{Q}\boldsymbol{v})\boldsymbol{\cdot}\boldsymbol{w} = \boldsymbol{v}\boldsymbol{\cdot}(\boldsymbol{Q}^T\boldsymbol{w})$, we obtain the simplified form

$$\nabla_{\boldsymbol{y}}\boldsymbol{\cdot}(\boldsymbol{\Lambda}\nabla_{\boldsymbol{y}}u) + (\boldsymbol{Q}^T\boldsymbol{b})\boldsymbol{\cdot}\nabla_{\boldsymbol{y}}u + cu = f. \tag{2.41}$$

Viewing $\nabla_{\boldsymbol{y}}u$ as a vector in $\mathbb{R}^n$, the quadratic form can now be classified as a (generalised)

(i) *ellipse* if all eigenvalues have the same sign,
(ii) *parabola* if at least one of the eigenvalues is zero, or
(iii) *hyperbola* if all eigenvalues are nonzero and have the same sign, except for one.

These geometric descriptions make sense in $\mathbb{R}^2$ at least. The corresponding PDEs are classified similarly: elliptic, parabolic, and hyperbolic. If there are at least two positive and negative eigenvalues (and the others are all nonzero), one sometimes calls the PDE ultrahyperbolic.

If we scale the variables $y_1, \ldots, y_n$ by multiplying them by $\sqrt{|\lambda_1|}, \ldots, \sqrt{|\lambda_n|}$, respectively (unless $\lambda_i = 0$), we will obtain a quadratic form with $\pm 1, 0$ as eigenvalues. Hence it is not restrictive to assume that this has already been done. It then follows that the multidimensional Laplace operator ∇^2 becomes an important symbol to describe second order PDEs in several dimensions. In particular we have the following:

the *(elliptic) Laplace equation*

$$\nabla^2 u = 0, \tag{2.42a}$$

the *(parabolic) heat equation*

$$\frac{\partial u}{\partial t} = \nabla^2 u, \tag{2.42b}$$

and the *(hyperbolic) wave equation*

$$\frac{\partial^2 u}{\partial t^2} = \nabla^2 u. \tag{2.42c}$$

2.5 Reduction to ODEs; Similarity Solutions

In the foregoing we have seen that it is sometimes possible to reformulate a (set of) PDE(s) as a set of ODEs that describe the solution along characteristics. In this section we will reiterate this in a slightly different way. To that purpose, consider the initial value problem

$$\frac{\partial u}{\partial t} + b(u)\frac{\partial u}{\partial x} = c(u), \qquad x \in \mathbb{R}, \quad t > 0, \tag{2.43a}$$

$$u(x,0) = v(x), \qquad x \in \mathbb{R}. \tag{2.43b}$$

For the special case $c(u) \equiv 0$ we have derived in Example 2.2 from the corresponding characteristic equations the following (implicit) representation of the solution:

$$u(x,t) = v(x - b(u(x,t))).$$

If, furthermore, $b(u) = b =$ constant, this solution reduces to $u(x,t) = v(x - bt)$, which is a wave propagating undisturbed with speed b.

Based on these observations, we try a solution of the form

$$u(x,t) = \hat{u}(\xi), \qquad \xi := x - st. \tag{2.44}$$

This solution is called a *travelling-wave* solution with *wave speed* s. Substituting (2.44) in (2.43a), we obtain the ODE

$$(b(\hat{u}) - s)\frac{\mathrm{d}\hat{u}}{\mathrm{d}\xi} = c(\hat{u}). \tag{2.45}$$

The travelling wave $\hat{u}(\xi)$ and its wave speed s have to be determined from (2.45) and (2.43b). We will illustrate this in the next example.

Example 2.8 An example from combustion theory reads [88]

$$\frac{\partial u}{\partial t} + u\frac{\partial u}{\partial x} = c(u) := u(1-u)(u-\beta)/\tau, \tag{$*$}$$

with parameters $\tau > 0$ and $0 < \beta < 1$. The corresponding ODE $u'(t) = c(u(t))$ has the stable equilibrium solutions $u(x,t) \equiv 0, 1$ and the unstable one $u(x,t) \equiv \beta$. This means that any initial value $v(x) \neq \beta$ approaches either one of the stable solutions for $t \to \infty$. Substituting (2.44) in $(*)$, we obtain the ODE

$$(\hat{u} - s)\frac{\mathrm{d}\hat{u}}{\mathrm{d}\xi} = \frac{1}{\tau}\hat{u}(1-\hat{u})(\hat{u}-\beta). \tag{$**$}$$

Suppose that the initial solution $v(x)$ increases monotonically from zero to one. Then the appropriate boundary conditions for $(**)$ read

$$\lim_{\xi\to-\infty} \hat{u}(\xi) = 0, \qquad \lim_{\xi\to\infty} \hat{u}(\xi) = 1.$$

Equation $(**)$ can only satisfy these boundary conditions if $s = \beta$; otherwise $\hat{u}(\xi)$ cannot cross the unstable solution $u(x,t) \equiv \beta$. The resulting solution then reads

$$\hat{u}(\xi) = \left(1 + \mathrm{e}^{-\xi/\tau}\right)^{-1}.$$

□

Travelling waves occur naturally in transport equations but also in a number of other equations. We return to this in Chapter 10, where we consider travelling waves for a parabolic equation.

Another class of solutions consists of the *similarity solutions*, which are functions of a (dimensionless) combination of x and t. We will introduce these solutions for the homogeneous transport equation (2.43a); i.e., $c(u) \equiv 0$. A more systematic discussion of similarity solutions, based on dimensional analysis, is presented in Chapter 7. Ignoring initial and boundary conditions, we see that if $u(x,t)$ is a solution of (2.43a), then $u_\alpha(x,t) :=$

$u(\alpha x, \alpha t)$ is a solution as well for any $\alpha > 0$. Therefore we may try a similarity solution of the form

$$u(x,t) = \hat{u}(\eta), \qquad \eta := \frac{x}{t}; \tag{2.46}$$

i.e., $u(x,t) =$ constant along rays $x/t =$ constant through the origin of the (x,t) plane. In Chapter 12 we will use this formulation to compute solutions of hyperbolic equations. Substituting (2.46) into (2.43a), we have

$$\big(b(\hat{u}) - \eta\big)\frac{\mathrm{d}\hat{u}}{\mathrm{d}\eta} = 0, \tag{2.47}$$

implying that either $\hat{u}(\eta) =$ constant, resulting in the trivial solution $u(x,t) \equiv$ constant, or $b(\hat{u}) = \eta$. In the latter case we obtain the solution

$$u(x,t) = b^{-1}(x/t), \tag{2.48}$$

with $b^{-1}(u)$ the inverse of $b(u)$, assuming it exists.

Example 2.9 Consider the traffic flow problem of Example 1.2 given by the transport equation

$$\frac{\partial n}{\partial t} + \frac{\partial f(n)}{\partial x} = 0.$$

A model for the flux $f(n)$ proposed in [88] reads

$$f(n) := u_{\mathrm{m}}\, n \left(1 - \frac{n}{n_{\mathrm{m}}}\right),$$

with u_{m} the maximum speed of vehicles and n_{m} the maximum density of cars. We may verify that $\hat{u}$ satisfies the equation

$$b(n) = f'(n) = u_{\mathrm{m}}\left(1 - \frac{2n}{n_{\mathrm{m}}}\right) = \eta,$$

resulting in the similarity solution

$$u(x,t) = \frac{1}{2} n_{\mathrm{m}} \left(1 - \frac{x}{u_{\mathrm{m}}\, t}\right).$$

See also Example 12.14. □

Similarity solutions are also of importance for parabolic equations. In Chapter 10 we give a full account; here we restrict ourselves to an example.

Example 2.10 Consider the heat equation

$$\frac{\partial u}{\partial t} = \frac{\partial^2 u}{\partial x^2}. \tag{$*$}$$

Note that if $u(x,t)$ is a solution of $(*)$, then so is $u_\alpha(x,t) := u(\alpha x, \alpha^2 t)$ for any $\alpha > 0$. Therefore an obvious choice for a similarity solution is

$$u(x,t) = \hat{u}(\eta), \qquad \eta := \frac{x}{\sqrt{t}}.$$

Substituting $\hat{u}(\eta)$ in $(*)$, we obtain the ODE

$$\frac{\mathrm{d}^2\hat{u}}{\mathrm{d}\eta^2} + \frac{1}{2}\eta\frac{\mathrm{d}\hat{u}}{\mathrm{d}\eta} = 0.$$

This equation can be solved, with solution

$$\hat{u}(\eta) = C_1 \int_0^{\frac{1}{2}\eta} \mathrm{e}^{-\tau^2}\,\mathrm{d}\tau + C_2,$$

where C_1, C_2 are constants to be determined from the initial and boundary conditions. □

2.6 Initial and Boundary Conditions; Well-Posedness

For any differential equation one needs to specify the solution somewhere and somehow. The actual problem is then to find a solution of the PDE subject to certain conditions. If the latter are given at a time point onward from which the evolution takes place, we call them *initial values*. For the space domain we have (possibly) *boundary conditions*. For time-dependent problems we usually have both *initial* and *boundary conditions*. If we have purely initial conditions, we call the problem a *Cauchy problem*.

To determine whether the problem is meaningful we use the following definition.

Definition 2.11 (Hadamard's well-posedness conditions). *A problem is well posed if*

(i) *a solution exists,*

(ii) *the solution is unique, and*

(iii) *the solution depends continuously on the data, in particular the initial and boundary values.*

Note that these conditions imply (i) that one should not have too many (conflicting) initial and boundary conditions, (ii) that we should not have too few, and (iii) that the effect of small perturbations is also small. The latter may be interpreted either in a *local* (small strip, interval, etc.) or *global* (infinite strip, interval, etc.) region.

In an elliptic problem the interaction between the coordinate points, as described by the equation, is in all directions. In time this is of course not possible. Therefore it is physically very unlikely that a time-dependent equation is of elliptic type. This is made precise by the following example.

Example 2.12 Consider the elliptic equation

$$\frac{\partial^2 u}{\partial x^2} + \frac{\partial^2 u}{\partial y^2} = 0, \qquad x \in \mathbb{R}, \quad y > 0,$$

where $u = u(x, y; n)$ is subject to

$$u(x, 0; n) = 0, \qquad x \in \mathbb{R},$$

$$\frac{\partial u}{\partial y}(x, 0; n) = \frac{1}{n}\sin nx, \qquad x \in \mathbb{R}, \quad n \in \mathbb{N}.$$

This initial value problem was used by Hadamard to show that a Cauchy problem setting is not appropriate for elliptic problems. This can be seen as follows. One can easily check that

$$u(x, y; n) := \frac{1}{n^2} \sin nx \sinh ny$$

is a solution on $\mathbb{R} \times [0, \infty) \times \mathbb{N}$. For $y > 0$, $x = \frac{1}{2}\pi$, and n odd we note that $|u(x, y; n)| \to \infty$, however small y is. On the other hand, the initial conditions $u(x, 0) = \frac{\partial}{\partial y} u(x, 0) = 0$ give the solution $u(x, y) = 0$. This shows that u does not depend continuously on the initial data; i.e., it violates criterion (iii). □

Elliptic PDEs typically give rise to boundary value problems. We remark that *elliptic operators* will play a role by themselves as well as being part of parabolic or hyperbolic problems. Now consider a hyperbolic problem. Let $\mathcal{D}$ be a smooth curve and $\boldsymbol{n}$ denote the normal direction. Then a Cauchy problem has as initial values

$$u = A(x, t), \quad \frac{\partial u}{\partial \boldsymbol{n}} = B(x, t), \quad (x, t)^T \in \mathcal{D}. \tag{2.49}$$

Theorem 2.13. *Let $\mathcal{D}$ be a curve in $\mathbb{R}^2$ such that $\mathcal{D}$ intersects the characteristics only once. Then the Cauchy problem consisting of equation* (2.16) *and initial conditions* (2.49) *is well posed.*

Proof. Let $\mathcal{D}$ have a representation $\varphi(x, t) = 0$. Let $\psi(x, t)$ be such that φ, ψ is a genuine coordinate transformation (i.e., the Jacobian is nonzero). Then we can reformulate (2.16) in terms of φ, ψ, giving

$$\alpha \frac{\partial^2 u}{\partial \varphi^2} + \beta \frac{\partial^2 u}{\partial \varphi \partial \psi} + \gamma \frac{\partial^2 u}{\partial \psi^2} + \delta \frac{\partial u}{\partial \varphi} + \varepsilon \frac{\partial u}{\partial \psi} = \zeta, \tag{$*$}$$

where

$$\alpha = a \left(\frac{\partial \varphi}{\partial x} \right)^2 + b \frac{\partial \varphi}{\partial x} \frac{\partial \psi}{\partial x} + c \left(\frac{\partial \psi}{\partial x} \right)^2, \quad \text{etc.}$$

For $u = u(\varphi, \psi)$ we then have the initial values $u(0, \psi) = \hat{A}(\varphi)$, $u_\varphi(0, \psi) = \hat{B}(\psi)$. Hence we can find $u_\psi(0, \psi)$, $u_{\psi\psi}(0, \psi)$, and $u_{\varphi\psi}(0, \psi)$ but not $u_{\varphi\varphi}(0, \psi)$ from the initial conditions. If $\alpha \neq 0$, we can find $u_{\varphi\varphi}(0, \psi)$ and thus all higher-order derivatives from the transformed PDE ($*$). A formal solution away from $\mathcal{D}$ can now be found through a Taylor expansion:

$$u(\tilde{\varphi}, \tilde{\psi}) = \sum_{i=0}^{\infty} \sum_{j=0}^{i} \frac{1}{j!(i-j)!} \frac{\partial^i u}{\partial \varphi^j \partial \psi^{i-j}} (\varphi, \psi)(\tilde{\varphi} - \varphi)^j (\tilde{\psi} - \psi)^{i-j}.$$

So at least the solution exists in a neighbourhood of $\mathcal{D}$; i.e., the problem is well posed. Thus we only need to prove $\alpha \neq 0$. However, this follows from the fact that if $\xi(t, x) = 0$ denotes a characteristic, then $\det(\boldsymbol{A}\xi_t + \boldsymbol{B}\xi_x) = a\xi_x^2 + b\xi_x\xi_t + c\xi_t^2 = 0$, while $\mathcal{D}$ was assumed not to be a characteristic. □

The preceding theorem also applies to parabolic problems.

Example 2.14 Consider the PDE

$$\frac{\partial u}{\partial t} = \frac{\partial^2 u}{\partial x^2}, \qquad x \in I \subset \mathbb{R}, \quad t > 0,$$

subject to the initial conditions

$$u(x, 0) = \alpha(x), \qquad \frac{\partial u}{\partial t}(x, 0) = \beta(x).$$

We can trivially see that $u_{xx}(x, 0)$ should be equal to $\beta(x)$, which is not true in general. If, on the other hand, we had a Cauchy problem in the x variable, i.e.,

$$u(0, t) = \bar{\alpha}(t), \qquad \frac{\partial u}{\partial x}(0, t) = \bar{\beta}(t),$$

then we could still show the solution to exist; cf. Theorem 2.13. □

Theorem 2.15. *If* (2.16) *is parabolic, i.e.,* $a = b = 0$, *and* $u(x, 0) = v(x)$ *is given, then this defines a well-posed Cauchy problem, at least locally.*

Proof. As in the proof of Theorem 2.13, one finds that $u_x(x, 0)$ and $u_{xx}(x, 0)$ are well defined. Hence $u_t(x, 0)$ is well defined and then so are their higher derivatives. Hence we conclude the existence of $u(x, t)$ in a neighbourhood of $t = 0$, whence we obtain local existence. □

2.7 Discussion

- Classifying equations as elliptic, parabolic, and hyperbolic is more tradition than natural. The most important distinction is between boundary value problems and initial (boundary) value problems, which have an evolutionary character. In the latter the information travels with a "finite speed," while for the former everything happens with "infinite speed." As remarked in the discussion of the previous chapter, the use of spatial coordinates does not preclude the problem from having an evolution. The boundary data are then typically needed on only a part of the boundary.
- A possible classification of evolutionary PDEs is whether or not they allow for wave-like solutions; by this we mean solutions of the type $f(x - st)$. Hyperbolic equations clearly have such solutions. But there also exist *dispersive waves*, which do not necessarily fit into the definition of hyperbolicity that was given here and will be used later in Chapter 12. For more details see [169].

Exercises

2.1. Consider the PDE

$$\frac{\partial^2 u}{\partial x^2} + 4\frac{\partial^2 u}{\partial x \partial y} + 3\frac{\partial^2 u}{\partial y^2} + 3\frac{\partial u}{\partial x} - \frac{\partial u}{\partial y} + 2u = 0.$$

(a) Show that it is hyperbolic.

(b) Find the characteristics and bring the PDE to normal form.

(c) Find a coordinate transformation such that the first order terms vanish in the resulting equation.

2.2. Consider the PDE

$$\frac{\partial^2 u}{\partial x^2} + 2\frac{\partial^2 u}{\partial x \partial y} + \frac{\partial^2 u}{\partial y^2} + 5\frac{\partial u}{\partial x} + 5\frac{\partial u}{\partial y} + u = 0.$$

(a) Show that it is parabolic.

(b) Find the normal form.

(c) Find a coordinate transformation such that the first order terms vanish in the resulting equation.

2.3. Consider the PDE

$$\frac{\partial^2 u}{\partial x^2} - 6\frac{\partial^2 u}{\partial x \partial y} + 12\frac{\partial^2 u}{\partial y^2} + 4\frac{\partial u}{\partial x} + \frac{\partial u}{\partial y} = 0.$$

(a) Show that it is elliptic.

(b) Find the normal form.

(c) Find a coordinate transformation such that the first order terms vanish in the resulting equation.

2.4. Classify the PDE

$$\frac{\partial^2 u}{\partial x^2} + 2\frac{\partial^2 u}{\partial y \partial z} + \cos x \frac{\partial u}{\partial z} - e^y u = \cosh z.$$

2.5. Show that in d-dimensional space any second order elliptic PDE with constant coefficients can be brought to the form

$$\sum_{i=1}^{d} \frac{\partial^2 u}{\partial {x_i}^2} + cu = f.$$

2.6. Show that in n-dimensional space any second order hyperbolic PDE with constant coefficients can be brought to the form

$$\frac{\partial^2 u}{\partial {x_n}^2} = \sum_{i=1}^{n-1} \frac{\partial^2 u}{\partial {x_i}^2} + cu + f.$$

2.7. Can you classify the PDE

$$x^\alpha y^\beta \frac{\partial^2 u}{\partial x^2} + x^\gamma y^\delta \frac{\partial^2 u}{\partial y^2} + \frac{\partial u}{\partial x} + \frac{\partial u}{\partial y} = 0?$$

2.8. Consider the hyperbolic equation

$$\frac{\partial^2 u}{\partial x \partial y} = 0$$

on the unit square, where u is given on the boundary. Show that this problem is not well posed.

2.9. Consider the parabolic equation

$$\frac{\partial u}{\partial t} = \frac{\partial^2 u}{\partial x^2}$$

for (x, t) in the positive (x, t) plane. Let $u(x, 0)$ be given. Show that this Cauchy problem is not well posed.

2.10. Determine the solution of

$$\frac{\partial u}{\partial t} + \frac{\partial u}{\partial x} = u, \qquad x \in \mathbb{R}, \quad t > 0,$$

$$u(x, 0) = \mathrm{e}^{x^2}, \qquad x \in \mathbb{R}.$$

2.11. Find the characteristics of the equation

$$\frac{\partial u}{\partial t} + x\frac{\partial u}{\partial x} = 0.$$

2.12. Show that all travelling-wave solutions of the wave equation

$$\frac{\partial^2 u}{\partial t^2} = c^2\frac{\partial^2 u}{\partial x^2}$$

are of the form $u(x, t) = \hat{u}_1(x - ct)$ and $u(x, t) = \hat{u}_2(x + ct)$. Consequently, the general solution (see Chapter 12) is given by

$$u(x, t) = \hat{u}_1(x - ct) + \hat{u}_2(x + ct).$$

Chapter 3

Fourier Theory

Fourier theory plays an important role in applied analysis. In this chapter we give an overview of the most important aspects needed in this book. First, we introduce an inner product and (orthonormal) basis functions in Section 3.1. Here we also define Fourier series, consider their convergence, and give Parseval's identity. We give both the complex and the trigonometric representations. Finally, the integral analogue of a series is introduced and exemplified. Next, in Section 3.3, the discrete form of the Fourier transform is considered, derived from the continuous version. Again, convergence and Parseval's identity are studied. Also, important phenomena, such as aliasing that show the restrictions of the discrete Fourier transform are treated. One very important application of this Fourier transform is in analysing linear equations with periodic boundary values. Despite the limitations of this problem class, it turns out that many physically meaningful concepts, such as stability, dissipation, and dispersion, can be studied quite fruitfully for the transformed equation, both in the continuous and in the discrete cases. In Section 3.4, therefore, the use of these transformations is demonstrated, leading to the important concept of the dispersion relation.

3.1 Fourier Series

A powerful tool in analysis is the expansion of a function f in terms of suitably chosen functions that form a *basis*. There are several ways to find such expansions. The most elegant way to describe the function mathematically is to use projections, for which we need the concept of "orthogonality." This is provided by using an inner product much like the one met in linear algebra (the "natural inner product"). Since this involves integration, we shall restrict ourselves to a finite interval, which is typically chosen as $(0, L)$. The functions we consider are *square integrable*, i.e., L_2-functions. Moreover, we shall assume that they are *periodically* extended to the full real axis. We then define the inner product for two such functions as

$$(f, g) := \int_0^L f(x)\bar{g}(x)\,\mathrm{d}x, \tag{3.1}$$

where the overbar $\bar{\ }$ denotes the complex conjugate (i.e., $\overline{a + \mathrm{i}b} = a - \mathrm{i}b$). Note that from periodicity it follows that any integral over an interval of length L is equivalent. As can simply be verified, the functions

$$p_j(x) := e^{i\alpha_j x}, \qquad j = 0, \pm 1, \pm 2, \dots , \tag{3.2a}$$

where

$$\alpha_j = \frac{2\pi j}{L}, \tag{3.2b}$$

are *orthogonal*. If we change (3.1) into

$$(f, g)_L := \frac{1}{L}\int_0^L f(x)\bar{g}(x)\,dx, \tag{3.1'}$$

the functions are even *orthonormal*; i.e., $(p_j, p_k) = \delta_{jk}$. We call α_j a *wave number* if x denotes a spatial variable and a *frequency* if the independent parameter is time. In physical texts a wave number is usually denoted by the letter k and a frequency is denoted by ω.

Introducing

$$c_j := \frac{1}{L}\int_0^L f(y)\,e^{-i\alpha_j y}\,dy, \tag{3.3}$$

we can form a *Fourier series*

$$\tilde{f}(x) := \sum_{j=-\infty}^{\infty} c_j\, e^{i\alpha_j x}. \tag{3.4}$$

The function $p_j(x) = e^{i\alpha_j x}$ is called a *Fourier mode* with wave number (frequency) α_j. The important question is whether $\tilde{f}$ can be identified with f (and of course whether $\tilde{f}$ makes sense at all). We then say that the Fourier series of f *converges* to f. In fact, no simple test is known that is both necessary and sufficient to relate a periodic function with its Fourier coefficients [25]. There is, however, a vast amount of partial results.

We have the following theorems.

Theorem 3.1. *If for all points $x \in (0, L)$ the left and right derivatives of f, i.e.,*

$$\lim_{h\downarrow 0}\frac{f(x+h)-f(x)}{h} \quad \text{and} \quad \lim_{h\downarrow 0}\frac{f(x)-f(x-h)}{h}$$

exist, then the Fourier series (3.4) *of f at x converges to $f(x)$.*

Theorem 3.2. *If f is continuous in x and $f(0) = f(L)$, then the Fourier series* (3.4) *of f converges uniformly to f (Section* C *in the appendix); i.e.,*

$$\lim_{N\to\infty}\sup_{x\in[0,L]}\left|\sum_{j=-N}^{N} c_j\, e^{i\alpha_j x} - f(x)\right| = 0.$$

Definition 3.3. *The function f is* piecewise continuous *on $[0, L]$ if there is a finite number of open subintervals $0 < x < x_1 \cdots x_{N-1} < x < L$ on which f is continuous, while the limits $f(0+), f(x_1\pm), \dots, f(L-)$ exist. The function f is* piecewise smooth *on $[0, L]$ if f and its derivative f' are both piecewise continuous.*

For such a function f we have the following theorem.

Theorem 3.4 (existence of Fourier series). *If a function f is piecewise smooth on the interval $[0, L]$, while $f(x) = \frac{1}{2}[f(x+) + f(x-)]$, then the Fourier series of f converges for every x to the L-periodic continuation of f.*

For a given Fourier series $\sum c_j \, \mathrm{e}^{\mathrm{i}\alpha_j x}$ we have the following theorem.

Theorem 3.5 (continuity of Fourier series). *If a Fourier series is absolutely convergent, i.e., $\sum |c_j| < \infty$, then it converges absolutely and uniformly to a continuous periodic function f such that c_j are just the Fourier coefficients of f.*

If f is a function only in L_2, then we still have at least the following identity.

Theorem 3.6 (Parseval's identity). *Let $f \in L_2(0, L)$ have Fourier coefficients c_j. Then*

$$(f, f)_L = \sum_{j=-\infty}^{\infty} c_j^2.$$

Proof. For a proof of Theorems 3.1, 3.2, and 3.6, see [25, 112]. □

Corollary 3.7. *If f and f' are piecewise smooth, then the Fourier coefficients c_j of f behave asymptotically for $j \to \infty$ like $c_j = \mathcal{O}(j^{-1})$.*

Proof. Integration by parts yields

$$c_j = \frac{1}{\mathrm{i}\alpha_j} \sum_{x=x_d} \Big[f(x) \, \mathrm{e}^{-\mathrm{i}\alpha_j x} \Big]_{x_d-}^{x_d+} - \frac{1}{\mathrm{i}\alpha_j} c_j',$$

where the summation runs over all points x_d of discontinuity of f (possibly including the endpoints) and c_j' is the jth Fourier coefficient of f'. As $c_j' \to 0$, the result follows. □

In a suitable context the inner product (f, f) may be interpreted as an *energy content* and thus $(f, f)_L$ as a *mean energy*. Therefore Theorem 3.6 is sometimes referred to as *Parseval's energy theorem*.

Often it is useful to rewrite the Fourier series in terms of trigonometric functions. Using the well-known relation

$$\mathrm{e}^{\mathrm{i}z} = \cos z + \mathrm{i} \sin z, \tag{3.5}$$

we thus find from (3.4) that

$$\tilde{f}(x) = a_0 + \sum_{j=1}^{\infty} a_j \cos(\alpha_j x) + b_j \sin(\alpha_j x), \tag{3.6a}$$

where

$$a_0 = c_0, \quad a_j = c_j + c_{-j}, \quad b_j = \mathrm{i}(c_j - c_{-j}). \tag{3.6b}$$

We can more directly write

$$
\begin{aligned}
a_0 &= \frac{1}{L}\int_0^L f(y)\,\mathrm{d}y, \qquad a_j = \frac{2}{L}\int_0^L f(y)\cos(\alpha_j y)\,\mathrm{d}y, \\
b_j &= \frac{2}{L}\int_0^L f(y)\sin(\alpha_j y)\,\mathrm{d}y, \qquad j = 1, 2, \dots .
\end{aligned}
\tag{3.7}
$$

Theorems 3.1 and 3.2 carry over to the trigonometric representations and for *Parseval's identity* (Theorem 3.6) we have

$$(f, f)_L = a_0^2 + \frac{1}{2}\sum_{j=1}^{\infty}(a_j^2 + b_j^2). \tag{3.8}$$

Example 3.8 Consider the function

$$f(x) = x, \qquad x \in (-\pi, \pi].$$

In order to make it periodic, we extend this function periodically beyond $(-\pi, \pi]$ to obtain the *sawtooth function*, as in Figure 3.1. It is then straightforward to see that $f \in L_2(-\pi, \pi)$. Since

$$a_j \propto \int_{-\pi}^{\pi} y\cos(jy)\,\mathrm{d}y = 0 \qquad \text{for all } j$$

and

$$b_j = \frac{1}{\pi}\int_{-\pi}^{\pi} y\sin(jy)\,\mathrm{d}y = \frac{1}{\pi}\left[\frac{-y\cos jy}{j} + \frac{\sin jy}{j^2}\right]_{-\pi}^{\pi} = \frac{2}{j}(-1)^{j+1},$$

we deduce, on account of Theorem 3.1 (we take $L = 2\pi$ and the interval $(-\pi, \pi]$ in (3.3)), that the resulting series

$$\tilde{f}(x) = 2\left[\sin x - \frac{\sin 2x}{2} + \frac{\sin 3x}{3} - \cdots\right]$$

converges to $f(x)$ for any $x \in (-\pi, \pi)$.

However, we cannot guarantee uniform convergence (see section C in the appendix) on the whole interval. At $x = \pi$ the series does not converge to $f(\pi) = \pi$ but rather to zero, the average of the left and right limits, as is shown graphically by Figure 3.2. We see an interesting phenomenon at the discontinuities of f: there is an overshoot to the left and an undershoot to the right. This is known as *Gibbs's phenomenon*. □

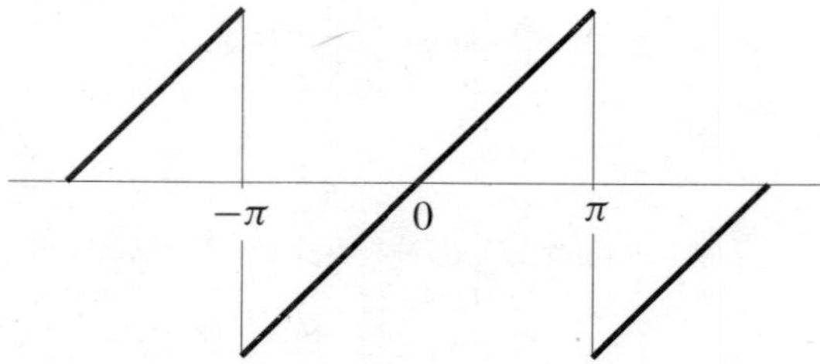

Figure 3.1. *Sawtooth function.*

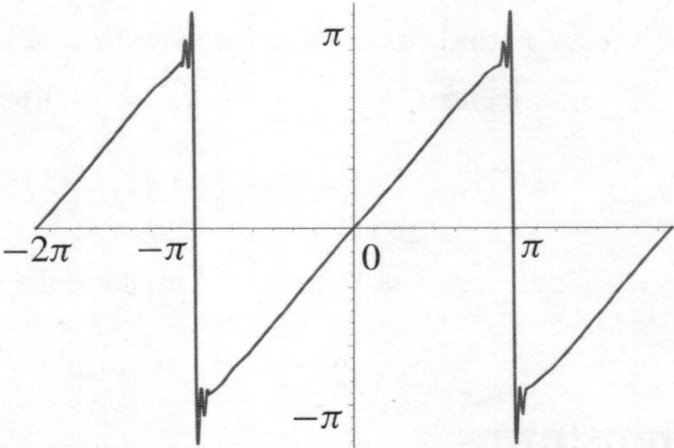

Figure 3.2. *Example of Gibbs's phenomenon (50 terms).*

The function in Example 3.8 was clearly odd and so were the Fourier terms. This is generally true. If f is odd (i.e., $f(-x) = -f(x)$), we have a *Fourier sine series*, and if f is even (i.e., $f(-x) = f(x)$), we have a *Fourier cosine series*.

Example 3.9 The following Fourier sine and cosine series define periodic functions with period one:

$$\sum_{n=1}^{\infty} \frac{\sin(2\pi nx)}{\pi n} = \left[\frac{1}{2} - x\right]_1, \qquad \sum_{n=1}^{\infty} \frac{\cos(2\pi nx)}{n} = -\log|2\sin\pi x|,$$

$$\sum_{n=1}^{\infty} \frac{\cos(2\pi nx)}{\pi^2 n^2} = \left[x^2 - x + \frac{1}{6}\right]_1, \qquad \sum_{n=1}^{\infty} \frac{\cos(2\pi nx)}{n^2 - \frac{1}{4}} = 2 - \pi|\sin\pi x|.$$

$[\cdot]_1$ denotes a function originally defined on $[0, 1]$ and continued periodically. Another interesting example is the *block-wave function*, defined along $[-1, 1]$ by

$$4\sum_{n=0}^{\infty} \frac{\sin(2n+1)\pi x}{(2n+1)\pi} = \mathrm{sign}(x)$$

and 2-periodically continued otherwise. □

In a straightforward way Fourier series are generalised to several dimensions. Suppose that the square-integrable function $f : \mathbb{R}^d \to \mathbb{R}$ is periodic in every argument x_j, $j = 1, \ldots, d$. Then

$$f(x_1, \ldots, x_d) := \sum_{m_1=-\infty}^{\infty} \cdots \sum_{m_d=-\infty}^{\infty} C_{m_1,\ldots,m_d}\, \mathrm{e}^{\mathrm{i}\alpha_{m_1}^{(1)} x_1 + \cdots + \mathrm{i}\alpha_{m_d}^{(d)} x_d}, \tag{3.9}$$

where

$$\alpha_{m_j}^{(j)} = \frac{2\pi}{L_j} m_j$$

and

$$C_{m_1,\ldots,m_d} = \frac{1}{L_1 \cdots L_d} \int_0^{L_1} \cdots \int_0^{L_d} f(y_1, \ldots, y_d)\, \mathrm{e}^{-\mathrm{i}\alpha_{m_1}^{(1)} y_1 - \cdots - \mathrm{i}\alpha_{m_d}^{(d)} y_d}\, \mathrm{d}y_1 \cdots \mathrm{d}y_d. \tag{3.10}$$

If f is scaled such that it is 2π-periodic in each independent variable, (3.9) may be written more compactly as

$$f(\boldsymbol{x}) := \sum_{\boldsymbol{m}=-\infty}^{\infty} C_{\boldsymbol{m}} \, \mathrm{e}^{\mathrm{i}\boldsymbol{m}\cdot\boldsymbol{x}}, \qquad C_{\boldsymbol{m}} = \frac{1}{(2\pi)^d} \int_0^{2\pi} f(\boldsymbol{y}) \, \mathrm{e}^{-\mathrm{i}\boldsymbol{m}\cdot\boldsymbol{y}} \, \mathrm{d}\boldsymbol{y}, \tag{3.11}$$

where $\boldsymbol{m} \in \mathbb{Z}^d$ denotes the index vector $\boldsymbol{m} = [m_1, \ldots, m_d]$ and the sum and integral signs are to be interpreted d-fold.

3.2 Fourier Transforms

There exists an integral analogue to the Fourier series. Recalling (3.2), we may let $L \to \infty$; i.e., $\frac{1}{L} \to 0$. Writing

$$\Delta\alpha := \frac{1}{L}, \tag{3.12}$$

we have for an L-periodic, square-integrable function f that

$$c_j = \Delta\alpha \int_0^{(\Delta\alpha)^{-1}} f(y) \, \mathrm{e}^{-2\pi \mathrm{i} j \Delta\alpha y} \, \mathrm{d}y, \tag{3.13}$$

whence

$$f(x) = \sum_{j=-\infty}^{\infty} \mathrm{e}^{2\pi \mathrm{i} j \Delta\alpha x} \, \Delta\alpha \int_0^{(\Delta\alpha)^{-1}} f(y) \, \mathrm{e}^{-2\pi \mathrm{i} j \Delta\alpha y} \, \mathrm{d}y. \tag{3.14}$$

Here we recognize a Riemann sum of the function $g(z)$, where

$$g(z) := \frac{1}{2\pi} \mathrm{e}^{\mathrm{i}zx} \int_0^{(\Delta\alpha)^{-1}} f(y) \, \mathrm{e}^{-\mathrm{i}zy} \, \mathrm{d}y. \tag{3.15}$$

One should just take piecewise constant approximations of g at the points $2\pi j \Delta\alpha$, $j = 0, \pm 1, \pm 2, \ldots$, multiplied by the interval width $2\pi\,\Delta\alpha$. Hence by a limit argument we find

$$f(x) = \frac{1}{2\pi} \int_{-\infty}^{\infty} \left[\int_{-\infty}^{\infty} f(y) \, \mathrm{e}^{-\mathrm{i}\alpha y} \, \mathrm{d}y \right] \mathrm{e}^{\mathrm{i}\alpha x} \, \mathrm{d}\alpha, \tag{3.16}$$

which leads to $\hat{f}$, the *Fourier transform* of f, which is, together with its inversion, given by

$$\hat{f}(\alpha) := \int_{-\infty}^{\infty} f(x) \, \mathrm{e}^{-\mathrm{i}\alpha x} \, \mathrm{d}x, \tag{3.17a}$$

$$f(x) = \frac{1}{2\pi} \int_{-\infty}^{\infty} \hat{f}(\alpha) \, \mathrm{e}^{\mathrm{i}\alpha x} \, \mathrm{d}\alpha. \tag{3.17b}$$

(Note that other, equivalent, definitions frequently occur, causing a lot of confusion.) The Fourier transform (also called the Fourier integral) plays an important role in analysing problems with a continuous spectrum of wave numbers (or frequencies). One can show that $|f|$ and $|\hat{f}|$ cannot vanish simultaneously outside a finite domain. Note that it is sufficient for the existence of $\hat{f}$ that $f \in L_1$, but if $f \in L_2$, then also $\hat{f} \in L_2$ [25] and therefore both (3.17a) and (3.17b) exist.

Example 3.10

(i) The "top hat" function, given by

$$f(x) = \begin{cases} 1, & -1 \le x \le 1, \\ 0, & |x| > 1, \end{cases}$$

has a Fourier transform that is closely related to the sinc-function:

$$\hat{f}(\alpha) = \int_{-1}^{1} \mathrm{e}^{-\mathrm{i}\alpha x}\, \mathrm{d}x = \frac{\mathrm{e}^{\mathrm{i}\alpha} - \mathrm{e}^{-\mathrm{i}\alpha}}{\mathrm{i}\alpha} = \frac{2\sin\alpha}{\alpha} = 2\,\mathrm{sinc}\left(\frac{\alpha}{\pi}\right).$$

(ii) The decaying exponential, vanishing for $x < 0$, defined by

$$f(x) = \begin{cases} \mathrm{e}^{-px}, & x > 0, \\ 0\ \ , & x < 0, \end{cases}$$

where $p > 0$, has a Fourier transform consisting of a single pole in the upper complex α plane:

$$\hat{f}(\alpha) = \int_{0}^{\infty} \mathrm{e}^{-px}\, \mathrm{e}^{-\mathrm{i}\alpha x}\, \mathrm{d}x = \left[-\frac{\mathrm{e}^{-(p+\mathrm{i}\alpha)x}}{p+\mathrm{i}\alpha}\right]_0^\infty = \frac{1}{p+\mathrm{i}\alpha}.$$

(iii) Another important example is $f(x) = p^{-1/2}\, \mathrm{e}^{-\frac{1}{2}(x/p)^2}$, where $p > 0$, with the similar

$$\hat{f}(\alpha) = \int_{-\infty}^{\infty} p^{-\frac{1}{2}}\, \mathrm{e}^{-\frac{1}{2}(x/p)^2}\, \mathrm{e}^{-\mathrm{i}\alpha x}\, \mathrm{d}x = (2\pi p)^{\frac{1}{2}}\, \mathrm{e}^{-\frac{1}{2}(p\alpha)^2}.$$

Note that where f depends on p, $\hat{f}$ depends on its reciprocal. □

If $\alpha^n \hat{f}(\alpha)$ is Fourier transformable, where $n \in \mathbb{N}$ and $\hat{f}(\alpha)$ is the Fourier transform of $f(x)$, then

$$\frac{\mathrm{d}^n}{\mathrm{d}x^n} f(x) = \frac{1}{2\pi} \int_{-\infty}^{\infty} (\mathrm{i}\alpha)^n \hat{f}(\alpha)\, \mathrm{e}^{\mathrm{i}\alpha x}\, \mathrm{d}\alpha. \tag{3.18}$$

Definition 3.11. *The convolution product $f * g$ of $f, g \in L_2$ is defined as*

$$(f * g)(x) = \int_{-\infty}^{\infty} f(x-y) g(y)\, \mathrm{d}y.$$

Theorem 3.12 (convolution theorem). *The Fourier-transformed convolution product of $f, g \in L_2$ is the product of their Fourier transforms:*

$$\widehat{f * g} = \hat{f}\hat{g}.$$

In other words, the inverse Fourier transform of $\hat{f}\hat{g}$ is equal to the convolution product:

$$\frac{1}{2\pi} \int_{-\infty}^{\infty} \hat{f}(\alpha)\hat{g}(\alpha)\, \mathrm{e}^{\mathrm{i}\alpha x}\, \mathrm{d}\alpha = \int_{-\infty}^{\infty} f(x-y) g(y) \mathrm{d}y.$$

Proof. As f and g are square integrable, we may change the order of integration to get

$$\frac{1}{2\pi}\int_{-\infty}^{\infty}\hat{f}(\alpha)\hat{g}(\alpha)\,\mathrm{e}^{\mathrm{i}\alpha x}\,\mathrm{d}\alpha = \frac{1}{2\pi}\int_{-\infty}^{\infty}\hat{f}(\alpha)\int_{-\infty}^{\infty} g(y)\,\mathrm{e}^{-\mathrm{i}\alpha y}\,\mathrm{d}y\,\mathrm{e}^{\mathrm{i}\alpha x}\,\mathrm{d}\alpha$$

$$= \frac{1}{2\pi}\int_{-\infty}^{\infty} g(y)\int_{-\infty}^{\infty}\hat{f}(\alpha)\,\mathrm{e}^{\mathrm{i}\alpha(x-y)}\,\mathrm{d}\alpha\,\mathrm{d}y = \int_{-\infty}^{\infty} f(x-y)g(y)\,\mathrm{d}y. \qquad \square$$

If we consider Theorem 3.12 for $x = 0$ and take $g(y) = \overline{f(-y)}$ with $\hat{g}(\alpha) = \overline{\hat{f}(\alpha)}$, we obtain the analogue of Parseval's identity (Theorem 3.6) for integrals:

$$\int_{-\infty}^{\infty}|f(x)|^2\,\mathrm{d}x = \frac{1}{2\pi}\int_{-\infty}^{\infty}|\hat{f}(\alpha)|^2\,\mathrm{d}\alpha. \tag{3.19}$$

This is sometimes referred to as the *energy theorem* for the continuous case.

Example 3.13 Consider Example 3.10(ii) again. We obtain indeed

$$\int_{-\infty}^{\infty}|f(x)|^2\,\mathrm{d}x = \int_{0}^{\infty}\mathrm{e}^{-2px}\,\mathrm{d}x = \frac{1}{2p},$$

$$\frac{1}{2\pi}\int_{-\infty}^{\infty}|\hat{f}(\alpha)|^2\,\mathrm{d}\alpha = \frac{1}{2\pi}\int_{-\infty}^{\infty}\frac{1}{|p+\mathrm{i}\alpha|^2}\,\mathrm{d}\alpha = \frac{1}{2\pi}\int_{-\infty}^{\infty}\frac{1}{p^2+\alpha^2}\,\mathrm{d}\alpha = \frac{1}{2p}. \qquad \square$$

Let f be a Fourier-transformable function on $\mathbb{R}$. Noting that the function

$$\sum_{m=-\infty}^{\infty} f(mL+x) \tag{3.20}$$

is periodic in x with period L, we can write (following [20])

$$\sum_{m=-\infty}^{\infty} f(mL+x) = \sum_{j=-\infty}^{\infty}\left[\frac{1}{L}\int_{0}^{L}\sum_{m=-\infty}^{\infty} f(mL+y)\,\mathrm{e}^{-\mathrm{i}\alpha_j y}\,\mathrm{d}y\right]\mathrm{e}^{\mathrm{i}\alpha_j x}$$

$$= \frac{1}{L}\sum_{j=-\infty}^{\infty}\left[\int_{-\infty}^{\infty} f(\eta)\,\mathrm{e}^{-\mathrm{i}\alpha_j\eta}\,\mathrm{d}\eta\right]\mathrm{e}^{\mathrm{i}\alpha_j x} = \frac{1}{L}\sum_{j=-\infty}^{\infty}\hat{f}(\alpha_j)\,\mathrm{e}^{\mathrm{i}\alpha_j x}$$

(with α_j as defined in (3.2b)), which leads, for $x = 0$, to Poisson's formula

$$\sum_{m=-\infty}^{\infty} f(mL) = \frac{1}{L}\sum_{m=-\infty}^{\infty}\hat{f}\left(\frac{2\pi}{L}m\right), \tag{3.21}$$

where the average of the left and right limit is to be taken at any discontinuities.

Example 3.14 Poisson's formula is an excellent tool for accelerating slowly converging series. From Example 3.10(ii) we find, for $p > 0$ and $L = 2\pi$, that

$$\frac{1}{2\pi p} + \frac{p}{\pi}\sum_{m=1}^{\infty}\frac{1}{m^2+p^2} = \frac{1}{2} + \sum_{m=1}^{\infty}\mathrm{e}^{-2\pi pm}.$$

The left-hand side converges algebraically slowly, in contrast to the fast exponential convergence of the right-hand side. As a bonus, we have in this case an explicit expression if we recognize the geometric series with common ratio $\mathrm{e}^{-2\pi p}$. $\square$

We are interested in knowing when a given $\hat{f}(\omega)$ is a time-Fourier transform of a physical signal $f(t)$. First, in order for f to be real, $\hat{f}$ has to satisfy the *reality condition*

$$\hat{f}(\omega) = \overline{\hat{f}(-\omega)}. \tag{3.22}$$

No physical process can exist for all time. A process $f(t)$ that starts by some cause at some finite time $t = t_0$ and vanishes before t_0 is called *causal*. The corresponding Fourier transform

$$\hat{f}(\omega) = \int_{t_0}^{\infty} f(t)\, \mathrm{e}^{-\mathrm{i}\omega t}\, \mathrm{d}t$$

has the property that $\hat{f}(\omega)$ is analytic in the lower complex half-plane $\mathrm{Im}(\omega) < 0$. So this is a necessary condition on $\hat{f}$ for f to be causal. A sufficient condition is the following *causality condition* [112].

Theorem 3.15 (causality condition). *If $\hat{f}(\omega)$ is analytic in* $\mathrm{Im}(\omega) \le 0$, *$|\hat{f}(\omega)|^2$ is integrable along the real axis, and there is a real t_0 such that $\hat{f}(\omega)\,\mathrm{e}^{\mathrm{i}\omega t_0} \to 0$ uniformly with regard to* $\arg(\omega)$ *for $|\omega| \to \infty$ in the lower complex half-plane, then $f(t)$ is causal and vanishes for $t < t_0$.*

Proof. It is no restriction for the proof to assume that $t_0 = 0$. Consider, for $t < 0$, the integral $\int \hat{f}(\omega)\,\mathrm{e}^{\mathrm{i}\omega t}\,\mathrm{d}\omega$ along the real contour $[-R, R]$ to be closed via a semicircle of radius R in the lower complex half-plane. As the integrand is analytic, the integral is zero. Let $R \to \infty$. The contribution I_R of the integral along the semicircle tends to zero because

$$|I_R| \le \int_0^{\pi} |\hat{f}(\omega)|\, \mathrm{e}^{-|t|R\sin\theta}\, R\,\mathrm{d}\theta \le 2R \max_{\theta} |\hat{f}(\omega)| \int_0^{\frac{1}{2}\pi} \mathrm{e}^{-|t|R2\theta/\pi}\,\mathrm{d}\theta \to 0,$$

where $\omega = R\,\mathrm{e}^{-\mathrm{i}\theta}$. So the contribution from the real axis, being equal to $2\pi f(t)$, is also zero. □

Note that the *lower* complex half-space becomes the *upper* half-space if the opposite Fourier sign convention is taken.

Example 3.16 The Fourier transform $\hat{f}(\omega) = (p + \mathrm{i}\omega)^{-1}$ is causal if $p > 0$, as may be confirmed by the inverse transform

$$f(t) = \frac{1}{2\pi} \int_{-\infty}^{\infty} \frac{\mathrm{e}^{\mathrm{i}\omega t}}{p + \mathrm{i}\omega}\,\mathrm{d}\omega = \begin{cases} \mathrm{e}^{-pt} & \text{if } t > 0, \\ 0 & \text{if } t < 0. \end{cases}$$

In the limit of no damping ($p \downarrow 0$) the singularity at $\omega = \mathrm{i}p$ moves to $\omega = 0$, which is on the real axis. The integral is to be interpreted via a suitable indentation of the contour *under* the pole in order to retain causality. □

In a straightforward way Fourier transforms can be generalised to several dimensions. For the square-integrable function $f\colon \mathbb{R}^d \to \mathbb{R}$ we have the couple

$$\hat{f}(\boldsymbol{\alpha}) := \int_{-\infty}^{\infty} f(\boldsymbol{x})\,\mathrm{e}^{-\mathrm{i}\boldsymbol{\alpha}\cdot\boldsymbol{x}}\,\mathrm{d}\boldsymbol{x}, \tag{3.23a}$$

$$f(\boldsymbol{x}) = \frac{1}{(2\pi)^d}\int_{-\infty}^{\infty} \hat{f}(\boldsymbol{\alpha})\,\mathrm{e}^{\mathrm{i}\boldsymbol{\alpha}\cdot\boldsymbol{x}}\,\mathrm{d}\boldsymbol{\alpha}, \tag{3.23b}$$

where $\boldsymbol{\alpha} \in \mathbb{R}^d$ denotes the Fourier wave number vector and the integral signs are to be interpreted d-fold.

3.3 Discrete Fourier Transforms

Often a function is known at a certain finite number of points only. For instance, f may be measured at a few values of the argument ("sampled"). If one uses a numerical approximation method based on discretising of the argument, as in the case of finite differences, there is also a natural set of points on which f is "monitored" only. Given the *grid points* x_k, $k = 0, \ldots, N-1$, we can give straightforward analogues of the continuous case by considering the *discrete inner product*

$$\langle f, g\rangle := \sum_{k=0}^{N-1} f(x_k)\,\overline{g(x_k)}, \tag{3.24}$$

where $x_k \in [0, L]$. Because of the special properties exhibited by $\mathrm{e}^{\mathrm{i}\alpha x}$ when the points x_k are chosen *equispaced*, i.e., $x_{k+1} - x_k$ is constant ($k = 0, \ldots, N-1$ and $x_N = x_0 + L$), we shall take them as

$$x_k := \frac{k}{N}L, \qquad k = 0, \ldots, N-1. \tag{3.25}$$

It can be verified that the functions

$$p_j(x) := \mathrm{e}^{\mathrm{i}\alpha_j x} \tag{3.26}$$

with

$$\alpha_j = \frac{2\pi j}{L}$$

are orthogonal. This is stated in the following theorem.

Theorem 3.17. *The polynomials p_j defined in* (3.26) *are orthogonal with respect to* (3.24) *in the sense that $\langle p_j, p_l\rangle = 0$ for all j, l, with $j - l$ not a multiple of N.*

Proof. Substituting (3.26) in (3.24), we have

$$\langle p_j, p_l\rangle = \sum_{k=0}^{N-1} \exp\left[2\pi\mathrm{i}(j-l)\frac{k}{N}\right] = \sum_{k=0}^{N-1}\sigma^k = \begin{cases} \dfrac{\sigma^N - 1}{\sigma - 1} & \text{if } \sigma \neq 1,\\ N & \text{if } \sigma = 1,\end{cases}$$

with $\sigma := \exp\left(2\pi\mathrm{i}\frac{j-l}{N}\right)$. Apparently, $\sigma^N = 1$, while $\sigma \neq 1$ if $(j-l) \bmod N \neq 0$, which proves the result. □

Corollary 3.18. $\langle p_j, p_j \rangle = N$.

We therefore may as well consider the inner product

$$\langle f, g \rangle_N := \frac{1}{N} \sum_{k=0}^{N-1} f(x_k)\, \overline{g(x_k)}, \tag{3.24$'$}$$

which makes $\{p_j(x)\}_{j=0}^{N-1}$ an orthonormal set of basis functions. We can now give a *discrete Fourier transform* (DFT) of a function f:

$$c_j := \langle f, p_j \rangle_N = \frac{1}{N} \sum_{k=0}^{N-1} f(x_k)\, \mathrm{e}^{-2\pi \mathrm{i} jk/N}. \tag{3.27}$$

Due to the special choice of the grid, we immediately find that $\exp(\mathrm{i}\alpha_j x_k) = 1$ when j is a multiple of N. Hence for $j \geq N$ we have

$$\mathrm{e}^{\mathrm{i}\alpha_j x_k} = \mathrm{e}^{2\pi \mathrm{i} \hat{j} k/N}, \qquad \hat{j} = j \bmod N. \tag{3.28}$$

This relation tells us that the basis functions $p_j(x)$, for $j \geq N$, will not provide additional information to represent (an approximation of) f. The phenomenon in which we cannot distinguish discrete Fourier components of p_j and p_{j+lN}, $l \in \mathbb{Z}$, is called *aliasing*. Hence we will be satisfied to have the finite series

$$\hat{f}(x) := \sum_{j=0}^{N-1} c_j\, \mathrm{e}^{2\pi \mathrm{i} jx}, \qquad x \in [0, L]. \tag{3.29}$$

Of particular interest is $\hat{f}(x)$ at the points $x = x_k$, because here the original values of f are exactly recovered; i.e., $f(x_k) = \hat{f}(x_k)$. So at these points f is completely defined by the coefficients c_j and vice versa. If we define

$$\hat{c}_k := \frac{1}{\sqrt{N}} f(x_k), \tag{3.30}$$

the following reciprocity relation between c_j and $\hat{c}_k$ may be shown.

Property 3.19 (reciprocity of DFT).

$$\hat{c}_k = \frac{1}{\sqrt{N}} \sum_{j=0}^{N-1} c_j\, \mathrm{e}^{2\pi \mathrm{i} jk/N}, \quad c_j = \frac{1}{\sqrt{N}} \sum_{k=0}^{N-1} \hat{c}_k\, \mathrm{e}^{-2\pi \mathrm{i} jk/N}.$$

We remark that the DFT has very important applications, in particular through its efficient implementation the *fast Fourier transform* (FFT; see [149]).

From an approximation point of view, aliasing implies that we cannot obtain more information about a function f than the *sampling rate* (the density of the grid) allows us; see Example 3.20. In particular, we thus conclude that the discrete wave numbers generated by the grid will limit the accuracy of the approximation $\hat{f}(x) - f(x)$.

Example 3.20 In Figure 3.3 we have drawn two sine functions, one with wave number 1, sampled with a rate of $20/2\pi$, and one with wave number 21. As $\sin x$ coincides with $\sin(21x)$ right at the sampling points, $\sin(21x)$ cannot be represented with this sampling rate. □

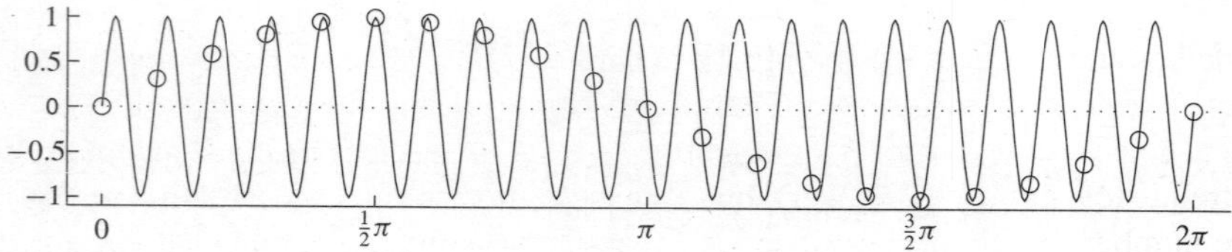

Figure 3.3. *Aliasing.*

3.4 Fourier Analysis Applied to PDEs

Consider now the linear PDE with constant coefficients

$$\frac{\partial u}{\partial t} = a\frac{\partial^2 u}{\partial x^2} + b\frac{\partial u}{\partial x} + cu. \tag{3.31}$$

We may look for solutions of this equation by using a Fourier series (on a finite domain) or Fourier integrals (on an infinite domain). Of course this does not make much sense in general if we do not specify the initial and boundary conditions. Yet one may hope that an analysis based on Fourier expansions—if successful—might give insight into the general case. If for instance such an analysis leads to prediction of instabilities or other undesirable phenomena, it may imply such a result for a more general case.

Since we have at least two variables, space and time, our ansatz will be

$$u(x,t) = \int_{-\infty}^{\infty}\int_{-\infty}^{\infty} \hat{u}(\kappa,\omega)\,\mathrm{e}^{\mathrm{i}\kappa x - \mathrm{i}\omega t}\,\mathrm{d}\kappa\,\mathrm{d}\omega, \tag{3.32}$$

where ω denotes frequency and κ is the wave number. For the natural definition of phase and group velocity below, these Fourier variables are defined with opposite signs; we could have used $e^{i\kappa x - i\omega t}$ equally well in (3.32). This expression is a very general one that is valid for any function $u(x,t)$. We are not dealing with any u, but with a solution of (3.31), and usually a simpler form is possible, e.g., if we require each Fourier mode to be a solution of the defining equation (3.31). In order to analyse the solution, we consider the *single mode*

$$u(x,t) = A\,\mathrm{e}^{\mathrm{i}\kappa x - \mathrm{i}\omega t}. \tag{3.33}$$

(Such a planar wave solution is also useful for analyzing nonlinear evolution equations; see [169].) Upon substituting (3.33) into (3.31), we obtain that the mode is a solution if

$$-\mathrm{i}\omega = -a\kappa^2 + \mathrm{i}b\kappa + c. \tag{3.34}$$

Since this relation gives information about the propagation properties of the various modes, it is often referred to as the *dispersion relation* (dispersion is explained below).

If we follow a modal wave crest, i.e., such that the phase $\kappa x - \omega t =$ constant, we move with the *phase velocity* or *wave speed*

$$v_{\mathrm{p}} := \frac{\omega}{\kappa}. \tag{3.35}$$

As a mode has an infinite extension in x, it is hard to tell by which velocity any associated properties, like energy, propagate. Therefore we consider a "localised mode," or wave packet, that decays slowly to zero for large $|x|$. This is not exactly one mode any more (with a single frequency and wave number), but a superposition of modes near a main frequency and wave number. To be more precise let $f(x)$ be an absolute integrable smooth function and ε (ε positive) be small compared to κ_0 such that $f(\varepsilon x)$ is the slowly varying envelope of the wave at $t = 0$:

$$u(x, 0) = \mathrm{e}^{\mathrm{i}\kappa_0 x} f(\varepsilon x). \tag{3.36a}$$

The Fourier transform of (3.36a) is given by

$$\int_{-\infty}^{\infty} \mathrm{e}^{-\mathrm{i}(\kappa-\kappa_0)x} f(\varepsilon x)\mathrm{d}x = \varepsilon^{-1} \hat{f}\left(\frac{\kappa - \kappa_0}{\varepsilon}\right). \tag{3.36b}$$

Let the modes be defined by a dispersion relation $\omega = \omega(\kappa)$. Hence we can write the wave packet as

$$\begin{aligned} u(x, t) :&= \frac{1}{2\pi} \int_{-\infty}^{\infty} \varepsilon^{-1} \hat{f}\left(\frac{\kappa - \kappa_0}{\varepsilon}\right) \mathrm{e}^{\mathrm{i}\kappa x - \mathrm{i}\omega(\kappa)t}\, \mathrm{d}\kappa \\ &= \frac{1}{2\pi} \int_{-\infty}^{\infty} \hat{f}(\eta)\, \mathrm{e}^{\mathrm{i}\kappa_0 x + \mathrm{i}\varepsilon\eta x - \mathrm{i}\omega(\kappa_0+\varepsilon\eta)t}\, \mathrm{d}\eta. \end{aligned}$$

By Taylor expanding $\omega(\kappa_0 + \varepsilon\eta) = \omega_0 + \varepsilon\eta\omega_0' + \mathcal{O}(\varepsilon^2)$ for small ε—where $\omega_0 = \omega(\kappa_0)$ and $\omega_0' = \omega'(\kappa_0)$—and rearranging terms, we find in good approximation that

$$\begin{aligned} u(x, t) &= \frac{1}{2\pi} \mathrm{e}^{\mathrm{i}\kappa_0 x - \mathrm{i}\omega_0 t} \int_{-\infty}^{\infty} \hat{f}(\eta)\, \mathrm{e}^{\mathrm{i}\varepsilon\eta[x - \omega_0' t]}\, \mathrm{d}\eta \\ &= \mathrm{e}^{\mathrm{i}\kappa_0 x - \mathrm{i}\omega_0 t} f\left(\varepsilon[x - \omega_0' t]\right). \end{aligned} \tag{3.37}$$

We see that the wave crests indeed propagate with the phase velocity ω_0/κ_0, whereas the group as a whole propagates with a velocity $\omega'(\kappa_0)$. This velocity is called the *group* or *energy* or *signal velocity* and is thus given by

$$v_{\mathrm{g}} := \frac{\mathrm{d}\omega}{\mathrm{d}\kappa}. \tag{3.38}$$

In general, all modes propagate with their own speed, and a group superposed of many different modes gets dispersed. The shape of the group remains intact, i.e., shows no dispersion, if all modes propagate with the same speed, i.e., if

$$\frac{\mathrm{d}v_{\mathrm{g}}}{\mathrm{d}\kappa} = \frac{\mathrm{d}^2\omega}{\mathrm{d}\kappa^2} = 0. \tag{3.39}$$

Note that a mode like $e^{-ict}\,e^{i\kappa(x-bt)}$ is clearly not dispersive, so condition (3.39) is not exactly equivalent to a condition of constant phase speed.

We return to (3.31) and consider two special cases. First, let b and c be zero; i.e., we have the standard heat equation. We then find upon substituting $-i\omega = -a\kappa^2$ that

$$u(x,t) = e^{i\kappa x}\, e^{-a\kappa^2 t}\,. \tag{3.40}$$

The first factor is just the spatial Fourier component, but the second is an exponentially growing/decaying quantity. Hence we see that we need $a > 0$ to have a stable mode, and we may call solution (3.40) *dissipative*.

The other special case is when $a = c = 0$. Then we have

$$\omega = -b\kappa. \tag{3.41}$$

This means that each mode is propagating with the same wave speed $-b$. In particular, we find

$$u(x,t) = e^{i(x+bt)\kappa}\,. \tag{3.42}$$

Hence on a line in the (x,t) plane where $x + bt$ is constant, i.e., on a *characteristic* (cf. Chapter 2), we note that u is constant. We may therefore call solutions like u in (3.41) *conservative*.

In the context of numerical methods for PDEs based on finite differences, analysing error propagation leads to typically linearized equations with slowly varying coefficients. Take, e.g., the "true" solution y satisfying

$$y_t = (A(y)y_x)_x + C(y), \tag{3.43}$$

the numerical solution $\tilde{y}$, and the error $u := y - \tilde{y}$. The error is by assumption small. Its typical fluctuations, due to the small grid and time steps, vary over a much shorter time and length scale than the "true" solution. As we are only interested in the error, i.e., the behaviour along the short time and length scales, the linearized equation for u in the neighbourhood of $x = x_0$ and $t = t_0$ may look like

$$\frac{\partial u}{\partial t} = a\frac{\partial^2 u}{\partial x^2} + b\frac{\partial u}{\partial x} + cu, \tag{3.44}$$

where a, b, and c are assumed constant, depending on $y(x_0, t_0)$. (Note that this equation is just an example and has no other purpose than to illustrate.) As before, the behaviour of u in x and t may be analysed by Fourier analysis. Assume for u a single mode of the form (3.33), i.e., with constant amplitude and phase. Substitution in (3.44) yields dispersion relation (3.34).

Fourier analysis in difference equations will turn out to be a powerful tool to determine necessary conditions for a numerical method to be practically useful. We shall defer details to the specific chapters where we assess the numerical methods. Refinements based on the method of multiple scales (Chapter 15, Section 15.4.2) allow us to include the variation of amplitude and phase with the (relatively) slowly varying x_0, t_0.

3.5 Discussion

- Fourier theory is an essential tool in many applications that are far beyond the goals of this book. Traditionally, the Fourier series have been used to approximate functions. There is a host of other choices for this, depending on the application; see, e.g., books on special functions, like [107]. Quite another application is the efficient solution of systems through the FFT; see [118].

- Dispersion is a very important concept for the theory of waves. As it will turn out in Chapters 12, 13, and 14 in both the analytical and the numerical study of hyperbolic problems, the actual behaviour of the solution critically depends on properties like dissipation and dispersion. In fact, one of the major problems in numerically solving hyperbolic problems is to capture the physical behaviour, i.e., not to introduce too much "numerical" dissipation or dispersion. Finally, we remark that there are many equations with wave-like solutions that are not of hyperbolic type. These are called *dispersive* waves [169].

Exercises

3.1. Let L be some positive number. Show that the functions $\frac{1}{2}\sqrt{2}\,s$, $s\cos(j\pi s^2 x)$, and $s\sin(j\pi s^2 x)$, with $s := L^{-\frac{1}{2}}$, $j = 1, 2, \ldots$, form an orthonormal basis on $(-L, L)$.

3.2. Show that an even function on $(-L, L)$ is orthogonal to an odd function.

3.3. Prove the reciprocity relation for DFT (Property 3.19).

3.4. (a) Find the Fourier series of $\cos^2 x$.

(b) Find the Fourier series of x^2 defined on $[0, 1]$.

(c) Consider the function $f(x) = x(x-1)$ defined on $[0, 1]$. Let $\sum_{j=1}^{\infty} \gamma_j \sin(j\pi x)$ be the Fourier sine series of f. Show that $\gamma_j = 0$ for j even and $\gamma_j = -8/(j\pi)^3$ for j odd.

3.5. Given the Fourier coefficients of $f(x)$, determine the Fourier coefficients of the first and second derivatives.

3.6. If $f(x)$ is a periodic L_2-function with a jump condition at a say, then the Fourier series at that point converges to $\frac{1}{2}(f(a+) + f(a-))$, where $a+$ denotes the right limit and $a-$ the left limit. Show this for $f(x) = x$ on $(-\pi, \pi)$.

3.7. Find the Fourier series of the following function, defined on $[0, 1]$:

$$f(x) = \frac{2}{3}x^3 - x^2 + \frac{1}{3}x.$$

3.8. Show that the functions given by

$$f_j(x) = \sin\left(\left(j + \frac{1}{2}\right)\pi x\right), \qquad j \in \mathbb{Z},$$

are orthonormal on $[-1, 1]$.

3.9. Show by utilizing Poisson's formula that

$$\sum_{m=1}^{\infty} \frac{\sin m}{m} = \frac{\pi - 1}{2}.$$

3.10. Determine

$$\sum_{j=0}^{\infty} \frac{(-1)^j}{2j+1}$$

from the Fourier series of $\frac{1}{2} - x$ on $[0, 1]$.

3.11. (a) Determine the dispersion relation for the *beam equation*

$$\frac{\partial^2 u}{\partial t^2} + c^2 \frac{\partial^4 u}{\partial x^4} = 0.$$

(b) Repeat part (a) for the *Korteweg–de Vries equation*

$$\frac{\partial u}{\partial t} + c\frac{\partial u}{\partial x} + d\frac{\partial^3 u}{\partial x^3} = 0.$$

(c) Repeat part (a) for the *Boussinesq equation*

$$\frac{\partial^2 u}{\partial t^2} - c^2 \frac{\partial^2 u}{\partial x^2} = b^2 \frac{\partial^4 u}{\partial x^2 \partial t^2}.$$

3.12. The dispersion relation for water waves is given by

$$\omega^2 = \left(1 + \frac{T}{\rho g}\kappa^2\right) g\kappa \tanh(\kappa h),$$

where ω is the frequency, κ is the wave number, h is the undisturbed water height, $g = 9.8$ m/s^2 is the gravitational acceleration, $\rho = 1000$ kg/m^3 is the water density, and $T = 0.074$ N/m is the surface tension. Waves controlled by surface tension ($T\kappa^2/\rho g$ is not small) are called ripples. Waves controlled by just gravity are called gravity waves.

(a) Verify that for deep water the phase velocity of gravity waves is twice the group velocity, so the waves are dispersive.

(b) Verify that for long waves ($\kappa \to 0$) the group and phase velocity become the same (and equal to $\sqrt{gh}$), so long waves propagate without dispersion.

(c) A practical parameter to maximize the wave number range of dispersionless waves is the water depth h. A device that uses these waves to physically model sound waves (which are dispersionless) is called a ripple tank [90].
Consider $V(z, \beta) = v_g/\sqrt{gh}$ as a function of $z = \kappa h$ and $\beta = T/\rho g h^2$. Our aim is to select β such that V remains close to one for a considerable interval $0 < z < z_0$. Verify that $V(0, \beta) = 1$ and $V_z(0, \beta) = V_{zzz}(0, \beta) = 0$. For what value of β is $V_{zz}(0, \beta) = 0$? This value produces practically dispersionless waves for z between 0 and 0.5, i.e., any wavelength larger than $4\pi h$.

By decreasing β slightly, the range in z may be increased with an acceptable deviation of V. In terms of h, suitable values are found for 5 mm to 8 mm at a wave speed of 22 cm/s to 28 cm/s.

3.13. Show that for higher-dimensional waves the group velocity, implied by the dispersion relation $\omega = \omega(\boldsymbol{\kappa})$, is given by

$$\boldsymbol{v}_{\mathrm{g}} = \frac{\partial \omega}{\partial \boldsymbol{\kappa}}.$$

$\frac{\partial}{\partial \boldsymbol{\kappa}}$ denotes the gradient with respect to $\boldsymbol{\kappa}$.

Chapter 4

Distributions and Fundamental Solutions

This chapter is devoted to rather fundamental concepts. In Section 4.1 we first sketch the idea behind a *fundamental solution*. For a number of properties and phenomena of PDEs the concept of *distribution* is needed. Important functions such as Dirac delta functions and Heaviside functions are an instance of this. In Section 4.2 we first consider distributions in one dimension and define what we mean by convergence in a distributional sense. The extension to higher dimensions, which is rather straightforward, is treated in Section 4.3. Distributions play a crucial role in problems that do not possess solutions "in a classical sense," i.e., that are nonsmooth. This then leads to a notion of solutions in a so-called weak form, which are solutions in distributional sense. They are discussed in Section 4.4. Another use of distributions is describing particular solutions of linear PDEs. These are called fundamental solutions; see Section 4.5. These fundamental solutions are defined on the whole $\mathbb{R}^d$. A special form of such a solution in the distributional sense is the Green's function, which moreover satisfies the homogeneous boundary condition. In fact the latter leads to expressions of the solution in terms of the source term of the equation. A more classical approach is to use a Duhamel integral, giving an expression for the solution of a PDE by superposing elementary solutions that represent the source term. These Duhamel integrals are discussed in Section 4.6. In fact, it turns out that there is a natural relationship between these two forms.

4.1 Introduction

Consider the Cauchy problem for the ODE

$$\frac{\mathrm{d}u}{\mathrm{d}t} = \lambda u + f(t), \qquad t > 0, \tag{4.1a}$$

$$u(0) = u_0. \tag{4.1b}$$

If $f(t) \equiv 0$, the solution, v say, is simply given by

$$v(t) = \mathrm{e}^{\lambda t}\, u_0. \tag{4.2}$$

In order to find the general solution of (4.1), one can use the *variation of constant* method, i.e., substitute

$$u(t) = c(t)v(t)$$

and determine $c(t)$ from (4.1a). The well-known result is

$$u(t) = \mathrm{e}^{\lambda t}\, u_0 + \int_0^t v(t)(v(\tau))^{-1} f(\tau)\,\mathrm{d}\tau. \tag{4.3}$$

Here $w(t;\tau)$, defined as

$$w(t;\tau) := v(t)(v(\tau))^{-1}, \qquad t \geq \tau, \tag{4.4}$$

is sometimes called a *fundamental solution.* For τ fixed it satisfies the homogeneous part of (4.1a). One can view the integral in (4.3) as a superposition of initial value solutions propagating values $f(\tau)\,\mathrm{d}\tau$ until t. One can also describe this as follows. Let

$$\xi := t - \tau, \tag{4.5a}$$

$$\overline{w}(\xi) := w(t;\tau). \tag{4.5b}$$

Clearly, $\overline{w}(\xi) = \mathrm{e}^{\lambda\xi}$ for $\xi \geq 0$. From (4.3) we conclude that it makes sense to define $\overline{w}(\xi) = 0$ for $\xi < 0$. But beyond the point $\xi = 0$, $\overline{w}(\xi)$ satisfies the ODE

$$\frac{\mathrm{d}\overline{w}}{\mathrm{d}\xi} = \lambda\overline{w}, \qquad \xi \neq 0, \tag{4.6}$$

with solution proportional to $\mathrm{e}^{\lambda\xi}$, so at the point $\xi = 0$ a jump occurs. This may be described by adding a (for the moment hypothetical) source term $\delta(\xi)$ to the right-hand side of (4.6) in such a way that

$$\overline{w}(\xi) = \int_{-\infty}^{\xi} \mathrm{e}^{\lambda(\xi - s)}\,\delta(s)\,\mathrm{d}s = \mathrm{e}^{\lambda\xi}, \qquad \xi > 0. \tag{4.7}$$

This is a nontrivial matter (even more: it is a condition impossible for any normal function to satisfy!), but in loose terms we apparently require $\delta(\xi)$ to be a spike-like function that integrates to zero along any interval except for an infinitesimally small interval around $\xi = 0$ where its integral contribution adds up to one; i.e.,

$$\int_{-\infty}^{\infty} \delta(\xi)\,\mathrm{d}\xi = 1. \tag{4.8}$$

This function δ is sometimes called the *(Dirac) delta function.* In the next section we shall give a more rigorous definition. Summarizing, we may view the fundamental solution $w(t;\tau)$ to satisfy the Cauchy problem

$$\frac{\mathrm{d}}{\mathrm{d}t} w(t;\tau) = \lambda w(t;\tau) + \delta(t - \tau), \quad t > \tau, \tag{4.9a}$$

$$w(\tau;\tau) = 0. \tag{4.9b}$$

This notion can simply be extended to vector-valued ODEs and to boundary value problems as well. We finally remark that for Cauchy problems the value $\tau = 0$ needs special consideration; here one just has a solution of the homogeneous problem satisfying $w(0,0) = 1$.

4.2 Distributions in One Variable

In Section 4.1 we saw that we needed a notion of a function of which some properties only made sense after integration. It took mathematicians awhile before they had the proper formulation for the delta function. It shows up as a special instance of a distribution, to be discussed below.

Let $\mathcal{D}$ be some class of functions $\mathbb{R} \to \mathbb{R}$, to be called *test functions*, and consider mappings, or functionals, from $\mathcal{D}$ to $\mathbb{R}$. A simple but important class of examples is generated by the "inner" product of a test function $\varphi \in \mathcal{D}$ and a given integrable real function $f : \mathbb{R} \to \mathbb{R}$; i.e.,

$$(f, \varphi) := \int_{-\infty}^{\infty} f(x)\varphi(x)\,\mathrm{d}x. \tag{4.10}$$

If the functional can be written in this way, it is identified with the function f and we call it "the functional f." We see that this functional is even *linear*; i.e.,

$$(f, \alpha\varphi_1 + \beta\varphi_2) = \alpha\,(f, \varphi_1) + \beta\,(f, \varphi_2). \tag{4.11}$$

A suitable test space can be found as follows. First we define the *support* of a function $\varphi : \mathbb{R} \to \mathbb{R}$ as the closure of the set of all points x such that $\varphi(x) \neq 0$:

$$\mathrm{supp}(\varphi) = \overline{\{x \in \mathbb{R} \mid \varphi(x) \neq 0\}}.$$

Example 4.1 The infinitely many times differentiable real function φ, defined by

$$\varphi(x) = \begin{cases} \exp\left[(x^2-1)^{-1}\right] & \text{for } |x| < 1, \\ 0 & \text{for } |x| \geq 1, \end{cases}$$

clearly has $\mathrm{supp}(\varphi) = [-1, 1]$. □

If $\mathrm{supp}(\varphi)$ is a bounded set, then φ is said to have a *compact support*. Now define the test function space

$$\mathcal{D} = C_0^\infty(\mathbb{R}) := \{\varphi \in C^\infty(\mathbb{R}) \mid \varphi \text{ has compact support}\}. \tag{4.12}$$

This space of test functions $\mathcal{D}$ will be used throughout unless indicated otherwise.

In order to have a practically meaningful linear functional, we would like it to be continuous. For this we need a convergence concept. In view of the compact support property the following makes sense.

Definition 4.2. *A sequence of test functions $\{\varphi_i\}_{i\geq 0}$, where $\varphi_i \in C_0^\infty(\mathbb{R})$, is called convergent to zero if*

(i) *there is a closed and finite interval I such that* $\mathrm{supp}(\varphi_i) \subset I$;

(ii) $\lim_{i\to\infty} \frac{\mathrm{d}^k}{\mathrm{d}x^k}\varphi_i(x) = 0$ *uniformly on $\mathbb{R}$ for $k = 0, 1, 2, \ldots$.*

From now on we shall identify $\mathcal{D}$ with $C_0^\infty(\mathbb{R})$, equipped with the convergence notion as defined in Definition 4.2.

Definition 4.3. *A linear functional f is continuous if for any sequence $\{\varphi_i\}_{i\geq 0} \subset \mathcal{D}$ that is convergent to zero, also $(f, \varphi_i) \to 0$.*

(Because of linearity it suffices to consider continuity at zero.) The space of continuous linear functionals on $\mathcal{D}$ is the dual space $\mathcal{D}'$. A *continuous linear functional* is called a *distribution*. Actually, we see that $\mathcal{D}'$ is a linear space by writing, for $f, g \in \mathcal{D}'$, that

$$(\alpha f + \beta g, \varphi) = \alpha(f, \varphi) + \beta(g, \varphi) \qquad \text{for all} \quad \varphi \in \mathcal{D}. \tag{4.13}$$

If f is locally integrable (i.e., the integral exists on any finite interval), then by (4.10) f generates a distribution. Such a distribution is called *regular* and may be identified with f. The delta function encountered in Section 4.1 can now be defined by

$$(\delta, \varphi) = \varphi(0) \qquad \text{for all} \quad \varphi \in \mathcal{D}. \tag{4.14}$$

This is not a regular distribution (see, e.g., [147]). However, we will write this, as tradition does, like (4.10); i.e.,

$$(\delta, \varphi) = \int_{-\infty}^{\infty} \delta(x)\varphi(x)\, \mathrm{d}x. \tag{4.15}$$

Take due note that this is just symbolism. The delta function $\delta(x)$ should be interpreted via its definition (4.14). It is not a function in the classical sense.

If $f, g \in \mathcal{D}'$ are regular distributions, it follows from (4.10) that if $(f, \varphi) = (g, \varphi)$ for all $\varphi \in \mathcal{D}$, then $f(x) = g(x)$ almost everywhere. In general we call two distributions f and g *identical* if

$$(f, \varphi) = (g, \varphi) \qquad \text{for all} \quad \varphi \in \mathcal{D}. \tag{4.16}$$

If we *shift* the argument in a distribution f by ξ, i.e.,

$$f_\xi(x) := f(x - \xi), \tag{4.17}$$

we find

$$(f_\xi, \varphi) = (f, \varphi_{-\xi}). \tag{4.18}$$

For a *shifted delta function* $\delta(x - \xi)$ we then define $\delta_\xi(x) = \delta(x - \xi)$ by

$$(\delta_\xi, \varphi) = (\delta, \varphi_{-\xi}) = \varphi(\xi) \qquad \text{for all} \quad \varphi \in \mathcal{D}. \tag{4.19}$$

We can also define a multiplication of $f \in \mathcal{D}'$ by an integrable function g through the relation

$$(gf, \varphi) := (f, g\varphi). \tag{4.20}$$

Note that gf is again a distribution. It follows that the product of a distribution and a regular distribution is defined. This is not necessarily the case for the product of two nonregular distributions.

Example 4.4 If we multiply the delta function by $g(x) = x$, we obtain

$$(g\delta, \varphi) = (\delta, g\varphi).$$

Hence the test functions $g\varphi$ are all zero at $x = 0$. Consequently, $(g\delta, \varphi) = 0$, or

$$x\delta(x) \equiv 0.$$

□

Using integration by parts, we find, for an ordinary differentiable function f, that

$$(f', \varphi) = \Big[f(x)\varphi(x)\Big]_{-\infty}^{\infty} - (f, \varphi') = -(f, \varphi'), \tag{4.21}$$

since φ has compact support. Therefore we define the derivative of a distribution f in general by

$$(f', \varphi) := -(f, \varphi') \qquad \text{for all} \quad \varphi \in \mathcal{D}. \tag{4.22}$$

Clearly, f' is a distribution again. Note that f is, in this sense, arbitrarily many times differentiable.

An important application is the following. Let $H(x)$ be defined by

$$H(x) = \begin{cases} 0 & \text{for } x < 0, \\ \dfrac{1}{2} & \text{for } x = 0, \\ 1 & \text{for } x > 0, \end{cases} \tag{4.23}$$

which is the *Heaviside function*. It generates the Heaviside distribution if we write, for any $\varphi \in \mathcal{D}$,

$$(H, \varphi) = \int_{-\infty}^{\infty} H(x)\varphi(x)\,\mathrm{d}x = \int_0^{\infty} \varphi(x)\,\mathrm{d}x. \tag{4.24a}$$

From (4.22) we obtain

$$(H', \varphi) = -(H, \varphi') = -\int_0^{\infty} \varphi'(x)\,\mathrm{d}x = \varphi(0). \tag{4.24b}$$

Hence we may identify the distribution $H'(x)$ with $\delta(x)$.

As it will turn out in later chapters, it is useful to have representations of the delta function as a convergent sequence of regular distributions. First, we introduce the following definition.

Definition 4.5. *Let $\{f_i\}_{i\geq 0}$ be a sequence of distributions. Then this sequence converges to f, denoted by $f_i \to f \in \mathcal{D}$, if*

$$\lim_{i\to\infty} (f_i, \varphi) = (f, \varphi) \qquad \textit{for all} \quad \varphi \in \mathcal{D}. \tag{4.25a}$$

Let I be some interval in $\mathbb{R}$. If $\{f_\lambda\}_{\lambda\in I}$ is a family of regular distributions, continuously parametrized by a parameter λ, then we say $f_\lambda \to f$ as $\lambda \to \lambda_0 \in \bar{I}$ if

$$\lim_{\lambda\to\lambda_0} (f_\lambda, \varphi) = (f, \varphi) \qquad \textit{for all} \quad \varphi \in \mathcal{D}. \tag{4.25b}$$

With these definitions we have the following theorem.

Theorem 4.6. *Let f_λ be a locally integrable function with*

(i) $f_\lambda(x) \geq 0$, $x \in \mathbb{R}$, $\lambda > 0$,

(ii) $\int_{-\infty}^{\infty} f_\lambda(x)\,\mathrm{d}x = 1$, $\lambda > 0$,

(iii) $\lim_{\lambda\to 0} \int_a^b f_\lambda(x)\,\mathrm{d}x = 1$ *for any* $a < 0 < b$.

Then $\{f_\lambda\}_{\lambda\in(0,\infty)}$ *is a family of regular distributions with*

$$f_\lambda(x) \to \delta(x) \qquad \textit{for} \quad \lambda \to 0. \tag{4.26}$$

Proof. Consider for any $\varphi \in \mathcal{D}$ the difference $I(\lambda) := (f_\lambda, \varphi) - \varphi(0)$. From the properties of f_λ it follows that $I(\lambda) = \int_{-\infty}^{\infty} f_\lambda(x)(\varphi(x) - \varphi(0))\,\mathrm{d}x$. We split the integration interval into three parts: $(-\infty, -a)$, $(-a, a)$, and (a, ∞) for some $a > 0$. Apparently φ is bounded on $\mathbb{R}$, say, $|\varphi(x)| < M$ for $x \in \mathbb{R}$. Hence $|\varphi(x) - \varphi(0)| \leq 2M$. Furthermore, φ is continuous, so for a given value of ε, choose a such that $|\varphi(x) - \varphi(0)| \leq \frac{1}{2}\varepsilon$, $x \in (-a, a)$. Finally, let λ_0 be such that $\int_{-a}^{a} f_\lambda(x)\,\mathrm{d}x > 1 - (4M)^{-1}\varepsilon$ for $0 < \lambda < \lambda_0$. Then $|I(\lambda)| \leq |\int_{-\infty}^{-a}(\;)\,\mathrm{d}x + \int_a^\infty(\;)\,\mathrm{d}x| + |\int_{-a}^{a}(\;)\,\mathrm{d}x| \leq 2M(4M)^{-1}\varepsilon + \frac{1}{2}\varepsilon = \varepsilon$. This shows (4.26). □

A sequence f_λ as introduced in Theorem 4.6 is sometimes referred to as a *delta sequence*. Such a delta sequence has both theoretical and practical significance; the latter in a situation where one needs, e.g., numerical approximations.

Example 4.7

(i) The top hat sequence (see Figure 4.1)

$$f_\lambda(x) := \begin{cases} \dfrac{1}{2}\lambda^{-1} & \text{for } x \in (-\lambda, \lambda), \\ 0 & \text{for } |x| > \lambda \end{cases}$$

satisfies all requirements of Theorem 4.6.

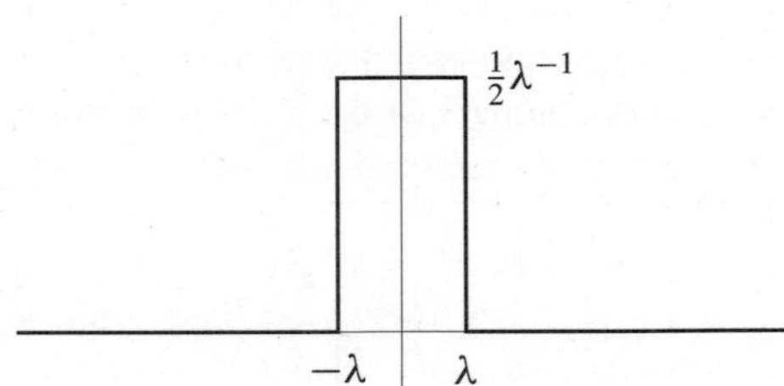

Figure 4.1. *A top hat delta sequence.*

(ii) Another well-known delta sequence is given by (see Figure 4.2)

$$f_\lambda(x) := \frac{1}{\sqrt{2\pi\lambda}} \exp\left(-\frac{x^2}{2\lambda}\right), \qquad x \in \mathbb{R}, \quad \lambda > 0,$$

which is the probability density function for a normal distribution (in the ordinary *statistical* sense!) with variance λ. The requirements from Theorem 4.6 can be verified by introducing the transformation $y^2 = x^2/2\lambda$ and the error function

$$\operatorname{erf}(z) := \frac{2}{\sqrt{\pi}} \int_0^z \mathrm{e}^{-y^2}\,\mathrm{d}y$$

(cf. [1]), which has the property $\operatorname{erf}(\infty) = 1$. This sequence has been important in probability theory, but also in parabolic problems; see Chapter 10. Finally, note that $f_\lambda \in C^\infty(\mathbb{R})$. □

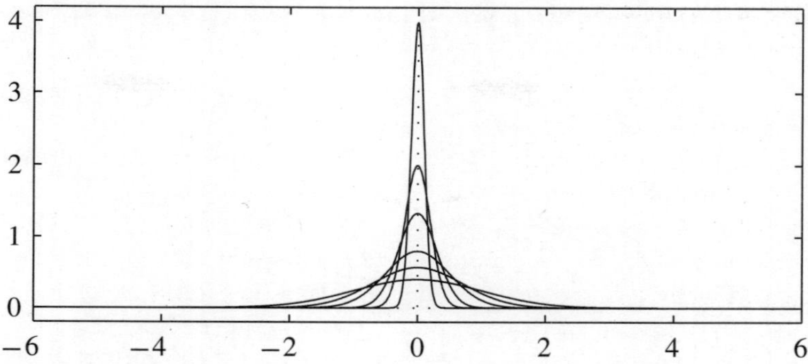

Figure 4.2. *An exponential delta sequence* ($\lambda = 0.01, 0.04, 0.09, 0.25, 0.49, 1.0$).

4.3 Distributions in Several Variables

For f a multivariate real function mapping $\mathbb{R}^n \to \mathbb{R}$, depending on, say, $\boldsymbol{x} = (x_1, x_2, \dots, x_n)^T$, we can give straightforward generalisations of the foregoing concepts. The test space $\mathcal{D}$ now consists of functions in $C^\infty(\mathbb{R}^n)$ defined by

$$C_0^\infty(\mathbb{R}^n) := \{\varphi \in C^\infty(\mathbb{R}^n) \mid \varphi \text{ has compact support}\}, \tag{4.27}$$

equipped with the following notion of convergence.

Definition 4.8. *A sequence* $\{\varphi_i\}_{i\geq 0}$, *with* $\varphi_i \in C_0^\infty(\mathbb{R}^n)$, *is called convergent to zero if*

(i) *there is a closed and bounded set* $\Omega \subset \mathbb{R}^n$ *such that for all* i $\operatorname{supp}(\varphi_i) \subset \Omega$,
(ii) $\lim_{i\to\infty} \varphi_i(x) = 0$ *uniformly in* $\mathbb{R}^n$, *and likewise all partial derivatives.*

A *distribution* is a continuous linear functional on $\mathcal{D}$. If $f(\boldsymbol{x})$ is locally integrable, then we can identify f with its distribution, where

$$(f, \varphi) := \int_{-\infty}^{\infty} \cdots \int_{-\infty}^{\infty} f(\boldsymbol{x})\varphi(\boldsymbol{x})\,\mathrm{d}x_1 \cdots \mathrm{d}x_n. \tag{4.28}$$

In particular, we can find the delta function $\delta(\boldsymbol{x})$ defined by

$$(\delta, \varphi) = \varphi(\mathbf{0}) \qquad \text{for all} \quad \varphi \in \mathcal{D}. \tag{4.29}$$

Differentiation is defined by the relation

$$\left(\frac{\partial f}{\partial x_i}, \varphi\right) = -\left(f, \frac{\partial \varphi}{\partial x_i}\right) \qquad \text{for all} \quad \varphi \in D\,. \tag{4.30}$$

Note that this "derivative" is again a distribution. Theorem 4.6 can be generalised straightforwardly.

Example 4.9 We can give simple generalisations of the top hat and normal distribution representations of the multivariate delta function as follows:

$$\text{(i)} \quad f_\lambda(\boldsymbol{x}) = \begin{cases} (2\lambda)^{-n} & \text{for } \|\boldsymbol{x}\|_\infty \le \lambda, \\ 0 & \text{for } \|\boldsymbol{x}\|_\infty \ge \lambda, \end{cases} \tag{4.31}$$

$$\text{(ii)} \quad f_\lambda(\boldsymbol{x}) = \left(\frac{1}{\sqrt{2\pi\lambda}}\right)^n \exp\left(-\frac{\|\boldsymbol{x}\|_2^2}{2\lambda}\right), \tag{4.32}$$

where $\boldsymbol{x} \in \mathbb{R}^n$, $\|\boldsymbol{x}\|_\infty = \max_{1\le i\le n}(|x_i|)$, and $\|\boldsymbol{x}\|_2^2 = x_1^2 + \cdots + x_n^2$. □

If we have two distributions of a single variable, we can form a direct product to obtain a multivariate distribution. Let f_1 and f_2 be distributions. Then

$$(f_1(x_1) f_2(x_2), \varphi(x_1, x_2)) := (f_1(x_1), (f_2(x_2), \varphi(x_1, x_2))), \tag{4.33}$$

with the inner products defined in an obvious way. We have retained the arguments in f_1, f_2, and φ for clarity.

Example 4.10

$$\delta(\boldsymbol{x}) = \delta(x_1)\delta(x_2)\cdots\delta(x_n).$$ □

4.4 Strong and Weak Solutions

In order to facilitate the following discussion, we shall denote a general linear second order PDE in d spatial variables, $x_1, \ldots, x_d$ say, as

$$\mathcal{L}[u] = f, \tag{4.34a}$$

where

$$\mathcal{L}[u] := \sum_{i,j=1}^{d} a_{ij} \frac{\partial^2 u}{\partial x_i \partial x_j} + \sum_{i=1}^{d} b_i \frac{\partial u}{\partial x_i} + cu\,. \tag{4.34b}$$

Let $\boldsymbol{x} := (x_1, x_2, \ldots, x_d)^T \in \Omega \subset \mathbb{R}^d$. If the *operator* $\mathcal{L}$ is really second order, i.e., $a_{ij} \neq 0$ for at least one pair (i, j), then we call u a *strong solution* of (4.34) if $u \in C^2(\Omega)$. If $\mathcal{L}$ is first order, i.e., all $a_{ij} = 0$, then u is a *strong solution* if $u \in C^1(\Omega)$.

Assuming the coefficients in (4.34b) to be C^∞, we may look for solutions in a distributional sense. We call u a *generalised* or *weak solution* if (4.34) is satisfied in a distributional sense, i.e., if

$$(\mathcal{L}[u], \varphi) = (f, \varphi) \qquad \text{for all} \quad \varphi \in \mathcal{D}. \tag{4.35}$$

In other words, $\mathcal{L}[u] - f$ is identical to the zero distribution. We may restrict $\mathcal{D}$ to test functions with support within Ω. Then u is a *weak solution in* Ω. A strong solution is also a weak solution, but not the other way around.

We next introduce the concept of an *adjoint operator*, defined in general by

$$(\mathcal{L}[u], v) =: (u, \mathcal{L}^*[v]). \tag{4.36}$$

If v is a test function, then using (4.30) we find in particular that

$$\mathcal{L}^*[v] := \sum_{i,j=1}^{d} \frac{\partial^2}{\partial x_i \partial x_j}(a_{ij}v) - \sum_{i=1}^{d} \frac{\partial}{\partial x_i}(b_i v) + cv. \tag{4.37}$$

For the operator $\mathcal{L}$ and its adjoint $\mathcal{L}^*$ the following property holds.

Property 4.11. *For $u, v \in C^2(\mathbb{R}^d)$ we have*

$$v\mathcal{L}[u] - u\mathcal{L}^*[v] = \nabla \cdot \boldsymbol{w}, \tag{4.38a}$$

where the vector function $\boldsymbol{w} := (w_1, \ldots, w_d)^T$ is given by

$$w_i = \sum_{j=1}^{d} \left(a_{ij} \frac{\partial u}{\partial x_j} v - u \frac{\partial}{\partial x_j}(a_{ij}v) \right) + b_i uv, \qquad i = 1, \ldots, d. \tag{4.38b}$$

Proof. The proof is trivial by direct substitution of (4.38b) into (4.38a). □

Next we have the following result.

Corollary 4.12. *If u is a distribution and $\varphi \in \mathcal{D}$, then*

$$\int_\Omega \left(\varphi\mathcal{L}[u] - u\mathcal{L}^*[\varphi]\right) \mathrm{d}V = \oint_{\partial\Omega} \boldsymbol{w} \cdot \boldsymbol{n} \,\mathrm{d}S, \tag{4.39}$$

where $\boldsymbol{n}$ is the unit outward normal of the boundary $\partial\Omega$ of Ω.

Proof. Apply the divergence theorem (J.13) Section J in the appendix to (4.38a). □

Example 4.13 Consider the Cauchy problem

$$\frac{\partial u}{\partial t} - \frac{\partial u}{\partial x} = 0, \qquad x \in \mathbb{R}, \quad t > 0,$$

$$u(x, 0) = H(x), \qquad x \in \mathbb{R}.$$

We see that the PDE has characteristics $x + t = c$, $c \in \mathbb{R}$. Hence we have a discontinuity along the line $x + t = 0$. So the solution appears to be

$$u(x, t) = H(x + t),$$

which is clearly not a classical solution. Let φ be a test function and consider (cf. (4.30))

$$I := -\int_0^\infty \int_{-\infty}^\infty \left(\frac{\partial u}{\partial t} - \frac{\partial u}{\partial x} \right) \varphi \,\mathrm{d}x \,\mathrm{d}t = \int_0^\infty \int_{-\infty}^\infty u \left(\frac{\partial \varphi}{\partial t} - \frac{\partial \varphi}{\partial x} \right) \mathrm{d}x \,\mathrm{d}t.$$

We split the domain into the regions $R_1 = \{x + t > 0\}$, where $u = 1$, and $R_2 = \{x + t < 0\}$, where $u = 0$. Then

$$
\begin{aligned}
I = \iint_{R_1} u \left(\frac{\partial \varphi}{\partial t} - \frac{\partial \varphi}{\partial x} \right) \mathrm{d}x\mathrm{d}t &= \int_0^\infty \int_{-t}^\infty \left(\frac{\partial \varphi}{\partial t} - \frac{\partial \varphi}{\partial x} \right) \mathrm{d}x\mathrm{d}t \\
&= \int_{-\infty}^0 \int_{-x}^\infty \frac{\partial \varphi}{\partial t} \mathrm{d}t\mathrm{d}x + \int_0^\infty \int_0^\infty \frac{\partial \varphi}{\partial t} \mathrm{d}t\mathrm{d}x - \int_0^\infty \int_{-t}^\infty \frac{\partial \varphi}{\partial x} \mathrm{d}x\mathrm{d}t \\
&= \int_{-\infty}^0 -\varphi(x, -x)\, \mathrm{d}x + \int_0^\infty -\varphi(x, 0)\, \mathrm{d}x + \int_0^\infty \varphi(-t, t)\, \mathrm{d}t = 0
\end{aligned}
$$

because φ vanishes along the borders of its domain, i.e. $\varphi(x, 0) = 0$. Hence $I = 0$ and so u is a weak solution. □

The example above shows that the weak solution concept nicely captures "shock structures." We shall encounter this in Chapter 12, where the "jump" condition in u will be met again as a Rankine–Hugoniot condition.

We finally show that a weak solution with sufficient smoothness is always a strong solution.

Theorem 4.14. *Let $u \in C^2(\Omega)$ be a weak solution of* (4.34a) *(or at least $C^1(\Omega)$ if $a_{ij} \equiv 0$). Then u is a strong solution on Ω.*

Proof. From (4.35) and (4.39) we find

$$
\int_\Omega \mathcal{L}[u]\varphi\, \mathrm{d}V = \int_\Omega u\mathcal{L}^*[\varphi]\, \mathrm{d}V = \int_\Omega f\varphi\, \mathrm{d}V,
$$

since $w \equiv 0$ on $\partial\Omega$. Hence

$$
\int_\Omega (\mathcal{L}[u] - f)\varphi\, \mathrm{d}V = 0.
$$

Since $\mathcal{L}[u] - f \in C^0(\Omega)$ we conclude that $\mathcal{L}[u] = f$. □

In the rest of this book we will often encounter systems of PDEs of the form

$$
\frac{\partial \boldsymbol{u}}{\partial t} = \boldsymbol{\mathcal{L}}[\boldsymbol{u}], \qquad x \in \mathbb{R}, \quad t > 0, \tag{4.40}
$$

where $\boldsymbol{\mathcal{L}}$ is a vector-valued differential operator containing only spatial derivatives that works on the components of the vector function $\boldsymbol{u} \in \mathbb{R}^m$. Analogously to (4.35), we can define a weak solution of (4.40) by the requirement

$$
\int_0^\infty \int_{-\infty}^\infty \left(\frac{\partial \boldsymbol{u}}{\partial t} - \boldsymbol{\mathcal{L}}[\boldsymbol{u}] \right) \cdot \boldsymbol{\varphi}\, \mathrm{d}x\mathrm{d}t = 0 \qquad \text{for all} \quad \boldsymbol{\varphi} \in \mathcal{D}^m. \tag{4.41}
$$

In this definition $\boldsymbol{\varphi}$ is a vector-valued test function with components in $\mathcal{D}$. We will apply this definition in Chapter 12 to systems of hyperbolic equations.

4.5 Fundamental Solutions

We now have the tools to generalise the findings of Section 4.1. The "solution" w found there will now be recognized as a weak solution.

Consider (4.34a). For a general linear differential operator $\mathcal{L}$ and for any $\boldsymbol{\xi}$, we call $w(\boldsymbol{x};\boldsymbol{\xi})$ a *fundamental solution* if

$$\mathcal{L}[w](\boldsymbol{x};\boldsymbol{\xi}) = \delta(\boldsymbol{x}-\boldsymbol{\xi}), \qquad \boldsymbol{x} \in \mathbb{R}^d. \tag{4.42}$$

Example 4.15 A fundamental solution of the Laplace equation in $\mathbb{R}^2$ satisfying

$$\nabla^2 w = \delta(\boldsymbol{x}-\boldsymbol{\xi})$$

is found to be given by

$$w(\boldsymbol{x};\boldsymbol{\xi}) := \frac{1}{2\pi}\ln(\|\boldsymbol{x}-\boldsymbol{\xi}\|_2).$$

This is not evident because, when considered as an ordinary function, $\nabla^2 w$ is identically zero, except at $\boldsymbol{x}=\boldsymbol{\xi}$, where it does not exist. As a result, any integral $\iint_{\mathbb{R}^2}\varphi\nabla^2 w\,\mathrm{d}\boldsymbol{x}$ is zero rather than equal to $\varphi(\boldsymbol{\xi})$. However, when we interpret $\nabla^2 w$ as a distribution, the gradient is defined by (4.30), and we have

$$\begin{aligned}\iint_{\mathbb{R}^2}\varphi\nabla^2 w\,\mathrm{d}\boldsymbol{x} &= -\iint_{\mathbb{R}^2}\nabla\varphi\cdot\nabla w\,d\boldsymbol{x} = -\int_0^{2\pi}\int_0^{\infty}\frac{\partial\varphi}{\partial r}\frac{\partial w}{\partial r}r\,\mathrm{d}r\mathrm{d}\theta\\ &= -\frac{1}{2\pi}\int_0^{2\pi}\int_0^{\infty}\frac{\partial\varphi}{\partial r}\,\mathrm{d}r\mathrm{d}\theta = \frac{1}{2\pi}\int_0^{2\pi}\varphi(\boldsymbol{\xi})\,\mathrm{d}\theta = \varphi(\boldsymbol{\xi}),\end{aligned}$$

with $\boldsymbol{x}-\boldsymbol{\xi}$ written in polar coordinates r and θ. □

Now consider a general linear inhomogeneous problem with linear boundary value operator $\mathcal{B}$, where we would like the solution u on a domain Ω; i.e.,

$$\mathcal{L}[u] = f, \qquad \boldsymbol{x} \in \Omega, \tag{4.43a}$$

$$\mathcal{B}[u] = \beta, \qquad \boldsymbol{x} \in \partial\Omega. \tag{4.43b}$$

Then we can construct u as follows. First, we seek a *particular* solution, $u_{\mathrm{p}}(\boldsymbol{x})$ say, such that

$$\mathcal{L}[u_{\mathrm{p}}] = f, \qquad \boldsymbol{x} \in \Omega, \tag{4.44a}$$

$$\mathcal{B}[u_{\mathrm{p}}] = 0, \qquad \boldsymbol{x} \in \partial\Omega. \tag{4.44b}$$

In order to find u_{p}, we employ special weak solutions derived from a fundamental solution by adding a suitable homogeneous (strong) solution of (4.44) (i.e., with $f=0$), giving rise to *Green's functions* $G(\boldsymbol{x};\boldsymbol{\xi})$:

$$\mathcal{L}[G](\boldsymbol{x};\boldsymbol{\xi}) = \delta(\boldsymbol{x}-\boldsymbol{\xi}), \qquad \boldsymbol{x} \in \Omega, \tag{4.45a}$$

$$\mathcal{B}[G](\boldsymbol{x};\boldsymbol{\xi}) = 0, \qquad \boldsymbol{x} \in \partial\Omega. \tag{4.45b}$$

(Note the difference between a fundamental solution and the Green's function.) Since we can write $f(\boldsymbol{x})$ formally as a superposition of delta functions in the form

$$f(\boldsymbol{x}) = \int_{\Omega} \delta(\boldsymbol{x}-\boldsymbol{\xi}) f(\boldsymbol{\xi})\, \mathrm{d}V_{\xi}, \tag{4.46}$$

where the integration is with respect to the $\boldsymbol{\xi}$ coordinates, we can construct u_{p} by convolution as

$$u_{\mathrm{p}}(\boldsymbol{x}) = \int_{\Omega} G(\boldsymbol{x};\boldsymbol{\xi}) f(\boldsymbol{\xi})\, \mathrm{d}\boldsymbol{\xi}. \tag{4.47}$$

The complete solution of (4.43) is now found by adding a (homogeneous) solution u_{h} of

$$\mathcal{L}[u_{\mathrm{h}}] = 0, \qquad \boldsymbol{x} \in \Omega, \tag{4.48a}$$

$$\mathcal{B}[u_{\mathrm{h}}] = \beta. \qquad \boldsymbol{x} \in \partial\Omega. \tag{4.48b}$$

Such adding of a suitable homogeneous solution to a particular solution (e.g., in order to make the boundary data fit) is called *superposition*.

Note that u_{h}, if it exists, need not be unique. This then gives rise to *Fredholm's alternative*: either u_{h} is unique on Ω or there exist infinitely many solutions for (4.48). The general solution of (4.43) is given by

$$u(\boldsymbol{x}) = \int_{\Omega} G(\boldsymbol{x};\boldsymbol{\xi}) f(\boldsymbol{\xi})\, \mathrm{d}V_{\xi} + u_{\mathrm{h}}(\boldsymbol{x}), \tag{4.49}$$

which is an analogue of (4.3).

We remark that the fundamental solution $w(\boldsymbol{x};\boldsymbol{\xi})$ can be found by translation of the argument from $w(\boldsymbol{x};\boldsymbol{0})$ if $\mathcal{L}$ has constant coefficients; i.e.,

$$w(\boldsymbol{x};\boldsymbol{\xi}) = w(\boldsymbol{x}-\boldsymbol{\xi};\boldsymbol{0}). \tag{4.50}$$

We can generalise this concept of the fundamental solution by taking t as an additional independent variable. So consider

$$\mathcal{L}_1[u] + \mathcal{L}_2[u] = f, \tag{4.51}$$

where $\mathcal{L}_1$ denotes a differential operator with respect to t and $\mathcal{L}_2$ is a differential operator with respect to $\boldsymbol{x}$. For each $(\boldsymbol{\xi},\tau)$ a fundamental solution $w(\boldsymbol{x},t;\xi,\tau)$ then satisfies the problem

$$\mathcal{L}_1[w](\boldsymbol{x},t;\boldsymbol{\xi},\tau) + \mathcal{L}_2[w](\boldsymbol{x},t;\boldsymbol{\xi},\tau) = \delta(\boldsymbol{x}-\boldsymbol{\xi})\,\delta(t-\tau), \qquad \boldsymbol{x} \in \mathbb{R}^d, \quad t > 0, \tag{4.52a}$$

$$w(\boldsymbol{x},t;\boldsymbol{\xi},\tau) = 0, \qquad \boldsymbol{x} \in \mathbb{R}^d, \quad t < \tau. \tag{4.52b}$$

Note that this formulation takes care of the evolutionary character of the problem (*causality principle*) with time intervals (τ,∞). We shall see later (Chapter 10) that solutions of these problems can be found using such fundamental solutions. An alternative way is to make use of Duhamel integrals; see the next section.

4.6 Initial (Boundary) Value Problems; Duhamel Integrals

Most problems we deal with in this book have both a space and a time dependence. In the *linear* case the equations are in fact either one of the following type:

$$\frac{\partial u}{\partial t} = \mathcal{L}_1[u] + f(\boldsymbol{x}, t), \qquad \mathbf{x} \in \Omega \subset \mathbb{R}^d, \quad t > 0, \tag{4.53}$$

$$\frac{\partial^2 u}{\partial t^2} = \mathcal{L}_2[u] + f(\boldsymbol{x}, t), \qquad \mathbf{x} \in \Omega \subset \mathbb{R}^d, \quad t > 0. \tag{4.54}$$

Here $\mathcal{L}_1$ and $\mathcal{L}_2$ are homogeneous second order differential operators in $\boldsymbol{x}$. In the chapters that will follow we shall investigate the solutions of the various types of PDEs. Generally speaking, we need to specify a condition on the *boundary* $\partial\Omega$ of the *spatial domain* Ω, say

$$\mathcal{B}[u] = \beta(t), \qquad \boldsymbol{x} \in \partial\Omega, \quad t > 0. \tag{4.55}$$

Moreover, we need initial conditions at $t = 0$. For (4.53) they have the form

$$u(\boldsymbol{x}, 0) = a(\boldsymbol{x}), \qquad \boldsymbol{x} \in \Omega. \tag{4.56}$$

For (4.54) we have, besides (4.56),

$$\frac{\partial u}{\partial t}(\boldsymbol{x}, 0) = b(\boldsymbol{x}), \qquad \boldsymbol{x} \in \Omega. \tag{4.57}$$

Now consider the following family of problems defined on $\Omega \times (\tau, \infty)$, associated with (4.53), (4.55), and (4.56):

$$\frac{\partial w}{\partial t}(\boldsymbol{x}, t; \tau) = \mathcal{L}_1[w](\boldsymbol{x}, t; \tau), \qquad \boldsymbol{x} \in \Omega, \quad t > \tau, \tag{4.58a}$$

$$\mathcal{B}[w](\boldsymbol{x}, t; \tau) = 0, \qquad \boldsymbol{x} \in \partial\Omega, \quad t > \tau, \tag{4.58b}$$

$$w(\boldsymbol{x}, \tau; \tau) = f(\boldsymbol{x}, \tau), \qquad \boldsymbol{x} \in \Omega. \tag{4.58c}$$

These initial boundary value problems together build the actual solution; see the following theorem.

Theorem 4.16 (Duhamel integral I). *Assume that the initial boundary value problem* (4.53), (4.55) *with* $\beta(t) \equiv 0$, *and* (4.56) *with* $a(\boldsymbol{x}) \equiv 0$ *has a unique solution* u. *Then it is given by the* Duhamel integral

$$u(\boldsymbol{x}, t) = \int_0^t w(\boldsymbol{x}, t; \tau)\, \mathrm{d}\tau, \tag{4.59}$$

where $w(\cdot\,, \cdot\,; \tau)$ *is a solution of* (4.58).

Proof. If we differentiate the right-hand side of (4.59) with respect to t, we get

$$\begin{aligned}\frac{\partial}{\partial t}\int_0^t w(\boldsymbol{x},t;\tau)\,\mathrm{d}\tau &= w(\boldsymbol{x},t;t)+\int_0^t \frac{\partial w}{\partial t}(\boldsymbol{x},t;\tau)\,\mathrm{d}\tau\\ &= f(\boldsymbol{x},t)+\int_0^t \mathcal{L}_1[w](\boldsymbol{x},t;\tau)\,\mathrm{d}\tau = f(\boldsymbol{x},t)+\mathcal{L}_1\left[\int_0^t w(\boldsymbol{x},t;\tau)\,\mathrm{d}\tau\right].\end{aligned}$$

Hence $\int_0^t w(\boldsymbol{x},t;\tau)\,\mathrm{d}\tau$ satisfies (4.53). Moreover, it is easy to see that (4.55), with $\beta(t)\equiv 0$, and (4.56), with $a(\boldsymbol{x})\equiv 0$, are satisfied, so that $\int_0^t w(\boldsymbol{x},t;\tau)\,\mathrm{d}\tau$ may be identified with the solution $u(\boldsymbol{x},t)$. □

Example 4.17 Consider the problem

$$\frac{\partial u}{\partial t}=\frac{\partial^2 u}{\partial x^2}+f(x), \qquad x\in\mathbb{R}, \quad t>0,$$
$$u(x,0)=0, \qquad x\in\mathbb{R}.$$

It can be checked that the solution of this initial value problem is given by (4.59), with

$$w(x,t;\tau):=\frac{1}{2\sqrt{\pi(t-\tau)}}\int_{-\infty}^{\infty}\exp\left(-\frac{(x-\xi)^2}{4(t-\tau)}\right)f(\xi)\,\mathrm{d}\xi.$$ □

Similarly, consider a problem associated with (4.54), and (4.55) to (4.57):

$$\frac{\partial^2 w}{\partial t^2}(\boldsymbol{x},t;\tau)=\mathcal{L}_2[w](\boldsymbol{x},t;\tau), \qquad \boldsymbol{x}\in\Omega, \quad t>\tau, \tag{4.60a}$$

$$\mathcal{B}[w](\boldsymbol{x},t;\tau)=0, \qquad \boldsymbol{x}\in\partial\Omega, \quad t>\tau, \tag{4.60b}$$

$$w(\boldsymbol{x},\tau;\tau)=0, \qquad \boldsymbol{x}\in\Omega, \tag{4.60c}$$

$$\frac{\partial w}{\partial t}(\boldsymbol{x},\tau;\tau)=f(\boldsymbol{x},\tau), \qquad \boldsymbol{x}\in\Omega. \tag{4.60d}$$

Then we find the following result.

Theorem 4.18 (Duhamel integral II). *Assume that the initial boundary value problem* (4.54), (4.55) *with* $\beta(t)\equiv 0$, (4.56) *with* $a(\boldsymbol{x})\equiv 0$, *and* (4.57) *with* $b(\boldsymbol{x})\equiv 0$ *has a unique solution. Then* u *is given by*

$$u(\boldsymbol{x},t)=\int_0^t w(\boldsymbol{x},t;\tau)\,\mathrm{d}\tau, \tag{4.61}$$

where $w(\cdot,\cdot;\tau)$ *is a solution of* (4.60).

Proof. If w is a solution of (4.60), we find from differentiating the right-hand side of (4.61) once that

$$\frac{\partial}{\partial t}\int_0^t w(\boldsymbol{x},t;\tau)\,\mathrm{d}\tau=\int_0^t \frac{\partial w}{\partial t}(\boldsymbol{x},t;\tau)\,\mathrm{d}\tau.$$

Differentiating this relation once more then gives

$$\frac{\partial^2}{\partial t^2}\int_0^t w(\boldsymbol{x},t;\tau)\,\mathrm{d}\tau = \frac{\partial w}{\partial t}(\boldsymbol{x},t;t) + \int_0^t \frac{\partial^2 w}{\partial t^2}(\boldsymbol{x},t;\tau)\,\mathrm{d}\tau$$

$$= f(\boldsymbol{x},t) + \mathcal{L}_2\left[\int_0^t w(\boldsymbol{x},t,\tau)\,\mathrm{d}\tau\right],$$

which shows that (4.61) is the solution of (4.54) to (4.57) with homogeneous conditions. □

If $a \neq 0$ in (4.56), we find, instead of (4.61), that

$$u(\boldsymbol{x},t) = w(\boldsymbol{x},t;0)\,a(\boldsymbol{x}) + \int_0^t w(\boldsymbol{x},t;\tau)\,\mathrm{d}\tau. \tag{4.62a}$$

Moreover, if $b \neq 0$ in (4.57), we obtain

$$u(\boldsymbol{x},t) = v(\boldsymbol{x},t)\,a(\boldsymbol{x}) + w(\boldsymbol{x},t;0)\,b(\boldsymbol{x}) + \int_0^t w(\boldsymbol{x},t;\tau)\,\mathrm{d}\tau, \tag{4.62b}$$

where w is a solution of (4.60) and v is a solution of

$$\frac{\partial^2 v}{\partial t^2}(\boldsymbol{x},t) = \mathcal{L}[v](\boldsymbol{x},t), \qquad \boldsymbol{x}\in\Omega, \quad t>0, \tag{4.63a}$$

$$\mathcal{B}[v](\boldsymbol{x},t) = 0, \qquad \boldsymbol{x}\in\partial\Omega, \quad t>0, \tag{4.63b}$$

$$v(\boldsymbol{x},0) = 1, \qquad \boldsymbol{x}\in\Omega, \tag{4.63c}$$

$$\frac{\partial v}{\partial t}(\boldsymbol{x},0) = 0, \qquad \boldsymbol{x}\in\Omega. \tag{4.63d}$$

The Duhamel integral is in fact nothing but the superposition of fundamental solutions (with respect to time), and we have the following theorem.

Theorem 4.19. *Let $\mathcal{D}'$ be the set of distributions in the parameter t.*

(i) *Then w satisfying* (4.58), *with $f(\boldsymbol{x},t)\equiv 1$, is a fundamental solution of*

$$\frac{\partial w}{\partial t}(\boldsymbol{x},t;\tau) = \mathcal{L}_1[w](\boldsymbol{x},t;\tau) + \delta(\tau). \tag{4.64a}$$

(ii) *Also, w satisfying* (4.60), *with $f(\boldsymbol{x},t)\equiv 1$, is a fundamental solution of*

$$\frac{\partial^2 w}{\partial t^2}(\boldsymbol{x},t;\tau) = \mathcal{L}_2[w](\boldsymbol{x},t;\tau) + \delta(\tau). \tag{4.64b}$$

Here we have defined $w(\boldsymbol{x},t)\equiv 0$ for $t<\tau$.

Proof.

(i) From (4.58) we obtain

$$\int_0^\infty \left(\frac{\partial w}{\partial t} - \mathcal{L}_1[w]\right)\varphi\,\mathrm{d}t = -\int_0^\infty \left(\frac{\partial \varphi}{\partial t}\,w + \varphi\mathcal{L}_1[w]\right)\,\mathrm{d}t$$

$$= -\int_\tau^\infty \left(\frac{\partial \varphi}{\partial t}\,w + \varphi\mathcal{L}_1[w]\right)\,\mathrm{d}t = \varphi(\tau) + \int_\tau^\infty \left(\frac{\partial w}{\partial t} - \mathcal{L}_1[w]\right)\varphi\,\mathrm{d}t = \varphi(\tau).$$

(ii) Likewise, from (4.60) we deduce

$$\int_0^\infty \left(\frac{\partial w}{\partial t} - \mathcal{L}_2[w]\right)\varphi\,\mathrm{d}t = \int_0^\infty \left(\frac{\partial^2 \varphi}{\partial t^2}\,w - \varphi\mathcal{L}_2[w]\right)\,\mathrm{d}t$$

$$= \int_\tau^\infty \left(\frac{\partial^2 \varphi}{\partial t^2}\,w - \varphi\mathcal{L}_2[w]\right)\,\mathrm{d}t$$

$$= \varphi(\tau) + \int_\tau^\infty \left(\frac{\partial^2 w}{\partial t^2} - \mathcal{L}_2[w]\right)\varphi\,\mathrm{d}t = \varphi(\tau).$$ □

Example 4.20 Consider the same problem as in Example 4.17. As one can easily see, a fundamental solution, i.e., a solution of

$$\frac{\partial u}{\partial t} = \frac{\partial^2 u}{\partial x^2} + \delta(x-\xi)\delta(t-\tau), \qquad x \in \mathbb{R}, \quad t > 0,$$

$$u(x,t) = 0, \qquad x \in \mathbb{R}, \quad t < \tau,$$

is given by

$$w(x,t;\xi,\tau) := \frac{1}{2\sqrt{\pi(t-\tau)}}\exp\left(-\frac{(x-\xi)^2}{4(t-\tau)}\right).$$

Hence $u(x,t;\tau) = \int_{-\infty}^{\infty} w(x,t;\xi,\tau)\mathrm{d}\xi$. □

As a final application of the Duhamel integral, consider initial boundary value problems on a one-dimensional semi-infinite domain, i.e., $\Omega = [0,\infty)$, with homogeneous source term and with homogeneous initial condition(s) (i.e., $a(x) \equiv 0$ in (4.56) for equation (4.53) and $a(x) = b(x) \equiv 0$ in (4.56) and (4.57) for equation (4.54)). Let the boundary condition (4.55) be given by

$$u(0,t) = \beta(t), \qquad t > 0. \tag{4.65}$$

Assume that the function $w(x,t;\tau)$ can actually be written as

$$\overline{w}(x,t-\tau) := w(x,t;\tau). \tag{4.66}$$

Let $\overline{w}$ satisfy the following initial boundary value problem corresponding to either (4.53) or (4.54), i.e.,

$$\frac{\partial^i \overline{w}}{\partial t^i} = \mathcal{L}_i[\overline{w}], \qquad x \in [0,\infty), \quad t > \tau \quad (i = 1, 2), \tag{4.67a}$$

$$\overline{w}(0,t) = 1, \qquad t > 0, \tag{4.67b}$$

$$\overline{w}(x,0) = 0, \qquad x \in [0,\infty) \text{ for (4.53)}. \tag{4.67c}$$

$$\overline{w}(x,0) = \frac{\partial \overline{w}}{\partial t}(x,0) = 0, \qquad x \in [0,\infty) \text{ for (4.54)} \tag{4.67d}$$

Then we have the following representation.

Property 4.21. *The solution of* (4.53) *or* (4.54) *with* $f(x,t) \equiv 0$, (4.65), (4.56) *with* $a(x) = 0$, *and* (for eq. (4.54)) (4.57) *with* $b(x) \equiv 0$, *is given by*

$$u(x,t) = \frac{\partial}{\partial t}\int_0^t \overline{w}(x,t-\tau)\beta(\tau)\,\mathrm{d}\tau. \tag{4.68}$$

Proof. Denote the right-hand side of (4.68) by $q(x,t)$. Then

$$q(x,t) = \overline{w}(x,t)\beta(0) + \int_0^t \overline{w}(x,t-\tau)\beta'(\tau)\,\mathrm{d}\tau,$$

and, using all properties of $\overline{w}$, we find

$$\frac{\partial q}{\partial t}(x,t) = \frac{\partial}{\partial t}\overline{w}(x,t)\beta(0) + \int_0^t \frac{\partial}{\partial t}\overline{w}(x,t-\tau)\beta'(\tau)\,\mathrm{d}\tau,$$

$$\frac{\partial^2 q}{\partial t^2}(x,t) = \frac{\partial^2}{\partial t^2}\overline{w}(x,t)\beta(0) + \int_0^t \frac{\partial^2}{\partial t^2}\overline{w}(x,t-\tau)\beta'(\tau)\,\mathrm{d}\tau,$$

$$\mathcal{L}_i[q](x,t) = \mathcal{L}_i[\overline{w}](x,t)\beta(0) + \int_0^t \mathcal{L}_i[\overline{w}](x,t-\tau)\beta'(\tau)\,\mathrm{d}\tau \qquad (i = 1, 2).$$

This shows that q satisfies the PDE (4.53) or (4.54). Furthermore, we have

$$q(0,t) = \beta(0) + \int_0^t \beta'(\tau)\,\mathrm{d}\tau = \beta(t),$$

so that it also satisfies the boundary condition. □

4.7 Discussion

- Distributions are a fundamental concept and a bit of an outsider in this book as far as rigour is concerned; cf. [119]. From a mathematical point of view they open entire new vistas in which concepts like differentiability can be treated for larger classes of problems, like in the theory pseudodifferential operators; see, e.g., [154]. They will turn out to be essential, though, when it comes to hyperbolic equations. Here one cannot solve actual problems without taking recourse to weak solutions.

- Fundamental solutions have a value in their own right, as it will turn out, e.g., in Chapters 8 and 10. Another interesting application is their use in the boundary element method, where a boundary value problem is first transformed into an integral equation by using Green's theorem, after which the latter is discretised. For boundary value problems on infinite domains, as well as for certain other problems, this method can be very attractive; see, e.g., [16, 51].

Exercises

4.1. Consider the set of functions

$$f(x;t) = \frac{t}{\pi(x^2+t^2)}, \qquad x \in \mathbb{R}, \quad t > 0.$$

Show that $\lim_{t\downarrow 0} f(x;t) = \delta(x)$.

4.2. Let the following set of (locally) integrable functions for $x \in \mathbb{R}$, $t > 0$, be given:

$$f(x;t) \geq 0,$$

They have the properties

$$\int_{-\infty}^{\infty} f(x;t)\,\mathrm{d}x = 1, \qquad t > 0,$$

$$\lim_{t\downarrow 0} \int_a^b f(x;t)\,\mathrm{d}x = 1 \quad \text{for} \quad a < 0 < b.$$

Show that $\lim_{t\downarrow 0} f(x;t) = \delta(x)$.

4.3. Show that a (fundamental) solution of the problem

$$\nabla^2 w(x;\xi) = \delta(\boldsymbol{\xi}), \qquad \boldsymbol{x} \in \mathbb{R}^3,$$

is given by

$$w(\boldsymbol{x};\boldsymbol{\xi}) = -\frac{1}{4\pi\,\|\boldsymbol{x}-\boldsymbol{\xi}\|_2}.$$

4.4. Determine the Green's function of the problem

$$\frac{\mathrm{d}u}{\mathrm{d}x} = 0, \qquad x \in (\alpha, \beta),$$

$$u(\alpha) = u_\alpha, \qquad u(\beta) = u_\beta.$$

4.5. Given the form of a fundamental solution as found in Exercise 4.3, determine a Green's function for the problem

$$\nabla^2 u = 0, \qquad \boldsymbol{x} \in \Omega := \mathcal{B}(\boldsymbol{0}; 1),$$
$$u = g(\boldsymbol{x}), \qquad \boldsymbol{x} \in \partial\Omega,$$

where $\mathcal{B}(\boldsymbol{0}; 1)$ denotes the unit ball of radius one centred at the origin. *Hint:* Use an appropriate mirror point.

4.6. Consider the problem

$$\frac{\partial u}{\partial t} = \frac{\partial^2 u}{\partial x^2}, \qquad x \in \mathbb{R}, \quad t > 0,$$
$$u(x, 0) = v(x), \qquad x \in \mathbb{R}.$$

Use a Duhamel integral to show that the solution is given by

$$\frac{1}{2\sqrt{\pi t}} \int_{-\infty}^{\infty} \exp\left(-\frac{(x-\xi)^2}{4t}\right) v(\xi)\,\mathrm{d}\xi.$$

4.7. Under the same assumptions as in Property 4.21, show that the following holds:

$$u(x, t) = \int_0^t \beta(t) \frac{\partial}{\partial t} \overline{w}(x, t - \tau)\,\mathrm{d}\tau.$$

Chapter 5

Approximation by Finite Differences

This chapter gives the basic methods that will be used in this book for numerical computations. In Section 5.1 we start with simple polynomial interpolation, which is then used to derive differentiation and integration formulas. These form the basis of the finite difference approximations treated in Section 5.2 for space variables, both for one- and higher-dimensional problems. Some attention is paid to curved boundaries and coordinate systems other than Cartesian ones. Reformulating a problem in terms of integrals leads to finite volume methods, dealt with in Section 5.3. These are akin to finite difference methods but have a "conservation property" that is useful in certain applications. In Section 5.4 the finite differences are used to define methods for solving initial value problems. We briefly overview both one-step methods and multistep methods. Also in this section we encounter the notion of stiffness, which plays an essential role in evolution equations. The methods described in this chapter all lead to *difference equations*, which are solved numerically to obtain an approximate solution. The last two sections are devoted to the stability and convergence of such numerical methods. In Section 5.5 we discuss the celebrated Lax equivalence theorem, stating that for consistent discretisations stability is equivalent to convergence. In Section 5.6 a practical criterion based on Fourier analysis is worked out to establish necessary conditions for stability.

5.1 Basic Methods

The method of choice for approximating the differential equations in this book is finite differences. In this chapter we therefore discuss the basic tools underlying this method. These tools are interpolation, numerical differentiation, and numerical quadrature, which are considered first in Section 5.1.1. The remainder of this section is devoted to difference methods proper.

5.1.1 Interpolation, Numerical Differentiation, and Quadrature

If we want to approximate a function $u = u(\xi)$, we can use a variety of methods. The most basic ones (appropriate for our goal) are those based on interpolation. In view of the available computer arithmetic we opt for polynomials since these can be evaluated directly by the simple algebraic operations ($+, -, \times, \div$) that a computer can carry out.

Suppose $u = u(\xi)$ is defined on an interval $[\alpha, \beta]$. Let $\xi_1, \xi_2, \ldots, \xi_m$ be some given points, ordered as $\alpha \le \xi_1 < \xi_2 < \cdots < \xi_m \le \beta$, and let p be the interpolation polynomial with

$$p(\xi_i) = u(\xi_i), \qquad i = 1, 2, \ldots, m. \tag{5.1}$$

The points ξ_i $(i = 1, 2, \ldots, m)$ are referred to as *interpolation points*; m is sometimes referred to as the interpolation order. Then $p(\xi)$, the *Lagrangian interpolation polynomial*, is given by

$$p(\xi) := \sum_{i=1}^{m} L_i(\xi) u(\xi_i), \qquad L_i(\xi) := \prod_{\substack{j=1 \\ j\neq i}}^{m} \frac{\xi - \xi_j}{\xi_i - \xi_j}, \tag{5.2a}$$

which is a polynomial of degree not more than $m - 1$. The terms $L_i(\xi)$ are called *Lagrangian basis polynomials*. For any $\xi \in (\alpha, \beta)$ the interpolation error of $p(\xi)$ is given by

$$u(\xi) - p(\xi) = \frac{u^{(m)}(\zeta)}{m!}\omega(\xi), \qquad \zeta \in (\alpha, \beta), \tag{5.2b}$$

where

$$\omega(\xi) := \prod_{j=1}^{m} (\xi - \xi_j). \tag{5.2c}$$

Note that ζ generally depends on ξ; i.e., $\zeta = \zeta(\xi)$. One can even prove [117] that

$$\frac{\mathrm{d}}{\mathrm{d}\xi}\left(\frac{u^{(m)}(\zeta)}{m!}\right) = \frac{u^{(m+1)}(\eta)}{(m+1)!}, \qquad \eta \in (\alpha, \beta). \tag{5.3}$$

Often the interpolation points ξ_i are *equispaced*; i.e., for some h we have $\xi_i - \xi_{i-1} =: h$ $(i = 2, 3, \ldots, m)$. Then the factor $\omega(\xi)$ should be scaled by a factor h^{-m}. More generally, it makes sense to use a scaling with

$$h := \max_{2\le i\le m} |\xi_i - \xi_{i-1}|,$$

resulting in the error term

$$u(\xi) - p(\xi) = h^m \frac{u^{(m)}(\zeta)}{m!}\left(h^{-m}\omega(\xi)\right). \tag{5.4}$$

Let ξ be any of the interpolation points, say $\xi = \xi_n$. We can approximate the derivative of u in ξ_n by differentiating the interpolation polynomial $p(\xi)$. This way, we find

$$\frac{\mathrm{d}u}{\mathrm{d}\xi}(\xi_n) \doteq \frac{\mathrm{d}p}{\mathrm{d}\xi}(\xi_n) = \sum_{i=1}^{m} \frac{\mathrm{d}L_i}{\mathrm{d}\xi}(\xi_n) u(\xi_i). \tag{5.5}$$

Applying the proper scaling, we can rewrite (5.5) as (suppressing the obvious dependence of the coefficients on ξ_n)

$$\frac{\mathrm{d}u}{\mathrm{d}\xi}(\xi_n) \doteq \frac{1}{h}\sum_{i=1}^{m} a_i u(\xi_i) \qquad (n = 1, 2, \ldots, m), \tag{5.6a}$$

where the *weights* a_i are given by

$$a_i := h\frac{\mathrm{d}L_i}{\mathrm{d}\xi}(\xi_n) \qquad (i = 1, 2, \ldots, m). \tag{5.6b}$$

Note that these weights are independent of the scale in which ξ is "measured." Such "scaled" (or from a physical point of view "dimensionless") variables have obvious advantages. Actually, in this book we will encounter many such "dimensionless" variables, often with a physical interpretation. A formula like the one on the right-hand side of (5.6a) is called a *finite difference*.

Differentiating (5.2b) and taking into account that $\omega(\xi_n) = 0$ for any interpolation point ξ_n, we find the following expression for the error in (5.6a):

$$\frac{\mathrm{d}u}{\mathrm{d}\xi}(\xi_n) - \frac{1}{h}\sum_{i=1}^{m} a_i u(\xi_i) = \frac{u^{(m)}(\zeta)}{m!}\frac{\mathrm{d}\omega}{\mathrm{d}\xi}(\xi_n), \qquad \zeta \in (\alpha, \beta). \tag{5.7}$$

In view of (5.2b) we thus conclude that this error can be written as

$$\frac{\mathrm{d}u}{\mathrm{d}\xi}(\xi_n) - \frac{1}{h}\sum_{i=1}^{m} a_i u(\xi_i) = c_m h^{m-1}\frac{u^{(m)}(\zeta)}{m!}, \qquad c_m \in \mathbb{R}, \quad \zeta \in (\alpha, \beta). \tag{5.8}$$

Example 5.1 For $m = 2$ (linear interpolation) we choose $\xi_1 = \alpha$ and $\xi_2 = \beta$, giving the interpolation polynomial

$$p(\xi) = L_1(\xi)u(\xi_1) + L_2(\xi)u(\xi_2), \qquad L_1(\xi) = -\frac{1}{h}(\xi - \xi_2), \quad L_2(\xi) = \frac{1}{h}(\xi - \xi_1),$$

where h is the interval length; i.e., $h = \xi_2 - \xi_1$. The error for $\xi \in (\xi_1, \xi_2)$ is given by

$$u(\xi) - p(\xi) = \frac{u^{(2)}(\zeta)}{2}(\xi - \xi_1)(\xi - \xi_2), \qquad \zeta \in (\xi_1, \xi_2).$$

Differentiation of $p(\xi)$ gives

$$\frac{\mathrm{d}u}{\mathrm{d}\xi}(\xi) \doteq \frac{\mathrm{d}p}{\mathrm{d}\xi}(\xi) = \frac{1}{h}(u(\xi_2) - u(\xi_1)).$$

The error here is given by

$$\frac{\mathrm{d}u}{\mathrm{d}\xi}(\xi_n) - \frac{1}{h}(u(\xi_2) - u(\xi_1)) = c_n h\frac{u^{(2)}(\zeta)}{2}, \qquad n = 1, 2, \quad \zeta \in (\alpha, \beta),$$

where $c_1 = -\frac{1}{2}$ and $c_2 = \frac{1}{2}$. □

Example 5.2 If $m = 3$ and $\alpha = \xi_1 < \xi_2 < \xi_3 = \beta$, the interpolation polynomial reads

$$p(\xi) = L_1(\xi)u(\xi_1) + L_2(\xi)u(\xi_2) + L_3(\xi)u(\xi_3),$$

$$L_1(\xi) = \frac{1}{h_1(h_1 + h_2)}(\xi - \xi_2)(\xi - \xi_3), \qquad L_2(\xi) = -\frac{1}{h_1 h_2}(\xi - \xi_1)(\xi - \xi_3),$$

$$L_3(\xi) = \frac{1}{(h_1 + h_2)h_2}(\xi - \xi_1)(\xi - \xi_2),$$

where $h_1 = \xi_2 - \xi_1$ and $h_2 = \xi_3 - \xi_2$. Hence we find for the central point ξ_2 that

$$\frac{\mathrm{d}u}{\mathrm{d}\xi}(\xi_2) \doteq \frac{\mathrm{d}p}{\mathrm{d}\xi}(\xi_2) = \frac{-h_2}{h_1(h_1+h_2)}u(\xi_1) + \frac{h_2-h_1}{h_1h_2}u(\xi_2) + \frac{h_1}{(h_1+h_2)h_2}u(\xi_3).$$

The error is given by

$$\frac{\mathrm{d}u}{\mathrm{d}\xi}(\xi_2) - \frac{\mathrm{d}p}{\mathrm{d}\xi}(\xi_2) = \frac{-h_1h_2}{6}u^{(3)}(\zeta), \qquad \zeta \in (\alpha, \beta).$$

If the interpolation points are equispaced, i.e., $h_1 = h_2 = h$, the approximation becomes

$$\frac{\mathrm{d}u}{\mathrm{d}\xi}(\xi_2) \doteq \frac{1}{2h}(u(\xi_3) - u(\xi_1)). \qquad \square$$

For higher-order derivatives we can follow the same idea. In particular, we obtain for the kth derivative, where $k \leq m$, that

$$\frac{\mathrm{d}^k u}{\mathrm{d}\xi^k}(\xi_n) \doteq \frac{1}{h^k}\sum_{i=1}^{m} a_i(k)u(\xi_i), \qquad a_i(k) := h^k\frac{\mathrm{d}^k L_i}{\mathrm{d}\xi^k}(\xi_n). \tag{5.9a}$$

Using Leibniz's rule and repeatedly applying (5.3), we find the error in the difference formula to be

$$\frac{\mathrm{d}^k u}{\mathrm{d}\xi^k}(\xi_n) - \frac{1}{h^k}\sum_{i=1}^{m} a_i(k)u(\xi_i) = \sum_{l=1}^{k}\binom{k}{l}\frac{u^{(m+k-l)}(\zeta_{k,l})}{(m+k-l)!}\frac{\mathrm{d}^l\omega}{\mathrm{d}\xi^l}(\xi_n), \qquad \zeta_{k,l} \in (\alpha, \beta). \tag{5.9b}$$

In the derivation of (5.9b) we have used that $\omega(\xi_n) = 0$.

Example 5.3 Consider the second order interpolation polynomial p as derived in Example 5.1. Differentiating p twice, we obtain

$$\frac{\mathrm{d}^2 u}{\mathrm{d}\xi^2}(\xi) \doteq \frac{\mathrm{d}^2 p}{\mathrm{d}\xi^2}(\xi) = \frac{2}{h_1(h_1+h_2)}u(\xi_1) - \frac{2}{h_1h_2}u(\xi_2) + \frac{2}{(h_1+h_2)h_2}u(\xi_3).$$

In particular, if $h_1 = h_2 = h$, then we find at ξ_2 that

$$\frac{\mathrm{d}^2 u}{\mathrm{d}\xi^2}(\xi_2) \doteq \frac{1}{h^2}\big(u(\xi_1) - 2u(\xi_2) + u(\xi_3)\big),$$

with an error

$$\frac{\mathrm{d}^2 u}{\mathrm{d}\xi^2}(\xi_2) - \frac{1}{h^2}\big(u(\xi_1) - 2u(\xi_2) + u(\xi_3)\big) = \frac{h^2}{12}u^{(4)}(\zeta_{2,2}). \qquad \square$$

We can also use Lagrangian interpolation to obtain approximations of integrals. Consider

$$\int_{\alpha}^{\beta} u(\xi)\,\mathrm{d}\xi. \tag{5.10}$$

Integrating the interpolation polynomial for $u(\xi)$, we find the approximation

$$\int_{\alpha}^{\beta} u(\xi)\mathrm{d}\xi \doteq h\sum_{i=1}^{m} b_i u(\xi_i), \tag{5.11a}$$

where the weights b_i are given by

$$b_i := \frac{1}{h}\int_\alpha^\beta L_i(\xi)\,\mathrm{d}\xi \qquad (i = 1, 2, \ldots, m). \tag{5.11b}$$

The right-hand side in (5.11a) is called a *quadrature formula* with *weights* b_i. Again, b_i is "dimensionless." The expression for the error is slightly more complicated to derive; it may, e.g., contain a derivative higher than the interpolation order would suggest. For all relevant methods it turns out to be of the form

$$\int_\alpha^\beta u(\xi)\,\mathrm{d}\xi - h\sum_{i=1}^m b_i u(\xi_i) = ch^{q+1}u^{(q)}(\zeta), \qquad c \in \mathbb{R}, \quad q \in \mathbb{N}, \quad \zeta \in (\alpha, \beta). \tag{5.12}$$

Clearly $q \geq m$, since the interpolation formula is exact for polynomials of degree not more than $m - 1$.

Example 5.4 Consider the linear interpolation polynomial introduced in Example 5.1. Then integration gives

$$\int_{\xi_1}^{\xi_2} u(\xi)\,\mathrm{d}\xi \doteq \int_{\xi_1}^{\xi_2} p(\xi)\,\mathrm{d}\xi = \frac{h}{2}\big(u(\xi_1) + u(\xi_2)\big),$$

which is called the *trapezoidal formula*. In order to find the error, we use that the method is exact for polynomials of degree zero and one. Employing the method for $u(\xi) = \xi^2$, we find that the error is not zero. This then easily shows that

$$\int_{\xi_1}^{\xi_2} u(\xi)\,\mathrm{d}\xi - \frac{h}{2}(u(\xi_1) + u(\xi_2)) = -\frac{1}{12}h^3u^{(2)}(\zeta), \qquad \zeta \in (\alpha, \beta). \qquad \square$$

Example 5.5 Let us consider the case $m = 1$, but with the point ξ_1 chosen in a smart way, i.e., the middle of the interval, or $\xi_1 := \frac{1}{2}(\alpha + \beta)$. We obtain

$$u(\xi) \doteq p(\xi) = u\left(\frac{1}{2}(\alpha + \beta)\right),$$

resulting in the approximation

$$\int_\alpha^\beta u(\xi)\,\mathrm{d}\xi \doteq \int_\alpha^\beta p(\xi)\,\mathrm{d}\xi = (\beta - \alpha)u\left(\frac{1}{2}(\alpha + \beta)\right).$$

This formula is called the *midpoint formula*. It is exact for both constant polynomials and linear functions, as one can easily check. It is not exact for $u(\xi) = \xi^2$. As in the previous example, one can thus deduce that the error is given by

$$\int_\alpha^\beta u(\xi)\,\mathrm{d}\xi - (\beta - \alpha)u\left(\frac{1}{2}(\alpha + \beta)\right) = \frac{1}{24}h^3u^{(2)}(\zeta), \qquad \zeta \in (\alpha, \beta). \qquad \square$$

There are special quadrature methods that can achieve high order by choosing the points ξ_n in a special way; see section D in the appendix.

5.1.2 Finite Difference Equations

We would now like to use these approximations in the context of differential equations. Typically, we then have to approximate an equation like

$$\frac{\mathrm{d}u}{\mathrm{d}\xi} = f(u). \tag{5.13}$$

Here f may still be a complicated function (in particular containing derivatives of other variables). If we replace the derivative $\frac{\mathrm{d}u}{\mathrm{d}\xi}$ at the points ξ_n $(n = 1, 2, \ldots, m)$ by the finite difference formula (5.6), we obtain, for the numerical approximation u_i of $u(\xi_i)$ (i.e., $u_i \doteq u(\xi_i)$), the *difference equation*

$$\frac{1}{h}\sum_{i=1}^{m} a_i u_i = f(u_n), \qquad n = 1, 2, \ldots, m, \tag{5.14}$$

with the coefficients a_i defined in (5.6b). If we have a kth order ODE

$$\frac{\mathrm{d}^k u}{\mathrm{d}\xi^k} = f(u), \tag{5.15}$$

the difference equation reads

$$\frac{1}{h^k}\sum_{i=1}^{m} a_i(k) u_i = f(u_n), \qquad n = 1, 2, \ldots, m. \tag{5.16}$$

Naturally, in order to make (5.16) nontrivial, we need to require $m \geq k + 1$.

Example 5.6 Let us summarise what we have derived in the previous examples. For (5.13) we obtain from Example 5.1 the difference equations

$$\frac{1}{h}\big(u(\xi_2) - u(\xi_1)\big) = f(u(\xi_1)) \qquad \textit{(explicit Euler difference equation)},$$

$$\frac{1}{h}\big(u(\xi_2) - u(\xi_1)\big) = f(u(\xi_2)) \qquad \textit{(implicit Euler difference equation)},$$

and from Example 5.2 we obtain

$$\frac{-h_2}{h_1(h_1+h_2)}u(\xi_1) + \frac{h_2 - h_1}{h_1 h_2}u(\xi_2) + \frac{h_1}{(h_1+h_2)h_2}u(\xi_3) = f(u(\xi_n)) \qquad (n = 1, 2, 3).$$

For (5.15) with $k = 2$ we obtain from Example 5.3 the difference scheme

$$\frac{1}{h^2}\big(u(\xi_1) - 2u(\xi_2) + u(\xi_3)\big) = f(u(\xi_2)). \qquad \square$$

There is an alternative way to obtain a difference equation for (5.13). We can integrate (5.13) over an interval (ξ_i, ξ_n) say; i.e.,

$$u(\xi_n) - u(\xi_i) = \int_{\xi_i}^{\xi_n} f(u(\xi))\,\mathrm{d}\xi. \tag{5.17}$$

So we have reformulated the differential equation as an integral equation. The integral in (5.17) is then approximated by a suitable quadrature rule; cf. (5.11). This leads to the relation

$$u(\xi_n) - u(\xi_i) \doteq h \sum_{j=1}^{m} b_j f(u(\xi_j)), \qquad b_j := \frac{1}{h} \int_{\xi_i}^{\xi_n} L_j(\xi)\, \mathrm{d}\xi. \tag{5.18}$$

The resulting difference equation can be written as (again with $u_i \doteq u(\xi_i)$ and $h := \max_{2 \le n \le m} |\xi_n - \xi_{n-1}|$)

$$u_n - u_i = h \sum_{j=1}^{m} b_j f(u_j), \qquad n = 2, 3, \ldots, m, \quad i = 1, 2, \ldots, n-1. \tag{5.19}$$

Difference equations are used to solve a differential equation on a domain. So if the differential equation (5.13) is defined for $\xi \in (\alpha, \beta)$, we have to choose specific points at which we would like to apply the difference equation. This process is called *discretisation* and the result is called a *grid* or *mesh*, which consists of a set of points $\{\eta_k\}_{k=1}^{N}$, say. If the points are equispaced, we call the grid *uniform*; in this case we see that the *grid size* is $h = (\beta - \alpha)/(N-1)$. At each grid point η_k we have to apply a difference equation like (5.14) or (5.19). If the number of interpolation points m equals one or two, this is trivial. If $m > 2$, two typical choices are often taken:

(i) The nodes $\xi_1, \xi_2, \ldots, \xi_m$ are chosen from (neighbouring) grid points η_n.

(ii) The nodes $\xi_2, \ldots, \xi_{m-1}$ are chosen as intermediate points on $(\xi_1, \xi_m) = (\eta_k, \eta_{k+1})$.

For the choice (i) we take the grid points $\eta_{k+1-r-(m-1)}, \ldots, \eta_{k+1-r}$ as interpolation points and apply either (5.14) or (5.19). We would like to write the latter difference equation in a form commensurate with the differential equation; i.e.,

$$\frac{1}{h}\big(u_{k+1} - u_k\big) = \sum_{j=r}^{r+m-1} b_j f(u_{k+1-j}), \qquad r = 0, 1, \ldots, m-1, \tag{5.20}$$

which is a *multistep* difference equation (because it involves more than one step, at least for $m \ge 1$). In an initial value context it makes sense to choose $r = 0$ or $r = 1$. We may consider more general difference equations, of which (5.14) and (5.20) are special cases:

$$\frac{1}{h} \sum_{j=0}^{m-1} a_j u_{k+1-j} = \sum_{j=r}^{r+m-1} b_j f(u_{k+1-j}), \qquad r = 0, 1. \tag{5.21}$$

For the choice (ii) the intermediate points need to be eliminated from the equation somehow because they do not correspond to grid points. Let us start with the relation (5.17), with $\xi_1 = \eta_k$ and $\xi_n = \eta_{k+1}$, so

$$u(\eta_{k+1}) - u(\eta_k) = \int_{\eta_k}^{\eta_{k+1}} f\big(u(\xi)\big)\, \mathrm{d}\xi \doteq h \sum_{j=1}^{m} b_j f\big(u(\xi_j)\big). \tag{5.22}$$

The idea now is to simply use (5.17) again in order to find extra relations for each of the function values $u(\xi_j)$ such that they can be eliminated. Of course we again have to approximate the introduced integrals by some quadrature formula. So we obtain

$$u(\xi_j) - u(\xi_1) = u(\xi_j) - u(\eta_k) = \int_{\eta_k}^{\xi_j} f\big(u(\xi)\big)\,\mathrm{d}\xi \doteq h \sum_{l=1}^{m} \gamma_{jl} f\big(u(\xi_l)\big), \qquad j = 2, 3, \ldots, m-1, \tag{5.23}$$

for some $\gamma_{j1}, \ldots, \gamma_{jm}$. From this we thus find

$$\frac{1}{h}\big(u_{k+1} - u_k\big) = \sum_{i=1}^{m} b_i k_i, \qquad k_i := f\left(u_k + h \sum_{j=1}^{m} \gamma_{ij} k_j\right). \tag{5.24}$$

This process is sometimes called *collocation*. Formula (5.24) is clearly a one-step difference method. The integer m is called the number of *stages* of the method.

Example 5.7 Consider the (first order) interpolation formula defined on the interval (η_k, η_{k+1}):

$$p(\xi) = f\big(u(\xi_1)\big), \qquad \xi_1 := \frac{1}{2}(\eta_k + \eta_{k+1}).$$

This way, we find the approximation

$$u(\eta_{k+1}) - u(\eta_k) = \int_{\eta_k}^{\eta_{k+1}} f(\xi)\,\mathrm{d}\xi \doteq hf\big(u(\xi_1)\big),$$

with $h = \eta_{k+1} - \eta_k$ the grid size, from which we have to eliminate $u(\xi_1)$. To this end we use the relation found from a simple first order quadrature formula

$$u(\xi_1) - u(\eta_k) \doteq \frac{1}{2}hf\big(u(\eta_k)\big).$$

Combining of these two relations gives

$$u_{k+1} - u_k = hf\left(u_k + \frac{1}{2}hf(u_k)\right).$$

□

Since we are replacing the differential equation by a difference equation, we are making an error in general. A useful concept to measure this is the *local discretisation error* $\delta(\eta_k)$, say. In order to let the difference equation be commensurate with the differential equation, we shall define the discretisation error for *the properly scaled form* (cf. (5.21)). We thus have

$$\delta(\eta_k) := \frac{1}{h} \sum_{j=0}^{m-1} a_j u(\eta_{k+1-j}) - \sum_{j=r}^{r+m-1} b_j f(u(\eta_{k+1-j})). \tag{5.25}$$

Hence $\delta(\eta_k)$ can be found by *substituting the exact solution and determining the residual*. Similarly, for (5.24) we define

$$\delta(\eta_k) := \frac{1}{h}(u(\eta_{k+1}) - u(\eta_k)) - \sum_{i=1}^{m} b_i k_i^*, \qquad k_i^* := f\left(u(\eta_k) + h \sum_{j=1}^{m} \gamma_{ij} k_j^*\right). \tag{5.26}$$

The local discretisation error $\delta(\eta_k)$ is always an expression of the form

$$\delta(\eta_k) = ch^p f^{(p+1)}(\vartheta), \qquad c \in \mathbb{R}, \quad p \in \mathbb{N}, \quad \vartheta \in (\eta_k, \eta_{k+1}), \tag{5.27}$$

and we say that $\delta(\eta_k)$ is of *order* p. The higher the order, the larger the value of m required in general. However, there is no simple rule relating these (cf. [21]).

Example 5.8 Consider the difference equation (cf. Example 5.6)

$$u_{k+1} - u_k = hf(u_k).$$

Substituting $f(u(\eta_k)) = u'(\eta_k)$, we find

$$\delta(\eta_k) = \frac{1}{h}\big(u(\eta_{k+1}) - u(\eta_k)\big) - u'(\eta_k),$$

where $h := \eta_{k+1} - \eta_k$. If we use a Taylor series expansion around η_k, we obtain

$$\begin{aligned}\delta(\eta_k) &= \frac{1}{h}\left(u(\eta_k) + hu'(\eta_k) + \frac{1}{2}h^2 u''(\theta) - u(\eta_k)\right) - u'(\eta_k)\\ &= \frac{1}{2}hu''(\vartheta), \qquad \vartheta \in (\eta_k, \eta_{k+1}).\end{aligned}$$

□

5.2 Finite Difference Methods for Spatial Variables

The difference equations we defined in the previous section can be applied to space and time derivatives separately. In this section we consider spatial derivatives first.

5.2.1 Finite Difference Approximations

By far the oldest method of approximating derivatives is by using finite differences. To start with, consider for a function $u = u(x)$ approximations of $\frac{\mathrm{d}}{\mathrm{d}x}u$ and $\frac{\mathrm{d}^2}{\mathrm{d}x^2}u$. Using the basic ideas outlined in Section 5.1, we find the difference formulas

$$\frac{\mathrm{d}u}{\mathrm{d}x}(x) \doteq \frac{1}{h}\big(u(x+h) - u(x)\big) \qquad \textit{(forward difference)}, \tag{5.28a}$$

$$\frac{\mathrm{d}u}{\mathrm{d}x}(x) \doteq \frac{1}{h}\big(u(x) - u(x-h)\big) \qquad \textit{(backward difference)}, \tag{5.28b}$$

$$\frac{\mathrm{d}u}{\mathrm{d}x}(x) \doteq \frac{1}{2h}\big(u(x+h) - u(x-h)\big) \qquad \textit{(central difference)}. \tag{5.28c}$$

Employing Taylor series expansions, we obtain the following expressions for the corresponding local discretisation errors:

$$\delta(x) = -\frac{1}{2}h\frac{\mathrm{d}^2u}{\mathrm{d}x^2}(\xi), \qquad \xi \in (x, x+h), \tag{5.29a}$$

$$\delta(x) = +\frac{1}{2}h\frac{\mathrm{d}^2u}{\mathrm{d}x^2}(\xi), \qquad \xi \in (x-h, x), \tag{5.29b}$$

$$\delta(x) = -\frac{1}{6}h^2\frac{\mathrm{d}^3u}{\mathrm{d}x^3}(\xi), \qquad \xi \in (x-h, x+h). \tag{5.29c}$$

For the second order derivative $\frac{\mathrm{d}^2}{\mathrm{d}x^2}u$ we find the (second order) *central difference*

$$\frac{\mathrm{d}^2u}{\mathrm{d}x^2}(x) \doteq \frac{2}{h_1(h_1+h_2)}u(x-h_1) - \frac{2}{h_1h_2}u(x) + \frac{2}{(h_1+h_2)h_2}u(x+h_2). \tag{5.30}$$

Expanding in a Taylor series around x, we find for the local discretisation error (defined in an obvious way, similar to the first order case)

$$\delta(x) = \frac{h_2-h_1}{3}\frac{\mathrm{d}^3u}{\mathrm{d}x^3}(x) - \frac{1}{12}\frac{h_1^3+h_2^3}{h_1+h_2}\frac{\mathrm{d}^4u}{\mathrm{d}x^4}(\xi), \qquad \xi \in (x-h_1, x+h_2). \tag{5.31}$$

Clearly, if h_1 is significantly different from h_2, we have $\delta(x) = \mathcal{O}(h)$, where $h := \max(h_1, h_2)$. On the other hand, if $h_1 = h_2 = h$, we have $\delta(x) = \mathcal{O}(h^2)$.

Often second order derivatives appear in a differential equation as $\frac{\mathrm{d}}{\mathrm{d}x}(a(x)\frac{\mathrm{d}}{\mathrm{d}x}u)$. One may approximate these by applying (5.28a) and (5.28b) as follows:

$$\begin{aligned}\frac{\mathrm{d}}{\mathrm{d}x}\left(a(x)\frac{\mathrm{d}u}{\mathrm{d}x}\right)(x) &\doteq \frac{1}{h}\frac{\mathrm{d}}{\mathrm{d}x}\Big(a(x)\big(u(x+h)-u(x)\big)\Big)\\ &\doteq \frac{1}{h^2}\Big(a(x)\big(u(x+h)-u(x)\big) - a(x-h)\big(u(x)-u(x-h)\big)\Big)\\ &= \frac{1}{h^2}\Big(a(x)u(x+h) - \big(a(x)+a(x-h)\big)u(x) + a(x-h)u(x-h)\Big). \end{aligned}\tag{5.32a}$$

One can do better by making the second derivative more symmetric through the use of central differences. For this we note that (5.28c) can also be seen as an average of (5.28a) and (5.28b). Applying this idea, we find

$$\begin{aligned}\frac{\mathrm{d}}{\mathrm{d}x}\left(a(x)\frac{\mathrm{d}u}{\mathrm{d}x}\right)(x) &\doteq \frac{1}{h}\frac{\mathrm{d}}{\mathrm{d}x}\left(a\left(x+\frac{1}{2}h\right)\big(u(x+h)-u(x)\big)\right)\\ &\doteq \frac{1}{h^2}\left[a\left(x+\frac{1}{2}h\right)\big(u(x+h)-u(x)\big) - a\left(x-\frac{1}{2}h\right)\big(u(x)-u(x-h)\big)\right]\\ &= \frac{1}{h^2}\left[a\left(x+\frac{1}{2}h\right)u(x+h) - \left(a\left(x+\frac{1}{2}h\right)+a\left(x-\frac{1}{2}h\right)\right)u(x)\right.\\ &\qquad \left.+ a\left(x-\frac{1}{2}h\right)u(x-h)\right]. \end{aligned}\tag{5.32b}$$

Both difference formulas can be shown to be second order; we leave this as an exercise.

The finite difference approximations mentioned in this section are often referred to as a *difference scheme*. Such a scheme is typically of a local nature and at a certain point describes the relation of the value of the unknown to that of its neighbours. The points involved in this are called the *stencil*. Since a numerical scheme will usually only give an approximation of the derivatives, we shall use a special notation for the approximate solution (see Section 5.1). Given the points x_k, $k = 1, 2, \ldots, M$, on the grid we shall write $u_k \doteq u(x_k)$.

5.2.2 Finite Difference Approximations in Several Dimensions

If we have several spatial dimensions, we can apply the difference schemes given in Section 5.2.1 to each of the variables separately. In particular, for the Laplace operator ∇^2 in two dimensions we thus find, with mesh widths Δx in the x direction and Δy in the y direction, that

$$\mathcal{L}[u] := \nabla^2 u = \frac{\partial^2 u}{\partial x^2} + \frac{\partial^2 u}{\partial y^2} \doteq \frac{1}{\Delta x^2}\big[u(x+\Delta x, y) - 2u(x, y) + u(x-\Delta x, y)\big] + \frac{1}{\Delta y^2}\big[u(x, y+\Delta y) - 2u(x, y) + u(x, y-\Delta y)\big]. \quad (5.33)$$

In this book we shall often use subscripts j and k for the approximate solution at the mesh points in the x and y directions, respectively; more precisely, we have $u_{j,k} \doteq u(x_j, y_k)$. If we denote the numerical approximation of the Laplace operator by $\mathcal{L}_\Delta$, we may write (5.33) as (cf. Figure 5.1)

$$\mathcal{L}_\Delta[u_{j,k}] := \frac{1}{\Delta x^2}\big(u_{j+1,k} - 2u_{j,k} + u_{j-1,k}\big) + \frac{1}{\Delta y^2}\big(u_{j,k+1} - 2u_{j,k} + u_{j,k-1}\big). \quad (5.34a)$$

Alternatively, to indicate the positions of points in a stencil one often employs directions on a compass, like in Figure 5.1, with respect to a central point C, say. The approximate solution at C is written as u_C. So (5.33) then becomes

$$\mathcal{L}_\Delta[u_C] = \frac{1}{\Delta x^2}\big(u_E - 2u_C + u_W\big) + \frac{1}{\Delta y^2}\big(u_N - 2u_C + u_S\big). \quad (5.34b)$$

Unlike the one-dimensional case, the two-dimensional case poses a problem related to the geometry of the domain, even for nonuniform grids. Indeed, a rectangular grid will rarely fit in with boundary curves. Hence we have to adapt the stencil appropriately for points neighbouring the boundaries. If we have, e.g., a problem like in Figure 5.2, with uniform mesh width h, then we can easily employ the difference formula (5.30). Defining $\alpha := \frac{1}{h}(x_{E'} - x_P)$, we simply find

$$\left(\frac{\partial^2 u}{\partial x^2}\right)_C \doteq \frac{2}{h^2}\left(\frac{1}{\alpha+1}u_W - \frac{1}{\alpha}u_P + \frac{1}{\alpha(\alpha+1)}u_{E'}\right), \quad (5.35)$$

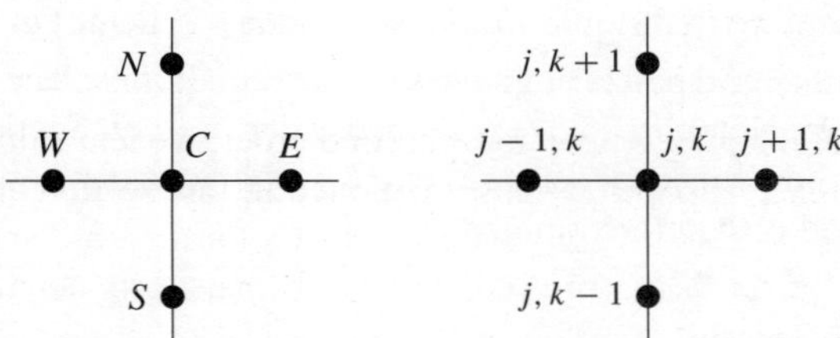

Figure 5.1. *Stencil of the Laplace operator in compass notation (left) and index notation (right).*

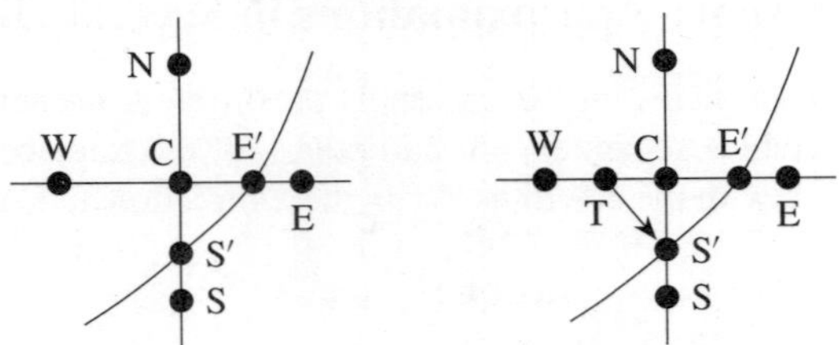

Figure 5.2. *Stencil of the Laplace operator near a boundary.*

and likewise for $\frac{\partial^2}{\partial y^2}u$. Note that this only makes sense if $u_{\mathrm{E}'}$ is given, i.e., if we know u at the boundary point E′. If we have a derivative in the boundary condition, the situation is more complicated. Consider Figure 5.2, where we have to employ the boundary condition $\frac{\partial u}{\partial n}$ given at S′. Let us define

$$\alpha := \frac{1}{h}(x_{\mathrm{C}} - x_{\mathrm{T}}), \quad \beta := \frac{1}{h}(y_{\mathrm{C}} - y_{\mathrm{S}'}), \quad \gamma := \sqrt{\alpha^2 + \beta^2}.$$

Then

$$\left(\frac{\partial u}{\partial n}\right)_{\mathrm{S}'} \doteq \frac{1}{\gamma h}\left(u_{\mathrm{S}'} - u_{\mathrm{T}}\right). \tag{5.36}$$

By linear interpolation we find

$$u_{\mathrm{T}} \doteq \alpha u_{\mathrm{W}} + (1-\alpha)u_{\mathrm{C}}. \tag{5.37}$$

Using (5.36), we can therefore express $u_{\mathrm{S}'}$ in values at grid points and the given boundary condition $\left(\frac{\partial u}{\partial n}\right)_{\mathrm{S}'}$:

$$u_{\mathrm{S}'} \doteq \alpha u_{\mathrm{W}} + (1-\alpha)u_{\mathrm{C}} + \gamma h\left(\frac{\partial u}{\partial n}\right)_{\mathrm{S}'}. \tag{5.38}$$

The value $u_{\mathrm{S}'}$ shows up in discretising $\frac{\partial^2 u}{\partial y^2}$ (cf. (5.30)):

$$\left(\frac{\partial^2 u}{\partial y^2}\right)_{\mathrm{C}} \doteq \frac{2}{h^2}\left(\frac{1}{\beta(\beta+1)}u_{\mathrm{S}'} - \frac{1}{\beta}u_{\mathrm{C}} + \frac{1}{\beta+1}u_{\mathrm{N}}\right). \tag{5.39}$$

Clearly, upon substituting (5.38) into (5.39), we obtain a scheme for discretising $\left(\frac{\partial^2}{\partial y^2}u\right)_{\mathrm{C}}$, with unknowns at either grid points or given boundary conditions. The treatment for $\left(\frac{\partial}{\partial x}u\right)_{\mathrm{E}'}$ goes in a similar way. Note that (5.37) is second order in h; cf. Section 5.1. Although we have considered two-dimensional schemes exclusively, the extension to three dimensions is straightforward and is therefore omitted.

5.2.3 Other Coordinate Systems

We conclude this section by considering some special coordinate systems, e.g., because these are better suited to the geometry of the problem or because we want to take advantage of certain symmetries. Let us start with the Poisson problem in polar coordinates for smooth source f and regular and continuous u:

$$\frac{1}{r}\frac{\partial}{\partial r}\left(r\frac{\partial u}{\partial r}\right)+\frac{1}{r^2}\frac{\partial^2 u}{\partial \phi^2}=f(r,\phi). \tag{5.40}$$

First we define a grid with radial mesh width Δr and angular mesh width $\Delta\phi$. The points on the circle with radius r have an angular coordinate $k\Delta\phi$ where k increases in a counterclockwise direction. Let $u_{j,k}$ be the approximation of $u(j\Delta r, k\Delta\phi)$. To discretise the various terms in the Laplace operator we can now use any of the difference methods we have seen so far; see, e.g., (5.30) and (5.32). We will not elaborate on this here in general. The main problem is the singularity of the coordinate system at the origin. One way to deal with this is to apply averaging of the values around $r=0$. Let $\Omega := \mathcal{D}(\mathbf{0}, \frac{1}{2}\Delta r)$ denote the disc with radius $\frac{1}{2}\Delta r$ around the origin $\mathbf{0}$. Then we find

$$\iint_\Omega \left(\frac{1}{r}\frac{\partial}{\partial r}\left(r\frac{\partial u}{\partial r}\right)+\frac{1}{r^2}\frac{\partial^2 u}{\partial \phi^2}\right) r\mathrm{d}r\mathrm{d}\phi=\frac{1}{2}\Delta r\int_0^{2\pi}\frac{\partial u}{\partial r}\left(\frac{1}{2}\Delta r,\phi\right)\mathrm{d}\phi. \tag{5.41}$$

We now use central differences to approximate $\frac{\partial}{\partial r}u(\frac{1}{2}\Delta r,\phi)$ in (5.41). We thus have

$$\frac{1}{2}\Delta r\int_0^{2\pi}\frac{\partial u}{\partial r}\left(\frac{1}{2}\Delta r,\phi\right)\mathrm{d}\phi \doteq \frac{1}{2}\Delta r\sum_{k=1}^{N}\frac{u_{1,k}-u_{0,k}}{\Delta r}\Delta\phi,$$

where $N=2\pi/\Delta\phi$ is the number of grid points in the ϕ direction. In general the source term f also needs to be integrated. To do this numerically we approximate it by its value at $\mathbf{0}$. So we have

$$\iint_\Omega f(r,\phi)r\,\mathrm{d}r\mathrm{d}\phi \doteq \frac{1}{4}\pi\Delta r^2 f(\mathbf{0}).$$

Combining these results, we thus find an equation involving $u_{0,0}$:

$$-u_{0,0}+\frac{1}{N}\sum_{k=1}^{N}u_{1,k}=\frac{1}{4}\Delta r^2 f(\mathbf{0}), \tag{5.42}$$

where we have used continuity so that $u_{0,k}=u_{0,0}$ for $k=0,1,\ldots,N$.

Another way to handle the singular origin is by applying Taylor series. We will demonstrate this approach in three dimensions as follows. In cylindrical coordinates r,ϕ,z (see Figure 5.3) we have

$$\nabla^2 u=\frac{1}{r}\frac{\partial}{\partial r}\left(r\frac{\partial u}{\partial r}\right)+\frac{1}{r^2}\frac{\partial^2 u}{\partial \phi^2}+\frac{\partial^2 u}{\partial z^2}, \tag{5.43}$$

and in spherical coordinates r,θ,ϕ we have

$$\nabla^2 u=\frac{1}{r^2}\frac{\partial}{\partial r}\left(r^2\frac{\partial u}{\partial r}\right)+\frac{1}{r^2\sin\theta}\frac{\partial}{\partial\theta}\left(\sin\theta\frac{\partial u}{\partial\theta}\right)+\frac{1}{r^2\sin^2\theta}\frac{\partial^2 u}{\partial\phi^2}. \tag{5.44}$$

With *cylindrical symmetry* the second term on the right-hand side of (5.43) vanishes, while all but the first term on the right-hand side of (5.44) vanish with *spherical symmetry*.

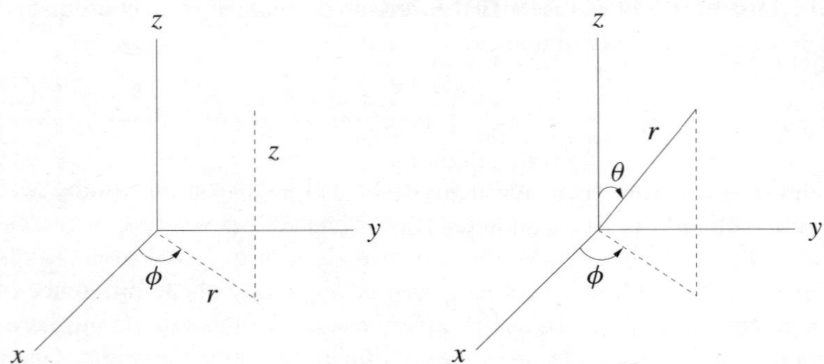

Figure 5.3. *Cylindrical (left) and spherical (right) coordinates.*

Suppressing any dependence on other variables, we expand $u(r)$ around $u(0)$ to obtain

$$u(r) = u(0) + \frac{1}{2}r^2\frac{\partial^2 u}{\partial r^2}(0) + \mathcal{O}(r^3), \tag{5.45}$$

where we have $\frac{\partial}{\partial r}u(0) = 0$ because of regularity. By inserting this expansion into (5.43) or (5.44), we find

$$\nabla^2 u = \frac{1}{r}\frac{\partial}{\partial r}\left(r\frac{\partial u}{\partial r}\right) + \frac{\partial^2 u}{\partial z^2} = 2\frac{\partial^2 u}{\partial r^2}(0) + \frac{\partial^2 u}{\partial z^2}(0) + \mathcal{O}(r) \tag{5.46a}$$

in cylindrical coordinates and

$$\nabla^2 u = \frac{1}{r^2}\frac{\partial}{\partial r}\left(r^2\frac{\partial u}{\partial r}\right) = 3\frac{\partial^2 u}{\partial r^2}(0) + \mathcal{O}(r) \tag{5.46b}$$

in spherical coordinates. Note that the θ and ϕ derivatives vanish at $r = 0$ because u is continuous. Once more employing the Taylor expansion (5.45) at $r = \Delta r$, we can derive the difference approximations

$$\frac{1}{r}\frac{\partial}{\partial r}\left(r\frac{\partial u}{\partial r}\right) \doteq \frac{4}{\Delta r^2}\big(u(\Delta r) - u(0)\big), \tag{5.47a}$$

$$\frac{1}{r^2}\frac{\partial}{\partial r}\left(r^2\frac{\partial u}{\partial r}\right) \doteq \frac{6}{\Delta r^2}\big(u(\Delta r) - u(0)\big) \tag{5.47b}$$

in cylindrical and spherical coordinates, respectively.

5.3 Finite Volume Methods

Rather than directly using finite differences to approximate spatial derivatives, we may first rewrite the PDE in its integral form. In later chapters we shall see that this is actually physically more natural, allowing the occurrence of discontinuities that are excluded in the differential formulation.

5.3.1 The One-Dimensional Case

Consider the ODE

$$\frac{\mathrm{d}f(u)}{\mathrm{d}x} = s(u), \tag{5.48}$$

where $f(u)$ is a (complicated) function of u and/or derivatives of u; however, for the sake of brevity we just write $f(u)$. We shall call $f(u)$ the *flux*, anticipating a concept that will be encountered later. In this section we consider the simple form

$$f(u) := a\frac{\mathrm{d}u}{\mathrm{d}x} + bu. \tag{5.49}$$

The separate terms $a\frac{\mathrm{d}}{\mathrm{d}x}u$ and bu are sometimes referred to as the *conductive flux* and the *convective flux*, respectively. Furthermore, $s(u)$ is a function of u only and represents a source term in physical applications.

On an arbitrary interval (α, β) we find by integration that

$$f\big(u(\beta)\big) - f\big(u(\alpha)\big) = \int_\alpha^\beta s(u(x))\,\mathrm{d}x. \tag{5.50}$$

If $a = 0$, we obtain a pure integral equation in (5.50), which we can solve numerically by applying suitable quadrature formulas, as was done in Section 5.1. If $a \neq 0$, we need to find a way to approximate the derivative term appropriately, e.g., through a finite difference formula. The finite volume method is based on this integral formulation of (5.48). In fact, we can distinguish between two variants of the finite volume method based on the choice of the interval (α, β).

Suppose we have a uniform grid $x_1, x_2, \ldots, x_M$ of grid size Δx on which we want to compute the solution $u(x)$ of (5.48). In the *cell-centered* method we apply the integral relation to intervals $(\alpha, \beta) = (x_{j-1/2}, x_{j+1/2}) =: V_j$, where the intermediate point $x_{j+1/2}$ is located halfway between the grid points x_j and x_{j+1}, i.e., $x_{j+1/2} := \frac{1}{2}(x_j + x_{j+1})$, and likewise for $x_{j-1/2}$. The interval V_j is called the *control volume*; see Figure 5.4. For this choice of the control volume it is obvious to approximate the integral of $s(u)$ in (5.50) by the midpoint rule; cf. Example 5.5. We find

$$\int_{x_{j-\frac{1}{2}}}^{x_{j+\frac{1}{2}}} s(u(x))\,\mathrm{d}x \doteq s(u(x_j))\Delta x. \tag{5.51}$$

Next we have to approximate the flux $f(u)$ at the cell boundaries $x_{j\pm1/2}$. For the conductive part it is natural to employ central differences, and for the convective part we simply use

Figure 5.4. *Control volume for a cell-centered (left) and a vertex-centered (right) finite volume method.*

linear interpolation, resulting in the approximation

$$f\big(u(x_{j+\frac{1}{2}})\big) \doteq \frac{a}{\Delta x}\big(u(x_{j+1}) - u(x_j)\big) + \frac{1}{2}b\big(u(x_j) + u(x_{j+1})\big). \tag{5.52}$$

A similar expression holds for $f(u(x_{j-1/2}))$. Combining the approximations in (5.51) and (5.52) with the integral formulation (5.50), we obtain the difference scheme

$$\frac{a}{\Delta x}\big(u_{j+1} - 2u_j + u_{j-1}\big) + \frac{b}{2}\big(u_{j+1} - u_{j-1}\big) = s(u_j)\Delta x, \tag{5.53}$$

where u_j denotes the numerical approximation of $u(x_j)$. Of course, in this case the method turns out to be exactly the same as would result from applying central differences (5.28c) and (5.30) directly. In Chapter 13 this finite volume approach will turn out to have a clear advantage over straightforward finite differences when solving conservation laws numerically.

Alternatively, in the *vertex-centered* method, we take for the jth control volume $V_j = (x_j, x_{j+1})$; i.e., the control volumes coincide with the intervals between grid points. The integral of $s(u)$ is in this case conveniently approximated by the trapezoidal rule; i.e.,

$$\int_{x_j}^{x_{j+1}} s(u(x))\,\mathrm{d}x \doteq \frac{1}{2}\Delta x\big[s\big(u(x_j)\big) + s\big(u(x_{j+1})\big)\big]. \tag{5.54}$$

To approximate the flux $f(u(x_j))$ at the left edge (vertex) of V_j we may take the backward difference formula (BDF) for the conductive part, whereas the convective part needs no approximation. This yields

$$f\big(u(x_j)\big) \doteq \frac{a}{\Delta x}\big(u(x_j) - u(x_{j-1})\big) + bu(x_j), \tag{5.55}$$

and likewise for $f(u(x_{j+1}))$. Hence we obtain the scheme

$$\frac{a}{\Delta x}\big(u_{j+1} - 2u_j + u_{j-1}\big) + b\big(u_{j+1} - u_j\big) = \frac{1}{2}\Delta x\big(s(u_j) + s(u_{j+1})\big). \tag{5.56}$$

Note that this is not symmetric with respect to the middle point x_j. Another possibility is to choose $V_j = (x_{j-1}, x_j)$ and employ the forward difference formula for the conductive part of the flux.

5.3.2 The Multidimensional Case

The derivation of the finite volume method for a differential equation in a two- or three-dimensional domain is similar to the one-dimensional case discussed in the previous section. The starting point is the generalisation of (5.48) to

$$\nabla\cdot \boldsymbol{f}(u) = s(u), \tag{5.57}$$

where $\boldsymbol{f}(u)$ is now the flux vector, generally depending on u and ∇u, and $s(u)$ is the source term, solely depending on u. Equation (5.57) is an example of a *conservation law*. For the flux vector we adopt the model

$$\boldsymbol{f}(u) = a\nabla u + \boldsymbol{b}u, \tag{5.58}$$

with $a\nabla u$ and $\boldsymbol{b}u$ the conductive and convective parts, respectively.

Integrating (5.57) over an arbitrary domain $\Omega \subset \mathbb{R}^d$ $(d = 2, 3)$ and using Gauss's theorem (see appendix section (J.13)), we obtain

$$\oint_{\partial\Omega} \boldsymbol{f}(u) \cdot \boldsymbol{n} \, \mathrm{d}S = \int_{\Omega} s(u) \, \mathrm{d}V, \tag{5.59}$$

where $\boldsymbol{n}$ is the unit outward normal on the boundary $\partial\Omega$ of Ω. The finite volume formulation is based on this integral formulation of (5.57). More precisely, in the finite volume method the computational domain is covered with a grid and a set of nonoverlapping control volumes, where a numerical approximation of (5.59) is applied. The arrangement of grid points and control volumes determines the method. In what follows we restrict ourselves to the cell-centred method, where grid points are located in the centre of the control volumes. Furthermore, for the sake of simplicity we restrict ourselves to a two-dimensional domain covered with a rectangular grid.

Consider the control volume V_{C} as shown in Figure 5.5. In the cell-centred finite volume method we apply relation (5.59) to this control volume and approximate the integrals involved by suitable quadrature rules. Naturally, the midpoint rule is again an obvious choice to approximate the integral of $s(u)$, and we find, by applying this in both variables, that

$$\int_{V_{\mathrm{C}}} s(u) \, \mathrm{d}V \doteq s(u(\boldsymbol{x}_{\mathrm{C}})) \Delta x \Delta y. \tag{5.60}$$

The contour integral of the flux in (5.59) has to be split into four parts corresponding with the four sides of $\Gamma = \partial\Omega$. The orientation of Γ is as indicated in Figure 5.5. Let $\boldsymbol{f} = (f_1, f_2)^T$. Then we have

$$\oint_{\partial V_{\mathrm{C}}} \boldsymbol{f}(u) \cdot \boldsymbol{n} \, \mathrm{d}S = \int_{\Gamma_{\mathrm{e}}} f_1(u(x_{\mathrm{e}}, \tau)) \, \mathrm{d}\tau + \int_{\Gamma_{\mathrm{n}}} f_2(u(\tau, y_{\mathrm{n}})) \, \mathrm{d}\tau - \int_{\Gamma_{\mathrm{w}}} f_1(u(x_{\mathrm{w}}, \tau)) \, \mathrm{d}\tau - \int_{\Gamma_{\mathrm{s}}} f_2(u(\tau, y_{\mathrm{s}})) \, \mathrm{d}\tau, \tag{5.61}$$

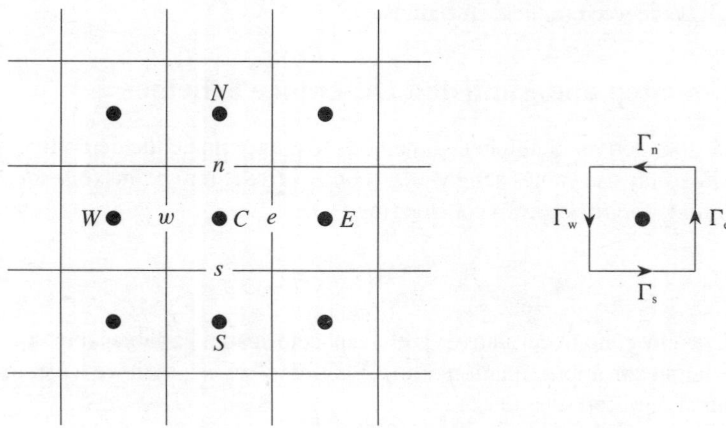

Figure 5.5. *Two-dimensional control volume for a cell-centered finite volume method.*

where τ is the parameter describing Γ. Next we approximate each of these integrals by the midpoint rule to find

$$\oint_{\partial V_C} \boldsymbol{f}(u)\cdot\boldsymbol{n}\,\mathrm{d}S \doteq \big[f_1(u(\boldsymbol{x}_e)) - f_1(u(\boldsymbol{x}_w))\big]\Delta y + \big[f_2(u(\boldsymbol{x}_n)) - f_2(u(\boldsymbol{x}_s))\big]\Delta x, \tag{5.62}$$

where $\boldsymbol{x}_e$ denotes the centre of the eastern cell face Γ_e, etc.; see Figure 5.5. Substituting the approximations (5.60) and (5.62) into (5.59), we obtain

$$\big(F_{1,e} - F_{1,w}\big)\Delta y + \big(F_{2,n} - F_{2,s}\big)\Delta x = s(u_C)\Delta x\Delta y, \tag{5.63}$$

with u_C and $F_{1,e}$ the numerical approximations of $u(\boldsymbol{x}_C)$ and $f_1(u(\boldsymbol{x}_e))$, respectively, etc. Finally, we have to determine the numerical fluxes. Like in (5.52) we have

$$F_{1,e} = \frac{a}{\Delta x}(u_E - u_C) + \frac{1}{2}b_1(u_C + u_E), \tag{5.64}$$

and similar expressions hold for the other numerical fluxes. Substituting these expressions in the discrete conservation law (5.63) and rearranging terms, we finally obtain

$$\frac{a}{\Delta x^2}(u_E - 2u_C + u_W) + \frac{b_1}{2\Delta x}(u_E - u_W) + \frac{a}{\Delta y^2}(u_N - 2u_C + u_S) + \frac{b_2}{2\Delta y}(u_N - u_S) = s(u_C). \tag{5.65}$$

5.4 Difference Methods for Initial Value Problems

When discretising the time derivative, we don't only encounter problems due to approximation. Indeed, evolution equations, continuous or discrete, involve questions relating to (non)growth. The growth behaviour of the discrete solution may be completely different from that of the continuous one. For obvious reasons we would like this behaviour to be the same; i.e., the numerical solution should mimic the physically relevant properties. Instability, in particular, will become apparent on longer time intervals. This section and the next two will be devoted to these questions.

5.4.1 One-Step and Multistep Difference Schemes

In Section 5.1 we derived a number of methods to approximate the derivative $\frac{\mathrm{d}}{\mathrm{d}t}u$. As was stated there, we may use values at either two consecutive (time) points (*one-step*) or a larger number of points (*multistep*). So consider the ODE

$$\frac{\mathrm{d}u}{\mathrm{d}t} = f(u,t), \tag{5.66}$$

where $f(u,t)$ may contain derivatives with respect to spatial variables (but not with respect to t). If we denote the approximation of $u(t^n)$ at time t^n by u^n, then we have, for a step size Δt, the general one-step scheme

$$\frac{1}{\Delta t}\big(u^{n+1} - u^n\big) = \Phi\big(u^n, u^{n+1}, t^n; \Delta t\big). \tag{5.67}$$

In particular, a collocation method with m stages is called a *Runge–Kutta method.* We have then (cf. (5.24))

$$\frac{1}{\Delta t}(u^{n+1} - u^n) = \sum_{j=1}^{m} \beta_j k_j^n, \tag{5.68a}$$

$$k_j^n := f\left(t^n + \rho_j \Delta t, u^n + \Delta t \sum_{l=1}^{m} \gamma_{j,l} k_l^n\right). \tag{5.68b}$$

Now recall that collocation was based on using quadrature formulas on certain intervals. Since any such quadrature formula is at least exact for the constant function (as it is based on polynomial interpolation), it follows that the sum of the weights must be equal to the interval length. Hence we find the following algebraic relations for the coefficients β_j, ρ_j, and $\gamma_{j,l}$:

$$\sum_{j=1}^{m} \beta_j = 1, \quad \rho_j = \sum_{l=1}^{m} \gamma_{j,l} \quad (j = 1, 2, \ldots, m). \tag{5.69}$$

Note that $0 \leq \rho_j \leq 1$ $(j = 1, 2, \ldots, n)$.

The five most widely used one-step methods are as follows:

$$\frac{1}{\Delta t}(u^{n+1} - u^n) = f(u^n, t^n) \qquad \textit{(explicit (or forward) Euler method)}, \tag{5.70}$$

$$\frac{1}{\Delta t}(u^{n+1} - u^n) = f(u^{n+1}, t^{n+1}) \qquad \textit{(implicit (or backward) Euler method)}, \tag{5.71}$$

$$\frac{1}{\Delta t}(u^{n+1} - u^n) = (1-\vartheta) f(u^n, t^n) + \vartheta f(u^{n+1}, t^{n+1}), \quad \vartheta \in [0, 1] \quad (\vartheta \textit{ method}), \tag{5.72}$$

$$\frac{1}{\Delta t}(u^{n+1} - u^n) = \frac{1}{2}\Big(f(u^n, t^n) + f(u^{n+1}, t^{n+1})\Big) \qquad \textit{(trapezoidal rule)}, \tag{5.73}$$

$$\frac{1}{\Delta t}(u^{n+1} - u^n) = f\left(\frac{1}{2}(u^n + u^{n+1}), t^n + \frac{1}{2}\Delta t\right) \qquad \textit{(midpoint rule)}. \tag{5.74}$$

Note that (5.72) is a linear combination of (5.70) and (5.71) and that (5.73) is a special case of (5.72), i.e., with $\vartheta = \frac{1}{2}$. From what we have found in Section 5.1 we conclude that (5.70) and (5.71) are first order, and thus we expect (5.72) to be first order in general as well. Since (5.73) is second order, this clearly also applies to (5.72) for $\vartheta = \frac{1}{2}$. Finally, we found in Section 5.1 that (5.74) is a second order method.

Quite another matter is the way the method produces the (approximate) value at the new time point t^{n+1}, given u^n. From the set above only (5.70) directly gives a value for u^{n+1}; it is therefore called an *explicit method.* All other methods (except for (5.72), with $\vartheta = 0$, which is of course just (5.70)) result in an equation that should be solved to find u^{n+1}. Therefore such methods are called *implicit*. We briefly discuss how to deal with the latter problem at the end of this section.

Example 5.9 The well known, *classical Runge–Kutta* formula, going back to Runge (1895) and Kutta (1901), is a fourth order method and reads

$$\frac{1}{\Delta t}(u^{n+1} - u^n) = \frac{1}{6}(k_1 + 2k_2 + 2k_3 + k_4),$$

where

$$\begin{aligned} k_1 &:= f(u^n, t^n), \\ k_2 &:= f\left(u^n + \frac{1}{2}\Delta t\, k_1,\ t^n + \frac{1}{2}\Delta t\right), \\ k_3 &:= f\left(u^n + \frac{1}{2}\Delta t\, k_2,\ t^n + \frac{1}{2}\Delta t\right), \\ k_4 &:= f(u^n + \Delta t\, k_3,\ t^n + \Delta t). \end{aligned}$$

Clearly, $\frac{1}{6} + \frac{1}{3} + \frac{1}{3} + \frac{1}{6} = 1$ (cf. (5.69)). Furthermore, we see that (cf. (5.69)) $\rho_1 = 0$, $\rho_2 = \frac{1}{2} = \gamma_{2,1}$, $\rho_3 = \frac{1}{2} = \gamma_{3,2}$, and $\rho_4 = 1 = \gamma_{4,3}$. □

A *multistep method* has the form (5.21) with $h = \Delta t$, $r = 0$, and k and m replaced by n and $k+1$, respectively:

$$\frac{1}{\Delta t}\sum_{j=0}^{k} a_j u^{n+1-j} = \sum_{j=0}^{k} b_j f\left(u^{n+1-j}, t^{n+1-j}\right). \tag{5.75}$$

Since k is equal to the number of steps, (5.75) is a k-step method. Two special cases of (5.75) are most widely used. The first one is the family of *Adams methods*; see section D in the appendix. They are based on direct integration of the interpolation polynomial of f and can be obtained from (5.20) by choosing $m = k+1$ and $r = 0$ or $r = 1$ and read

$$\frac{1}{\Delta t}\left(u^{n+1} - u^n\right) = \sum_{j=0}^{k} b_j f\left(u^{n+1-j}, t^{n+1-j}\right). \tag{5.76}$$

If $b_0 = 0$, the methods are clearly explicit (*Adams–Bashforth*); if $b_0 \neq 0$, they are *implicit* (*Adams–Moulton*); see appendix section D. The latter are based on finite differences using previous time points and are called *backward differences*. This results in the *BDF* family:

$$\frac{1}{\Delta t}\sum_{j=0}^{k} a_j u^{n+1-j} = f\left(u^{n+1}, t^{n+1}\right). \tag{5.77}$$

It is important to note that k-step methods require more than one initial value. This reflects the fact that their solution space is higher dimensional. One may use a one-step method as a start-up procedure. Yet there is still another problem. To see this consider their use for the model problem

$$\frac{\mathrm{d}u}{\mathrm{d}t} = \lambda u. \tag{5.78}$$

Then (5.75) yields the difference equation

$$\sum_{j=0}^{k} (a_j - \Delta t \lambda b_j) u^{n+1-j} = 0. \tag{5.79}$$

This difference equation with constant coefficients has basis solutions of the form

$$u^n = \mu^n(\Delta t \lambda),$$

where μ is a root of the characteristic polynomial

$$\sum_{j=0}^{k}(a_j - \Delta t\lambda b_j)\mu^{k-j} = 0. \tag{5.80}$$

If μ is a double root, then $u^n = n\mu^n(\Delta t\lambda)$ is also a solution and so are ones with higher powers of n as coefficient if the root has even higher multiplicity. For the exact solution of (5.78) one has a mode $u(t)$ satisfying $u(t^n) = \exp(\Delta t\lambda)u(t^{n-1})$. Hence one should expect at least one root of (5.80) to be approximately equal to $\exp(\Delta t\lambda)$ for $\Delta t\lambda$ small. In particular, this means that for this root we have $\mu(\Delta t\lambda) \to 1$ for $\Delta t\lambda \to 0$. In fact, this is a necessary condition for the method to be meaningful and is exactly equal to saying that the polynomial

$$\sum_{j=0}^{k} a_j\mu^{k-j} = 0$$

has a root equal to one. The corresponding root of (5.80) is called the *essential root*. All other roots of (5.80) are called *parasitic roots*. The corresponding solutions of the difference equation are then referred to as the *essential solution* and the *parasitic solutions*, respectively. The latter do not have any practical significance and may blur the numerically relevant one if no precautions are taken. This is a stability problem that is absent in (implicit) Adams methods and BDF methods. For a more general discussion one may consult [97]. We demonstrate a typical problem in the following example.

Example 5.10 Consider the *explicit midpoint rule* for (5.66), i.e.,

$$\frac{1}{\Delta t}(u^{n+1} - u^{n-1}) = 2f(u^n, t^n).$$

The characteristic polynomial for the model problem reads

$$\mu^2 - 2\lambda\Delta t\mu - 1 = 0.$$

Hence we find $\mu = \pm 1 + \lambda\Delta t + \mathcal{O}((\lambda\Delta t)^2)$. So to leading order we have the general solution

$$u^n \doteq A(1+\lambda\Delta t)^n + B(-1+\lambda\Delta t)^n, \quad A, B \in \mathbb{R}.$$

The constants A, B follow from the initial conditions. We may approximate $1+\lambda\Delta t$ by $\exp(\lambda\Delta t)$. Therefore we have $u^n \doteq A\exp(\lambda t^n) + B(-1)^n\exp(-\lambda t^n)$. If $\lambda < 0$, as typically holds for stable problems, we have an oscillatory (absolutely) growing parasitic mode; hence the problem is unstable. □

Both one-step and k-step methods have advantages and disadvantages. An advantage of one-step methods is their flexibility in choosing the time step Δt (which is preferably not necessarily constant in order to have an efficient time integration). A disadvantage is that each stage requires the evaluation of f (which is often a boundary value problem in our setting!). The latter problem is not present for a k-step method, as it can simply use previously computed "f values." The disadvantage of a k-step method is the change of the time step Δt. One often does this by using interpolation of f values (not u values). This process, together with using an entire family (e.g., Adams–Moulton or BDF), has resulted in *variable step, variable order methods*. For more details see, e.g., [97].

If we have an implicit method, we are faced with the problem of "retrieving" the value u^{n+1}. There are basically two ways to solve such equations: (i) successive substitution and (ii) root finding. We consider them both briefly.

(i) We can symbolically write an implicit method as

$$u^{n+1} = \varphi\left(u^{n+1}\right). \tag{5.81}$$

If we have some reasonable estimate of u^{n+1}, say v_0, from some explicit method, the "predictor," then we can perform successive substitution to produce a (we hope converging) sequence $v_j,\ j = 0, 1, 2, \ldots$, via the "corrector"

$$v_{j+1} = \varphi(v_j).$$

In order to let this converge, we not only need v_0 to be close enough to u^{n+1} but also should have $|\varphi'(v)| < 1$. The latter condition typically implies that

$$\Delta t \left| b_0 \frac{\partial f}{\partial u} \right| < 1$$

in a neighbourhood of u^{n+1}; cf. (5.76). Unfortunately, problems are often such that this requirement implies Δt be taken unduly small. Another problem caused by the latter behaviour of f is discussed in the next section.

(ii) We can also rewrite the implicit method as

$$\psi\left(u^{n+1}\right) = 0. \tag{5.82}$$

Again, given some initial guess v_0 for u^{n+1}, e.g., found from extrapolating values at previous time points, we can carry out a Newton algorithm for $j = 0, 1, 2, \ldots$:

$$v_{j+1} = v_j - \frac{\psi(v_j)}{\psi'(v_j)}.$$

One can show that this procedure does not suffer from the restriction that we have with (i). For more details one may consult [97].

5.4.2 Time Scales and Stiffness

Many time-dependent problems have solutions that are fairly smooth, except for some (smaller) regions where they exhibit rapid motion. This phenomenon is reflected by the PDE, which then has *more than one time scale*. The relatively smooth part is related to the larger time scales and the highly active part is related to the smaller time scales. These high-activity zones are called *layers*; they often occur at boundaries of the domain (*boundary layers*) but sometimes also at the interior of the domain (*interior layers*). A typical, although academic, example is the initial value problem

$$\frac{\mathrm{d}u}{\mathrm{d}t} = -100\left(u - \cos t + \frac{1}{100}\sin t\right), \tag{5.83a}$$

$$u(0) = 0. \tag{5.83b}$$

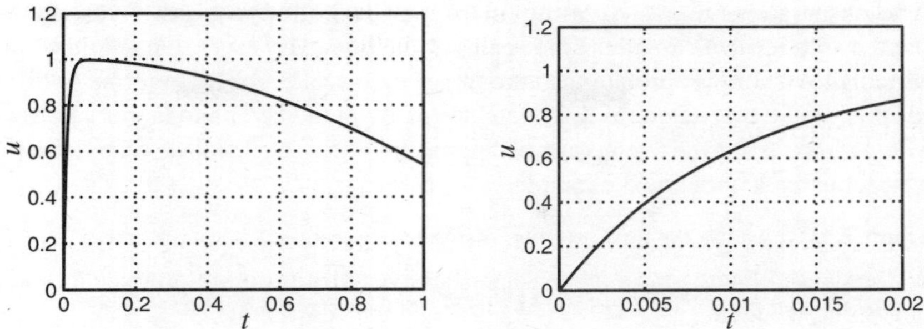

Figure 5.6. *The solution* (5.84) *on the intervals* (0, 1) *(left) and* (0, 0.02) *(right).*

Its solution is easily seen to be

$$u(t) = \cos t - \mathrm{e}^{-100t}\,. \tag{5.84}$$

In Figure 5.6 we give a graph of u and an enlarged part of the interval (0, 0.02).

Clearly, u exhibits a boundary layer of about 10^{-2} thickness; the time scales differ by a factor of 100. It is important to realize that the notion of time scale is a relative one. For if we *stretched* the time via

$$\tau := 100t, \tag{5.85}$$

then the differential equation (5.83a) could be rewritten as

$$\frac{\mathrm{d}u}{\mathrm{d}\tau} = -u + \cos(0.01\tau) - 0.01\sin(0.01\tau). \tag{5.86}$$

Its solution,

$$u(\tau) = \cos(0.01\tau) - \mathrm{e}^{-\tau},$$

does not look like one with a highly active region. However, it is clear that the actual behaviour of the solution for larger values of τ should not be measured on the unit scale, but rather on one a factor of 100 larger. In this book we will often encounter such problems after nondimensionalising the equations, as parameters like the factor 100 in (5.83a) show up naturally. In fact, one may even rewrite (5.83a) as

$$\frac{1}{100}\frac{\mathrm{d}u}{\mathrm{d}t} = -u + \cos t - \frac{1}{100}\sin t$$

to see that $\frac{1}{100}\frac{\mathrm{d}u}{\mathrm{d}t} \doteq 0$ outside the boundary layer. Taking this idea even further, we can rewrite the equation as

$$\varepsilon\left(\frac{\mathrm{d}u}{\mathrm{d}t} + \sin t\right) = -u + \cos t, \tag{5.87}$$

where $\varepsilon = 0.01$. If we take $\varepsilon = 0$ in (5.87), we obtain the *reduced equation*

$$-u(t) + \cos t = 0. \tag{5.88}$$

Outside the layer the solution is accurate within a bound ε.

The observations above have important numerical implications. We only need to compute the solution with smaller time scales in the boundary layer. Outside this layer the solution satisfies a simpler problem, so the larger time scales apply. In terms of difference methods this can be translated to smaller values of Δt in the layer and larger values outside the layer, at least from the viewpoint of approximation (i.e., local discretisation) errors. However, let us look at the next example.

Example 5.11 If we use the forward Euler method to discretise (5.83a), we find

$$\begin{aligned} u^{n+1} &= u^n - 100\Delta t\left(u^n - \cos t^n + \frac{1}{100}\sin t^n\right) \\ &= (1 - 100\Delta t)u^n + \Delta t(100\cos t^n - \sin t^n). \end{aligned}$$

Suppose we want to have a local error smaller than 10^{-2} in absolute value. Since the local error is given by $\frac{1}{2}\Delta t u''(\tau^n)$ (cf. Example 5.8), we can estimate a value for Δt, sufficiently small to resolve the layer, from the inequality

$$\frac{1}{2}\Delta t 10^4 \,\mathrm{e}^{-100t} \le 10^{-2}.$$

This implies a time step restriction

$$\Delta t \le 2 \times 10^{-6}\,\mathrm{e}^{100t}.$$

Hence for t close to zero this would require steps as small as 2×10^{-6}. Outside the layer we have $u(t) \approx \cos t$, whence a guaranteed step size Δt there is estimated from

$$\frac{1}{2}\Delta t \le 10^{-2} \Rightarrow \Delta t \le 2 \times 10^{-2}.$$

In order to see if the latter finding is correct, we consider a variant of problem (5.83), i.e., where we replace (5.83b) by the initial value $u(0) = 1$. As one can easily see, there is no boundary layer then. So a step size $\Delta t = 2 \times 10^{-2}$ seems reasonable to find an approximation of the solution with an absolute accuracy of 10^{-2}. However, applying of the explicit Euler method to this problem might lead to unstable numerical solutions, as shown in Figure 5.7. □

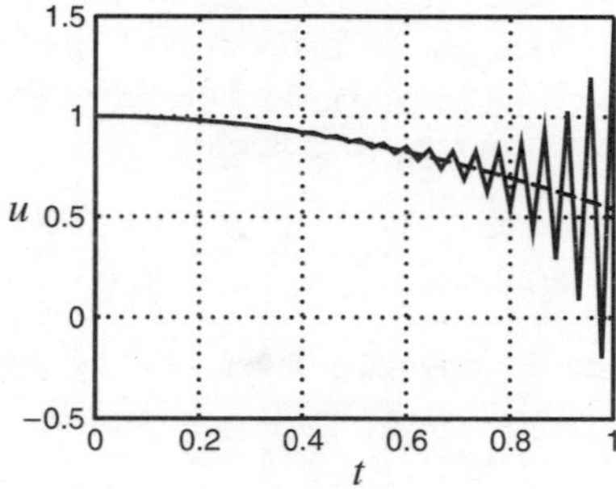

Figure 5.7. *Exact and numerical solutions of* (5.83a) *with initial condition* $u(0) = 1$. *The numerical solution is computed with the forward Euler method with* $\Delta t = 1/45$.

As the previous example shows, accuracy of the approximation is not enough for the method to be effective. It is easy to see that there are other requirements. To find this out let us consider two distinct initial values for the forward Euler method discussed in Example 5.11:

$$u^0 = 1, \tag{5.89a}$$

$$\widetilde{u}^0 = 1 + \delta, \tag{5.89b}$$

with $0 < \delta \ll 1$. Clearly, (5.89a) and (5.89b) generate sequences $\{u^n\}$ and $\{\widetilde{u}^n\}$, respectively. We now find

$$\widetilde{u}^{n+1} - u^{n+1} = (1 - 100\Delta t)(\widetilde{u}^n - u^n). \tag{5.90}$$

Hence

$$\widetilde{u}^n - u^n = (1 - 100\Delta t)^n \delta.$$

If $\Delta t > 2 \times 10^{-2}$, we see that $\{\widetilde{u}^n - u^n\}$ becomes unbounded. A necessary stability requirement is therefore

$$|1 - 100\Delta t| \le 1. \tag{5.91}$$

The phenomenon that *accuracy requirements* following from smoothness of the solutions allow for larger time steps, whereas *stability arguments*, such as in the forward Euler method, require smaller time steps, relates to the *stiffness* of the problem. In general, a problem is called *stiff* if the actual time scale is (much) larger than small time scales potentially existing in the problem. In this sense the problem is not stiff in the boundary layer, only outside the layer.

Fortunately, not all difference methods have a problem with stiffness. Indeed, consider the implicit Euler method applied to (5.83a):

$$u^{n+1} = u^n - 100\Delta t\left(u^{n+1} - \cos t^{n+1} + \frac{1}{100}\sin t^{n+1}\right). \tag{5.92}$$

If we use the same initial values as in (5.89), we now find

$$\widetilde{u}^{n+1} - u^{n+1} = \widetilde{u}^n - u^n - 100\Delta t(\widetilde{u}^{n+1} - u^{n+1}), \tag{5.93}$$

whence

$$\widetilde{u}^{n+1} - u^{n+1} = \frac{1}{1 + 100\Delta t}(\widetilde{u}^n - u^n). \tag{5.94}$$

Obviously, $\left|\frac{1}{1+100\Delta t}\right| < 1$ for all $\Delta t > 0$, so no stability constraint applies.

Example 5.12 If we employ the ϑ method (5.72), we find that the stability requirement for (5.83a) is

$$\left|\frac{1 - 100(1 - \vartheta)\Delta t}{1 + 100\vartheta\Delta t}\right| \le 1.$$

From this inequality we can easily derive the time step restriction $(1 - 2\vartheta)\Delta t \le 0.02$, implying that the method is unconditionally stable for $0.5 \le \vartheta \le 1$. □

To conclude this section we remark that only implicit schemes (like backward Euler) can handle stiffness. Yet not all implicit schemes have this property. There exist special Runge–Kutta schemes, but most widely used are still the BDF (multistep) methods. For more details see [58].

5.5 Discretisation, Convergence, and Stability

The difference methods outlined so far for space and time variables separately have to be combined to obtain a numerical method for approximating solutions of time-dependent PDEs. This brings up the question of how accurate the actual approximations are. This is not only a matter of choosing a proper mesh, i.e., commensurate with the "activity" of the solution. Indeed, for time-dependent problems there is also the question of ensuring that the method has suitable numerical stability properties.

5.5.1 The Method of Lines

A general time-dependent (scalar) PDE can be written as

$$\mathcal{A}(t)[u] = \mathcal{B}(\boldsymbol{x})[u] + c(\boldsymbol{x}, t, u), \tag{5.95}$$

where $\mathcal{A}(t)$ is a differential operator in t and $\mathcal{B}(\boldsymbol{x})$ is a differential operator in $\boldsymbol{x}$, whereas c does not contain any derivatives. Typically, $\mathcal{A}(t)$ is either $\frac{\partial}{\partial t}$ or $\frac{\partial^2}{\partial t^2}$. If we denote the relevant space-time domain (where we would like to compute the solution) by $\Omega \times I$ with $\Omega \subset \mathbb{R}^d$ and $I \subset [0, \infty)$, we basically have two options. The first one is to discretise $\mathcal{B}(\boldsymbol{x})$ by finite differences (directly or through a finite volume method). To do this we have to find a grid on Ω; this discretisation makes the resulting (semidiscrete) problem finite dimensional. From this we obtain a relation between (numerical) solution values on that grid. If we consider this grid as fixed on Ω, we thus find from (5.95) an *initial value* ODE for solution values on the grid, coupled through the finite difference scheme. In a one-dimensional setting we typically have a situation like that in Figure 5.8.

The lines there represent the domains of the thus obtained (time-dependent) variables. This discretisation is therefore called the *method of lines* (MOL); more accurately this is actually the *longitudinal* MOL. The second option is to discretise the time derivative first, and this approach is called the *transversal* MOL; see Figure 5.8. Here we have a boundary value problem (PDE or ODE) at discrete time levels. These are coupled through the finite

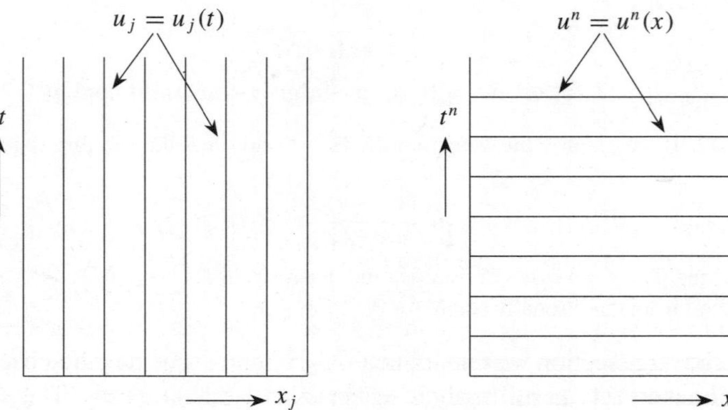

Figure 5.8. *Grids for the longitudinal (left) and transversal (right) method of lines.*

differences of the time derivative(s). In particular, for parabolic problems this transversal MOL leads to a sequence of elliptic problems.

Both types of MOL have advantages and disadvantages. The longitudinal is the simplest and therefore most attractive, as it transforms the PDE to an ODE, which lends itself to easy computation. However, one should realize that a fixed (let alone a uniform) mesh for all $t \in I$ is not likely to give good results if the high activity of the solution is moving in time. If a good boundary value problem solver is available, then the transversal MOL seems attractive for the latter type of problems. However, changing meshes from one time step to the next is not trivial here either. In practice a hybrid MOL is often used. We shall base our derivations in this book mainly on uniform fixed meshes, used in the longitudinal MOL. When needed we shall use more sophisticated gridding.

5.5.2 Consistency, Stability, and Convergence of the MOL

After discretising the spatial derivatives in (5.95), we end up with a vectorial system of equations

$$\mathcal{A}(t)[\mathbf{u}] = \mathbf{B}\mathbf{u} + \mathbf{f}, \tag{5.96}$$

where $\mathbf{u}(t) \in \mathbb{R}^M$, with M the number of grid points, is the vector of unknown values and $\mathbf{B}\mathbf{u}$ is the result of discretising $\mathcal{B}(\boldsymbol{x})[u]$. We consider the case where $\mathcal{A}(t)$ is either $\frac{\partial}{\partial t}$ or $\frac{\partial^2}{\partial t^2}$, so that we actually have the ODE

$$\frac{\mathrm{d}\mathbf{u}}{\mathrm{d}t} = \mathbf{B}\mathbf{u} + \mathbf{f} \tag{5.97}$$

or

$$\frac{\mathrm{d}^2\mathbf{u}}{\mathrm{d}t^2} = \mathbf{B}\mathbf{u} + \mathbf{f}. \tag{5.98}$$

As is standard procedure for ODEs, one may reduce (5.98) to one of type (5.97) by introducing a new variable $\mathbf{v}$, defined as

$$\mathbf{v} := \begin{pmatrix} \mathbf{u} \\ \dfrac{\mathrm{d}\mathbf{u}}{\mathrm{d}t} \end{pmatrix}. \tag{5.99}$$

In the remainder of this section we will consider (5.97). Although this looks like an innocuous system of ODEs, one has to bear in mind that for a convergence analysis the dimension M goes to infinity as the step size Δx approaches zero. A rigorous analysis of the numerical integration of (5.97) will therefore be complicated. Below we shall be satisfied with a finite-dimensional convergence analysis nevertheless, keeping an eye on the infinite-dimensional limit case (for a more functional analytic approach see, e.g., [120]).

First we need a way to measure errors such that they are independent of the dimension M. This can be achieved by using Hölder norms; see the appendix, section G. In particular, we shall use for $\mathbf{u} = (u_1, u_2, \dots, u_M)^T$ one of the norms

$$\|\mathbf{u}\|_\infty := \max_{1\le i\le M} |u_i| \qquad \textit{(infinity norm)}, \tag{5.100a}$$

$$\|\mathbf{u}\|_2 := \left(\sum_{i=1}^{M} |u_i|^2\right)^{\frac{1}{2}} \qquad \textit{(2-norm)}. \tag{5.100b}$$

One easily sees that the length of a vector depends on the norm used. Rather than (5.100b), we may use the norm

$$\|\mathbf{u}\|_{2,\mathrm{m}} := \frac{1}{\sqrt{M}}\left(\sum_{i=1}^{M} |u_i|^2\right)^{\frac{1}{2}} \qquad \textit{(median 2-norm)}. \tag{5.100c}$$

Note that, e.g., for $\mathrm{e} := (1, 1, \dots, 1)^T$ we have $\|\mathrm{e}\|_2 = \sqrt{M}$ and $\|\mathrm{e}\|_{2,\mathrm{m}} = 1$.

Let us now assume the PDE to be linear with $c = c(\boldsymbol{x}, t)$. Then a typical one-step method results in, say,

$$\frac{1}{\Delta t}(\mathbf{u}^{n+1} - \mathbf{u}^n) = \mathbf{R}^n\mathbf{u}^{n+1} + \mathbf{S}^n\mathbf{u}^n + \mathbf{f}^{n,n+1}. \tag{5.101a}$$

Assuming that $\mathbf{I} - \Delta t\,\mathbf{R}^n$ is nonsingular, we have

$$\mathbf{u}^{n+1} = \left(\mathbf{I} - \Delta t\,\mathbf{R}^n\right)^{-1}\Big((\mathbf{I} + \Delta t\,\mathbf{S}^n)\mathbf{u}^n + \mathbf{f}^{n,n+1}\Big). \tag{5.101b}$$

Here $\mathbf{R}^n$ and $\mathbf{S}^n$ are $M \times M$ matrices and $\mathbf{f}^{n,n+1} \in \mathbb{R}^M$. Note that, as before, the superscript n denotes a time index, not a power, whereas the double superscript is a reminder that the source term arose from two time level contributions. If we denote the exact solution of (5.95) restricted to the grid by $\mathbf{u}(\cdot, t^n)$, i.e., $\mathbf{u}(\cdot, t^n) := \big(u(\boldsymbol{x}_1, t^n), u(\boldsymbol{x}_2, t^n), \dots, u(\boldsymbol{x}_M, t^n)\big)^T$, then we can define the *local discretisation error (vector)* $\mathbf{d}(\cdot, t^n)$ by

$$\mathbf{d}(\cdot, t^n) := \frac{1}{\Delta t}\Big(\mathbf{u}(\cdot, t^{n+1}) - \mathbf{u}(\cdot, t^n)\Big) - \mathbf{R}^n\mathbf{u}(\cdot, t^{n+1}) - \mathbf{S}^n\mathbf{u}(\cdot, t^n) - \mathbf{f}^{n,n+1}. \tag{5.102}$$

Now let the time step Δt be related to Δx by $\Delta t = \varphi(\Delta x)$, where we require $\varphi(\Delta x) \to 0$ for $\Delta x \to 0$. We shall refer to the latter procedure as a *refinement path*. The analysis below will be valid on a finite time interval $[0, T]$, say. Then we have the following definition.

Definition 5.13. *The difference scheme (5.101) is consistent for $\Delta t = \varphi(\Delta x)$ if $\|\mathbf{d}(\cdot, t^n)\| \to 0$ for $\Delta x \to 0$ and for all time levels $t^n \le T$.*

One should note that this consistency definition requires the (scalar) local discretisation errors to go to zero if $\Delta t, \Delta x \to 0$ (the finite difference methods we have constructed so far have this property automatically); however, it also requires a proper handling of the initial and boundary conditions.

Example 5.14 If we have to solve a second order problem like in (5.98), then we clearly have to specify two initial conditions at $t = 0$. Often this is a given value of the solution and its derivative. The latter should be discretised with a similar accuracy as the time-stepping method to prevent the initial errors from being persistent. □

The next property we need deals with appropriate error propagation. Recalling the scalar situation (5.90), we therefore consider (cf. (5.101))

$$\left(\mathbf{I} - \Delta t\,\mathbf{R}^n\right)^{-1}\left(\mathbf{I} + \Delta t\,\mathbf{S}^n\right).$$

Let us now consider system (5.97) defined on some time interval $[0, T]$. We then have the following definition.

Definition 5.15. *The difference scheme* (5.101) *is called* stable *for $\Delta t = \varphi(\Delta x)$ if, for some ε and $\Delta x < \varepsilon$, there exists a constant $C(\varepsilon)$ such that, for all k, n with $k \le n$ and $n\Delta t \le T$,*

$$\left\| \prod_{i=k}^{n} \left(\mathbf{I} - \Delta t\,\mathbf{R}^i\right)^{-1}\left(\mathbf{I} + \Delta t\,\mathbf{S}^i\right) \right\| \le C(\varepsilon).$$

The latter concept needs some further explanation. We give this through an example.

Example 5.16 If $\mathbf{R}^i = \mathbf{R}$ and $\mathbf{S}^i = \mathbf{S}$ for all i, we may write

$$\left(\mathbf{I} - \Delta t\,\mathbf{R}\right)^{-1}\left(\mathbf{I} + \Delta t\,\mathbf{S}\right) = \mathbf{T}\mathbf{J}(\Delta t)\mathbf{T}^{-1},$$

where $\mathbf{J}(\Delta t)$ is the Jordan normal form; see the appendix, section H. Hence

$$\prod_{i=k}^{n} \left(\mathbf{I} - \Delta t\,\mathbf{R}\right)^{-1}\left(\mathbf{I} + \Delta t\,\mathbf{S}\right) = \mathbf{T}\left(\mathbf{J}(\Delta t)\right)^{n-k+1}\mathbf{T}^{-1}.$$

So the required bound exists if and only if $\|(\mathbf{J}(\Delta t))^{n-k+1}\|$ is uniformly bounded; this is the case if and only if $\mathbf{J}(\Delta t)$ has eigenvalues smaller than one in modulus and/or unimodular eigenvalues that are geometrically simple; see the appendix, section I. Note that such a property may hold, even if

$$\left\|\left(\mathbf{I} - \Delta t\,\mathbf{R}\right)^{-1}\left(\mathbf{I} + \Delta t\,\mathbf{S}\right)\right\| > 1,$$

for the 2-norm $\|\cdot\|_2$ or the infinity norm $\|\cdot\|_\infty$. □

Finally, we introduce a convergence concept. To this end we define the *global discretisation error (vector)* $\mathbf{e}(\cdot, t^n)$ as

$$\mathbf{e}(\cdot, t^n) := \mathbf{u}(\cdot, t^n) - \mathbf{u}^n. \tag{5.103}$$

This has to be understood with $\mathbf{e}$ and $\mathbf{u}$ being vectors with the appropriate dimension. For a first order time-dependent problem it is trivial to assume that $\mathbf{e}(\cdot, t^0) = \mathbf{0}$. For second order problems this is generally not true. We simply assume the starting error to be small enough and disregard it from now on. As a further (but trivial) simplification we let $t^0 = 0$. We now have the following definition.

Definition 5.17. *The difference scheme* (5.101) *is convergent for $\Delta t = \varphi(\Delta x)$ if the following holds: for any $c(\boldsymbol{x}, t, u)$ and initial value $u(\boldsymbol{x}, t^*)$ $(0 \le t^* \le T)$ for which* (5.95) *has a bounded solution, $\|\mathbf{e}(\cdot, t^n)\| \to 0$ for $\Delta x \to 0$ and $n\Delta t \le T$.*

Again we remark that convergence for a PDE is much more complicated than for an ODE, the major problem being the dimension M approaching infinity. The concepts of consistency and stability are, for many problem classes, equivalent. This is the famous Lax theorem [120], of which we give a proof for the restricted setting sketched above.

Theorem 5.18 (Lax equivalence theorem). *Let* (5.101) *be consistent. Then it is convergent if and only if it is stable.*

Proof. Suppose (5.101) is stable. Then we first have to prove that it is convergent. From (5.101) and (5.102) we obtain

$$(\mathbf{I} - \Delta t\, \mathbf{R}^n)\, \mathbf{e}(\cdot, t^{n+1}) = (\mathbf{I} + \Delta t\, \mathbf{S}^n)\, \mathbf{e}(\cdot, t^n) + \Delta t\, \mathbf{d}(\cdot, t^n).$$

Since $\mathbf{e}(\cdot, t^0) = \mathbf{0}$, we thus find

$$\mathbf{e}(\cdot, t^n) = \Delta t \sum_{k=0}^{n-1} \left[\prod_{i=k+1}^{n-1} (\mathbf{I} - \Delta t\, \mathbf{R}^i)^{-1} (\mathbf{I} + \Delta t\, \mathbf{S}^i) \right] (\mathbf{I} - \Delta t\, \mathbf{R}^k)^{-1} \mathbf{d}(\cdot, t^k),$$

where the matrix product is defined as $\prod_{i=k+1}^{n-1} \mathbf{C}^i := \mathbf{C}^{n-1} \cdots \mathbf{C}^{k+2} \mathbf{C}^{k+1}$. Hence we obtain

$$\|\mathbf{e}(\cdot, t^n)\| \le \Delta t\, C(\varepsilon) \sum_{k=0}^{n-1} \|(\mathbf{I} - \Delta t\, \mathbf{R}^k)^{-1}\| \|\mathbf{d}(\cdot, t^k)\| \le C(\varepsilon) D T \max_k \|\mathbf{d}(\cdot, t^k)\|.$$

Here $C(\varepsilon)$ is the stability bound as implied by Definition 5.15 and $D := \max_k \|(\mathbf{I} - \Delta t \mathbf{R}^k)^{-1}\|$. In the applications that follow the latter bound is trivial; often even $D = 1$. The convergence then follows from the consistency.

Conversely, suppose (5.101) is convergent. Then we have to prove that it is stable. We first remark that it is not restrictive to assume $c(\boldsymbol{x}, t, u) \equiv 0$, and consequently $\mathbf{f} = \mathbf{0}$, and moreover we assume that $\mathbf{u}^k = \mathbf{u}(\cdot, t^k)$ for all k with $k \le n \le T/\Delta t$. The latter assumption implies that we may look at problem (5.95) on the interval $[t^k, T]$ with $k\Delta t \le T$. Under these assumptions we have for any "initial" value $\mathbf{u}(\cdot, t^k)$ that

$$\|\mathbf{u}(\cdot, t^n) - \mathbf{u}^n\| = \left\| \left[\prod_{i=k}^{n-1} (\mathbf{I} - \Delta t\, \mathbf{R}^i)^{-1} (\mathbf{I} + \Delta t\, \mathbf{S}^i) \right] \mathbf{u}(\cdot, t^k) - \mathbf{u}(\cdot, t^n) \right\|.$$

Since (5.101) is convergent, the left-hand side is bounded by some value $\epsilon(\Delta x)$, with $\epsilon(\Delta x) \to 0$ if $\Delta x < \delta x$ and $\delta x \to 0$. Moreover, on a finite interval $\subseteq [0, T]$, $\|\mathbf{u}(\cdot, t^n)\|$ must be bounded by a constant E, say. Applying the triangle inequality, we thus obtain

$$\left\| \left[\prod_{i=k}^{n-1} (\mathbf{I} - \Delta t\, \mathbf{R}^i)^{-1} (\mathbf{I} + \Delta t\, \mathbf{S}^i) \right] \mathbf{u}(\cdot, t^k) \right\| \le E + \varepsilon$$

for $\Delta x \le \delta x$. Hence for M (dimension of vector space) fixed we find that the set

$$\left\{ \left\| \prod_{i=k}^{n-1} (\mathbf{I} - \Delta t\, \mathbf{R}^i)^{-1} (\mathbf{I} + \Delta t\, \mathbf{S}^i)\, \mathbf{u}(\cdot, t^k) \right\| \right\}$$

is bounded for $k, n \leq T/\Delta t$ and thus

$$\left\{ \left\| \prod_{i=k}^{n-1} (\mathbf{I} - \Delta t\, \mathbf{R}^i)^{-1} (\mathbf{I} + \Delta t\, \mathbf{S}^i) \right\| \right\}$$

is bounded for $k, n \leq T/\Delta t$; i.e., we have stability. □

Example 5.19 Let $\mathbf{R}^i \equiv \mathbf{R}$ and $\mathbf{S}^i \equiv \mathbf{S}$, where $\mathbf{R}$ and $\mathbf{S}$ commute and have simple eigenvalues, say equal to $r_1, r_2, \ldots, r_M$ and $s_1, s_2, \ldots, s_M$, respectively. Then it is easily verified that

$$(\mathbf{I} - \Delta t\, \mathbf{R}^i)^{-1} (\mathbf{I} + \Delta t\, \mathbf{S}^i) = \mathbf{T}\boldsymbol{\Lambda}(\Delta t)\mathbf{T}^{-1},$$

$$\boldsymbol{\Lambda}(\Delta t) := \operatorname{diag}\left(\frac{1 + \Delta t\, s_1}{1 - \Delta t\, r_1}, \frac{1 + \Delta t\, s_2}{1 - \Delta t\, r_2}, \ldots, \frac{1 + \Delta t\, s_M}{1 - \Delta t\, r_M} \right) \mathbf{T}^{-1},$$

for some diagonalising matrix $\mathbf{T}$. Thus we have

$$\prod_{i=k}^{n} (\mathbf{I} - \Delta t\, \mathbf{R})^{-1} (\mathbf{I} + \Delta t\, \mathbf{S}) = \mathbf{T}(\boldsymbol{\Lambda}(\Delta t))^{n+1-k}\mathbf{T}^{-1},$$

and, consequently, stability (and thus convergence) follows if

$$\max_l \left| \frac{1 + \Delta t\, s_l}{1 - \Delta t\, r_l} \right| \leq 1.$$

Clearly, the eigenvalues $\{s_l\}$ and $\{r_l\}$ depend on the spatial discretisation. □

5.6 Fourier Mode Analysis

The method outlined in the previous section is often too heavy a tool for establishing stability results. Indeed, as we saw in Example 5.19, one may need quantities like eigenvalues of a matrix, which are hard to get. One has to remember that the bulk of the matrix elements in $\mathbf{R}^i$ and S^i are formed by applying a difference scheme locally. Hence it is tempting to look at the latter exclusively in order to find out about stability (and how the refinement path $\Delta t = \varphi(\Delta x)$ should be chosen, for that matter). This should then give at least necessary conditions. In order to mathematically justify this approach, we may restrict ourselves to one-dimensional problems with periodic boundary conditions. The latter is at least convenient for the operator notation that will follow; however, we shall not aim at a complete derivation, given the restrictive (yet extremely useful!) nature of the setting. The basic idea is to view the difference method as an operator defined on the discrete set of points $\{x_1, x_2, \ldots, x_M\}$. So, rather than deriving a matrix system like (5.96) (which includes the boundary conditions as well) we investigate the discrete analogue of (5.95). To do this let us introduce the notation for a function defined on the grid $x_1, x_2, \ldots, x_M$:

$$u_\Delta^n := \{u_j^n\}_{j=1}^{M}, \tag{5.104}$$

where $u_j^n \doteq u(x_j, t^n)$. Apparently, $u_\Delta^n : \{1, 2, \ldots, M\} \to \mathbb{R}$, which is the analogue of $u(\cdot, t^n) : \Omega \to \mathbb{R}$. If we set $c(\boldsymbol{x}, t, u) \equiv 0$ in (5.95) and apply a one-step method, we may write the result as

$$\mathcal{A}_\Delta[u_j^{n+1}] = \mathcal{B}_\Delta[u_j^n], \tag{5.105}$$

where $\mathcal{A}_\Delta$ and $\mathcal{B}_\Delta$ denote linear (difference) operators.

Example 5.20 Consider the equation $\frac{\partial}{\partial t}u - \frac{\partial}{\partial x}u = 0$. Obviously, $\mathcal{A}(t) = \frac{\partial}{\partial t}$, $\mathcal{B}(x) = \frac{\partial}{\partial x}$, and $c(x, t, u) = 0$. Applying forward differences yields

$$\frac{\partial u}{\partial t}(x_j, t^n) \doteq \frac{1}{\Delta t}(u_j^{n+1} - u_j^n), \qquad \frac{\partial u}{\partial x}(x_j, t^n) \doteq \frac{1}{\Delta x}(u_{j+1}^n - u_j^n).$$

Hence

$$\mathcal{A}_\Delta[u_j^{n+1}] := \frac{1}{\Delta t}u_j^{n+1}, \qquad \mathcal{B}_\Delta[u_j^n] := \frac{1}{\Delta t}u_j^n + \frac{1}{\Delta x}(u_{j+1}^n - u_j^n).$$

The indices j and n are generic, just like x and t. Note that $\mathcal{A}_\Delta$ and $\mathcal{B}_\Delta$ are determined up to a scaling factor. We could, for example, have multiplied both operators by Δt. However, the above representation has the same dimension as the corresponding PDE. □

We shall need norms for operators like $\mathcal{A}_\Delta$ and $\mathcal{B}_\Delta$. Anticipating the use of Fourier series, we first recall the 2-norm for $u_\Delta = \{u_j\}_{j=1}^M$ as

$$\|u_\Delta\|_2 := \left(\sum_{j=1}^M u_j^2\right)^{\frac{1}{2}}. \tag{5.106}$$

We can then define the 2-norm for an operator like $\mathcal{A}_\Delta$ as

$$\left\|\mathcal{A}_\Delta\right\|_2 := \max_{u_\Delta \neq 0} \frac{\|\mathcal{A}_\Delta u_\Delta\|_2}{\|u_\Delta\|_2} = \max_{u_\Delta \neq 0} \frac{\left(\sum_{j=1}^M \left(\mathcal{A}_\Delta[u_j]\right)^2\right)^{\frac{1}{2}}}{\left(\sum_{j=1}^M u_j^2\right)^{\frac{1}{2}}}. \tag{5.107}$$

Example 5.21 For the choices in Example 5.20 we find

$$\|\mathcal{A}_\Delta^{-1}\mathcal{B}_\Delta\|_2^2 = \max_{u_\Delta \neq 0} \frac{\|(\mathcal{A}_\Delta^{-1}\mathcal{B}_\Delta)u_\Delta\|_2^2}{\|u_\Delta\|_2^2} = \max_{u_\Delta \neq 0} \frac{\sum_{j=1}^M \left(u_j + \frac{\Delta t}{\Delta x}(u_{j+1} - u_j)\right)^2}{\sum_{j=1}^M u_j^2}.$$

Define $\sigma := \frac{\Delta t}{\Delta x}$ and let $0 < \sigma \leq 1$. Then we have, from the triangle inequality for the 2-norm applied to the numerator, that

$$\sum_{j=1}^M \left((1-\sigma)u_j + \sigma u_{j+1}\right)^2 \leq \left[\left(\sum_{j=1}^M (1-\sigma)^2 u_j^2\right)^{\frac{1}{2}} + \left(\sum_{j=1}^M \sigma^2 u_{j+1}^2\right)^{\frac{1}{2}}\right]^2$$
$$= \left[(1-\sigma+\sigma)\left(\sum_{j=1}^M u_j^2\right)^{\frac{1}{2}}\right]^2 = \sum_{j=1}^M u_j^2.$$

The equality sign is taken for $u_j \equiv 1$. Hence we obtain $\|\mathcal{A}_\Delta^{-1}\mathcal{B}_\Delta\|_2 = 1$. Note that we have used periodicity. □

We now obtain from (5.105) that

$$u_j^{n+l} = \left(\mathcal{A}_\Delta^{-1}\mathcal{B}_\Delta\right)^l [u_j^n].$$

Hence

$$\frac{\|u_\Delta^{n+l}\|_2}{\|u_\Delta^n\|_2} \leq \left\| \left(\mathcal{A}_\Delta^{-1}\mathcal{B}_\Delta\right)^l \right\|_2. \tag{5.108}$$

As a consequence, we have

$$\sup_{u_\Delta^n \neq 0} \frac{\|u_\Delta^{n+1}\|}{\|u_\Delta^n\|} \leq \left\| \mathcal{A}_\Delta^{-1}\mathcal{B}_\Delta \right\|_2. \tag{5.109}$$

Stability of (5.105) is therefore seen to be equivalent to

$$\left\{ \left\| \left(\mathcal{A}_\Delta^{-1}\mathcal{B}_\Delta\right)^l \right\|_2 \right\}_{l\geq 1}$$

being bounded. Like in Section 5.5.2, this has to be interpreted in terms of a suitable refinement path, say $\Delta t = \varphi(\Delta x)$, and for Δx small enough.

A useful tool in estimating growth is the discrete Fourier transform (DFT) (see Chapter 3). Recall that $\mathrm{e}^{2\pi\mathrm{i}j/(M-1)}$, where $j/(M-1) = x_j$, represents a typical mode. We shall apply the DFT to the solution u_Δ^n, which is defined at the grid points $x_1, x_2, \ldots, x_M$. Then we can write

$$u_j^n = \sum_{k=1}^{M} \hat{u}_k^n \, \mathrm{e}^{2\pi\mathrm{i}kj/(M-1)}, \tag{5.110a}$$

where

$$\hat{u}_k^n := \sum_{j=1}^{M} u_j^n \, \mathrm{e}^{-2\pi\mathrm{i}kj/(M-1)}. \tag{5.110b}$$

Clearly, in (5.110a) and (5.110b) the periodicity of u_Δ^n is tacitly assumed (which should carry over from $u(\cdot, t^n)$). We can use the expansion (5.110a) in (5.105), giving

$$\sum_{k=1}^{M} \hat{u}_k^{n+1} \, \mathrm{e}^{2\pi\mathrm{i}kj/(M-1)} = \sum_{k=1}^{M} \hat{u}_k^n (\mathcal{A}_\Delta^{-1}\mathcal{B}_\Delta)\left[\mathrm{e}^{2\pi\mathrm{i}kj/(M-1)}\right]. \tag{5.111}$$

If we collect all terms containing the factor $\mathrm{e}^{2\pi\mathrm{i}kj/(M-1)}$ in (5.111), we can write this as

$$\hat{u}_k^{n+1} = \mathcal{G}_k \hat{u}_k^n. \tag{5.112}$$

The quantity $\mathcal{G}$ is called the *growth factor* (which does not depend on n, apparently).

Example 5.22 If we apply this idea to the operators $\mathcal{A}_\Delta$ and $\mathcal{B}_\Delta$ from Example 5.20, we obtain

$$\mathcal{A}_\Delta\left[\mathrm{e}^{2\pi\mathrm{i}kj/(M-1)}\right] = \frac{1}{\Delta t}\,\mathrm{e}^{2\pi\mathrm{i}kj/(M-1)},$$
$$\mathcal{B}_\Delta\left[\mathrm{e}^{2\pi\mathrm{i}kj/(M-1)}\right] = \frac{1}{\Delta t}\,\mathrm{e}^{2\pi\mathrm{i}kj/(M-1)} + \frac{1}{\Delta x}\left(\mathrm{e}^{2\pi\mathrm{i}k(j+1)/(M-1)} - \mathrm{e}^{2\pi\mathrm{i}kj/(M-1)}\right).$$

Consequently, we find for the growth factor

$$\mathcal{G}_k = \mathrm{e}^{-2\pi\mathrm{i}kj/(M-1)}\,(\mathcal{A}_\Delta^{-1}\mathcal{B}_\Delta)\left[\mathrm{e}^{2\pi\mathrm{i}kj/(M-1)}\right] = 1 + \frac{\Delta t}{\Delta x}\left(\mathrm{e}^{2\pi\mathrm{i}k/(M-1)} - 1\right). \qquad \square$$

The usefulness of this growth factor is shown by the following. We apparently have

$$\begin{aligned} u_j^{n+l} &= \sum_{k=1}^{M} \hat{u}_k^{n+l} \, \mathrm{e}^{2\pi \mathrm{i} kj/(M-1)} \\ &= \sum_{k=1}^{M} \mathcal{G}_k \hat{u}_k^{n+l-1} \, \mathrm{e}^{2\pi \mathrm{i} kj/(M-1)} \\ &\vdots \\ &= \sum_{k=1}^{M} \hat{u}_k^{n} \big(\mathcal{G}_k\big)^l \, \mathrm{e}^{2\pi \mathrm{i} kj/(M-1)} . \end{aligned} \tag{5.113}$$

Hence we deduce from Parseval's identity (cf. Chapter 3) that

$$\|u_\Delta^{n+l}\|_2^2 = \sum_{k=1}^{M} \Big(|\mathcal{G}_k|^l |\hat{u}_k^n| \Big)^2 \leq \max_k |\mathcal{G}_k|^{2l} \sum_{k=1}^{M} |\hat{u}_k^n|^2 = \max_k |\mathcal{G}_k|^{2l} \|u_\Delta^n\|_2^2. \tag{5.114}$$

Equality in (5.114) can be obtained if all $|\mathcal{G}_k|$ are equal. Hence we conclude that

$$\sup_{u_\Delta^n \neq 0} \frac{\|u_\Delta^{n+l}\|}{\|u_\Delta^n\|} = \max_k |\mathcal{G}_k|^l. \tag{5.115}$$

Comparing (5.108) with (5.115), we conclude that $\max_k |\mathcal{G}_k|^l$ is the required upper bound for

$$\big\| \mathcal{A}_\Delta^{-1} \mathcal{B}_\Delta \big\|_2.$$

Hence stability of (5.104) is established by requiring $\sup_l \{\max_k |\mathcal{G}_k|^l\}$ to be bounded. This leads to the following property.

Property 5.23 (von Neumann). *A necessary condition for* (5.104) *to be stable on* $[0, T]$ *is that*

$$|\mathcal{G}_k| \leq 1 + \mathcal{O}(\Delta t) \tag{5.116}$$

uniformly in k.

Proof. Let $|\mathcal{G}_k| \leq 1 + c\Delta t$ for some $c > 0$. Then

$$|\mathcal{G}_k|^l \leq \big(1 + c\Delta t\big)^l \leq \mathrm{e}^{cl\Delta t} \leq \mathrm{e}^{cT},$$

which is bounded for every finite T. □

Property 5.23 will often turn out to be the most convenient way to find at least necessary conditions for (5.105) to be stable. One should recall, however, that our setting was quite restricted anyway (linear, constant coefficients, periodic boundary conditions). Hence we may use this property as a test to reject certain methods (and refinement paths $\Delta t = \varphi(\Delta x)$), as a method should at least be meaningful for a member of the above-mentioned restricted class of problems.

5.7 Discussion

- Finite differences are the oldest and simplest methods of finding approximate solutions of PDEs. Therefore there exist a host of books on the subject. It is interesting to see that finite volume methods lead to finite difference methods as well. Finite element methods (FEMs) can often be formulated such that they lead to difference schemes; see [82]. In particular, for complicated domains (in two or three space dimensions), however, the finite difference schemes may result in complicated bookkeeping that FEMs do not encounter. Bearing in mind that there is much similarity, finite differences are the most transparent methods of understanding typical phenomena like stability, etc., of numerical methods, and therefore they are the method of choice.

- Implicit time integrators that address stiffness may look very attractive at first sight. There exist many such methods, in particular implicit Runge–Kutta methods that are not dealt with here; see, e.g., [58, 21]. They come with a price, however. Indeed, implicit methods require the solution of a (possibly nonlinear) system. This may lead to such complexity that despite the relatively small step required by an explicit method, the overall complexity is still lower.

- An important question is how to find an appropriate mesh in general, both for spatial and for temporal discretisation. The simple approach of a uniform mesh requires it to be small enough to cope with the "activity" (in terms of magnitude of derivatives) of the solution. Clearly, for nonuniform behaviour of the solution this is not efficient. There are all sorts of adaptive methods to cope with this problem. For time discretisation this is relatively easy; see, e.g., [97, 57]. For spatial discretisation one often employs equidistribution techniques; see [3, 11].

- The Lax theorem proved in this chapter has a more universal character. In many instances one can show that for a consistent scheme stability is equivalent to convergence; see [148]. The statement and proof in this chapter are for scalar one-step methods only. For multistep methods the amplification factor becomes an amplification matrix. The estimates then typically should hold for eigenvalues of the matrix.

Exercises

5.1. Although numerical derivatives are more accurate the smaller the grid size chosen, this is only true in exact arithmetic. In practice we have to deal with rounding errors as well. This exercise deals with this problem.

(a) Let the function $f(\xi)$ be perturbed by some $\delta(\xi)$, where $|\delta(\xi)| \leq \varepsilon$, for some ε and $\xi \in I$, I being some relevant interval. Show that a finite difference approximation of $f'(\xi_1)$, $\xi_1 \in I$, has a rounding error approximately proportional to ε/h, where h is the grid size.

(b) If we use a kth order approximation for f', then we have an approximately optimal grid size (i.e., the sum of the absolute approximation error and the rounding error is minimal) if $h \approx \varepsilon^{1/(k+1)}$. Note that if, e.g., $k = 1$ (as we have for a forward difference), this result means that the last half of the digits in the mantissa of the result may not be significant.

5.2. Derive a Runge–Kutta method by taking the trapezoidal rule as the method of integration and the Euler forward method to approximate the implicit part on the right-hand side.

5.3. A way to determine a finite difference approximation is to impose that it is exact for polynomials of a degree as large as possible. As an example we consider the two-step backward difference

$$a_1 u(\xi_1) + a_2 u(\xi_2) + a_3 u(\xi_3) \qquad (\xi_1 > \xi_2 > \xi_3)$$

as an approximation of $\frac{\mathrm{d}u}{\mathrm{d}\xi}(\xi_1)$.

(a) Why can you ensure this to be exact for $u(\xi) := 1$, $u(\xi) := \xi$, and $u(\xi) := \xi^2$?

(b) Find the coefficients a_1, a_2, and a_3.

(c) Try the formula for $u(\xi) := \xi^3$ and see that it is not differentiated exactly. Assume then that the error is of the form $cu^{(q)}(\zeta)$ and determine c, q, and ζ.

5.4. To see that the error term in integration is of the form (5.12), we consider the case of the trapezoidal rule; see Example 5.4. So consider the interval (ξ_1, ξ_2). The error is given by

$$\int_{\xi_1}^{\xi_2} \frac{1}{2} u^{(2)}(\zeta(\xi))(\xi - \xi_1)(\xi - \xi_2)\,\mathrm{d}\xi.$$

The mean value theorem says that $\int_{\xi_1}^{\xi_2} f(\xi)g(\xi)\,\mathrm{d}\xi = f(\eta)\int_{\xi_1}^{\xi_2} g(\xi)\,\mathrm{d}\xi$ for some $\eta \in (\xi_1, \xi_2)$ if g has a constant sign. Now prove the error term given in Example 5.4.

5.5. The weights and error term of a quadrature formula can be determined by requiring that it be exact for polynomials up to a certain degree. As an example we consider the integral

$$\int_{-h}^{h} u(\xi)\,\mathrm{d}\xi,$$

which we would like to approximate by the three-point formula

$$b_1 u(-h) + b_2 u(0) + b_3 u(h).$$

We assume that the error term is of the form $cu^{(q)}(\zeta)$.

(a) Show that b_1, b_2, and b_3 can be chosen such that the formula is exact for polynomials of degree up to two.

(b) Show that this formula is also exact for any odd function. Thus it is also exact for $u(\xi) = \xi^3$.

(c) Determine the error term.

5.6. The approximations of the second derivatives in (5.32a) and (5.32b) are second order. Show this by using Taylor expansions.

5.7. Let Ω be the trapezium defined by the corner points $(0, 0)$, $(0, 2)$, $(2, 1)$, and $(2, 0)$. We want to discretise the Poisson equation

$$\nabla^2 u = f, \qquad (x, y) \in \Omega.$$

Let the grid size be given by $h = \Delta x = \Delta y = 2/M$ for some M.

(a) Assume that Dirichlet boundary conditions are given. Derive the scheme at a point neighbouring the part of the boundary connecting (0, 2) and (2, 1).

(b) Repeat part (a) for Neumann boundary conditions.

5.8. Consider the difference scheme (5.42).

(a) Give the stencil.

(b) Show that it is second order accurate.

5.9. Show that the finite volume scheme (5.53) is second order accurate.

5.10. The multistep method (5.75) is called *root stable* if the scheme with $f(u, t) \equiv 0$ is a stable recursion, i.e., all roots of the *characteristic polynomial*

$$\sum_{j=0}^{k} a_j \xi^{k-j}$$

are smaller than one (in absolute value) or equal to one and simple. Now let the multistep method not be root stable. If we apply this method to the ODE

$$\frac{\mathrm{d}x}{\mathrm{d}t} = f(t),$$

then show that the difference equation

$$\frac{1}{\Delta t} \sum_{j=0}^{k} a_j x^{n+1-j} = \sum_{j=0}^{k} b_j f\left(t^{n+1-j}\right)$$

grows unstable.

Hint: Any solution is a linear combination of basis solutions; a typical basis function corresponding to the root λ is given by λ^n, or $n^j \lambda^n$, $j = 1, \ldots, l-1$, if λ is an l-fold root.

5.11. Consider the *explicit midpoint rule*

$$u^{n+1} = u^{n-1} + 2\Delta t f(u^n, t^n).$$

(a) Show that this method is root stable (see Exercise 5.10).

(b) Apply this method to the initial value problem

$$\frac{\mathrm{d}u}{\mathrm{d}t} = \lambda u, \qquad t > 0,$$

$$u(0) = 1.$$

5.12. Consider the ODE

$$\frac{\mathrm{d}u}{\mathrm{d}t} = \mathrm{i}\, u,$$

where i is the imaginary unit.

(a) Show that the Euler forward method is unstable (i.e., the solution grows unbounded) for any step size (and the number of steps large enough).

(b) Show that the solution of the Euler backward method becomes arbitrarily small for any step size (and the number of steps large enough).

(c) Show that the midpoint rule gives a solution that remains constant in absolute value.

5.13. The BDF-2 method reads

$$\frac{3}{2}u^{n+1} - 2u^n + \frac{1}{2}u^{n-1} = \Delta t f(u^{n+1}, t^{n+1}).$$

Show that this method applied to the initial value problem

$$\begin{aligned} \frac{\mathrm{d}u}{\mathrm{d}t} &= \lambda u, \qquad t > 0, \\ u(0) &= 1, \end{aligned}$$

is unconditionally stable for $\lambda \leq 0$.
Hint: Note that the recursion you obtain has constant coefficients and thus the problem boils down to finding the characteristic roots.

5.14. Define the matrix

$$\mathbf{J} := \begin{pmatrix} \frac{1}{2} & 1 \\ 0 & \frac{1}{2} \end{pmatrix}.$$

Show that $\rho(\mathbf{J}) < 1$ but that $\|\mathbf{J}\|_\infty > 1$.

5.15. Consider the following PDE

$$\frac{\partial u}{\partial t} = a\frac{\partial^2 u}{\partial x^2} + b\frac{\partial u}{\partial x} + cu, \qquad a > 0, \quad b, c \in \mathbb{R}.$$

Discretise this equation by means of the explicit Euler method for the time derivative and with central differences for the space derivatives.

(a) Give the operators $\mathcal{A}_\Delta$ and $\mathcal{B}_\Delta$ (cf. (5.105)) explicitly.

(b) Determine the growth factor $\mathcal{G}_k$.

5.16. Consider the PDE

$$\frac{\partial u}{\partial t} + b\frac{\partial u}{\partial x} = 0, \qquad b \in \mathbb{R}.$$

Discretise this equation in space and time with central differences. Show that this leads to an unconditionally unstable difference equation.

Chapter 6

The Equations of Continuum Mechanics and Electromagnetics

Important areas of application of the methods described in this book are found in continuum physics. They are based on an almost axiomatic footing of conservation laws, completed with problem-dependent constitutive relations. Some care is needed when the well-established laws for mass, momentum, and energy conservation are reformulated for a continuum. These laws will therefore be given in detail. The equations for compressible viscous flow and for linear elastic deformations will be written out in full. In view of their importance the equations for electromagnetic fields will also be derived, but without taking full account of possible coupling between electromagnetic forces and stresses in the material.

6.1 Introduction

The major areas of application of the problems and methods considered in this book are found in the physics of continua, which encompasses the theories of fluid dynamics, deformation of elastic media, and electromagnetic phenomena. Therefore we will summarise the respective theories here. In the context of this book this is necessarily very brief and concise. We will, for example, not consider the combined effect of electromagnetic and inertial forces. The reader is advised to consult the extensive existing literature for this and other details and for further background information.

Continuum physics considers the deformation and motion of matter under internal and external forces (inertia, stresses, gravity, or electromagnetic fields) at a macroscopic level, disregarding the molecular structure other than its integrated effects. The prevailing equations are based on the postulates that mass, momentum, and energy are conserved. Therefore these equations are called *conservation equations*. They are universal and do not contain the properties of the material in question. The number of unknowns is larger than the number of equations. Therefore the equations are not sufficient to determine the problem, and we need in addition *constitutive* and *thermodynamic relations*. These relations represent the properties of the material considered, and their choice is part of the modeling process (see Chapter 7).

We will start with the equations for fluid flow and elastic deformation, leaving aside any electromagnetic effects. Then we will present the equations for electromagnetic fields, without dilating upon any mechanical coupling (except the production of heat).

6.2 Eulerian and Lagrangian Coordinates

Consider the deformation and motion of material continuously distributed in some physical domain. We adopt the *continuum hypothesis* to consider the material as being made up of a coherent collection of "particles," each consisting of sufficiently many, but not too many, molecules in order to allow us to speak pointwise of velocity, pressure, temperature, etc. Thus we study the matter at a macroscopic level and do not consider explicitly the interaction between the individual molecules. The particles interact with each other via contact forces (stresses) that depend on the material considered.

There are two approaches to describing a deformation or motion. In the *Lagrangian* description we follow particles that move with the continuum. In the alternative *Eulerian* description we maintain a fixed position and consider particles that pass this position. Let the position $\boldsymbol{x}$ of a particle at time t be written as

$$\boldsymbol{x} = \boldsymbol{\varphi}(\boldsymbol{x}^*, t), \tag{6.1}$$

where $\boldsymbol{x}^*$ is a reference position for which we take the position of this particle at an initial time t_0; i.e., $\boldsymbol{x}^* = \boldsymbol{\varphi}(\boldsymbol{x}^*, t_0)$. Thus a particle can be specified by its Lagrangian coordinates $(\boldsymbol{x}^*, t)$, when following the motion, or by its Eulerian coordinates $(\boldsymbol{x}, t)$. A generic variable f can be expressed in terms of the Lagrangian or the Eulerian coordinates, and we write $f = f(\boldsymbol{x}, t) = f^*(\boldsymbol{x}^*, t)$. The velocity $\boldsymbol{v}^*(\boldsymbol{x}^*, t)$ of a particle is the time rate of change of the position of this particle expressed in Lagrangian coordinates:

$$\boldsymbol{v}^* := \frac{\partial \boldsymbol{\varphi}}{\partial t}(\boldsymbol{x}^*, t). \tag{6.2}$$

The time rate of change of a variable $f^*(\boldsymbol{x}^*, t)$ may be expressed in Eulerian coordinates as the *material* or *convective derivative* of f and is given by

$$\frac{\partial f^*}{\partial t}(\boldsymbol{x}^*, t) = \frac{\mathrm{d}f}{\mathrm{d}t}(\boldsymbol{\varphi}(\boldsymbol{x}^*, t), t) = \frac{\partial f}{\partial t}(\boldsymbol{x}, t) + \boldsymbol{v}(\boldsymbol{x}, t)\cdot\nabla f(\boldsymbol{x}, t). \tag{6.3}$$

The first term on the right-hand side of (6.3) is the local time derivative and the second term is the convective derivative, which is in fact the directional derivative of f in the direction of velocity $\boldsymbol{v}$.

The *displacement vector* is defined by

$$\boldsymbol{u} = \boldsymbol{x} - \boldsymbol{x}^* = \boldsymbol{\varphi}(\boldsymbol{x}^*, t) - \boldsymbol{x}^*. \tag{6.4}$$

In the theory of small elastic deformations we need the *linear deformation tensor* or *linear strain tensor*

$$\mathcal{E} := \frac{1}{2}\nabla \boldsymbol{u} + \frac{1}{2}(\nabla \boldsymbol{u})^T. \tag{6.5}$$

Under the assumption of the deformations being small, the gradient $\nabla \boldsymbol{u}$ may be interpreted with respect to either $\boldsymbol{x}$ or $\boldsymbol{x}^*$.

In the theory of Newtonian viscous fluid flow (Section 6.8.2) we need the *deformation velocity tensor* or the *rate of deformation tensor*

$$\mathcal{D} := \frac{1}{2}\nabla \boldsymbol{v} + \frac{1}{2}(\nabla \boldsymbol{v})^T. \tag{6.6}$$

(Note that the definition of the tensor $\nabla \boldsymbol{v}$ is ambiguous. Both $(\frac{\partial}{\partial x_j} v_i)$ and $(\frac{\partial}{\partial x_i} v_j)$ occur. Due to the symmetry of $\mathcal{D}$ this ambiguity is not important here. The same is true for $\mathcal{E}$.)

Example 6.1 A piece of rubber is stretched uniformly in all directions such that its local coordinate system $\boldsymbol{x}$ is related to its rest position $\boldsymbol{x}^*$ by $\boldsymbol{x} = (1+\alpha)\boldsymbol{x}^*$. The displacement vector is then

$$\boldsymbol{u} = \boldsymbol{x} - \boldsymbol{x}^* = \alpha \boldsymbol{x}^* = \frac{\alpha}{1+\alpha}\boldsymbol{x}.$$

The linear deformation tensor is thus given by $\mathcal{E} = \alpha\mathcal{I}$ for small α. □

Example 6.2 A velocity field directed along the x axis with linear shear in the y direction is given by $\boldsymbol{v} = \alpha y \boldsymbol{e}_x$. The deformation velocity tensor is then

$$\mathcal{D} = \frac{1}{2}\alpha \begin{pmatrix} 0 & 1 & 0 \\ 1 & 0 & 0 \\ 0 & 0 & 0 \end{pmatrix}.$$ □

6.3 The Transport Theorem

A rigorous derivation of the conservation laws is based on the *transport theorem*, which we will derive first. The theorem is most relevant for a moving fluid, although it remains equally valid for a deforming solid.

Let $q(\boldsymbol{x}, t)$ be a quantity per unit volume of the material (a fluid, say). Consider a control volume $\Omega(t)$ moving with the flow. Ω is called a material volume and its surface $\partial\Omega$ is called a material surface. Define

$$F(t) := \int_{\Omega(t)} q(\boldsymbol{x}, t)\,\mathrm{d}V. \tag{6.7}$$

For example, if $q(\boldsymbol{x}, t) = \rho(\boldsymbol{x}, t)$ is the mass density of the fluid, then $F(t)$ is the total mass of fluid contained in the control volume. The time rate of change of $F(t)$ is given by

$$\frac{\mathrm{d}F(t)}{\mathrm{d}t} = \int_{\Omega(t)} \Big(\frac{\partial q}{\partial t}(\boldsymbol{x}, t) + \nabla\boldsymbol{\cdot}(q\boldsymbol{v})(\boldsymbol{x}, t)\Big)\,\mathrm{d}V. \tag{6.8}$$

Equation (6.8) is called the *transport theorem* (see also (J.20) in Appendix J).

A concise proof is as follows. Consider $F(t+h) - F(t)$ asymptotically for small h. We write symbolically $\Omega(t+h) = \Omega(t) + \mathrm{d}\Omega$. Then we have

$$\begin{aligned} F(t+h) - F(t) &\doteq \int_{\Omega(t)+\mathrm{d}\Omega} \Big(q(\boldsymbol{x}, t) + h\frac{\partial q}{\partial t}(\boldsymbol{x}, t)\Big)\,\mathrm{d}V - \int_{\Omega(t)} q(\boldsymbol{x}, t)\,\mathrm{d}V \\ &\doteq \int_{\mathrm{d}\Omega} q(\boldsymbol{x}, t)\,\mathrm{d}V + h\int_{\Omega(t)} \frac{\partial q}{\partial t}(\boldsymbol{x}, t)\,\mathrm{d}V. \end{aligned}$$

We apply locally near the surface $\partial\Omega$ of Ω an orthogonal coordinate system $\boldsymbol{x} = \boldsymbol{\sigma} + \lambda\boldsymbol{n}$, where $\lambda = 0$ denotes the surface $\partial\Omega$ and $\boldsymbol{n}$ is the unit outward normal of $\partial\Omega$. As the volume Ω is a material volume, it moves with the fluid velocity $\boldsymbol{v}$, and therefore the surface $\partial\Omega$ has moved a distance $h\boldsymbol{v}$ between times t and $t+h$. So we have

$$\int_{\mathrm{d}\Omega} q(\boldsymbol{x}, t)\,\mathrm{d}V = \oint_{\partial\Omega}\int_0^{h(\boldsymbol{v}\cdot\boldsymbol{n})} q(\boldsymbol{\sigma} + \lambda\boldsymbol{n}, t)\,\mathrm{d}\lambda\mathrm{d}S \doteq \oint_{\partial\Omega} q(\boldsymbol{\sigma})h(\boldsymbol{v}\boldsymbol{\cdot}\boldsymbol{n})\,\mathrm{d}S = h\int_{\Omega} \nabla\boldsymbol{\cdot}(q\boldsymbol{v})\,\mathrm{d}V,$$

where we used Gauss's divergence theorem for the last step. After dividing by h and taking the limit $h \to 0$, we obtain (6.8).

Example 6.3 A flow field $\boldsymbol{v}$ that keeps the content of any convected volume $\Omega(t)$ constant is called incompressible. It satisfies

$$\frac{\mathrm{d}}{\mathrm{d}t}\int_{\Omega(t)} 1\,\mathrm{d}V = \int_{\Omega(t)} \nabla\cdot\boldsymbol{v}\,\mathrm{d}V = 0$$

and is therefore divergence free; i.e., $\nabla\cdot\boldsymbol{v} = 0$. □

6.4 Conservation Equations

Consider the deformation and motion of matter in some domain $\Omega(t)$ that is moving with the material (i.e., Ω is a material volume). From physics we know that the matter is subject to some very strict limitations called conservation laws. These laws postulate that, without a source, certain properties like mass, momentum, and energy remain unchanged. Any such property P can be described by a density E (the amount of P per unit volume) and an associated flux density $\boldsymbol{F}$ (the amount of P that flows per unit time through a unit material surface normal to $\boldsymbol{F}$) such that the change of P of a given material volume Ω must be exactly equal to the sum of the net influx through the volume's surface $\partial\Omega$ and any possible production from a source distribution Q:

$$\frac{\mathrm{d}}{\mathrm{d}t}\int_{\Omega} E\,\mathrm{d}V = -\oint_{\partial\Omega} \boldsymbol{F}\cdot\boldsymbol{n}\,\mathrm{d}S + \int_{\Omega} Q\,\mathrm{d}V, \tag{6.9}$$

where $\boldsymbol{n}$ denotes the unit outward normal of $\partial\Omega$. As Ω moves with the flow with velocity $\boldsymbol{v}$, this is, according to the transport theorem and Gauss's divergence theorem, equivalent to

$$\int_{\Omega}\left(\frac{\partial E}{\partial t} + \nabla\cdot(E\boldsymbol{v} + \boldsymbol{F}) - Q\right)\mathrm{d}V = 0. \tag{6.10}$$

If this is true for any material volume in our region of interest, the integrand itself must vanish, so we have

$$\frac{\partial E}{\partial t} + \nabla\cdot(E\boldsymbol{v} + \boldsymbol{F}) = Q. \tag{6.11}$$

It is clear that while $\boldsymbol{F}$ denotes the flux density through a material surface, $E\boldsymbol{v} + \boldsymbol{F}$ is the flux through a fixed unit surface. Relation (6.11) is the general form of a conservation law for a continuum. In the following we will derive specific versions of this general format. Note that we will often use just "flux" instead of "flux density."

6.5 Conservation of Mass

As there exist in the present context no mass sources, the mass of any material volume $\Omega(t)$ is constant, so

$$\frac{\mathrm{d}}{\mathrm{d}t}\int_{\Omega(t)} \rho\,\mathrm{d}V = 0. \tag{6.12}$$

Applying the transport theorem to (6.12) with $q(\boldsymbol{x}, t) = \rho$ gives

$$\int_{\Omega(t)} \left(\frac{\partial \rho}{\partial t} + \nabla \cdot (\rho \boldsymbol{v}) \right) \mathrm{d}V = 0. \tag{6.13}$$

Since this conservation law holds for any $\Omega(t)$, the differential form of the mass conservation law

$$\frac{\partial \rho}{\partial t} + \nabla \cdot (\rho \boldsymbol{v}) = 0 \tag{6.14}$$

must be satisfied. Equation (6.14) is called the *continuity equation*, written in conservative form. By using the material derivative (6.3), we can rewrite (6.14) into convective form:

$$\frac{\mathrm{d}\rho}{\mathrm{d}t} + \rho \nabla \cdot \boldsymbol{v} = 0. \tag{6.15}$$

6.6 Conservation of Momentum

The equations of motion of a continuum describe conservation of linear momentum and angular momentum. First, the law of *conservation of linear momentum* reads

$$\frac{\mathrm{d}}{\mathrm{d}t} \int_{\Omega(t)} \rho \boldsymbol{v} \, \mathrm{d}V = \oint_{\partial\Omega(t)} \boldsymbol{t}(\boldsymbol{n}) \, \mathrm{d}S + \int_{\Omega(t)} \rho \boldsymbol{f} \, \mathrm{d}V, \tag{6.16}$$

with $\boldsymbol{t}(\boldsymbol{n})$ the *stress vector*, i.e., the *internal* or *contact force* field per unit area, acting on the boundary $\partial\Omega$ of the material volume Ω, and $\boldsymbol{f}$ the specific (i.e., per unit mass) *external* or *volume force* field acting on the material contained by Ω. In (6.16) we have explicitly written $\boldsymbol{t}(\boldsymbol{n})$ to denote the dependence of the stress vector on the outward unit normal $\boldsymbol{n}$ on $\partial\Omega$. The conservation law in (6.16) states that the rate of change of momentum of the material contained in Ω, due to the movement of Ω with velocity $\boldsymbol{v}$, is equal to the sum of contact forces and volume forces acting on the material.

Second, the law of *conservation of angular momentum* is expressed by

$$\frac{\mathrm{d}}{\mathrm{d}t} \int_{\Omega(t)} \boldsymbol{x} \times \rho \boldsymbol{v} \, \mathrm{d}V = \oint_{\partial\Omega(t)} \boldsymbol{x} \times \boldsymbol{t}(\boldsymbol{n}) \, \mathrm{d}S + \int_{\Omega(t)} \boldsymbol{x} \times \rho \boldsymbol{f} \, \mathrm{d}V, \tag{6.17}$$

meaning that the rate of change of angular momentum of the material in $\Omega(t)$, when the control volume moves with the continuum, is equal to the sum of the moment of the contact forces and the moment of the volume forces acting on it.

In the following the conservation law of linear momentum will be used to develop the equations of motion for a continuum. From the conservation law of angular momentum a certain symmetry in the stress vector will be derived.

Consider the ith ($i = 1, 2, 3$) component of the conservation law of linear momentum (6.16). Applying the transport theorem (6.8) to its left-hand side with $q(\boldsymbol{x}, t) = \rho(\boldsymbol{x}, t) v_i(\boldsymbol{x}, t)$ yields

$$\frac{\mathrm{d}}{\mathrm{d}t} \int_{\Omega(t)} \rho v_i \, \mathrm{d}V = \int_{\Omega(t)} \left(\frac{\partial}{\partial t}(\rho v_i) + \nabla \cdot (\rho \boldsymbol{v} v_i) \right) \mathrm{d}V. \tag{6.18}$$

The stress vector $\boldsymbol{t}(\boldsymbol{n})$ acting on a surface with normal $\boldsymbol{n}$ is completely determined by the stress vectors $\boldsymbol{t}(\boldsymbol{e}_j)$ for the unit vectors $\boldsymbol{e}_j$ $(j = 1, 2, 3)$. The *stress tensor* $\mathcal{T}$ is defined by

$$\mathcal{T} = (\mathcal{T}_{ij}) = \big(\boldsymbol{t}(\boldsymbol{e}_1), \boldsymbol{t}(\boldsymbol{e}_2), \boldsymbol{t}(\boldsymbol{e}_3)\big) \quad \text{with} \quad \mathcal{T}_{ij} := t_i(\boldsymbol{e}_j); \tag{6.19}$$

i.e., $\mathcal{T}_{ij}$ is the ith component of the stress vector $\boldsymbol{t}$ acting on a surface with unit normal $\boldsymbol{e}_j$. (Note that there is no uniformity in the nomenclature. Some authors define the stress tensor as $\mathcal{T}^T$. Since $\mathcal{T}$ will in general be symmetric, this is usually of no concern.) Applying the principle of local equilibrium, it follows that [2, 24, 41]

$$t_i(\boldsymbol{n}) = \sum_{j=1}^{3} \mathcal{T}_{ij} n_j, \quad i = 1, 2, 3, \tag{6.20a}$$

or

$$\boldsymbol{t}(\boldsymbol{n}) = \mathcal{T}\boldsymbol{n}. \tag{6.20b}$$

The surface integral in (6.16) can now be replaced by a volume integral using Gauss's theorem (J.13), and we find

$$\oint_{\partial\Omega(t)} \boldsymbol{t}(\boldsymbol{n})\,\mathrm{d}S = \oint_{\partial\Omega(t)} \mathcal{T}\boldsymbol{n}\,\mathrm{d}S = \int_{\Omega(t)} \nabla\cdot\mathcal{T}^T\,\mathrm{d}V, \tag{6.21}$$

where $\nabla\cdot\mathcal{T}^T$ is defined by

$$\nabla\cdot\mathcal{T}^T := \begin{pmatrix} \nabla\cdot\mathcal{T}_{1*} \\ \nabla\cdot\mathcal{T}_{2*} \\ \nabla\cdot\mathcal{T}_{3*} \end{pmatrix} \tag{6.22}$$

(see the appendix (L.3)) and $\mathcal{T}_{i*}$ denotes the ith row of $\mathcal{T}$. Combining (6.16), (6.18), and (6.21), we get the ith component of the conservation law of momentum:

$$\int_{\Omega(t)} \left(\frac{\partial}{\partial t}(\rho v_i) + \nabla\cdot(\rho\boldsymbol{v}v_i)\right) \mathrm{d}V = \int_{\Omega(t)} \nabla\cdot\mathcal{T}_{i*}\,\mathrm{d}V + \int_{\Omega(t)} \rho f_i\,\mathrm{d}V. \tag{6.23}$$

This becomes, in differential, conservative form,

$$\frac{\partial}{\partial t}(\rho v_i) + \nabla\cdot(\rho\boldsymbol{v}v_i) = \nabla\cdot\mathcal{T}_{i*} + \rho f_i, \quad i = 1, 2, 3, \tag{6.24a}$$

or, in vectorial notation,

$$\frac{\partial}{\partial t}(\rho\boldsymbol{v}) + \nabla\cdot(\rho\boldsymbol{v}\boldsymbol{v}^T) = \nabla\cdot\mathcal{T}^T + \rho\boldsymbol{f}. \tag{6.24b}$$

The divergence of the dyadic product $\rho\boldsymbol{v}\boldsymbol{v}^T$ is defined analogously to (6.22).

From the conservation of angular momentum (6.17) it happens (see below) that $\mathcal{T} = \mathcal{T}^T$; in other words, the stress tensor $\mathcal{T}$ is symmetric (for nonpolar media). Therefore in (6.24b) we can safely change $\nabla\cdot\mathcal{T}^T$ into $\nabla\cdot\mathcal{T}$. If, in addition, we use the continuity equation (6.14), we finally get the equation in its best-known, convective, form

$$\rho\frac{\mathrm{d}}{\mathrm{d}t}\boldsymbol{v} = \nabla\cdot\mathcal{T} + \rho\boldsymbol{f}. \tag{6.25}$$

This is known as *Cauchy's equation of motion* [24].

A brief proof of the symmetry of $\mathcal{T}$ is as follows [2]. By employing the transport theorem (6.8) and the continuity equation (6.14), we can derive for the left-hand side of (6.17)

$$\frac{\mathrm{d}}{\mathrm{d}t}\int_{\Omega(t)} \boldsymbol{x}\times\rho\boldsymbol{v}\,\mathrm{d}V = \int_{\Omega(t)} \rho\frac{\mathrm{d}}{\mathrm{d}t}(\boldsymbol{x}\times\boldsymbol{v})\,\mathrm{d}V = \int_{\Omega(t)} \rho\boldsymbol{x}\times\frac{\mathrm{d}\boldsymbol{v}}{\mathrm{d}t}\,\mathrm{d}V. \tag{6.26}$$

The second integral in (6.17) can be rewritten by means of Gauss's theorem (J.13) as

$$\oint_{\partial\Omega(t)} \boldsymbol{x}\times\boldsymbol{t}(\boldsymbol{n})\,\mathrm{d}S = \oint_{\partial\Omega(t)} (\boldsymbol{x}\times\mathcal{T})\cdot\boldsymbol{n}\,\mathrm{d}S = \int_{\Omega(t)} \left(\boldsymbol{x}\times(\nabla\cdot\mathcal{T}^T)+\boldsymbol{t}^*\right)\mathrm{d}V, \tag{6.27}$$

where $\boldsymbol{t}^*$ denotes a vector, related to the antisymmetric part of $\mathcal{T}$, given by

$$\boldsymbol{t}^* := \begin{pmatrix} \mathcal{T}_{32}-\mathcal{T}_{23} \\ \mathcal{T}_{13}-\mathcal{T}_{31} \\ \mathcal{T}_{12}-\mathcal{T}_{21} \end{pmatrix}. \tag{6.28}$$

By combining (6.17) with (6.26) and (6.27), we get

$$\int_{\Omega(t)} \boldsymbol{x}\times\left(\rho\frac{\mathrm{d}}{\mathrm{d}t}\boldsymbol{v}-\nabla\cdot\mathcal{T}^T-\rho\boldsymbol{f}\right)\mathrm{d}V = \int_{\Omega(t)} \boldsymbol{t}^*\,\mathrm{d}V. \tag{6.29}$$

Together with (6.14) and (6.24b) this implies that $\boldsymbol{t}^*$ vanishes; i.e., $\mathcal{T}$ is symmetric.

6.7 Conservation of Energy

Finally, we have the law of *conservation of energy*. The rate of change of kinetic energy ($\frac{1}{2}\rho|\boldsymbol{v}|^2$) and internal energy ($\rho e$) should equal the mechanical power of the stresses ($\boldsymbol{t}(\boldsymbol{n})\cdot\boldsymbol{v}$) and volume forces ($\rho\boldsymbol{f}\cdot\boldsymbol{v}$) acting on the material, plus the heat supplied by internal sources (ρr) and exchanged across the border ($-\boldsymbol{q}\cdot\boldsymbol{n}$). This is given by

$$\frac{\mathrm{d}}{\mathrm{d}t}\int_{\Omega(t)} \rho E\,\mathrm{d}V = \oint_{\partial\Omega(t)} \boldsymbol{t}(\boldsymbol{n})\cdot\boldsymbol{v}\,\mathrm{d}S + \int_{\Omega(t)} \rho\boldsymbol{f}\cdot\boldsymbol{v}\,\mathrm{d}V - \oint_{\partial\Omega(t)} \boldsymbol{q}\cdot\boldsymbol{n}\,\mathrm{d}S + \int_{\Omega(t)} \rho r\,\mathrm{d}V, \tag{6.30}$$

with $\boldsymbol{q}$ the heat flux vector and r the specific heat supply, e.g., by a distribution of radioactive sources or electric (Joule) heating (see (6.64) below). The *specific energy* E is defined by

$$E := e + \frac{1}{2}|\boldsymbol{v}|^2, \tag{6.31}$$

with e the *specific internal energy* (a thermodynamic property) of the material and $\frac{1}{2}|\boldsymbol{v}|^2$ the *specific kinetic energy* of the continuum. The term $-\boldsymbol{q}\cdot\boldsymbol{n}$ is the amount of energy per unit area and per unit time that is transmitted through $\partial\Omega(t)$ to the fluid in material volume $\Omega(t)$.

By using the transport theorem (6.8) and Gauss's theorem (J.13), we can convert (6.30) into the volume integral

$$\int_{\Omega(t)} \left(\frac{\partial}{\partial t}(\rho E)+\nabla\cdot(\rho\boldsymbol{v}E)\right)\mathrm{d}V = \int_{\Omega(t)} (\nabla\cdot(\mathcal{T}\boldsymbol{v})+\rho\boldsymbol{f}\cdot\boldsymbol{v}-\nabla\cdot\boldsymbol{q}+\rho r))\,\mathrm{d}V, \tag{6.32}$$

which holds for any volume that moves with the material. Therefore we have the *energy equation* in differential form:

$$\frac{\partial}{\partial t}(\rho E) + \nabla\cdot(\rho \boldsymbol{v} E) = \nabla\cdot(\mathcal{T}\boldsymbol{v}) + \rho \boldsymbol{f}\cdot\boldsymbol{v} - \nabla\cdot\boldsymbol{q} + \rho r. \tag{6.33}$$

This may be further simplified by taking the inner product between the momentum equations (6.24b) and the velocity $\boldsymbol{v}$ and subtracting the result from (6.33). This yields, using the symmetry of $\mathcal{T}$, the equation in the conservative form

$$\frac{\partial}{\partial t}(\rho e) + \nabla\cdot(\rho \boldsymbol{v} e) = \mathcal{T}:\nabla\boldsymbol{v} - \nabla\cdot\boldsymbol{q} + \rho r, \tag{6.34}$$

where the double inner product $\mathcal{T}:\nabla\boldsymbol{v}$ is given by (see (L.5) in Appendix L)

$$\mathcal{T}:\nabla\boldsymbol{v} := \sum_{i=1}^{3}\sum_{j=1}^{3}\mathcal{T}_{ij}\frac{\partial v_i}{\partial x_j}. \tag{6.35}$$

If we use the equation of mass conservation (6.12), we obtain the equation in convective form:

$$\rho\frac{\mathrm{d}}{\mathrm{d}t}e = \mathcal{T}:\nabla\boldsymbol{v} - \nabla\cdot\boldsymbol{q} + \rho r. \tag{6.36}$$

6.8 Constitutive Equations and Thermodynamic Relations

The conservation laws for deformable media are the continuity equation (6.14), the momentum equations (6.24b), and the energy equation (6.34). These 5 equations have 14 unknowns. Therefore the mathematical description is not complete without some closure relations describing properties of the matter and certain instantaneous (in both time and space) interactions between material parameters. These relations are called the *constitutive equations* and *thermodynamic relations*. In particular, the constitutive equations are not as basic as the conservation laws, and their form will depend on the model adopted. In fact, this part of the model, the constitutive equations, may be called the *physical model*.

The conservation laws have to be completed with models for the stress tensor $\mathcal{T}$ and the heat flux vector $\boldsymbol{q}$, thermodynamic relations, internal heat sources r, and the volume force field $\boldsymbol{f}$. Apart from possible apparent forces in an accelerating coordinate system (called Coriolis forces), gravity is usually the only external force field working on the matter, so $\boldsymbol{f} = g\boldsymbol{e}_g$, with g the acceleration of gravity and $\boldsymbol{e}_g$ the unit vector in the direction of gravitation. Except when indicated otherwise, we will further assume no internal heat sources, so $r = 0$.

It should be noted that these constitutive relations between $\mathcal{T}$, $\mathcal{D}$, and $\mathcal{E}$ are based on the physics, and are therefore subject to certain compatibility conditions. The material may be isotropic [24], i.e., have no preferred directions, and the relation must be independent of orientation. The material may be homogeneous, in which case the relation is independent of position. A very important condition is the compatibility of the relations with the principle of objectivity. This means that the relation should be equivalent for any observer, i.e., in any frame of reference [24]. This may be formalized as follows. The transformation that connects two frames of reference by the relations

$$\boldsymbol{x}' = \boldsymbol{x}'_0(t) + \boldsymbol{Q}(t)\boldsymbol{x}, \qquad t' = t - t_0, \tag{6.37a}$$

where $\boldsymbol{Q}$ is a rotation, i.e., an orthogonal tensor with $\det \boldsymbol{Q} = 1$, is called an *observer transformation*. Suppose such an observer transformation identifies a scalar field ϕ, vector field $\boldsymbol{u}$, and tensor field $\mathcal{T}$ with corresponding fields ϕ', $\boldsymbol{u}'$, and $\mathcal{T}'$ in the following way:

$$\phi'(\boldsymbol{x}', t') = \phi(\boldsymbol{x}, t), \qquad \boldsymbol{u}'(\boldsymbol{x}', t') = \boldsymbol{Q}(t)\boldsymbol{u}(\boldsymbol{x}, t), \qquad \mathcal{T}'(\boldsymbol{x}', t') = \boldsymbol{Q}(t)\mathcal{T}(\boldsymbol{x}, t)\,\boldsymbol{Q}^T(t). \tag{6.37b}$$

Then these fields are called *objective*. For a further discussion on this important restriction we refer to the literature.

In the remainder of this section we will consider some examples of physical models, defined by constitutive equations and, where relevant, supplemented by suitable thermodynamic relations.

A simple but important example is the problem of heat conduction in rigid material. This assumption of rigidity implies the absence of any deformation or any response to external forcing. The stress tensor, therefore, plays no role. The resulting heat equation is similar to the equation that describes mass diffusion.

We will further consider two important types of material that are characterized by their stress tensor: viscous fluids and elastic material. A fluid at rest (i.e., in equilibrium) sustains normal stresses by compression, but cannot sustain any shear stress, and $\mathcal{T}$ only depends on the rate of deformation tensor $\mathcal{D}$ (6.6). Elastic material, on the other hand, is characterized by a response to any deformation and for small deformations the stress tensor depends on the deformation tensor $\mathcal{E}$ (6.5). If the material is purely elastic, the stress depends on $\mathcal{E}$ only and vanishes when $\mathcal{E}$ vanishes (the material is in its undeformed reference state).

6.8.1 Heat Conduction and Mass Diffusion

If the material is rigid, allowing no motion or deformation, all conservation laws are trivially satisfied except for the heat balance in the energy equation. If the internal energy e is a function of the temperature T only, we can define $C = \frac{\mathrm{d}}{\mathrm{d}T}e$ as the heat capacity or specific heat of the material. Furthermore, if the heat flux satisfies *Fourier's law* of heat conduction

$$\boldsymbol{q} = -\kappa(T)\nabla T, \tag{6.38}$$

with κ the coefficient of heat conduction, (6.36) reduces to

$$\rho C \frac{\partial}{\partial t} T = \nabla\cdot(\kappa \nabla T) + \rho r. \tag{6.39}$$

It is worth mentioning here that diffusion of heat and diffusion of molecules of a solute in a liquid are essentially similar processes. If the concentration of the solute is c and its flux is $\boldsymbol{j}$, a widely used constitutive law that relates $\boldsymbol{j}$ to c in the absence of a temperature gradient is given by [83]

$$\boldsymbol{j} = -D(c)\nabla c, \tag{6.40}$$

which is called *Fick's law*, the analogue of Fourier's law. From the equation of mass conservation for the solute, i.e.,

$$\frac{\partial c}{\partial t} + \nabla\cdot \boldsymbol{j} = Q,$$

where Q is a source of solute, we obtain the diffusion equation

$$\frac{\partial c}{\partial t} = \nabla\cdot\big(D\nabla c\big) + Q. \tag{6.41}$$

6.8.2 Newtonian Viscous Fluid

For fluids it is useful to split the stress tensor into a part depending on the thermodynamic pressure p, representing the stationary normal components, and a viscous part $\boldsymbol{\tau}$, related to the velocity gradients:

$$\mathcal{T} = -p\mathcal{I} + \boldsymbol{\tau}, \tag{6.42}$$

where $\boldsymbol{\tau}$ is called the *viscous stress tensor*. Let us recall the just-derived conservation equations:

$$\text{mass:} \quad \frac{\partial}{\partial t}\rho \quad + \nabla\cdot(\rho\boldsymbol{v}) \quad = 0, \tag{6.43a}$$

$$\text{momentum:} \quad \frac{\partial}{\partial t}(\rho\boldsymbol{v}) + \nabla\cdot(\rho\boldsymbol{v}\boldsymbol{v}^T) = -\nabla p + \nabla\cdot\boldsymbol{\tau} + \rho\boldsymbol{f}, \tag{6.43b}$$

$$\text{energy:} \quad \frac{\partial}{\partial t}(\rho e) + \nabla\cdot(\rho\boldsymbol{v}e) \quad = -\nabla\cdot\boldsymbol{q} + \nabla\cdot(-p\boldsymbol{v} + \boldsymbol{\tau}\boldsymbol{v}). \tag{6.43c}$$

Depending on the application, it is often convenient to introduce the specific *enthalpy* $h := e + p/\rho$, or *entropy* s and absolute temperature T, via the fundamental law of thermodynamics for a reversible process:

$$T\mathrm{d}s = \mathrm{d}e + p\mathrm{d}\rho^{-1} = \mathrm{d}h - \rho^{-1}\mathrm{d}p. \tag{6.44}$$

With the convective derivative (6.3), and noting that $\boldsymbol{\tau}:\nabla\boldsymbol{v} = \nabla\cdot(\boldsymbol{\tau}\boldsymbol{v}) - \boldsymbol{v}\cdot(\nabla\cdot\boldsymbol{\tau})$ since $\boldsymbol{\tau}$ is symmetric, the above conservation laws may be rewritten in their convective form:

$$\text{mass:} \quad \frac{\mathrm{d}}{\mathrm{d}t}\rho = -\rho\nabla\cdot\boldsymbol{v}, \tag{6.45a}$$

$$\text{momentum:} \quad \rho\frac{\mathrm{d}}{\mathrm{d}t}\boldsymbol{v} = -\nabla p + \nabla\cdot\boldsymbol{\tau} + \rho\boldsymbol{f}, \tag{6.45b}$$

$$\text{energy:} \quad \rho\frac{\mathrm{d}}{\mathrm{d}t}e = -\nabla\cdot\boldsymbol{q} - p\nabla\cdot\boldsymbol{v} + \boldsymbol{\tau}:\nabla\boldsymbol{v}, \tag{6.45c}$$

$$\rho\frac{\mathrm{d}}{\mathrm{d}t}h = \frac{\mathrm{d}}{\mathrm{d}t}p - \nabla\cdot\boldsymbol{q} + \boldsymbol{\tau}:\nabla\boldsymbol{v}, \tag{6.45d}$$

$$\rho T\frac{\mathrm{d}}{\mathrm{d}t}s = -\nabla\cdot\boldsymbol{q} + \boldsymbol{\tau}:\nabla\boldsymbol{v}. \tag{6.45e}$$

Equations (6.45c), (6.45d), and (6.45e) are three equivalent forms of the energy equation. For an ideal fluid, e and h depend on T only and we may introduce $\mathrm{d}e = C_V \mathrm{d}T$ and $\mathrm{d}h = C_P \mathrm{d}T$ such that we get

$$\text{energy 1:}\quad \rho C_V \frac{\mathrm{d}}{\mathrm{d}t} T = -\nabla\cdot\boldsymbol{q} - p\nabla\cdot\boldsymbol{v} + \boldsymbol{\tau}:\nabla\boldsymbol{v}, \tag{6.45f}$$

$$\text{energy 2:}\quad \rho C_P \frac{\mathrm{d}}{\mathrm{d}t} T = \frac{\mathrm{d}}{\mathrm{d}t} p - \nabla\cdot\boldsymbol{q} + \boldsymbol{\tau}:\nabla\boldsymbol{v}. \tag{6.45g}$$

For an incompressible fluid we have $T\mathrm{d}s = C_V \mathrm{d}T$ with

$$\text{energy 3:}\quad \rho C_V \frac{\mathrm{d}}{\mathrm{d}t} T = -\nabla\cdot\boldsymbol{q} + \boldsymbol{\tau}:\nabla\boldsymbol{v}. \tag{6.45h}$$

$C_V = T\left(\frac{\partial s}{\partial T}\right)_\rho$ is the *heat capacity* or *specific heat at constant volume*. $C_P = T\left(\frac{\partial s}{\partial T}\right)_p$ is the *heat capacity* or *specific heat at constant pressure* [83, 26]. In general (for an ideal gas) they may depend on temperature. For a *perfect gas* they are constants. (There is no uniformity in nomenclature. Some authors use perfect gas for what we define as ideal gas.) For an incompressible fluid like a liquid, C_V is practically equal to C_P. A notable exception is matter during phase transition (fusion, vaporization), where all the added heat (enthalpy) is used for the solid-liquid or liquid-vapor transition, rather than an increase of temperature. The amount of heat involved with the phase transition is usually called *latent heat*.

We will consider here an ideal, heat-conducting, and viscous fluid, which is described by the following relations:

$$\text{ideal gas relation:}\quad p = \rho \mathcal{R} T, \tag{6.46a}$$

$$\text{Fourier's heat flux model:}\quad \boldsymbol{q} = -\kappa \nabla T, \tag{6.46b}$$

$$\text{Newton's viscous stress tensor:}\quad \boldsymbol{\tau} = 2\mu\mathcal{D} + \lambda(\nabla\cdot\boldsymbol{v})\mathcal{I}, \tag{6.46c}$$

where λ and μ are viscosity coefficients and $\mathcal{D}$ is the deformation velocity tensor, defined in (6.6), while $\mathcal{R}$ is the specific gas constant. For air $\mathcal{R} = 286.73\ \mathrm{J/(kg\cdot K)}$. For an ideal gas we also have the relationship $\mathcal{R} = C_P - C_V$. Under these assumptions, (6.45a) and (6.45b), usually supplemented by an energy equation (6.45f) or (6.45g), are called the *Navier–Stokes equations*. (There is no uniformity in the nomenclature. Sometimes this name is only given to the momentum equation (6.45b), and sometimes only to the equations for incompressible flow.)

For an isentropic process we have

$$C_P \mathrm{d}T - \rho^{-1}\mathrm{d}p = C_V \mathrm{d}T - p\rho^{-2}\mathrm{d}\rho = 0, \qquad \text{so} \quad \frac{\mathrm{d}p}{\mathrm{d}\rho} = \gamma\frac{p}{\rho}, \tag{6.47}$$

where $\gamma = C_P/C_V$ is the *specific heat ratio* (= 1.4 for air). From the definition of the speed of sound $c^2 := (\partial p/\partial\rho)_s$ we have

$$c = (\gamma p/\rho)^{1/2} \quad \text{or} \quad c = (\gamma R T)^{1/2}. \tag{6.48}$$

It is instructive to introduce (see (L.2) in Appendix L) the deviatoric deformation velocity tensor $\mathcal{D}' := \mathcal{D} - \frac{1}{3}(\nabla\cdot\boldsymbol{v})\mathcal{I}$ such that

$$\mathcal{T} = \left(-p + \left(\lambda + \frac{2}{3}\mu\right)(\nabla\cdot\boldsymbol{v})\right)\mathcal{I} + 2\mu\mathcal{D}'. \tag{6.49}$$

The first part of the stress represents the fluid's resistance against dynamic compression; the second part is its resistance against shear. It shows that the mechanical pressure $p_m := -\frac{1}{3}\,\mathrm{tr}(\mathcal{T})$ is not equivalent to the thermodynamic pressure p. From the continuity equation (6.14) the difference is found to be proportional to relative changes of density:

$$p - p_m = \left(\lambda + \frac{2}{3}\mu\right)(\nabla\cdot\boldsymbol{v}) = \left(\lambda + \frac{2}{3}\mu\right)\frac{1}{\rho}\frac{\mathrm{d}}{\mathrm{d}t}\rho,$$

where $\lambda + \frac{2}{3}\mu$ is called the coefficient of *bulk viscosity* (or expansion viscosity or second viscosity). If, according to Stokes's hypothesis, the fluid is in local thermodynamic equilibrium and both pressures are the same, this coefficient vanishes (see [2, 15, 41]). The coefficient μ is sometimes called the coefficient of *dynamic viscosity*, in contrast to the ratio $\nu = \mu/\rho$, which is called the coefficient of *kinematic viscosity*. It should be noted that the viscosity coefficients in general depend on the temperature.

Important simplifications are obtained (see Chapter 7) if we may neglect viscosity (the fluid is called a gas) or if the fluid is incompressible (the fluid is called a liquid). In the latter case the energy equation is decoupled from the mass and momentum equations and may be solved separately.

Finally, we note that at a free surface $\mathcal{S}$ of a fluid the *surface tension* produces a pressure jump across $\mathcal{S}$, which is proportional to the sum of the principal curvatures of the surface. The factor of proportionality σ (say) is usually called "surface tension," but this is really a force per unit length. If $\boldsymbol{n}$ denotes a unit vector field, (outward) normal to $\mathcal{S}$, then the pressure jump $[p]_{\mathcal{S}} := p_{\text{inside}} - p_{\text{outside}}$ is given by [46]

$$[p]_{\mathcal{S}} = \sigma\nabla\cdot\boldsymbol{n} \quad \text{at} \quad \mathcal{S}. \tag{6.50}$$

It can be proved that any smooth $\boldsymbol{n}$ yields the same $\nabla\cdot\boldsymbol{n}$ at $\mathcal{S}$. The *contact angle* θ (Figure 6.1) between the fluid-free surface and the wetted solid surface (e.g., of the container of the fluid) is—in equilibrium—a material property that does not depend on the shape of the fluid.

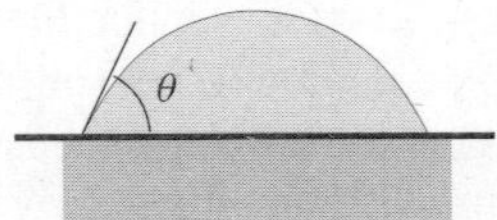

Figure 6.1. *Contact angle.*

6.8.3 Linear Elastic and Viscoelastic Deformations

It can be shown that for deformations small enough to allow linearization of homogeneous isotropic elastic material the stress tensor satisfies *Hooke's law*, i.e., the linear relation

$$\mathcal{T} = \lambda\,\mathrm{tr}(\mathcal{E})\mathcal{I} + 2\mu\mathcal{E}, \tag{6.51a}$$

or, written out componentwise,

$$t_{ij} = \lambda\delta_{ij}\big(e_{11} + e_{22} + e_{33}\big) + 2\mu e_{ij}, \tag{6.51b}$$

where we followed the tradition to write the Cartesian components of $\mathcal{T}$ as t_{ij} and of $\mathcal{E}$ as e_{ij}. The material parameters λ and μ are called Lamé coefficients. The material is called linear elastic.

From the observation that with small deformations density changes of the material are negligible we may consider ρ constant and we can ignore the continuity equation (6.14). Similarly, in the absence of heat sources the energy equation (6.36) may usually be decoupled from the elastic deformation problem. Another important simplification implied by the assumption of small displacements is the fact that the acceleration in the equation of motion (6.25) simplifies to a double time-derivative of displacement $\boldsymbol{u}$. By eliminating $e_{ij} = \frac{1}{2}(\frac{\partial}{\partial x_j}u_i + \frac{\partial}{\partial x_i}u_j)$, we may finally obtain Navier's equations

$$\rho \frac{\partial^2}{\partial t^2}\boldsymbol{u} = (\lambda + \mu)\nabla\big(\nabla\cdot\boldsymbol{u}\big) + \mu\nabla^2\boldsymbol{u} + \rho\boldsymbol{f}. \tag{6.52}$$

If a piece of elastic material is uniformly stretched in one direction, we have, say, $t_{11} > 0$, while the other t_{ij} are zero. This does not imply that only $e_{11} > 0$. The material is usually contracted in the transverse directions. From the inverted relation (6.51), $2\mu\mathcal{E} = \mathcal{T} - \frac{\lambda}{3\lambda+2\mu}\operatorname{tr}(\mathcal{T})\mathcal{I}$, we find that

$$e_{11} = \frac{\lambda+\mu}{\mu(3\lambda+2\mu)}t_{11}, \quad e_{22} = e_{33} = -\frac{\lambda}{2\mu(3\lambda+2\mu)}t_{11}, \quad e_{ij} = 0 \quad (i \neq j).$$

From the ratios t_{11}/e_{11} and $-e_{22}/e_{11}$ two constants naturally appear:

$$E := \frac{\mu(3\lambda+2\mu)}{\lambda+\mu} \quad \text{(Young's modulus)}, \qquad \nu := \frac{\lambda}{2(\lambda+\mu)} \quad \text{(Poisson's ratio)}.$$

E is positive; ν is less than 0.5 and usually positive. The inverted equation (6.51) is thus

$$E\mathcal{E} = (1+\nu)\mathcal{T} - \nu\operatorname{tr}(\mathcal{T})\mathcal{I}.$$

Since $\int_\Omega \nabla\cdot\boldsymbol{u}\,\mathrm{d}V = \oint_{\partial\Omega}\boldsymbol{u}\cdot\boldsymbol{n}\,\mathrm{d}S$ describes (linearised) the increase of a volume Ω by a displacement field $\boldsymbol{u}$, the material is incompressible (no volume changes) if $\nabla\cdot\boldsymbol{u} = \operatorname{tr}(\mathcal{E}) = 0$. As $\operatorname{tr}(\mathcal{T}) = \frac{E}{1-2\nu}\operatorname{tr}(\mathcal{E})$ is to be finite, while E is finite, we have necessarily $\nu = 0.5$ for incompressible elastic material.

Viscoelastic material is characterised by a stress-strain relation that depends on the history of the deformation and allows a certain amount of dissipation. The simplest and most common models are as follows:

- *Kelvin–Voigt model:*

$$\mathcal{T} = \lambda\big(\operatorname{tr}(\mathcal{E}) + \theta_1\operatorname{tr}(\dot{\mathcal{E}})\big)\mathcal{I} + 2\mu\big(\mathcal{E} + \theta_2\dot{\mathcal{E}}\big), \tag{6.53a}$$

  where the dot denotes a derivative with respect to time. λ and μ are equivalent Lamé coefficients, while θ_1 and θ_2 are time parameters. Since there exists an equilibrium with constant stress, this material is sometimes called a viscoelastic solid. Note that $\operatorname{tr}(\mathcal{T}) = (3\lambda+2\mu)\operatorname{tr}(\mathcal{E}) + (3\lambda\theta_1 + 2\mu\theta_2)\operatorname{tr}(\dot{\mathcal{E}})$. The coefficient of $\operatorname{tr}(\dot{\mathcal{E}})$, $3\lambda\theta_1 + 2\mu\theta_2$, is usually small enough to be neglected.

- *Maxwell model:*

$$E\dot{\mathcal{E}} = (1+\nu)(\dot{\mathcal{T}} + \gamma_1 \mathcal{T}) - \nu(\mathrm{tr}(\dot{\mathcal{T}}) + \gamma_2\,\mathrm{tr}(\mathcal{T}))\mathcal{I}, \tag{6.53b}$$

where E and ν are equivalents of Young's modulus and Poisson's ratio, while γ_1 and γ_2 are time parameters. As a constant stress produces a constant flow rate, this material is sometimes called a viscoelastic fluid.

Both models may be considered as special cases of further generalisations.

Example 6.4 In the case of pure shear stress in the 1,2-direction (i.e., t_{12} is constant), we have with Kelvin–Voigt material the equation $t_{12} = 2\mu(e_{12} + \theta_2 \dot{e}_{12})$, yielding an exponential decay to a limiting deformation gradient $e_{12}(t) = \frac{1}{2\mu} t_{12} + C\,\mathrm{e}^{-t/\theta_2}$.

In the case of pure shear in the 1,2-direction with a fixed velocity gradient (i.e., $\dot{e}_{12}$ is constant), we have with Maxwell material the equation $E\dot{e}_{12} = (1+\nu)(\dot{t}_{12} + \gamma_1 t_{12})$, yielding an exponential decay to a limiting shear stress $t_{12}(t) = \frac{E}{1+\nu}\dot{e}_{12} + C\,\mathrm{e}^{-t/\gamma_1}$. □

6.9 Maxwell Equations

The above discussion was aimed at a derivation from first principles of the Navier–Stokes equations, describing the motion of fluids, and the equations of linear elastic deformations. Although this relates to the main area of application considered in this book, we cannot leave unmentioned another monument in applied mathematics, the Maxwell equations for electromagnetic fields.

Electric charge is described by a charge density Q and a current density $\boldsymbol{J}$, corresponding to changes in motion. These charges and currents produce electromagnetic fields described by (i) the electric field intensity $\boldsymbol{E}$, which applies a force $q\boldsymbol{E}$ to a point charge q, and (ii) the magnetic-flux density or magnetic induction $\boldsymbol{B}$, which applies a torque $\boldsymbol{m} \times \boldsymbol{B}$ to a magnetic dipole with magnetic moment $\boldsymbol{m}$. Further, we introduce the derived fields (iii) $\boldsymbol{D}$, the electric displacement, and (iv) $\boldsymbol{H}$, the magnetic field intensity.

For these fields we have the following equations.

- *Coulomb's law.* The net effect of a charge distribution in a fixed volume Ω is equivalent to the total flux of electric displacement $\boldsymbol{D}$ out through Ω's surface $\partial\Omega$ (with $\boldsymbol{n}$ the outward unit normal):

$$\int_\Omega Q\,\mathrm{d}V = \oint_{\partial\Omega} \boldsymbol{D}\cdot\boldsymbol{n}\,\mathrm{d}S. \tag{6.54}$$

- *Ampère–Maxwell's law.* A current or a changing electric displacement causes a magnetic field as follows. The work done by a magnetic field $\boldsymbol{H}$ along a closed contour C is equal to the flux of the total current $\boldsymbol{J} + \frac{\partial}{\partial t}\boldsymbol{D}$ through the enclosed surface S:

$$\oint_C \boldsymbol{H}\cdot\mathrm{d}\boldsymbol{\ell} = \int_S \left(\boldsymbol{J} + \frac{\partial}{\partial t}\boldsymbol{D}\right)\cdot\boldsymbol{n}\,\mathrm{d}S. \tag{6.55}$$

- *Faraday–Henry's law.* A changing magnetic field causes an electric field as follows. The electromotive force induced by $\boldsymbol{E}$ around a circuit C is equal but opposite to the rate of change of the magnetic flux $\boldsymbol{B}$ through the enclosed surface $\mathcal{S}$:

$$\oint_C \boldsymbol{E} \cdot \mathrm{d}\boldsymbol{\ell} = -\frac{\mathrm{d}}{\mathrm{d}t} \int_S \boldsymbol{B} \cdot \boldsymbol{n} \, \mathrm{d}S. \tag{6.56}$$

- *Absence of free magnetic poles.* This yields a vanishing total flux of magnetic induction $\boldsymbol{B}$ through the surface of a volume Ω:

$$\oint_{\partial\Omega} \boldsymbol{B} \cdot \boldsymbol{n} \, \mathrm{d}S = 0. \tag{6.57}$$

- *Maxwell equations.* As these equations are valid for any volume Ω or contour $\mathcal{C}$, the identities are valid locally and may be reformulated into the differential equations

$$\begin{aligned} \nabla\cdot\boldsymbol{D} = Q, \qquad \nabla\times\boldsymbol{H} = \boldsymbol{J} + \frac{\partial}{\partial t}\boldsymbol{D}, \\ \nabla\times\boldsymbol{E} + \frac{\partial}{\partial t}\boldsymbol{B} = 0, \qquad \nabla\cdot\boldsymbol{B} = 0, \end{aligned} \tag{6.58}$$

known as the Maxwell equations.

By taking the divergence of Ampère–Maxwell's law (in differential form), we obtain the law of *conservation of electric charge*:

$$\frac{\partial}{\partial t} Q + \nabla\cdot\boldsymbol{J} = 0. \tag{6.59}$$

In integral form this says that the variation of the total charge of a fixed volume Ω (in the absence of sources) is equal to the net flux of the current into Ω's surface $\partial\Omega$:

$$\frac{\mathrm{d}}{\mathrm{d}t} \int_\Omega Q \, \mathrm{d}V = -\oint_{\partial\Omega} (\boldsymbol{J} \cdot \boldsymbol{n}) \, \mathrm{d}S. \tag{6.60}$$

Taking the divergence of Faraday–Henry's law yields that $\nabla\cdot\boldsymbol{B}$ is stationary (and therefore equals zero if it starts that way). So the Maxwell equations consist of only 7 independent relations for 16 unknowns. Therefore we need additional relations to describe the field uniquely.

6.9.1 Constitutive Relations

The fields $\boldsymbol{D}$, $\boldsymbol{H}$, and $\boldsymbol{J}$ are related to $\boldsymbol{E}$ and $\boldsymbol{B}$ via constitutive equations that depend on the problem considered. For example, in vacuum and isotropic diamagnetic and paramagnetic media we have the simple linear relationship

$$\boldsymbol{B} = \mu \boldsymbol{H}, \tag{6.61}$$

where μ is the magnetic permeability and is denoted by μ_0 for vacuum. Its numerical value is $\mu_0 = 1.2566371 \times 10^{-6}$ H/m. In vacuum and isotropic dielectric media we have the linear relation

$$\boldsymbol{D} = \epsilon \boldsymbol{E}, \tag{6.62}$$

where ϵ is the electric permittivity and is denoted by ϵ_0 for vacuum. Its numerical value is $\epsilon_0 = 8.8541853 \times 10^{-12}$ F/m. Note that $\mu_0\epsilon_0 c^2 = 1$, where $c = 2.99792458 \times 10^8$ m/s is the speed of light in vacuum and $\mu_0 = 4\pi \times 10^{-7}$ H/m.

The relation between the current and the electric field, the generalized *Ohm's law*, is, for a wide range of conditions, linear, and is given by

$$\boldsymbol{J} = \sigma \boldsymbol{E}, \tag{6.63}$$

where σ is the conductivity of the medium.

Example 6.5 A stationary point charge has a charge distribution $Q(\boldsymbol{x}) = q\delta(\boldsymbol{x})$. As the field is stationary, we have $\nabla \times \boldsymbol{E} = 0$. Hence $\boldsymbol{E}$ is conservative and may be written as the gradient of a potential $\boldsymbol{E} = -\nabla\phi$. If the field is in vacuum, we have $\boldsymbol{D} = \epsilon_0 \boldsymbol{E}$, and so ϕ satisfies Poisson's equation $\nabla^2\phi = -\frac{q}{\epsilon_0}\delta(\boldsymbol{x})$. In infinite space this has solution $\phi = \frac{q}{4\pi\epsilon_0 r}$, where $r = |\boldsymbol{x}|$. □

6.9.2 Energy Conservation and Poynting's Theorem

Energy is dissipated in conducting media (otherwise $\boldsymbol{J} = 0$) by *Joule heating*, given (per unit volume) by the power density $\boldsymbol{J} \cdot \boldsymbol{E}$. With Ohm's law (6.63) this simplifies to $\boldsymbol{J} \cdot \boldsymbol{E} = \sigma |\boldsymbol{E}|^2$.

This is illustrated by the following energy conservation law. If we define the rate of change of energy density $\frac{\partial}{\partial t}u := \boldsymbol{E} \cdot \frac{\partial}{\partial t}\boldsymbol{D} + \boldsymbol{B} \cdot \frac{\partial}{\partial t}\boldsymbol{H}$ and the energy flux vector $\boldsymbol{S} := \boldsymbol{E} \times \boldsymbol{H}$ (also called the Poynting vector), the Maxwell equations may be recast into the identity

$$\frac{\partial u}{\partial t} + \nabla \cdot \boldsymbol{S} = -\boldsymbol{J} \cdot \boldsymbol{E}, \tag{6.64}$$

which is known as *Poynting's theorem*. It shows that the rate of change of electromagnetic energy within a certain volume plus the energy flowing out through the boundaries per unit time is equal to minus the work done by the field inside the volume.

6.9.3 Electromagnetic Waves and Lorentz's Force

Disturbances of an electromagnetic field propagate like waves. For linear material, where $\boldsymbol{B} = \mu \boldsymbol{H}$ and $\boldsymbol{D} = \epsilon \boldsymbol{E}$, the Maxwell equations can be recast into a set of wave equations as follows.

As $\boldsymbol{B}$ is solenoidal (divergence free), it can be written as the curl of a vector potential $\boldsymbol{A}$. Since $\nabla \times \nabla = 0$, this vector potential is defined up to a scalar potential α, so we write

$$\boldsymbol{B} = \nabla \times (\boldsymbol{A} + \nabla\alpha).$$

From $\nabla \times (\boldsymbol{E} + \frac{\partial}{\partial t}\boldsymbol{A} + \frac{\partial}{\partial t}\nabla\alpha) = \boldsymbol{0}$ it follows that there is a scalar potential ψ such that

$$\boldsymbol{E} = -\nabla\psi - \frac{\partial}{\partial t}\boldsymbol{A}$$

because we can absorb $\frac{\partial}{\partial t}\alpha$ into ψ. This yields

$$\nabla^2\psi + \frac{\partial}{\partial t}(\nabla \cdot \boldsymbol{A}) = -\frac{1}{\epsilon}Q.$$

With the vector identity $\nabla\times(\nabla\times\boldsymbol{A}) = \nabla(\nabla\cdot\boldsymbol{A}) - \nabla^2\boldsymbol{A}$, we have

$$\nabla^2\boldsymbol{A} - \frac{1}{c^2}\frac{\partial^2\boldsymbol{A}}{\partial t^2} - \nabla\left(\nabla\cdot\boldsymbol{A} + \frac{1}{c^2}\frac{\partial\psi}{\partial t}\right) = -\mu\boldsymbol{J},$$

where we introduced the speed of light c in the medium considered, with $\mu\epsilon = c^{-2}$. If we choose ψ and $\boldsymbol{A}$ such that $\nabla\cdot\boldsymbol{A} + \frac{1}{c^2}\frac{\partial}{\partial t}\psi = 0$, the *Lorenz gauge condition*, we have finally the set of wave equations

$$\nabla^2\boldsymbol{A} - \frac{1}{c^2}\frac{\partial^2\boldsymbol{A}}{\partial t^2} = -\mu\boldsymbol{J}, \tag{6.65a}$$

$$\nabla^2\psi - \frac{1}{c^2}\frac{\partial^2\psi}{\partial t^2} = -\frac{1}{\epsilon}Q. \tag{6.65b}$$

Note that the freedom in α provides us with a fairly large class of possible potentials ψ.

Other gauge conditions are possible, e.g., the *Coulomb gauge condition*, where $\nabla\cdot\boldsymbol{A} = 0$, leading to a Poisson, rather than a wave, equation for ψ:

$$\nabla^2\psi = -\frac{1}{\epsilon}Q.$$

We finally remark that the *Lorentz force* $q(\boldsymbol{E} + \boldsymbol{v}\times\boldsymbol{B})$ describes the force acting on a point charge q moving with velocity $\boldsymbol{v}$ in the presence of an electromagnetic field. The path of the particle may be determined by recalling Newton's equations describing the change of momentum due to an external force.

Example 6.6 In free space, in the absence of charge Q or current $\boldsymbol{J}$, in a medium satisfying the linear relations (6.61) and (6.62), we have the time-harmonic plane-wave solution (see Chapter 3) given by the real part of

$$\boldsymbol{E} = \boldsymbol{E}_0\,\mathrm{e}^{\mathrm{i}\omega t - \mathrm{i}\boldsymbol{k}\cdot\boldsymbol{x}}, \qquad \boldsymbol{H} = \boldsymbol{H}_0\,\mathrm{e}^{\mathrm{i}\omega t - \mathrm{i}\boldsymbol{k}\cdot\boldsymbol{x}},$$

where $\boldsymbol{k} = k\boldsymbol{\kappa}$ is the wave vector and $\boldsymbol{E}_0 = E_0\boldsymbol{e}$ and $\boldsymbol{H}_0 = H_0\boldsymbol{h}$ are the vectorial amplitudes. Unit vectors $\boldsymbol{\kappa}$ and $\boldsymbol{e}$ and $\boldsymbol{h}$ denote the direction of propagation and the polarization of $\boldsymbol{E}$ and $\boldsymbol{H}$, respectively. They form an orthonormal triple with $\boldsymbol{\kappa} = \boldsymbol{e}\times\boldsymbol{h}$, $\boldsymbol{e} = \boldsymbol{h}\times\boldsymbol{\kappa}$, and $\boldsymbol{h} = \boldsymbol{\kappa}\times\boldsymbol{e}$. The modulus of the wave vector is $k = \omega/c$, where $c = (\epsilon\mu)^{-1/2}$, while the moduli of the vectorial amplitudes satisfy $E_0 = ZH_0$, where $Z = (\mu/\epsilon)^{1/2}$ is the impedance of the medium. □

6.10 Discussion

- As myriad applications of mathematics are found in physics, in particular in continuum physics, it is useful to have the pertaining equations concisely summarized. It is, however, very rare that the equations are used in the very comprehensive way presented. Usually, the problem is much more limited, and it is wise to simplify the equations first before attempting to tackle them mathematically. This highly nontrivial step is called "modeling," and the next chapter will be devoted to it.
- The section on conservation laws is not only of interest purely physically. These laws form the basis of many numerical methods based on the integral formulations of the conservation laws and are known as finite volume methods.

Exercises

6.1. Verify that (6.45), with relations (6.46), for inviscid nonconducting fluids simplify to the Euler equations

$$\frac{\mathrm{d}}{\mathrm{d}t}\rho = -\rho\nabla\cdot\boldsymbol{v}, \quad \rho\frac{\mathrm{d}}{\mathrm{d}t}\boldsymbol{v} = -\nabla p + \rho\boldsymbol{f}, \quad \frac{\mathrm{d}}{\mathrm{d}t}s = 0.$$

6.2. Verify that (6.45), with relations (6.46), for incompressible fluids ($\rho = \rho_0$) of constant viscosity simplify to

$$\nabla\cdot\boldsymbol{v} = 0, \quad \rho_0\frac{\mathrm{d}}{\mathrm{d}t}\boldsymbol{v} = -\nabla p + \mu\nabla^2\boldsymbol{v} + \rho_0\boldsymbol{f}, \quad \rho_0 T\frac{\mathrm{d}}{\mathrm{d}t}s = -\nabla\cdot\boldsymbol{q} + \boldsymbol{\tau}:\nabla\boldsymbol{v}.$$

6.3. (a) Derive for an ideal gas the relation

$$\mathrm{d}s = C_V\frac{\mathrm{d}p}{p} - C_P\frac{\mathrm{d}\rho}{\rho}.$$

When is it possible, and when is it beneficial, to integrate this expression?

(b) Show that a homentropic (s is constant) perfect gas is a polytropic gas (p/ρ^n is constant).

(c) Show that for a homentropic perfect gas flow the following relation holds:

$$\frac{1}{\rho}\nabla p = \frac{C_P}{\mathcal{R}}\nabla\frac{p}{\rho} = \frac{\gamma}{\gamma - 1}\nabla\frac{p}{\rho}.$$

(d) Prove *Kelvin's theorem* of the conservation of circulation: in an inviscid homentropic flow field any circulation

$$\Gamma := \oint_{\partial\mathcal{A}} \boldsymbol{v}\cdot\mathrm{d}\boldsymbol{\ell},$$

where $\partial\mathcal{A}$ is a closed material contour (the boundary of a material surface $\mathcal{A}$, i.e., moving with the flow), is a constant while it is convected with the flow.

6.4. Show, by using (6.52), that $\nabla^2(\nabla\cdot\boldsymbol{u}) = 0$ and $\nabla^4\boldsymbol{u} = \boldsymbol{0}$ if $\boldsymbol{f} = \frac{\partial^2}{\partial t^2}\boldsymbol{u} = \boldsymbol{0}$.

6.5. Verify the identity

$$\frac{\partial^2 e_{ij}}{\partial x_k \partial x_\ell} + \frac{\partial^2 e_{k\ell}}{\partial x_i \partial x_j} = \frac{\partial^2 e_{ik}}{\partial x_j \partial x_\ell} + \frac{\partial^2 e_{j\ell}}{\partial x_i \partial x_k},$$

and show that it yields six compatibility relations for the components of $\boldsymbol{\mathcal{E}}$.

6.6. Show that the harmonic plane wave $\boldsymbol{u} = U\boldsymbol{v}\,\mathrm{e}^{\mathrm{i}\omega t - \mathrm{i}\boldsymbol{k}\cdot\boldsymbol{x}}$ with frequency ω, wave vector $\boldsymbol{k}$, amplitude U, and polarization vector $\boldsymbol{v}$ that satisfies Navier's equation (6.52) with $\boldsymbol{f} = \boldsymbol{0}$ has the dispersion relation

$$\rho\omega^2 = (\lambda + 2\mu)(\boldsymbol{k}\cdot\boldsymbol{k})$$

if the wave is longitudinal ($\boldsymbol{v} \parallel \boldsymbol{k}$) and

$$\rho\omega^2 = \mu(\boldsymbol{k}\cdot\boldsymbol{k})$$

if the wave is transverse ($\boldsymbol{v} \perp \boldsymbol{k}$).

6.7. Consider two electromagnetic media satisfying relations (6.61) and (6.62) with constants ϵ_1, μ_1 and ϵ_2, μ_2. The media are separated from each other by a plane interface with unit normal vector $\boldsymbol{\kappa}$ directed from medium 1 to medium 2. The interface is given by $\boldsymbol{\kappa} \cdot \boldsymbol{x} = 0$. In medium 1 a plane wave propagates in direction $\boldsymbol{\kappa}$ (see Example 6.6) and reflects and transmits at the interface. In medium 1 the wave is given by

$$\boldsymbol{E} = \boldsymbol{E}_0(\mathrm{e}^{-\mathrm{i}\boldsymbol{k}_1 \cdot \boldsymbol{x}} + R\,\mathrm{e}^{\mathrm{i}\boldsymbol{k}_1 \cdot \boldsymbol{x}})\,\mathrm{e}^{\mathrm{i}\omega t}, \qquad \boldsymbol{H} = Z_1^{-1}\boldsymbol{E}_0(\mathrm{e}^{-\mathrm{i}\boldsymbol{k}_1 \cdot \boldsymbol{x}} - R\,\mathrm{e}^{\mathrm{i}\boldsymbol{k}_1 \cdot \boldsymbol{x}})\,\mathrm{e}^{\mathrm{i}\omega t},$$

while in medium 2 it is given by

$$\boldsymbol{E} = \boldsymbol{E}_0 T\,\mathrm{e}^{\mathrm{i}\omega t - \mathrm{i}\boldsymbol{k}_2 \cdot \boldsymbol{x}}, \qquad \boldsymbol{H} = Z_2^{-1}\boldsymbol{E}_0 T\,\mathrm{e}^{\mathrm{i}\omega t - \mathrm{i}\boldsymbol{k}_2 \cdot \boldsymbol{x}}.$$

Given that at an interface the normal components of D and B, and the tangential components of $\boldsymbol{E}$ and $\boldsymbol{H}$, are continuous, determine the reflection and transmission coefficients R and T.

Chapter 7

The Art of Modeling

This chapter describes the ideas and principles of modeling a real-life problem. It is necessarily a bit contemplative in nature. In Section 7.1 we discuss how modeling may be defined. There are various ways to model real-life situations. We then discuss the kinds of models that can be distinguished in Section 7.2. They all relate to the (thrifty) way a real-world problem description is translated to a mathematical formulation and the usefulness of the conclusions drawn from the latter for the actual problem at hand. In order to decide which part of the resulting equations really matters, it is necessary to make them dimensionless. To do this one has to scale the parameters and variables first, which is treated in Section 7.3. We show that by Buckingham's theorem there is always a subset of relevant (maybe redefined) variables in which the problem can be formulated. As a result, one can often indicate large or small coefficients for some of the contributions in the equation. If they are small enough, we can neglect them, at least in most of the domain. This kind of treating the equations of mass, momentum, and energy for fluid flow is discussed to some extent in Section 7.4. It is shown which simpler equations (models) may result from them.

7.1 Introduction

Mathematics has, historically, its major sources of inspiration in applications [33, 47, 91]. It is just the unexpected question from practice that forces one to go off the beaten track. Also it is usually easier to portray properties of a mathematical abstraction with a concrete example at hand. Therefore it is safe to say that most mathematics is applied, applicable, or emerges from applications.

Before mathematics can be applied to a real problem, the problem must be described mathematically. We need a mathematical representation of its primitive elements and their relations, and the problem must be formulated in equations and formulas to render it amenable to formal manipulation and to clarify the inherent structure. This is called mathematical modeling. An informal definition could be as follows:

> *Mathematically modeling is describing a real-world problem in a mathematical way by what is called a* model, *such that it becomes possible to deploy mathematical tools for its solution. The model should be based on first principles*

and elementary relations and it should be accurate enough to have reasonable claims to predict both quantitative and qualitative aspects of the original problem. The accuracy of the description should be limited in order that the model not be unnecessarily complex.

This is evidently a very loose definition. Apart from the question of what is meant by a problem being described in a mathematical way, there is the confusing paradox that we only know the precision of our model if we can compare it with a better model, but this better model is exactly what we try to avoid as it is usually unnecessarily complex! In general we do not know a problem and its accompanying model well enough to be absolutely sure that the sought description is consistent, complete, and sufficiently accurate for the purpose, and not too formidable for any treatment. A model is, therefore, to a certain extent a vague concept. Nevertheless, modeling plays a key role in applied mathematics since mathematics cannot be applied to any real-world problem without the intermediate steps of modeling. Therefore a more structured approach is necessary, which is the aim of this chapter.

Some people define *modeling* as the process of translating a real-world problem into mathematical terms. We will not do so, as this definition is too wide to include the subtle aspects of "limited precision" (to be discussed in a minute). Therefore we will introduce the word *mathematising*, defined as the process of translating a real-world problem into mathematical terms. It is a translation in the sense that we translate from the inaccurate, verbose "everyday" language to the language of mathematics. For example, the geometrical presence and evolution of objects in space and time may be described parametrically in a suitable coordinate system. Any properties or fields that are expected to play a role may be formulated by functions in time and space, explicitly or implicitly, e.g., as a differential equation.

Mathematising is an elementary but not trivial step. In fact, it forms probably the single most important step in the progress of science. It requires the distinction, naming, and exact specification of the essential relevant elementary objects and their interrelations, where mathematics acts as a language in which the problem is described. If theory is available for the mathematical problem obtained in this way, the problem considered may be subjected to the strict logic of mathematics, and reasoning in this language will transcend the limited and inaccurate ordinary language. Mathematising is, therefore, as well as being the link between the mathematical world and the real world, also important for science in general.

A very important point to note is the fact that such a mathematised formulation is *always* at some level simplified. Earth can be modeled by a point or a sphere in astronomical applications, or by an infinite half-space or not at all in problems of human scale. Based on the level of simplification, sophistication, or accuracy, we can associate an inherent hierarchy with the set of possible descriptions. A model may be too crude, but also it may be too refined. It is too crude if it just doesn't describe the problem considered, or if the numbers it produces are not accurate enough to be acceptable. It is too refined if it includes irrelevant effects that make the problem untreatable, or make the model so complicated that important relations or trends remain hidden. According to Barenblatt [14], every mathematical model is based on "intermediate asymptotics."

The ultimate goal of mathematising a problem is a deeper understanding and a more profound analysis and solution of the problem. Usually, a more refined problem translation is more accurate but also more complicated and more difficult—if not impossible!—to analyse and solve than a simpler one. Therefore not every mathematical translation is a good one. We will call a good mathematical translation a *model* or *mathematical model* if it is *lean* or *thrifty* in the sense that it describes our problem quantitatively or qualitatively to a suitable or required accuracy with a *minimal* number of essentially different parameters and variables. (We say "essentially different" in view of a reduction that is always possible by writing the problem in dimensionless form. See Buckingham's Theorem 7.11 below.) Again, this definition is rather subjective, as it greatly depends on the context of the problem considered and our knowledge and resources. So there will rarely be one "best" model. At the same time, it shows that modeling, even if relying significantly on intuition, is part of the mathematical analysis.

7.2 Models

We will distinguish the following three classes of models.

7.2.1 Systematic Models

Other possible names are *asymptotic models* or *reducing models*. The starting point here is to use available complete models, which are adequate, but overcomplete, so that effects are included that are irrelevant, uninteresting, or negligibly small, thereby making the mathematical problem unnecessarily complex. By using available additional information (order of magnitude of the parameters), we can make assumptions that minimize in a systematic way the *overcomplete* model into a *good* model by taking a parameter that is already large or small to its asymptotic limit: small parameters are taken zero; large parameters become infinite; an almost symmetry becomes a full symmetry.

Examples of systematic models are found in particular in the well-established fields of continuum physics, considered in Chapter 6. An ordinary flow is usually described by a model that is reduced from the full, i.e., compressible and viscous, Navier–Stokes equations. This will be elaborated on in detail in Section 7.4.

Example 7.1 (Convection-diffusion) Consider the following simple convection-diffusion problem. For a temperature field T and given velocity field $\boldsymbol{v}$ in space $\boldsymbol{x}$ and time t and thermal diffusion coefficient α we have the assumed "overcomplete" model

$$\frac{\partial T}{\partial t} + \boldsymbol{v} \cdot \nabla T = \alpha \nabla^2 T,$$

which is, however, difficult to solve. If we have reasons to assume that the diffusion term may be ignored, we obtain the reduced problem

$$\frac{\partial T}{\partial t} + \boldsymbol{v} \cdot \nabla T = 0, \qquad (*)$$

which is far more attractive than the full problem, as it may be solved exactly. Along the streamlines $\boldsymbol{x} = \boldsymbol{\xi}(t)$ given by

$$\frac{\mathrm{d}\boldsymbol{\xi}}{\mathrm{d}t} = \boldsymbol{v},$$

equation (∗) simplifies to

$$\frac{\mathrm{d}}{\mathrm{d}t}T(\boldsymbol{\xi}(t),t)=0$$

with solution $T=T(\boldsymbol{\xi}(t),t)=$ constant. □

Example 7.2 (the pendulum) The dynamics of an idealised frictionless undriven pendulum consisting of a point mass suspended by a weightless cord of length L may be described by the difficult nonlinear pendulum equation

$$\frac{\mathrm{d}^2\theta}{\mathrm{d}t^2}=-\frac{g}{L}\sin\theta,\qquad \theta(0)=\theta_0,\qquad \frac{\mathrm{d}}{\mathrm{d}t}\theta(0)=0,$$

where g is the gravitational acceleration and $\theta=\theta(t)$ is the angle at time t of the cord with the vertical. Since $\frac{\mathrm{d}}{\mathrm{d}t}\theta(0)=0$ and the pendulum is undriven, $\theta(t)$ will never be larger than θ_0. So if $|\theta_0|\ll 1$, we may assume that θ is always much smaller than one, and we may approximate, at least for some time, the nonlinear term $\sin\theta$ by θ. This yields the much simpler model

$$\frac{\mathrm{d}^2\theta}{\mathrm{d}t^2}=-\frac{g}{L}\theta,\qquad \theta(0)=\theta_0,\qquad \frac{\mathrm{d}}{\mathrm{d}t}\theta(0)=0,$$

which has solution $\theta(t)=\theta_0\cos(\omega t)$ with $\omega=\sqrt{g/L}$. □

7.2.2 Constructing Models

Other possible names are *building block models* or *lumped-parameter models*. Here we build our problem description step by step from low to high, from simple to more complex, by adding effects and elements lumped together in building blocks, until the required accuracy or adequacy is obtained.

Example 7.3 (the air pump) The following problem of air release by a simple air pump may be an example of a building block model. Consider a pump of cross section S and length $a(t)$ that depends on the piston position. Initially, $a(0)=L$. Under pressure, the enclosed volume of air $Sa(t)$ leaves the pump through a small hole, forming a jet of cross section S_j and (mean) velocity v_j. From time $t=0$ a spring pushes against the piston with a force $F=\lambda a(t)$. Assuming any inertia effects of the piston to be much smaller than the inertia of the flow, the piston force is balanced by a pressure increase from atmospheric pressure p_∞ outside to the value p_0 inside the pump. So $F=S(p_0-p_\infty)$. Assuming incompressible air of density ρ_0

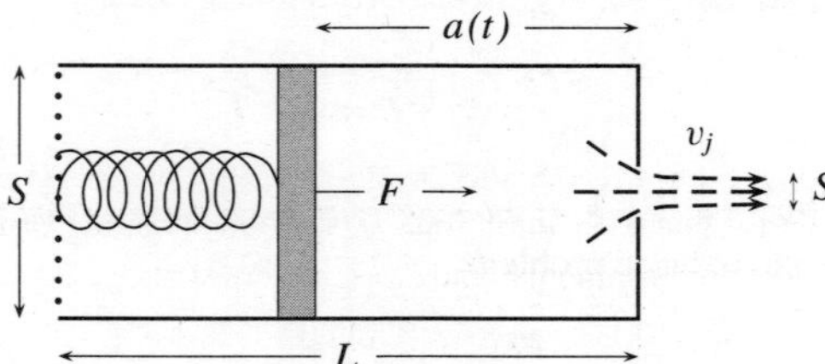

and a practically vanishing velocity inside the pump, conservation of mass (see (6.12)) tells us that {the variation inside} = −{what goes out}. Written as a formula this is

$$\frac{\mathrm{d}}{\mathrm{d}t}\big(\rho_0 Sa(t)\big)=-\rho_0 v_j S_j,$$

leading to

$$v_j = -\frac{S}{S_j}\frac{\mathrm{d}a}{\mathrm{d}t}.$$

From Bernoulli's law (to be introduced below; see (7.10)), relating pressure p and velocity v by $p + \frac{1}{2}\rho_0 v^2 = \text{constant}$, and noting that the pressure inside the jet is equal to the atmospheric pressure p_∞ (the jet cannot support a pressure difference), we can deduce that

$$p_0 = p_j + \frac{1}{2}\rho_0 v_j^2 = p_\infty + \frac{1}{2}\rho_0 v_j^2,$$

resulting in

$$\frac{1}{2}\rho_0 v_j^2 = \frac{\lambda a}{S}.$$

Putting these together, we have the model

$$\frac{\mathrm{d}a}{\mathrm{d}t} = -K\sqrt{a} \qquad \text{with} \quad K = \frac{S_j}{S}\sqrt{\frac{2\lambda}{\rho_0 S}},$$

which is easily solved by

$$a(t) = L\left(1 - \frac{1}{2}\frac{Kt}{\sqrt{L}}\right)^2 \qquad \text{along} \quad 0 \le t \le \frac{2\sqrt{L}}{K}. \qquad \square$$

Example 7.4 **(the flexible bar)** A brilliant example of a constructive model is the Bernoulli–Euler model of elastic deformation of slender bars, in which case the bar is described by a flexible line of vanishing cross section. The essentials of the theory were developed long before the general results described by (6.51) and (6.52) were available. In principle, the equations for the line should be implied by the general three-dimensional theory by utilizing the slenderness of the bar in the limit to zero. This, however, is not straightforward, Therefore the classical derivation is still important [48, 151].

For simplicity we will restrict the analysis to the case of deformation and motion in the vertical plane. Torsion and friction with any surroundings are neglected. The two-dimensional equations of motion are written as a differential equation for the position vector.

The line is described by the position vector $\boldsymbol{x}(s,t)$ as a function of curve length s and time t, with natural local coordinate s such that $|\boldsymbol{x}'| = 1$, where $\{\ \}' = \frac{\partial}{\partial s}\{\ \}$ and $\dot{\{\ \}} = \frac{\partial}{\partial t}\{\ \}$ (see, e.g., [92]). Introduce the right-hand orthogonal basis $[\boldsymbol{t}, \boldsymbol{n}, \boldsymbol{b}]$ consisting of the tangential unit vector $\boldsymbol{t} = \boldsymbol{x}'$, the principal normal unit vector $\boldsymbol{n}$, and the binormal unit vector $\boldsymbol{b}$ such that $\boldsymbol{b} = \boldsymbol{t} \times \boldsymbol{n}$, $\boldsymbol{n} = \boldsymbol{b} \times \boldsymbol{t}$, and $\boldsymbol{t} = \boldsymbol{n} \times \boldsymbol{b}$. The curvature vector is $\boldsymbol{k} = \boldsymbol{t}' = \boldsymbol{x}''$, with curvature $|\kappa| = |\boldsymbol{k}|$ defined such that $\boldsymbol{k} = \kappa\boldsymbol{n}$. The torsion or second curvature vector is $\boldsymbol{b}' = -\tau\boldsymbol{n}$, with torsion τ. Note that $\boldsymbol{n}' = -\kappa\boldsymbol{t} + \tau\boldsymbol{b}$.

Introduce a bar element of length $\mathrm{d}s$, loaded by an external line load $\boldsymbol{q}$ and internal forces $\boldsymbol{F}$ and moments $\boldsymbol{M}$ at both ends. The basic equations are derived from the equilibrium of the dynamic forces, the equilibrium of the moments, and from the constitutive equations as follows [84].

For a beam there is a moment around $\boldsymbol{b}$ (bending) and around $\boldsymbol{t}$ (torsion), so $\boldsymbol{M} = M_B\boldsymbol{b} + M_T\boldsymbol{t}$. Torsion will be assumed to be zero, and M_B is given by the following Bernoulli hypothesis. See Figure 7.1.

Consider a small bar of length ℓ and cross section A, bent over an angle ψ. From $R\psi = \ell$ and $(R+\xi)\psi = \ell + \mathrm{d}\ell$ it follows that $\mathrm{d}\ell/\ell = \xi/R = \xi\kappa$. The residual force $\mathrm{d}f$ at a cross-sectional slice $\mathrm{d}A$ that causes the bar to bend is by Hooke's law given by $\mathrm{d}f = E\kappa\xi\,\mathrm{d}A$. The

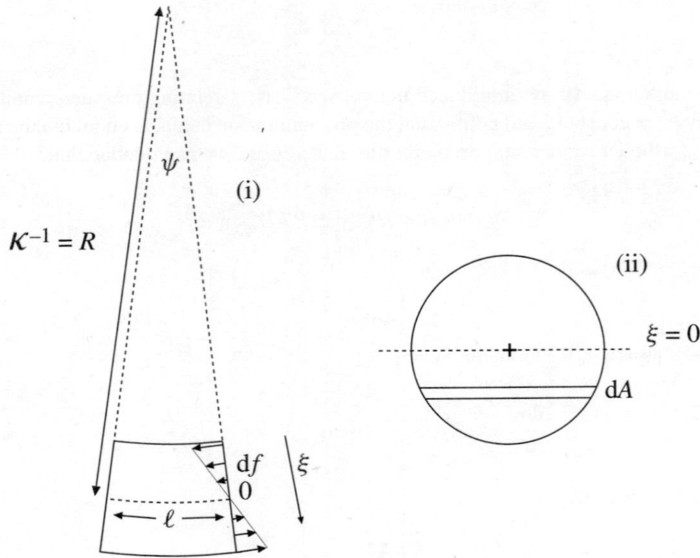

Figure 7.1. *Sketch of bar element. Side view* (i) *and cross section* (ii).

moment applied by $\mathrm{d}f$ is then

$$M_B = \int_A \xi \,\mathrm{d}f = E\kappa \int_A \xi^2 \mathrm{d}A = EI\kappa, \tag{$*$}$$

where the bending stiffness EI is the product of Young's modulus E and the second moment of cross-sectional area I.

Since the force $\boldsymbol{F}$ is the only cause of the deformation, $\boldsymbol{F}$ lies in the plane of the tangent and the principal normal, so $\boldsymbol{F} = T\boldsymbol{t} + S\boldsymbol{n}$, where T is called the normal force and S is called the shearing force. The dynamic force equilibrium $\mathrm{d}\boldsymbol{F} + \boldsymbol{q}\,\mathrm{d}s = m_0\ddot{\boldsymbol{x}}\,\mathrm{d}s$ (where m_0 is the mass per unit length) and the moment equilibrium $\mathrm{d}\boldsymbol{M} + \mathrm{d}\boldsymbol{x} \times \boldsymbol{F} = \boldsymbol{0}$ yield

$$\boldsymbol{F}' + \boldsymbol{q} = m_0\ddot{\boldsymbol{x}}, \qquad \boldsymbol{M}' + \boldsymbol{t} \times \boldsymbol{F} = \boldsymbol{0}.$$

From the vector identity $\boldsymbol{t} \times (\boldsymbol{M}' + \boldsymbol{t} \times \boldsymbol{F}) = \boldsymbol{t} \times \boldsymbol{M}' + T\boldsymbol{t} - \boldsymbol{F} = \boldsymbol{0}$ we obtain $(\boldsymbol{t} \times \boldsymbol{M}')' + (T\boldsymbol{t})' + \boldsymbol{q} = m_0\ddot{\boldsymbol{x}}$. With ($*$) and $\tau = M_T = 0$, we have $\boldsymbol{t} \times \boldsymbol{M}' = [\boldsymbol{t} \times (M_B\boldsymbol{b})]' - \kappa\boldsymbol{n} \times (M_B\boldsymbol{b}) = -(M_B\boldsymbol{n})' - M_B\kappa\boldsymbol{t} = -EI\boldsymbol{k}' - EI\kappa^2\boldsymbol{t}$, which yields the following differential equation for the position vector:

$$\left[-EI(\boldsymbol{x}''' + \kappa^2\boldsymbol{x}') + T\boldsymbol{x}'\right]' + \boldsymbol{q} = m_0\ddot{\boldsymbol{x}}.$$

With $\boldsymbol{x}(s,t) = (x(s,t), y(s,t), 0)$, we denote the angle between horizon and tangent $\boldsymbol{t}(s,t)$ as $\phi(s,t)$, so we have

$$x(s,t) = x(0,t) + \int_0^s \cos\phi(\varsigma,t)\,\mathrm{d}\varsigma, \;\; y(s,t) = y(0,t) + \int_0^s \sin\phi(\varsigma,t)\,\mathrm{d}\varsigma.$$

Note that $\kappa^2 = |\boldsymbol{x}''|^2 = (\phi')^2$. If the line is loaded by its own weight only, we have $\boldsymbol{q} = (0, -Q, 0)$, where $Q = m_0 g$ is the weight per unit length and g is the acceleration of gravity.

In x and y coordinates, we have finally

$$\frac{\partial}{\partial s}\left(EI\,\frac{\partial^2\phi}{\partial s^2}\sin\phi + T\cos\phi\right) = m_0\frac{\partial^2 x}{\partial t^2},$$

$$\frac{\partial}{\partial s}\left(-EI\,\frac{\partial^2\phi}{\partial s^2}\cos\phi + T\sin\phi\right) - Q = m_0\frac{\partial^2 y}{\partial t^2}.$$

In the stationary state we can integrate the equations (with integration constants H_0 and V_0, say), and by eliminating T we get the single equation

$$EI\frac{\mathrm{d}^2\phi}{\mathrm{d}s^2} = H_0\sin\phi - (Qs + V_0)\cos\phi. \qquad (\dagger)$$

Note that the present theory allows large deflections of the bar, although the elastic compression and extension at each element $\mathrm{d}s$ are small enough to apply linear elastic theory. If the deflections are small, we may write $x = s$, $y' = \phi$, and $T =$ constant and derive the linear beam equation

$$EI\frac{\partial^4 y}{\partial x^4} - T\frac{\partial^2 y}{\partial x^2} + Q + m_0\frac{\partial^2 y}{\partial t^2} = 0.$$ □

7.2.3 Canonical Models

Another possible name is *characteristic models* or *quintessential models*. Here an existing model is further reduced to describe only a certain aspect of the problem considered. These models are particularly important if the mathematical analysis of a model from one of the other categories lacks available theory. The development of such a theory is usually hindered by too many irrelevant details. These models are useful for understanding but are usually far away from the original full problem setting and therefore not suitable for direct industrial application.

Example 7.5 (Burgers equation) The Navier–Stokes equations for incompressible viscous flow (see Chapter 7), given in terms of a velocity and pressure field, are

$$\frac{\partial}{\partial t}\boldsymbol{v} + \boldsymbol{v}\cdot\nabla\boldsymbol{v} = -\frac{1}{\rho}\nabla p + \nu\nabla^2\boldsymbol{v}, \qquad \nabla\cdot\boldsymbol{v} = 0.$$

These equations are in general very complex. In particular, the coupling between the nonlinear and viscous terms, yielding instabilities and turbulence, is complicated and difficult to analyse. Therefore Burgers proposed to consider the following very simplified version of the equations, where the pressure gradient has been neglected and only behaviour in one dimension is taken into account. This equation,

$$\frac{\partial u}{\partial t} + u\frac{\partial u}{\partial x} = \nu\frac{\partial^2 u}{\partial x^2},$$

is called the Burgers equation [169]. A great deal of insight is obtained by the remarkable transformation

$$u = -2\nu\varphi^{-1}\frac{\partial\varphi}{\partial x}$$

found independently by Cole (1951) and Hopf (1950), by which the nonlinear equation is reduced to a linear equation

$$\frac{\partial\varphi}{\partial t} - \nu\frac{\partial^2\varphi}{\partial x^2} = C(t)\varphi,$$

related to the heat equation, where $C(t)$ is an arbitrary function of t. This equation is well understood and allows many exact solutions. □

Example 7.6 **(Sommerfeld's diffraction problem)** An important model problem for the understanding of acoustic or electromagnetic wave scattering at sharp edges is Sommerfeld's flat plate diffraction problem [145]. Consider a flat plate of vanishing thickness at $y = 0$, $x < 0$. Time-harmonic acoustic plane waves, with circular frequency ω, propagate in the direction $\boldsymbol{e}_x \cos\theta_i + \boldsymbol{e}_y \sin\theta_i$ and scatter at the solid plate. The velocity potential is independent

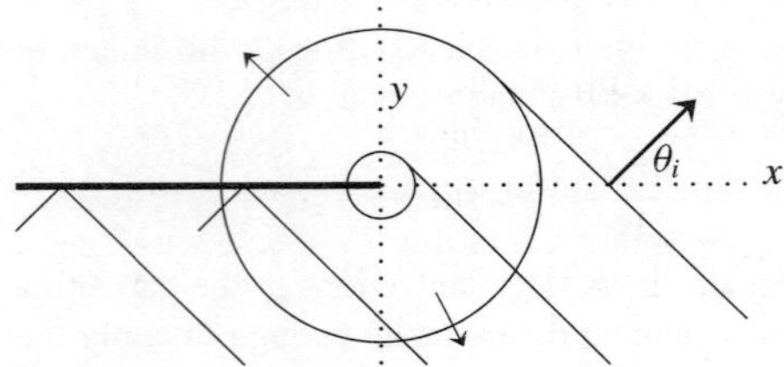

of z and may be written in the usual complex notation $\phi^*(x, y, t) = \mathrm{Re}(\phi(x, y)\,\mathrm{e}^{\mathrm{i}\omega t})$. As ϕ^* satisfies the wave equation, we have for ϕ the Helmholtz equation with boundary conditions

$$\nabla^2\phi + \kappa^2\phi = 0, \qquad \text{where} \quad \kappa = \frac{\omega}{c},$$
$$\frac{\partial\phi}{\partial y} = 0 \qquad \text{at} \quad y = 0,\ x < 0,$$

while the incident wave is given by $\phi_i(x, y) = \mathrm{e}^{-\mathrm{i}\kappa r\cos(\theta-\theta_i)}$, where $x = r\cos\theta$, $y = r\sin\theta$. It is important to note that the problem as stated does not have a unique solution. This is caused by three modeling simplifications that we made. We assumed that

1. the field is time harmonic, i.e., exists for all time;
2. the medium is nondissipative;
3. the plate edge is infinitely sharp.

From assumptions 1 and 2 we lost information about the propagation direction of the field, which is (except from the incident part) supposed to radiate away from the scattering edge. This is to be compared with the (acceptable) outward radiating solution $f(t - r/c)/r$ and (unacceptable) inward radiating solution $g(t + r/c)/r$ of the three-dimensional wave equation; see Example 12.31 in Section 12.6.2. We have to add radiation conditions in the form of a causality condition (assume the field to be switched on at some time long ago) or allow a small amount of dissipation, usually by giving c a vanishingly small imaginary part. If the source region is finite, a third option may be to prescribe the direction of the energy flux vector in the far field. In the present problem it comes down to the condition that for large κr the radiated part of the solution should behave like $\exp(\mathrm{i}\omega t - \mathrm{i}\kappa r)$. Therefore it is of utmost importance to note the sign convention $+\mathrm{i}\omega t$ in the exponential. Assumption 3 is of another kind. To represent the solid wall we prescribed the normal component of the gradient to vanish at the plate. At any discontinuous change of the wall's normal vector (i.e., at any sharp edge) the boundary condition is not defined any more, and a line source (delta function and its derivatives) may "hide" itself, producing a false additional field. In the present problem we would have the following false solutions (decaying to zero sufficiently fast for $r \to \infty$) [18]:

$$\psi(x, y) = H_\nu^{(2)}(\kappa r)\sin(\nu\theta), \qquad \nu = n + \frac{1}{2}, \quad n = 1, 2, \ldots,$$

where $H_\nu^{(2)}$ is the Hankel function of the second kind [1] (chosen in compliance with the radiation condition) of order ν. To exclude these solutions we have to add the *edge condition* of an integrable energy $\sim |\nabla\phi|^2$ in any neighbourhood of the edge. This tells us that, e.g., $\nu = \frac{1}{2}$

is allowed, but any higher order is not. When we take these extra conditions into account, the solution is found to be

$$\phi(x, y) = \mathrm{e}^{-\mathrm{i}\kappa r}\big(F(\gamma_i^{+}) + F(\gamma_i^{-})\big), \qquad \gamma_i^{\pm} = (2\kappa r)^{1/2} \sin \frac{1}{2}(\theta \mp \theta_i),$$

where F denotes a version of Fresnel's integral $F(z) = \pi^{-\frac{1}{2}}\,\mathrm{e}^{\frac{1}{4}\pi\mathrm{i}+\mathrm{i}z^2}\int_z^{\infty}\mathrm{e}^{-\mathrm{i}t^2}\,\mathrm{d}t$. □

Note that an asymptotic model may start as a building block model, which is only found at a later stage to be too comprehensive. Similarly, a canonical model may reduce from an asymptotic model if the latter appears to contain a particular, not yet understood, effect, which should be investigated in isolation before any progress with the original model can be made.

7.3 Nondimensionalisation and Scaling

Modeling means that one has to decide which effects are relevant and should be included and which are irrelevant and can be ignored. More in general, we may expect a hierarchy in relevance from most dominant, via less relevant and locally irrelevant, to absolutely unimportant effects or contributions. Relevant and irrelevant are rather vague qualifications. To make this operational we will relate them to small and large terms in our mathematical description (equations, etc.).

7.3.1 General Concepts

Small and large have no absolute meaning as long as we have not defined our "measuring stick." To illustrate this we may imagine the following science fiction scenario. Suppose we are lost in outer space, with all planets, stars, and galaxies so far away that they are only seen as sizeless spots on our retina. Then a rock drifts slowly into our field of vision. As long as we are not close enough for a stereoscopic view with both our eyes, we are not able to compare its size or distance with anything we know. There is no way to estimate if it is big and far away, or small and nearby. Only the rock itself is our scale of reference. A similar experience is found when we look into a microscope of unknown amplification. An object, visible but not recognizable, may be as big as an amoeba, or as small as a virus or a molecule. Reinterpreting the famous saying of Protagoras, "Man is the measure of all things," nothing we observe is small or large, fast or slow, in any absolute sense. It is only by comparison that these qualifications have a meaning.

The next question is, what do we use for comparing? We can use an absolute or universal measuring stick, like a meter or a kilogram, to archive the observations and be able to reproduce them exactly. However, we use a natural scale, like typical sizes in the problem itself, if we want to classify the type of phenomenon.

The following concepts are important in this respect.

When we model, we need to understand the problem in advance to a certain degree, such that we are able to formulate the relevant physical laws and relations. Therefore, in modeling, the natural scaling is the appropriate one to use. We introduce for all our dependent and independent variables typical values taken from the problem in question, e.g., a length L for the independent spatial variable x, and a velocity V for the dependent variable v, and thus an intrinsic time L/V for time coordinate t. We refer to this as *inherent scaling*.

When more than one problem parameter in the same units is available, e.g., a length L and a width D, or a time L/V and an inverse frequency ω^{-1}, it is inevitable that if one is selected for the scaling, the combination with the others gives us new parameters, like D/L or $\omega L/V$. These are now independent of the units (meters, seconds) and are therefore called *dimensionless parameters*. Incidentally, this meaning of the word "dimension" has nothing to do with the mathematical meaning of the number of independent basis vectors in a vector space. Dimensionless parameters are very important for a systematic classification of types of problems. They measure the relative importance of certain effects in an absolute way.

Consider a model depending on n physical quantities $q_1, q_2, \ldots, q_n$. Each quantity q has a *dimension* (unit of scale, dimensional unit) denoted by $[q]$, such that q can be written as

$$q = u[q].$$

The dimension is derived from a set of r independent base units d_i, e.g., the SI base units [m, kg, s, A, K, mol, cd] [153]. If the model is a proper one, reflecting the intrinsic relations between the variables, it should not depend on the arbitrary use of meters or inches, etc. Let the model be formally given by the relation

$$f(q_1, q_2, \ldots, q_n) = 0.$$

This relation should be equivalent for all choices of sets of independent base units. In other words, it should be *dimensionally homogeneous.* We refer to this as the *principle of dimensional homogeneity.* In order to achieve this, the dimension function has to satisfy the following conditions.

- Terms that are added, like $q_1 + q_2$, should have the same dimensions; i.e., $[q_1] = [q_2]$.
- The dimension of a product should be the product of the dimensions; i.e., if $q_0 = q_1 q_2$, then $[q_0] = [q_1][q_2]$.
- Terms that occur as the argument of a dimensionless function like sin or exp should have dimension one, i.e., be dimensionless. So if $\sin(q_1 q_2)$ occurs, then $[q_1][q_2] = 1$.

It can be shown (see [91, 13, 14]) that this is only possible if the dimension function is written as a monomial of powers of d_i, so

$$[q_j] = d_1^{\mu_{1j}} d_2^{\mu_{2j}} \cdots d_r^{\mu_{rj}} = \prod_{i=1}^{r} d_i^{\mu_{ij}}.$$

Example 7.7 **(a simple scaling problem)** Consider the following model of a quantity x satisfying the equation

$$ax^2 + bx + c = 0.$$

Assume that x denotes a length, with units in meters, denoted by $[x] = \mathrm{m}$, and c is a velocity with units in meters per second, or $[c] = \mathrm{m/s}$. If the equation is dimensionally homogeneous with $[ax^2] = [bx] = [c]$, the units of the other parameters a and b cannot be anything but $[a] = [c]/[x^2] = 1/(\mathrm{m} \cdot \mathrm{s})$ and $[b] = [c]/[x] = 1/\mathrm{s}$. Therefore we can scale time and length in several combinations to obtain a reduced problem as follows.

$$\text{If } b, c \neq 0: \quad x := \frac{c}{b}X, \quad a := \frac{b^2}{c}\alpha, \quad \alpha X^2 + X + 1 = 0.$$

$$\text{If } ac > 0: \quad x := \sqrt{\frac{c}{a}}X, \quad b := \sqrt{ac}\,\beta, \quad X^2 + \beta X + 1 = 0.$$

$$\text{If } a \neq 0: \quad x := \frac{b}{a}X, \quad c := \frac{b^2}{a}\gamma, \quad X^2 + X + \gamma = 0.$$

The constants α, β, and γ are dimensionless, parametrising the respective reduced problem. It should be noted that all of these scalings are equivalent (no information is lost), but they are not equally useful. The preferred reduction is the one in which x is scaled on a value typically occurring in the situation considered, and X is henceforth of order unity. So a careful inspection of the range of numerical values of x and the parameters a, b, c is essential. Only then can the dimensionless parameter α, β, or γ tell us more about the behaviour of X. □

Example 7.8 (a cooling problem) Consider an object Ω of typical size L that initially has a temperature distribution $T(\boldsymbol{x}, 0) = T_0(\boldsymbol{x})$. The temperature T satisfies the following heat diffusion equation with thermal diffusion constant α:

$$\begin{aligned}
\frac{\partial T}{\partial t} &= \alpha \nabla^2 T, && \boldsymbol{x} \in \Omega, \quad t > 0,\\
T(\boldsymbol{x}, t) &= 0, && \boldsymbol{x} \in \partial\Omega, \quad t > 0,\\
T(\boldsymbol{x}, 0) &= T_0(\boldsymbol{x}), && \boldsymbol{x} \in \Omega.
\end{aligned}$$

The edges of the object are kept at a constant temperature $T(\partial\Omega, t) = 0$ (Figure 7.2). Note that the steady state solution is $T(\boldsymbol{x}, t) \equiv 0$. So any gradient of T is always coupled to a variation in time. We scale $\boldsymbol{x}$ on L, the only length scale in the problem. As the problem is linear, it is not really necessary to scale T, but we could use the mean or maximum value of T_0. There is no explicit time scale, t_0 say, in the problem, e.g., from an external source. If we leave it unspecified for the moment and write $\boldsymbol{x} = L\boldsymbol{\xi}$ and $t = t_0\tau$, then we obtain

$$\frac{1}{t_0}\frac{\partial T}{\partial \tau} = \frac{\alpha}{L^2}\nabla_\xi^2 T.$$

As is also clear from the equation, the only parameter with the dimension of time is the number L^2/α. Therefore, as long as no steady state is achieved, the balance between decay and diffusion implies that the typical decay time (the half-life, say) is given, in order of magnitude, by this number. It is thus the natural time to scale on and so we have $t_0 = L^2/\alpha$. □

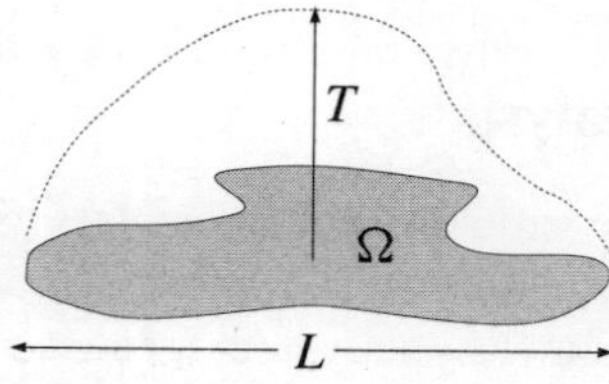

Figure 7.2. *A temperature distribution.*

Example 7.9 (electrically heated metal) A piece of metal Ω of size L is heated from an initial state $T(\boldsymbol{x},t) \equiv 0$ to a temperature distribution T by applying an electric field with potential ψ and typical voltage V (Figure 7.3). This heat source, amounting to the energy dissipation of the electric field (see Section 6.9.2), is given by the inhomogeneous term $\sigma|\nabla\psi|^2$ in the inhomogeneous heat equation

$$\begin{aligned} \rho C \frac{\partial T}{\partial t} &= \kappa \nabla^2 T + \sigma |\nabla \psi|^2, && \boldsymbol{x} \in \Omega, \quad t > 0, \\ T(\boldsymbol{x}, t) &= 0, && \boldsymbol{x} \in \partial\Omega, \quad t > 0, \\ T(\boldsymbol{x}, 0) &= 0, && \boldsymbol{x} \in \Omega. \end{aligned}$$

The edges are kept at $T = 0$. If we introduce the formal scaling $T = T_0 u$, $t = t_0 \tau$, $\boldsymbol{x} = L\boldsymbol{\xi}$, and $\psi = V\Psi$, then we get

$$\frac{\rho C T_0}{t_0} \frac{\partial u}{\partial \tau} = \frac{\kappa T_0}{L^2} \nabla_\xi^2 u + \frac{\sigma V^2}{L^2} |\nabla_\xi \Psi|^2.$$

Assuming a balance between the storage (first) and dissipation (second) terms during the initial phase of the process (although details may vary with the applied field ψ), it follows that the generated heat is dissipated through the metal with a typical decay time of $\mathcal{O}(\rho C L^2/\kappa)$, which is therefore a natural choice for the scaling time t_0.

Assuming a balance between the dissipation and source (third) terms in the stationary state, it follows that the final temperature of the stationary state is typically $\mathcal{O}(\sigma V^2/\kappa)$, which is therefore a suitable choice for T_0, the scaling temperature.

Note that the boundary conditions are rather important. If the edges were thermally isolated, we would, at least initially, have no temperature gradients scaling on L. Only the storage term would balance the source term, and there would be no other temperature to scale on than $\sigma V^2 t_0/\rho C L^2$. In other words, the temperature would rise approximately linearly in time.

See Example 7.19 for an extensive description. □

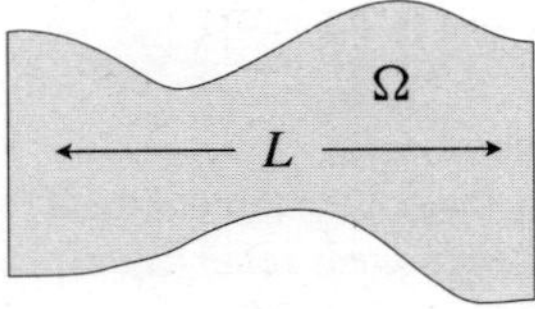

Figure 7.3. *A piece of metal heated by an electric field.*

7.3.2 Dimensional Analysis

In any description of reality the variables and parameters have physical dimensions and are therefore dimensionally related. However, any physical law is only universally valid if the expression or expressions are dimensionally homogeneous and independent of the physical dimensions used. We will see below that this fact alone implies that the set of variables and parameters can always be reduced to a smaller set of essentially independent, dimensionless, variables and parameters.

Definition 7.10. *Dimensionless groups consisting of a combination of problem parameters only are called* dimensionless numbers. *Groups consisting of a combination of parameters and several variables are called* similarity variables.

As the number of dimensionless groups tells us something about the complexity of the model involved (and for a good model this should not be too high; see Sections 7.1 and 7.2), we are interested in determining how many dimensionless groups are possible at most.

A fundamental result quantifying this is known as *Buckingham's Π theorem* (Π stands for the products).

Theorem 7.11 (Buckingham's Π theorem). *If a physical problem is described by n variables and parameters in r dimensions, the number of essentially different problem parameters (dimensionless groups) is at most $n - r$.*

Proof. Evidently, $n > r$. Suppose we have n physical variables and parameters $q_1, \ldots, q_n$ in r independent physical dimensions $d_1, \ldots, d_r$. All d_i occur at least once. So we postulate a relation

$$f(q_1, q_2, \ldots, q_n) = 0,$$

while the dimension $[q_j]$ of each q_j may be written as

$$[q_j] = d_1^{\mu_{1j}} d_2^{\mu_{2j}} \cdots d_r^{\mu_{rj}} = \prod_{i=1}^{r} d_i^{\mu_{ij}}.$$

From the principle of dimensional homogeneity (this relation should be equivalent for all base units) and the fact that any dimension can be written as a power law monomial, it can be shown [91, 13, 14] that f can be written as

$$\Phi(R_1, \ldots, R_m) = 0,$$

where Φ depends on m dimensionless groups of $q_1, \ldots, q_n$ of the form $R_k = q_1^{\alpha_{1k}} \cdots q_n^{\alpha_{nk}}$. The smallest possible m is the number to be determined.

Since each group is dimensionless, we have the dimension of R_k given by

$$\begin{aligned}[R_k] &= [q_1^{\alpha_{1k}} q_2^{\alpha_{2k}} \cdots q_n^{\alpha_{nk}}] = [q_1^{\alpha_{1k}}][q_2^{\alpha_{2k}}] \cdots [q_n^{\alpha_{nk}}] \\ &= \prod_{j=1}^{n} [q_j^{\alpha_{jk}}] = \prod_{j=1}^{n} \prod_{i=1}^{r} d_i^{\mu_{ij}\alpha_{jk}} = \prod_{i=1}^{r} d_i^{\sum_{j=1}^{n} \mu_{ij}\alpha_{jk}} = 1\end{aligned}$$

for $k = 1, \ldots, m$. This is only possible if any of the exponents of d_i is zero. In other words, m, the number of possible groups, is the number of independent (nontrivial) solutions $\boldsymbol{\xi} = (\xi_1, \ldots, \xi_n)^T$ of

$$\sum_{j=1}^{n} \mu_{ij}\xi_j = 0 \qquad \text{for} \quad i = 1, 2, \ldots, r,$$

or, in matrix notation,

$$\begin{pmatrix} \mu_{11} & \mu_{12} & \cdots & \mu_{1n} \\ \mu_{21} & & & \vdots \\ \vdots & & \ddots & \vdots \\ \mu_{r1} & \cdots & \cdots & \mu_{rn} \end{pmatrix} \begin{pmatrix} \xi_1 \\ \xi_2 \\ \vdots \\ \xi_n \end{pmatrix} = \begin{pmatrix} 0 \\ 0 \\ \vdots \\ 0 \end{pmatrix}.$$

Hence we have at least $n - r$ nontrivial solutions (i.e., nonzero and apart from a multiplicative factor) because the number of independent solutions is equal to the dimension n of the solution vector minus the rank of the matrix, which is at most r, the number of equations. However, the base units $d_1, \ldots, d_r$ are by assumption independent and occur at least once, so all r equations are independent and the rank of the matrix is exactly r, and we have $n - r$ nontrivial solutions.

On the other hand, as long as we have not solved the problem in detail, we may not be certain that all $R_1, \ldots, R_{n-r}$ are indeed necessary to describe the problem. Therefore the number of dimensionless groups is at most $n - r$. □

Corollary 7.12. *If a physical quantity q_0 is expressed by n quantities $q_1, \ldots, q_n$ in r dimensional base units, it depends on (at most) $n - r$ dimensionless parameters.*

Proof. From Theorem 7.11 it follows directly that if $q_0 = f(q_1, \ldots, q_n)$, it can be written as $q_0 = q_1^{\gamma_1} \cdots q_n^{\gamma_n} F(R_1, \ldots, R_m)$, where $m = n - r$. □

Example 7.13 (viscous drag) Consider the drag D—the reaction force due to the surrounding flow—of a sphere of radius a moving with velocity V in a viscous fluid with viscosity μ and density ρ. We assume, as our model, that the drag D depends only on ρ, V, μ, and a. (This is true for relatively low velocities, an infinite medium, and a relatively large sphere.)

Now we verify the dimensions of the parameters $[D] = \text{kg} \cdot \text{m/s}^2$, $[\rho] = \text{kg/m}^3$, $[V] = \text{m/s}$, $[\mu] = \text{kg/(m} \cdot \text{s)}$, and $[a] = \text{m}$. Presented in the form of a table, with at each entry the corresponding exponent of the base units kg, m, and s, this is as follows:

	ρ	V	μ	a	D
kg	1	0	1	0	1
m	−3	1	−1	1	1
s	0	−1	−1	0	−2

We conclude that we can express D in a functional relationship with at most $4 - 3 = 1$ dimensionless constant. The best-known form is the Reynolds number $Re = \rho V L/\mu$. The drag is usually scaled either on the pressure difference $\frac{1}{2}\rho V^2 a^2$ between front and back of the sphere or on the viscous friction $a V \mu$ due to wall shear stress. This leads to the functional relations

$$D = \frac{1}{2}\rho V^2 a^2 F(Re) = a V \mu\, G(Re), \qquad \text{where} \qquad Re = \frac{\rho V L}{\mu}.$$

The first expression is the proper scaling for nearly inviscid flow (*Re* large) and the second is for very viscous flow (*Re* small). □

Example 7.14 (an intense explosion) A famous example, originally due to G.I. Taylor [13, 12, 14], is the analysis of the shock front propagation of a very intense (e.g., nuclear bomb) explosion. From physical considerations the radius of the shock wave front R depends, during the early stages of the explosion when the pressure inside the shock wave is much higher than outside, only on the time interval t since the explosion, the initial energy E, and the initial air density ρ_0. Since $[R] = \text{m}$, $[t] = \text{s}$, $E = \text{kg} \cdot \text{m}^2/\text{s}^2$, and $\rho_0 = \text{kg/m}^3$, we have $3 - 3 = 0$ dimensionless groups. In other words, we can express R as

$$R = \text{constant} \left(\frac{E}{\rho_0}\right)^{1/5} t^{2/5}.$$

The full solution to the appropriate gas dynamical problem showed that the constant has a value close to unity. □

Example 7.15 (membrane resonance) The resonance frequency ω of a freely suspended membrane (no resonance cavity) is determined by the air density ρ_a and sound speed c_a, and the membrane tension T, density σ, and diameter a. As $[\rho_a] = \text{kg/m}^3$, $[c_a] = \text{m/s}$, $[T] = \text{kg/s}^2$, $[\sigma] = \text{kg/m}^2$, and $[a] = \text{m}$, we have $5 - 3 = 2$ dimensionless numbers determining ω. A possible choice is

$$\omega = \frac{c_M}{a} F\left(\frac{c_M}{c_a}, \frac{\rho_a a}{\sigma}\right),$$

where for convenience we introduced $c_M = (T/\sigma)^{\frac{1}{2}}$, the propagation speed of transversal waves in the membrane in the absence of air loading. □

Example 7.16 (a sessile drop with surface tension) The height h of a drop of liquid at rest on a horizontal surface with the effect of gravity balanced by surface tension is a function of liquid density ρ, volume L^3, acceleration of gravity g, surface tension γ, and contact angle θ. As $[h] = \text{m}$, $[\rho] = \text{kg/m}^3$, $[L] = \text{m}$, $[g] = \text{m/s}^2$, $[\gamma] = \text{kg/s}^2$, and $[\theta] = 1$, we have $5 - 3 = 2$ dimensionless numbers. Possible choices are (θ is already dimensionless)

$$h = LF(Bo, \theta) = \left(\frac{\gamma}{\rho g}\right)^{\frac{1}{2}} G(Bo, \theta), \qquad \text{where} \quad Bo = \frac{\rho g L^2}{\gamma}.$$

Bo is known as the Bond number. The first form is suitable when Bo is small (high relative surface tension). The drop becomes spherical and h is comparable with L. The second form is the proper scaling when Bo is large (low relative surface tension). The drop is flat as a pancake and h is comparable with $\sqrt{\gamma/\rho g}$ [124]. □

7.3.3 Similarity Solutions

If the problem contains no other length scale than the spatial variable x itself and no other time scale than the time variable t itself, dimensionless groups can only occur by combinations of x and t. As a result, the spatial distribution of the solution develops in time but remains geometrically self-similar [14]. We call this a similarity solution. Dimensional analysis of the independent variables naturally suggests the possibility of such solutions, although not always the form. Following Zeldovich and Barenblatt [13], we call self-similar solutions that can be constructed using dimensional analysis alone *self-similar solutions of the first kind*. There is complete similarity in all the parameters and variables, independent as well as dependent. *Self-similar solutions of the second kind* are self-similar solutions with incomplete similarity in the dependent variables. They are connected to an eigenvalue problem. We refer to [13] for an extensive discussion.

We will illustrate the concept of similarity solutions by the following examples.

Example 7.17 (the heated bar) Consider the following heat conduction problem. A very long thermally isolated bar, initially at uniform temperature zero, is heated at one end by a constant flux. There is no source at the other end. The bar is modeled as semi-infinite (Figure 7.4), with a crosswise constant temperature distribution, while the temperature T is described by the following one-dimensional equation for heat conduction, with constant heat diffusion coefficient α:

$$\frac{\partial T}{\partial t} = \alpha \frac{\partial^2 T}{\partial x^2}, \qquad x \in [0, \infty), \quad t > 0,$$

with initial and boundary conditions

$$\begin{aligned} \frac{\partial}{\partial x} T(0,t) &= -Q_0, & & t > 0, \\ T(x,0) &= 0, & x &\in [0,\infty), \\ 0 \leq T(x,t) &< \infty, & x &\to \infty, \quad t > 0. \end{aligned}$$

If we try to scale, we find that we have modeled any explicit length, time, or temperature scale out of our problem. So we can only make the problem dimensionless on the available implicit scales:

- As there is no length scale in x or t, the intrinsic length scale can only be $\sqrt{\alpha t}$.
- The only temperature in the problem is $Q_0 x$ or $Q_0\sqrt{\alpha t}$.

Therefore we assume

$$T(x,t) = Q_0 x g(\eta),$$

where the similarity variable η is given by

$$\eta = \frac{x}{\sqrt{4\alpha t}}.$$

It follows that g satisfies the reduced ODE

$$\frac{1}{2}\eta g'' + (1 + \eta^2) g' = 0$$

with boundary conditions

$$\lim_{\eta\to\infty} g(\eta) = 0, \qquad \lim_{\eta\downarrow 0}\big(g(\eta) + \eta g'(\eta)\big) = -1.$$

This has the solution

$$g(\eta) = \frac{1}{\eta\sqrt{\pi}} \exp(-\eta^2) - \operatorname{erfc}(\eta),$$

where $\operatorname{erfc}(x) := \frac{2}{\sqrt{\pi}} \int_x^\infty \mathrm{e}^{-\xi^2}\, \mathrm{d}\xi$ is the complementary error function. Hence we have

$$T(x,t) = Q_0 \left[\sqrt{\frac{4\alpha t}{\pi}} \exp(-\eta^2) - x \operatorname{erfc}(\eta) \right].$$

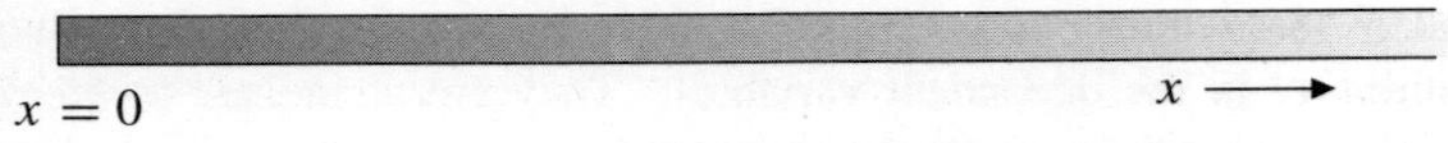

$x = 0$ $x \longrightarrow$

Figure 7.4. *A semi-infinite bar heated from* $x = 0$.

(Note that there exists no stationary solution!) The found solution is completely similar, both in the independent and in the dependent variables. Therefore it is a similarity solution of the first kind. □

Example 7.18 (convection) In the convection problem

$$\frac{\partial u}{\partial t} + U_0 \frac{\partial u}{\partial x} = 0, \qquad x \in \mathbb{R}, \quad t > 0,$$

$$u(x,0) = H(x), \qquad x \in \mathbb{R},$$

there is the length scale given by $U_0 t$ and a length scale, say L, hidden in the initial profile $H(x)$, as x cannot occur on its own. The dimensions of u and H are the same, say H_0, and we write $H(x) = H_0 h(\frac{x}{L})$. We scale $x = L\xi$, $t = \frac{L}{U_0}\tau$, $u = H_0 v$, and $v(\xi, 0) = h(\xi)$ to get

$$\frac{\partial v}{\partial \tau} + \frac{\partial v}{\partial \xi} = 0,$$

with solution $v(\xi, \tau) = h(\xi - \tau)$. □

Example 7.19 (Ohmic heating at a corner) Consider the following canonical model, originating from the problem of Joule heating in miniature soldering [126], for the edge singularity of the time-dependent temperature field generated in a homogeneous and isotropic conductor by an electric field (see (6.58)). The electric current density $\boldsymbol{J}$ and the electric field $\boldsymbol{E}$ satisfy Ohm's law (6.63) $\boldsymbol{J} = \sigma \boldsymbol{E}$, where σ is the electric conductivity, i.e., the inverse of the specific electric resistance. For an effectively stationary current flow the conservation of electric charge (6.59) leads to a vanishing divergence of the electric current density, $\nabla \cdot \boldsymbol{J} = 0$. The electric field $\boldsymbol{E}$ satisfies $\nabla \times \boldsymbol{E} = \boldsymbol{0}$ and therefore has a potential ψ, with $\boldsymbol{E} = -\nabla\psi$, satisfying $\nabla \cdot (\sigma \nabla \psi) = 0$. The electric conductivity σ is a material parameter that depends quite strongly on temperature. Nevertheless, to make progress we will assume a constant σ, independent of T. This, then, leads to the Laplace equation for ψ:

$$\nabla^2 \psi = 0. \qquad (*)$$

The heat dissipated as a result of the work done by the field per unit time and volume (Ohmic heating) is given by Joule's law $\boldsymbol{J} \cdot \boldsymbol{E}$ (see Section 6.9.2) and leads to the heat source distribution

$$\sigma |\nabla \psi|^2.$$

Since energy is conserved, the net rate of heat conduction and the rate of increase of internal energy are balanced by the heat source (6.39) with ρr the above heat source), which yields the equation for temperature T:

$$\rho C \frac{\partial T}{\partial t} = \kappa \nabla^2 T + \sigma |\nabla \psi|^2.$$

The thermal conductivity κ, the density ρ, and the specific heat of the material C depend mildly on temperature, but we assume these parameters constant.

Since we are interested in the role of the edge only, the conductor is modeled, in cylindrical (r, ϕ) coordinates, as an electrically and thermally isolated infinite wedge-shaped two-dimensional region (without any geometrical length scale; Figure 7.5) $0 \le \phi \le \nu$ with an electric field with potential

$$\psi(x, y) = (\nu/\pi) A\, r^{\pi/\nu} \cos(\phi\pi/\nu)$$

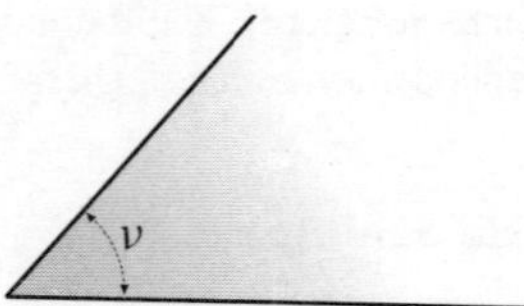

Figure 7.5. *A wedge-shaped conductor heated by an electric field.*

(a similarity solution of the second kind of $(*)$), while the temperature distribution T due to the heat generated by this source is given by

$$\rho C \frac{\partial T}{\partial t} = \kappa \nabla^2 T + \sigma A^2 r^{2\pi/\nu - 2}, \qquad (**)$$

with boundary and initial conditions

$$\frac{\partial T}{\partial \phi} = 0 \quad \text{at} \quad \phi = 0,\ \phi = \nu, \qquad T(x, y, 0) \equiv 0.$$

Since there are no other (point) sources at $r = 0$, we have the additional condition of a finite field at the origin: $0 \le T(0, 0, t) < \infty$. Boundary conditions and the symmetric source imply that T is a function of t and r only, so that $(**)$ reduces to

$$\rho C \frac{\partial T}{\partial t} = \kappa \left(\frac{\partial^2 T}{\partial r^2} + \frac{1}{r} \frac{\partial T}{\partial r} \right) + \sigma A^2 r^{2\pi/\nu - 2}. \qquad (\dagger)$$

Owing to the homogeneous initial and boundary conditions, the infinite geometry, and the fact that the source is a monomial in r, homogeneous of order $2\pi/\nu - 2$, there is no length scale in the problem other than $\sqrt{(\frac{\kappa t}{\rho C})}$, while the temperature T can only scale on $\frac{\sigma A^2}{\kappa} r^{2\pi/\nu}$. This indicates that a similarity solution (of the first kind) is possible of the form

$$T(r, t) = \frac{\sigma}{4\kappa} A^2 r^{2\pi/\nu} h(X), \qquad X = \frac{\rho C r^2}{4\kappa t},$$

where X is a similarity variable, reducing $(\dagger)$ to

$$X^2 h'' + X \left(1 + \frac{2\pi}{\nu} + X \right) h' + \frac{\pi^2}{\nu^2} h = -1,$$

with boundary conditions, corresponding to the behaviour near $r = 0$ and $t = 0$, of

$$0 \le X^{\pi/\nu} h(X) < \infty \quad \text{if} \quad X \downarrow 0, \qquad h(X) \to 0 \quad \text{if} \quad X \to \infty.$$

This equation may be recognized as related to the confluent hypergeometric equation in $-X$. It has solution, with the required behaviour in $X = 0$, given by

$$h(X) = \text{constant} \times X^{-\pi/\nu} M\left(-\frac{\pi}{\nu}; 1; -X \right) - \frac{\nu^2}{\pi^2},$$

where $M(a; b; z)$ is Kummer's function or the regular confluent hypergeometric function [1, Ch. 13] . From the asymptotic expansion of $M(-\pi/\nu; 1; -X)$ and the condition for $X \to \infty$

the unknown constant is found to be $(\nu/\pi)\Gamma(\pi/\nu)$. Putting everything together, we have the solution

$$T(r,t) = \frac{\sigma\nu^2 A^2}{4\pi^2\kappa} r^{2\pi/\nu} \left[\Gamma\left(1+\frac{\pi}{\nu}\right)\left(\frac{\rho C r^2}{4\kappa t}\right)^{-\pi/\nu} M\left(-\frac{\pi}{\nu}; 1; -\frac{\rho C r^2}{4\kappa t}\right) - 1 \right].$$

At the edge we have $T(0,t) \sim t^{\pi/\nu}$. This shows, together with the radial temperature distribution given in Figure 7.6, a marked difference in behaviour between outward ($\nu < \pi$) and inward ($\nu > \pi$) pointed wedges. For the first category the temperature at the corner rises smoothly and so slowly that it always remains behind the temperatures for larger r. For the other category it is the other way around. The corner temperature rises abruptly and so quickly that the values for larger r are always lower. □

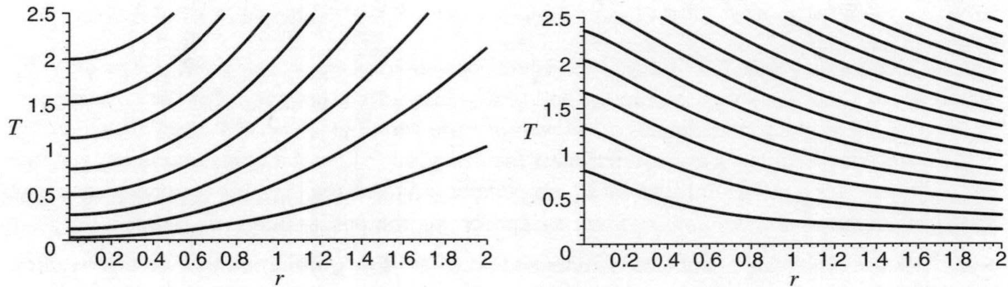

Figure 7.6. *Radial temperature distribution in wedge of* $\nu = \frac{1}{2}\pi$ *(left) and* $\nu = \frac{3}{2}\pi$ *(right) for* $\sigma A^2/4\kappa = 1$ *and* $4\kappa t/\rho C = \frac{1}{4}, \frac{2}{4}, \frac{3}{4}, \dots$.

Example 7.20 (decibels) Since the range of our human audible sensitivity is incredibly large (10^{14} in energy), the loudest and quietest levels are practically infinitely far away. Therefore we have no reference or scaling level to compare with other than the sound itself we are hearing. As a result, variations in loudness $\mathrm{d}L$ are perceived proportional to relative variations of the physical sound intensity $\mathrm{d}I/I$ and thus L varies logarithmically in I. As the intensity (the time-averaged energy flux) I is, for a single tone, proportional to the mean squared acoustic pressure p_{rms}^2, we have for suitable constants K and L_0 the relation $L = K\log(p_{\mathrm{rms}}) + L_0$. When

$$L = 2\log_{10}(p_{\mathrm{rms}}/p_0)$$

for a reference value $p_0 = 2 \times 10^{-5}$ Pa, we call L the sound pressure level in bels. The usual unit is one tenth of this, the decibel. □

Example 7.21 (duct modes) When the geometrical restrictions of a problem are invariant in one direction (say z), the absence of any length scale in z leads naturally to a trial solution independent of z. Usually, however, there are infinitely many solutions that are self-similar in z. These solutions are called *modes*. They are self-similar of the second kind and indeed related to an eigenvalue problem in (x, y). The missing length scale in the z direction is inherited from the available length scale in the crosswise (x, y) plane by a dispersion relation.

Consider, as an example, the acoustic wave problem in the hard-walled duct given by

$$\mathcal{D} = \{\boldsymbol{x} = (x, y, z) | (x, y) \in \mathcal{A}\},$$

where $\mathcal{A}$ is a simply connected two-dimensional area in the (x, y) plane. If the time-harmonic potential field is given by $\mathrm{Re}[\phi(\boldsymbol{x})\,\mathrm{e}^{\mathrm{i}\omega t}]$, then $\phi(\boldsymbol{x})$ satisfies the Helmholtz equation with boundary condition

$$\nabla^2\phi + \kappa^2\phi = 0 \quad \text{for} \quad \boldsymbol{x} \in \mathcal{D}, \qquad \boldsymbol{n}\cdot\nabla\phi = 0 \quad \text{for} \quad \boldsymbol{x} \in \partial\mathcal{D},$$

where $\kappa = \omega/c$ and $\boldsymbol{n}$ is the normal of the duct wall. In view of the invariance in z of the geometry and boundary conditions, we try solutions of the form

$$\phi(\boldsymbol{x}) = \psi(x, y)F(z)$$

to find that $\nabla^2\psi + (\kappa^2 + F''/F)\psi = 0$. This is only possible if $F''/F = -\gamma^2$, a constant. So $F(z) = F_0 \exp(\pm\mathrm{i}\gamma z)$. If we introduce the eigenvalue $\alpha^2 = \kappa^2 - \gamma^2$, we have for ψ the eigenvalue problem

$$-\nabla^2\psi = \alpha^2\psi \quad \text{for} \quad (x, y) \in \mathcal{A}, \qquad \boldsymbol{n}\cdot\nabla\psi = 0 \quad \text{for} \quad (x, y) \in \partial\mathcal{A}.$$

For the rectangular duct $\mathcal{A} = [0, a] \times [0, b]$ we have solutions $\psi = \cos(p\pi x/a)\cos(q\pi y/b)$, where $\alpha^2 = (p\pi/a)^2 + (q\pi/b)^2$ and p and q are nonnegative integers. For the circular duct $\mathcal{A} = \{r < R\}$ in polar coordinates (r, ϕ) we have the solutions $\psi = J_m(\alpha r)\,\mathrm{e}^{\pm\mathrm{i}m\phi}$, where J_m is the mth order ordinary Bessel function of the first kind [1], m is a nonnegative integer, and $\alpha R = j'_{m\mu} \geq 0$ is a nontrivial zero of J'_m. In general, it is true that for this boundary condition the eigenvalues α_n^2 are real and positive, except for the first one, which is $\alpha_1 = 0$.

Note that the eigenvalue problem is independent of κ. For a finite number of eigenvalues, i.e., with $0 \leq \alpha_n < \kappa$, γ_n is real and the mode propagates in z. For the infinitely many others, i.e., with $\alpha_n > \kappa$, γ_n is imaginary and the mode is evanescent, i.e., exponentially decaying in z. If $\alpha_n = \kappa$, then $\gamma_n = 0$ and the mode is in resonance, i.e., independent of z. The set of modes forms an L_2-complete set of orthogonal basis functions to represent any solution of the problem. □

7.4 Scaling and Reduction of the Navier–Stokes Equations

One of the most fruitful uses of scaling is the hierarchy it provides in comprehensive and rich models. In most applications such overcomplete models are not truly adapted to the problem, and therefore they are not true "lean" models, as we discussed above. From a suitable inherent scaling the order of magnitude of the various contributions or effects can be estimated, leading to a hierarchy on which we can base an asymptotic modeling.

We will give here one of the most important examples, the scaling and reduction of the compressible Navier–Stokes equations. Note that a typical, but not universal, approach will be worked out, since sometimes problem-dependent choices have to be made.

We recall the mass, momentum, and energy conservation equations (6.45) with constitutive relations (6.46) as derived in Chapter 6.

7.4.1 Scaling; Nondimensionalisation

Assume that we have the following typical scales for velocity, pressure, density, and temperature:

$\boldsymbol{v}$ with v_0, Δv; $\quad p$ with p_0, Δp; $\quad \rho$ with ρ_0, $\Delta\rho$; $\quad T$ with ΔT; $\quad \boldsymbol{x}$ with L; $\quad t$ with f^{-1}, where a subscript 0 refers to the primary variable and Δ denotes a typical difference. The typical time is written as the inverse of a frequency f. Note that sometimes we have more candidates for a scaling parameter. For example, the typical frequency may be enforced by an external source or instability, or it may be inherited from the intrinsic hydrodynamic frequency v_0/L or the inverse diffusion time $\kappa/\rho_0 C_P L^2$.

Assuming that the relations (6.46) hold, the mass, momentum, and energy equations

$$\frac{\mathrm{d}}{\mathrm{d}t}\rho = -\rho\nabla\cdot\boldsymbol{v}, \tag{6.45a}$$

$$\rho\frac{\mathrm{d}}{\mathrm{d}t}\boldsymbol{v} = -\nabla p + \nabla\cdot\boldsymbol{\tau}, \tag{6.45b}$$

$$\rho C_P\frac{\mathrm{d}}{\mathrm{d}t}T = \frac{\mathrm{d}}{\mathrm{d}t}p - \nabla\cdot\boldsymbol{q} + \boldsymbol{\tau}:\nabla\boldsymbol{v} \tag{6.45g}$$

become, symbolically, after scaling,

$$\begin{aligned}
f\Delta\rho\,,\ \frac{v_0\Delta\rho}{L} &= \frac{\rho_0\Delta v}{L},\\
f\rho_0\Delta v\,,\ \frac{\rho_0 v_0\Delta v}{L} &= \frac{\Delta p}{L}\,,\ \frac{\mu\Delta v}{L^2},\\
f\rho_0 C_P\Delta T\,,\ \frac{\rho_0 C_P v_0\Delta T}{L} &= f\Delta p\,,\ \frac{v_0\Delta p}{L}\,,\ \frac{\kappa\Delta T}{L^2}\,,\ \frac{\mu(\Delta v)^2}{L^2}.
\end{aligned}$$

From the gas laws we know that the typical sound speed squared c_0^2 scales with both p_0/ρ_0 and $\Delta p/\Delta\rho$ (not exactly but in order of magnitude), so we can take $p_0 = \rho_0 c_0^2$ and $\Delta\rho = \Delta p/c_0^2$. Furthermore, we will take here the common situation where $\Delta v = \mathcal{O}(v_0)$ and κ and μ are constants. (This may not always be the case!) With a rescaling such that all coefficients become dimensionless, we have

$$\begin{aligned}
\frac{fL\Delta p}{v_0\rho_0 c_0^2}\,,\ \frac{\Delta p}{\rho_0 c_0^2} &= 1,\\
\frac{fL}{v_0}\,,\ 1 &= \frac{\Delta p}{\rho_0 v_0^2}\,,\ \frac{\mu}{\rho_0 v_0 L},\\
\frac{fL}{v_0}\,,\ 1 &= \frac{fL\Delta p}{\rho_0 C_P v_0\Delta T}\,,\ \frac{\Delta p}{\rho_0 C_P\Delta T}\,,\ \frac{\kappa}{\rho_0 C_P v_0 L}\,,\ \frac{v_0\mu}{\rho_0 C_P L\Delta T}.
\end{aligned}$$

From these tables we can collect by inspection all the potentially relevant dimensionless numbers that may occur in problems described by the Navier–Stokes equations.

7.4.2 Some Dimensionless Groups with Their Common Names

In a natural way a large number of dimensionless numbers or groups, characterizing a flow, have appeared. Each represents a certain balance between two or more effects. If the typical values used for the scaling are well chosen, these numbers already provide an enormous amount of information about the physical problem considered. A more complete list, derived also from other fields, is given in the appendix, Section M.

$$\text{Reynolds:}\quad Re = \frac{\rho_0 v_0 L}{\mu},\qquad \text{Prandtl:}\quad Pr = \frac{C_P \mu}{\kappa},\qquad \text{Mach:}\quad Ma = \frac{v_0}{c_0},$$

$$\text{Strouhal:}\quad Sr = \frac{fL}{v_0},\qquad \text{Fourier:}\quad Fo = \frac{\kappa f^{-1}}{\rho_0 C_P L^2},\qquad \text{Euler:}\quad Eu = \frac{\Delta p}{\rho_0 v_0^2},$$

$$\text{Helmholtz:}\quad He = \frac{fL}{c_0},\qquad \text{Eckert:}\quad Ec = \frac{v_0^2}{C_P \Delta T},\qquad \text{Péclet:}\quad Pe = \frac{\rho_0 C_P v_0 L}{\kappa}.$$

These dimensionless numbers have the following physical interpretations.

- The Reynolds number *Re* describes how important viscous forces are compared to inertial forces and tells us if either viscosity or inertia may be neglected.
- The Prandtl number *Pr* describing the relative importance of viscous against heat diffusion, depends only on the material, and is for most gases and fluids of order one.
- The Mach number *Ma* compares the occurring velocities with the speed of sound and tells us whether the stationary velocity is so high that compressibility effects should be taken into account.
- The Strouhal number *Sr* compares an externally enforced frequency f with the hydrodynamically induced frequency v_0/L.
- The Fourier number *Fo* compares a time scale f^{-1} with the typical time necessary for the diffusion of heat along a distance L.
- The Euler number *Eu* compares the available pressure difference with the typical pressure difference that can be expected from hydrodynamical effects alone.
- The Helmholtz number *He* compares the typical wavelength of sound with the size of a scattering object or a source, which tells us a lot about the effectiveness of the scatterer or source.
- The Eckert number *Ec* compares the kinetic energy of the flow with available differences in enthalpy.
- The Péclet number *Pe* compares forced convection of heat with heat conduction.

Note that a dimensionless number is not always best interpreted by a balance between two effects. In that case the problem at hand may be better described by another selection of numbers. This is not difficult, as many dimensionless numbers are related. For example, $Pe = Pr\,Re$, $He = Sr\,Ma$, and $Sr\,Fo\,Pr\,Re = 1$.

The scaled equations are now, in dimensionless variables,

$$Eu\,Ma^2\left(Sr\frac{\partial \rho}{\partial t} + \boldsymbol{v}\cdot\nabla\rho\right) = -\rho\nabla\cdot\boldsymbol{v}, \tag{7.1a}$$

$$Sr\rho\frac{\partial \boldsymbol{v}}{\partial t} + \rho\boldsymbol{v}\cdot\nabla\boldsymbol{v} = -Eu\nabla p + \frac{1}{Re}\nabla\cdot\boldsymbol{\tau}, \tag{7.1b}$$

$$Sr\rho\frac{\partial T}{\partial t} + \rho\boldsymbol{v}\cdot\nabla T = Eu\,Ec\left(Sr\frac{\partial p}{\partial t} + \boldsymbol{v}\cdot\nabla p\right) + \frac{1}{Pe}\nabla^2 T + \frac{Ec}{Re}\boldsymbol{\tau}:\nabla\boldsymbol{v}. \tag{7.1c}$$

7.4.3 Asymptotic Reductions of the Navier–Stokes Equations

When Ma, Re, Sr, Ec, etc., become small or large, we may derive from (7.1) various reduced models by assuming the respective terms to vanish or dominate completely. Other terms are sometimes best taken equal to one, in particular if one of the involved scales is inherent (i.e., a result) rather than externally enforced.

A Fourier number $Fo = 1$ corresponds to a time scale $\rho_0 C_P L^2/\kappa$ determined by heat diffusion. A Strouhal number $Sr = 1$ corresponds to a frequency $f = v_0/L$, which is connected to hydrodynamical convection processes, e.g., the shedding of vortices from a blunt trailing edge. For a cylinder of diameter D, experiments indeed show that the shedding frequency is practically equal to $f_S = 0.2v_0/D$. The Euler number $Eu = 1$ corresponds to a present pressure gradient comparable to hydrodynamic pressure variations, $\rho_0 v_0^2$, which is the usual situation in incompressible inviscid (separated) flow around nonstreamlined bodies. We have to select $Eu\,Re = \Delta pL/v_0\mu = 1$ if the pressure gradient is mainly balanced by viscous forces, e.g., in very slow or very viscous flow. The combination $Eu\,Ma^2 = \Delta p/\rho_0 c_0^2 = 1$ corresponds to a pressure scaled on pressure variations due to compressible effects, which is common for acoustic problems (pressure-density coupling while Ma is small) or high-speed problems (v_0 comparable with c_0). We select $Eu\,Ec = \Delta p/C_P\Delta T = 1$ if enthalpy changes are mainly coupled to pressure variations and the flow is at least for the larger part isentropic (cf. (6.44)).

In the following we will derive some important examples of asymptotic models, systematically deduced from the Navier–Stokes equations.

- **Incompressible flow.** When the Mach number Ma tends to zero, while we take $Eu = Sr = 1$, we have incompressible flow described by

$$\nabla\cdot\boldsymbol{v} = 0, \qquad \rho\left(\frac{\partial\boldsymbol{v}}{\partial t} + \boldsymbol{v}\cdot\nabla\boldsymbol{v}\right) = -\nabla p + \frac{1}{Re}\nabla\cdot\boldsymbol{\tau}. \tag{7.2}$$

Very often an incompressible flow will have a uniform constant density, but this is not necessary. Note that the energy equation does not disappear but is decoupled from the other equations if viscosity and density may be taken independent of the temperature. The pressure does not play a role thermodynamically, as only its gradient occurs as a reaction force. If the fluid is Newtonian and ρ and μ are constant (they may be taken equal to one), we obtain for (7.2) the form

$$\nabla\cdot\boldsymbol{v} = 0, \qquad \frac{\partial\boldsymbol{v}}{\partial t} + \boldsymbol{v}\cdot\nabla\boldsymbol{v} = -\nabla p + \frac{1}{Re}\nabla^2\boldsymbol{v}. \tag{7.3}$$

A fully developed laminar flow in a circular pipe, called a Poiseuille flow, satisfies

$$\boldsymbol{v} = U(r)\boldsymbol{e}_x, \quad \nabla p = -K\boldsymbol{e}_x, \quad \frac{1}{r}\frac{\mathrm{d}}{\mathrm{d}r}\left(r\frac{\mathrm{d}}{\mathrm{d}r}U\right) = -ReK.$$

If the duct radius is unity, the solution is described by the parabolic profile

$$U(r) = \frac{1}{4}(1-r^2)ReK.$$

The equation for mass conservation of (7.3) can be solved by introducing a stream function $\boldsymbol{\Psi}$ defined by $\boldsymbol{v} = \nabla\times\boldsymbol{\Psi}$. This is particularly useful in two-dimensional flow when $\boldsymbol{\Psi} =$

$(0, 0, \psi)^T$ and $\boldsymbol{v} = (\psi_y, -\psi_x, 0)^T$. By taking the curl of the momentum equation, we remove the dependence on pressure and obtain (using (J.9)–(J.12)) the following equation for the transport of vorticity $\boldsymbol{\omega} := \nabla \times \boldsymbol{v}$:

$$\frac{\partial \boldsymbol{\omega}}{\partial t} + \boldsymbol{v} \cdot \nabla \boldsymbol{\omega} = \boldsymbol{\omega} \cdot \nabla \boldsymbol{v} + \frac{1}{Re} \nabla^2 \boldsymbol{\omega}. \tag{7.4}$$

Note that in two dimensions $\boldsymbol{\omega} \cdot \nabla \boldsymbol{v} = \mathbf{0}$, so (7.4) becomes a convection-diffusion equation in ω.

• **Inviscid compressible flow.** If the Reynolds number tends to infinity, usually the Péclet number does too, because $Pe = Pr\,Re$ and the Prandtl number is for most fluids and gases of order one. If we further take $Eu = Sr = Ma = Ec = 1$, we obtain a compressible inviscid flow described by

$$\frac{\partial \rho}{\partial t} + \boldsymbol{v} \cdot \nabla \rho = -\rho \nabla \cdot \boldsymbol{v}, \tag{7.5a}$$

$$\rho \left(\frac{\partial \boldsymbol{v}}{\partial t} + \boldsymbol{v} \cdot \nabla \boldsymbol{v} \right) = -\nabla p, \tag{7.5b}$$

$$\rho \left(\frac{\partial T}{\partial t} + \boldsymbol{v} \cdot \nabla T \right) = \frac{\partial p}{\partial t} + \boldsymbol{v} \cdot \nabla p. \tag{7.5c}$$

In terms of entropy, the last equation is equivalent to $\frac{\mathrm{d}}{\mathrm{d}t} s = 0$, so the flow is isentropic everywhere the assumptions hold.

• **Stokes flow.** If the velocities of a viscous flow are so low that Re tends to zero, while the time scales are all determined by the flow itself (no external forcing) such that Sr remains finite, we need to scale the pressure gradients on the inverse Reynolds number in order to have flow at all; i.e., Eu tends to infinity such that $Eu\,Re = 1$. If, in addition, the velocities remain so small compared to the sound speed, i.e., $Ma \ll 1$, that $Eu\,Ma^2$ tends to zero, we obtain the very viscous incompressible *Stokes flow* given by the Stokes equations

$$\nabla \cdot \boldsymbol{v} = 0, \qquad -\nabla p + \nabla \cdot \boldsymbol{\tau} = \mathbf{0}. \tag{7.6}$$

Again, it should be noted that the energy equation is not negligible but only decoupled from the other equations (provided that the viscosity is not temperature dependent).

• **Sound waves.** Consider small pressure-density perturbations in an atmosphere of uniform mean pressure. Assume that the frequencies are relatively high and the typical velocities are small but large enough to neglect viscosity, such that $Ma \to 0$ while we choose $Eu = Sr = Ma^{-1}$ and $Ec = Ma$, and Re and Pe remain finite or large. We then retain

$$\frac{\partial \rho}{\partial t} = -\rho \nabla \cdot \boldsymbol{v}, \quad \rho \frac{\partial \boldsymbol{v}}{\partial t} = -\nabla p, \quad \rho \frac{\partial T}{\partial t} = \frac{\partial p}{\partial t}.$$

Written in terms of entropy s, the last equation is equivalent to $\frac{\partial}{\partial t} s = 0$. This means that pressure and density perturbations are coupled isentropically by $\frac{\partial}{\partial t} p = c^2 \frac{\partial}{\partial t} \rho$. Noting that

ρc^2 is proportional to the mean pressure, which is constant, we can now eliminate ρ and $\boldsymbol{v}$ to obtain finally the wave equation

$$\frac{\partial^2 p}{\partial t^2} = \nabla\cdot\left(c^2 \nabla p\right). \tag{7.7}$$

If ρ and c are constant, we have the usual wave equation with constant coefficients.

• **Convection-diffusion.** In a given flow field with $Sr = 1$, Pe finite, and $EuEc$ and Ec/Re negligible, we get for the temperature the convection-diffusion problem

$$\rho\left(\frac{\partial T}{\partial t} + \boldsymbol{v}\cdot\nabla T\right) = \frac{1}{Pe}\nabla^2 T. \tag{7.8}$$

We end with two important reductions, not immediately obtainable from small-parameter considerations.

• **Potential flow.** If the flow is irrotational, i.e., the vorticity vector is $\boldsymbol{\omega} = \nabla\times\boldsymbol{v} = \boldsymbol{0}$, a scalar velocity potential φ may be introduced with

$$\boldsymbol{v} = \nabla\varphi.$$

For example, in inviscid homentropic flow, any vorticity is convected with the flow (see Kelvin's theorem, Exercise 6.3d), and if the flow starts out irrotational, it stays that way. In incompressible flow this potential is independent of pressure (except indirectly via boundary conditions) and satisfies Laplace's equation

$$\nabla^2\varphi = 0. \tag{7.9}$$

In two dimensions an important class of solutions may be generated by using the property of analytic functions $F(z)$ in the complex variable $z = x + \mathrm{i}y$ that both their real and imaginary parts satisfy Laplace's equation. If we introduce the complex potential $F(z) = \phi(x, y) + \mathrm{i}\psi(x, y)$, then the velocity $\boldsymbol{v} = (u, v)$ is given by $u - \mathrm{i}v = F'(z)$. Note that solutions may be constructed by superposition of elementary solutions (the problem is only nonlinear in pressure). For example, a uniform flow Uz and a dipole source flow R^2Uz^{-1} yield together the flow past a cylinder of radius R. As the flow is inviscid, this solution is not unique and any multiple of a line vortex flow $-\mathrm{i}\Gamma\ln(z)/2\pi$ may be added to get

$$F(z) = Uz + \frac{R^2U}{z} - \mathrm{i}\frac{\Gamma}{2\pi}\ln(z).$$

By itself this solution is not very practical because no high-Reynolds-number flow will pass a cylinder without separating and creating a turbulent wake. It may, however, be a starting point for a larger family of solutions $F(\zeta(z))$ to be obtained by conformal mappings $\zeta \mapsto z$. For example, the Joukowski transformation

$$z = \left(\zeta + \frac{\lambda^2}{\zeta}\right)\mathrm{e}^{-\mathrm{i}\alpha}$$

maps the circle $|\zeta - \zeta_c| = R$ in ζ domain to an airfoil in z domain if $\lambda = \xi_c + \sqrt{R^2 - \eta_c^2}$, where $\zeta_c = \xi_c + \mathrm{i}\eta_c$ (see Figure 7.7). (Take e.g., $\xi_c = -0.03$, $\eta_c = 0.03$, $R = 1$, and $\alpha = 0.05$.) The corresponding flow around the airfoil is given by

$$F(z) = U\zeta\,\mathrm{e}^{-\mathrm{i}\alpha} + \frac{R^2 U\,\mathrm{e}^{\mathrm{i}\alpha}}{\zeta - \zeta_c} - \mathrm{i}\frac{\Gamma}{2\pi}\ln(\zeta - \zeta_c).$$

The undetermined circulation Γ is found by requiring the flow to be nonsingular at the trailing edge $\zeta = \lambda$ or $z = 2\lambda\,\mathrm{e}^{-\mathrm{i}\alpha}$ (the so-called Kutta condition) and we obtain $\Gamma = -4\pi RU\sin(\alpha + \beta)$, where $\beta = \arcsin(\eta_c/R)$. This condition is a remainder of the effect of viscosity near the trailing edge. Note that when we drop viscosity in our model the no-slip boundary condition cannot be maintained as no solution would exist. However, dropping the no-slip condition altogether is too much and will produce a nonunique solution. It can be shown that for small angles of incidence the inviscid limit yields a condition between slip and no-slip: the no-slip condition can be dropped almost everywhere, except near the trailing edge, where it degenerates to the Kutta condition of nonsingular velocity.

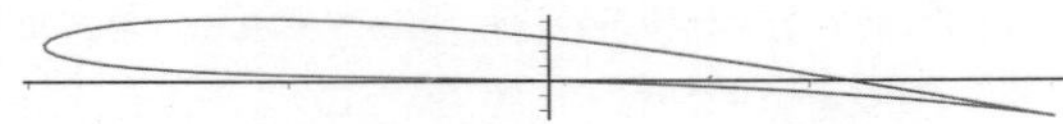

Figure 7.7. *A Joukowski airfoil.*

- **Bernoulli's law.** In stationary, incompressible and inviscid flow of constant density ρ_0 the momentum equation

$$\rho_0\left(\frac{\partial}{\partial t}\boldsymbol{v} + \boldsymbol{v}\cdot\nabla\boldsymbol{v}\right) = -\nabla p$$

can be integrated along a streamline, leading to an equation known as Bernoulli's law that describes conservation of mechanical energy density. By using the vector identity

$$\boldsymbol{v}\cdot\nabla\boldsymbol{v} = \frac{1}{2}\nabla|\boldsymbol{v}|^2 + \boldsymbol{\omega}\times\boldsymbol{v}$$

and noting that $(\boldsymbol{\omega}\times\boldsymbol{v})\cdot\mathrm{d}\boldsymbol{\ell} = 0$, we can integrate along a streamline

$$\int\left[\frac{1}{2}\rho_0\nabla|\boldsymbol{v}|^2 + \rho_0\boldsymbol{\omega}\times\boldsymbol{v} + \nabla p\right]\cdot\mathrm{d}\boldsymbol{\ell} = \int\nabla\left[\frac{1}{2}\rho_0|\boldsymbol{v}|^2 + p\right]\cdot\mathrm{d}\boldsymbol{\ell} = 0$$

to get the famous and useful identity

$$\frac{1}{2}\rho_0|\boldsymbol{v}|^2 + p = \text{constant}. \tag{7.10}$$

Bernoulli's equation in irrotational flow is valid everywhere rather than only along a streamline. By introducing a potential, we can generalise to unsteady flow with gravity in the z direction as follows:

$$\frac{\partial\varphi}{\partial t} + \frac{1}{2}|\nabla\varphi|^2 + \frac{p}{\rho_0} + gz = C(t), \tag{7.11}$$

where C is an arbitrary function of time. A typical example is water. Bernoulli's equation may be generalised to include compressibility if the flow is *barotropic*, i.e., the pressure is a function of density alone. For example, if we have an irrotational inviscid homentropic perfect gas flow, we obtain, with the result of Exercise 6.3c, that

$$\frac{\partial \varphi}{\partial t} + \frac{1}{2}|\nabla \varphi|^2 + \frac{\gamma}{\gamma - 1}\frac{p}{\rho} = C(t), \tag{7.12}$$

where γ is the specific heat ratio and C is an arbitrary function of time.

7.5 Discussion

- The mathematical solution of a real-world problem starts with the modeling phase, where the problem is described in a mathematical representation of its primitive elements and their relations.

- As the solution is not served by unnecessary complexity, we are interested in an adequate mathematical description with the lowest number of essentially independent parameters and variables. We will call this a model.

- A model may be constructed ad hoc by collecting as many pieces as possible of information and combining them into what we call a building block model. Usually, this is the first type of model if a new area of research is being explored.

  Once the foundations are laid, it is not necessary any more to start from scratch, but we can use the complete and comprehensive descriptions that have become available. By removing any unnecessary details, we can simplify these usually overcomplete models to the model required. This is what we call a systematic or asymptotic model.

- A very important aspect in modeling is the identification of a hierarchy of importance to distinguish between the important, the less important, and the unimportant effects. From this hierarchy we may decide which aspects can be included and which can be neglected in the model. An important tool in this respect is the notion of scaling and dimensionless numbers. If this is done properly, it is possible to characterize a problem without any calculation.

- A natural consequence of dimensional analysis is the possibility of similarity solutions. Self-similar solutions that can be constructed using dimensional analysis alone are called self-similar of the first kind. Self-similar solutions of the second kind have incomplete similarity in the dependent variables. They are connected to an eigenvalue problem.

- A sometimes underrated aspect is the question of existence and uniqueness. While making modeling assumptions for a reduced or asymptotic model, we have to make sure that the boundary and related conditions are always consistent with the other equations and accurately reflect the physics of the problem. Examples have been given in the above. When we neglect the effects of viscosity, the no-slip boundary condition of no-slip has to be dropped, but sometimes this is too drastic a simplification, and amendments (like the Kutta condition or the causality condition) have to be made that were not necessary in the higher-level model. The same is true for waves scattered at edges that are simplified to sharp edges. In problems of sound with mean flow the edge condition and the Kutta condition may occur in combination [121, 122]. There is no doubt that many more examples exist from other fields of application.

- Although modeling is the basis of any application of mathematics, there are not really many books that deal with this aspect from a more general point of view. A few representative texts are [152, 47, 33, 91].

Exercises

7.1. Simplify, by suitable scaling of the variables, the Korteweg–de Vries equation

$$\frac{\partial u}{\partial t} + c\left(1 + \frac{3}{2}u\right)\frac{\partial u}{\partial x} + \frac{1}{6}ch^2\frac{\partial^3 u}{\partial x^3} = 0$$

and the linearized Boussinesq equation

$$\frac{\partial^2 u}{\partial t^2} - c^2\frac{\partial^2 u}{\partial x^2} - \frac{1}{3}h^2\frac{\partial^4 u}{\partial x^2 \partial t^2} = 0$$

such that the coefficients become equal to one.

7.2. Nondimensionalise the telegraph equation (∗) in Example 1.4,

$$\frac{\partial^2 u}{\partial t^2} + (a+b)\frac{\partial u}{\partial t} + abu - c^2\frac{\partial^2 u}{\partial x^2} = 0,$$

such that the resulting problem depends on one dimensionless number only.

7.3. The drag D of a moving ship, due to viscous effects and wave generation, depends on its length L and velocity V, water viscosity μ, gravity acceleration g and water density ρ. The dimensional units are $[D] = \text{kg} \cdot \text{m/s}^2$, $[L] = \text{m}$, $[V] = \text{m/s}$, $[\mu] = \text{kg/(m} \cdot \text{s)}$, $[g] = \text{m/s}^2$, and $[\rho] = \text{kg/m}^3$. By how many dimensionless groups is the problem completely described? Give an example of such a set of dimensionless groups (these are not uniquely defined).

7.4. Is the self-similar solution of Example 7.18 of the first or of the second kind?

7.5. Analyse Example 7.19 in terms of Buckingham's theorem. Verify the dimensional groups, including their number. Note that $[\sigma] = \text{A}^2 \cdot \text{s}^3/(\text{m}^3 \cdot \text{kg})$, $[\rho C] = \text{kg/(m} \cdot \text{s}^2 \cdot \text{K)}$, $[\psi] = \text{kg} \cdot \text{m}^2\text{/(s}^3 \cdot \text{A)}$, and $[\kappa] = \text{kg} \cdot \text{m}/(\text{s}^3 \cdot \text{K})$.

7.6. Reconsider Example 7.1. Make the problem complete by adding boundary and initial conditions. Make the problem dimensionless by scaling on the inherent time and length scales. Determine conditions, in terms of a dimensionless number, for which the diffusion term can be neglected. Note that the order of the differential equation is then reduced from two to one. What are the consequences for the boundary and/or initial conditions?

7.7. Consider (†) of Example 7.4 to describe a stationary suspended flexible bar of length L. Make the problem complete by adding suitable boundary conditions at $x = 0$ and

$x = D$, where $1 - D/L$ is positive and not small. Note the two integration constants H_0 and V_0, so we need four conditions. Make the problem dimensionless by scaling lengths on L and forces on QL. Under what condition, in terms of a dimensionless parameter, can we neglect bending stiffness (i.e., the term multiplying ϕ_{ss})? The result describes a cable with vanishing bending stiffness, or a catenary. What are the consequences for the boundary conditions? Solve this equation.

7.8. Material of concentration c is diffused from a container located at $|x| \leq a$ through a membrane at $|x| = a$ into the outer medium $|x| > a$. In the container and in the medium the diffusion is described by

$$\frac{\partial c}{\partial t} = \frac{\partial}{\partial x}\left(D\frac{\partial c}{\partial x}\right).$$

The diffusion coefficient in the container is $D = D_i$; in the outer medium it is $D = D_o$. Initially, $c = 0$ for $|x| > a$ and $c = c_0$ in $|x| \leq a$. At the interfaces $x = \pm a$, c and Dc_x are continuous.

(a) Describe the problem in dimensionless variables.

(b) Approximate the solution for t large (which is the same as for $a \to 0$). Hint: Use the fact that eventually the majority of the material is diffused to the outer medium, while conservation of mass requires that $\int_{-\infty}^{\infty} c(x,t)\,\mathrm{d}x = \text{constant}$. Then derive a similarity solution. Note the symmetry in x.

(c) Do the same for the analogous problems in two and three dimensions by utilising Chapter 10, Section 10.3.

7.9. We are interested in knowing at what distance D a boat of height H is still visible above the horizon. Criticize the following (incorrect) model.
The height of a person is negligible compared to Earth's radius R. So our field of visibility is just in the tangent plane of Earth's at the position of the observer. A boat is visible in this plane if $\cos(\frac{D}{R}) \geq \frac{R}{R+H}$. Since D/R and H/R are small, this is equivalent to $D^2 \leq 2RH$.

7.10. A simple model for the temperature T in the ground, at time t and depth z, is

$$\rho C\frac{\partial T}{\partial t} = \kappa\frac{\partial^2 T}{\partial z^2}, \qquad -\infty < z \leq 0,$$

where $T(0,t) = T_0(t)$ is the given air temperature at the surface $z = 0$. ρ denotes the density, C the specific heat, and κ the thermal conductivity of the soil. For dry sand $\rho = 1600\,\text{kg/m}^3$, $C = 800\,\text{J/(kg}\cdot\text{K)}$, and $\kappa = 0.3\,\text{W/(m}\cdot\text{K)}$, and for saturated sand $\rho = 2000\,\text{kg/m}^3$, $C = 1480\,\text{J/(kg}\cdot\text{K)}$, $\kappa = 2.2\,\text{W/(m}\cdot\text{K)}$. Estimate, by scaling, the typical penetration depth of the yearly temperature variations (steady state). Compare this with an exact solution if T_0 varies harmonically like $T_0(t) = A + B\cos(\omega t)$, where $\omega = 2\pi(1\text{ year})^{-1}$, and T is in steady state. If the temperature for $z \to -\infty$ is below 0°C, we call this soil *permafrost*.

7.11. A reasonably complete model of a kettle drum consists of a cavity in free space, with acoustic pressure perturbations $p(\boldsymbol{x},t) = \mathrm{Re}(\hat{p}(\boldsymbol{x})\,\mathrm{e}^{\mathrm{i}\omega t})$ inside and outside the cavity given by

$$\nabla^2\hat{p} + \kappa^2\hat{p} = 0, \qquad \mathrm{i}\omega\rho_0\hat{\boldsymbol{v}} + \nabla\hat{p} = 0,$$

where $\kappa = \omega/c_0$, ρ_0 denotes the mean air density, and $\hat{\boldsymbol{v}}$ represents the corresponding velocity perturbations. The cavity is hard walled on all sides ($\hat{\boldsymbol{v}} \cdot \boldsymbol{n} = 0$) except one, which is closed by an elastic membrane (tension T, mass density σ). The membrane displacement $\hat{\eta}$ is driven by, and drives, the pressure difference across the membrane:

$$T\nabla^2\hat{\eta} + \omega^2\sigma\hat{\eta} = \hat{p}_{\text{upper}} - \hat{p}_{\text{lower}}.$$

As the air follows the membrane, the normal velocity $\hat{\boldsymbol{v}} \cdot \boldsymbol{n}$ at both sides of the membrane is equal to $\mathrm{i}\omega\hat{\eta}$. The basic question of interest is to determine the spectrum of (complex) resonance frequencies.

Consider the following canonical model of the kettle drum problem. A semi-infinite pipe ($0 \leq x < \infty$) of radius a, closed at $x = 0$, contains a piston-like element at $x = L$ (modeling the membrane), which is driven by the pressure difference across $x = L$ and kept in position by a spring:

$$\frac{\partial^2 \hat{p}}{\partial x^2} + \kappa^2 \hat{p} = 0 \qquad \text{for} \quad x \in (0, L) \cup (L, \infty)$$

$$-\frac{8}{a^2} T\hat{\eta} + \omega^2\sigma\hat{\eta} = \hat{p}(L+) - \hat{p}(L-) \qquad \text{at} \quad x = L$$

$$\frac{\partial \hat{p}}{\partial x} = 0 \ \text{at}\ x = 0, \qquad \frac{\partial \hat{p}}{\partial x} = \omega^2\rho_0\hat{\eta} \ \text{at}\ x = L, \quad \text{outgoing waves for}\ x \to \infty.$$

Determine the equation for ω, solve this for some simple cases, and try to indicate the general solution graphically in the complex ω plane for dimensionless groups of parameters. Are there solutions with $\mathrm{Im}(\omega) = 0$? How are these to be interpreted physically?

Chapter 8

The Analysis of Elliptic Equations

In this chapter we discuss analytical methods for elliptic equations. We define in Section 8.1 several boundary value problems for elliptic equations and investigate the uniqueness of the solution. The concepts of eigenvalues and eigenfunctions of an elliptic boundary value problem are introduced in Section 8.2. An important analytical solution method for linear elliptic equations is separation of variables, which is discussed in Section 8.3. In Section 8.4 we introduce the fundamental solution of the Poisson equation, which is a solution of the Poisson equation with a Dirac delta function as source term. If, furthermore, the fundamental solution satisfies certain homogeneous boundary conditions, it is called a Green's function. Next, in Section 8.5, we derive integral representations for the solution of elliptic boundary value problems using these Green's functions. A qualitative description of the solution of the Poisson equation is based on the maximum principle and is presented in Section 8.6. As an application of elliptic equations, we study in Section 8.7 the Stokes equations, which describe creeping flow. In particular, we compute the (slow) flow around a sphere.

8.1 The Laplace Operator

Elliptic differential equations typically occur in problems that describe stationary situations; i.e., time has no explicit role. The simplest and most well known elliptic equation is the *Laplace equation*, defined on a domain $\Omega \subset \mathbb{R}^d$ $(d = 1, 2, 3)$, say:

$$\mathcal{L}[u] := \nabla^2 u = 0, \qquad \boldsymbol{x} \in \Omega. \tag{8.1}$$

In the inhomogeneous case we have the *Poisson equation*

$$\mathcal{L}[u] := \nabla^2 u = f(\boldsymbol{x}), \qquad \boldsymbol{x} \in \Omega. \tag{8.2}$$

A further type that is often encountered is the *Helmholtz equation*, which is actually related to the eigenvalue problem of (8.1):

$$\mathcal{L}[u] := \nabla^2 u - \lambda u = 0, \qquad \boldsymbol{x} \in \Omega, \qquad \lambda \in \mathbb{R}. \tag{8.3}$$

The Laplace operator ∇^2 also occurs quite often in time-dependent problems like the heat equation or the wave equation. This provides for additional interest in investigating problems like (8.1), (8.2), and (8.3).

8.1.1 Problem Types

The Laplace operator typically occurs in situations where the flux f of a variable is proportional to its gradient. As in Example 1.1 we may consider a concentration c that causes a flow to areas with lower concentration; i.e.,

$$f = -D\nabla c, \tag{8.4}$$

with $D > 0$ the diffusion coefficient. If we apply Gauss's divergence theorem (see (J.13) in the appendix) to an arbitrary volume $W \subset \Omega$, we find

$$-\int_W \nabla\cdot(D\nabla c)\,\mathrm{d}V = -\oint_{\partial W} (D\nabla c)\cdot\boldsymbol{n}\,\mathrm{d}S = \oint_{\partial W} \boldsymbol{f}\cdot\boldsymbol{n}\,\mathrm{d}S. \tag{8.5}$$

Without sources or sinks, the net flux through ∂W should be 0. If, furthermore, the diffusion coefficient D is constant, we obtain (8.1). Any solution in $C^2(\Omega)$ of (8.1) is called a *harmonic function*.

In order to define a solution more precisely, we have to specify a *boundary condition*. Three common cases are distinguished for $\boldsymbol{x} \in \partial\Omega$:

$$u(\boldsymbol{x}) = a(\boldsymbol{x}) \qquad \textit{(Dirichlet)}, \tag{8.6a}$$

$$\frac{\partial u}{\partial \boldsymbol{n}}(\boldsymbol{x}) = b(\boldsymbol{x}) \qquad \textit{(Neumann)}, \tag{8.6b}$$

$$\alpha u(\boldsymbol{x}) + \beta\frac{\partial u}{\partial \boldsymbol{n}}(\boldsymbol{x}) = c(\boldsymbol{x}), \qquad \alpha, \beta \neq 0 \qquad \textit{(Robin)}, \tag{8.6c}$$

with $\boldsymbol{n}$ the outward unit normal on $\partial\Omega$. $\frac{\partial}{\partial n}$ denotes the normal derivative; i.e., $\frac{\partial}{\partial n}u := \boldsymbol{n}\cdot\nabla u$. Condition (8.6a) is called a *Dirichlet boundary condition*, (8.6b) is called a *Neumann boundary condition*, and (8.6c) is called a *Robin boundary condition*. We can easily establish the uniqueness of a solution of (8.1) and either one of the two boundary conditions (8.6a) or (8.6c). It is instructive to illustrate these problems for the one-dimensional case, where the Poisson equation reduces to an ODE.

Example 8.1 Consider the two-point Dirichlet boundary value problem

$$\mathcal{L}[u] := \frac{\mathrm{d}^2 u}{\mathrm{d}x^2} = f(x), \qquad x \in (0, \pi),$$

$$u(0) = u(\pi) = 0,$$

where f is piecewise smooth on $(0, \pi)$. We now use Fourier sine series to find the solution in a formal way. We take the ansatz

$$u(x) = \sum_{k=1}^{\infty} a_k \sin(kx),$$

which has the advantage that u satisfies the boundary conditions identically *if* the series converges uniformly and the found solution is continuous at the endpoints (appendix, Section C).This is to be verified afterward. We also expand $f(x)$ in a Fourier sine series

$$f(x) = \sum_{k=1}^{\infty} f_k \sin(kx),$$

where

$$f_k = \frac{2}{\pi} \int_0^{\pi} f(x) \sin(kx)\, \mathrm{d}x, \qquad k = 1, 2, \ldots,$$

and $f_k \to 0$ for $k \to \infty$. We then find

$$\sum_{k=1}^{\infty} \left(k^2 a_k + f_k\right) \sin(kx) = 0.$$

From the uniqueness of the Fourier coefficients of the null function it follows that $k^2 a_k = -f_k$, yielding the solution

$$u(x) = -\sum_{k=1}^{\infty} \frac{f_k}{k^2} \sin(kx), \qquad x \in (0, \pi),$$

which is indeed uniformly convergent. □

Example 8.2 Consider the two-point Neumann boundary value problem

$$\mathcal{L}[u] := \frac{\mathrm{d}^2 u}{\mathrm{d}x^2} = f(x), \qquad x \in (0, \pi),$$

$$\frac{\mathrm{d}u}{\mathrm{d}x}(0) = \frac{\mathrm{d}u}{\mathrm{d}x}(1) = 0,$$

where f and f' are both piecewise smooth on $(0, \pi)$. Because of the form of the boundary conditions, it seems advantageous to expand the solution $u(x)$ in a Fourier cosine series as

$$u(x) = b_0 + \sum_{k=1}^{\infty} b_k \cos(kx).$$

If u' converges uniformly (to be verified afterward), the boundary conditions are automatically satisfied. Likewise, we have

$$f(x) = f_0 + \sum_{k=1}^{\infty} f_k \cos(kx),$$

with coefficients f_k given by

$$f_0 = \frac{1}{\pi} \int_0^{\pi} f(x)\, \mathrm{d}x, \qquad f_k = \frac{2}{\pi} \int_0^{\pi} f(x) \cos(kx)\, \mathrm{d}x, \qquad k = 1, 2, \ldots.$$

Note that $f_k = \mathcal{O}(k^{-1})$ for $k \to \infty$ (see Corollary 3.7). Substituting these expansions into the differential equation, we find the relation

$$\sum_{k=0}^{\infty} \left(k^2 b_k + f_k\right) \cos(kx) = 0,$$

so that $k^2 b_k + f_k = 0$. In particular, $f_0 = 0$, implying that $f(x)$ should satisfy the consistency condition

$$\int_0^\pi f(x)\,\mathrm{d}x = 0$$

stating that the average value of $f(x)$ over $(0, \pi)$ vanishes. As a consequence, b_0 is undetermined and the solution reads

$$u(x) = b_0 - \sum_{k=1}^{\infty} \frac{f_k}{k^2} \cos(kx),$$

with b_0 arbitrary. Thus $u(x)$ is determined up to an additive constant. Since $f_k/k = \mathcal{O}(k^{-2})$, it is readily verified that u' converges uniformly and satisfies the boundary conditions. □

Example 8.3 Consider the two-point Helmholtz boundary value problem

$$\mathcal{L}[u] := \frac{\mathrm{d}^2 u}{\mathrm{d}x^2} - \lambda\, u = 0, \qquad x \in (0, \pi),$$
$$u(0) = u(\pi) = 0.$$

To solve this linear ODE in a formal way, we look for solutions of the form $u(x) = \mathrm{e}^{\mu x}$. Substituting this solution into the differential equation, we find that μ should satisfy the characteristic equation

$$\mu^2 - \lambda = 0.$$

So as a general solution we obtain

$$u(x) = A\,\mathrm{e}^{\sqrt{\lambda} x} + B\,\mathrm{e}^{-\sqrt{\lambda} x}, \qquad A, B \in \mathbb{R}.$$

Using the boundary conditions, we find

$$A + B = 0, \qquad A\,\mathrm{e}^{\sqrt{\lambda}\pi} + B\,\mathrm{e}^{-\sqrt{\lambda}\pi} = 0.$$

For arbitrary λ this system has only the trivial solution $A = B = 0$. Nontrivial solutions (the eigenvalue problem) exist if the determinant of the system is zero:

$$\begin{vmatrix} 1 & 1 \\ \mathrm{e}^{\sqrt{\lambda}\pi} & \mathrm{e}^{-\sqrt{\lambda}\pi} \end{vmatrix} = \mathrm{e}^{-\sqrt{\lambda}\pi} - \mathrm{e}^{\sqrt{\lambda}\pi} = 0.$$

This relation implies that $\mathrm{e}^{2\sqrt{\lambda}\pi} = 1$, which has the solutions $2\sqrt{\lambda}\pi = k2\pi\mathrm{i}$ $(k = 0, 1, 2, \dots)$. Apparently, the only possible values of λ that allow solutions are given by

$$\lambda = \lambda_k = -k^2, \qquad k = 0, 1, 2, \dots.$$

Choosing $A = -B = -\frac{1}{2}\mathrm{i}$, we find the corresponding solutions

$$u(x) = u_k(x) = \sin(kx), \qquad k = 0, 1, 2, \dots.$$

So we have either no solutions or infinitely many solutions. Note that $k = 0$ corresponds to the trivial solution $u_0(x) \equiv 0$ and should therefore be discarded. □

8.1.2 Uniqueness

One can simply investigate the uniqueness of the Laplace equation (8.1) for a solution satisfying either one of the boundary conditions (8.6). This is done in the next theorem.

Theorem 8.4. *A harmonic function satisfying the Dirichlet boundary condition* (8.6a) *is unique. A harmonic function satisfying the Neumann boundary condition* (8.6b) *is unique up to an additive constant. If* $\operatorname{sign}(\alpha) = \operatorname{sign}(\beta)$, *then a harmonic function satisfying the Robin boundary condition* (8.6c) *is unique.*

Proof. First, consider the boundary condition (8.6a) with $a(\boldsymbol{x}) \equiv 0$. Using the first identity of Green (J.16), we obtain

$$\int_\Omega \left(u\nabla^2 u + \nabla u \cdot \nabla u\right) \mathrm{d}V = \oint_{\partial\Omega} u \frac{\partial u}{\partial \boldsymbol{n}} \,\mathrm{d}S,$$

so that

$$\int_\Omega |\nabla u|^2 \,\mathrm{d}V = 0, \tag{$*$}$$

whence u is constant in Ω. Because of continuity we conclude that $u(\boldsymbol{x}) \equiv 0$. If we now have two harmonic solutions, $u_1(\boldsymbol{x})$ and $u_2(\boldsymbol{x})$ say, both satisfying (8.6a), then obviously $u_1(\boldsymbol{x}) - u_2(\boldsymbol{x}) \equiv 0$ in Ω, which implies uniqueness. For the boundary condition (8.6b) with $b(\boldsymbol{x}) \equiv 0$ we derive ($*$) for a harmonic function $u(\boldsymbol{x})$ in a similar manner. This trivially implies $u(\boldsymbol{x})$ to be constant in Ω. If $u_1(\boldsymbol{x})$ and $u_2(\boldsymbol{x})$ are two harmonic functions satisfying the boundary condition (8.6b), then $u_1(\boldsymbol{x}) - u_2(\boldsymbol{x})$ can be identified with $u(\boldsymbol{x})$, as before. Hence a solution is unique, apart from an additive constant. Finally, for a harmonic function $u(\boldsymbol{x})$ satisfying the boundary condition (8.6c) we obtain

$$\int_\Omega |\nabla u|^2 \,\mathrm{d}V = -\oint_{\partial\Omega} \frac{\alpha}{\beta} u^2 \,\mathrm{d}S \le 0.$$

Obviously this can only be true if $u(\boldsymbol{x}) \equiv 0$. Uniqueness then follows in the same fashion as for the first case. □

8.2 Eigenvalues and Eigenfunctions

Eigenvalue problems play an important role, either directly, like the Helmholtz equation, or indirectly, when we characterise a PDE. In this section we will mainly consider the latter aspect. We first consider the one-dimensional case.

8.2.1 The One-Dimensional Eigenvalue Problem

Probably the simplest eigenvalue problem is given by

$$\frac{\mathrm{d}^2 u}{\mathrm{d}x^2} = \lambda u, \qquad x \in (0, a), \tag{8.7a}$$

$$u(0) = u(a) = 0. \tag{8.7b}$$

Note that (8.7a) is just the one-dimensional Helmholtz equation. Parameter λ is not prescribed but is part of the solution. It is called an eigenvalue and its value is determined by the condition that (8.7) has a nontrivial solution. The general solution of (8.7a) can be written like

$$u(x) = A\,\mathrm{e}^{\sqrt{\lambda}x} + B\,\mathrm{e}^{-\sqrt{\lambda}x}, \tag{8.8}$$

where A and B follow from the boundary conditions (8.7b). We find the equations

$$A + B = 0, \qquad A\,\mathrm{e}^{\sqrt{\lambda}a} + B\,\mathrm{e}^{-\sqrt{\lambda}a} = 0. \tag{8.9}$$

This homogeneous system only has a nontrivial solution if the following determinant is zero:

$$\begin{vmatrix} 1 & 1 \\ \mathrm{e}^{\sqrt{\lambda}a} & \mathrm{e}^{-\sqrt{\lambda}a} \end{vmatrix} = \mathrm{e}^{-\sqrt{\lambda}a} - \mathrm{e}^{\sqrt{\lambda}a} = 0. \tag{8.10}$$

We conclude that $\mathrm{e}^{2\sqrt{\lambda}a} = 1 = \mathrm{e}^{2k\pi\mathrm{i}}$, and the eigenvalues are found to be

$$\lambda = \lambda_k = -\left(\frac{k\pi}{a}\right)^2, \qquad k = 0, 1, 2, \ldots. \tag{8.11}$$

Since $B = -A$, we find for the corresponding eigenfunctions

$$u(x) = u_k(x) = \sin\left(\frac{k\pi x}{a}\right), \qquad k = 0, 1, 2, \ldots. \tag{8.12}$$

Any multiple of $u_k(x)$ is an eigenfunction as well. Note that $u_0(x) \equiv 0$ and therefore λ_0 and $u_0(x)$ should be discarded as trivial solutions.

The inner product (f, g), defined in (3.1), reduces for real-valued functions to

$$(f, g) := \int_0^a f(x)g(x)\,\mathrm{d}x. \tag{8.13}$$

It is readily seen that $(u_k, u_l) = 0$ for $k \neq l$. In other words, the eigenfunctions u_k are orthogonal with respect to the inner product (8.13). We can put this in a more formal setting by considering the operator

$$\mathcal{L}[u] := \frac{\mathrm{d}^2}{\mathrm{d}x^2}u. \tag{8.14}$$

If we apply integration by parts twice, we find for a suitable "test function" $v = v(x)$ with $v(0) = v(a) = 0$ the relation

$$\big(\mathcal{L}[u], v\big) = \int_0^a \frac{\mathrm{d}^2u}{\mathrm{d}x^2}(x)v(x)\,\mathrm{d}x = \int_0^a u(x)\frac{\mathrm{d}^2v}{\mathrm{d}x^2}(x)\,\mathrm{d}x = \big(u, \mathcal{L}[v]\big). \tag{8.15}$$

This is a special property of this operator. More specifically, we define the adjoint operator $\mathcal{A}^*$ of an operator $\mathcal{A}$ by (cf. Chapter 4)

$$\big(u, \mathcal{A}[v]\big) = \big(\mathcal{A}^*[u], v\big). \tag{8.16}$$

Then we see from (8.15) that $\mathcal{L}$ defined in (8.14) has the symmetry property

$$\big(\mathcal{L}[u], v\big) = \big(u, \mathcal{L}[v]\big) = \big(\mathcal{L}^*[u], v\big). \tag{8.17}$$

We therefore call this $\mathcal{L}$ *self-adjoint*. More generally, if we have the operator $\mathcal{L}$ defined by

$$\mathcal{L}[u] := \frac{\mathrm{d}}{\mathrm{d}x}\left(p(x)\frac{\mathrm{d}u}{\mathrm{d}x}\right) + q(x)u(x), \tag{8.18}$$

where p and q are real, then $\mathcal{L}$ is self-adjoint. We leave it as an exercise to see that an operator with an explicit first derivative term is not self-adjoint. There is a remarkable similarity with symmetric matrices. Indeed, symmetric matrices have an orthogonal system of eigenvectors that correspond to real eigenvalues. This similarity is often exploited in numerical approaches that try to preserve the self-adjointness of (8.18) in a discrete form; see also Chapter 9.

We can also derive the following (cf. (8.15)). Let λ be an eigenvalue of (8.18) with homogeneous boundary conditions and u be the corresponding eigenfunction. Then

$$\lambda(u, u) = (\mathcal{L}[u], u) = \int_0^a \left(-p(x)\left|\frac{\mathrm{d}u}{\mathrm{d}x}\right|^2 + q(x)|u|^2\right)\mathrm{d}x. \tag{8.19}$$

Hence λ is real, which is in agreement with the eigenvalues of (8.7) and is in line with what we know for symmetric matrices. Moreover, we see that $\lambda < 0$ for $p > 0$, and $q \le 0$. With these requirements on p and q we have found the analogue of what we call negative definite symmetric matrices. (A matrix $\mathbf{C}$ is negative definite if $\mathbf{z}^T\mathbf{C}\,\mathbf{z} < 0$ for any $\mathbf{z} > \mathbf{0}$.)

Remark. Sometimes one rather prefers to use the operator $-\mathcal{L}$ in order to have strictly positive eigenvalues.

Example 8.5 Consider the operator $\mathcal{L}$, with

$$\mathcal{L}[u] := \frac{\mathrm{d}^2u}{\mathrm{d}x^2} + qu, \qquad q \in \mathbb{R},$$

and let $u(0) = u(a) = 0$. If we try to find the eigenvalues as we did for (8.7), we obtain

$$\lambda_k = -\left(\frac{k\pi}{a}\right)^2 + q, \qquad k = 1, 2, \ldots.$$

Clearly, all eigenvalues are negative if $q < (\pi/a)^2$, which is slightly more relaxed than the requirement $q \le 0$ that we used before. □

Example 8.6 If we have purely Neumann boundary conditions, we get eigensolutions different from (8.11) and (8.12). Thus consider the boundary value problem

$$\frac{\mathrm{d}^2u}{\mathrm{d}x^2} = \lambda u, \qquad x \in (0, a),$$

$$\frac{\mathrm{d}u}{\mathrm{d}x}(0) = \frac{\mathrm{d}u}{\mathrm{d}x}(a) = 0.$$

Analogously to the derivation of the eigensolution (8.11) and (8.12), we find a general solution of the form (8.8), where the coefficients A and B have to satisfy the equations

$$A - B = 0, \qquad A\,\mathrm{e}^{\sqrt{\lambda}a} - B\,\mathrm{e}^{-\sqrt{\lambda}a} = 0.$$

We again find the eigenvalues

$$\lambda_k = -\left(\frac{k\pi}{a}\right)^2, \qquad k = 0, 1, 2, \ldots,$$

but now they correspond to the eigenfunctions

$$u_k(x) = \cos\left(\frac{k\pi x}{a}\right), \qquad k = 0, 1, 2, \ldots.$$

Note that $u_0(x) \equiv 1$ corresponds to any constant that can be added to the solution of the Neumann problem for the Poisson equation, illustrating that this solution is not unique. □

There is a close relationship between eigenfunctions and Fourier series (cf. Chapter 3). If we expand a function f with $f(0) = f(a) = 0$ on $[0, a]$, then we obtain the Fourier sine series

$$f(x) = \sum_{k=1}^{\infty} f_k \sin\left(\frac{k\pi x}{a}\right), \tag{8.20}$$

so $f(x)$ is actually expanded in terms of eigenfunctions of (8.7). This property holds more generally for eigenfunctions and has a host of consequences. An appropriate setting for this is variational calculus, which is outside the scope of this text.

8.2.2 Eigenvalue Problems in Several Dimensions

The analysis of eigenvalue problems can be extended to higher dimensions. The starting point is a self-adjoint generalization of the Helmholtz equation defined on a domain Ω with homogeneous Dirichlet boundary condition; i.e.,

$$\mathcal{L}[u] := \nabla^2 u + qu = \lambda\, u, \qquad \boldsymbol{x} \in \Omega, \tag{8.21a}$$

$$u(\boldsymbol{x}) = 0, \qquad \boldsymbol{x} \in \partial\Omega. \tag{8.21b}$$

We first introduce an inner product on Ω. Let u and v be defined on Ω and satisfy the homogeneous Dirichlet boundary condition as in (8.21b). Then

$$(u, v) := \int_\Omega uv \, \mathrm{d}V. \tag{8.22}$$

Our first goal is to show that $\mathcal{L}$ is self-adjoint. We have, using the second identity of Green, that

$$\big(\mathcal{L}[u], v\big) = \int_\Omega (\nabla^2 u + qu)v \, \mathrm{d}V = \int_\Omega u\big(\nabla^2 v + qv\big) \, \mathrm{d}V = \big(u, \mathcal{L}[v]\big) = \big(\mathcal{L}^*[u], v\big). \tag{8.23}$$

Note that we have used that $v = u = 0$ on $\partial\Omega$. If we now let u be an eigensolution corresponding to λ (presupposing its existence), then we find by Green's first identity that

$$\lambda(u, u) = (\mathcal{L}[u], u) = \int_\Omega (\nabla^2 u + qu)u \, \mathrm{d}V = -\int_\Omega \big(|\nabla u|^2 - qu^2\big) \, \mathrm{d}V. \tag{8.24}$$

Hence we see that $\lambda \in \mathbb{R}$. Moreover, if $q < 0$, then $\lambda < 0$.

Now let u_k and u_l be eigenfunctions corresponding to the eigenvalues λ_k and λ_l, respectively, with $\lambda_k \neq \lambda_l$. Then

$$u_l\mathcal{L}[u_k] - u_k\mathcal{L}[u_l] = u_l\nabla^2 u_k - u_k\nabla^2 u_l = \big(\lambda_k - \lambda_l\big)u_k u_l. \tag{8.25}$$

After integration we obtain from (8.25) that

$$\int_\Omega \big(u_l\nabla^2 u_k - u_k\nabla^2 u_l\big)\,\mathrm{d}V = \big(\lambda_k - \lambda_l\big)\int_\Omega u_k u_l\,\mathrm{d}V. \tag{8.26}$$

For the left-hand side in (8.26), we can apply Green's second identity, which then results in

$$\oint_{\partial\Omega}\left(u_l\frac{\partial u_k}{\partial \boldsymbol{n}} - u_k\frac{\partial u_l}{\partial \boldsymbol{n}}\right)\mathrm{d}S = \big(\lambda_k - \lambda_l\big)\int_\Omega u_k u_l\,\mathrm{d}V. \tag{8.27}$$

Clearly, the integral on the left in (8.27) is zero, from which we derive

$$\big(u_k, u_l\big) = \int_\Omega u_k u_l\,\mathrm{d}V = 0; \tag{8.28}$$

i.e., the eigensolutions form an orthogonal set.

For a general domain Ω one has to use numerical methods to find (an approximation of) the eigenvalues. If we have a rectangular domain, say $\Omega = \{(x, y) \in \mathbb{R}^2 \mid 0 \le x \le a, 0 \le y \le b\}$, we can simply obtain the eigensolutions from what we found for the one-dimensional case when $q(x) = 0$. It is left as an exercise to show that

$$\lambda_{k,l} = -\pi^2\left(\frac{k^2}{a^2} + \frac{l^2}{b^2}\right), \tag{8.29a}$$

$$u_{k,l}(x, y) = \sin\left(\frac{k\pi x}{a}\right)\sin\left(\frac{l\pi y}{b}\right) \qquad (k, l = 1, 2, \dots) \tag{8.29b}$$

are eigenvalues and eigenfunctions, respectively. We conclude this section with a general property, which we give without proof (see, e.g., [36]).

Property 8.7. *Let the area of a two-dimensional domain be A. Then there exists an infinite sequence of eigenvalues of the Laplace equation. They can be ordered such that the mth eigenvalue λ_m satisfies, for large m,*

$$\lambda_m \sim -\frac{4\pi m}{A}. \tag{8.30}$$

Example 8.8 Consider a Dirichlet problem on the unit square $\Omega := \{(x, y) \in \mathbb{R}^2 \mid 0 < x < 1, 0 < y < 1\}$. From (8.29a) we see that $\lambda_{k,l} = -\pi^2(k^2 + l^2)$. Let $m(\lambda)$ be the number of eigenvalues, still larger than a fixed negative λ, or, equivalently, the number of index pairs (k, l) satisfying $k^2 + l^2 < |\lambda|/\pi^2$. We can relabel the eigenvalues $\{\lambda_{k,l}\}$ as $\{\mu_m\}$ by letting $\lambda_{k,l}$ gradually decrease, starting from $\lambda_{1,1}$. In Figure 8.1 we have indicated the set of index pairs $\{(k, l)\}$ that lie in the first quadrant of a circle with radius $\frac{1}{\pi}\sqrt{|\lambda|}$. The area of this quarter circle is equal to $|\lambda|/4\pi$. Hence

$$m(\lambda) \sim \frac{|\lambda|}{4\pi}.$$

Reverting the argument, we obtain $\mu_m \sim -4\pi m$, in agreement with (8.30). □

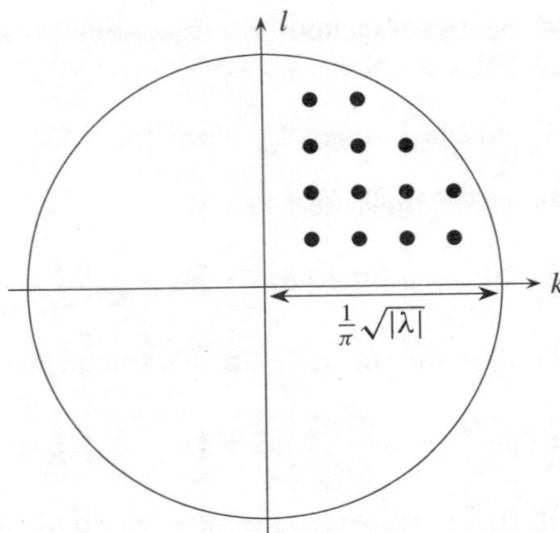

Figure 8.1. *The eigenvalues contained in the first quadrant of the circle* $k^2 + l^2 < |\lambda|/\pi$.

8.3 Separation of Variables

The method of separation of variables is useful for linear problems with constant coefficients and homogeneous boundary conditions. In this section we will apply the method to the two-dimensional Laplace equation (8.1). The method is based on the assumption that the solution $u = u(x, y)$ can be written as a product of a function v, say, depending solely on x, and a function w, say, depending solely on y; i.e.,

$$u(x, y) = v(x)w(y). \tag{8.31}$$

If we substitute this in (8.1), we obtain

$$\frac{\mathrm{d}^2 v}{\mathrm{d}x^2}(x)w(y) = -v(x)\frac{\mathrm{d}^2 w}{\mathrm{d}y^2}(y).$$

Dividing by $v(x)w(y)$, we can write this as

$$\frac{1}{v(x)}\frac{\mathrm{d}^2 v}{\mathrm{d}x^2}(x) = -\frac{1}{w(y)}\frac{\mathrm{d}^2 w}{\mathrm{d}y^2}(y). \tag{8.32}$$

Since the left-hand side in (8.32) is a function of x only and the right-hand side is a function of y only, they must be independent of both, i.e., constant. Let us denote this constant by λ. Then we apparently have the two *eigenvalue problems*

$$\frac{\mathrm{d}^2 v}{\mathrm{d}x^2}(x) = \lambda v(x), \tag{8.33a}$$

$$\frac{\mathrm{d}^2 w}{\mathrm{d}y^2}(y) = -\lambda w(y). \tag{8.33b}$$

The constant λ is called the separation constant. Let us now assume that Ω is the unit square, i.e., $0 < x, y < 1$, and that, e.g., the following boundary conditions are given:

$$u(0, y) = u(1, y) = 0, \qquad u(x, 0) = u(x, 1) = x(1 - x). \tag{8.34}$$

From the homogeneous boundary conditions we conclude that $v(0) = v(1) = 0$ and so (8.33a) is a genuine eigenvalue problem. We then find that the eigenvalues λ_k are given by

$$\lambda_k = -k^2\pi^2, \qquad k = 1, 2, \ldots, \tag{8.35a}$$

and the corresponding eigenfunctions are given by

$$v_k(x) = \sin(k\pi x), \qquad k = 1, 2, \ldots; \tag{8.35b}$$

cf. Section 8.2.1. The functions $w(y) = w_k(y)$ corresponding to λ_k can be determined from the ODE (8.33b) and we find

$$w_k(y) = \alpha_k \, \mathrm{e}^{k\pi y} + \beta_k \, \mathrm{e}^{-k\pi y}, \qquad k = 1, 2, \ldots, \tag{8.36}$$

for some α_k, β_k. In order to determine the desired solution $u(x, y)$, we apply the *superposition principle*; i.e., we assume that

$$u(x, t) = \sum_{k=1}^{\infty} v_k(x) w_k(y) = \sum_{k=1}^{\infty} \left(\alpha_k \, \mathrm{e}^{k\pi y} + \beta_k \, \mathrm{e}^{-k\pi y}\right) \sin(k\pi x). \tag{8.37}$$

This superposition is possible since the PDE is linear. Note that (8.37) is in fact a Fourier-sine series in the x variable, so that the homogeneous boundary conditions are automatically satisfied *if* the series converges uniformly. This is to be verified afterward. The coefficients α_k and β_k follow from the remaining boundary conditions. First we have to find a Fourier-sine series for the function $x(x - 1)$. We obtain

$$x(x-1) = \sum_{k=1}^{\infty} \gamma_k \sin(k\pi x), \tag{8.38a}$$

where

$$\gamma_k = \begin{cases} 0 & \text{for } k \text{ even}, \\ -\dfrac{8}{(k\pi)^3} & \text{for } k \text{ odd}. \end{cases} \tag{8.38b}$$

The coefficients now follow from comparing (8.38a) and (8.37) for $x = 0$ and $x = 1$. This gives the set of equations

$$\alpha_k + \beta_k = \gamma_k, \tag{8.39a}$$

$$\alpha_k \, \mathrm{e}^{k\pi} + \beta \, \mathrm{e}^{-k\pi} = \gamma_k. \tag{8.39b}$$

With some straightforward arithmetic one finds from (8.39) the coefficients resulting in the final (indeed uniformly converging) series solution

$$u(x, y) = \sum_{k=1}^{\infty} \gamma_k \sin(k\pi x) \frac{\cosh(k\pi(y - \frac{1}{2}))}{\cosh(\frac{1}{2}k\pi)}. \tag{8.40}$$

The method of separation of variables apparently requires knowledge of the eigenvalues and eigenfunctions of the separate systems. This means that the boundary conditions should allow the separation to result in homogeneous boundary conditions for either one of the subproblems. The method is not restricted to Cartesian coordinates. Indeed, consider the PDE

$$\nabla^2 u = \frac{\partial^2 u}{\partial r^2} + \frac{1}{r}\frac{\partial u}{\partial r} + \frac{1}{r^2}\frac{\partial^2 u}{\partial \phi^2} = 0 \tag{8.41}$$

defined on a finite wedge $\{(r, \phi) \mid 0 < r < R, 0 < \phi < \Phi\}$, where u satisfies the boundary conditions

$$u(r, 0) = u(r, \Phi) = 0, \qquad u(R, \phi) = g(\phi) \tag{8.42}$$

for some $g(\phi)$. Writing

$$u(r, \phi) = v(r)w(\phi),$$

we obtain from (8.41) that

$$\frac{r^2}{v}\frac{\mathrm{d}^2 v}{\mathrm{d}r^2} + \frac{r}{v}\frac{\mathrm{d}v}{\mathrm{d}r} = -\frac{1}{w}\frac{\mathrm{d}^2 w}{\mathrm{d}\phi^2} = -\lambda. \tag{8.43}$$

We now have the two eigenvalue problems

$$\frac{\mathrm{d}^2 w}{\mathrm{d}\phi^2} = \lambda w, \tag{8.44a}$$

$$r^2\frac{\mathrm{d}^2 v}{\mathrm{d}r^2} + r\frac{\mathrm{d}v}{\mathrm{d}r} = -\lambda v. \tag{8.44b}$$

From the homogeneous boundary conditions in (8.42) we see that $w(0) = w(\Phi) = 0$ and consequently w is the solution of a genuine eigenproblem with the typical eigenvalues

$$\lambda_k = -\omega_k^2, \qquad \omega_k := \frac{k\pi}{\Phi} \qquad (k = 1, 2, \dots) \tag{8.45}$$

and corresponding eigenfunctions

$$w_k(\phi) = \sin(\omega_k \phi) \qquad (k = 1, 2, \dots). \tag{8.46}$$

For each λ_k the equation for v can now be solved by simple substitution of this λ_k in (8.44b). Substituting a solution of the form $v(r) = r^\mu$, we can easily see that $v_k(r)$ has the form

$$v(r) = \alpha_k r^{\omega_k} + \beta_k r^{-\omega_k}. \tag{8.47}$$

By requiring v_k to be bounded, we see that $\beta_k = 0$. The resulting solution of the boundary value problem (8.41), (8.42) is then given by

$$u(r, \phi) = \sum_{k=1}^{\infty} \alpha_k r^{\omega_k} \sin(\omega_k \phi). \tag{8.48}$$

Like in the previous case, if we have a Fourier-sine expansion of the function $g(\phi)$, we can formally determine the coefficients α_k.

Example 8.9 Denote by γ_k the coefficients of the Fourier-sine expansion of $g(\phi)$. Then we have

$$\gamma_k = \frac{2}{\Phi}\int_0^{\Phi} g(\phi)\sin(\omega_k\phi)\,\mathrm{d}\phi.$$

Hence from the boundary condition at $r = R$ we obtain

$$\alpha_k = \gamma_k R^{-\omega_k},$$

so the final solution reads

$$u(r,\phi) = \sum_{k=1}^{\infty} \gamma_k \left(\frac{r}{R}\right)^{\omega_k} \sin(\omega_k\phi).$$

This result is interesting as it shows the smoothness of the solution on a wedge in the neighbourhood of the corner point. Indeed, if we consider, e.g., $\frac{\partial u}{\partial r}$, we obtain

$$\frac{\partial u}{\partial r} = \frac{1}{R}\sum_{k=1}^{\infty} \gamma_k\omega_k \left(\frac{r}{R}\right)^{\omega_k-1} \sin(\omega_k\phi).$$

We see already that for $\Phi > \pi$ the first term ($k = 1$) is not bounded. The conclusion more generally, therefore, is that corners in a domain imply less smoothness. In particular, for reentrant corners (i.e., those with angles larger than π) this already holds true for the first derivative. This corner problem has, of course, consequences when solving a problem numerically. □

8.4 Fundamental Solutions

Before looking at general Poisson problems, as we shall do in Section 8.5, it is meaningful to investigate the fundamental solution w of the Poisson equation, i.e., the solution of the equation

$$\nabla^2 w(\boldsymbol{x};\boldsymbol{\xi}) = \delta(\boldsymbol{x}-\boldsymbol{\xi}). \tag{8.49}$$

We look for symmetry solutions depending only on the distance $r = \|\boldsymbol{x}-\boldsymbol{\xi}\|_2$, i.e., circular and spherically symmetric solutions in the two-dimensional and three-dimensional spaces, respectively. This simplifies the problem to an ODE. Indeed, let us denote this solution by $\bar{w}(r)$. Then it satisfies the differential equation

$$\nabla^2 \bar{w} = \frac{\mathrm{d}^2\bar{w}}{\mathrm{d}r^2} + \frac{d-1}{r}\frac{\mathrm{d}\bar{w}}{\mathrm{d}r} = \delta(r) \qquad (d = 2, 3). \tag{8.50}$$

Clearly, (8.50) has a general solution, for $r > 0$, of

$$\bar{w}(r) = \begin{cases} A\ln r + B & \text{if } d = 2, \\ \dfrac{A}{r} + B & \text{if } d = 3, \end{cases} \tag{8.51}$$

with $A, B \in \mathbb{R}$. For $d = 3$ we may choose $\bar{w}(r) \to 0$ for $r \to \infty$, implying that $B = 0$. We shall also choose $B = 0$ for $d = 2$. Therefore we are left to find A such that (8.50) is satisfied. To this end we define a ball $\mathcal{B}(\mathbf{0};\rho)$ around $\mathbf{0}$ with radius ρ and denote by $\partial\mathcal{B}(\mathbf{0};\rho)$ its sphere. From Gauss's theorem we then formally obtain

$$\int_{\mathcal{B}(\mathbf{0};\rho)} \nabla^2\bar{w}(r)\,\mathrm{d}V = \oint_{\partial\mathcal{B}(\mathbf{0};\rho)} \frac{\partial\bar{w}}{\partial\boldsymbol{n}}(r)\,\mathrm{d}S = \oint_{\partial\mathcal{B}(\mathbf{0};\rho)} \frac{\mathrm{d}\bar{w}}{\mathrm{d}r}(r)\,\mathrm{d}S. \tag{8.52}$$

Since $\int_{\mathcal{B}(\mathbf{0};\rho)} \nabla^2 \bar{w}(r)\,\mathrm{d}V = 1$, we can determine A. For $d = 2$ we find

$$\oint_{\partial\mathcal{B}(\mathbf{0};\rho)} \frac{\mathrm{d}\bar{w}}{\mathrm{d}r}(r)\,\mathrm{d}S = \oint_{\partial\mathcal{B}(\mathbf{0};\rho)} \frac{A}{\rho}\,\mathrm{d}S = 2\pi A,$$

from which we conclude that $A = 1/(2\pi)$. Likewise, for $d = 3$ we obtain

$$\oint_{\partial\mathcal{B}(\mathbf{0};\rho)} \frac{\mathrm{d}\bar{w}}{\mathrm{d}r}(r)\,\mathrm{d}S = \oint_{\partial\mathcal{B}(\mathbf{0};\rho)} \frac{-A}{\rho^2}\,\mathrm{d}S = -4\pi A,$$

so that in this case $A = -1/(4\pi)$. Using this in (8.51), we obtain for the fundamental solution

$$w(\boldsymbol{x};\boldsymbol{\xi}) = \begin{cases} \dfrac{1}{2\pi} \ln \|\boldsymbol{x} - \boldsymbol{\xi}\|_2 & \text{if } d = 2, \\[2ex] \dfrac{-1}{4\pi \|\boldsymbol{x} - \boldsymbol{\xi}\|_2} & \text{if } d = 3. \end{cases} \tag{8.53}$$

We leave it as an exercise to show that $w(\boldsymbol{x};\boldsymbol{\xi})$ is the (weak) solution of (8.49).

Example 8.10 Consider the scalar problem

$$\mathcal{L}[u] := \frac{\mathrm{d}^2 u}{\mathrm{d}x^2} = 0.$$

We would like to find a fundamental solution $w(x;\xi)$ satisfying

$$\frac{\mathrm{d}^2 w}{\mathrm{d}x^2}(x;\xi) = \delta(x - \xi).$$

Since the general solution of the ODE for $x \neq \xi$ can be written as $A + Bx$, we may take as ansatz for w

$$w(x;\xi) = \begin{cases} A_1 + B_1 x & \text{if } x < \xi, \\ A_2 + B_2 x & \text{if } x > \xi. \end{cases}$$

Using integration, we obtain for sufficiently small ρ that

$$\frac{\mathrm{d}w}{\mathrm{d}x}(\xi + \rho;\xi) - \frac{\mathrm{d}w}{\mathrm{d}x}(\xi - \rho;\xi) = \int_{\xi-\rho}^{\xi+\rho} \delta(x - \xi)\,\mathrm{d}x = 1.$$

Applying this condition to the solution above, we find

$$B_2 - B_1 = 1.$$

Since w is apparently continuous at $x = \xi$, we also find

$$A_1 + B_1\xi = A_2 + B_2\xi.$$

Hence we obtain the fundamental solution

$$w(x;\xi) = \begin{cases} A_1 + B_1 x & \text{if } x \leq \xi, \\ A_1 - \xi + (B_1 + 1)x & \text{if } x > \xi, \end{cases}$$

where A_1 and B_1 are arbitrary constants. □

Of course the fundamental solution relates to the (homogeneous) operator only. In the next section we shall construct solutions of (8.49) that satisfy the homogeneous form of the boundary condition as well. It is constructive to illustrate this here for the one-dimensional case.

Example 8.11 Consider the fundamental solution derived in Example 8.5. We now want to solve the boundary value problem

$$\mathcal{L}[w](x;\xi) := \frac{\mathrm{d}^2 w}{\mathrm{d}x^2}(x;\xi) = \delta(x-\xi), \qquad x \in (0,1),$$

$$w(0;\xi) = w(1;\xi) = 0.$$

Applying the boundary conditions to the general form in Example 8.5, we find

$$A_1 = 0, \qquad A_1 - \xi + B_1 + 1 = 0.$$

Therefore the fundamental solution is given by

$$w(x,\xi) = \begin{cases} x(\xi - 1) & \text{if } x \le \xi, \\ (x-1)\xi & \text{if } x > \xi. \end{cases}$$ □

8.5 Green's Functions; Superposition

Consider the Poisson problem (8.2). We can formally write the right-hand side $f(\boldsymbol{x})$ as

$$f(\boldsymbol{x}) = \int_\Omega \delta(\boldsymbol{x} - \boldsymbol{\xi}) f(\boldsymbol{\xi})\,\mathrm{d}V_{\boldsymbol{\xi}}, \qquad \boldsymbol{x} \in \Omega, \tag{8.54}$$

where the subscript $\boldsymbol{\xi}$ denotes integration with respect to $\boldsymbol{\xi}$. Using this representation of $f(\boldsymbol{x})$, we can derive a *particular solution* $u_\mathrm{p}(\boldsymbol{x})$ in the following way. We multiply (8.49) by $f(\boldsymbol{\xi})$ and integrate with respect to $\boldsymbol{\xi}$ over the domain Ω. This way, we find

$$u_\mathrm{p}(\boldsymbol{x}) = \int_\Omega w(\boldsymbol{x};\boldsymbol{\xi}) f(\boldsymbol{\xi})\,\mathrm{d}V_{\boldsymbol{\xi}}. \tag{8.55}$$

Inserting the fundamental solution (8.53), we obtain

$$u_\mathrm{p}(\boldsymbol{x}) = \begin{cases} \dfrac{1}{2\pi}\displaystyle\int_\Omega \ln \|\boldsymbol{x} - \boldsymbol{\xi}\|_2 f(\boldsymbol{\xi})\,\mathrm{d}V_{\boldsymbol{\xi}} & \text{if } d = 2, \\ -\dfrac{1}{4\pi}\displaystyle\int_\Omega \dfrac{1}{\|\boldsymbol{x} - \boldsymbol{\xi}\|_2} f(\boldsymbol{\xi})\,\mathrm{d}V_{\boldsymbol{\xi}} & \text{if } d = 3. \end{cases} \tag{8.56}$$

In general $u_\mathrm{p}(\boldsymbol{x})$ does not satisfy the prescribed boundary condition (see (8.6)). However, it is clear that $u(\boldsymbol{x}) - u_\mathrm{p}(\boldsymbol{x})$ satisfies the Laplace equation (8.1). Hence there exists a harmonic function $u_\mathrm{h}(\boldsymbol{x})$ such that

$$u(\boldsymbol{x}) = u_\mathrm{p}(\boldsymbol{x}) + u_\mathrm{h}(\boldsymbol{x}). \tag{8.57}$$

This is called *superposition*. The harmonic function $u_h(\boldsymbol{x})$ has to be determined such that $u(\boldsymbol{x})$ satisfies the prescribed boundary condition. Suppose we have a Dirichlet problem. Then it is obvious that $u_h(\boldsymbol{x})$ is the solution of the boundary value problem

$$\nabla^2 u_h = 0, \qquad \boldsymbol{x} \in \Omega, \tag{8.58a}$$

$$u_h(\boldsymbol{x}) = u(\boldsymbol{x}) - u_p(\boldsymbol{x}), \qquad \boldsymbol{x} \in \partial\Omega. \tag{8.58b}$$

In a similar way, we can derive a solution for a Neumann or a Robin boundary value problem.

Example 8.12 Consider a two-dimensional Dirichlet problem on the half-space $\Omega := \{(x, y) \in \mathbb{R}^2 \mid y > 0\}$ (Figure 8.2). The fundamental solution (8.53), and consequently also the particular solution (8.56), does not satisfy the homogeneous boundary condition on the line $y = 0$. In order to overcome this problem, we have to modify the fundamental solution. This can be done in the following way. Let $\boldsymbol{\xi} = (\xi, \eta)$ be an arbitrary point in Ω. Then we take a mirror point $\boldsymbol{\xi}^* := (\xi, -\eta)$ with respect to the line $y = 0$ and modify the fundamental solution as follows:

$$w(\boldsymbol{x}; \boldsymbol{\xi}) = \frac{1}{2\pi} \ln \|\boldsymbol{x} - \boldsymbol{\xi}\|_2 - \frac{1}{2\pi} \ln \|\boldsymbol{x} - \boldsymbol{\xi}^*\|_2 = \frac{1}{2\pi} \ln \left(\frac{\|\boldsymbol{x} - \boldsymbol{\xi}\|_2}{\|\boldsymbol{x} - \boldsymbol{\xi}^*\|_2} \right).$$

It is obvious that this fundamental solution satisfies the homogeneous boundary condition $w(\boldsymbol{x}; \boldsymbol{\xi}) = 0$ on the line $y = 0$. Moreover, since $\boldsymbol{\xi}^* \notin \Omega$, $w(\boldsymbol{x}; \boldsymbol{\xi})$ is also a solution of (8.49). The solution of the Dirichlet problem is given by

$$u(\boldsymbol{x}) = \frac{1}{2\pi} \int_\Omega \ln \left(\frac{\|\boldsymbol{x} - \boldsymbol{\xi}\|_2}{\|\boldsymbol{x} - \boldsymbol{\xi}^*\|_2} \right) f(\boldsymbol{\xi})\, \mathrm{d}V_{\boldsymbol{\xi}}. \qquad \square$$

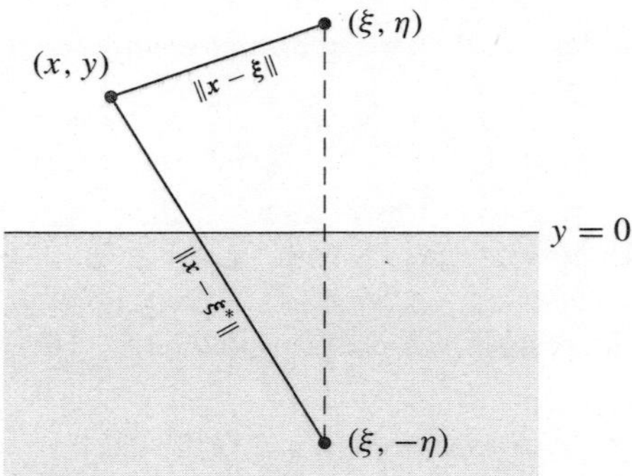

Figure 8.2. *A point* $\boldsymbol{\xi} \in \Omega$ *and its mirror point* $\boldsymbol{\xi}^*$.

Another way to use this superposition principle is to find a particular solution $u_p(\boldsymbol{x})$ satisfying the inhomogeneous equation with homogeneous boundary conditions and a harmonic function $u_h(\boldsymbol{x})$ satisfying the Laplace equation with the appropriate boundary conditions. This leads to the introduction of the *Green's function* $G(\boldsymbol{x}; \boldsymbol{\xi})$. The Green's function $G(\boldsymbol{x}; \boldsymbol{\xi})$ for the Dirichlet problem (8.2) and (8.6a) is by definition the solution of the boundary value problem

$$\nabla^2 G(\boldsymbol{x}; \boldsymbol{\xi}) = \delta(\boldsymbol{x} - \boldsymbol{\xi}), \qquad \boldsymbol{x} \in \Omega, \tag{8.59a}$$

$$G(\boldsymbol{x}; \boldsymbol{\xi}) = 0, \qquad \boldsymbol{x} \in \partial\Omega. \tag{8.59b}$$

Using superposition, we immediately see that the Green's function $G(\boldsymbol{x};\boldsymbol{\xi})$ equals the fundamental solution $w(\boldsymbol{x};\boldsymbol{\xi})$ (see Section 8.4), apart from a harmonic function. Using Theorem 8.4, we therefore conclude that the solution is unique. From the second identity of Green

$$\int_\Omega \left(u\nabla^2 v - v\nabla^2 u\right)\mathrm{d}V = \oint_{\partial\Omega}\left(u\frac{\partial v}{\partial \boldsymbol{n}} - v\frac{\partial u}{\partial \boldsymbol{n}}\right)\mathrm{d}S, \tag{8.60}$$

we can find a representation of $u(\boldsymbol{x})$ as announced. Indeed, if we take $v = G(\boldsymbol{x};\boldsymbol{\xi})$, then we obtain

$$u(\boldsymbol{\xi}) = \int_\Omega G(\boldsymbol{x};\boldsymbol{\xi}) f(\boldsymbol{x})\,\mathrm{d}V + \oint_{\partial\Omega} a(\boldsymbol{x})\frac{\partial G}{\partial \boldsymbol{n}}(\boldsymbol{x};\boldsymbol{\xi})\,\mathrm{d}S. \tag{8.61}$$

Also, for Neumann problems, we can construct a solution by superposition. First, we remark that the divergence theorem provides for a constraint relating the source term $f(\boldsymbol{x})$ and the boundary function $b(\boldsymbol{x})$ (see (8.6b)). Indeed, we have

$$\int_\Omega f(\boldsymbol{x})\,\mathrm{d}V = \int_\Omega \nabla^2 u\,\mathrm{d}V = \oint_{\partial\Omega}\frac{\partial u}{\partial \boldsymbol{n}}\,\mathrm{d}S = \oint_{\partial\Omega} b(\boldsymbol{x})\,\mathrm{d}S. \tag{8.62}$$

As a candidate for the Green's function $G(\boldsymbol{x};\boldsymbol{\xi})$ we might take the function satisfying (8.49) and the homogeneous Neumann boundary condition. However, this choice of $G(\boldsymbol{x};\boldsymbol{\xi})$ obviously does not satisfy the constraint (8.62). Another possibility is to define the Green's function as the solution of the boundary value problem

$$\nabla^2 G(\boldsymbol{x};\boldsymbol{\xi}) = \delta(\boldsymbol{x}-\boldsymbol{\xi}) - \frac{1}{\kappa}, \qquad \kappa := \int_\Omega \mathrm{d}V, \qquad \boldsymbol{x}\in\Omega, \tag{8.63a}$$

$$\frac{\partial G}{\partial \boldsymbol{n}}(\boldsymbol{x};\boldsymbol{\xi}) = 0, \qquad \boldsymbol{x}\in\partial\Omega, \tag{8.63b}$$

where the right-hand side of (8.63a) has been modified to enforce (8.62). We then obtain from (8.60) with $v = G(\boldsymbol{x};\boldsymbol{\xi})$ that

$$u(\boldsymbol{\xi}) = \int_\Omega G(\boldsymbol{x};\boldsymbol{\xi}) f(\boldsymbol{x})\,\mathrm{d}V + \frac{1}{\kappa}\int_\Omega u(\boldsymbol{x})\,\mathrm{d}V - \oint_{\partial\Omega} G(\boldsymbol{x};\boldsymbol{\xi}) b(\boldsymbol{\xi})\,\mathrm{d}S. \tag{8.64}$$

First, note that for this Green's function the constraint (8.62) is satisfied. Indeed, we have

$$\oint_{\partial\Omega}\frac{\partial G}{\partial \boldsymbol{n}}(\boldsymbol{x};\boldsymbol{\xi})\,\mathrm{d}S = \int_\Omega \nabla^2 G(\boldsymbol{x};\boldsymbol{\xi})\,\mathrm{d}V = 1 - \frac{1}{\kappa}\int_\Omega \mathrm{d}V = 0.$$

Second, we remark that the first integral in (8.64) represents the particular solution and the last one represents the harmonic function. The term $\frac{1}{\kappa}\int_\Omega u(\boldsymbol{x})\,\mathrm{d}V$ is just an additive constant.

Finally, for the boundary value problem with Robin boundary condition (8.6c), we define the Green's function $G(\boldsymbol{x};\boldsymbol{\xi})$ as solution of

$$\nabla^2 G(\boldsymbol{x};\boldsymbol{\xi}) = \delta(\boldsymbol{x}-\boldsymbol{\xi}), \qquad \boldsymbol{x}\in\Omega, \tag{8.65a}$$

$$\alpha G(\boldsymbol{x};\boldsymbol{\xi}) + \beta\frac{\partial G}{\partial \boldsymbol{n}}(\boldsymbol{x};\boldsymbol{\xi}) = 0, \qquad \boldsymbol{x}\in\partial\Omega. \tag{8.65b}$$

Analogously to the previous derivations, we obtain the following integral representation of $u(\boldsymbol{\xi})$:

$$u(\boldsymbol{\xi}) = \int_{\Omega} G(\boldsymbol{x};\boldsymbol{\xi}) f(\boldsymbol{x})\,\mathrm{d}V - \oint_{\partial\Omega} \left(\frac{\alpha}{\beta} u(\boldsymbol{x}) + \frac{\partial u}{\partial \boldsymbol{n}}(\boldsymbol{x})\right) G(\boldsymbol{x};\boldsymbol{\xi})\,\mathrm{d}S$$
$$= \int_{\Omega} G(\boldsymbol{x};\boldsymbol{\xi}) f(\boldsymbol{x})\,\mathrm{d}V - \frac{1}{\beta}\oint_{\partial\Omega} G(\boldsymbol{x};\boldsymbol{\xi}) c(\boldsymbol{x})\,\mathrm{d}S, \tag{8.66}$$

provided that $\beta \neq 0$.

A particular solution of (8.63a) can be found by subtracting the following solution $v_{\mathrm{p}}(\boldsymbol{x})$ (which is independent of $\boldsymbol{\xi}$) of the fundamental solution $w(\boldsymbol{x};\boldsymbol{\xi})$:

$$v_{\mathrm{p}}(\boldsymbol{x}) = \frac{1}{4\kappa}\|\boldsymbol{x}\|_2^2 \quad \text{if} \quad d = 2, \tag{8.67a}$$

$$v_{\mathrm{p}}(\boldsymbol{x}) = \frac{1}{6\kappa}\|\boldsymbol{x}\|_2^2 \quad \text{if} \quad d = 3. \tag{8.67b}$$

The Green's function $G(\boldsymbol{x};\boldsymbol{\xi})$ then differs from $w(\boldsymbol{x};\boldsymbol{\xi}) - v_{\mathrm{p}}(\boldsymbol{x})$ by a harmonic function, which can be found formally by solving a (homogeneous) Neumann problem with known boundary conditions.

Example 8.13 Consider the following Dirichlet boundary value problem on the half-space:

$$\nabla^2 u = f(\boldsymbol{x}), \qquad \boldsymbol{x} \in \Omega := \{(x,y) \in \mathbb{R}^2 \mid y > 0\},$$
$$u(x,0) = a(x), \quad u(x,y) \to 0 \quad \text{for} \quad y \to \infty,$$

where the boundary function $a(x) \to 0$ as $|x| \to \infty$. From Example 8.12 we derive for the Green's function with $\boldsymbol{\xi} = (\xi, \eta)$ that

$$G(x,y;\xi,\eta) = \frac{1}{2\pi}\ln\left(\frac{(x-\xi)^2 + (y-\eta)^2}{(x-\xi)^2 + (y+\eta)^2}\right)^{\frac{1}{2}}.$$

Hence on $\partial\Omega$ we find

$$\frac{\partial G}{\partial \boldsymbol{n}}(x,0;\xi,\eta) = \frac{1}{\pi}\frac{\eta}{(x-\xi)^2 + \eta^2}.$$

So, formally, the solution of the boundary value problem above is given by

$$u(\xi,\eta) = \frac{1}{2\pi}\int_0^{\infty}\int_{-\infty}^{\infty} \ln\left(\frac{(x-\xi)^2 + (y-\eta)^2}{(x-\xi)^2 + (y+\eta)^2}\right)^{\frac{1}{2}} f(x,y)\,\mathrm{d}x\mathrm{d}y$$
$$+ \frac{\eta}{\pi}\int_{-\infty}^{\infty} \frac{a(x)}{(x-\xi)^2 + \eta^2}\,\mathrm{d}x. \qquad \square$$

The last example shows the power and weakness of an analytical approach. On the one hand, by mere construction, one can show that a solution exists, and one can give estimates for it. On the other hand, the expressions one obtains are often complicated and usually not directly solvable in closed form.

We would like to point out another important fact. The expressions we have derived in this section for the solutions can be seen as inverting the operator form for the original problem; i.e., both the equation and the boundary condition. In particular, the Green's function can be interpreted as the ("constrained") inverse of the Laplace operator. In the next chapter we will investigate numerical methods for such problems, where this "inversion" will be met again, now in terms of matrices.

8.6 The Maximum Principle

The Laplace operator has a nice property that gives rise to a number of useful results. If the second derivative of a scalar function $u(x)$ is zero, it is linear. On a finite interval the absolute maximum is attained at either end of the interval. For a Poisson problem this is described in the following property.

Property 8.14 (maximum principle). *Let $\nabla^2 u(\boldsymbol{x}) = 0$ for all $\boldsymbol{x} \in \Omega$. Then $u(\boldsymbol{x})$ satisfies the inequalities*

$$m := \min_{\boldsymbol{\xi}\in\partial\Omega} u(\boldsymbol{\xi}) \leq u(\boldsymbol{x}) \leq \max_{\boldsymbol{\xi}\in\partial\Omega} u(\boldsymbol{\xi}) =: M. \tag{8.68}$$

Proof. Define the function $v_\varepsilon(\boldsymbol{x}) := u(\boldsymbol{x}) + \varepsilon \, \|\boldsymbol{x}\|_2^2$ for $\varepsilon > 0$. Clearly, we have

$$\nabla^2 v_\varepsilon(\boldsymbol{x}) = \nabla^2 u(\boldsymbol{x}) + 2d\varepsilon > 0, \qquad \boldsymbol{x} \in \Omega \quad (d = 2, 3). \tag{$*$}$$

Suppose $v_\varepsilon(\boldsymbol{x})$ has a maximum in the interior of Ω, say at $\boldsymbol{x}_0$. Then

$$\nabla^2 v_\varepsilon(\boldsymbol{x}_0) \leq 0.$$

Since this contradicts the inequality in $(*)$, we conclude that v_ε can only attain its maximum at the boundary $\partial\Omega$. If we denote the latter by M_ε, we derive the required upper bound by letting $\varepsilon \downarrow 0$. The lower bound follows from a similar argument, now using $-u(\boldsymbol{x})$ and $v_\varepsilon(\boldsymbol{x}) := -u(\boldsymbol{x}) + \varepsilon \, \|\boldsymbol{x}\|_2^2$ instead. □

The theorem above is sometimes also referred to as the minimum-maximum principle. In particular, the maximum principle can be extended to Poisson problems.

Property 8.15. *Let $\nabla^2 u(\boldsymbol{x}) \geq 0$ for all $\boldsymbol{x} \in \Omega$. Then $u(\boldsymbol{x})$ attains its maximum at the boundary; i.e.,*

$$u(\boldsymbol{x}) \leq \max_{\boldsymbol{x}\in\partial\Omega} u(\boldsymbol{\xi}). \tag{8.69}$$

Proof. It is easy to see that the arguments in the proof of Property 8.14 still apply for this case. □

The latter property is quite powerful in that it gives the possibility of comparing solutions of two Poisson problems. The following property states this more precisely.

Property 8.16 (comparison theorem). *Consider the two Poisson equations $\nabla^2 u_1(\boldsymbol{x}) = f_1(\boldsymbol{x})$ and $\nabla^2 u_2(\boldsymbol{x}) = f_2(\boldsymbol{x})$ with $f_1(\boldsymbol{x}) \geq f_2(\boldsymbol{x})$ for all $\boldsymbol{x} \in \Omega$. Then*

$$u_1(\boldsymbol{x}) \leq u_2(\boldsymbol{x}) + \max_{\boldsymbol{x}\in\partial\Omega} \big(u_1(\boldsymbol{x}) - u_2(\boldsymbol{x})\big). \tag{8.70}$$

Proof. Since $\nabla^2(u_1(\boldsymbol{x}) - u_2(\boldsymbol{x})) \geq 0$, the result follows directly from Property 8.15. □

Corollary 8.17. *A Dirichlet problem has a unique solution that depends continuously on the boundary data.*

Proof. If there are two solutions, then the difference is a harmonic function satisfying the homogeneous boundary condition. We can apply Property 8.14 to conclude that this difference must be zero. The continuous dependence is a consequence of the comparison theorem. □

8.7 The Stokes Equations

The problems met in practice are often more complex than the ones discussed thus far. A typical example of these are the Navier–Stokes equations in fluid dynamics and their creeping flow simplification, the Stokes equations. Here we shall not dwell on either their derivation or their application. This is done in Chapters 7 and 16. The first complication we meet is that a flow in one dimension is not of interest and thus a nontrivial formulation requires a Laplacian operating on a vector. Let $\boldsymbol{u}$ and p be a velocity and a pressure field. Then the Stokes equations for a domain $\Omega \subset \mathbb{R}^d$ $(d = 2, 3)$ read

$$\nabla^2 \boldsymbol{u} - \nabla p = 0, \tag{8.71a}$$

$$\nabla \cdot \boldsymbol{u} = 0. \tag{8.71b}$$

Note that $\nabla^2 \boldsymbol{u}$ is to be taken componentwise. From (8.71a) it is clear that ∇p, rather than p, is a (dependent) variable, so we always need to specify p somewhere in the domain. We shall moreover prescribe the Dirichlet boundary condition

$$\boldsymbol{u} \cdot \boldsymbol{n} = b(\boldsymbol{x}), \qquad \boldsymbol{x} \in \partial\Omega. \tag{8.72}$$

First, consider the case where $b(\boldsymbol{x}) \equiv 0$. We now want to show that the solution of boundary value problem (8.71) and (8.72) is unique. We shall need a bit of vector calculus to do this. First, we use the relation (see appendix J)

$$\nabla \cdot (p\boldsymbol{u}) = p \nabla \cdot \boldsymbol{u} + \boldsymbol{u} \cdot \nabla p. \tag{8.73}$$

Applying the divergence theorem to this relation and taking into account (8.71b) and (8.72), we find

$$\int_\Omega \boldsymbol{u} \cdot \nabla p \, \mathrm{d}V = \int_\Omega \nabla \cdot (p\boldsymbol{u}) \, \mathrm{d}V = \oint_{\partial\Omega} (p\boldsymbol{u}) \cdot \boldsymbol{n} \, \mathrm{d}S = \oint_{\partial\Omega} p b(\boldsymbol{x}) \, \mathrm{d}S. \tag{8.74}$$

Since $b(\boldsymbol{x}) \equiv 0$, we obtain

$$\int_\Omega \boldsymbol{u} \cdot \nabla p \, \mathrm{d}V = 0, \tag{8.75}$$

which states that the velocity field and the pressure gradient are orthogonal. Furthermore, we have the following relation for the velocity field $\boldsymbol{u} := (u, v, w)^T$:

$$\nabla \cdot (\boldsymbol{u} \cdot \nabla \boldsymbol{u}) = |\nabla \boldsymbol{u}|^2 + \boldsymbol{u} \cdot \nabla^2 \boldsymbol{u}, \tag{8.76}$$

where $\boldsymbol{u}\cdot\nabla\boldsymbol{u} := (\boldsymbol{u}\cdot\boldsymbol{u}_x, \boldsymbol{u}\cdot\boldsymbol{u}_y, \boldsymbol{u}\cdot\boldsymbol{u}_z)^T$ and $|\nabla\boldsymbol{u}|^2 := |\nabla u|^2 + |\nabla v|^2 + |\nabla w|^2$. Integrating this relation yields

$$\int_\Omega \nabla\cdot\big(\boldsymbol{u}\cdot\nabla\boldsymbol{u}\big)\,\mathrm{d}V = \int_\Omega |\nabla\boldsymbol{u}|^2\,\mathrm{d}V + \int_\Omega \boldsymbol{u}\cdot\nabla^2\boldsymbol{u}\,\mathrm{d}V. \tag{8.77}$$

The divergence theorem now implies for the left-hand side of (8.77) that

$$\int_\Omega \nabla\cdot\big(\boldsymbol{u}\cdot\nabla\boldsymbol{u}\big)\,\mathrm{d}V = \oint_{\partial\Omega} \big(\boldsymbol{u}\cdot\nabla\boldsymbol{u}\big)\cdot\boldsymbol{n}\,\mathrm{d}S = 0. \tag{8.78}$$

The zero value follows from (8.72). Let us now take the inner product of $\nabla^2\boldsymbol{u} - \nabla p$ and $\boldsymbol{u}$; this trivially gives

$$\int_\Omega \boldsymbol{u}\cdot\nabla^2\boldsymbol{u}\,\mathrm{d}V - \int_\Omega \boldsymbol{u}\cdot\nabla p\,\mathrm{d}V = 0. \tag{8.79}$$

The second term in (8.79) is zero on account of (8.75). So the first integral in (8.79) is zero as well. Combining (8.77), (8.78), and (8.79), we therefore conclude that

$$\int_\Omega |\nabla\boldsymbol{u}|^2\,\mathrm{d}V = 0. \tag{8.80}$$

Clearly $|\nabla\boldsymbol{u}|^2 = 0$ only if $\nabla\boldsymbol{u} = \boldsymbol{0}$, so $\boldsymbol{u}$ is constant. On account of the boundary condition (8.72), with $b(\boldsymbol{x}) \equiv 0$, we conclude that $\boldsymbol{u}(\boldsymbol{x}) \equiv 0$. It is trivial now to see that this implies uniqueness of the general problem (8.71), (8.72).

The linearity of the problem allows us to construct fundamental solutions $\boldsymbol{u}_i$, p_i called the *Stokeslets*; cf. Section 8.4. They are named after G.G. Stokes but were really first discovered by H.A. Lorentz in 1896; see [93]. In two dimensions they are the solution of

$$\nabla^2\boldsymbol{u}_i - \nabla p_i = \delta(\boldsymbol{x}-\boldsymbol{\xi})\boldsymbol{e}_i \qquad (i = 1, 2), \tag{8.81a}$$

$$\nabla\cdot\boldsymbol{u}_i = 0. \tag{8.81b}$$

Here $\boldsymbol{e}_1 = (1, 0)^T$ and $\boldsymbol{e}_2 = (0, 1)^T$. We have for $\boldsymbol{u}_i = (u_i^1, u_i^2)^T$ that

$$u_i^j(\boldsymbol{x};\boldsymbol{\xi}) = \frac{1}{4\pi}\left(\delta_{j,i}\ln\|\boldsymbol{x}-\boldsymbol{\xi}\|_2 - \frac{(x_j-\xi_j)(x_i-\xi_i)}{\|\boldsymbol{x}-\boldsymbol{\xi}\|_2^2}\right) \qquad (i, j = 1, 2), \tag{8.82a}$$

$$p_i(\boldsymbol{x};\boldsymbol{\xi}) = -\frac{x_i-\xi_i}{2\pi\|\boldsymbol{x}-\boldsymbol{\xi}\|_2^2}. \tag{8.82b}$$

These Stokeslets can be used to derive an integral formulation for the Stokes equations.

We conclude this section with an example of one of the few Stokes problems that can be solved analytically. It regards the flow past a sphere of radius one. We assume that the velocity $\boldsymbol{u}$ is equal to zero at the sphere and equal to one in the direction of the flow (say the z direction) at infinity. The pressure p approaches a limit value there. This example will give us an opportunity to show some more vector calculus. To start with we shall use the spherical coordinates (r, θ, ϕ), and we obtain, with $\boldsymbol{u} = (u_r, u_\theta, u_\phi)^T$, that

$$\nabla\cdot\boldsymbol{u} = \frac{1}{r^2}\frac{\partial}{\partial r}\big(r^2u_r\big) + \frac{1}{r\sin\theta}\frac{\partial}{\partial\theta}\big(\sin\theta\, u_\theta\big) + \frac{1}{r\sin\theta}\frac{\partial}{\partial\phi}u_\phi = 0. \tag{8.83}$$

We assume axial symmetry, implying that $\frac{\partial}{\partial\phi} = 0$ and $u_\phi = 0$. The construction of the solution now employs the notions of a stream function, $\psi = \psi(r,\theta)$ say. In particular, we require

$$u_r = \frac{1}{r^2 \sin\theta}\frac{\partial\psi}{\partial\theta}, \qquad u_\theta = -\frac{1}{r\sin\theta}\frac{\partial\psi}{\partial r}. \tag{8.84}$$

Substituting these relations into (8.83) (with $u_\phi = 0$) gives

$$\frac{1}{r^2\sin\theta}\frac{\partial^2\psi}{\partial r\partial\theta} - \frac{1}{r^2\sin\theta}\frac{\partial^2\psi}{\partial\theta\partial r} = 0.$$

So we have $\nabla\cdot\boldsymbol{u} = 0$, indeed. The relations in (8.84) imply that the velocity can be written as

$$\boldsymbol{u} = \nabla\times\widetilde{\boldsymbol{\psi}}, \qquad \widetilde{\boldsymbol{\psi}} := \frac{\psi}{r\sin\theta}\boldsymbol{e}_\phi. \tag{8.85}$$

Next it is convenient to introduce the vorticity vector $\boldsymbol{\omega} = \nabla\times\boldsymbol{u}$. Obviously, $\widetilde{\boldsymbol{\psi}}$ and $\boldsymbol{\omega}$ are related by

$$\nabla\times\left(\nabla\times\widetilde{\boldsymbol{\psi}}\right) = \boldsymbol{\omega}. \tag{8.86}$$

Applying the curl operator twice to the vector $\widetilde{\boldsymbol{\psi}}$ defined in (8.85) results in a vector in the same direction. More precisely, we have

$$\boldsymbol{\omega} = \frac{-1}{r\sin\theta}\mathcal{D}[\psi]\boldsymbol{e}_\phi, \tag{8.87}$$

where the differential operator $\mathcal{D}$ is defined by

$$\mathcal{D}[\psi] := \frac{\partial^2\psi}{\partial r^2} + \frac{1}{r^2}\frac{\partial^2\psi}{\partial\theta^2} - \frac{\cot\theta}{r^2}\frac{\partial\psi}{\partial\theta}. \tag{8.88}$$

Using the relation $\nabla^2\boldsymbol{u} = \nabla(\nabla\cdot\boldsymbol{u}) - \nabla\times\boldsymbol{\omega}$ (see the appendix, Section J), we can rewrite (8.71a) as

$$\nabla\times\boldsymbol{\omega} = -\nabla p. \tag{8.89}$$

In order to eliminate the pressure gradient, we apply the curl operator to (8.89) and find

$$\nabla\times\left(\nabla\times\boldsymbol{\omega}\right) = \boldsymbol{0}. \tag{8.90}$$

Analogously to the derivation of (8.87), we also find

$$\nabla\times\left(\nabla\times\boldsymbol{\omega}\right) = \frac{1}{r\sin\theta}\mathcal{D}^2[\psi]\boldsymbol{e}_\phi. \tag{8.91}$$

Hence we conclude from (8.90) that

$$\mathcal{D}^2[\psi] = 0. \tag{8.92}$$

At the sphere we have $\boldsymbol{u} = \boldsymbol{0}$, so we obtain as boundary conditions for ψ that

$$\frac{\partial\psi}{\partial r} = \frac{\partial\psi}{\partial\theta} = 0 \qquad \text{at } r = 1, \tag{8.93}$$

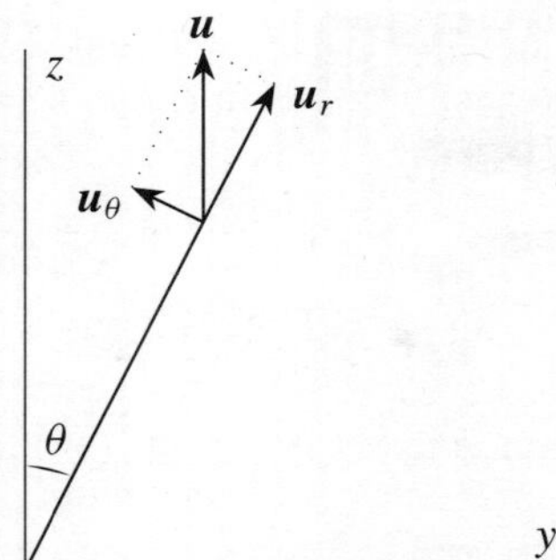

Figure 8.3. *Asymptotic boundary condition for the velocity* **u**.

so apparently $\psi(r,\theta) =$ constant at $r = 1$ and therefore we may as well take $\psi(1,\theta) = 0$. As for the asymptotic boundary condition, we see from Figure 8.3 that $u_r \sim \cos\theta$ and $u_\theta \sim -\sin\theta$. Using this in (8.84) gives $\frac{\partial\psi}{\partial\theta} \sim r^2 \sin\theta\cos\theta$, so

$$\psi(r,\theta) \sim \frac{1}{2}r^2\sin^2\theta \quad \text{for } r \to \infty. \tag{8.94}$$

Note that an integration constant is immaterial.

Returning now to (8.92), we apply separation of variables, i.e.,

$$\psi(r,\theta) = f(r)\,g(\theta). \tag{8.95}$$

Actually, with some further (tedious) analysis it turns out that $g(\theta)$ may be taken equal to $\sin^2\theta$. Straightforward computation then reveals

$$\mathcal{D}[\psi] = \left(\frac{\mathrm{d}^2 f}{\mathrm{d}r^2} - \frac{2f}{r^2}\right)\sin^2\theta. \tag{8.96}$$

Once more applying the rule (8.96) we find

$$\mathcal{D}^2[\psi] = \left(\frac{\mathrm{d}^4 f}{\mathrm{d}r^4} - \frac{4}{r^2}\frac{\mathrm{d}^2 f}{\mathrm{d}r^2} + \frac{8}{r^3}\frac{\mathrm{d}f}{\mathrm{d}r} - \frac{8}{r^4}f\right)\sin^2\theta = 0. \tag{8.97}$$

We have finally arrived at an ODE whose general solution is

$$f(r) = \alpha r^4 + \beta r^2 + \gamma r + \frac{\delta}{r}. \tag{8.98}$$

Using the condition at infinity (8.94), we find $\alpha = 0$, $\beta = \frac{1}{2}$, and using the boundary condition at the sphere (8.93), we then find $\gamma = -\frac{3}{4}$, $\delta = \frac{1}{4}$. We thus find for the stream function

$$\psi(r,\theta) = \left(\frac{1}{2}r^2 - \frac{3}{4}r + \frac{1}{4r}\right)\sin^2\theta. \tag{8.99}$$

The actual sought flow field $\boldsymbol{u}$ is then given by

$$u_r = \left(1 - \frac{3}{2r} + \frac{1}{2r^3}\right)\cos\theta, \qquad u_\theta = \left(-1 + \frac{3}{4r} + \frac{1}{4r^3}\right)\sin\theta, \tag{8.100}$$

while for the pressure p we obtain

$$p = -\frac{3\cos\theta}{2r^2} + p_0, \tag{8.101}$$

with p_0 the constant pressure at infinity.

8.8 Discussion

- Problems of elliptic type are probably the most common. At the least, elliptic operators appear in a large number of problems in mathematical physics. As we saw in Chapter 2, we encounter the Laplace operator in parabolic problems as well as hyperbolic problems. A hyperbolic problem like the wave equation reduces to an elliptic problem when we consider only time-harmonic solutions. When the equation is of parabolic type, elliptic problems appear as steady state problems, letting the time go to infinity. Thus we may consider situations like the temperature distribution in a room with heat sources and sinks or the concentration in a vessel that is draining with constant replenishing [30, 31]. The name "potential problem" comes from electrostatics. It refers to the potential from electrical charges, following from the Maxwell equations. In fact, the Maxwell equations are a good example of a hyperbolic problem [67].

- In fluid mechanics the velocity potential and stream function of an incompressible irrotational flow satisfy an elliptic equation [15]. Typical mechanical problems involve the deflection of a thin membrane in two dimensions. For deflection of beams one encounters the *biharmonic equation*. This is still elliptic but contains the ∇^4 operator. There are many similarities with harmonic analysis [44].

- In Chapter 16 we shall encounter a number of elliptic problems arising in a practical context. To start with, in Section 16.6 the groundwater table is modeled as a nonlinear boundary value problem in one dimension. This nonlinearity arises in the second derivative term, which contains a small coefficient. It is shown when and how an asymptotic expansion can "solve" this problem. Section 16.8 deals with chemical reactions in small pellets. The problem is described as the steady state of a reaction-diffusion equation of a spherically symmetric problem. Here too, solutions can be obtained in terms of an asymptotic expansion. An example of a viscous flow problem leading to the Stokes equation is the forming of glass products; see Section 16.11. The problem discussed here can be solved analytically using thin-layer approximations.

Exercises

8.1. In this exercise we determine the spherically symmetric fundamental solution $w(\boldsymbol{x})$ of the d-dimensional Poisson equation, i.e., $w(\boldsymbol{x}) = \bar{w}(r)$ with $r = \|\boldsymbol{x}\|_2$.

(a) Show that $\bar{w}(r)$ is a solution of the ODE

$$\frac{\mathrm{d}^2\bar{w}}{\mathrm{d}r^2} + \frac{d-1}{r}\frac{\mathrm{d}\bar{w}}{\mathrm{d}r} = \delta(r).$$

(b) For $d > 2$ and the requirement that $\lim_{r\to\infty} \bar{w}(r) = 0$, show that

$$\bar{w}(r) = \frac{A_d}{r^{d-2}},$$

where the integration constants A_d satisfy

$$A_d \int_{\mathcal{S}(\mathbf{0};\rho)} \frac{\mathrm{d}}{\mathrm{d}r}\left(r^{2-d}\right) \mathrm{d}S = 1,$$

with $\mathcal{S}(\mathbf{0};\rho)$ the sphere with centre $\mathbf{0}$ and radius ρ. One can prove that the constants A_d can be found from

$$A_d = \frac{1}{(2-d)\, B_d},$$

where B_d is the surface area of the d-dimensional unit sphere $\mathcal{S}(\mathbf{0};1)$. Note that $B_d = 2\pi^{d/2}\Gamma(d/2)$.

8.2. Let $\nabla^2 u(\boldsymbol{x}) = 0$ for all $\boldsymbol{x} \in \Omega$ and $u(\boldsymbol{x}) > 0$ for all $\boldsymbol{x} \in \partial\Omega$. Prove that $u(\boldsymbol{x}) > 0$ for all $\boldsymbol{x} \in \Omega$.

8.3. Consider the Neumann problem

$$\begin{aligned} \nabla^2 u &= c, && \boldsymbol{x} \in \Omega := \left\{(x, y) \in \mathbb{R}^2 \mid x^2 + y^2 < 1\right\}, \\ \frac{\partial u}{\partial \boldsymbol{n}} &= 1, && \boldsymbol{x} \in \delta\Omega, \end{aligned}$$

where c is a constant. Show that this boundary value problem only has solutions for $c = 2$. Determine all of these solutions.

8.4. Let $\nabla^2 u_i(\boldsymbol{x}) = 0$ $(i = 1, 2, \ldots, N)$ for all $\boldsymbol{x} \in \Omega$. Let $u_{i-1}(\boldsymbol{x}) \le u_i(\boldsymbol{x}) \le u_{i+1}(\boldsymbol{x})$ $(i = 2, 3, \ldots, N-1)$ for all $\boldsymbol{x} \in \partial\Omega$. Show that $u_{i-1}(\boldsymbol{x}) \le u_i(\boldsymbol{x}) \le u_{i+1}(\boldsymbol{x})$ $(i = 2, 3, \ldots, N-1)$ for all $\boldsymbol{x} \in \Omega$.

8.5. Let $\nabla^2 u(\boldsymbol{x}) = 0$ for all $\boldsymbol{x} \in \Omega$. Split the boundary $\partial\Omega$ into two simply connected parts, $\partial\Omega_1$ and $\partial\Omega_2$, such that $\partial\Omega_1 \cup \partial\Omega_2 = \partial\Omega$. Let $u(\boldsymbol{x})$ satisfy the boundary conditions

$$\begin{aligned} u(\boldsymbol{x}) &= \alpha(\boldsymbol{x}), && \boldsymbol{x} \in \partial\Omega_1, \\ \frac{\partial u}{\partial \boldsymbol{n}}(\boldsymbol{x}) &= \beta(\boldsymbol{x}), && \boldsymbol{x} \in \partial\Omega_2. \end{aligned}$$

Show that the solution $u(\boldsymbol{x})$ is unique.

8.6. Consider the domain $\Omega := \{(x, y) \in \mathbb{R}^2 \mid x > 0,\ y > 0\}$. Construct a Green's function for the Dirichlet and Neumann problem using mirror points.

8.7. Let $a, b \in \mathbb{R}$ with $a < b$. Define the domain $\Omega := \{(x, y) \in \mathbb{R}^2 \mid a < y < b\}$. Construct a Green's function for the Dirichlet problem on Ω using mirror points. Show that the Green's function is symmetric; i.e.,

$$G(\boldsymbol{x}; \boldsymbol{\xi}) = G(\boldsymbol{\xi}; \boldsymbol{x}).$$

8.8. Consider on the unit disc $\Omega := \{(r, \phi) \in \mathbb{R}^2 \mid 0 \le r < 1, -\pi \le \phi < \pi\}$ the Neumann problem

$$\nabla^2 u = \frac{1}{r}\frac{\partial}{\partial r}\left(r\frac{\partial u}{\partial r}\right) + \frac{1}{r^2}\frac{\partial^2 u}{\partial \phi^2} = 0$$

satisfying the boundary condition

$$\frac{\partial u}{\partial \boldsymbol{n}}(1, \phi) = g(\phi),$$

where the function $g(\phi)$ satisfies the constraint

$$\int_{-\pi}^{\pi} g(\phi)\,\mathrm{d}\phi = 0.$$

Find a formal solution.

8.9. Prove the *mean value theorem*: If $u(\boldsymbol{x})$ is harmonic in $\Omega \subset \mathbb{R}^3$ and $\mathcal{S}(\boldsymbol{x}; \rho)$ denotes the sphere in Ω with centre $\boldsymbol{x}$ and radius ρ, then

$$u(\boldsymbol{x}) = \frac{1}{4\pi\rho^2}\int_{\mathcal{S}(\boldsymbol{x};\rho)} u(\boldsymbol{y})\,\mathrm{d}S_y,$$

where the subscript $\boldsymbol{y}$ denotes that the integration is carried out in the $\boldsymbol{y}$ variables.

8.10. Consider on the strip $\Omega := \{(x, y) \in \mathbb{R}^2 \mid 0 < x < 1,\ 0 < y\}$ the Dirichlet problem

$$\begin{aligned} \nabla^2 u &= 0, & \boldsymbol{x} &\in \Omega, \\ u(0, y) &= u(1, y) = 0, & 0 &\le y < \infty, \\ u(x, 0) &= x(1 - x), & 0 &\le x \le 1, \\ u &\text{ is bounded in } \Omega. \end{aligned}$$

Show by separation of variables that the solution is given by

$$u(x, y) = \frac{8}{\pi^3}\sum_{k=0}^{\infty}\frac{1}{(2k+1)^3}\,\mathrm{e}^{-(2k+1)\pi y}\sin(2k+1)\pi x.$$

8.11. The Helmholtz equation may not allow for the maximum principle. Prove this for the one-dimensional case. Consider the equation

$$\frac{\mathrm{d}^2 u}{\mathrm{d}x^2} + \lambda u = 0, \qquad x \in \Omega := (0, \pi),$$

where $\lambda > 1$.

8.12. The maximum principle may also apply to nonlinear problems. For this consider the boundary value problem

$$\begin{aligned} \nabla^2 u &= \lambda\, \mathrm{e}^u, & \boldsymbol{x} &\in \Omega, \\ u &= 0, & \boldsymbol{x} &\in \partial\Omega. \end{aligned}$$

(a) Prove that $u(\boldsymbol{x}) > 0$ for all $\boldsymbol{x} \in \Omega$ if $\lambda > 0$.

(b) Construct a comparison function to estimate $u(\boldsymbol{x})$.

8.13. If we apply an affine coordinate transformation (corresponding to rigid body motions), the Laplace operator remains invariant. Prove this for $\boldsymbol{x} \in \mathbb{R}^2$.

8.14. Consider the Helmholtz equation on a unit square, i.e.,

$$\begin{aligned} \nabla^2 u + ku &= 0, & \boldsymbol{x} &\in \Omega := \left\{(x, y) \in \mathbb{R}^2 \,\middle|\, 0 < x,\ y < 1\right\}, \\ u &= 0, & \boldsymbol{x} &\in \partial\Omega, \end{aligned}$$

where $k < 0$. Compute the eigenvalues and eigenfunctions.

8.15. Show that the operator

$$\mathcal{L}[u] := p(x)\frac{\mathrm{d}^2 u}{\mathrm{d}x^2} + q(x)\frac{\mathrm{d}u}{\mathrm{d}x}$$

is not self-adjoint unless $q(x) = \frac{\mathrm{d}}{\mathrm{d}x} p(x)$.

8.16. Consider the eigenvalue problem

$$\begin{aligned} \mathcal{L}[u] := \frac{\mathrm{d}^2 u}{\mathrm{d}x^2} &= \lambda u, & x &\in (0, 1), \\ u(0) &= 0, & \frac{\mathrm{d}u}{\mathrm{d}x}(1) &= 0. \end{aligned}$$

Determine the eigenvalues and eigenfunctions.

8.17. Consider the eigenvalue problem

$$\begin{aligned} \mathcal{L}[u] := \frac{\mathrm{d}^2 u}{\mathrm{d}x^2} &= \lambda u, & x &\in (0, 1), \\ u(0) - \frac{\mathrm{d}u}{\mathrm{d}x}(0) &= 0, & u(1) &= 0. \end{aligned}$$

Show that the eigenvalues are negative.

8.18. Consider the equation

$$\mathcal{L}[u] := \nabla^2 u + qu = \lambda u, \qquad \boldsymbol{x} \in \Omega,$$

where $q < 0$. Show that the eigenvalues λ_k are nonpositive if we impose homogeneous Dirichlet or Neumann boundary conditions. Also show that $\lambda_k \leq 0$ if we impose the homogeneous Robin boundary condition (8.6c) with $\mathrm{sign}(\alpha) = \mathrm{sign}(\beta)$.

8.19. Show that for the operator

$$\mathcal{L}[u] := \nabla^2 u + a(\boldsymbol{x})\frac{\partial u}{\partial x} + b(\boldsymbol{x})\frac{\partial u}{\partial y}$$

the adjoint operator is given by

$$\mathcal{L}^*[v] = \nabla^2 v - \frac{\partial}{\partial x}\big(a(\boldsymbol{x})v\big) - \frac{\partial}{\partial y}\big(b(\boldsymbol{x})v\big).$$

Chapter 9

Numerical Methods for Elliptic Equations

This chapter is devoted to the numerical solution of elliptic equations. In Section 9.1 we recall how the difference methods that were derived in Chapter 5 are used. Moreover, the discretisation of derivative boundary conditions is described. In Section 9.2 the maximum principle for discrete problems is discussed. As it turns out, there are many analogies with the continuous case. The discrete maximum principle is then applied in Section 9.3 to estimate the global discretisation error. As an alternative, the global discretisation error is estimated using the matrix method. Both methods enable one to assess the global effect of discretisation errors that are exclusively made at the boundary. In Section 9.4 some basic solution methods for linear systems resulting from discretising an elliptic equation are discussed. In particular, Gauss-type iterative methods are analyzed. These iterative methods are evolutionary in character and can therefore be interpreted in terms of a parabolic equation. This latter equation is called the closure of the iterative method. The discrete analogue of Green's functions is presented in Section 9.5. This provides a method of explicitly computing the inverse of the discretisation matrix. Nonlinear problems are dealt with in Section 9.6 and the Newton and Gauss–Jacobi iteration and transient methods of solution are discussed. Finally, in Section 9.7 we consider the numerical solution of the Stokes equations, in particular a special iteration method called pressure correction.

9.1 Discretisation and Boundary Conditions

In this chapter we shall restrict ourselves to problems with at most two independent variables. First, let us consider the linear equation

$$a\frac{\partial^2 u}{\partial x^2} + b\frac{\partial^2 u}{\partial y^2} + c\frac{\partial u}{\partial x} + d\frac{\partial u}{\partial y} + eu = f, \qquad (x, y) \in \Omega \subset \mathbb{R}^2, \quad a, b > 0, \ e \leq 0, \tag{9.1}$$

where the coefficients a, b, c, d, and e are constant. In Chapter 5 we introduced second order central difference approximations for both first and second order derivatives. Applying these central differences to (9.1) on a uniform grid with grid sizes Δx and Δy in the x and

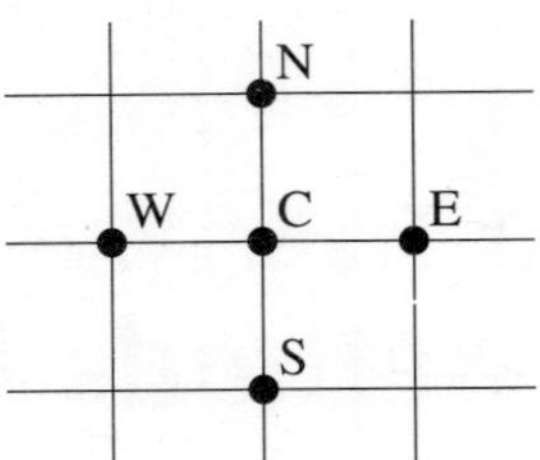

Figure 9.1. *Five-point stencil for the difference equation* (9.2).

y directions respectively, we obtain the difference scheme (cf. (5.33))

$$\begin{aligned}\frac{1}{\Delta x^2}\left(a+\frac{1}{2}c\Delta x\right)u_{\mathrm{E}}+\frac{1}{\Delta x^2}\left(a-\frac{1}{2}c\Delta x\right)u_{\mathrm{W}}+\frac{1}{\Delta y^2}\left(b+\frac{1}{2}d\Delta y\right)u_{\mathrm{N}}\\ +\frac{1}{\Delta y^2}\left(b-\frac{1}{2}d\Delta y\right)u_{\mathrm{S}}+\left(e-\frac{2a}{\Delta x^2}-\frac{2b}{\Delta y^2}\right)u_{\mathrm{C}}=f(\boldsymbol{x}_{\mathrm{C}}),\end{aligned} \tag{9.2}$$

where u_{C} denotes the numerical approximation of $u(\boldsymbol{x}_{\mathrm{C}})$, etc. The stencil of (9.2) is depicted in Figure 9.1. The difference equation (9.2) can also formally be written as

$$\alpha_0 u_{\mathrm{C}}+\alpha_1 u_{\mathrm{N}}+\alpha_2 u_{\mathrm{E}}+\alpha_3 u_{\mathrm{S}}+\alpha_4 u_{\mathrm{W}}=f(\boldsymbol{x}_{\mathrm{C}}). \tag{9.3}$$

From our assumptions we immediately see that the following property holds.

Property 9.1. *The coefficients in the difference scheme* (9.3) *satisfy the inequalities*

$$\alpha_0<0, \tag{9.4a}$$

$$\alpha_0+\sum_{i=1}^{4}\alpha_i\le 0, \tag{9.4b}$$

$$\alpha_i>0 \quad (i=1,2,3,4) \quad \textit{if } \Delta x \textit{ and } \Delta y \textit{ are small enough.} \tag{9.4c}$$

The grid points considered thus far are interior grid points. If we have a Dirichlet boundary condition, we can simply substitute the given data into the scheme to eliminate any one of the points $\boldsymbol{x}_{\mathrm{E}}, \boldsymbol{x}_{\mathrm{S}}, \boldsymbol{x}_{\mathrm{W}}$ or $\boldsymbol{x}_{\mathrm{N}}$ or (at corners) two of them. In Section 5.2.2 we considered how to deal with curved boundaries. Below we shall deal with straight lines being the edges of a rectangular domain. In particular, let us consider the treatment of a Robin boundary condition, which automatically includes the case of Neumann boundary conditions. Consider a situation like in Figure 9.2. Suppose that along a left vertical boundary we have

$$\frac{\partial u}{\partial x}=qu+g(y). \tag{9.5}$$

A straightforward, though only first order, discrete approximation of (9.5) would be the forward difference

$$\frac{1}{\Delta x}(u_{\mathrm{E}}-u_{\mathrm{C}})=qu_{\mathrm{C}}+g(y_{\mathrm{C}}). \tag{9.6}$$

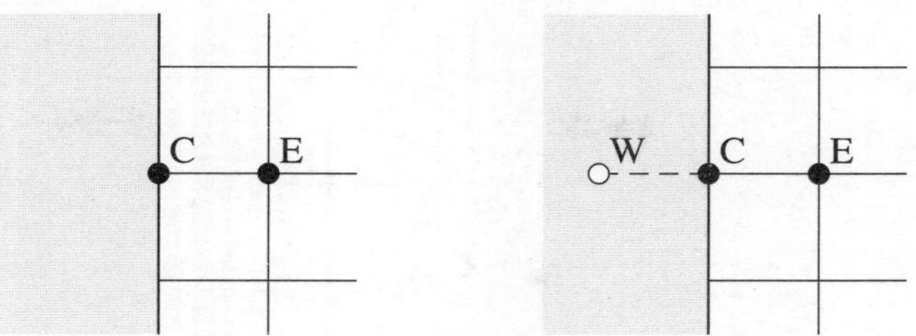

Figure 9.2. *Stencil for the Robin boundary condition* (9.5) *without (left) and with (right) a virtual grid point W.*

This then gives

$$u_{\text{C}} = \frac{u_{\text{E}} - \Delta x g(y_{\text{C}})}{1 + q\Delta x}. \tag{9.7}$$

Using this value for a boundary point $\boldsymbol{x}_{\text{C}}$ differs from using a Dirichlet boundary value in two ways. First, it changes the scheme, as the coefficient of u_{E} appearing in the difference equation at $\boldsymbol{x}_{\text{E}}$ is clearly updated with a value $\alpha/(1 + q\Delta x)$, where α is the coefficient of u_{C}. Second, it also affects the *local discretisation error* of such a point. Indeed, since (9.6) leaves an error of $\mathcal{O}(\Delta x)$, using u_{C} computed from (9.7) induces an additional error of $\mathcal{O}(\Delta x^2)$ in any scheme where it is substituted. In (9.2) this therefore introduces local errors of $\mathcal{O}(1)$. A better way is to employ *virtual grid points*. For example, if we use an extra grid point $\boldsymbol{x}_{\text{W}}$ like in Figure 9.2 and apply central differences to (9.5), we obtain

$$\frac{1}{2\Delta x}(u_{\text{E}} - u_{\text{W}}) = qu_{\text{C}} + g(y_{\text{C}}). \tag{9.8}$$

This leads to

$$u_{\text{W}} = u_{\text{E}} - 2\Delta x(qu_{\text{C}} + g(y_{\text{C}})). \tag{9.9}$$

Substitution into (9.2) then gives a stencil containing grid points within the domain only. Since (9.8) is second order, this then induces an extra local error of $\mathcal{O}(\Delta x)$. This order is still less than desired. However, we shall show in Example 9.14 that this does not affect the error globally.

Example 9.2 An alternative method defines a grid such that the boundary is at the middle of the first interval. We can then employ a *box-scheme* approximation of (9.5); i.e.,

$$\frac{1}{\Delta x}(u_{\text{E}} - u_{\text{C}}) = \frac{1}{2}q(u_{\text{C}} + u_{\text{E}}) + g(y_{\text{C}}).$$

From this relation we can eliminate u_{C} and find

$$u_{\text{C}} = \frac{(1 - \frac{1}{2}q\Delta x)u_{\text{E}} - \Delta x g(y_{\text{C}})}{1 + \frac{1}{2}q\Delta x}. \qquad \square$$

9.2 The Maximum Principle

In the previous section we constructed the central difference scheme, which may be written as a discrete operator $\mathcal{L}_\Delta$ acting on a grid function u_Δ at a generic point of a grid Ω_Δ

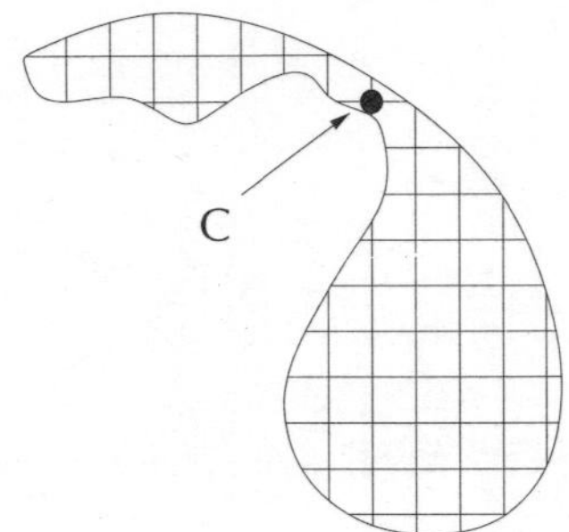

Figure 9.3. *An unconnected grid; grid point $\boldsymbol{x}_C$ is not connected to any other grid point.*

covering the domain. So for an interior grid point $\boldsymbol{x}_C$ we have

$$\mathcal{L}_\Delta[u_C] = f_C, \tag{9.10}$$

where $u_C = u_\Delta(\boldsymbol{x}_C)$ is the numerical approximation of $u(\boldsymbol{x}_C)$ and f_C is the source term, including contributions from the boundary conditions when applicable. If $\boldsymbol{x}_C$ is a *genuine interior grid point*, i.e., it has no neighbouring grid points on the boundary, then $f_C = f(\boldsymbol{x}_C)$. For the theory that follows it is important that such a point be a generic interior point. We have illustrated this in Figure 9.3. Apparently, the grid point $\boldsymbol{x}_C$ is not connected to any of its neighbours, so the total set of difference equations is uncoupled. We now require that any two interior grid points can be connected by a curve that follows grid lines in the interior of the domain. Such a grid will be called *connected.* One should note that a grid can always be assumed connected for Δx, Δy small enough. We remark that an interior grid point corresponds to an unknown value. Also, a grid point on that part of the boundary where either a Neumann or a Robin condition is specified contains an unknown value. For such points we proceed as follows. Consider, e.g., the discrete Robin condition (9.6), which we rewrite as

$$\frac{1}{\Delta x}\big(0u_W - (1 + q\Delta x)u_C + u_E\big) = g(y_C). \tag{9.11}$$

Then we can view $\boldsymbol{x}_C$ as an interior grid point. Note that $\boldsymbol{x}_W$ is a virtual point that does not play a role as such; neither does the value u_W in the difference equation. By doing this for all such points, we see that we can interpret our discrete problem as one with Dirichlet boundary conditions exclusively; the actual solution value at such points may be taken as zero so that the points correspond to (virtual) homogeneous Dirichlet boundary conditions. In more complex situations the difference equation, arising after having eliminated the virtual point(s) from a discrete Robin or Neumann boundary condition, can be treated likewise: the values at virtual points are incorporated in the difference equation with a zero coefficient. In what follows we need at least one real Dirichlet or Robin boundary point. This corresponds of course to the well-posedness of the continuous problem, which requires Robin or Dirichlet data, at least on part of the boundary. So for a two-dimensional problem with central differences and with $\boldsymbol{x}_C$ as any interior point in the general sense as described above, we write the difference equation as

$$\mathcal{L}_\Delta[u_C] := \alpha_0 u_C + \sum_{i=1}^{4} \alpha_i u_{P_i}. \tag{9.12}$$

Here we have written $u_{P_1}, \ldots, u_{P_4}$ for u_N, u_E, u_S, and u_W in a more generic way. Note that the coefficients in (9.12) coincide with those in (9.3) if $\boldsymbol{x}_C$ is a genuine interior point. We first deal with the case where we have a strict equality in (9.4b), i.e., $\alpha_0 + \sum_{i=1}^{4} \alpha_i = 0$ (which is, e.g., the case for a Dirichlet problem on a rectangle with a uniform grid).

Theorem 9.3. *Let the grid Ω_Δ be connected and let $\mathcal{L}_\Delta[u_C] = 0$ for all interior grid points $\boldsymbol{x}_C$, where $\mathcal{L}_\Delta$ is defined in (9.12). Moreover, the coefficients $\alpha_0, \ldots, \alpha_4$ satisfy the conditions*

$$\alpha_0 + \sum_{i=1}^{4} \alpha_i = 0, \qquad \alpha_i > 0 \qquad (i = 1, 2, 3, 4).$$

Then for any grid point $\boldsymbol{x}_C$ we have

$$m := \min_{\boldsymbol{x}_P \in \partial\Omega_\Delta} u_P \le u_C \le \max_{\boldsymbol{x}_P \in \partial\Omega_\Delta} u_P =: M,$$

where $\partial\Omega_\Delta$ denotes the set of grid points on the boundary.

Proof. We prove this by contradiction. Suppose that $u_C > M$ at some interior point $\boldsymbol{x}_C$. Since $\alpha_0 \neq 0$ (because $\alpha_0 < 0$), we find from (9.3) that

$$u_C = -\sum_{i=1}^{4} \frac{\alpha_i}{\alpha_0} u_{P_i};$$

i.e., u_C is a weighted average of its neighbouring values u_{P_i}. This equality only makes sense if $u_{P_i} = u_C$ for all i. Repeating this argument, we conclude that all interior points must have a value equal to u_C. At least one is connected to a boundary point where the value M is taken. This leads to a contradiction. A similar argument can be applied to see that $u_C \ge m$ at all interior grid points. The case that $\boldsymbol{x}_C$ is a boundary point is trivial. □

Property 9.4 (maximum principle). *Let the grid Ω_Δ be connected and let $\mathcal{L}_\Delta[u_C] \ge 0$ for all interior grid points. Then for any interior grid point $\boldsymbol{x}_C$ we have*

$$u_C \le M := \max_{\boldsymbol{x}_P \in \partial\Omega_\Delta} u_P.$$

Proof. Suppose that $u_C > M$ for some interior grid point $\boldsymbol{x}_C$. From $\mathcal{L}_\Delta[u_C] \ge 0$ and $\alpha_0 < 0$ we deduce that

$$u_C \le -\sum_{i=1}^{4} \frac{\alpha_i}{\alpha_0} u_{P_i} \le u_C.$$

This leads to the same conclusion; i.e., $u_{P_i} = u_C$, etc. □

Example 9.5 Consider the two-point boundary value problem

$$\mathcal{L}[u] := \frac{\mathrm{d}^2 u}{\mathrm{d}x^2} = 6x, \qquad x \in (0,1),$$
$$u(0) = u(1) = 0,$$

which has the solution $u(x) = x^3 - x$. If we use central differences, we obtain the difference equation

$$\mathcal{L}_\Delta[u_\mathrm{C}] := \frac{1}{\Delta x^2}\left(u_\mathrm{E} + u_\mathrm{W} - 2u_\mathrm{C}\right) = 6x_\mathrm{C}.$$

Since central differences are exact for polynomials up to degree three, we see that $u_\mathrm{C} = u(x_\mathrm{C}) < 0$. This is in agreement with Property 9.4, from which we conclude that $u_\mathrm{C} < \max(u(0), u(1)) = 0$, indeed, □

Like in the continuous case there exists a comparison theorem as well.

Theorem 9.6 (comparison theorem). *Consider the two operator equations*

$$\mathcal{L}_\Delta[u_\mathrm{C}] = f_\mathrm{C}, \qquad \mathcal{L}_\Delta[v_\mathrm{C}] = g_\mathrm{C},$$

with $f \geq g$ pointwise (i.e., at every grid point). Then

$$u_\mathrm{C} \leq v_\mathrm{C} + \max_{x_\mathrm{P} \in \partial\Omega_\Delta} \left(u_\mathrm{P} - v_\mathrm{P}\right).$$

Proof. Since $\mathcal{L}_\Delta[u_\mathrm{C} - v_\mathrm{C}] \geq 0$, we can deduce this directly from Property 9.4. □

Example 9.7 If we have the two-point boundary value problem

$$\mathcal{L}[u] := \frac{\mathrm{d}^2 u}{\mathrm{d}x^2} = \mathrm{e}^x, \qquad x \in (0,1),$$
$$u(0) = u(1) = 0,$$

we can use the comparison theorem and estimate the solution of

$$\mathcal{L}_\Delta[u_\mathrm{C}] = \exp(x_\mathrm{C})$$

in terms of the solution given in Example 9.5. Indeed, we have $\mathrm{e}^x \geq (\mathrm{e}/6)\,6x$ for $x \in (0,1)$. If we use the auxiliary function $v(x) := \frac{1}{6}\mathrm{e}(x^3 - x)$, which is the solution of

$$\mathcal{L}_\Delta[v_\mathrm{C}] = \mathrm{e}x_\mathrm{C}, \qquad v(0) = v(1) = 0;$$

as the comparison function, we obtain from the comparison theorem that

$$u_\mathrm{C} \leq v_\mathrm{C} + \max(u(0) - v(0), u(1) - v(1)) = \frac{\mathrm{e}}{6}({x_\mathrm{C}}^3 - x_\mathrm{C}).$$

□

Again, like in the continuous case, we can easily show uniqueness of a solution on a connected grid.

Theorem 9.8. *On a connected grid the solution of* (9.3) *with suitable boundary data is unique.*

Proof. Let u_Δ and v_Δ be two such solutions. Then $u_\Delta - v_\Delta$ satisfies the homogeneous problem with homogeneous data. From Theorem 9.4 it then follows that $u_P - v_P \leq 0$. By interchanging u_Δ and v_Δ, we similarly find $v_P - u_P \leq 0$. This shows that $u_P = v_P$ for any interior point $\boldsymbol{x}_P$. □

9.3 Estimates of the Global Error

The actual global discretisation error for an elliptic problem is typically found by solving a system with the local error as the known right-hand side. In this section we shall consider two approaches to estimating this error. In Section 9.3.1 our results will use the maximum principle to obtain suitable bounds. Although this method is not totally unrelated to the matrix method (actually solving the aforementioned system), we consider the latter separately in Section 9.3.2. Finally, in Section 9.3.3, we investigate the effect of errors originating from the boundary.

9.3.1 Estimates Based on the Maximum Principle

Let us denote by $\mathbf{A}$ the matrix with coefficients as appearing in the difference equations, i.e., from collecting all difference equations (9.3) for all interior grid points, including possible boundary points where u is not given. In the following we will denote this set of grid points by Ω_Δ. We can order the unknowns in many ways. Although this is a nontrivial matter, we leave this question of ordering for the time being. It is extremely important, however, when computing the numerical solution. Writing the vector of unknowns as $\mathbf{u}$ and the known forcing and boundary data as $\mathbf{f}$, we have

$$\mathbf{A}\mathbf{u} = \mathbf{f}. \tag{9.13}$$

So we have two different representations for the numerical solution: viz. the grid function u_Δ satisfying the operator equation (9.10) and the vector of unknowns $\mathbf{u}$ satisfying the linear system (9.13). We will use both formulations for the numerical solution and also for the discretisation errors to be defined next.

The local discretisation error is actually the residual or defect when the exact solution of (9.1) is substituted in (9.13). Let us denote the vector of values at the (similarly ordered!) grid points of the exact solution in a similar fashion by $\mathbf{u}^*$. Then the *local discretisation error* $\mathbf{d}$ is defined by

$$\mathbf{A}\mathbf{u}^* =: \mathbf{f} + \mathbf{d}. \tag{9.14}$$

Since the matrix $\mathbf{A}$ is constant, the *global discretisation error*

$$\mathbf{e} := \mathbf{u}^* - \mathbf{u} \tag{9.15}$$

simply follows from (9.13) and (9.14); i.e.,

$$\mathbf{A}\mathbf{e} = \mathbf{d}. \tag{9.16}$$

We see that we have to solve a system to find the global error. Employing the maximum principle for this is conceptually simple. We first look for a solution of the problem with a right-hand side bounded below by one. We then use an appropriate scaling and use the resulting solution as the comparison function. This is borne out in the following theorems. Inequalities for vectors are meant to hold elementwise. To start with, we restrict ourselves to Dirichlet boundary data on a suitable domain so that there are no approximation errors coming from the boundary (one may think of a domain where the boundary coincides with grid lines). It is important to realise that a difference equation for the global error holds at any interior point (where the solution is unknown); here the coefficient corresponding to the boundary point(s) is taken as zero. As remarked before, we may as well take the boundary value for the error to be zero. Thus we have a homogeneous Dirichlet problem for the error.

Theorem 9.9. *Assume that there exists a vector* $\mathbf{v}$ *with nonnegative entries such that*

$$\mathbf{A}\mathbf{v} \geq \mathrm{e} := (1, 1, \ldots, 1)^T.$$

Then the global discretisation error is bounded by

$$\|\mathbf{e}\|_\infty \leq \max_{x_P \in \partial\Omega_\Delta} (v_P)\, \|\mathbf{d}\|_\infty.$$

Proof. Since $\mathcal{L}_\Delta[v_C] \geq 1$ for all grid points $x_C \in \Omega_\Delta$, we clearly have

$$\mathcal{L}_\Delta[\varphi v_C - e_C] \geq \varphi - d_C \geq 0,$$

where $\varphi := \|\mathbf{d}\|_\infty$. Applying the maximum principle (see Property 9.4), we obtain

$$-e_C \leq \varphi v_C - e_C \leq \max_{x_P \in \partial\Omega_\Delta} \big(\varphi v_P - e_P\big) = \varphi \max_{x_P \in \partial\Omega_\Delta} v_P. \qquad (*)$$

In the derivation of $(*)$ we have used that $e_P = 0$ at the boundary grid points. Likewise, from

$$\mathcal{L}_\Delta[\varphi v_C + e_C] \geq \varphi + d_C \geq 0,$$

we can derive the inequalities

$$e_C \leq \varphi v_C + e_C \leq \max_{x_P \in \partial\Omega_\Delta} \big(\varphi v_P + e_P\big) = \varphi \max_{x_P \in \partial\Omega_\Delta} v_P. \qquad (**)$$

Combining the inequalities in $(*)$ and $(**)$, we find

$$|e_C| \leq \varphi \max_{x_P \in \partial\Omega_\Delta} v_P$$

for all $x_C \in \Omega_\Delta$. □

We remark that a comparison function v_Δ, like in the theorem above, is often chosen as the solution of a suitable continuous problem restricted to the grid Ω_Δ. Since the right-hand side is not so important (although its maximum at $\partial\Omega$ should not be too large of course), one can comfortably restrict oneself to polynomial solutions of sufficiently low degree.

Example 9.10 Consider the Poisson problem

$$\begin{aligned} \mathcal{L}[u] &:= \nabla^2 u = f(x, y), \qquad && (x, y) \in \Omega := (0, 1) \times (0, 1), \\ u(x, y) &= 0, && (x, y) \in \partial\Omega. \end{aligned}$$

For a constant grid size $h = \Delta x = \Delta y$ we obtain the difference scheme

$$\mathcal{L}_\Delta[u_\mathrm{C}] := \frac{1}{h^2}\big(u_\mathrm{N} + u_\mathrm{E} + u_\mathrm{S} + u_\mathrm{W} - 4u_\mathrm{C}\big) = f(\boldsymbol{x}_\mathrm{C}).$$

As comparison function v_Δ we use the restriction of

$$v(x, y) := \frac{1}{4}(x^2 + y^2).$$

Clearly we have $\mathcal{L}_\Delta[v_\mathrm{C}] = 1$ and $\max\limits_{\boldsymbol{x}_\mathrm{P}\in\partial\Omega_\Delta}(v_\mathrm{P}) = \frac{1}{2}$. Applying Theorem 9.9, we conclude that

$$|u_\mathrm{P}| \le \frac{1}{2}\max_{\boldsymbol{x}_\mathrm{P}\in\Omega_\Delta}|f(\boldsymbol{x}_\mathrm{P})|.$$

Applying this result to (9.16), we can determine an upper bound for the global discretisation error. □

Next we can prove convergence of the scheme in the following sense.

Theorem 9.11. *If the local discretisation error of* (9.10) *is $\mathcal{O}(h^p)$, then for a well-posed problem we also have an $\mathcal{O}(h^p)$ global discretisation error (pointwise).*

Proof. For a well-posed problem we have at least one point where a Dirichlet/Robin condition is given, so that a maximum principle holds. The result then follows from Theorem 9.9, applied to (9.16), noting that we should have zero data for the error at the boundary. □

Example 9.12 Consider the boundary value problem

$$\mathcal{L}[u] := \nabla^2 u = -2\sin x \sin y, \qquad (x, y) \in \Omega := (0, 1) \times (0, 1),$$

$$u(x, y) = \begin{cases} 0 & \text{if } x = 0,\ y \in (0, 1), \\ 0 & \text{if } y = 0,\ x \in (0, 1), \\ \sin 1 \sin y & \text{if } x = 1,\ y \in (0, 1), \\ \sin x \sin 1 & \text{if } y = 1,\ x \in (0, 1). \end{cases}$$

Clearly, this problem has the solution $u(x, y) = \sin x \sin y$. Discretising with central differences and uniform grid size h gives a local discretisation error equal to (cf. (5.30))

$$d(x_j, y_k) = -\frac{1}{12}h^2\left(\frac{\partial^4 u}{\partial x^4}(\xi, y_k) + \frac{\partial^4 u}{\partial y^4}(x_j, \eta)\right),$$

with $\xi \in (x_j - h, x_j + h)$ and $\eta \in (y_k - h, y_k + h)$. This can be bounded by

$$|d(x_j, y_k)| \le \frac{1}{6}h^2.$$

Using the comparison function v_Δ as defined in Example 9.10, we obtain

$$\max_{\boldsymbol{x}_\mathrm{P}\in\Omega_\Delta}|e_\mathrm{P}| \le \frac{1}{2}\frac{1}{6}h^2 = \frac{1}{12}h^2.$$ □

We can sometimes improve our error estimates by distinguishing between two subgrids and considering estimates there separately. In particular, this may be useful when some of the boundary data are corrupted with errors. The latter may be due to the fact that the grid does not coincide with a boundary (cf. (5.35)), or it may be due to discretisation errors arising from Robin or Neumann boundary conditions. The next theorem deals with this question.

Theorem 9.13. *Let Ω^*_Δ be a subgrid of Ω_Δ containing a part of the boundary. Denote by $\boldsymbol{x}_{P^*}$ a grid point in Ω^*_Δ and by $\boldsymbol{x}_P$ an interior grid point in $\Omega_\Delta \setminus \Omega^*_\Delta$. Assume there exists a nonnegative grid function v_Δ such that*

$$\mathcal{L}_\Delta[v_P] \geq 1, \qquad \boldsymbol{x}_P \in \Omega_\Delta \setminus \Omega^*_\Delta,$$
$$\mathcal{L}_\Delta[v_{P^*}] \geq \beta > 0, \quad \boldsymbol{x}_{P^*} \in \Omega^*_\Delta.$$

Then the grid function e_Δ corresponding to the global discretisation error **e** *(with homogeneous boundary conditions) is estimated by*

$$|e_C| \leq \max_{\boldsymbol{x}_P \in \partial\Omega_\Delta} (v_P) \max(\varphi, \varphi^*/\beta),$$

where

$$\varphi := \max_{\boldsymbol{x}_P \in \Omega_\Delta \setminus \Omega^*_\Delta} |d_P|, \qquad \varphi^* := \max_{\boldsymbol{x}_{P^*} \in \Omega^*_\Delta} |d_{P^*}|.$$

Proof. For $\boldsymbol{x}_P \in \Omega_\Delta \setminus \Omega^*_\Delta$ and $\boldsymbol{x}_{P^*} \in \Omega^*_\Delta$ we have, respectively,

$$\mathcal{L}_\Delta[\varphi v_P \pm e_P] = \varphi \mathcal{L}_\Delta[v_P] \pm \mathcal{L}_\Delta[e_P] \geq \varphi \pm d_P \geq 0,$$

$$\mathcal{L}_\Delta\left[\tfrac{1}{\beta}\varphi^* v_{P^*} \pm e_{P^*}\right] = \frac{1}{\beta}\varphi^* \mathcal{L}_\Delta[v_{P^*}] \pm \mathcal{L}_\Delta[e_{P^*}] \geq \varphi^* \pm d_{P^*} \geq 0.$$

Combining both formulas, we conclude that

$$\mathcal{L}_\Delta[\tilde{\varphi} v_P \pm e_P] \geq 0$$

for all $\boldsymbol{x}_P \in \Omega_\Delta$, where $\tilde{\varphi} := \max(\varphi, \varphi^*/\beta)$. The proof then follows the same formal reasoning as that of Theorem 9.9. □

To show how Theorem 9.13 may be used we have the following example.

Example 9.14 Consider the two-point boundary value problem

$$\mathcal{L}[u] := \frac{\mathrm{d}^2 u}{\mathrm{d}x^2} = \mathrm{e}^x, \qquad x \in (0, 1),$$

$$\frac{\mathrm{d}u}{\mathrm{d}x}(0) = 1, \qquad u(1) = \mathrm{e}.$$

If we discretise the problem on the grid $\Omega_\Delta = \{x_0, x_1, \ldots, x_M\}$, where $x_j = j\Delta x$, $\Delta x = 1/(M+1)$, we obtain the scheme

$$\mathcal{L}_\Delta[u_j] := \frac{1}{\Delta x^2}\left(u_{j+1} - 2u_j + u_{j-1}\right) = \exp(x_j), \qquad j = 0, 1, 2, \ldots, M.$$

The right boundary condition is straightforward. In order to deal with the left boundary condition, we introduce a virtual point $x_{-1} = -\Delta x$ and apply central differences at x_0, giving

$$u_1 - u_{-1} = 2\Delta x.$$

Eliminating u_{-1} from this relation and substituting into the difference equation for $j = 0$, we obtain

$$\mathcal{L}_\Delta[u_0] := \frac{2}{\Delta x^2}(u_1 - u_0) = 1 + \frac{2}{\Delta x}. \qquad (*)$$

We may view $(*)$ as a difference equation at the points x_{-1}, x_0, x_1 with a (Dirichlet) boundary condition; i.e.,

$$\mathcal{L}_\Delta[u_0] = \frac{1}{\Delta x^2}(2u_1 - 2u_0 + 0\,u_{-1}).$$

As a (discrete) comparison function we now take

$$v_j := \frac{1}{2}x_j(x_j - p)$$

for some p. We can choose p such that the comparison function satisfies our requirements and even such that it minimises the global error. We have

$$\mathcal{L}_\Delta[v_j] = 1, \qquad j = 1, 2, \ldots, M,$$

and

$$\mathcal{L}_\Delta[v_0] = 1 - \frac{p}{\Delta x}.$$

Obviously, $\Omega^*_\Delta = \{x_0\}$. So a possible choice is $p = -1$, implying that $\beta = 1 + 1/\Delta x$. Since the problem is smooth we see that $\varphi = \mathcal{O}(\Delta x^2)$ and $\varphi^* = \mathcal{O}(\Delta x)$. Moreover, $\max_{x_\mathrm{P} \in \partial\Omega_\Delta}(v_\mathrm{P}) = 1$. Application of Theorem 9.13 then yields $\|\mathbf{e}\|_\infty = \mathcal{O}(\Delta x^2)$, which is sharper than $\mathcal{O}(\Delta x)$, which one would have obtained by direct application of the comparison theorem (and might have expected because the local error arising from the boundary condition is first order). Optimising for p will not give a higher order (as second order is the best one can get, of course). One may, however, try to optimise p such that the coefficient in the error estimate is as small as possible. We omit this. □

9.3.2 Error Estimates Using the Matrix Method

Although the maximum principle is a very powerful tool, there are situations where employing the matrix representation may be at least as useful, as will be illustrated here and in Chapter 11. At this point it is important to remark that some useful properties only hold if the grid points are ordered in a special way. For our purposes (and two-dimensional problems) it is sufficient to assume that they are ordered lexicographically (in reading order). Another useful ordering is red-black, i.e., first the red squares of a chess board and then the black ones (again in reading order).

To start we introduce a number of concepts for matrices arising from discretised PDEs with suitable orderings. The first one is positivity of vectors and matrices.

Definition 9.15. *A vector* $\mathbf{v} = (v_1, v_2, \ldots, v_n)^T$ *is called positive and denoted* $\mathbf{v} > \mathbf{0}$ *if* $v_i > 0$ *for* $i = 1, 2, \ldots, n$. *A matrix* $\mathbf{A} = (a_{i,j})$ *is called positive and denoted* $\mathbf{A} > \mathbf{O}$ *if* $a_{i,j} > 0$ *for* $i, j = 1, 2, \ldots, n$. *It is called semipositive and denoted* $\mathbf{A} \geq \mathbf{O}$ *if* $a_{i,j} \geq 0$ *for* $i, j = 1, 2, \ldots, n$.

Definition 9.16. *An $n \times n$ matrix $\mathbf{A} = (a_{i,j})$ is called an* L*-matrix if $a_{i,i} > 0$ for $i = 1, 2, \ldots, n$ and $a_{i,j} \leq 0$ for $i, j = 1, 2, \ldots, n,\ j \neq i$. An $n \times n$ matrix $\mathbf{A} = (a_{i,j})$ is called an* M*-matrix if $a_{i,j} \leq 0$ for $i, j = 1, 2, \ldots, n,\ j \neq i$, $\mathbf{A}$ is nonsingular, and its inverse $\mathbf{A}^{-1}$ is semipositive.*

Matrix and vector inequalities like $\mathbf{A} \leq \mathbf{B}$ follow in an obvious way from Definition 9.15. It is not difficult to see that an M-matrix is an L-matrix. The following theorem gives a sufficient condition for the converse.

Theorem 9.17. *Let $\mathbf{A}$ be an* L*-matrix. Let $\mathbf{D}$ denote the diagonal of $\mathbf{A}$, $\mathbf{C} := \mathbf{D} - \mathbf{A}$, and $\mathbf{B} := \mathbf{D}^{-1}\mathbf{C}$. Then $\mathbf{A}$ is an* M*-matrix if and only if $\rho(\mathbf{B}) < 1$.*

Proof. Suppose $\rho(\mathbf{B}) < 1$. Then the matrix $\mathbf{I} - \mathbf{B}$ is nonsingular and the series $\mathbf{I} + \mathbf{B} + \mathbf{B}^2 + \cdots$ converges to $(\mathbf{I} - \mathbf{B})^{-1}$. Since $\mathbf{D}, \mathbf{C} \geq \mathbf{O}$, we conclude that $\mathbf{B}$ and $(\mathbf{I} - \mathbf{B})^{-1}$ are semipositive. Moreover, since $\mathbf{D}$ and $\mathbf{I} - \mathbf{B}$ are nonsingular, it follows that $\mathbf{A} = \mathbf{D}(\mathbf{I} - \mathbf{B})$ is also nonsingular and $\mathbf{A}^{-1} = (\mathbf{I} - \mathbf{B})^{-1}\mathbf{D}^{-1}$ is semipositive. Hence $\mathbf{A}$ is an M-matrix. Conversely, let $\mathbf{A}$ be an M-matrix. Define

$$\hat{\mathbf{A}} := \mathbf{D}^{-1/2}\,\mathbf{A}\,\mathbf{D}^{-1/2}.$$

Then $\hat{\mathbf{A}} = \mathbf{I} - \mathbf{D}^{1/2}\,\mathbf{B}\,\mathbf{D}^{-1/2}$. Clearly, $\hat{\mathbf{A}}$ is an M-matrix. If we finally define

$$\hat{\mathbf{B}} := \mathbf{D}^{1/2}\,\mathbf{B}\,\mathbf{D}^{-1/2},$$

which is clearly similar to $\mathbf{B}$, we can write

$$\hat{\mathbf{A}}^{-1} = \left(\mathbf{I} - \hat{\mathbf{B}}\right)^{-1} = \left(\mathbf{I} + \hat{\mathbf{B}} + \hat{\mathbf{B}}^2 + \cdots + \hat{\mathbf{B}}^m\right) + \left(\mathbf{I} - \hat{\mathbf{B}}\right)^{-1}\hat{\mathbf{B}}^{m+1}.$$

Since $(\mathbf{I} - \hat{\mathbf{B}})^{-1}$ and $\hat{\mathbf{B}}$ are semipositive as well, we conclude that $\mathbf{I} + \hat{\mathbf{B}} + \hat{\mathbf{B}}^2 + \cdots + \hat{\mathbf{B}}^m$ is bounded above elementwise by the corresponding elements of $(\mathbf{I} - \hat{\mathbf{B}})^{-1}$, which implies that the series $\mathbf{I} + \hat{\mathbf{B}} + \hat{\mathbf{B}}^2 + \cdots$ converges; i.e., $\rho(\mathbf{B}) = \rho(\hat{\mathbf{B}}) < 1$. □

One can give a useful property that gives a pointwise estimate of a matrix and thus can be seen as a matrix analogue of the comparison theorem.

Property 9.18. *Let $\mathbf{B}$ be an* M*-matrix. Let $\mathbf{A} := \mathbf{B} + \mathbf{F}$, where we have the diagonal matrix $\mathbf{F} \geq \mathbf{O}$. Then $\mathbf{A}^{-1} \leq \mathbf{B}^{-1}$.*

Proof. Write $\mathbf{B} =: \mathbf{D}(\mathbf{I} - \mathbf{J})$ and $\mathbf{A} =: \mathbf{E}(\mathbf{I} - \mathbf{K})$, with $\mathbf{D}$ and $\mathbf{E}$ the diagonal parts of $\mathbf{B}$ and $\mathbf{A}$, respectively. Since $\rho(\mathbf{J}) < 1$ (see Theorem 9.17), $(\mathbf{I} - \mathbf{J})^{-1}$ exists. By assumption we moreover have $\mathbf{K} \leq \mathbf{J}$, which implies that $\mathbf{K}^i \to 0$ for $i \to \infty$; i.e., $(\mathbf{I} - \mathbf{K})^{-1}$ exists. Actually, we can even deduce from $\mathbf{K}^i \leq \mathbf{J}^i$ that $(\mathbf{I} - \mathbf{K})^{-1} \leq (\mathbf{I} - \mathbf{J})^{-1}$. Combining results, we see that $(\mathbf{I} - \mathbf{K})^{-1}\mathbf{E}^{-1} \leq (\mathbf{I} - \mathbf{J})^{-1}\mathbf{D}^{-1}$; i.e., $\mathbf{A}^{-1} \leq \mathbf{B}^{-1}$. □

Typically, the matrices we have encountered will give rise to the negative of M-matrices. However, rather than considering the problem

$$\nabla^2 u = f,$$

one may write

$$-\nabla^2 u = \tilde{f} := -f,$$

as we shall do henceforth in this chapter. The usual central differences then give rise to an M-matrix, as is borne out in the next example.

Example 9.19 If we use central differences to discretise the boundary value problem

$$-\frac{\mathrm{d}^2 u}{\mathrm{d}x^2} = -bu, \qquad x \in (0,1),$$

$$u(0) = u(1) = 0,$$

where $b > 0$, we obtain the matrix

$$\mathbf{A} = \mathbf{B} + b\mathbf{I}, \qquad \mathbf{B} := \frac{1}{\Delta x^2}\begin{pmatrix} 2 & -1 & & & \\ -1 & 2 & -1 & & \\ & \ddots & \ddots & \ddots & \\ & & -1 & 2 & -1 \\ & & & -1 & 2 \end{pmatrix},$$

where $\Delta x = 1/(M+1)$ is the grid size. Clearly, $\mathbf{B}$ is an L-matrix. The matrix $\mathbf{D}^{-1}\mathbf{C}$, where $\mathbf{D}$ is the diagonal part of $\mathbf{B}$ and $\mathbf{C} = \mathbf{D} - \mathbf{B}$, is symmetric and has Gershgorin circles with radius one, centred around zero. The only possible eigenvalues λ with $|\lambda| = 1$ might be $+1, -1$. It is trivial to see that neither value is an eigenvalue, which implies that $\rho(\mathbf{D}^{-1}\mathbf{C}) < 1$. So, according to Theorem 9.17, $\mathbf{B}$ is an M-matrix. In fact, we can give the explicit inverse of $\mathbf{B}^{-1}$ as

$$\mathbf{B}^{-1} := \frac{1}{(M+1)^3}\begin{pmatrix} M & M-1 & M-2 & \cdots & 1 \\ M-1 & (M-1)2 & (M-2)2 & \cdots & 2 \\ M-2 & (M-2)2 & (M-2)3 & \cdots & 3 \\ \vdots & \vdots & \vdots & & \vdots \\ 1 & 2 & 3 & \cdots & M \end{pmatrix};$$

see also Example 9.28. We find $\|\mathbf{B}^{-1}\|_\infty = \frac{1}{8}$. Following the same reasoning as above, we can prove that $\mathbf{A}$ is also an M-matrix, and, moreover, $\mathbf{A}^{-1} \leq \mathbf{B}^{-1}$. From this inequality we see that $\|\mathbf{A}^{-1}\|_\infty \leq \|\mathbf{B}^{-1}\|_\infty$. □

We will now apply these properties of M-matrices to estimate the global discretisation error. Consider the linear system (9.13). The corresponding global discretisation error $\mathbf{e}$ satisfies the system (9.16). If the matrix $\mathbf{A}$ is an M-matrix, we find

$$|\mathbf{e}| \leq \mathbf{A}^{-1}|\mathbf{d}|, \tag{9.17}$$

where the absolute value of the vectors has to be taken elementwise; i.e., $|\mathbf{e}| := (|e_1|, |e_2|, \ldots, |e_N|)^T$, with e_j the global discretisation error in the jth grid point and N the number of points in Ω_Δ, and $|\mathbf{d}|$ is defined likewise. This immediately gives, for the problem in Example 9.19, that

$$\|\mathbf{e}\|_\infty \leq \|\mathbf{A}^{-1}\|_\infty \max_j |d_j| \leq \frac{C}{8}\Delta x^2, \tag{9.18}$$

where C is a bound on the fourth derivatives of the exact solution $u(\boldsymbol{x})$. Of course it is crucial to have an estimate of $\|\mathbf{A}^{-1}\|_\infty$. For more general problems it is often not so easy to obtain an estimate for $\|\mathbf{e}\|_\infty$ and we proceed as follows. Introducing the vector

$$\mathrm{e} := (1, 1, \ldots, 1)^T,$$

we can write, instead of (9.17),

$$|\mathbf{e}| \leq \mathbf{A}^{-1}\mathrm{e} \max_j |d_j| = \mathrm{v} \max_j |d_j|, \tag{9.19}$$

where the vector v is the solution of the system

$$\mathbf{A}\mathrm{v} = \mathrm{e}. \tag{9.20}$$

In many situations the (positive) vector v can be estimated rather sharply from above, thus making it possibile to estimate the global discretisation error through (9.19). We demonstrate this by an example. One should note the similarity with the use of a comparison function in Section 9.3.1.

Example 9.20 For the problem in Example 9.19 with $b = 0$ we define the vector v to be such that its jth coordinate is $\mathrm{v}_j = \frac{1}{2}x_j(1 - x_j)$. This is based on the fact that the difference equation is exact for polynomial solutions of degree up to two. It is easy to see that (9.20) holds. Clearly, $\|\mathrm{v}\|_\infty = \frac{1}{8}$, confirming the estimate in (9.18). □

Usually v can only be estimated from above by some $\tilde{\mathrm{v}}$, say, such that

$$\mathrm{e} = \mathbf{A}\mathrm{v} \leq \mathbf{A}\tilde{\mathrm{v}}.$$

The next example demonstrates this for a problem that has a Robin boundary condition at one side. In the next subsection we will consider the global error due to discretising this. Here we will be concerned with estimating the effect of the local discretisation errors arising from the interior.

Example 9.21 Consider the boundary value problem

$$-\frac{\mathrm{d}^2u}{\mathrm{d}x^2} = f(x), \qquad x \in (0, 1),$$
$$\frac{\mathrm{d}u}{\mathrm{d}x}(0) = u(0), \qquad u(1) = 0.$$

To discretise this problem we employ central differences on the grid $\Omega_\Delta = \{x_0, x_1, \ldots, x_M\}$ with $x_j = j\Delta x$, $\Delta x = 1/(M+1)$. We introduce the virtual point x_{-1} and apply a central difference to discretise the Robin boundary condition at the left boundary as

$$\frac{u_1 - u_{-1}}{2\Delta x} = u_0.$$

After eliminating u_{-1} using the difference scheme at the points x_{-1}, x_0, x_1, we obtain the global system

$$\mathbf{A} := \frac{1}{\Delta x^2}\begin{pmatrix} 2(1+\Delta x) & -2 & & & \\ -1 & 2 & -1 & & \\ & \ddots & \ddots & \ddots & \\ & & -1 & 2 & -1 \\ & & & -1 & 2 \end{pmatrix};$$

cf. Example 9.14. As one can easily see, the vector $\tilde{\mathrm{v}} = (\tilde{\mathrm{v}}_0, \tilde{\mathrm{v}}_1, \ldots, \tilde{\mathrm{v}}_M)^T$ with $\tilde{\mathrm{v}}_j = \frac{1}{2}(1 - x_j^2)$ satisfies $\mathbf{A}\tilde{\mathrm{v}} \geq \mathrm{e}$. Since $\|\tilde{\mathrm{v}}\|_\infty \leq \frac{1}{2}$ we find the estimate $\|\mathbf{e}\|_\infty \leq \frac{1}{2}C\Delta x^2$, with C a bound on the third and fourth derivatives of $u(x)$. □

We conclude this subsection by demonstrating the analysis for a problem in two dimensions.

Example 9.22 Consider the boundary value problem

$$\begin{aligned}
-\nabla^2 u &= f(\boldsymbol{x}), & \boldsymbol{x} &\in \Omega := (0,1)\times(0,1),\\
\frac{\partial u}{\partial y}(x,0) &= u(x,0), & x &\in (0,1),\\
u(x,1) &= g_2(x), & x &\in (0,1),\\
\frac{\partial u}{\partial x}(0,y) &= u(0,y), & y &\in (0,1),\\
u(1,y) &= g_4(y), & x &\in (0,1).
\end{aligned}$$

To discretise this problem we use central differences on a uniform grid with grid spacing $h = \Delta x = \Delta y = 1/(M+1)$ for some M. In order to discretise the boundary conditions at $y = 0$ and $x = 0$, we have to introduce virtual grid points at the lines $y = -h$ and $x = -h$, respectively. The virtual points arising in the difference equation are then eliminated via the discrete boundary condition, like in Example 9.14. Typically, a stencil at the left boundary then looks like Figure 9.4. So we have (away from the corners)

$$\frac{1}{h^2}\big((4+2h)u_{\mathrm{C}} - u_{\mathrm{N}} - u_{\mathrm{S}} - 2u_{\mathrm{E}}\big) = f(\boldsymbol{x}_{\mathrm{C}}).$$

One thus obtains $M+1$ grid lines in the y direction (at $x_j,\ j = 0, 1, \ldots, M$) and $M+1$ grid lines in the x direction (at $y_k,\ k = 0, 1, \ldots, M$). In analogy with the previous example, we choose

$$\tilde{\mathrm{v}}_{\mathrm{P}} = \frac{1}{4}(1 - x_j^2) + \frac{1}{4}(1 - y_k^2),$$

with $\tilde{\mathrm{v}}_{\mathrm{P}}$ the component of $\tilde{\mathrm{v}}$ corresponding to the grid point $\boldsymbol{x}_{\mathrm{P}} = (x_j, y_k) = (jh, kh)$. We can again easily check that $\tilde{\mathrm{v}}$ satisfies $\mathbf{A}\tilde{\mathrm{v}} \geq \mathrm{e}$. Moreover, $\|\tilde{\mathrm{v}}\|_\infty = \frac{1}{2}$, implying an error estimate of the form $\|\mathbf{e}\|_\infty \leq \frac{1}{2}Ch^2$ for some constant C. □

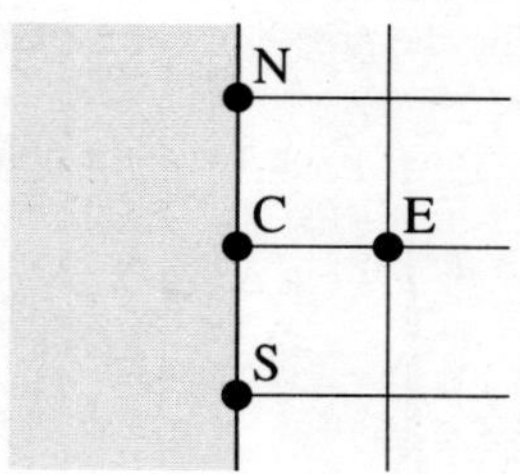

Figure 9.4. *Stencil at the left boundary.*

9.3.3 The Effect of Errors Originating from the Boundary

The boundary data need special attention. For the sake of simplicity, we restrict ourselves to a uniform grid of size h in either one or two space dimensions. Recalling the vector e introduced in Section 9.3.2, we define a vector $\mathrm{e}_{\mathrm{bnd}}$ to be found from e by making all components zero if the corresponding grid point is a genuine interior point, i.e., has no neighbours on the boundary. The genuine interior points induce a vector $\mathrm{e}_{\mathrm{int}}$ defined by

$$\mathrm{e}_{\mathrm{int}} := \mathrm{e} - \mathrm{e}_{\mathrm{bnd}}. \tag{9.21}$$

From (9.17) we thus find

$$|\mathbf{e}| \leq \mathbf{A}^{-1}\big(\mathrm{e}_{\mathrm{int}} \max_{\mathrm{int}} |d_{\mathrm{int}}| + \mathrm{e}_{\mathrm{bnd}} \max_{\mathrm{bnd}} |d_{\mathrm{bnd}}|\big). \tag{9.22}$$

Here d_{int} and d_{bnd} denote the local discretisation errors at the internal and boundary points, respectively. The crucial point is now to realise that $\mathbf{A}$ is usually *weakly diagonally dominant*; i.e., $|a_{i,i}| \geq \sum_{j\neq i} |a_{i,j}|$ for all i, with an inequality for at least one i, corresponding to a grid point that is not a genuine interior point. Hence, for some constant $C_{\mathrm{bnd}} > 0$, which depends on the problem at hand, we have $\mathbf{A}\mathrm{e} \geq C_{\mathrm{bnd}}\mathrm{e}_{\mathrm{bnd}}/h^2$, giving

$$\mathbf{A}^{-1}\mathrm{e}_{\mathrm{bnd}} \leq \frac{h^2}{C_{\mathrm{bnd}}}\mathrm{e}. \tag{9.23}$$

Since $\mathbf{A}^{-1}\mathrm{e}_{\mathrm{int}} \leq \mathbf{A}^{-1}\mathrm{e}$, we estimate the global effect of the interior local errors, say $\|\mathbf{A}^{-1}\mathrm{e}\|_\infty \leq C_{\mathrm{int}}$, for some C_{int}. Combining this with (9.22), we then obtain

$$\|\mathbf{e}\|_\infty \leq C_{\mathrm{int}} \max_{\mathrm{int}} |d_{\mathrm{int}}| + \frac{h^2}{C_{\mathrm{bnd}}} \max_{\mathrm{bnd}} |d_{\mathrm{bnd}}|. \tag{9.24}$$

The result in (9.24) implies that local errors at the boundary may be two orders less than the local errors made in the interior without affecting the order of the global error. This phenomenon explains why the often thoughtless treatment of derivative boundary conditions may not affect the overall result dramatically.

Example 9.23 Consider the two-point boundary value problem

$$-\frac{\mathrm{d}^2u}{\mathrm{d}x^2} = f(x), \qquad x \in (0,1),$$

$$u(0) = 0, \qquad u(1) = R,$$

where R stands for a representation that can only be approximated; say we use $\bar{R}$ instead of R, with $\bar{R} = R + \delta$. After discretising the problem with central differences with grid size h, we obtain the matrix

$$\mathbf{A} := \frac{1}{h^2}\begin{pmatrix} 2 & -1 & & & \\ -1 & 2 & -1 & & \\ & \ddots & \ddots & \ddots & \\ & & -1 & 2 & -1 \\ & & & -1 & 2 \end{pmatrix}.$$

From this we see that we can take $C_{\mathrm{bnd}} = 1$ in (9.24). The effect of the approximation error in the boundary condition is only globally felt as $h^2\,\delta$! □

We can generalise the preceding analysis to problems where boundary condition errors occur at several places on the boundary. To start with let us split the boundary $\Gamma_\Delta := \partial\Omega_\Delta$ into a number of subboundaries, $\Gamma_{\Delta,k}$, $k = 1, 2, \ldots, K$, for some K, say, and introduce the vectors (cf. (9.22))

$$\mathrm{e} = \mathrm{e}_{\text{int}} + \sum_{k=1}^{K} \mathrm{e}_{\mathrm{b},k}, \tag{9.25}$$

where the vector $\mathrm{e}_{\mathrm{b},k}$ contains zero entries if the corresponding grid point has no neighbours on $\Gamma_{\Delta,k}$. Still assuming that $\mathbf{A}^{-1}$ is positive, we then obtain, in an obvious notation (cf. (9.22)), that

$$|\mathbf{e}| \leq \max_{\text{int}} |d_{\text{int}}|\,\mathbf{A}^{-1}\mathrm{e}_{\text{int}} + \sum_{k=1}^{K} \max_{\Gamma_{\Delta,k}} |d_{\mathrm{b},k}|\,\mathbf{A}^{-1}\mathrm{e}_{\mathrm{b},k}. \tag{9.26}$$

For problems with derivative boundary conditions we now try to find positive vectors $\mathrm{w}_{\mathrm{b},k}$ and positive constants $C_{\mathrm{b},k}$ such that

$$\mathbf{A}\mathrm{w}_{\mathrm{b},k} \geq \frac{C_{\mathrm{b},k}}{h^2}\mathrm{e}_{\mathrm{b},k}, \tag{9.27}$$

so that we have (cf. (9.23))

$$\mathbf{A}^{-1}\mathrm{e}_{\mathrm{b},k} \leq \frac{h^2}{C_{\mathrm{b},k}}\mathrm{w}_{\mathrm{b},k}, \qquad k = 1, 2, \ldots, K. \tag{9.28}$$

The eventual estimate then reads

$$\|\mathbf{e}\|_\infty \leq C_{\text{int}} \max_{\text{int}} |d_{\text{int}}| + h^2 \sum_{k=1}^{K} \max_{\Gamma_{\Delta,k}} |d_{\mathrm{b},k}| \frac{\|\mathrm{w}_{\mathrm{b},k}\|_\infty}{C_{\mathrm{b},k}}. \tag{9.29}$$

If some part of the boundary, say with index k, has a Dirichlet boundary condition, the corresponding estimates above are trivial. Indeed, one can take $\mathrm{w}_{\mathrm{b},k} = \mathrm{e}$ and $C_{\mathrm{b},k} = 1$, from which it follows that the contribution to the global error is just $h^2 \max_{\Gamma_{\Delta,k}} |d_{\mathrm{b},k}|$. The next two examples show how these estimates may be used.

Example 9.24 Let us replace the boundary condition at $x = 1$ in Example 9.23 by the Robin condition

$$\frac{\mathrm{d}u}{\mathrm{d}x}(1) + u(1) = g.$$

We can approximate this boundary condition by the *box scheme*:

$$\frac{1}{h}\left(u\left(1+\frac{1}{2}h\right) - u\left(1-\frac{1}{2}h\right)\right) + \frac{1}{2}\left(u\left(1-\frac{1}{2}h\right) + u\left(1+\frac{1}{2}h\right)\right) \doteq g.$$

As one can easily see this scheme has a discretisation error $\mathcal{O}(h^2)$. To discretise this problem we employ central differences on the interior grid $\{x_1, x_2, \ldots, x_M\}$, with $x_j = (j - \frac{1}{2})h$, $h = 1/M$. This way, we obtain a standard problem as in the previous example, but for the last equation. Introducing the virtual grid point $x_{M+1} = 1 + \frac{1}{2}h$, we can apply the box scheme and obtain a relation between u_M and u_{M+1}. We can use this latter relation to eliminate u_{M+1} from the difference equation at x_M. This gives

$$\frac{1}{h^2}\left(\left(2-\frac{1-\frac{1}{2}h}{1+\frac{1}{2}h}\right)u_M - u_{M-1}\right) = f_M + \frac{1}{h}\frac{g}{1+\frac{1}{2}h}.$$

As a result, the local discretisation error due to discretising the boundary condition is $\mathcal{O}(h)$. It is clear that we should split the boundary into a left part $\Gamma_{\Delta,1}$, say, and a right part $\Gamma_{\Delta,2}$, the latter being the only troublesome one. If we take $\mathrm{w}_{b,2} = (1, 2, \ldots, M)^T$, we find that $\mathbf{A}\mathrm{w}_{b,2}$ has zeros everywhere except for the last element, which is given by

$$\frac{1}{h^2}\left(\left(2-\frac{1-\frac{1}{2}h}{1+\frac{1}{2}h}\right)M - M + 1\right) = \frac{1}{h^2}\frac{2+\frac{1}{2}h}{1+\frac{1}{2}h} \geq \frac{1}{h^2}.$$

Thus the constant $C_{b,2}$ can be chosen equal to one and we find, for the contribution of the discretisation error of the Robin boundary condition,

$$h^2\frac{\|\mathrm{w}_{b,2}\|_\infty}{C_{b,2}}|d_{b,2}| \leq \frac{1}{2}Mh^2\mathcal{O}(h) = \mathcal{O}(h^2).$$

This result is similar to what we found from applying the maximum principle when using a virtual point and the second order central difference scheme in Example 9.14. □

The next example deals with a two-dimensional problem already encountered in Example 9.22.

Example 9.25 Consider the boundary value problem of Example 9.22. The grid points on the edges $y = 0$, $y = 1$, $x = 0$, and $x = 1$ will be denoted by $\Gamma_{\Delta,1}$, $\Gamma_{\Delta,2}$, $\Gamma_{\Delta,3}$, and $\Gamma_{\Delta,4}$, respectively. As in Example 9.22 we use central differences to discretise the Poisson equation and the boundary conditions. An appropriate vector $\mathrm{w}_{b,3}$ then has entries

$$\left(\mathrm{w}_{b,3}\right)_{\mathrm{P}} = M + 1 - j,$$

corresponding to grid point $\boldsymbol{x}_{\mathrm{P}} = (x_j, y_k)$. It is easy to see that

$$\mathbf{A}\mathrm{w}_{b,3} \geq \frac{2}{h^2}\mathrm{e}_{b,3}.$$

Thus we find

$$\mathbf{A}^{-1}\mathrm{e}_{b,3} \leq \frac{h^2}{2}\mathrm{w}_{b,3} \leq \frac{h}{2}\mathrm{e}.$$

We can treat the lower boundary in a similar way to obtain

$$\mathbf{A}^{-1}\mathrm{e}_{b,1} \leq \frac{h}{2}\mathrm{e}.$$

Since the local errors from discretising the Neumann boundary conditions are first order (though the central difference is second order), we can conclude again that they have a global effect of second order only. The errors arising form discretising internal points were discussed in Example 9.22. □

9.4 Solution of Linear Systems

After discretising a linear elliptic equation, we obtain a system of equations as in (9.13). In Section 9.4.1 we first consider the one-dimensional case, for which this system can most efficiently be solved by standard LU-decomposition. For problems in higher dimensions this becomes inefficient and therefore iterative methods are often employed. This topic is briefly introduced in Section 9.4.2. To conclude we discuss in Section 9.4.3 the *closure* of iterative methods.

9.4.1 One-Dimensional Dirichlet Problems

Before we deal with problems in higher dimensions, it is instructive to start with a one-dimensional problem. So consider the boundary value problem

$$a\,\frac{\mathrm{d}^2u}{\mathrm{d}x^2}+c\,\frac{\mathrm{d}u}{\mathrm{d}x}+e\,u=f,\qquad x\in(0,1),\tag{9.30a}$$

$$u(0)=A,\qquad u(1)=B.\tag{9.30b}$$

If the coefficient functions a, c, and e are constant, we obtain the central difference scheme (cf. (9.2))

$$\frac{1}{\Delta x^2}\left(\left(a+\frac{1}{2}c\Delta x\right)u_E+\left(a-\frac{1}{2}c\Delta x\right)u_W-(2a-\Delta x^2 e)\,u_C\right)=f(\boldsymbol{x}_{\mathrm{c}}).\tag{9.31}$$

If we label the grid points x_j, with the index j running from 0 to $M+1$, so

$$x_j=j\Delta x,\qquad j=0,1,\ldots,M+1,\qquad \Delta x:=1/(M+1),$$

and denote the numerical approximation of $u(x_j)$ by u_j, then (9.31) can be written as a second order difference equation of the form

$$\alpha_j u_{j-1}+\beta_j u_j+\gamma_j u_{j+1}=f_j,\qquad j=1,2,\ldots,M.\tag{9.32a}$$

Moreover, from the boundary conditions (9.30b) we conclude that

$$u_0=A,\qquad u_{M+1}=B.\tag{9.32b}$$

The equations (9.32a) and (9.32b) result in the linear system

$$\mathbf{A}\mathbf{u}=\mathbf{f},\tag{9.33}$$

where

$$\mathbf{A}:=\begin{pmatrix}\beta_1 & \gamma_1 & & & \\ \alpha_2 & \beta_2 & \gamma_2 & & \\ & \ddots & \ddots & \ddots & \\ & & \alpha_{M-1} & \beta_{M-1} & \gamma_{M-1}\\ & & & \alpha_M & \beta_M\end{pmatrix},\quad \mathbf{u}:=\begin{pmatrix}u_1\\ u_2\\ \vdots\\ u_{M-1}\\ u_M\end{pmatrix},\quad \mathbf{b}:=\begin{pmatrix}f_1-\alpha_1 A\\ f_2\\ \vdots\\ f_{M-1}\\ f_M-\gamma_M B\end{pmatrix}.\tag{9.34}$$

Clearly, **A** is *sparse*; i.e., has a large number of systematic zero entries. In this (still rather simple) case we may use a direct solver based on LU-decomposition [52]. It is instructive to work this out. Let us anticipate the existence of this LU-decomposition without pivoting, i.e., without interchanging rows. Say

$$\mathbf{A} = \mathbf{LU}, \tag{9.35}$$

where the matrices **L** and **U** are given by

$$\mathbf{L} := \begin{pmatrix} 1 & & & & \\ l_2 & 1 & & & \\ & \ddots & \ddots & & \\ & & l_{M-1} & 1 & \\ & & & l_M & 1 \end{pmatrix}, \quad \mathbf{U} := \begin{pmatrix} v_1 & w_1 & & & \\ & v_2 & w_2 & & \\ & & \ddots & \ddots & \\ & & & v_{M-1} & w_{M-1} \\ & & & & v_M \end{pmatrix}. \tag{9.36}$$

It is easy to see that the matrices **L** and **U** maintain the same sparsity structure as **A**. Simply comparing the coefficients in **LU** and **A** reveals

$$\begin{aligned} w_j &= \gamma_j, & j &= 1, 2, \ldots, M-1, \\ v_1 &= \beta_1, & & \\ l_j &= \alpha_j / v_{j-1}, & j &= 2, 3, \ldots, M, \\ v_j &= \beta_j - l_j \gamma_{j-1}, & j &= 2, 3, \ldots, M. \end{aligned} \tag{9.37}$$

In the simplest case where $a(\boldsymbol{x}) \equiv 1$, $c(\boldsymbol{x}) \equiv 0$, and $e(\boldsymbol{x}) \equiv 0$, we can easily show that $| v_j |> 1$, so that the algorithm (9.37), without partial pivoting, does not break down.

The matrix **A** in the latter case is symmetric. One can show that its eigenvalues λ_j are given by

$$\lambda_j = -\frac{4}{\Delta x^2} \sin^2\left(\frac{1}{2}\pi x_j\right), \qquad j = 1, 2, \ldots, M, \tag{9.38}$$

which should be approximations of the eigenvalues $-(j\pi)^2$ $(j = 1, 2, \ldots)$ of the exact operator, i.e., the differential operator $\frac{\mathrm{d}^2}{\mathrm{d}x^2}$ combined with (homogeneous) Dirichlet boundary conditions; cf. (8.11). As an example, we have plotted in Figure 9.5 both the eigenvalues of **A** for $M = 20$ and the exact eigenvalues. In particular, we see that the smallest eigenvalue in absolute value, λ_1, equals

$$\lambda_1 = -\pi^2 + \mathcal{O}(\Delta x^2). \tag{9.39}$$

Hence λ_1 is a sharp approximation of the exact eigenvalue $-\pi^2$. However, the largest eigenvalue, in absolute value, λ_M, is approximately equal to $-4/\Delta x^2$, which is not a good approximation of the Mth eigenvalue $-(M\pi)^2$ of the exact operator.

9.4.2 Basic Iterative Methods

For two-dimensional problems the situation is similar, but more complicated. Below we shall only discuss the simple case of a Poisson equation on the unit square with homogeneous

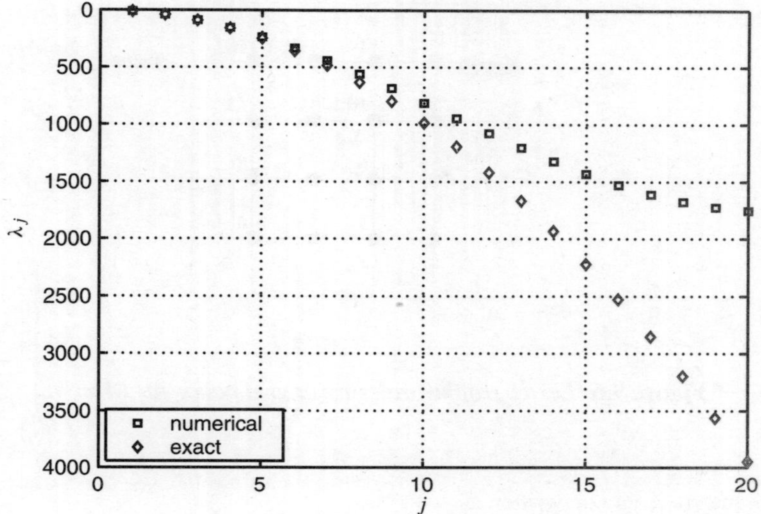

Figure 9.5. *The eigenvalues of* **A** *for* $M = 20$ *compared with the exact eigenvalues.*

Dirichlet boundary conditions, i.e.,

$$\nabla^2 u = f, \qquad (x, y) \in \Omega := (0,1) \times (0,1), \tag{9.40a}$$

$$u = 0, \qquad (x, y) \in \partial\Omega, \tag{9.40b}$$

discretised with central differences on the uniform grid

$$(x_j, y_k) := (jh, kh) \qquad j, k = 0, 1, \ldots, M+1, \quad h := 1/(M+1). \tag{9.41}$$

More general PDEs and discretisations can be dealt with in a straightforward way. Note that there is no natural ordering of the (internal) grid points. A common way to label these is in lexicographical order; i.e., we number the grid points along horizontal grid lines, as indicated in Figure 9.6, starting in the lower left corner. If we use the notation $u_{j,k}$ for the numerical approximation of $u(x_j, y_k)$, the central difference scheme for (9.40a) reads

$$\frac{1}{h^2}\left(u_{j+1,k} + u_{j-1,k} + u_{j,k+1} + u_{j,k-1} - 4u_{j,k}\right) = f(x_j, y_k) \qquad j, k = 1, 2, \ldots, M; \tag{9.42a}$$

cf. (9.2). These equations have to be completed with the numerical boundary conditions

$$u_{j,k} = 0 \qquad \text{if} \quad j, k = 0, M+1. \tag{9.42b}$$

Introducing the vector of unknowns $\mathbf{u}$

$$\mathbf{u} := \left(u_{1,1}, u_{2,1}, \ldots, u_{M,1} \mid u_{1,2}, u_{2,2}, \ldots, u_{M,2} \mid \ldots \mid u_{1,M}, u_{2,M}, \ldots, u_{M,M}\right)^T, \tag{9.43}$$

we can write (9.42a) and (9.42b) as the linear system

$$\mathbf{A}\mathbf{u} = \mathbf{f}, \tag{9.44}$$

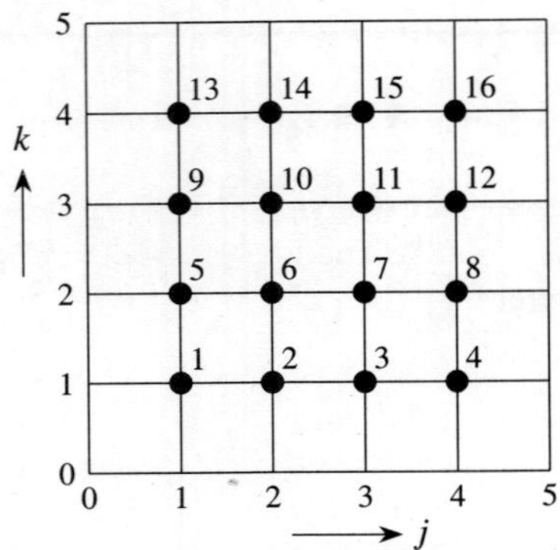

Figure 9.6. *Lexicographic ordering of grid points for $M = 4$.*

where the matrix $\mathbf{A}$ can be written as

$$\mathbf{A} = \frac{1}{h^2}\begin{pmatrix} \mathbf{B} & \mathbf{I} & & & \\ \mathbf{I} & \mathbf{B} & \mathbf{I} & & \\ & \ddots & \ddots & \ddots & \\ & & \mathbf{I} & \mathbf{B} & \mathbf{I} \\ & & & \mathbf{I} & \mathbf{B} \end{pmatrix}, \qquad \mathbf{B} := \begin{pmatrix} -4 & 1 & & & \\ 1 & -4 & 1 & & \\ & \ddots & \ddots & \ddots & \\ & & 1 & -4 & 1 \\ & & & 1 & -4 \end{pmatrix}. \tag{9.45}$$

The partitioning of $\mathbf{u}$ and $\mathbf{A}$ is based on the numbering along horizontal grid lines.

The matrix $\mathbf{A}$ is called *block tridiagonal*. Since all elements below the Mth lower codiagonal as well as above the Mth upper codiagonal are zero, it is also referred to as a *band matrix*, with *bandwidth* $2M + 1$. Note that we only need to store five elements per row (or rather five linear arrays). LU-decomposition will give a fill-in within this band, however, as one can easily check. This is typical for systems arising from elliptic (higher-dimensional) PDEs. A host of alternatives, in particular *iterative methods*, have been designed. They are all meant to reduce memory usage and to be at least as fast as LU-decomposition. Most of them are faster because one does not need full accuracy of the solution, which is perturbed by discretisation errors anyway. This important branch of numerical linear algebra is, however, too vast to even be superficially covered by this book; one may consult, e.g., [52, 135]. Nevertheless, we shall briefly discuss two classical iterative methods, i.e., Gauss–Jacobi and Gauss–Seidel, which have a merit of their own and lend themselves to a simple description.

The basic idea is to obtain an iteration of the form

$$\begin{aligned} &\mathbf{u}^0 \text{ given}, \\ &\mathbf{u}^{l+1} = \mathbf{M}\,\mathbf{u}^l + \mathbf{c}^l \qquad (l = 0, 1, 2, \ldots). \end{aligned} \tag{9.46}$$

The matrix $\mathbf{M}$ is called the *iteration matrix*. Gauss-type methods try to update the vector elementwise in sequential order (i.e., components with increasing index). To see how this is done let us split the matrix $\mathbf{A} = (a_{i,j})$ into three parts as follows:

$$\mathbf{A} =: \mathbf{L} + \mathbf{D} + \mathbf{U}, \tag{9.47}$$

where $\mathbf{L}$ is the lower triangular part $((a_{i,j})_{i>j})$, $\mathbf{D}$ is the diagonal $\mathrm{diag}(a_{i,i})$, and $\mathbf{U}$ is the upper triangular part $((a_{i,j})_{i<j})$ of $\mathbf{A}$. The first method we discuss is *Gauss–Jacobi*. Since $\mathbf{D}$ is nonsingular, we can define

$$\frac{4}{h^2}u_{j,k}^{l+1} = \frac{1}{h^2}\left(u_{j+1,k}^{l} + u_{j-1,k}^{l} + u_{j,k+1}^{l} + u_{j,k-1}^{l}\right) - f(x_j, y_k). \tag{9.48}$$

This corresponds to

$$-\mathbf{D}\mathbf{u}^{l+1} = \left(\mathbf{L} + \mathbf{U}\right)\mathbf{u}^{l} - \mathbf{f}. \tag{9.49}$$

We see that we can identify the iteration matrix $\mathbf{M} = \mathbf{M}_{\mathrm{J}}$, say, and the vector $\mathbf{c}^{l}$ above with

$$\mathbf{M}_{\mathrm{J}} := -\mathbf{D}^{-1}\left(\mathbf{L} + \mathbf{U}\right), \qquad \mathbf{c}^{l} = \mathbf{D}^{-1}\mathbf{f}. \tag{9.50}$$

We will analyse this method now. To start with we note that the eigenvalues $\lambda_{j,k}$ of $\mathbf{A}$ are given by (see the appendix, Section F)

$$\lambda_{j,k} = -\frac{4}{h^2}\left(\sin^2\left(\frac{1}{2}\pi x_j\right) + \sin^2\left(\frac{1}{2}\pi y_k\right)\right). \tag{9.51}$$

Hence $\mathbf{A}$ is symmetric and negative definite. We have convergence if $\rho(\mathbf{M}_{\mathrm{J}}) < 1$; see Appendix I. Since

$$\mathbf{M}_{\mathrm{J}} = \frac{h^2}{4}\mathbf{A} + \mathbf{I}, \tag{9.52}$$

we find that the eigenvalues of $\mathbf{M}_{\mathrm{J}}$ $\mu_{j,k}$ say, are given by

$$\mu_{j,k} = 1 - \sin^2\left(\frac{1}{2}\pi x_j\right) - \sin^2\left(\frac{1}{2}\pi y_k\right). \tag{9.53}$$

Apparently, $\rho(\mathbf{M}_{\mathrm{J}}) \approx 1 - (\pi h)^2/2$. In order to assess this, we remark that we should be content if, after m iterations, the error has shrunk to $\mathcal{O}(h^2)$, being the global discretisation error of this problem. Since $(1 - \frac{1}{2}(\pi h)^2)^m \approx \mathrm{e}^{-\frac{1}{2}m(\pi h)^2}$, we thus find that

$$m \approx \frac{4}{\pi^2}M^2 \ln M.$$

An improvement is to employ all updated values per row, instead of just the diagonal element. This gives the *Gauss–Seidel* method, which reads

$$-\frac{1}{h^2}\left(u_{j-1,k}^{l+1} + u_{j,k-1}^{l+1}\right) + \frac{4}{h^2}u_{j,k}^{l+1} = \frac{1}{h^2}\left(u_{j+1,k}^{l} + u_{j,k+1}^{l}\right) - f(x_j, y_k), \tag{9.54}$$

or, in matrix-vector formulation,

$$-\left(\mathbf{L} + \mathbf{D}\right)\mathbf{u}^{l+1} = \mathbf{U}\mathbf{u}^{l} - \mathbf{f}. \tag{9.55}$$

Note that the system (9.55) can be "solved" directly (back substitution). The iteration matrix, $\mathbf{M}_\mathrm{S}$ say, and the vector $\mathbf{c}^l$ are now given by

$$\mathbf{M}_\mathrm{S} := -\left(\mathbf{L}+\mathbf{D}\right)^{-1}\mathbf{U}, \quad \mathbf{c}^l := \left(\mathbf{L}+\mathbf{D}\right)^{-1}\mathbf{f}. \tag{9.56}$$

In order to analyse the spectral radius of the system, we need the following technical property.

Property 9.26. *For the block-tridiagonal matrix* $\mathbf{A}$ *in* (9.45) *the following holds:* λ_J *is an eigenvalue of* $\mathbf{M}_\mathrm{J}$ *in* (9.50) *if and only if* $\lambda_\mathrm{S} = \lambda_\mathrm{J}^2$ *is an eigenvalue of* $\mathbf{M}_\mathrm{S}$ *in* (9.56).

Proof. Consider the diagonal matrix $\mathbf{E}$

$$\mathbf{E} := \operatorname{diag}\left(\alpha, \alpha^2, \ldots, \alpha^M\right)$$

and the block diagonal matrix

$$\hat{\mathbf{E}} := \operatorname{diag}\left(\alpha\mathbf{E}, \alpha^2\mathbf{E}, \ldots, \alpha^M\mathbf{E}\right).$$

Then

$$\hat{\mathbf{E}}(\mathbf{L}+\mathbf{U})\hat{\mathbf{E}}^{-1} = \alpha\mathbf{L} + \frac{1}{\alpha}\mathbf{U}.$$

So $\alpha\mathbf{L} + \frac{1}{\alpha}\mathbf{U}$ has the same eigenvalues as $\mathbf{L}+\mathbf{U}$. Now we first note that $\det(\mathbf{I}+\mathbf{D}^{-1}\mathbf{L}) = 1$, whence

$$\begin{aligned}
\det\left(\lambda\mathbf{I}-\mathbf{M}_\mathrm{S}\right) &= \det\left(\left(\mathbf{I}+\mathbf{D}^{-1}\mathbf{L}\right)\left(\lambda\mathbf{I}-\mathbf{M}_\mathrm{S}\right)\right) \\
&= \det\left(\lambda\mathbf{I}+\lambda\mathbf{D}^{-1}\mathbf{L}+\mathbf{D}^{-1}\mathbf{U}\right) \\
&= \det\left(\lambda\mathbf{I}-\frac{1}{4}h^2(\lambda\mathbf{L}+\mathbf{U})\right).
\end{aligned}$$

If $\lambda = \lambda_\mathrm{S} = \lambda_\mathrm{J}^2$, we find

$$\begin{aligned}
\det\left(\lambda_\mathrm{S}\mathbf{I}-\mathbf{M}_\mathrm{S}\right) &= \lambda_\mathrm{J}^{M^2}\det\left(\lambda_\mathrm{J}\mathbf{I}-\frac{1}{4}h^2\left(\lambda_\mathrm{J}\mathbf{L}+\frac{1}{\lambda_\mathrm{J}}\mathbf{U}\right)\right) \\
&= \lambda_\mathrm{J}^{M^2}\det\left(\lambda_\mathrm{J}\mathbf{I}-\frac{1}{4}h^2(\mathbf{L}+\mathbf{U})\right) \\
&= \lambda_\mathrm{J}^{M^2}\det\left(\lambda_\mathrm{J}\mathbf{I}-\mathbf{M}_\mathrm{J}\right) = 0.
\end{aligned}$$

For the second equality we used $\alpha = \lambda_\mathrm{J}$ in the assertion above. The derivation can also be carried out starting with λ_J as an eigenvalue of $\mathbf{M}_\mathrm{J}$. □

From this property we immediately see that

$$\rho(\mathbf{M}_\mathrm{S}) \approx \left(1-\frac{1}{2}(\pi h)^2\right)^2 \approx 1-(\pi h)^2.$$

The number of iterations m to obtain an iteration error of $\mathcal{O}(h^2)$ is therefore approximately equal to

$$m \approx \frac{2}{\pi^2} M^2 \ln M,$$

which is about a factor of two faster than Gauss–Jacobi, at least asymptotically. There are many improvements to the Gauss-type methods. For example, one may apply relaxation; i.e., the contribution of old "values" is damped by a factor, say ω. We shall not discuss this any further. Another class of methods is related to solving a variational form of the equation. An example is the conjugate gradient method. Together with preconditioning techniques this gives a very powerful method. For further details we refer to the literature [135].

Example 9.27 In this example we compute a numerical solution of the boundary value problem (9.40) with $f(x, y) = 2(x(x-1) + y(y-1))$. The exact solution is in this case given by $u(x, y) = x(1-x)y(1-y)$. We choose $M = 20$ and apply the Gauss–Jacobi and Gauss–Seidel methods to system (9.44). A convergence history of both methods is given in Figure 9.7; i.e., $\|\mathbf{r}^l\|_2/\|\mathbf{r}^0\|_2$ is given as a function of the iteration index l, where $\mathbf{r}^l := \mathbf{A}\mathbf{u}^l - \mathbf{f}$ is the residual after l iterations. From this figure we see, e.g., that approximately 800 Gauss–Jacobi iterations are necessary to reduce the 2-norm of the initial residue $\mathbf{r}^0$ by a factor of 10^4, whereas approximately 400 Gauss–Seidel iterations suffice. This nicely confirms that Gauss–Seidel is twice as fast as Gauss–Jacobi. □

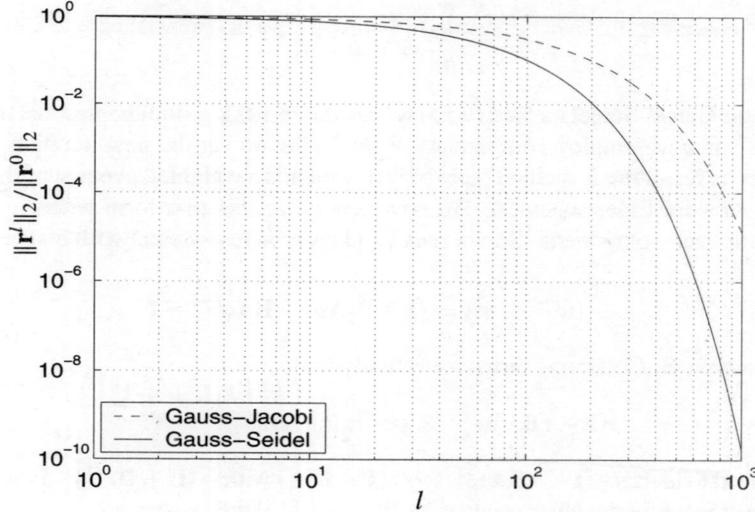

Figure 9.7. *Convergence history of the Gauss–Jacobi and Gauss–Seidel iterative methods for the Dirichlet problem* (9.40) *with* $M = 20$.

9.4.3 Closure of Iterative Methods; Operator Splitting

Iterative methods change for increasing l and are therefore evolutionary in character. It is sometimes possible to write an iterative scheme as a time integration method for a set of

ODEs. This set of ODEs is then called the *time closure* of the method. To start with we show this for the Gauss–Jacobi method. Recall from (9.49), with $\mathbf{D} = -\frac{4}{h^2}\mathbf{I}$, that

$$\mathbf{u}^{l+1} = \frac{h^2}{4}\left(\mathbf{L}+\mathbf{U}\right)\mathbf{u}^l - \frac{h^2}{4}\mathbf{f}. \tag{9.57}$$

Hence we find the increment

$$\mathbf{u}^{l+1} - \mathbf{u}^l = \frac{h^2}{4}\left(\mathbf{L}+\mathbf{U}+\mathbf{D}\right)\mathbf{u}^l - \frac{h^2}{4}\mathbf{f} = \frac{h^2}{4}\left(\mathbf{A}\mathbf{u}^l - \mathbf{f}\right). \tag{9.58}$$

Introducing the artificial time step $\tau := h^2/4$, we see that (9.58) actually amounts to

$$\frac{1}{\tau}\left(\mathbf{u}^{l+1} - \mathbf{u}^l\right) = \mathbf{A}\mathbf{u}^l - \mathbf{f}. \tag{9.59}$$

This is nothing but an explicit Euler discretisation with step size τ of the ODE

$$\frac{d\mathbf{u}}{dt} = \mathbf{A}\mathbf{u} - \mathbf{f}. \tag{9.60}$$

Hence (9.60) may be called the time closure of (9.49), i.e., the continuous version of the iterative method. In fact, the *closure* of the method as such is the parabolic PDE

$$\frac{\partial u}{\partial t} = \nabla^2 u - f.$$

For the Gauss–Seidel method (9.55) we can do something similar; note that the method is implicit since we employ components of the "solution" at the new iteration level; see (9.54). Let us recall the ϑ method (see (5.72)), which is a weighted average of the forward and the backward Euler schemes. We now generalize this to a form where ϑ might be different for each component. So we seek to identify Gauss–Seidel with a scheme of the form

$$\frac{1}{\tau}\left(\mathbf{u}^{l+1} - \mathbf{u}^l\right) = \left(\mathbf{I}-\mathbf{B}\right)\mathbf{A}\mathbf{u}^l + \mathbf{B}\mathbf{A}\mathbf{u}^{l+1} - \mathbf{f} \tag{9.61}$$

for some matrix $\mathbf{B}$. Collecting terms we actually find

$$\left(\mathbf{I} - \tau\mathbf{B}\mathbf{A}\right)\mathbf{u}^{l+1} = \left(\mathbf{I} + \tau(\mathbf{I}-\mathbf{B})\mathbf{A}\right)\mathbf{u}^l - \tau\mathbf{f}. \tag{9.62}$$

We now need to identify $(\mathbf{I} - \tau\mathbf{B}\mathbf{A})^{-1}(\mathbf{I} + \tau(\mathbf{I}-\mathbf{B})\mathbf{A})$ with $-(\mathbf{L}+\mathbf{D})^{-1}\mathbf{U}$. It makes sense to scale the factors in the latter product by $\mathbf{D} = -\frac{4}{h^2}\mathbf{I}$. Hence we try

$$\mathbf{I} - \tau\mathbf{B}\mathbf{A} = -\frac{1}{4}h^2\mathbf{L} + \mathbf{I}, \tag{9.63a}$$

$$\mathbf{I} + \tau(\mathbf{I}-\mathbf{B})\mathbf{A} = \frac{1}{4}h^2\mathbf{U}. \tag{9.63b}$$

From this we obtain the expression

$$\mathbf{B} = \mathbf{L}\mathbf{A}^{-1} = \mathbf{I} - \left(\mathbf{D}+\mathbf{U}\right)\mathbf{A}^{-1}. \tag{9.64}$$

We can then rewrite (9.62) as an (implicit) time step scheme for solving (9.60) as

$$\frac{1}{\tau}(\mathbf{u}^{l+1} - \mathbf{u}^{l}) = (\mathbf{D} + \mathbf{U})\mathbf{u}^{l} + \mathbf{L}\mathbf{u}^{l+1} - \mathbf{f}, \tag{9.65}$$

from which we find the time closure of the Gauss–Seidel method

$$\frac{\mathrm{d}\mathbf{u}}{\mathrm{d}t} = (\mathbf{D} + \mathbf{U})\mathbf{u} + \mathbf{L}\mathbf{u} - \mathbf{f} = \mathbf{A}\mathbf{u} - \mathbf{f}.$$

Apparently, we have split the operator into two terms. This type of scheme, of which (9.65) is just an example, is also referred to as *operator splitting*.

It is possible to split the operator in many other ways. We shall consider here one of these, i.e., *alternating direction implicit (ADI)*. The idea is to split the matrix $\mathbf{A}$ into a matrix derived from discretising $\frac{\partial^2}{\partial x^2}u$ and $\frac{\partial^2}{\partial y^2}u$. In an obvious notation we have

$$\mathbf{A} = \mathbf{A}_x + \mathbf{A}_y. \tag{9.66}$$

The scheme then reads

$$\mathbf{u}^{l+1} = \mathbf{u}^{l} + \tau\mathbf{A}_x\mathbf{u}^{l} + \tau\mathbf{A}_y\mathbf{u}^{l+1} - \tau\mathbf{f}, \tag{9.67a}$$

$$\mathbf{u}^{l+2} = \mathbf{u}^{l+1} + \tau\mathbf{A}_x\boldsymbol{u}^{l+2} + \tau\mathbf{A}_y\boldsymbol{u}^{l+1} - \tau\mathbf{f}. \tag{9.67b}$$

The "implicit" direction is used for the discretisation in the y step first (and only!) and then for the discretisation in the x step next. We need to perform two steps per ADI iteration. We clearly have two discretisations of (9.60). A more detailed convergence analysis is therefore deferred to Section 11.8.1. The two-step iteration is

$$(\mathbf{I} - \tau\mathbf{A}_y)\mathbf{u}^{l+1} = (\mathbf{I} + \tau\mathbf{A}_x)\boldsymbol{u}^{l} - \tau\mathbf{f}, \tag{9.68a}$$

$$(\mathbf{I} - \tau\mathbf{A}_x)\mathbf{u}^{l+2} = (\mathbf{I} + \tau\mathbf{A}_y)\mathbf{u}^{l+1} - \tau\mathbf{f}. \tag{9.68b}$$

In contrast to Gauss–Jacobi and Gauss–Seidel, scheme (9.68) contains a matrix on the left that does not allow for a direct solve. However, both $\mathbf{I} - \tau\mathbf{A}_y$ and $\mathbf{I} - \tau\mathbf{A}_x$ are tridiagonal, so that each system allows for a simple (and cheap) LU-decomposition.

9.5 Green's Functions

The linear systems in (9.33) and (9.44) are analogues of the continuous (partial) differential equation and the boundary conditions combined. It is therefore no surprise that one can find a discrete analogue to the Green's function. Indeed, consider the system of equations

$$\mathbf{A}\,\mathbf{v}^j = \mathbf{e}^j, \qquad \mathbf{e}^j := (\,0, \ldots, 0, \underbrace{1}_{j\text{th position}}, 0, \ldots, 0\,)^T. \tag{9.69}$$

The vector $\mathbf{e}^j$ can be seen as a discrete "Dirac function." By construction the corresponding grid function v^j_Δ satisfies the difference equation (9.2) (or (9.31) in the one-dimensional case) with *homogeneous* source term, except for one point, where this source term equals one; furthermore, it satisfies homogeneous boundary conditions. In operator form we have

$$\mathcal{L}_\Delta[v^j_\Delta](\boldsymbol{x}_k) = \delta_{j,k}, \qquad \boldsymbol{x}_k \in \Omega_\Delta, \tag{9.70a}$$

$$v^j_\Delta(\boldsymbol{x}_k) = 0, \qquad \boldsymbol{x}_k \in \partial\Omega_\Delta, \tag{9.70b}$$

where $\boldsymbol{x}_k$ is the kth grid point in Ω_Δ, assuming some ordering of the grid points. Therefore v^j_Δ and its vector representation $\mathbf{v}^j$ are called a *discrete Green's function.*

Apparently, the Green's functions $\mathbf{v}^j$ are the columns of the inverse of $\mathbf{A}$; i.e.,

$$\mathbf{A}^{-1} = \left(\mathbf{v}^1 | \, \mathbf{v}^2 | \cdots | \, \mathbf{v}^N\right), \tag{9.71}$$

with N the number of grid points. It is interesting to see that these discrete Green's functions can be used to construct a particular solution of the following inhomogeneous problem with homogeneous boundary conditions:

$$\mathcal{L}_\Delta[v_\Delta](\boldsymbol{x}_k) = f(\boldsymbol{x}_k), \qquad \boldsymbol{x}_k \in \Omega_\Delta, \tag{9.72a}$$

$$v_\Delta(\boldsymbol{x}_k) = 0, \qquad \boldsymbol{x}_k \in \partial\Omega_\Delta. \tag{9.72b}$$

Comparing (9.70) and (9.72), we conclude that

$$v(\boldsymbol{x}_k) = \sum_{j=1}^{N} f(\boldsymbol{x}_j) v^j_\Delta(\boldsymbol{x}_k), \qquad \boldsymbol{x}_k \in \Omega_\Delta, \tag{9.73}$$

or, in matrix-vector notation,

$$\mathbf{v} = \mathbf{A}^{-1}\mathbf{f}. \tag{9.74}$$

There is a close analogy with the continuous case, dealt with in Section 8.5. Note that Green's functions and fundamental solutions differ only by a ("discrete") harmonic function. The complete solution u_Δ of the inhomogeneous problem

$$\mathcal{L}_\Delta[u_\Delta](\boldsymbol{x}_k) = f(\boldsymbol{x}_k), \qquad \boldsymbol{x}_k \in \Omega_\Delta, \tag{9.75a}$$

$$u_\Delta(\boldsymbol{x}_k) = g(\boldsymbol{x}_k), \qquad \boldsymbol{x}_k \in \partial\Omega_\Delta, \tag{9.75b}$$

is given by

$$u_\Delta(\boldsymbol{x}_k) = v_\Delta(\boldsymbol{x}_k) + w_\Delta(\boldsymbol{x}_k), \qquad \boldsymbol{x}_k \in \Omega_\Delta, \tag{9.76}$$

with v_Δ the solution of (9.72) and w_Δ the solution of the following homogeneous problem with inhomogeneous boundary conditions:

$$\mathcal{L}_\Delta[w_\Delta](\boldsymbol{x}_k) = 0, \qquad \boldsymbol{x}_k \in \Omega_\Delta, \tag{9.77a}$$

$$w_\Delta(\boldsymbol{x}_k) = g(\boldsymbol{x}_k), \qquad \boldsymbol{x}_k \in \partial\Omega_\Delta. \tag{9.77b}$$

In matrix-vector formulation we have

$$\mathbf{A}\mathbf{w} = \mathbf{g}, \tag{9.78}$$

where the vector $\mathbf{g}$ has zero components except for points where the scheme includes boundary points; at such a point the component is equal to minus the coefficient in the difference equation times the value at that boundary (or, when there are more, a sum of such contributions). As before, (9.76) is an instance of the *superposition principle*; cf. (8.60). We thus have

$$\mathbf{u} = \mathbf{v} + \mathbf{w}, \qquad \mathbf{A}\mathbf{u} = \mathbf{f} + \mathbf{g}. \tag{9.79}$$

Example 9.28 Consider the boundary value problem

$$\mathcal{L}[u] := \frac{\mathrm{d}^2 u}{\mathrm{d}x^2} = f(x), \qquad x \in (0,1),$$
$$x(0) = A, \qquad x(1) = B.$$

The fundamental solution of this differential equation is given by $u(x) = A + Bx$. Since this is linear it will be integrated exactly by the central difference scheme and hence it is also the solution of the discrete problem

$$\mathcal{L}_\Delta[u_\Delta](x_i) = f(x_i), \qquad i = 1, 2, \ldots, M$$
$$u_0 = A, \qquad u_{M+1} = B,$$

where $\mathcal{L}_\Delta$ is the central difference operator defined on the grid

$$x_j := j\Delta x, \qquad j = 0, 1, \ldots, M+1, \quad \Delta x = 1/(M+1).$$

The resulting coefficient matrix of the linear system $\mathbf{A}\,\mathbf{u} = \mathbf{f}$ reads

$$\mathbf{A} := \frac{1}{\Delta x^2}\begin{pmatrix} -2 & 1 & & & \\ 1 & -2 & 1 & & \\ & \ddots & \ddots & \ddots & \\ & & 1 & -2 & 1 \\ & & & 1 & -2 \end{pmatrix}.$$

The solution of $\mathbf{A}\,\mathbf{v}^j = \mathbf{e}^j$ can now be constructed as follows: For $i = 0, 1, \ldots, j$ it must have the form

$$v_i^j = B_1^j x_i$$

for some B_1^j. For $i = j, j+1, \ldots, M+1$ it must have the form

$$v_i^j = B_2^j (1 - x_i)$$

for some B_2^j. Note that this function v^j satisfies the homogeneous boundary conditions $v^j(x_0) = v^j(x_{M+1}) = 0$. At $i = j$ we have

$$v_{j-1}^j - 2v_j^j + v_{j+1}^j = \Delta x^2, \qquad B_1^j x_j = B_2^j (1 - x_j).$$

From these relations we can determine B_1^j and B_2^j. We obtain

$$B_1^j \left(x_{j-1} - 2x_j + \frac{x_j}{1 - x_j}(1 - x_{j+1}) \right) = \Delta x^2.$$

Hence

$$B_1^j = -\Delta x\,(1 - x_j), \qquad B_2^j = -\Delta x\, x_j.$$

So we find for the discrete Green's function

$$v_i^j = \begin{cases} \Delta x\,(x_j - 1)x_i & \text{for } i = 0, 1, \ldots, j, \\ \Delta x\, x_j(x_i - 1) & \text{for } i = j+1, j+2, \ldots, M+1. \end{cases}$$

We immediately find, for the inverse of the matrix $\mathbf{A}$,

$$\mathbf{A}^{-1} = -\frac{1}{(M+1)^3}\begin{pmatrix} M & M-1 & M-2 & \ldots & 1 \\ M-1 & (M-1)2 & (M-2)2 & \ldots & 2 \\ M-2 & (M-2)2 & (M-2)3 & \ldots & 3 \\ M-3 & (M-3)2 & (M-3)3 & \ldots & 4 \\ \vdots & \vdots & \vdots & & \vdots \\ 1 & 2 & 3 & \ldots & M \end{pmatrix}. \qquad \square$$

9.6 Nonlinear Problems

For nonlinear problems we have to adapt our solution procedure. We shall only consider PDEs with a linear differential operator. So we consider

$$\mathcal{L}[u] = f(u), \qquad \boldsymbol{x} \in \Omega, \tag{9.80a}$$

$$u = g(\boldsymbol{x}), \qquad \boldsymbol{x} \in \partial\Omega, \tag{9.80b}$$

where $f(u)$ is nonlinear in u. After discretisation, we obtain a nonlinear system of equations for the unknowns at the grid points, say $\mathbf{u} \in \mathbb{R}^N$:

$$\mathbf{N}(\mathbf{u}) = \mathbf{0}. \tag{9.81}$$

Since nonlinear problems occur quite frequently, we shall discuss three often used approaches: Newton iteration in Section 9.6.1, Gauss–Jacobi iteration in Section 9.6.2, and transient methods in Section 9.6.3.

9.6.1 Newton Iteration

The rationale behind Newton's method is the linearisation of $\mathbf{N}(\mathbf{u})$ at an approximation of the solution, $\tilde{\mathbf{u}}$ say:

$$\mathbf{N}(\mathbf{v}) \doteq \mathbf{N}(\tilde{\mathbf{u}}) + \mathbf{J}(\tilde{\mathbf{u}})\,(\mathbf{v} - \tilde{\mathbf{u}}). \tag{9.82}$$

Here $\mathbf{J}(\mathbf{u})$ is the Jacobi matrix of $\mathbf{N}(\mathbf{u})$, defined by

$$\mathbf{J}(\mathbf{u}) := \frac{\partial \mathbf{N}(\mathbf{u})}{\partial \mathbf{u}} = \left(\frac{\partial N_i(\mathbf{u})}{\partial u_j}\right). \tag{9.83}$$

Anticipating $\mathbf{v}$ to be a better approximation of the exact solution $\mathbf{u}$ than $\tilde{\mathbf{u}}$, we take $\mathbf{N}(\mathbf{v}) = \mathbf{0}$, and so we obtain, assuming $\mathbf{J}(\tilde{\mathbf{u}})$ is nonsingular,

$$\mathbf{v} - \tilde{\mathbf{u}} = -\,\mathbf{J}^{-1}(\tilde{\mathbf{u}})\,\mathbf{N}(\tilde{\mathbf{u}}). \tag{9.84}$$

Let $\mathbf{u}^0$ be some known approximate value. Then we have *Newton's method*:

$$\begin{aligned}
&\text{given,}\quad \mathbf{u}^0,\\
&\text{solve}\ \ \mathbf{J}(\mathbf{u}^l)\mathbf{s}^l = -\mathbf{N}(\mathbf{u}^l),\\
&\mathbf{u}^{l+1} = \mathbf{u}^l + \mathbf{s}^l.
\end{aligned} \tag{9.85}$$

The quantity $\mathbf{s}^l$ is called the *Newton update*. Note that $\mathbf{s}^l$ satisfies a linear equation, which can therefore be solved by a method described in Section 9.4. The convergence of the Newton method (9.85) is assessed in the following famous theorem, which we give without proof (see [68]).

Theorem 9.29 (Newton–Kantorovich). *Let* $\mathbf{N}(\mathbf{u})$ *be continuously differentiable in a neighbourhood* $\mathcal{B}(\mathbf{u}^0;\rho)$ *of* $\mathbf{u}^0$ *(i.e., a ball with centre* $\mathbf{u}^0$ *and radius* ρ*) and let the constants* C_1, C_2, *and* C_3 *be such that*

$$\begin{aligned}
\left\|\mathbf{J}^{-1}(\mathbf{u}^0)\mathbf{N}(\mathbf{u}^0)\right\| &\le C_1,\\
\left\|\mathbf{J}^{-1}(\mathbf{u}^0)\right\| &\le C_2,\\
\left\|\mathbf{J}(\mathbf{u}_1)-\mathbf{J}(\mathbf{u}_2)\right\| &\le C_3\left\|\mathbf{u}_1-\mathbf{u}_2\right\|, \qquad \mathbf{u}_1,\mathbf{u}_2\in\mathcal{B}(\mathbf{u}^0;\rho).
\end{aligned}$$

Define $\sigma := C_1C_2C_3$. *If* $\sigma \le \frac{1}{2}$ *and* $\rho \ge \rho_- := (1-\sqrt{1-2\sigma})/(C_2C_3)$, *then the following hold:*

(i) *There is a unique root* $\mathbf{u}^* \in \mathcal{B}(\mathbf{u}^0;\rho_-)$.

(ii) *The sequence of Newton iterates* $\{\mathbf{u}^l\}$ *converges to* $\mathbf{u}^*$.

If $\sigma < \frac{1}{2}$, *then, moreover,*

(iii) $\mathbf{u}^*$ *is the unique root in* $\mathcal{B}(\mathbf{u}^0;\min(\rho,\rho_+))$, *where* $\rho_+ := (1+\sqrt{1-2\sigma})/(C_2C_3)$.

(iv) $\left\|\mathbf{u}^l-\mathbf{u}^*\right\| \le (2\sigma)^{2^m}\frac{1}{C_2C_3}$.

Unfortunately, this theorem does not reveal that the method is actually asymptotically quadratically convergent. The main problem, however, is to find a suitable initial guess $\mathbf{u}^0$. The Newton–Kantorovich theorem at least shows that this will be less problematic the smoother f is.

The Jacobi matrix $\mathbf{J}(\mathbf{u})$ has a special form for our problem (9.80). Let $\mathbf{A}$ be the matrix that would have been found from discretising $\mathcal{L}[u]=0$. Then we have

$$\mathbf{J}(\mathbf{u}) = \mathbf{A} - \operatorname{diag}\left(f'(u_i)\right). \tag{9.86}$$

If we assume that $f'(u) > 0$, then the important requirement (9.4a) in Property 9.1 is certainly satisfied (even in a strict inequality sense).

Example 9.30 Consider the one-dimensional problem

$$\begin{aligned}
&\mathcal{L}[u] := \frac{\mathrm{d}^2u}{\mathrm{d}x^2} = -\alpha\mathrm{e}^u, \qquad x\in(0,1),\\
&\frac{\mathrm{d}u}{\mathrm{d}x}(0) = 0, \qquad u(1)=0,
\end{aligned}$$

where $\alpha > 0$ is some parameter. This boundary value problem only has a solution for $0 < \alpha < \alpha^* := 0.878457764$ [3]. If we use central differences on the grid

$$x_j := j\Delta x, \qquad j = 0, 1, \ldots, M+1, \quad \Delta x = 1/(M+1),$$

we obtain the following nonlinear system for $\mathbf{u} := (u_1, u_2, \ldots, u_M)^T$

$$\mathbf{N}(\mathbf{u}) := \mathbf{A}\mathbf{u} + \alpha \operatorname{diag}\left(\mathrm{e}^{u_1}, \mathrm{e}^{u_2}, \ldots, \mathrm{e}^{u_M}\right) = \mathbf{0}, \tag{$*$}$$

with

$$\mathbf{A} = \frac{1}{\Delta x^2}\begin{pmatrix} -2 & 2 & & & \\ 1 & -2 & 1 & & \\ & \ddots & \ddots & \ddots & \\ & & 1 & -2 & 1 \\ & & & 1 & -2 \end{pmatrix}.$$

The Jacobi matrix for this problem is obviously

$$\mathbf{J}(\mathbf{u}) = \mathbf{A} + \alpha \operatorname{diag}\left(\mathrm{e}^{u_1}, \mathrm{e}^{u_2}, \ldots, \mathrm{e}^{u_M}\right).$$

We have solved this boundary value problem for several values of α between 0.5 and α^* on a grid with $\Delta x = 0.01$. The nonlinear system $(*)$ is solved with Newton's method, and it is continued until $\|\mathbf{N}(\mathbf{u}^l)\|_2 <$ tol and $\|\mathbf{s}^l\|_2 <$ tol for some tolerance tol. Convergence is very fast; typically 5 to 10 iterations are needed for tol $= 10^{-4}$. This value is of the same order of magnitude as the discretisation error; a smaller value is not needed since the numerical solution contains a discretisation error anyway. □

In practice the Newton process does not converge without additional effort. To understand how to improve the method it is useful to define the object function

$$q(\mathbf{u}) := \|\mathbf{N}(\mathbf{u})\|_2^2 = \mathbf{N}(\mathbf{u})^T\,\mathbf{N}(\mathbf{u}). \tag{9.87}$$

The Newton update $\mathbf{s}$ is a *descent direction*; i.e., it tends to smaller values of $q(\mathbf{u})$. Since $\nabla q(\mathbf{u}) = 2\mathbf{J}^T(\mathbf{u})\mathbf{N}(\mathbf{u})$, we readily see that

$$\mathbf{s}^T\,\nabla q(\mathbf{u}) = 2\,\mathbf{s}^T\,\mathbf{J}^T\,\mathbf{N} = -2q(\mathbf{u}) < 0. \tag{9.88}$$

If a certain Newton step does not give a smaller value of the object function, only a fraction may do a better job. Indeed, from a Taylor expansion we see

$$q(\mathbf{u} + \lambda\,\mathbf{s}) = q(\mathbf{u}) + \lambda\,\mathbf{s}^T\nabla q(\mathbf{u}) + \mathcal{O}\big(\lambda^2\|\mathbf{s}\|_2^2\big). \tag{9.89}$$

If $\lambda\|\mathbf{s}\|_2$ is small enough, we see that ($\lambda > 0$!)

$$q(\mathbf{u} + \lambda\mathbf{s}) < q(\mathbf{u}).$$

Such a λ is called a *damping factor* and the variant is called *damped Newton*:

$$\begin{aligned} &\text{given,}\ \ \mathbf{u}^0, \\ &\text{solve}\ \ \mathbf{J}(\mathbf{u}^l)\mathbf{s}^l = -\mathbf{N}(\mathbf{u}^l), \\ &\mathbf{u}^{l+1} = \mathbf{u}^l + \lambda_l\,\mathbf{s}^l. \end{aligned} \tag{9.90}$$

Although one may try to find "optimal" values for λ_l from the model above, an often used method is to halve the damping factor if "convergence" fails (and repeat this until a smaller value of the object function is reached); if such a decrease is found, one then only cautiously increases the damping parameter; for more details see, e.g., [3].

Another obvious modification is to keep the Jacobi matrix fixed during a few iterations. Only at the very end, when quadratic convergence is becoming visible, might a new Jacobi evaluation at every step be meaningful.

9.6.2 Gauss–Jacobi Iteration

Rather than using the complete operator in Newton's method, we can use a Gauss–Jacobi or Gauss–Seidel splitting first. So let us rewrite the discrete nonlinear equation (9.81) as

$$\mathbf{A}\mathbf{u} = \mathbf{f}(\mathbf{u}), \tag{9.91}$$

with $\mathbf{A}$ the matrix resulting from discretising the homogeneous problem $\mathcal{L}[u] = 0$. In the rest of this section we take for $\mathcal{L}$ the two-dimensional Laplacian. If we split the matrix $\mathbf{A}$ as in (9.47), we obtain the nonlinear variant of the Gauss–Jacobi iteration

$$\mathbf{D}\mathbf{u}^{l+1} - \mathbf{f}(\mathbf{u}^{l+1}) = -(\mathbf{L} + \mathbf{U})\mathbf{u}^{l}. \tag{9.92}$$

An alternative to Newton's method is employing some form of operator splitting; see Section 9.4.3. Since $f(u)$ is nonlinear, it is convenient to leave this out of the implicit part. So, alternatively, we may use the following form, which is linear in $\mathbf{u}^{l+1}$:

$$\mathbf{D}\mathbf{u}^{l+1} = -(\mathbf{L} + \mathbf{U})\mathbf{u}^{l} + \mathbf{f}(\mathbf{u}^{l}). \tag{9.93}$$

Likewise, a nonlinear Gauss–Seidel variant is

$$(\mathbf{L} + \mathbf{D})\mathbf{u}^{l+1} = -\mathbf{U}\mathbf{u}^{l} + \mathbf{f}(\mathbf{u}^{l}). \tag{9.94}$$

In order to prove convergence of (9.93), we need the following lemma. Note that symmetric matrices have real eigenvalues.

Lemma 9.31. *Let* $\mathbf{B}$ *and* $\mathbf{C}$ *be symmetric matrices. Denote by* $\lambda_1, \ldots, \lambda_M$, $\mu_1, \ldots, \mu_M$, *and* $\nu_1, \ldots, \nu_M$ *the eigenvalues of* $\mathbf{B}$, $\mathbf{B} - \mathbf{C}$, *and* $\mathbf{C}$, *respectively. Then*

$$\begin{aligned} \max_i \lambda_i &\geq \max_j \mu_j + \min_k \nu_k, \\ \min_i \lambda_i &\leq \min_j \mu_j + \max_k \nu_k. \end{aligned} \tag{9.95}$$

Proof. The maximum eigenvalue of $\mathbf{B}$ satisfies

$$\begin{aligned} \max_i \lambda_i &= \max_{\|\mathbf{u}\|_2=1} \mathbf{u}^T\mathbf{B}\mathbf{u} \\ &= \max_{\|\mathbf{u}\|_2=1} \left(\mathbf{u}^T(\mathbf{B} - \mathbf{C})\mathbf{u} + \mathbf{u}^T\mathbf{C}\mathbf{u}\right) \\ &\geq \mathbf{v}^T(\mathbf{B} - \mathbf{C})\mathbf{v} + \mathbf{v}^T\mathbf{C}\mathbf{v} \\ &\geq \mathbf{v}^T(\mathbf{B} - \mathbf{C})\mathbf{v} + \min_k \nu_k, \end{aligned}$$

where $\mathbf{v}$ is the eigenvector, with $\|\mathbf{v}\|_2 = 1$, belonging to the largest eigenvalue of $\mathbf{B} - \mathbf{C}$, so

$$\mathbf{v}^T(\mathbf{B} - \mathbf{C})\mathbf{v} = \max_j \mu_j.$$

If we consider the minimum eigenvalue of $\mathbf{B}$, we obtain

$$\begin{aligned}\min_i \lambda_i &= \min_{\|\mathbf{u}\|_2=1} \mathbf{u}^T\mathbf{B}\mathbf{u} \\ &\leq \mathbf{w}^T(\mathbf{B} - \mathbf{C})\mathbf{w} + \mathbf{w}^T\mathbf{C}\mathbf{w} \\ &\leq \mathbf{w}^T(\mathbf{B} - \mathbf{C})\mathbf{w} + \max_k \nu_k,\end{aligned}$$

where $\mathbf{w}$ is the eigenvector, with $\|\mathbf{w}\|_2 = 1$, belonging to the smallest eigenvalue of $\mathbf{B} - \mathbf{C}$, so

$$\mathbf{v}^T(\mathbf{B} - \mathbf{C})\mathbf{v} = \min_j \mu_j.$$

From this the result immediately follows. □

We can now show the following property.

Property 9.32. *Assume we use central differences to discretise boundary value problem* (9.80). *Let C be a constant such that $f'(u) \geq C > 0$ for all u. Then the Gauss–Jacobi method* (9.93) *converges and the error $\mathbf{e}^l := \mathbf{u}^l - \mathbf{u}$ satisfies*

$$\|\mathbf{e}^{l+1}\|_2 \leq \left(1 - h^2\left(\frac{1}{2}\pi^2 + \frac{1}{4}C\right)\right)\|\mathbf{e}^l\|_2.$$

Proof. For the error vector $\mathbf{e}^l$ in the Gauss–Jacobi iteration we obtain

$$\mathbf{D}\mathbf{e}^{l+1} = -(\mathbf{L} + \mathbf{U})\mathbf{e}^l + \mathbf{f}(\mathbf{u}^l) - \mathbf{f}(\mathbf{u}) = -\left(\mathbf{L} + \mathbf{U} - \operatorname{diag}\left(f'(v_i^l)\right)\right)\mathbf{e}^l,$$

where v_i^l is some intermediate value between u_i and u_i^l. We write this simply as

$$-\mathbf{D}\mathbf{e}^{l+1} = (\mathbf{L} + \mathbf{U} - \mathbf{E}^l)\mathbf{e}^l, \qquad \mathbf{E}^l := \operatorname{diag}\left(f'(v_i^l)\right).$$

Note that $\mathbf{E}^l$ is a diagonal matrix with positive entries since $f'(u) > 0$. Hence

$$\mathbf{e}^{l+1} = \left(-\mathbf{D}^{-1}(\mathbf{L} + \mathbf{U}) + \mathbf{D}^{-1}\mathbf{E}^l\right)\mathbf{e}^l.$$

Taking the 2-norm, we obtain the upper bound

$$\|\mathbf{e}^{l+1}\|_2 \leq \left\|\mathbf{D}^{-1}(\mathbf{L} + \mathbf{U}) - \mathbf{D}^{-1}\mathbf{E}^l\right\|_2 \|\mathbf{e}^l\|_2.$$

We can now invoke Lemma 9.31 to estimate this matrix norm. Let us identify $\mathbf{D}^{-1}(\mathbf{L} + \mathbf{U})$ with $\mathbf{B}$ and $\mathbf{D}^{-1}\mathbf{E}^l$ with $\mathbf{C}$. Moreover, we remark that the eigenvalues of $\mathbf{D}^{-1}(\mathbf{L} + \mathbf{U})$ can be found from (9.53); in particular, we obtain in good approximation $\lambda_1 := \min_j \lambda_j \doteq -1 + \frac{(\pi h)^2}{2}$ and $\lambda_M := \max_j \lambda_j \doteq 1 - \frac{(\pi h^2)}{2}$. We thus obtain in good approximation

$$-1 + \frac{(\pi h)^2}{2} - \max_k \nu_k \leq \mu_j \leq 1 - \frac{(\pi h)^2}{2} - \min_k \nu_k.$$

Apparently the matrix $\mathbf{D}^{-1}(\mathbf{L}+\mathbf{U}) - \mathbf{D}^{-1}\mathbf{E}^l$ is symmetric and positive definite. Since $\nu_k \geq \frac{h^2}{4}C$, we see that

$$\left\|\mathbf{D}^{-1}(\mathbf{L}+\mathbf{U}) - \mathbf{D}^{-1}\mathbf{E}^l\right\|_2 \leq 1 - h^2\left(\frac{1}{2}\pi^2 + \frac{1}{4}C\right). \qquad \square$$

Note that the convergence is slightly enhanced by a larger value of C in the property above. Actually, the result is not essentially different from that of a linear problem where the PDE has a term proportional to u. For Gauss–Seidel a similar result holds.

9.6.3 Transient Methods

Quite another approach is to invoke the time closure of (9.91) directly, i.e.,

$$\frac{\mathrm{d}\mathbf{u}}{\mathrm{d}t} = \mathbf{A}\mathbf{u} - \mathbf{f}(\mathbf{u}). \tag{9.96}$$

The time closure automatically gives the proper sign, i.e., not $-\mathbf{A}\mathbf{u} + \mathbf{f}(\mathbf{u})$, which could have been chosen as well from a purely theoretical point of view. The basic idea is to compute the steady state solution of (9.96) by applying a suitable time integration technique; cf. Section 5.4. If we have an inaccurate initial guess $\mathbf{u}^0$, we expect to need many time steps to arrive at the steady state where $\frac{\mathrm{d}\mathbf{u}}{\mathrm{d}t} = \mathbf{0}$. This kind of approach is often referred to as the (false) *transient method*. Note that (9.96) is the semidiscretisation of the parabolic equation

$$\frac{\partial u}{\partial t} = \mathcal{L}[u] - f(u). \tag{9.97}$$

Solution methods for this equation are deferred to Chapter 11.

Since we are only interested in the steady state solution of (9.96), the stability of a time integrator is of more importance than high accuracy. Therefore the implicit Euler method is a good candidate, resulting in the following nonlinear system for the unknown $\mathbf{v} = \mathbf{u}^{n+1}$:

$$\mathbf{F}(\mathbf{v}) := \left(\mathbf{A} - \frac{1}{\Delta t}\mathbf{I}\right)\mathbf{v} - \mathbf{f}(\mathbf{v}) + \frac{1}{\Delta t}\mathbf{u}^n = \mathbf{0}, \tag{9.98}$$

with Δt the time step and $\mathbf{u}^n$ the approximation of $\mathbf{u}$ at time level $n\Delta t$. To solve (9.96) we can suitably use Newton iteration. Note that the Jacobi matrix $\mathbf{J}_F$ for (9.98) is given by

$$\mathbf{J}_F = \mathbf{A} - \frac{1}{\Delta t}\mathbf{I} - \operatorname{diag}\left(f'(v_i)\right). \tag{9.99}$$

It contains the extra term $(1/\Delta t)\mathbf{I}$ compared to the Jacobi matrix in (9.86), thus enhancing condition (9.4a) in Property 9.1. As a result, $\operatorname{cond}(\mathbf{J}_F)$ is often smaller than $\operatorname{cond}(\mathbf{J})$ and Newton iteration for (9.98) usually converges well. We have to solve (9.98) repeatedly until $\|\mathbf{u}^{n+1} - \mathbf{u}^n\| < \Delta t\,\mathrm{tol}$ for some tolerance tol.

We conclude this section by noting that Newton's method also has a time closure. Consider the damped Newton iteration

$$\mathbf{u}^{l+1} - \mathbf{u}^l = -\lambda_l \mathbf{J}^{-1}(\mathbf{u}^l)\mathbf{N}(\mathbf{u}^l). \tag{9.100}$$

Then, letting $\lambda_l \to 0$, we arrive at

$$\frac{\mathrm{d}\mathbf{u}}{\mathrm{d}t} = -\mathbf{J}^{-1}(\mathbf{u})\mathbf{N}(\mathbf{u}), \tag{9.101}$$

the *Davidenko equation*. Here, too, one can use a transient method as a practical way to determine the damping factor. Note that apparently λ_l should not be larger than one. This is intuitively more restrictive than a stability requirement for the Euler forward method would give.

9.7 A Pressure Correction Method for the Stokes Equations

In Section 8.7 we encountered the Stokes equations

$$\nabla^2 \boldsymbol{u} - \nabla p = \mathbf{0}, \tag{9.102a}$$

$$\nabla \cdot \boldsymbol{u} = 0, \tag{9.102b}$$

which describe creeping, steady flow of an incompressible fluid in a domain $\Omega \subset \mathbb{R}^d$ ($d = 2, 3$). In this section we restrict ourselves to two-dimensional problems; the extension to three dimensions is straightforward. In system (9.102) $\boldsymbol{u} = (u(x, y), v(x, y))^T$ and $p = p(x, y)$ are the velocity and the pressure field, respectively. We assume that we have Dirichlet boundary conditions for $\boldsymbol{u}$; i.e.,

$$\boldsymbol{u} = \boldsymbol{b}, \qquad \boldsymbol{x} \in \partial\Omega. \tag{9.103}$$

These boundary conditions have to satisfy the constraint (cf. (9.102b))

$$\oint_{\partial\Omega} \boldsymbol{b} \cdot \boldsymbol{n} \,\mathrm{d}s = 0, \tag{9.104}$$

with $\boldsymbol{n}$ the unit outward normal on $\partial\Omega$.

If we apply the divergence operator to the momentum equations (9.102a) and invoke the incompressibility condition (9.102b), we obtain the *pressure Poisson equation*

$$\nabla^2 p = 0; \tag{9.105}$$

i.e., the pressure is a harmonic function. Conversely, from (9.102a) and (9.105) we can simply deduce that $\nabla^2 d = 0$ with $d := \nabla \cdot \boldsymbol{u}$, and this equation has only the trivial solution $d(\boldsymbol{x}) \equiv 0$ if $d = \nabla \cdot \boldsymbol{u} = 0$ on the boundary $\partial\Omega$. This is then the proper boundary condition for (9.105). It can be shown that $\nabla \cdot \boldsymbol{u} = 0$ on the boundary implies the Neumann condition

$$\frac{\partial p}{\partial \boldsymbol{n}} = \boldsymbol{n} \cdot \nabla^2 \boldsymbol{u}, \qquad \boldsymbol{x} \in \partial\Omega, \tag{9.106}$$

which is just the normal component of (9.102a) applied at the boundary; cf. [55]. So we have two alternative formulations for the Stokes problem: either (9.102) with boundary condition (9.103) or (9.102a) and (9.105) together with the boundary conditions (9.103) and (9.106). In both formulations p is determined up to an additive constant. In the rest of this section we will derive a numerical scheme for (9.102).

In order to compute a numerical solution for (9.102), we cover the domain Ω with a uniform grid with grid sizes Δx and Δy in the x and y directions, respectively, and apply a finite difference scheme. First, we consider central differences. Let N denote the number of (unknown) internal grid points. Introducing the vectors

$$\begin{aligned} \mathbf{u} &:= (u_1, v_1, u_2, v_2, \dots, u_N, v_N)^T, \\ \mathbf{p} &:= (p_1, p_2, \dots, p_N)^T \end{aligned} \tag{9.107}$$

containing the velocity components and pressure at the grid points, ordered lexicographically, we obtain an algebraic system of the form (cf. (9.44))

$$\mathbf{Au} - \mathbf{Bp} = \mathbf{f}_1, \tag{9.108a}$$

$$\mathbf{B}^T\mathbf{u} = \mathbf{f}_2. \tag{9.108b}$$

In (9.108), $\mathbf{A}$, $\mathbf{B}$, and $\mathbf{B}^T$ are the central difference approximations of the Laplace, gradient, and divergence operators, respectively. Nonzero boundary conditions are taken into account through the vectors $\mathbf{f}_1$ and $\mathbf{f}_2$. The direct numerical solution of (9.108) is tedious because the pressure is not included in (9.108b), which precludes the straightforward application of standard iterative methods.

For the solution of (9.108) we consider the *pressure correction method*. The basic idea is to decouple the computation of $\mathbf{u}$ and $\mathbf{p}$ in a predictor-corrector fashion in combination with a (false) transient method. The transient method is obvious here, since (9.102a) is the time limit solution (i.e., for $t \to \infty$) of

$$\frac{\partial \boldsymbol{u}}{\partial t} = \nabla^2 \boldsymbol{u} - \nabla p. \tag{9.109}$$

Applying the method of lines to (9.109), we obtain the semidiscrete system

$$\frac{\mathrm{d}\mathbf{u}}{\mathrm{d}t} = \mathbf{Au} - \mathbf{Bp} - \mathbf{f}_1, \tag{9.110}$$

which should replace (9.108a). Since we are interested only in the time limit solution of (9.110) and (9.108b), we look for an unconditionally stable time integration method. As noted in Section 9.6.3, we are not interested in the accuracy of this integration as such. Therefore the implicit Euler method is an obvious choice. We find

$$\mathbf{u}^{n+1} = \mathbf{u}^n + \Delta t\big(\mathbf{A}\mathbf{u}^{n+1} - \mathbf{B}\mathbf{p}^{n+1} - \mathbf{f}_1\big), \tag{9.111a}$$

$$\mathbf{B}^T\mathbf{u}^{n+1} = \mathbf{f}_2. \tag{9.111b}$$

A predictor-corrector method for (9.111) then runs as follows. First, compute a predictor $\mathbf{u}^*$ for the velocity by simply replacing $\mathbf{p}^{n+1}$ in (9.111a) by $\mathbf{p}^n$; i.e.,

$$\mathbf{u}^* = \mathbf{u}^n + \Delta t\big(\mathbf{A}\mathbf{u}^* - \mathbf{B}\mathbf{p}^n - \mathbf{f}_1\big). \tag{9.112}$$

Note that this predictor does not satisfy (9.108b), a condition we will impose on the corrector. Indeed, the corrector is evaluated via (9.111a) with $\mathbf{u}^{n+1}$ on the right-hand side replaced by $\mathbf{u}^*$. Thus the corrected value, also denoted by $\mathbf{u}^{n+1}$, is given by

$$\mathbf{u}^{n+1} = \mathbf{u}^n + \Delta t\big(\mathbf{A}\mathbf{u}^* - \mathbf{B}\mathbf{p}^{n+1} - \mathbf{f}_1\big). \tag{9.113}$$

Subtracting (9.112) from (9.113), we obtain

$$\mathbf{u}^{n+1} - \mathbf{u}^* = -\Delta t \mathbf{B}\mathbf{q}^n, \qquad \mathbf{q}^n := \mathbf{p}^{n+1} - \mathbf{p}^n. \tag{9.114}$$

Since we require the corrector $\mathbf{u}^{n+1}$ to satisfy the discrete incompressibility condition (9.111b), we obtain, upon substituting (9.114) in (9.111b),

$$\mathbf{L}\mathbf{q}^n = \frac{1}{\Delta t}\big(\mathbf{B}^T\mathbf{u}^* - \mathbf{f}_2\big), \qquad \mathbf{L} := \mathbf{B}^T\mathbf{B}. \tag{9.115}$$

Equation (9.115) is a "discrete" Poisson equation, from which we can compute the pressure update $\mathbf{q}^n$. The velocity update simply follows from (9.114). This time integration procedure should be repeated until $\|\mathbf{u}^{n+1} - \mathbf{u}^n\| < \text{tol}\Delta t$ for some tolerance tol.

Collecting the various steps, we arrive at the *pressure correction algorithm*

$$\big(\mathbf{I} - \Delta t\mathbf{A}\big)\mathbf{u}^* = \mathbf{u}^n - \Delta t\big(\mathbf{B}\mathbf{p}^n + \mathbf{f}_1\big) \qquad \textit{(predictor)}, \tag{9.116a}$$

$$\mathbf{L}\mathbf{q}^n = \frac{1}{\Delta t}\big(\mathbf{B}^T\mathbf{u}^* - \mathbf{f}_2\big), \quad \mathbf{p}^{n+1} = \mathbf{p}^n + \mathbf{q}^n \qquad \textit{(pressure update)}, \tag{9.116b}$$

$$\mathbf{u}^{n+1} = \mathbf{u}^* - \Delta t\mathbf{B}\mathbf{q}^n \qquad \textit{(velocity update)}. \tag{9.116c}$$

There are a few difficulties associated with the discrete Poisson equation (9.115). First, the matrix $\mathbf{L}$ is singular, since $\mathbf{L}\mathrm{e} = \mathbf{0}$, with $\mathrm{e} := (1, 1, \dots, 1)^T$. This is in agreement with the fact that the pressure is determined up to an additive constant. In order to determine $\mathbf{p}$, we have to specify the pressure at one single point of the domain, e.g., somewhere near the boundary. The drawback is that the incompressibility condition (9.111b) is not satisfied at this point. Alternatively, we might solve the modified equation

$$\big(\mathbf{L} - \alpha\mathbf{I}\big)\mathbf{q}^n = \frac{1}{\Delta t}\big(\mathbf{B}^T\mathbf{u}^* - \mathbf{f}_2\big) - \alpha\mathbf{q}^{n-1}, \tag{9.117}$$

where $0 < \alpha \ll 1$. In this way we achieve that the coefficient matrix $\mathbf{L} - \alpha\mathbf{I}$ has an acceptable condition number. Condition (9.111b) then only holds in the limit $n \to \infty$.

Second, combining the central difference approximations of ∇p and $\nabla\cdot\mathbf{u}$ at an interior grid point, we see that the matrix $\mathbf{L}$ is the standard five-point discretisation of the Laplace operator, except on a grid with grid sizes $2\Delta x$ and $2\Delta y$, as indicated in Figure 9.8. Clearly, there is no direct coupling between neighbouring values of $\mathbf{q}$, and consequently also not between neighbouring values of the pressure. This means that the pressure field is decoupled in the following way. Ignoring boundary conditions, let $p_{j,k}$ denote the numerical approximation of the pressure at grid point $\boldsymbol{x}_\mathrm{c} = (x_j, y_k)$. Then $p_{j,k} + (-1)^j C$, $C \in \mathbb{R}$, is also a solution of (9.116). Apparently, the grid function $(-1)^j$ is a *spurious mode* of the numerical pressure. Likewise, $(-1)^k$ and $(-1)^{j+k}$ are spurious modes.

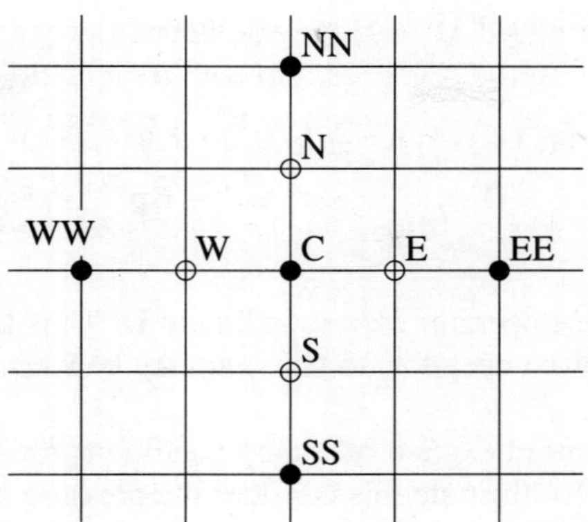

Figure 9.8. *Five-point stencil for* (9.115).

A way to avoid spurious modes is to use a *staggered grid*. In a staggered grid the nodes for the velocity components u and v and for the pressure p are displaced as indicated in Figure 9.9. The momentum equations (9.102a) are discretised at the u, v points and the continuity equation (9.102b) is discretised at the p points. In a staggered grid arrangement all first derivatives can be conveniently discretised by central differences on a grid with grid sizes $\Delta x/2$ and $\Delta y/2$. Indeed, we have for ∇p

$$\frac{\partial p}{\partial x}(x_{j+\frac{1}{2}}, y_k) \doteq \frac{1}{\Delta x}\big(p(x_{j+1}, y_k) - p(x_j, y_k)\big), \tag{9.118a}$$

$$\frac{\partial p}{\partial y}(x_j, y_{k+\frac{1}{2}}) \doteq \frac{1}{\Delta y}\big(p(x_j, y_{k+1}) - p(x_j, y_k)\big) \tag{9.118b}$$

and for $\nabla\cdot\boldsymbol{u}$

$$\nabla\cdot\boldsymbol{u}(x_j, y_k) \doteq \frac{1}{\Delta x}\big(u(x_{j+\frac{1}{2}}, y_k) - u(x_{j-\frac{1}{2}}, y_k)\big) + \frac{1}{\Delta y}\big(v(x_j, y_{k+\frac{1}{2}}) - v(x_j, y_{k-\frac{1}{2}})\big). \tag{9.118c}$$

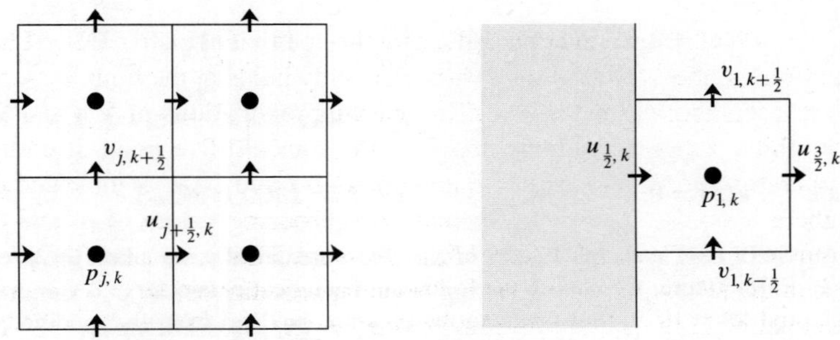

Figure 9.9. *A patch of a staggered grid in the interior (left) and near a boundary (right).*

Combining the discretisations in (9.118), we see that

$$\mathcal{L}_\Delta[q](x_j, y_k) = \frac{1}{\Delta x^2}\big(q(x_{j+1}, y_k) - 2q(x_j, y_k) + q(x_{j-1}, y_k)\big) + \frac{1}{\Delta y^2}\big(q(x_j, y_{k+1}) - 2q(x_j, y_k) + q(x_j, y_{k-1})\big), \tag{9.119}$$

where $\mathcal{L}_\Delta$ is the difference operator corresponding to $\mathbf{L}$. Thus $\mathbf{L}$ is the standard five-point discretisation of the Laplace operator, so that coupling between neighbouring p values is guaranteed.

Finally, we do not need explicit boundary conditions for q since there are no nodes for q on the boundary. To illustrate this consider the pressure node (x_1, y_k) adjacent to a vertical boundary; see Figure 9.9. For this node (9.111b) can be written as

$$\frac{1}{\Delta x}u^{n+1}_{\frac{3}{2},k} + \frac{1}{\Delta y}\left(v^{n+1}_{1,k+\frac{1}{2}} - v^{n+1}_{1,k-\frac{1}{2}}\right) = \frac{1}{\Delta x}u_{\frac{1}{2},k}, \tag{9.120}$$

where $u_{1/2,k}$ is the normal component of the velocity at the boundary. Using (9.114) in (9.120), we obtain the equation

$$\frac{1}{\Delta x^2}\left(q^n_{2,k} - q^n_{1,k}\right) + \frac{1}{\Delta y^2}\left(q^n_{1,k+1} - 2q^n_{1,k} + q^n_{1,k-1}\right) = \frac{1}{\Delta t}\left(\frac{1}{\Delta x}\left(u^*_{\frac{3}{2},k} - u_{\frac{1}{2},k}\right) + \frac{1}{\Delta y}\left(v^*_{1,k+1} - v^*_{1,k-1}\right)\right), \tag{9.121}$$

which only contains q values in the interior domain. Other boundaries can be treated in a similar way.

Example 9.33 In this example we compute a numerical solution of the Stokes equations (9.102) on a rectangular domain $\Omega := (0, a) \times (0, b)$ subject to the following boundary conditions:

inflow: $u(\boldsymbol{x}) = 16U_{\max}\dfrac{y}{b}\left(\dfrac{1}{2} - \dfrac{y}{b}\right)$, $v(\boldsymbol{x}) = 0$,

$$\boldsymbol{x} \in \partial\Omega_{\text{in}} := \left\{\boldsymbol{x} \,\middle|\, x = 0, 0 \le y \le \frac{1}{2}b\right\},$$

outflow: $u(\boldsymbol{x}) = 0$, $v(\boldsymbol{x}) = 16V_{\max}\left(\dfrac{x}{a} - \dfrac{1}{2}\right)\left(1 - \dfrac{x}{a}\right)$,

$$\boldsymbol{x} \in \partial\Omega_{\text{out}} := \left\{\boldsymbol{x} \,\middle|\, \frac{1}{2}a \le x \le a, y = b\right\},$$

solid wall: $\boldsymbol{u}(\boldsymbol{x}) = \boldsymbol{0}$, $\boldsymbol{x} \in \partial\Omega \setminus \left(\partial\Omega_{\text{in}} \cup \partial\Omega_{\text{out}}\right)$.

The constraint (9.104) leads to $aV_{\max} = bU_{\max}$. More specifically, we take $a = 2$, $b = 1$, and $U_{\max} = 1$. Furthermore, we choose the following numerical parameters: $\Delta x = \Delta y = 0.05$, $\Delta t = 0.1$, and tol $= 10^{-6}$. We compute the pressure (update) from the modified equation (9.117) with $\alpha = 10^{-6}$. The steady solution we obtained after 886 time steps satisfies $\|\mathbf{B}^T\mathbf{u} - \mathbf{f}_2\|_2 \le 6.25 \times 10^{-4}$ and is given in Figure 9.10. □

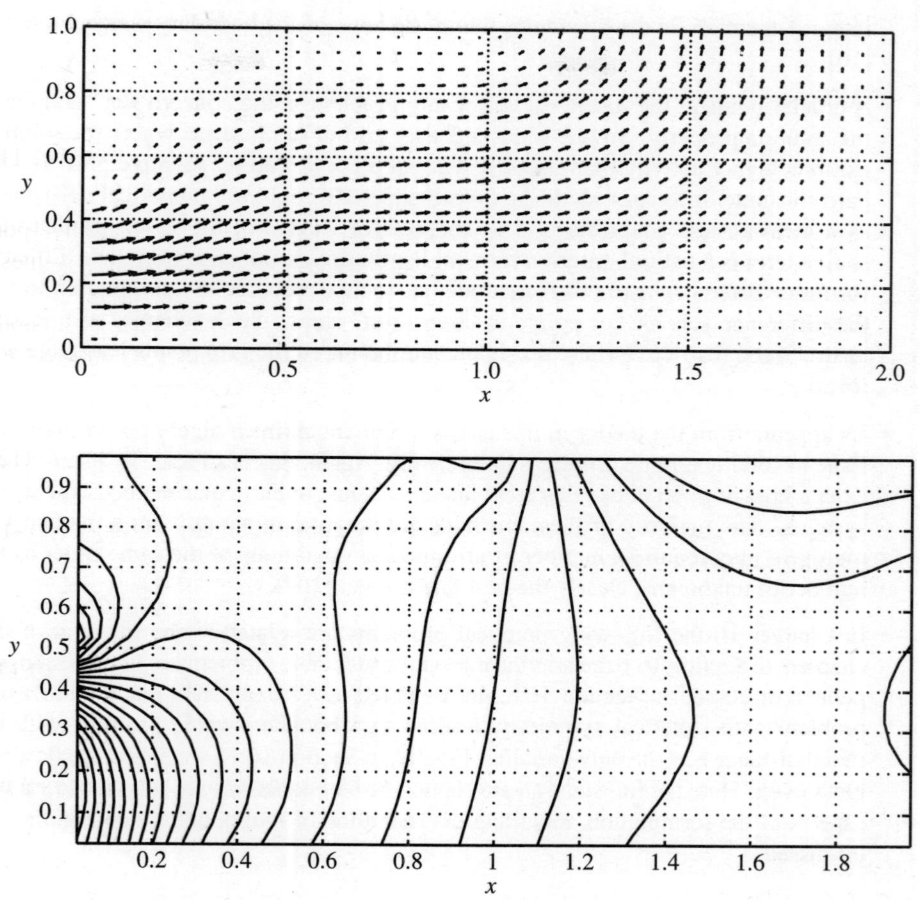

Figure 9.10. *The velocity (top) and the pressure field (bottom) of a Stokes flow in a rectangular domain.*

9.8 Discussion

- Elliptic problems can be numerically treated in a variety of ways. Although we considered finite difference methods here, this appears to be the area where finite element methods turn out to be very powerful [9]. The derivation and analysis of these methods is, however, so much different from the finite differences or finite volumes we use that it would require too much space for this book. These methods are based on the weak form of a differential equation, which, after applying Gauss's formula, results in a bilinear form, which is discretised in turn. The domain is then divided into small triangles (or quadrilaterals), which is the basis for the large flexibility as well as the possibility of working fairly easily for complex geometries. Another approach is to use Green's identity and employ knowledge of fundamental solutions to find an integral equation on the boundary; this is the basis of the boundary element method [16].

- A crucial point in the discretisation of any problem is the concern for "sampling" the solution properly. So one needs a denser grid in subdomains where the solution exhibits higher activity than outside. The problem of finding adaptive grids is very hard for finite difference methods. In principle one could use numerical derivatives as a measure and adapt the grid in an iterative way. For a Poisson problem the source term is a natural criterion for equidistribution of the error [11]. The difficulty for finite difference methods is that one has to take recourse to interpolation to find the difference scheme for points in "transition" areas. Such schemes also lead to unstructured matrices (where no simple numbering of the grid points leads to a nice form).

- As appears from the theory in this chapter, numerical linear algebra is an important issue in solving elliptic problems numerically. Again, this is an issue by itself. There exist a host of good books that treat linear systems, in particular methods for sparse systems. The majority of these methods are iterative in nature. Yet a method like multigrid may require a number of arithmetical operations of the same order as the number of unknowns, clearly the best one can get [167].

- In Chapter 16 the following practical problems are related to the material of this chapter. In Section 16.6 the nonlinear groundwater table problem is solved. Catalyzer pellets, discussed in Section 16.8, are modeled as a steady state reaction diffusion problem with spherical symmetry, leading to a boundary value problem with the radial distance r as the only variable. Finally, in Section 16.15 we model the flow in a glass oven. Here the full steady state Navier–Stokes equations are solved and various aspects of the formulation, including the (handling of the) boundary conditions, are discussed.

Exercises

9.1. One can use higher-order difference approximations for the Laplace operator. One possibility is to use five points per direction. Show the stencil of this scheme and determine the coefficients. Finally, show that this has fourth order accuracy.

9.2. Suppose we have the differential operator

$$\mathcal{L}[u] := \nabla^2 u + \frac{\partial^2 u}{\partial x \partial y}.$$

Use central differences on a grid with grid sizes $\Delta x = \Delta y = h$ to obtain a nine-point scheme on a square stencil whose coefficients are given by

$$\frac{1}{h^2}\begin{bmatrix} 0 & 1 & 0 \\ 1 & -4 & 1 \\ 0 & 1 & 0 \end{bmatrix} + \frac{1}{4h^2}\begin{bmatrix} -1 & 0 & 1 \\ 0 & -4 & 0 \\ 1 & 0 & -1 \end{bmatrix}.$$

9.3. Consider the Poisson equation on the unit square, i.e.,

$$\mathcal{L}[u] := \nabla^2 u = f, \qquad \boldsymbol{x} \in \Omega := (0, 1) \times (0, 1).$$

Let the boundary condition be given by $u(\boldsymbol{x}) = 0$ at the lower, left, and right boundaries, while $u(\boldsymbol{x}) = \sin(\pi x)$ at the upper boundary. Assume that we use central differences to approximate the solution.

(a) Show that the numerical solution at a grid point $\boldsymbol{x}_{\text{P}} = (x, y)$ is bounded by $x(1-x)y + \sin \pi$.

(b) Show that the global error vector $\mathbf{e}$ is bounded by

$$\|\mathbf{e}\|_\infty \leq \left(\frac{1}{8} + \sin \pi\right) \|\mathbf{d}\|_\infty,$$

where $\mathbf{d}$ is the local error vector.

9.4. Consider the problem

$$\mathcal{L}[u] := \nabla^2 u - \frac{\partial u}{\partial x} = 0$$

on the unit square. We use central differences to discretise the problem.

(a) Determine a maximum value for Δx such that the scheme satisfies the requirements of Property 9.1.

(b) Use a comparison function to give an estimate for the norm of the global error in terms of the norm of the local error.
Hint: You may use, e.g., $v(x, y) = x^2 + y^2$.

9.5. Consider the mixed problem

$$\mathcal{L}[u] := \nabla^2 u = f$$

defined on the unit square with homogeneous Dirichlet boundary conditions except at the right boundary, where $\frac{\partial u}{\partial n} = 0$.

(a) Use a one-sided difference to discretise the Neumann boundary conditions. Show how the stencil looks at a point with a right boundary point as neighbour. Show that this gives errors like $\mathcal{O}(\Delta x)^{-1}$.

(b) Show that the global error is at least first order accurate in Δx.

9.6. In this exercise we want to determine the coefficients in the LU-decomposition of the matrix resulting from the central difference discretisation of

$$\mathcal{L}[u] := \frac{\mathrm{d}^2 u}{\mathrm{d}x^2} = f, \qquad x \in (0, 1),$$

with Dirichlet and with mixed boundary conditions.

(a) Do this for the case where u is given at $x = 0$ and $x = 1$.

(b) Next consider the case where u is given at $x = 0$ and $\frac{\mathrm{d}u}{\mathrm{d}x}$ is given at $x = 1$. Determine the LU-decomposition when $\frac{\mathrm{d}u}{\mathrm{d}x}(1)$ is approximated by a backward difference.

(c) Repeat part (b) with Neumann and Dirichlet conditions interchanged. Which difference approximation would you use for $\frac{du}{dx}(0)$?

9.7. Show that LU-decomposition of the matrix **A** in (9.45) gives **L** and **U** matrices whose M diagonals immediately below and above the main diagonal (the *codiagonals*) will fill in (at least eventually).

9.8. Consider the matrix

$$\mathbf{A} = \begin{pmatrix} 2 & -1 & 0 \\ -1 & 2 & -1 \\ 0 & -1 & 2 \end{pmatrix}.$$

Determine the spectral radii $\rho(\mathbf{M}_J)$ and $\rho(\mathbf{M}_S)$.

9.9. Assume that a matrix $\mathbf{A} = (a_{i,j})$ is *strongly diagonally dominant*; i.e.,

$$\sum_{j \neq i} a_{i,j} < a_{i,i} \qquad \text{for all} \quad i.$$

Show that both the Gauss–Jacobi and the Gauss–Seidel methods converge.

9.10. Assume that the matrices $\mathbf{A}_x$ and $\mathbf{A}_y$ in (9.66) commute. Let us denote their eigenvalues by λ_j and μ_k, respectively.

(a) Show that the convergence of ADI is determined by the quotient

$$\max_{\lambda_j, \mu_k} \left| \frac{1 + \tau\lambda_j}{1 - \tau\lambda_j} \frac{1 + \tau\mu_k}{1 - \tau\mu_k} \right|.$$

(b) Show that convergence is assured if $\mathbf{A}_x$ and $\mathbf{A}_y$ have negative eigenvalues.

(c) Check the requirement under part (b) for a simple case.

9.11. Consider the problem

$$\mathcal{L}[u] := \frac{\mathrm{d}^2 u}{\mathrm{d}x^2} = \mathrm{e}^u, \qquad x \in (0, 1),$$

with u given at the boundary points $x = 0$ and $x = 1$. We discretise this boundary value problem with central differences. Show that the Newton algorithm for the resulting nonlinear system converges, at least locally.

9.12. Let Ω be the trapezium defined by the corner points (0, 0), (0, 2), (2, 1), (2, 0). We want to discretise the problem

$$\nabla^2 u = f, \qquad \boldsymbol{x} \in \Omega.$$

Let the grid size be given by $h = \Delta x = \Delta y = 2/M$ for some M.

(a) Assume that Dirichlet boundary conditions are given. Use central differences to discretise this problem. Pay special attention to points with a neighbour at the right boundary; see also Exercise 7 in Chapter 5.

(b) Show that the global discretisation error is second order.

9.13. Consider the Laplace equation in polar coordinates

$$\nabla^2 u := \frac{\partial^2 u}{\partial r^2} + \frac{1}{r}\frac{\partial u}{\partial r} + \frac{1}{r^2}\frac{\partial^2 u}{\partial \phi^2} = 0,$$

defined on the unit circle, with boundary value

$$u(1, \phi) = 1.$$

Let Δr and $\Delta\phi$ be the grid size in the r and ϕ directions, respectively, and use central differences.

(a) Find the system of equations for the unknowns at the grid points, ordered by taking them counterclockwise and from the innermost circle to the outermost circle of grid points.

(b) Find the equation at (0, 0); cf. Section 5.2.3.

9.14. Consider the Laplace equation (cf. (8.41)).

$$\nabla^2 u := \frac{\partial^2 u}{\partial r^2} + \frac{1}{r}\frac{\partial u}{\partial r} + \frac{1}{r^2}\frac{\partial^2 u}{\partial \phi^2} = 0$$

defined on a finite wedge $\{(r, \phi) \mid 0 < r < 1,\ 0 < \phi < \frac{1}{4}\pi\}$, where u satisfies the boundary conditions

$$u(r, 0) = u(r, \pi/4) = 0, \qquad u(1, \phi) = \sin 4\phi.$$

(a) Use central differences to discretise the problem. Give the explicit form of the system for the unknowns at the grid points.

(b) Solve the system for various values of the grid sizes Δr, $\Delta\phi$ and make a table of the error at a number of points (on the intersection of the grids) close to $\mathbf{0}$. Note: You can find the exact solution from (8.41).

9.15. In (9.108) we have a linear system that we hope has a unique solution. To see this we do the following: Instead of (9.108b), we take

$$-\mathbf{B}^T\mathbf{u} = \mathbf{f}_2.$$

Show that (9.108a) together with this equation leads to a positive definite symmetric system. Note: A matrix $\mathbf{C}$ is positive definite if $\mathbf{z}^T\mathbf{C}\,\mathbf{z} > 0$ for $\mathbf{z} \neq 0$. Complete the proof.

Chapter 10

Analysis of Parabolic Equations

The analytical theory of parabolic equations in this chapter starts (Section 10.1) with deriving solutions for Cauchy problems, i.e., pure initial value problems. The dependence of (fundamental) solutions on the time and space variables turns out to respect certain symmetries (Section 10.2). We study similarity solutions in Section 10.3, where we also derive formulas for the diffusion operator in other than Cartesian coordinates. Next we investigate the role of boundary conditions in Section 10.4. First we analyse problems on finite (spatial) domains. A special class of problems is formed by PDEs with moving boundaries called Stefan problems (Section 10.5). The last section, Section 10.6, is devoted to steady state solutions and travelling-wave solutions. A point of interest here is that stationary solutions constitute proper solutions of corresponding elliptic boundary value problems (which are the subject of Chapter 8).

10.1 Cauchy Problems

Parabolic equations arise in a variety of applications, mainly associated with diffusive processes. In Section 1.1 we saw an example of (chemical) *diffusion*. Heat flow also has a diffusive character, often called *conduction*. This may be combined with convection if the medium is a fluid in motion. In mechanics of fluids, internal friction, called *viscosity*, produces diffusion of momentum. In this section we shall consider the simplest form of a diffusive problem: the linear heat equation on infinite domains.

10.1.1 The Heat Equation in One Space Dimension

In order to investigate solutions of parabolic equations, we consider the following initial value problem for the simplest (nondimensional) form of the *heat equation*:

$$\frac{\partial u}{\partial t} = \frac{\partial^2 u}{\partial x^2}, \qquad x \in \mathbb{R}, \quad t > 0, \tag{10.1a}$$

$$u(x, 0) = v(x), \qquad x \in \mathbb{R}, \tag{10.1b}$$

where $|v|$ and $|u|$ are integrable over $\mathbb{R}$. (Note that limiting conditions for $x \to \pm\infty$ are to be expected if we portray the present initial value problem as the limit of an initial boundary value problem. See Example 10.2.) We seek a solution of the form

$$u(x,t) = p(x)q(t); \tag{10.2}$$

i.e., we *separate* the independent variables so that the functions p and q in the product depend on either one of the variables x and t exclusively. Upon substitution we find

$$p\frac{\mathrm{d}q}{\mathrm{d}t} = \frac{\mathrm{d}^2 p}{\mathrm{d}x^2}q,$$

or, assuming $p(x), q(t) \neq 0$,

$$\frac{1}{q}\frac{\mathrm{d}q}{\mathrm{d}t} = \frac{1}{p}\frac{\mathrm{d}^2 p}{\mathrm{d}x^2}. \tag{10.3}$$

The left-hand side of (10.3) depends solely on t and the right-hand side depends solely on x. This is only possible if both sides are equal to some constant λ, the separation constant. (As will appear below, it is sufficient to assume that λ is real.) So we obtain the two *eigenvalue* problems

$$\frac{\mathrm{d}^2 p}{\mathrm{d}x^2} = \lambda p, \tag{10.4a}$$

$$\frac{\mathrm{d}q}{\mathrm{d}t} = \lambda q. \tag{10.4b}$$

From (10.4b) we conclude that only nonpositive values of λ are allowed for a stable solution $q(t)$. Indeed, if $\lambda > 0$, then (10.4a) will have solutions $p(x)$ that are exponentially increasing for either $x \to -\infty$ or $x \to \infty$. Also, for $\lambda = 0$, (10.4a) has an unbounded solution given by $p(x) = C_1 x + C_2$. Therefore, in order for the solution $u(x,t)$ of (10.1a) to be bounded, we assume (with κ real) that

$$\lambda = -\kappa^2 < 0. \tag{10.5}$$

A general solution of (10.4) is then given by

$$p(x)q(t) = C\,\mathrm{e}^{\mathrm{i}\kappa x - \kappa^2 t}, \tag{10.6}$$

where C is a constant. Here we recognize a planar wave from the dispersion relation for the heat equation (10.1a). Apparently, we can view this planar wave as a Fourier mode. By Fourier analysis (Chapter 3), a general bounded solution of (10.1) can now be found by superposition over all possible κ. The suggested solution is now given by

$$u(x,t) = \int_{-\infty}^{\infty} \hat{v}(\kappa)\,\mathrm{e}^{\mathrm{i}\kappa x - \kappa^2 t}\,\mathrm{d}\kappa, \tag{10.7}$$

where, using the initial value $u(x,0) = v(x)$,

$$\hat{v}(\kappa) := \frac{1}{2\pi}\int_{-\infty}^{\infty} v(\xi)\,\mathrm{e}^{-\mathrm{i}\kappa\xi}\,\mathrm{d}\xi. \tag{10.8}$$

By substituting (10.8) in (10.7) and interchanging the order of integration, we obtain

$$u(x,t) = \int_{-\infty}^{\infty} v(\xi)\psi(x-\xi,t)\,\mathrm{d}\xi, \tag{10.9a}$$

where

$$\begin{aligned}\psi(\zeta,t) := \frac{1}{2\pi}\int_{-\infty}^{\infty} \mathrm{e}^{\mathrm{i}\kappa\zeta}\,\mathrm{e}^{-\kappa^2 t}\,\mathrm{d}\kappa &= \frac{1}{2\pi}\int_{-\infty}^{\infty} \mathrm{e}^{-(\kappa-\mathrm{i}\zeta/2t)^2 t-\zeta^2/4t}\,\mathrm{d}\kappa \\ &= \frac{\mathrm{e}^{-\zeta^2/4t}}{2\pi}\int_{-\infty}^{\infty} \mathrm{e}^{-\kappa^2 t}\,\mathrm{d}\kappa = \frac{\mathrm{e}^{-\zeta^2/4t}}{2\pi\sqrt{t}}\int_{-\infty}^{\infty} \mathrm{e}^{-s^2}\,\mathrm{d}s = \frac{\mathrm{e}^{-\zeta^2/4t}}{2\sqrt{\pi t}}.\end{aligned} \tag{10.9b}$$

We shifted the κ contour by an amount $\mathrm{i}\zeta/2t$ into the complex plane. Altogether we obtain the solution

$$u(x,t) = \frac{1}{2\sqrt{\pi t}}\int_{-\infty}^{\infty} v(\xi)\exp\left(-\frac{(x-\xi)^2}{4t}\right)\mathrm{d}\xi. \tag{10.10a}$$

We remark that the derivation closely follows that of the Duhamel integral in Section 4.6. One can easily verify that this solution satisfies (10.1a). By changing the coordinate of integration to $\xi = x + 2\eta\sqrt{t}$, we find

$$u(x,t) = \frac{1}{\sqrt{\pi}}\int_{-\infty}^{\infty} v(x+2\sqrt{t}\eta)\,\mathrm{e}^{-\eta^2}\,\mathrm{d}\eta. \tag{10.10b}$$

The initial condition is now readily verified to be

$$u(x,0) = \frac{1}{\sqrt{\pi}}\int_{-\infty}^{\infty} v(x)\,\mathrm{e}^{-\eta^2}\,\mathrm{d}\eta = v(x)\frac{1}{\sqrt{\pi}}\int_{-\infty}^{\infty} \mathrm{e}^{-\eta^2}\,\mathrm{d}\eta = v(x).$$

In Section 10.1.2 we shall show that the solution is also unique.

Next we show that $\psi(\zeta,t)$ is equivalent to a fundamental solution; see (4.5). By taking $v(x) = \delta(x)$ in (10.10a), we immediately see that ψ satisfies the initial value problem

$$\frac{\partial\psi}{\partial t} - \frac{\partial^2\psi}{\partial\zeta^2} = 0, \qquad \zeta\in\mathbb{R}, \quad t>0, \tag{10.11a}$$

$$\psi(\zeta,0) = \delta(\zeta), \qquad \zeta\in\mathbb{R}. \tag{10.11b}$$

A fundamental solution $w(x,t;\xi,\tau)$ of (10.1) satisfies

$$\frac{\partial w}{\partial t} - \frac{\partial^2 w}{\partial x^2} = \delta(x-\xi)\delta(t-\tau), \qquad x\in\mathbb{R}, \quad t>\tau, \tag{10.12a}$$

$$w(x,\xi,t,\tau) = 0, \qquad t<\tau. \tag{10.12b}$$

Property 10.1. *The function $w(x,t;\xi,\tau) = \psi(x-\xi,t-\tau)H(t-\tau)$, where H is the Heaviside step function, is a fundamental solution of* (10.1).

Proof. $H(t-\tau) \equiv 0$ for $t < \tau$. Furthermore, the difference between

$$w_t = \psi_t H + \psi\delta = \psi_t H + \delta(x-\xi)\delta(t-\tau) \text{ and } w_{xx} = \psi_{\zeta\zeta} H$$

is just $\delta(x-\xi)\delta(t-\tau)$. □

It follows that for $t > \tau$ we can identify $w(x,t;\xi,\tau)$ with $\psi(x-\xi, t-\tau)$. We remark that the ψ found in (10.9b) apparently satisfies

$$\int_{-\infty}^{\infty} \psi(\zeta,t)\,\mathrm{d}\zeta = \frac{1}{2\sqrt{\pi t}} \int_{-\infty}^{\infty} \mathrm{e}^{-\zeta^2/4t}\,\mathrm{d}\zeta = \frac{1}{\sqrt{\pi}} \int_{-\infty}^{\infty} \mathrm{e}^{-\mu^2}\,\mathrm{d}\mu = 1. \tag{10.13a}$$

Hence ψ can be seen as a probability distribution. In fact, it is the famous *Gaussian* or *normal* distribution. From the rightmost side of (10.13a) we see that

$$\operatorname{erf}(x) := \frac{2}{\sqrt{\pi}} \int_0^x \mathrm{e}^{-\mu^2}\,\mathrm{d}\mu, \tag{10.13b}$$

the *error function*, which plays such an important role in statistics, is related to the problem here. The same is true for the *complementary error function* defined by

$$\operatorname{erfc}(x) := \frac{2}{\sqrt{\pi}} \int_x^{\infty} \mathrm{e}^{-\mu^2}\,\mathrm{d}\mu = 1 - \operatorname{erf}(x). \tag{10.13c}$$

We will frequently encounter this function further on.

Example 10.2 If we do not restrict ourselves, in problems on infinite domains, to bounded solutions, we may easily encounter solutions that grow rather than decay in time. This growth may be associated with possible sources at infinity. Consider the examples

$$u(x,0) = \mathrm{e}^x \quad \text{with } u(x,t) = \mathrm{e}^{x+t},$$
$$u(x,0) = \frac{1}{2}x^2 \quad \text{with } u(x,t) = \frac{1}{2}x^2 - t.$$
□

Example 10.3 Rather than (10.1a), one may have

$$\frac{\partial u}{\partial t} = k\frac{\partial^2 u}{\partial x^2},$$

where k is a positive constant called the (thermal) diffusivity. We may apply a change of variables, $s := tk$ or $z := x/\sqrt{k}$, to obtain either $u_s = u_{xx}$ or $u_t = u_{zz}$. □

10.1.2 The Heat Equation in *d* Space Dimensions

Our formulation in Section 10.1.1 was chosen such that it lends itself easily to heat equations in more than one space variable. So consider

$$\frac{\partial u}{\partial t} = \nabla^2 u, \qquad \boldsymbol{x} \in \mathbb{R}^d, \quad t > 0, \tag{10.14a}$$

$$u(\boldsymbol{x},0) = v(\boldsymbol{x}), \qquad \boldsymbol{x} \in \mathbb{R}^d, \tag{10.14b}$$

where $|v|$ and $|u|$ are integrable over $\mathbb{R}^d$. A fundamental solution w (cf. Section 4.5) then satisfies (cf. (10.12))

$$\frac{\partial w}{\partial t} - \nabla^2 w = \delta(\boldsymbol{x} - \boldsymbol{\xi})\delta(t - \tau), \qquad \boldsymbol{x} \in \mathbb{R}^d, \quad t > 0 \tag{10.15a}$$

$$w(\boldsymbol{x}, t; \boldsymbol{\xi}, \tau) = 0, \qquad t < \tau. \tag{10.15b}$$

By analogy with Property 10.1, we have

$$w(\boldsymbol{x}, t; \boldsymbol{\xi}, \tau) = \left(\frac{1}{2\sqrt{\pi(t-\tau)}}\right)^d \exp\left(-\frac{\|\boldsymbol{x} - \boldsymbol{\xi}\|_2^2}{4(t-\tau)}\right) H(t - \tau). \tag{10.16}$$

Superposition then gives (cf. Chapter 4)

$$u(\boldsymbol{x}, t) = \left(\frac{1}{2\sqrt{\pi t}}\right)^d \int_{\mathbb{R}^d} v(\boldsymbol{\xi}) \exp\left(-\frac{\|\boldsymbol{x} - \boldsymbol{\xi}\|_2^2}{4t}\right) \mathrm{d}\boldsymbol{\xi}. \tag{10.17}$$

This solution exists and is unique.

Theorem 10.4. *Let v be continuous and integrable on $\mathbb{R}^d$. Then the function given by* (10.17) *satisfies initial value problem* (10.14) *uniquely.*

Proof. By virtue of the exponential and the fact that v is bounded, the defining integral of u is uniformly integrable for all $\boldsymbol{x}$ and $t > 0$. So u and likewise $\int w_t\, v(\boldsymbol{\xi})\mathrm{d}\boldsymbol{\xi}$ and $\int \nabla^2 w\, v(\boldsymbol{\xi})\mathrm{d}\boldsymbol{\xi}$ exist, and we can interchange differentiation and integration. Since $w_t = \nabla^2 w$, (10.14a) readily follows. To show that u also satisfies the initial condition we change variables $\boldsymbol{\xi} = \boldsymbol{x} + 2\boldsymbol{\eta}\sqrt{t}$ to obtain

$$u(\boldsymbol{x}, t) = \left(\frac{1}{\sqrt{\pi}}\right)^d \int_{\mathbb{R}^d} v(\boldsymbol{x} + 2\boldsymbol{\eta}\sqrt{t})\, \mathrm{e}^{-\|\boldsymbol{\eta}\|_2^2}\, \mathrm{d}\boldsymbol{\eta}.$$

With identity (10.13a), this is clearly equal to $v(\boldsymbol{x})$ for $t = 0$.

To prove that the solution is unique we assume that u_1 and u_2 are two solutions. Then the difference $w(\boldsymbol{x}, t) := u_1(\boldsymbol{x}, t) - u_2(\boldsymbol{x}, t)$ satisfies the initial value problem with $v(\boldsymbol{x}, t) \equiv 0$. From Green's first identity we have

$$\frac{\mathrm{d}}{\mathrm{d}t} \int_{\mathbb{R}^d} \frac{1}{2} w^2 \,\mathrm{d}V = \int_{\mathbb{R}^d} w \frac{\partial w}{\partial t} \,\mathrm{d}V = \int_{\mathbb{R}^d} w \nabla^2 w \,\mathrm{d}V = -\int_{\mathbb{R}^d} |\nabla w|^2 \,\mathrm{d}V < 0.$$

Hence

$$0 \le \int_{\mathbb{R}^d} w^2(\boldsymbol{x}, t)\,\mathrm{d}V \le \int_{\mathbb{R}^d} w^2(\boldsymbol{x}, 0)\,\mathrm{d}V = 0,$$

so $w \equiv 0$. □

10.1.3 Problems on Half-Spaces

It is of interest to study the solution of problems defined on a semi-infinite domain with values prescribed at the boundary. Although such a setting can also be handled in higher dimensions, we restrict ourselves to the one-dimensional case, with the domain $[0, \infty)$. We are looking for the solution of

$$\frac{\partial u}{\partial t} = \frac{\partial^2 u}{\partial x^2}, \qquad x \in [0, \infty), \qquad t > 0, \tag{10.18}$$

satisfying either

$$u(x, 0) = v(x), \qquad x \in [0, \infty), \tag{10.19a}$$

$$u(0, t) = 0, \qquad t > 0, \tag{10.19b}$$

for some given initial value $v(x)$ or

$$u(x, 0) = 0, \qquad x \in [0, \infty), \tag{10.20a}$$

$$u(0, t) = \beta(t), \qquad t > 0, \tag{10.20b}$$

for some boundary value $\beta(t)$. Clearly, the conditions (10.19) are different from but related to the one we encountered in (10.1b).

First consider initial boundary value problem (10.18) and (10.19). Formally, we may try to use a fundamental solution approach by employing a continuation of the initial condition on the left interval $(-\infty, 0]$. Let $\tilde{v}(x)$ be equal to $v(x)$ for $x \geq 0$ and some continuous extension for $x < 0$ but such that the representation

$$u(x, t) = \frac{1}{2\sqrt{\pi t}} \int_{-\infty}^{\infty} \tilde{v}(\xi) \exp\left(-\frac{(x-\xi)^2}{4t}\right) \mathrm{d}\xi \tag{10.21}$$

still satisfies boundary condition (10.19b). Simple substitution then reveals

$$\frac{1}{2\sqrt{\pi t}} \int_{-\infty}^{\infty} \tilde{v}(\xi) \exp\left(-\frac{\xi^2}{4t}\right) \mathrm{d}\xi = \frac{1}{2\sqrt{\pi t}} \int_{0}^{\infty} (\tilde{v}(\xi) + \tilde{v}(-\xi)) \exp\left(-\frac{\xi^2}{4t}\right) \mathrm{d}\xi = 0, \tag{10.22}$$

which is true for all t if

$$\tilde{v}(\xi) = -v(-\xi), \qquad \xi < 0. \tag{10.23}$$

As a consequence we find the representation

$$u(x, t) = \frac{1}{2\sqrt{\pi t}} \int_{0}^{\infty} v(\xi) \left[\exp\left(-\frac{(x-\xi)^2}{4t}\right) - \exp\left(-\frac{(x+\xi)^2}{4t}\right)\right] \mathrm{d}\xi. \tag{10.24}$$

Example 10.5 An important application is the heat shock. Suppose an object with a certain given temperature is suddenly exposed to a heat source (or sink). If this has a temperature different from the object, this means that the temperature profile initially exhibits a discontinuity. The simplest model is given by assuming that the initial temperature is constant; i.e.,

$$v(x) = v_0 > 0.$$

(See Example 7.17 for the related problem of a given heat flux.) The solution of initial boundary value problem (10.18) and (10.19) is then given by

$$u(x, t) = \frac{v_0}{2\sqrt{\pi t}} \left[\int_{0}^{\infty} \exp\left(-\frac{(x-\xi)^2}{4t}\right) \mathrm{d}\xi - \int_{0}^{\infty} \exp\left(-\frac{(x+\xi)^2}{4t}\right) \mathrm{d}\xi\right].$$

Substituting $z := -(x-\xi)/2\sqrt{t}$ in the first integral and $z := (x+\xi)/2\sqrt{t}$ in the second one gives

$$u(x,t) = \frac{v_0}{\sqrt{\pi}}\left[\int_{-x/2\sqrt{t}}^{\infty} \mathrm{e}^{-z^2}\,\mathrm{d}z - \int_{x/2\sqrt{t}}^{\infty} \mathrm{e}^{-z^2}\,\mathrm{d}z\right] = v_0\,\mathrm{erf}\left(\frac{x}{2\sqrt{t}}\right).$$

In Figure 10.1 we have displayed various stages of the solution for $v_0 = 1$. □

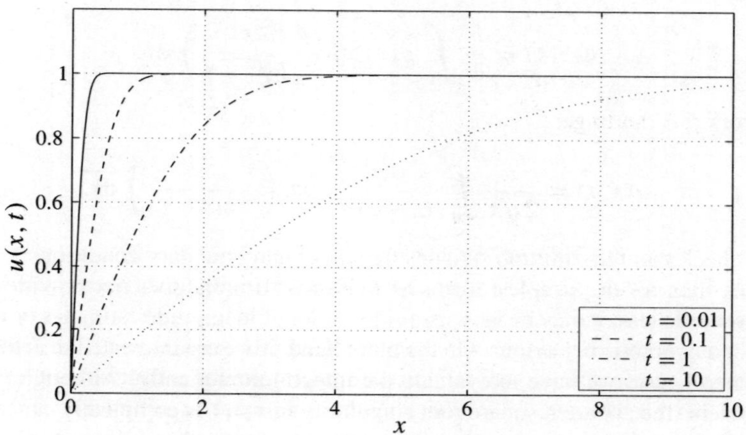

Figure 10.1. *Solution of the heat shock problem at various time levels.*

Example 10.6 We can also apply the analysis to a two-sided heat shock. Indeed, consider an object that is suddenly heated in a finite region, $[-1, 1]$ say. A very simple model is then given by the piecewise constant initial condition

$$v(x) = \begin{cases} 0 & \text{if} \quad x \in (-\infty, -1), \\ v_0 & \text{if} \quad x \in [-1, 1], \\ 0 & \text{if} \quad x \in (1, \infty), \end{cases}$$

where $v_0 \neq 0$, yielding the solution

$$u(x,t) = \frac{v_0}{2\sqrt{\pi t}}\int_{-1}^{1} \exp\left(-\frac{(x-\xi)^2}{4t}\right)\mathrm{d}\xi,$$

or, upon substituting $z := (x-\xi)/2\sqrt{t}$,

$$u(x,t) = \frac{v_0}{\sqrt{\pi}}\int_{(x-1)/2\sqrt{t}}^{(x+1)/2\sqrt{t}} \mathrm{e}^{-z^2}\,\mathrm{d}z = \frac{v_0}{2}\left(\mathrm{erf}\left(\frac{x+1}{2\sqrt{t}}\right) - \mathrm{erf}\left(\frac{x-1}{2\sqrt{t}}\right)\right),$$

which can also be found by employing the solution of the previous example. □

The other situation, i.e., initial boundary value problem (10.18) and (10.20), can be solved using the solution of the initial boundary value problem (10.18) and (10.19). First, consider the problem where $\beta(t) \equiv 1$. It is simple to see that

$$u(x,t) = 1 - \mathrm{erf}\left(\frac{x}{2\sqrt{t}}\right) = \mathrm{erfc}\left(\frac{x}{2\sqrt{t}}\right) \tag{10.25}$$

satisfies (10.18) and (10.20). We can now invoke the Duhamel integral (see Theorem 4.16) to obtain the solution. Note that we can define solutions $w(x, t-\tau)$ by

$$w(x, t-\tau) := \operatorname{erfc}\left(\frac{x}{2\sqrt{t-\tau}}\right). \tag{10.26}$$

Hence we obtain as the formal solution of (10.18) and (10.20) from (4.68) that

$$u(x,t) = \frac{\partial}{\partial t}\int_0^t \beta(\tau)\operatorname{erfc}\left(\frac{x}{2\sqrt{t-\tau}}\right)\mathrm{d}\tau. \tag{10.27}$$

We can work this out to get

$$u(x,t) = \frac{x}{2\sqrt{\pi}}\int_0^t \frac{\beta(\tau)}{(t-\tau)^{3/2}}\exp\left(-\frac{x^2}{4(t-\tau)}\right)\mathrm{d}\tau. \tag{10.28}$$

One may check that this solution satisfies the initial and boundary conditions.

Other than for the simplest forms of $\beta(\tau)$ this formula does not provide an explicit answer. Nevertheless, it may be very useful for order of magnitude estimates or the analysis of trends or asymptotic behaviour. On the other hand, if we are interested in actual numbers for the general case, we have to evaluate the integral numerically. Although this is not a major problem (the apparent square root singularity in $\tau = t$ is completely cancelled by the exponential), we have to compare the effort with other, more direct, numerical methods for solving the problem.

10.2 The Heat Equation with Spatial Symmetries

In Section 10.1.2 we formulated the solution of the d-dimensional heat equation

$$\frac{\partial u}{\partial t} = \nabla^2 u, \qquad \boldsymbol{x} \in \mathbb{R}^d, \quad t > 0, \tag{10.29}$$

for a given initial value for u. If boundary and initial conditions are cylindrically or spherically symmetric, so is the solution, and it is beneficial to rewrite the Laplacian ∇^2 in the corresponding coordinates (see the appendix, Section K) and suppress any of the nonradial derivatives. With circular (in two dimensions) or cylindrical (in three dimensions) symmetry, with $r^2 = x^2 + y^2$, we have

$$\frac{\partial u}{\partial t} = \frac{1}{r}\frac{\partial}{\partial r}\left(r\frac{\partial u}{\partial r}\right) \quad \text{or} \quad \frac{\partial u}{\partial t} = \frac{1}{r}\frac{\partial}{\partial r}\left(r\frac{\partial u}{\partial r}\right) + \frac{\partial^2}{\partial z^2}u. \tag{10.30a}$$

With spherical symmetry, with $r^2 = x^2 + y^2 + z^2$, we have

$$\frac{\partial u}{\partial t} = \frac{1}{r^2}\frac{\partial}{\partial r}\left(r^2\frac{\partial u}{\partial r}\right). \tag{10.30b}$$

In general we have thus for a radially symmetric field in $d = 1, 2, 3$-dimensional space that

$$\frac{\partial u}{\partial t} = \frac{1}{r^{d-1}}\frac{\partial}{\partial r}\left(r^{d-1}\frac{\partial u}{\partial r}\right) \qquad \text{for} \quad r = |\boldsymbol{x}|, \quad \boldsymbol{x} \in \mathbb{R}^d, \quad t > 0. \tag{10.30c}$$

Evidently, these PDEs are defined on the half-space $r \geq 0$. If u is smooth at $r = 0$, the radial symmetry induces the boundary condition

$$\frac{\partial u}{\partial r} = 0. \tag{10.31}$$

Finally, we remark that if for a circular or cylindrically symmetric problem the r domain consists of values $r \geq R_0 > 0$, for some R_0, we can use the transformation

$$v(\rho, t) := u(r, t), \qquad \rho := \ln r, \tag{10.32a}$$

resulting, for (10.30a), in

$$\frac{\partial v}{\partial t} = \mathrm{e}^{-2\rho} \frac{\partial^2 v}{\partial \rho^2} \quad \text{or} \quad \frac{\partial v}{\partial t} = \mathrm{e}^{-2\rho} \frac{\partial^2 v}{\partial \rho^2} + \frac{\partial^2 v}{\partial z^2}. \tag{10.32b}$$

A three-dimensional spherically symmetric problem can be transformed, for $r \geq R_0$, for some $R_0 > 0$, by

$$v(r, t) := u(r, t) r \tag{10.33a}$$

to a form equivalent to the one-dimensional standard heat equation

$$\frac{\partial v}{\partial t} = \frac{\partial^2 v}{\partial r^2}. \tag{10.33b}$$

This shows that any of the foregoing one-dimensional solutions corresponds immediately to a similar three-dimensional solution.

10.3 Similarity Solutions

In Section 7.3 we found by dimensional arguments that, in the absence of explicit length and time scales other than x and t themselves, dimensionless groups can only occur by mutual combinations of x and t. In the present problems related to the heat equation the prevailing combination appears to be $x/\sqrt{t}$. In the same way the dependent variable (the temperature, say) will scale on a given value if this is available, but will have to depend on a combination of x and t, and the flux if no value but only a flux is given. These observations give rise to similarity solutions (see also Section 2.5).

As these symmetry properties should be present in the problem independently of the physics, it seems useful to look for similarity solutions as soon as the spatial and time coordinates have an infinite extension. Assume that the problem is radially symmetric, so $u = u(r, t)$ where $d = 1$, 2, or 3, and r denotes the Euclidean distance to the origin. Consider, therefore,

$$\frac{\partial u}{\partial t} = \frac{1}{r^{d-1}} \frac{\partial}{\partial r} \left(r^{d-1} \frac{\partial u}{\partial r} \right) \quad \text{for} \quad r = |\boldsymbol{x}|, \quad \boldsymbol{x} \in \mathbb{R}^d, \quad t > 0. \tag{10.30c}$$

We will look for solutions of the form

$$u(r, t) = r^m f(\eta), \qquad \eta = \frac{r}{t^n}. \tag{10.34}$$

As $r = t^n\eta$, there is no need to include any factor of the type t^p. Differentiation with respect to t and r yields

$$\frac{\partial u}{\partial t} = -nr^{m-2}\eta^3 t^{2n-1} f'(\eta), \tag{10.35a}$$

$$\frac{\partial u}{\partial r} = r^{m-1}\big(mf(\eta) + \eta f'(\eta)\big), \tag{10.35b}$$

$$\frac{\partial^2 u}{\partial r^2} = r^{m-2}\big(m(m-1)f(\eta) + 2m\eta f'(\eta) + \eta^2 f''(\eta)\big) \tag{10.35c}$$

(the prime denotes differentiation with respect to η). After substitution in (10.30c), we find

$$\eta^2 f'' + \eta(2m + d - 1)f' + m(m + d - 2)f + n\eta^3 t^{2n-1} f' = 0, \tag{10.36}$$

which only has proper solutions if $n = \frac{1}{2}$. So we finally have the equation

$$\eta^2 f'' + \eta\left(\frac{1}{2}\eta^2 + 2m + d - 1\right)f' + m(m + d - 2)f = 0. \tag{10.37}$$

A solution of this equation for arbitrary m and d can be given in terms of hypergeometric or similar functions (see [1]), but this is too general to be of interest. It is more expedient to look for specific solutions once d and m are known. The value of m has to follow from the available boundary conditions and is sometimes immediately clear from dimensional arguments. A typical case may be found in Example 7.17.

Example 10.7 An important example of a similarity solution is the field of a steady point source. This is most efficiently found from the fundamental solution given in (10.16). This gives us instantly the field of a stroke of heat from the point source $\delta(\boldsymbol{x})\delta(t)$, i.e.,

$$u(\boldsymbol{x}, t) = \left(\frac{1}{2\sqrt{\pi t}}\right)^d \exp\left(-\frac{r^2}{4t}\right).$$

By integrating this solution with respect to time, we find the field of a steady point source $\delta(\boldsymbol{x})H(t)$ at the origin, switched on at $t = 0$, thus satisfying

$$\frac{\partial U}{\partial t} - \nabla^2 U = \delta(\boldsymbol{x})H(t).$$

In one dimension this is

$$U_1(x, t) = \int_0^t \frac{\mathrm{e}^{-x^2/4\tau}}{\sqrt{4\pi\tau}}\mathrm{d}\tau = \sqrt{\frac{t}{\pi}}\exp\left(-\frac{x^2}{4t}\right) - \frac{1}{2}|x|\operatorname{erfc}\left(\frac{|x|}{\sqrt{4t}}\right),$$

which we already encountered in Example 7.17. In two dimensions we have

$$U_2(r, t) = \int_0^t \frac{\mathrm{e}^{-r^2/4\tau}}{4\pi\tau}\mathrm{d}\tau = \frac{1}{4\pi}E_1\left(\frac{r^2}{4t}\right), \qquad \text{where } E_1(z) = \int_z^\infty \frac{\mathrm{e}^{-t}}{t}\,\mathrm{d}t.$$

$E_1(z)$ is known as the exponential integral [1]. In three dimensions we obtain

$$U_3(r, t) = \int_0^t \frac{\mathrm{e}^{-r^2/4\tau}}{(4\pi\tau)^{3/2}}\mathrm{d}\tau = \frac{1}{4\pi r}\operatorname{erfc}\left(\frac{r}{\sqrt{4t}}\right).$$

It may be verified that indeed the respective fluxes out of the source are given by

$$\lim_{h\downarrow 0}\left[-\frac{\partial U_1}{\partial x}\right]_{x=-h}^{x=h} = \lim_{h\downarrow 0}\left[\frac{1}{2}\,\text{sign}(x)\,\text{erfc}\left(\frac{|x|}{\sqrt{4t}}\right)\right]_{x=-h}^{x=h} = 1,$$

$$\lim_{h\downarrow 0}\left[-2\pi r\frac{\partial U_2}{\partial r}\right]_{r=h} = \lim_{h\downarrow 0}\left[\exp\left(-\frac{r^2}{4t}\right)\right]_{r=h} = 1,$$

$$\lim_{h\downarrow 0}\left[-4\pi r^2\frac{\partial U_3}{\partial r}\right]_{r=h} = \lim_{h\downarrow 0}\left[\frac{r}{\sqrt{\pi t}}\exp\left(-\frac{r^2}{4t}\right) + \text{erfc}\left(\frac{r}{\sqrt{4t}}\right)\right]_{r=h} = 1.$$

It is remarkable that the field of a point source of constant output sometimes (if $d = 3$) converges to a stationary state, but not always (if $d = 1, 2$). This depends on d, the number of spatial dimensions of the problem. For large t we find asymptotically [1]

$$U_1(x,t) = \sqrt{\frac{t}{\pi}} - \frac{1}{2}|x| + \mathcal{O}(t^{-1/2}),$$

$$U_2(r,t) = \frac{1}{4\pi}\left[\ln(t) - \ln\left(\frac{1}{4}r^2\right) - \gamma\right] + \mathcal{O}(t^{-1}),$$

$$U_3(r,t) = \frac{1}{4\pi r} + \mathcal{O}(t^{-1/2}),$$

where $\gamma = 0.5772\ldots$. Apparently, only the capacity of three-dimensional space is big enough to absorb all the heat of a stationary source! □

10.4 Initial Boundary Value Problems

When the problem is defined on a domain with boundaries, we need to provide *boundary conditions*. Sometimes these boundaries are part of the evolution process; we shall address the latter in the next section. In order to fix our thoughts, we think of u as a temperature here. Consider, as before, a PDE with given initial value:

$$\frac{\partial u}{\partial t} = \nabla\cdot(k\nabla u) = k\nabla^2 u, \qquad \boldsymbol{x}\in\Omega\subset\mathbb{R}^d, \quad t>0, \tag{10.38a}$$

$$u(x,0) = v(x), \tag{10.38b}$$

where k denotes the thermal diffusion coefficient and $-k\nabla u$ is the heat flux vector. The heat flux across a surface with unit normal vector $\boldsymbol{n}$ is thus given by $-k\frac{\partial}{\partial \boldsymbol{n}}u$. The first form of (10.38) is more general and is valid if k depends on $\boldsymbol{x}$ or u (or both). We assume that Ω is a compact simply connected domain with boundary $\partial\Omega$. Apart from a prescribed initial value, the following boundary conditions are usually considered:

(i) If the temperature at the boundary is prescribed, we call this a boundary condition of *Dirichlet type*:

$$u(\boldsymbol{x},t) = g(\boldsymbol{x},t), \qquad \boldsymbol{x}\in\partial\Omega, \quad t>0. \tag{10.39}$$

(ii) If the heat flux across the boundary is prescribed, we have a boundary condition of *Neumann type*:

$$-k\frac{\partial u}{\partial \boldsymbol{n}}(\boldsymbol{x},t) = g(\boldsymbol{x},t), \qquad \boldsymbol{x}\in\partial\Omega, \quad t>0. \tag{10.40}$$

Here $\frac{\partial u}{\partial \boldsymbol{n}} := \boldsymbol{n}\cdot\nabla u$, where $\boldsymbol{n}$ is the unit outward normal vector on $\partial\Omega$.

(iii) A linear combination of Dirichlet and Neumannn type boundary conditions is called a *Robin* or a *Newton* boundary condition:

$$-k\frac{\partial u}{\partial \boldsymbol{n}}(\boldsymbol{x},t) = s\big(u(\boldsymbol{x},t) - g(\boldsymbol{x},t)\big), \qquad \boldsymbol{x} \in \partial\Omega, \quad t > 0, \tag{10.41}$$

with the heat transfer coefficient $s > 0$. Put in physical terms, the heat flux is proportional to the difference between the temperature of the medium and the ambient.

(iv) If the loss of heat is caused by radiation, we have a *radiative* boundary condition:

$$-k\frac{\partial u}{\partial \boldsymbol{n}}(\boldsymbol{x},t) = \sigma\big(u(\boldsymbol{x},t)^4 - u_\infty^4\big), \qquad \boldsymbol{x} \in \partial\Omega, \quad t > 0, \tag{10.42}$$

where u_∞ is the specified ambient temperature of the surrounding medium and the parameter σ quantifies the emissivity properties of the surface.

We remark that boundary conditions without external forcing, i.e., where $g(\boldsymbol{x},t) \equiv 0$, are called homogeneous.

Example 10.8 Consider the initial boundary value problem

$$\begin{aligned}
&\frac{\partial u}{\partial t} = \frac{\partial^2 u}{\partial x^2}, && x \in (0,1), \quad t > 0,\\
&u(x,0) = v(x), && x \in (0,1),\\
&u(0,t) = u(1,t) = 0, && t > 0.
\end{aligned}$$

Clearly the boundary conditions are of Dirichlet type. Assume that v and v' are piecewise smooth such that the coefficients of the Fourier-sine expansion

$$v(x) = \sum_{k=1}^{\infty} v_k \sin(k\pi x)$$

converge according to $v_k = \mathcal{O}(1/k)$ as $k \to \infty$ (Corollary 3.7). Consider the formal series

$$u(x,t) = \sum_{k=1}^{\infty} v_k \,\mathrm{e}^{-k^2\pi^2 t} \sin(k\pi x).$$

Due to the exponential, the coefficients behave as $v_k\,\mathrm{e}^{-k^2\pi^2 t} = \mathcal{O}(k^{-n})$ as $k \to \infty$ for any n if $t > 0$, so the series and all of its derivatives converge uniformly for $t > 0$ (appendix, Section C). As a result, u is continuous, while differentiation and summation may be exchanged. It is then easily verified that u satisfies the differential equation, initial profile v, and boundary values $u(0,t) = u(1,t) = 0$. Note in particular that in the heat equation no discontinuity in any derivative can be sustained: (practically) any initial profile v yields immediately, for any $t > 0$, an infinitely differentiable temperature distribution u. □

Example 10.9 Consider the initial boundary value problem

$$\begin{aligned}
&\frac{\partial u}{\partial t} = \frac{\partial^2 u}{\partial x^2}, && x \in (0,1), \quad t > 0,\\
&u(x,0) = v(x), && x \in (0,1),\\
&\frac{\partial u}{\partial x}(0,t) = \frac{\partial u}{\partial x}(1,t) = 0, && t > 0.
\end{aligned}$$

This is a boundary condition of Neumann type. Analogously to the previous example, we assume that v and v' are piecewise smooth and expand v into the Fourier-cosine series

$$v(x) = \sum_{k=0}^{\infty} v_k \cos(k\pi x),$$

whose coefficients converge like $v_k = \mathcal{O}(1/k)$ as $k \to \infty$. The formal solution

$$u(x,t) = \sum_{k=0}^{\infty} v_k \, \mathrm{e}^{-k^2\pi^2 t} \cos(k\pi x)$$

and any of its derivatives converge uniformly for $t > 0$ to a continuous function. So this solution satisfies the equation and initial condition. In particular, it satisfies the boundary conditions. □

Uniqueness of solutions can be shown by integral arguments.

Theorem 10.10. *The solution of the linear initial boundary value problem* (10.38), *i.e., k is independent of u, with any linear boundary condition* (10.39), (10.40), *or* (10.41), *is unique.*

Proof. Let u_1 and u_2 be two solutions. Then $w(\boldsymbol{x},t) := u_1(\boldsymbol{x},t) - u_2(\boldsymbol{x},t)$ satisfies either initial boundary value problem with $g(\boldsymbol{x},t) \equiv 0$ and $v(\boldsymbol{x},t) \equiv 0$. From a variation of Green's first identity we have (for a fixed Ω)

$$\begin{aligned}
\frac{\mathrm{d}}{\mathrm{d}t}\int_\Omega \frac{1}{2} w^2 \,\mathrm{d}V &= \int_\Omega w \frac{\partial w}{\partial t}\,\mathrm{d}V = \int_\Omega w \nabla\cdot(k\nabla w)\,\mathrm{d}V = \int_\Omega (\nabla\cdot(kw\nabla w) - k|\nabla w|^2)\,\mathrm{d}V \\
&= -\int_\Omega k|\nabla w|^2\,\mathrm{d}V + \oint_{\partial\Omega} kw\frac{\partial w}{\partial \boldsymbol{n}}\,\mathrm{d}S. \\
&= \left\{\begin{array}{l} -\displaystyle\int_\Omega k|\nabla w|^2\,\mathrm{d}V \text{ for (10.39) or (10.40)} \\ -\displaystyle\int_\Omega k|\nabla w|^2\,\mathrm{d}V - \oint_{\partial\Omega} sw^2\,\mathrm{d}S \text{ for (10.41)} \end{array}\right\} < 0.
\end{aligned}$$

The first option corresponds to a Dirichlet or Neumann boundary condition, the other to a Robin boundary condition. Hence

$$0 \le \int_\Omega w^2(\boldsymbol{x},t)\,\mathrm{d}V \le \int_\Omega w^2(\boldsymbol{x},0)\,\mathrm{d}V = 0,$$

so $w \equiv 0$. □

10.5 Moving Boundaries; Stefan Problems

Diffusion processes often involve a moving boundary. Typical examples are contact problems between two materials like two fluids or a solid material and a dissolvent, between two phases of the same material like water and ice, and between wet and dry materials. In these problems the boundary is determined by some additional kinematic constraint. In Chapter 11 we shall consider one such problem in more detail. Here we will restrict ourselves to a class of problems referred to as Stefan problems [172]. One may typically think of ice of 0°C that is melting at a surface, the *front*, denoted by S. Apparently $S = S(t)$. In one dimension, its simplest form, the (dimensionless) temperature $u(x,t)$ of the water satisfies the heat equation

$$\frac{\partial u}{\partial t} = \frac{\partial^2 u}{\partial x^2}, \qquad 0 < x < S(t), \quad t > 0. \tag{10.43}$$

Before $t = 0$, the temperature is everywhere $u(x,t) = 0$, but at $t = 0$ the temperature at one end, $x = 0$, is suddenly increased to unity:

$$u(0,t) = 1, \qquad t \geq 0. \tag{10.44a}$$

This enforces the input of energy because the temperature elsewhere is lower. At $x = S(t)$, the interface with the ice, we have a continuous temperature, so

$$u(S,t) = 0, \qquad t > 0. \tag{10.44b}$$

As the position of the interface is unknown, we need an additional condition. This extra condition is found from the physics of the phase transition. The amount of specific latent heat released during the water-to-ice phase transition is again to be added when the ice melts. As the front travels with a certain speed, the produced heat travels with the same speed and is therefore equal to the heat flux. As we assumed that the ice is of constant temperature, i.e., $u_{\text{ic}} \equiv 0$, the flux into the ice vanishes, and we get (nondimensionally)

$$-\frac{\partial u}{\partial x}(S(t),t) = \alpha\frac{\mathrm{d}S}{\mathrm{d}t}, \qquad S(0) = 0. \tag{10.45}$$

The constant α in (10.45) is called the *Stefan constant*. Let us try to solve problem (10.43–10.45) by introducing, following (10.34) to (10.37), the similarity solution

$$u(x,t) = x^m f(\eta), \qquad \eta := \frac{x}{2\sqrt{t}}.$$

From (10.44a) it follows that $m = 0$, while from (10.44b) we can write for some constant γ that

$$S = 2\gamma\sqrt{t}.$$

Equation (10.43) and boundary conditions turn into

$$f'' + 2\eta f' = 0, \qquad f(0) = 1, \quad f(\gamma) = 0, \quad f'(\gamma) = -2\alpha\gamma, \tag{10.46}$$

along $0 \leq \eta \leq \gamma$, with solution

$$f(\eta) = 1 - \alpha\gamma\, \mathrm{e}^{\gamma^2}\sqrt{\pi}\,\mathrm{erf}\,(\eta), \tag{10.47}$$

while the unknown γ is given by the real positive root of

$$\alpha\gamma\sqrt{\pi}\,\mathrm{e}^{\gamma^2}\,\mathrm{erf}\,(\gamma) = 1. \tag{10.48}$$

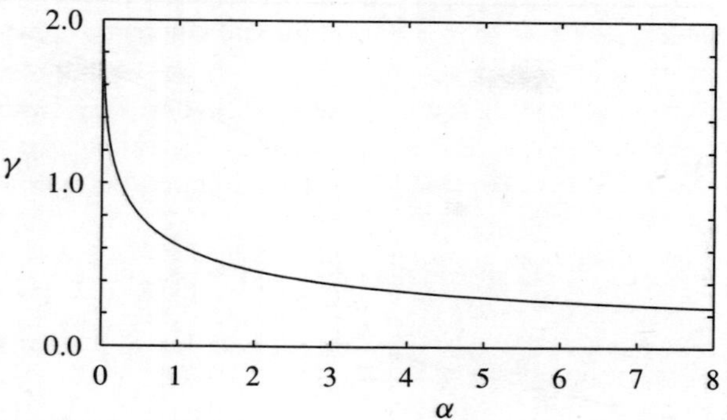

Figure 10.2. *Similarity coordinate γ of the interface as a function of the Stefan problem parameter α.*

Altogether we thus have

$$u(x,t) = 1 - \frac{\operatorname{erf}(x/2\sqrt{t})}{\operatorname{erf}(\gamma)}. \tag{10.49}$$

We may solve (10.48) numerically to obtain γ as a function of the Stefan parameter α. See, e.g., Figure 10.2. For small γ we can approximate $\gamma\sqrt{\pi}\,e^{\gamma^2}\operatorname{erf}(\gamma)$ by $2\gamma^2 + \cdots$, so for large α we have $\gamma \simeq 1/\sqrt{2\alpha}$. For large γ we have a dominating exponential that produces for small α a $\gamma \simeq (-\frac{1}{2}\ln\alpha)^{\frac{1}{2}}$.

Example 10.11 This single-phase problem, where the temperature varies only in the water, is easily generalised to a double-phase problem, where the temperature varies in both the ice and the water. In this case we need the more general Stefan condition

$$\rho L \frac{\mathrm{d}}{\mathrm{d}t} S = \left[-k_{\mathrm{wa}} \frac{\partial}{\partial x} T_{\mathrm{wa}} + k_{\mathrm{ic}} \frac{\partial}{\partial x} T_{\mathrm{ic}} \right]_{x=S(t)} \tag{10.50}$$

(now dimensional), where L denotes the amount of specific latent heat released during the water-to-ice phase transition.

Consider a Stefan problem of melting ice in the semi-infinite domain $x \in [0,\infty)$. Let the phase change be at $x = S$. Let the temperature of the ice at $t = 0$ be given by $T_0 < 0$. It follows that the temperature for $x \to \infty$ also equals T_0. At time $t = 0$ the temperature at $x = 0$ is suddenly increased to $T_1 > 0$. We have

$$\frac{\partial T_{\mathrm{wa}}}{\partial t} = k_{\mathrm{wa}} \frac{\partial^2 T_{\mathrm{wa}}}{\partial x^2}, \qquad 0 < x < S(t), \quad t > 0, \qquad T_{\mathrm{wa}}(0,t) = T_1,$$

$$\frac{\partial T_{\mathrm{ic}}}{\partial t} = k_{\mathrm{ic}} \frac{\partial^2 T_{\mathrm{ic}}}{\partial x^2}, \qquad S(t) < x < \infty, \quad t > 0, \qquad T_{\mathrm{ic}}(x,0) = T_{\mathrm{ic}}(\infty,t) = T_0,$$

$$\alpha \frac{\mathrm{d}}{\mathrm{d}t} S = \left[-k_{\mathrm{wa}} \frac{\partial}{\partial x} T_{\mathrm{wa}} + k_{\mathrm{ic}} \frac{\partial}{\partial x} T_{\mathrm{ic}} \right]_{x=S(t)}, \qquad \left[T_{\mathrm{ic}}(x,t) = T_{\mathrm{wa}}(x,t) = 0 \right]_{x=S(t)}.$$

Inspired by the previous solution, we find solutions of the form

$$S = 2\gamma\sqrt{t}, \quad T_{\mathrm{wa}} = T_1 \left(1 - \frac{\operatorname{erf}(x/2\sqrt{k_{\mathrm{wa}}t})}{\operatorname{erf}(\gamma_{\mathrm{wa}})} \right), \quad T_{\mathrm{ic}} = T_0 \left(1 - \frac{\operatorname{erfc}(x/2\sqrt{k_{\mathrm{ic}}t})}{\operatorname{erfc}(\gamma_{\mathrm{ic}})} \right),$$

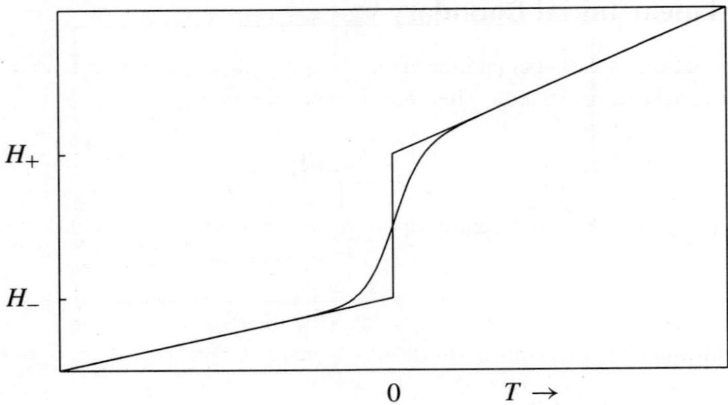

Figure 10.3. *Enthalpy function $H(T)$.*

where $\gamma_{\mathrm{wa}} = \gamma/\sqrt{k_{\mathrm{wa}}}$ and $\gamma_{\mathrm{ic}} = \gamma/\sqrt{k_{\mathrm{ic}}}$. The constant γ is to be determined from the equation

$$\alpha\sqrt{\pi} = T_1 \frac{\exp(-\gamma_{\mathrm{wa}}^2)}{\gamma_{\mathrm{wa}}\,\mathrm{erf}\,(\gamma_{\mathrm{wa}})} + T_0 \frac{\exp(-\gamma_{\mathrm{ic}}^2)}{\gamma_{\mathrm{ic}}\,\mathrm{erfc}(\gamma_{\mathrm{ic}})}. \qquad \square$$

From a physical point of view it is more sensible to use the *enthalpy* rather than the temperature as the dependent variable for the melting ice problem, as we have both "sensible" temperature and latent heat. The enthalpy, being the sum of both, is given as a function of the temperature. This then closes the equations. Let us denote this enthalpy by H. Then we have for given $H = H(T)$

$$\frac{\partial H}{\partial t} = \frac{\partial^2 T}{\partial x^2}, \qquad 0 < x < S(t), \quad t > 0, \tag{10.51a}$$

$$T(x,0) = 0, \tag{10.51b}$$

$$T(0,t) - \beta\,\frac{\partial T}{\partial x}(0,t) = 1. \tag{10.51c}$$

To start with we know the temperature at $t = 0$, say, and so we know the initial enthalpy as well. Two typical graphs of $H(T)$ are given in Figure 10.3. One shows a simple discontinuous H, with a discontinuity at $T = 0$ between H_- and H_+, whereas the other one shows an H of a material with a "mushy region," i.e., where the phase change is more gradually taking place (as happens in melting of alloys).

10.6 Long-Time Behaviour of Solutions

The solutions of parabolic problems are typically diffusive and smoothing steep gradients. As a result, any initial value tends to be "forgotten" and is therefore sometimes not as important as the long-time behaviour. We will consider some occasions where this is the case.

10.6.1 Linear Initial Boundary Value Problem

Parabolic equations are of special interest in studying phenomena that tend to a *steady state* situation or *equilibrium*. In fact, when considering the PDE

$$\frac{\partial u}{\partial t} = \mathcal{L}[u], \tag{10.52}$$

where $\mathcal{L}$ is a second order differential operator, it is natural to ask for solutions for which

$$\frac{\partial u}{\partial t} \to 0;$$

i.e., u will ultimately satisfy the *stationary elliptic equation*

$$\mathcal{L}[u] = 0.$$

This shows that there is an intimate relation between some parabolic and elliptic PDEs. Indeed, one often even tries to solve a difficult elliptic boundary value problem numerically by embedding it in a corresponding parabolic problem like (10.52) and solving the latter by implicit time stepping. This *false transient* approach is the basis of quite a few numerical methods. As we saw in Example 10.7, this will not always work and the problems should be analyzed carefully. At least the elliptic problem should have a solution, and the solution of the associated parabolic problem should converge for large time to this solution.

We shall illustrate the long-time behaviour of solutions of parabolic problems by a simple linear one-dimensional case. Consider for $U(x)$ the steady elliptic problem

$$\mathcal{L}[U] := a\frac{\partial^2 U}{\partial x^2} + b\frac{\partial U}{\partial x} + cU = 0, \qquad x \in (0,1), \tag{10.53a}$$

$$U(0) = p, \qquad U(1) = q, \tag{10.53b}$$

where $a > 0$, b, c, p, and q are constants. Suppose that we try to approach this solution by the auxiliary parabolic problem

$$\frac{\partial y}{\partial t} = a\frac{\partial^2 y}{\partial x^2} + b\frac{\partial y}{\partial x} + cy, \qquad x \in (0,1), \qquad t > 0, \tag{10.54a}$$

$$y(x,0) = v(x) + U(x), \qquad x \in (0,1), \tag{10.54b}$$

$$y(0,t) = p,\ y(1,t) = q, \qquad t > 0, \tag{10.54c}$$

where v is an arbitrary, in general unknown, but reasonably smooth, initial error to the finally sought solution. We are interested in knowing under which conditions solution y asymptotically tends to U. In other words, when does the difference $u = y - U$, satisfying

$$\frac{\partial u}{\partial t} = a\frac{\partial^2 u}{\partial x^2} + b\frac{\partial u}{\partial x} + cu, \qquad x \in (0,1), \quad t > 0, \tag{10.55a}$$

$$u(x,0) = v(x), \qquad x \in (0,1), \tag{10.55b}$$

$$u(0,t) = u(1,t) = 0, \qquad t > 0, \tag{10.55c}$$

tend to zero for $t \to \infty$?

After trying a transformation of the type $u(x,t) = \mathrm{e}^{\alpha t - \beta x}\,\tilde{u}(x,\gamma t)$, we find that the transformation $u(x,t) = \exp[(c - \frac{b^2}{4a})t - \frac{b}{2a}x]\tilde{u}(x,at)$ reduces the problem to the one in Example 10.8. So we deduce a Fourier series solution of (10.55) of the form

$$u(x,t) = \mathrm{e}^{(c-\frac{b^2}{4a})t - \frac{b}{2a}x} \sum_{k=1}^{\infty} A_k\,\mathrm{e}^{-\pi^2 k^2 at}\,\sin(k\pi x), \tag{10.56}$$

where the coefficients A_k are found from the Fourier-sine series expansion

$$v(x)\,\mathrm{e}^{\frac{b}{2a}x} = \sum_{k=1}^{\infty} A_k \sin(k\pi x) \tag{10.57}$$

in order to satisfy condition (10.55b). Although in general (10.57) is not uniformly convergent, the coefficients will at least decay like $A_k = \mathcal{O}(k^{-1})$ as $k \to \infty$. So, with the help of the exponential, series (10.56) will converge uniformly for any $t > 0$, and indeed $u(0,t) = u(1,t) = 0$ (appendix, Section C).

We conclude that $u(x,t) \to 0$ for $t \to \infty$ if the least attenuated mode tends to zero, i.e., if $c < \frac{b^2}{4a} + \pi^2 a$. In this case solution $y(x,t)$ will approach the stationary solution $U(x)$.

The corresponding problem with boundary conditions of Neumann type is more involved. Consider the initial boundary value problem

$$\frac{\partial u}{\partial t} = a\frac{\partial^2 u}{\partial x^2} + b\frac{\partial u}{\partial x} + cu, \qquad x \in (0,1), \quad t > 0, \tag{10.58a}$$

$$u(x,0) = v(x), \qquad x \in (0,1), \tag{10.58b}$$

$$\frac{\partial u}{\partial x}(0,t) = \frac{\partial u}{\partial x}(1,t) = 0, \qquad t > 0. \tag{10.58c}$$

After a similar transformation to before and separation of variables, we can construct a Fourier-type series expansion

$$u(x,t) = A_0\,\mathrm{e}^{ct} + \mathrm{e}^{(c-\frac{b^2}{4a})t - \frac{b}{2a}x} \sum_{k=1}^{\infty} A_k\,\mathrm{e}^{-\pi^2 k^2 at}\, f_k(x), \tag{10.59}$$

where $\{f_k(x)\}$ form the orthonormal set

$$f_k(x) = \left(\frac{2}{4a^2\pi^2k^2 + b^2}\right)^{\frac{1}{2}} \big(2a\pi k\cos k\pi x + b\sin k\pi x\big). \tag{10.60}$$

Except for A_0 (note that the solution is *not* unique), the coefficients A_k are determined by

$$A_k = \int_0^1 \big(v(x) - A_0\big)\,\mathrm{e}^{\frac{b}{2a}x}\, f_k(x)\,\mathrm{d}x. \tag{10.61}$$

We conclude that the restrictions on c for a decaying u are slightly more stringent than in the previous case. Unless we are able to make sure that $A_0 = 0$, we must have $c < 0$ for u to tend to zero for large t.

Example 10.12 Let $\Omega := [0,1] \times [0,1]$. Consider the initial boundary value problem

$$\begin{aligned} \frac{\partial u}{\partial t} &= \nabla^2 u + 1, && \boldsymbol{x} \in \Omega, \quad t > 0 \\ u(x,y,0) &= 0, && \boldsymbol{x} \in \Omega, \\ u(x,y,t) &= 0, && \boldsymbol{x} \in \partial\Omega, \quad t > 0. \end{aligned}$$

We shall use the Duhamel integral (see Theorem 4.16). To this end we have to find a solution of

$$\begin{aligned} \frac{\partial w}{\partial t}(\boldsymbol{x},t;\tau) &= \nabla^2 w(\boldsymbol{x},t;\tau), && \boldsymbol{x} \in \Omega, \quad t > \tau, \\ w(\boldsymbol{x},\tau;\tau) &= 1, && \boldsymbol{x} \in \Omega, \\ w(\boldsymbol{x},t;\tau) &= 0, && \boldsymbol{x} \in \partial\Omega, \quad t > \tau. \end{aligned}$$

Because of the boundary conditions it is convenient to have Fourier basis modes of the form $\sin(j\pi x)\sin(k\pi y)$ only. We find

$$1 = 16 \sum_{j,k \geq 0} \frac{\sin \alpha_j x}{\alpha_j} \frac{\sin \alpha_k y}{\alpha_k}, \qquad \text{where} \quad \alpha_j := (2j+1)\pi.$$

Using, e.g., the dispersion relation, we find the (uniformly converging) expansion

$$w(x,y,t;\tau) = 16 \sum_{j,k \geq 0} \frac{\sin \alpha_j x}{\alpha_j} \frac{\sin \alpha_k y}{\alpha_k} \, \mathrm{e}^{-(\alpha_j^2 + \alpha_k^2)(t-\tau)}.$$

Hence

$$u(x,y,t) = 16 \sum_{j,k \geq 0} \frac{\sin \alpha_j x}{\alpha_j} \frac{\sin \alpha_k y}{\alpha_k} \int_0^t \mathrm{e}^{-(\alpha_j^2 + \alpha_k^2)(t-\tau)} \, \mathrm{d}\tau.$$

If $t \to \infty$, we find

$$u(x,y,t) \to 16 \sum_{j,k \geq 0} \frac{\sin(\alpha_j x)\sin(\alpha_k y)}{\alpha_j \alpha_k (\alpha_j^2 + \alpha_k^2)} =: v(x,y).$$

By direct substitution and noting the uniform convergence, we may verify immediately that this is the solution of the Poisson problem

$$\begin{aligned} \nabla^2 v &= 1, && \boldsymbol{x} \in \Omega, \\ v &= 0, && \boldsymbol{x} \in \partial\Omega. \end{aligned}$$

□

10.6.2 Equilibrium and Travelling-Wave Solutions for Nonlinear Problems

Quite often diffusion equations arise in the analysis of chemical reactions or population dynamics or in the modeling of epidemics. Typically, the equation includes a nonlinear source or reaction term and takes on the form

$$\frac{\partial u}{\partial t} = \nabla \cdot (D \nabla u) + r(u). \tag{10.62}$$

The reaction term r arises from the application. For example, u may be a temperature and r the energy release from the chemical reaction; or u may be a concentration of some species and r a source or sink. A common model for r is the polynomial

$$r(u) = \gamma(u - \alpha)(\beta - u), \tag{10.63}$$

where $\alpha < \beta$ and $\gamma > 0$. If we further assume that D is a positive constant, we may rescale this equation by the transformation

$$\tilde{u} := \frac{u - \alpha}{\beta - \alpha}, \quad \tilde{\boldsymbol{x}} := \left(\frac{\gamma}{D}(\beta - \alpha)\right)^{1/2} \boldsymbol{x}, \quad \tilde{t} := \gamma(\beta - \alpha)t \tag{10.64}$$

(where we will omit the tilde henceforth) into the standard form

$$\frac{\partial u}{\partial t} = \nabla^2 u + u(1 - u). \tag{10.65}$$

This is known as *Fisher's equation.* It was introduced originally to model the spread of a gene in a population [173].

To demonstrate the basic ideas we restrict our discussion of (10.62) to the one-dimensional case

$$\frac{\partial u}{\partial t} = \frac{\partial^2 u}{\partial x^2} + r(u). \tag{10.66}$$

Clearly, $r(u)$ represents a source when it is positive and a sink when it is negative. Any zero of r, say $u = u_0$, is evidently also a solution of the equation if the boundaries are compatible with this solution, e.g., if the walls are isolated such that no heat or matter is lost. These stationary solutions are called *equilibrium solutions.* It is not clear in advance if any such solution will be attained for $t \to \infty$ if we start in its neighbourhood. This will be seen to be critically dependent on the sign of $r'(u_0)$. Let us therefore study the initial boundary value problem—with isolated boundaries and arbitrary initial condition—for a perturbation $\delta e(x, t)$ of $u = u_0$ given by

$$u(x, t) := u_0 + \delta e(x, t), \tag{10.67}$$

where δ is small and $r_0' := r'(u_0) \neq 0$. After linearisation of $r(u) = \delta r_0' e + \cdots$ (where we neglect any term smaller than $\mathcal{O}(\delta)$), we have the following problem for $e(x, t)$:

$$\frac{\partial e}{\partial t} = \frac{\partial^2 e}{\partial x^2} + r_0' e, \qquad x \in (0, 1), \qquad t > 0, \tag{10.68a}$$

$$e(x, 0) = v(x), \qquad x \in (0, 1), \tag{10.68b}$$

$$\frac{\partial e}{\partial x}(0, t) = \frac{\partial e}{\partial x}(1, t) = 0, \qquad t > 0. \tag{10.68c}$$

This has the same form as the problem we studied in (10.58). We can immediately infer that $u = u_0$ is an asymptotically stable equilibrium, i.e., $e \to 0$ as $t \to \infty$, if $r_0' < 0$. Similarly, it is asymptotically unstable if $r_0' > 0$. In particular, for the Fisher equation (10.65) we obtain that $u(x,t) \equiv 1$ is stable, while $u(x,t) \equiv 0$ is unstable.

It should be noted that other equilibrium solutions than the zeros of $r(u)$ are in principle also possible. They are, however, in general more difficult to analyse.

Example 10.13 The stability of solutions is an important issue in many reaction-diffusion equations. A typical question is whether the solution becomes unbounded in finite time, i.e., "blow up" occurs. We shall give a simple example of such a problem. Consider the following PDE with initial and boundary conditions:

$$\begin{aligned} \frac{\partial u}{\partial t} &= \frac{\partial^2 u}{\partial x^2} + u^3, && x \in (0,1), \quad t > 0, \\ u(x,0) &= 1, && x \in (0,1), \\ \frac{\partial u}{\partial x}(0,t) &= \frac{\partial u}{\partial x}(1,t) = 0, && t > 0. \end{aligned}$$

One easily verifies that a solution exists that is independent of x and hence is given by

$$u(x,t) = \frac{1}{\sqrt{1-2t}}.$$

It has the remarkable property that $u \to \infty$ for $t \to \frac{1}{2}$. □

Reaction-diffusion equations allow for further analysis by looking for *travelling-wave solutions* (see Section 2.5). These may be considered as similarity solutions of a particular type (Section 7.3.3, Example 7.18). As an example we consider Fisher's equation

$$\frac{\partial u}{\partial t} = \frac{\partial^2 u}{\partial x^2} + u(1-u), \qquad x \in \mathbb{R}, \quad t > 0. \tag{10.69}$$

We now try to find similarity solutions of the form

$$U(\xi) := U(x/c - t) = u(x,t).$$

By substituting this in (10.69), we obtain the autonomous ODE

$$U'' + c^2 U' + c^2 U(1-U) = 0, \qquad \xi \in \mathbb{R}, \tag{10.70}$$

where the prime denotes differentiation with respect to ξ. The physically interesting solutions are those that remain finite for $\xi \to \pm\infty$. Note that U depends on c^2, so the behaviour for left- and right-running waves is the same.

To facilitate the analysis it is useful to rewrite (10.70) as a first order system, i.e., with $y_1 := U$ and $y_2 := U'$:

$$\begin{aligned} y_1' &= y_2, \\ y_2' &= -c^2 y_2 - c^2 y_1(1-y_1). \end{aligned} \tag{10.71}$$

Insight into the behaviour of possible solutions is obtained by using the phase plane (see, e.g., [97]), i.e., graphs (trajectories) of y_2 as a function of y_1; see Figure 10.4. Of particular importance are the stationary points $(0,0)^T$ and $(1,0)^T$. The Jacobi matrix of the system

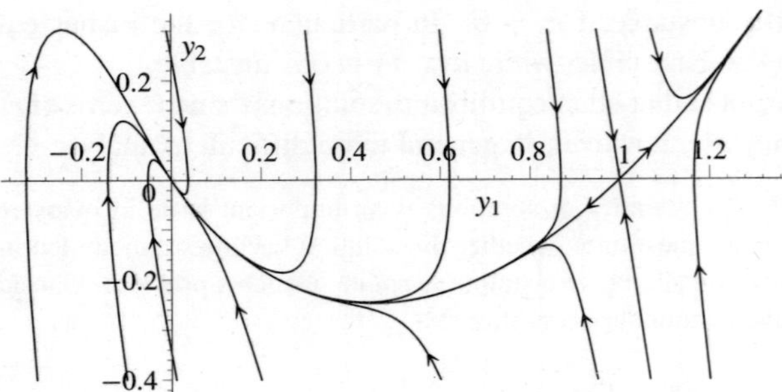

Figure 10.4. *Phase portrait of Fisher's travelling-wave problem* (10.71) *with* $c = 2.25$. *Arrows indicate the positive* ξ *direction. Note the trajectory that connects saddle point* $(1, 0)^T$ *with stable node* $(0, 0)^T$.

(10.71), linearised around $(0, 0)^T$, is given by

$$\begin{pmatrix} 0 & 1 \\ -c^2 & -c^2 \end{pmatrix}.$$

The corresponding eigenvalues

$$\lambda_{1,2} = -\frac{1}{2}c^2 \pm \frac{1}{2}\sqrt{c^4 - 4c^2}$$

are negative if $|c| \geq 2$ and complex with a negative real part otherwise. Hence the point $(0, 0)^T$ is a stable spiral point for $|c| < 2$ and a stable node for $|c| \geq 2$. At the other stationary point $(1, 0)^T$ one finds the Jacobi matrix

$$\begin{pmatrix} 0 & 1 \\ c^2 & -c^2 \end{pmatrix}$$

with the real eigenvalues

$$\lambda_{1,2} = -\frac{1}{2}c^2 \pm \frac{1}{2}\sqrt{c^4 + 4c^2},$$

one of which is always negative and the other always positive. So this point is an unstable saddle point (see Figure 10.4). Note that conclusions for both $(0, 0)^T$ and $(1, 0)^T$ are entirely in agreement with the results of (10.68).

Most of the trajectories are physically not interesting because they tend to infinity for $\xi \to \infty$ or $-\infty$. For each c there is one trajectory, however, that starts at $(1, 0)^T$ and ends at $(0, 0)^T$. This trajectory (see Figure 10.4) corresponds to the solution we are looking for. This solution may be understood as a transition of the system from the unstable stationary state $u(x, t) \equiv 0$ to the stable stationary state $u(x, t) \equiv 1$.

For models where u cannot be negative we have to assume additionally that $(0, 0)^T$ is not a spiral point; i.e., $|c| \geq 2$. In Figure 10.5 we have depicted the right-running wave for

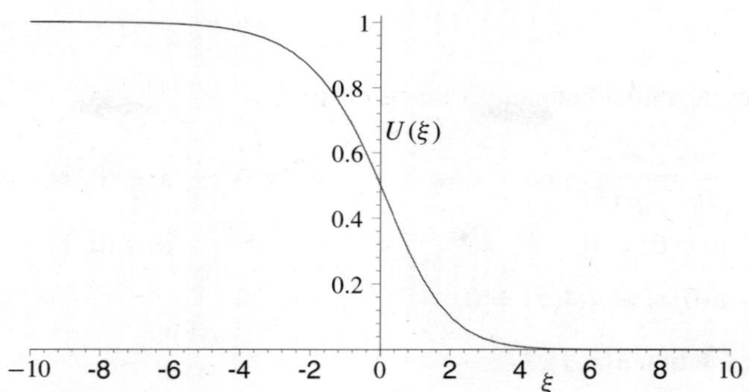

Figure 10.5. *The (numerically obtained) travelling-wave solution of Fisher's problem* (10.69), (10.70) *with* $c = 2.25$ *and* $U(0) = \frac{1}{2}$.

the value $c = 2.25$, normalized by $U(0) = \frac{1}{2}$. The wave's activity is exponentially localized around $x - ct = 0$.

10.7 Discussion

- Parabolic equations describe phenomena like conduction and diffusion. They are diffusive in nature; i.e., an initial profile is spread out rather than propagated in a certain direction. Conduction typically occurs in heat transfer, whereas diffusion refers to the "spreading" of a material, often in a chemistry context.

- Since solutions of elliptic equations can be seen as a steady state of solutions of parabolic equations, there are a number of correspondences between the two classes. An interesting one is that there is also a maximum principle for parabolic equations. The theory and statements follow similar lines as those outlined in Section 8.7. We may refer to [115].

- There exists an extensive literature on nonlinear problems. They occur in a multitude of problems, like population dynamics, chemical reactions, etc. In particular, the study of solitary waves has brought many interesting phenomena to light. For further reading we recommend [35].

- In Chapter 16 we will encounter parabolic problems in a number of practical situations. An example is that of thermal explosion in vessels in Section 16.3, which deals with chemical species that may spontaneously ignite at a certain temperature. An example of a Stefan problem is given in Section 16.12, where the melting of a solid by a laser beam is studied.

Exercises

10.1. Consider the initial boundary value problem

$$\begin{aligned} \frac{\partial u}{\partial t} &= \frac{\partial^2 u}{\partial x^2} - cu + \sin \pi x + \frac{1}{8}\sin 3\pi x, & x \in (0,1), \quad & t > 0, \\ u(x,0) &= 0, & x \in (0,1), \quad & \\ u(0,t) &= u(1,t) = 0, & & t > 0. \end{aligned}$$

Find its solution $u(x,t)$.

10.2. Consider the initial boundary value problem

$$\begin{aligned} \frac{\partial u}{\partial t} &= \frac{\partial^2 u}{\partial x^2}, & x \in (0,1), \quad & t > 0, \\ u(x,0) &= v(x), & x \in (0,1), \quad & \\ u(0,t) &= u(1,t) = 0, & & t > 0. \end{aligned}$$

(a) Find the Green's function $w(x,t;\tau)$ such that

$$u(x,t) = \int_0^t w(x,t;\tau)v(\tau)\,d\tau.$$

(b) Determine $u(x,t)$ for $v(x) = \sin^2(\pi x)$.

10.3. Consider the problem of Exercise 10.2, now with boundary conditions

$$\frac{\partial u}{\partial x}(0,t) = \frac{\partial u}{\partial x}(1,t) = 0, \qquad t > 0.$$

(a) Compute the solution $u(x,t)$.

(b) Determine $\lim_{t\to\infty} u(x,t)$ both from the expression under (a) and by direct computation.

10.4. Solve the initial boundary value problem

$$\begin{aligned} \frac{\partial u}{\partial t} &= \frac{\partial^2 u}{\partial x^2} - cu + \frac{1}{2}(1 - \cos \pi x), & x \in (0,1), \quad & t > 0, \\ u(x,0) &= 2(1 - \cos \pi x), & x \in (0,1), \quad & \\ \frac{\partial u}{\partial x}(0,t) &= \frac{\partial u}{\partial x}(1,t) = 0, & & t > 0. \end{aligned}$$

10.5. Solve the initial boundary value problem

$$\begin{aligned} \frac{\partial u}{\partial t} &= \frac{\partial^2 u}{\partial x^2}, & x \in (-1,1), \quad & t > 0, \\ u(x,0) &= 1, & x \in (-1,1), \quad & \\ u(-1,t) &= u(1,t) = 0, & & t > 0. \end{aligned}$$

10.6. A sphere with radius R has a constant initial temperature u_0. At $t = 0$ the sphere is cooled at $u = 0$.

(a) Show that the temperature $u(r, t)$ satisfies the initial boundary value problem

$$\frac{\partial u}{\partial t} = k\left(\frac{\partial^2 u}{\partial r^2} + \frac{2}{r}\frac{\partial u}{\partial r}\right), \qquad r \in (0, R), \quad t > 0,$$

$$u(r, 0) = u_0, \qquad r \in (0, R),$$

$$\frac{\partial u}{\partial r}(0, t) = u(R, t) = 0, \qquad t > 0.$$

(b) Determine $u(r, t)$.

10.7. Use Green's functions to solve the following problem, defined for $0 < x < L$ (for some L) and $t > 0$:

$$\frac{\partial u}{\partial t} = \frac{\partial^2 u}{\partial x^2} + g(x, t), \qquad x \in (0, L), \quad t > 0,$$

$$u(x, 0) = v(x), \qquad x \in (0, L),$$

$$u(0, t) = 0, \qquad \frac{\partial u}{\partial x}(L, t) + u(L, t) = 0, \qquad t > 0.$$

10.8. Consider the following initial boundary value problem for $x > 0, t > 0$ for some parameter $\alpha > 0$:

$$\frac{\partial u}{\partial t} = \frac{\partial^2 u}{\partial x^2}, \qquad x \in (0, \infty), \quad t > 0,$$

$$u(x, 0) = 0, \qquad x \in (0, \infty),$$

$$u(0, t) = t^\alpha, \qquad t > 0.$$

Find a similarity solution $\hat{u}(\eta)$ of the form

$$u(x, t) = t^\alpha \hat{u}(\eta) \text{ with } \eta = x/\sqrt{t}.$$

10.9. Find the solution of the following initial boundary value problem with periodic boundary conditions:

$$\frac{\partial u}{\partial t} = \frac{\partial^2 u}{\partial x^2}, \qquad x \in (-1, 1), \quad t > 0,$$

$$u(x, 0) = v(x), \qquad x \in (-1, 1),$$

$$u(-1, t) = u(1, t), \quad \frac{\partial u}{\partial x}(-1, t) = \frac{\partial u}{\partial x}(1, t), \qquad t > 0.$$

10.10. Find a similarity solution, for $x > 0,\ t > 0$, of

$$\frac{\partial u}{\partial t} = \frac{\partial^2 u}{\partial x^2} + 1, \qquad x \in (0, \infty), \quad t > 0,$$

$$u(x, 0) = 0, \qquad x \in (0, \infty),$$

$$u(0, t) = 0, \qquad t > 0.$$

10.11. Let $0 < a < \frac{1}{2}$.

(a) Find an ODE for the travelling-wave solution $\hat{u}(\xi) := \hat{u}(x - bt)$ of

$$\frac{\partial u}{\partial t} = \frac{\partial^2 u}{\partial x^2} - u(1-u)(a-u).$$

(b) Draw a phase plane for $\hat{u}$ and investigate when a solution exists subject to the conditions

$$\hat{u}(\xi) \to 1 \quad \text{for} \quad \xi \to \infty, \qquad \hat{u}(\xi) \to a \quad \text{for} \quad \xi \to -\infty.$$

10.12. Use separation of variables to compute the solution of

$$\begin{aligned} &\frac{\partial u}{\partial t} = \frac{\partial^2 u}{\partial x^2}, && x \in (0,1), \quad t > 0, \\ &u(x,0) = \sin(\pi x)\cos(\pi x), && x \in (0,1), \\ &u(0,t) = u(1,t) = 0, && t > 0. \end{aligned}$$

10.13. Use separation of variables to compute the solution of

$$\begin{aligned} &\frac{\partial u}{\partial t} = \frac{\partial^2 u}{\partial x^2}, && x \in (0,1), \quad t > 0, \\ &u(x,0) = \sin(\pi x)\cos(\pi x), && x \in (0,1), \\ &u(0,t) = 0, \ \frac{\partial u}{\partial x}(1,t) = 0, && t > 0. \end{aligned}$$

10.14. Consider the initial boundary value problem

$$\begin{aligned} &\frac{\partial u}{\partial t} = \frac{\partial^2 u}{\partial x^2} + \alpha u, && x \in (0,1), \quad t > 0, \\ &u(x,0) = v(x), && x \in (0,1), \\ &u(0,t) = u(1,t) = 0, && t > 0. \end{aligned}$$

(a) Determine the formal solution.

(b) Show that this solution is stable for $\alpha < \pi^2$.

10.15. Consider the problem

$$\begin{aligned} &\frac{\partial u}{\partial t} = \frac{\partial^2 u}{\partial x^2} + \alpha u, && x \in (0,1), \quad t > 0, \\ &u(x,0) = 0, && x \in (0,1), \\ &u(0,t) = u(1,t) = 0, && t > 0. \end{aligned}$$

(a) Determine the stationary solution, i.e., the solution for $t \to \infty$.

(b) Use the result in part (a) to compute the solution.

Chapter 11

Numerical Methods for Parabolic Equations

This chapter gives an overview of a few numerical techniques for solving parabolic problems, as well as their analysis for accuracy and stability. It starts with an introduction of the simple explicit Euler scheme in Section 11.1. A matrix-vector formulation appears to enhance the analysis of this scheme a great deal. This analysis is carried out in Section 11.2, where semidiscretisations, both of the spatial derivatives and of the time derivative, are considered. The former results in a vector ODE. Due to the inherent stiffness, it makes sense to consider implicit schemes, which is done in Section 11.3. Both the implicit Euler scheme and a weighted average of the Euler schemes (the ϑ method) are analysed. In Section 11.4 the matrix method, based on the semidiscretisation with respect to the spatial derivatives, is used to analyse the stability and accuracy of the one-step schemes of the earlier sections. A case where the ϑ scheme may fail is for an initial value problem with discontinuous data. This is investigated in detail in Section 11.5. Mixed boundary conditions require a special treatment as they may have an impact on stability; they are studied in Section 11.6. Problems in two space dimensions are briefly introduced in Section 11.7. It is shown that the explicit Euler scheme has a similar time step restriction as in the one-dimensional case. Implicit methods, however, involve the solution of linear systems that may lose their sparseness in direct methods (LU-decomposition). Hence we derive in Section 11.8 the alternating direction implicit (ADI) method, which has both favorable stability properties and good efficiency. We also present some mixed explicit/implicit methods for the advection-diffusion equation in this section. In Section 11.9 we turn to nonlinear parabolic equations. We discuss a number of options for a variety of nonlinear terms. Finally, in Section 11.10, we treat some numerical questions related to the Stefan problem, which is a parabolic problem with moving boundaries.

11.1 The Explicit Euler Scheme

In Chapter 10 we considered various ways to find an analytic solution of a parabolic PDE. Sometimes this is the method of choice, but most often the analytical expressions are less useful for actually finding numerical values. As an example, a series solution might converge very slowly, or an integral expression might still require a nontrivial numerical quadrature procedure. The major advantage of analytical methods is to obtain qualitative insight. If more precise (numerical) information is needed, one has to resort to numerical methods. To start with we consider the heat equation in one space variable. In fact, most of the ideas and methods can be applied to more complex problems, yet this fairly simple case demonstrates most of the algorithmic and analysis aspects of parabolic problems in general. Higher-dimensional as well as nonlinear cases are specifically treated in later sections. So let us study the problem

$$\frac{\partial u}{\partial t} = \frac{\partial^2 u}{\partial x^2} + f(x,t), \qquad x \in (0,1), \quad t > 0, \tag{11.1a}$$

$$u(x,0) = v(x), \qquad x \in (0,1), \tag{11.1b}$$

$$u(0,t) = g_\ell(t), \qquad t > 0, \tag{11.1c}$$

$$u(1,t) = g_\mathrm{r}(t), \qquad t > 0. \tag{11.1d}$$

For the source term $f(x,t) \equiv 0$, (11.1a) is often called the *heat equation*.

Recalling what we did in Chapter 5, we now introduce a grid for the domain $\Omega := \{(x,t) \in \mathbb{R}^2 \mid 0 < x < 1,\ t > 0\}$; see Figure 11.1. If Δx denotes the spatial grid size and Δt the time step, the grid points (x_j, t^n) can be chosen as

$$x_j := j\Delta x \qquad (j = 0,1,2,\ldots,M+1), \qquad \Delta x = 1/(M+1), \tag{11.2a}$$

$$t^n := n\Delta t \qquad (n = 0,1,2,\ldots). \tag{11.2b}$$

We *discretise* (11.1a) by replacing the time derivative $\frac{\partial}{\partial t}u(x_j,t^n)$ by the forward difference approximation defined in (5.28c) and the space derivative $\frac{\partial^2}{\partial x^2}u(x_j,t^n)$ by the central difference approximation given in (5.30). Writing u_j^n for the numerical approximation of $u(x_j,t^n)$ and defining $f_j^n := f(x_j,t^n)$, we obtain the *difference scheme*

$$\frac{1}{\Delta t}\left(u_j^{n+1} - u_j^n\right) = \frac{1}{\Delta x^2}\left(u_{j+1}^n - 2u_j^n + u_{j-1}^n\right) + f_j^n, \tag{11.3}$$

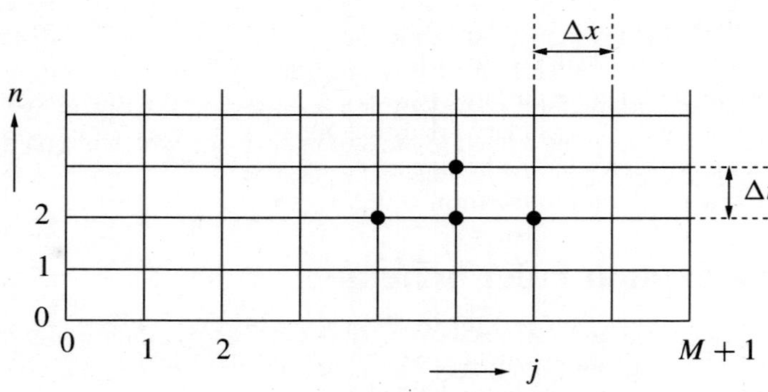

Figure 11.1. *Stencil for the explicit scheme* (11.4).

which holds at all internal grid points (x_j, t^n). The scheme (11.3) can simply be rewritten as

$$u_j^{n+1} = u_j^n + d\big(u_{j+1}^n - 2u_j^n + u_{j-1}^n\big) + \Delta t\, f_j^n, \tag{11.4}$$

where d is defined by

$$d := \frac{\Delta t}{\Delta x^2}. \tag{11.5}$$

The coefficient d is sometimes called the *diffusion number*. The dependence of the numerical value u_j^{n+1} at the new time level t^{n+1} on the numerical values u_k^n $(k = j \pm 1, j)$ at the old time level t^n is depicted in the stencil in Figure 11.1 (black dots). Clearly, each value at time level t^{n+1} can be computed independently from the other values at the same time level; i.e., the difference scheme (11.4) is explicit; it is often referred to as the *explicit Euler scheme*. The initial condition (11.1b) and the boundary conditions (11.1c), (11.1d) also have to be prescribed for the numerical solution. For the initial boundary value problem (11.1) this is trivial and we find

$$u_j^0 = v(x_j) \qquad (j = 1, 2, \ldots, M) \tag{11.6}$$

for the initial condition and

$$u_0^n = g_\ell(t^n), \quad u_{M+1}^n = g_r(t^n) \qquad (n = 0, 1, 2, \ldots) \tag{11.7}$$

for the boundary conditions.

Example 11.1 In this example we compute the numerical solution of the initial boundary value problem (11.1) for the initial function

$$v(x) = 4x(1 - x), \qquad 0 < x < 1,$$

and homogeneous boundary conditions (i.e., $g_\ell(t)$, $g_r(t) \equiv 0$). Furthermore, we take $f(x, t) \equiv 0$. We have employed the difference scheme (11.4) with grid size $\Delta x = 5 \times 10^{-2}$ and time steps $\Delta t = 1.2 \times 10^{-3}$ and $\Delta t = 1.3 \times 10^{-3}$. The resulting numerical solutions are shown in Figure 11.2. In the left column the solution is computed with $\Delta t = 1.2 \times 10^{-3}$, which is a good approximation of the exact solution. However, the column on the right shows the numerical solution for $\Delta t = 1.3 \times 10^{-3}$, which is obviously incorrect. When the number of time steps increases further, the numerical scheme eventually results in floating-point overflow. This example shows that the difference scheme (11.4) has to be applied with care. □

We can investigate the stability of this scheme in various ways. The simplest way is often to resort to Fourier analysis; see Chapter 5. As we saw there, such an analysis can detect when a scheme may be unstable (it rather gives necessary conditions for stability). The prerequisite is that the boundary conditions be periodic (like in the example), which we therefore assume. Consider the homogeneous scheme (11.4); i.e., we take $f_j^n \equiv 0$. We look for a numerical solution of the form

$$u_j^n = \mathrm{e}^{\mathrm{i}\kappa x_j - \mathrm{i}\omega t^n}, \tag{11.8}$$

where κ is the *wave number* and ω is the *frequency*; see Section 3.4. Suppressing the dependencies on wave number and frequency, we use two other variables, i.e., the *amplification factor* λ and *phase angle* φ, which are defined as

$$\lambda := \mathrm{e}^{-\mathrm{i}\omega\Delta t}, \qquad \varphi := \kappa\Delta x. \tag{11.9}$$

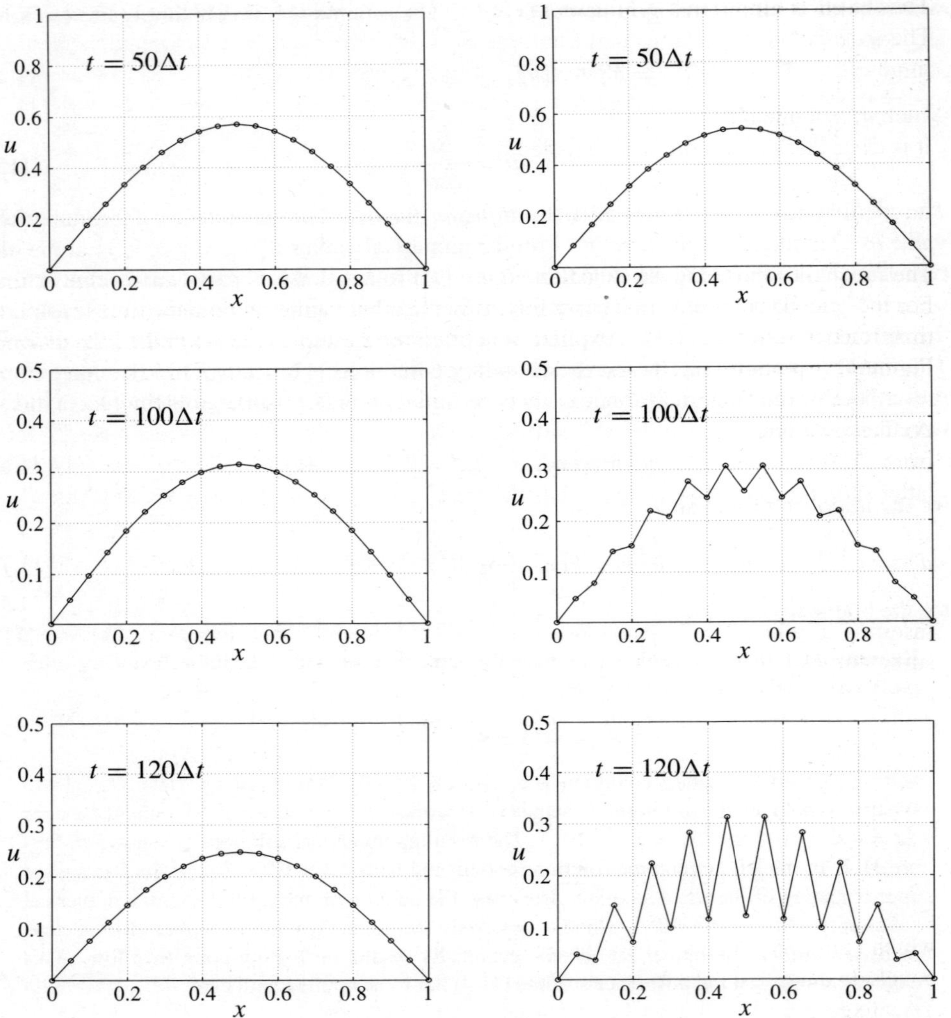

Figure 11.2. *Numerical solution of the heat equation computed with the explicit Euler scheme with time steps $\Delta t = 1.2 \times 10^{-3}$ (left) and $\Delta t = 1.3 \times 10^{-3}$ (right).*

This results in the representation

$$u_j^n = \lambda^n \, e^{i\kappa x_j} \, . \tag{11.10}$$

Substituting (11.10) in the homogeneous scheme and using the relations

$$u_j^{n+1} = \lambda u_j^n, \qquad u_{j\pm 1}^n = e^{\pm i\varphi} \, u_j^n, \tag{11.11}$$

we find the amplification factor

$$\lambda(\varphi) = 1 + d(e^{i\varphi} - 2 + e^{-i\varphi}) = 1 - 4d \sin^2\left(\frac{1}{2}\varphi\right). \tag{11.12}$$

Thus at each time step the Fourier mode $e^{i\kappa x_j}$ is amplified by the factor $\lambda(\varphi)$ in (11.12). The undesirable oscillations in the numerical solution in Figure 11.2 can be prevented if we impose the condition

$$|\lambda(\varphi)| \leq 1 \qquad \text{for} \quad 0 \leq \varphi \leq \pi. \tag{11.13}$$

It is clear that this condition is satisfied if

$$d = \frac{\Delta t}{\Delta x^2} \leq \frac{1}{2}. \tag{11.14}$$

The inequality in (11.14) is called the *stability condition* of the numerical scheme (11.4). For the grid size $\Delta x = 5 \times 10^{-2}$ in Example 11.1 the stability condition (11.14) gives the time step restriction $\Delta t \leq 1.25 \times 10^{-3}$. Hence the numerical solution in the left column in Figure 11.2 is stable and the numerical solution in the right column is not.

The next question is the accuracy of the method, which is related to the choice of both Δx and Δt. In (5.25) the *local discretisation error* was defined as the residual, d_j^n, found from substituting the exact solution $u(x, t)$ of (11.1a) in the difference scheme (11.3); the latter should be scaled such that it has the same dimensions as (11.1a). We thus find

$$d_j^n := \frac{1}{\Delta t}\big(u(x_j, t^{n+1}) - u(x_j, t^n)\big) - \frac{1}{\Delta x^2}\big(u(x_{j+1}, t^n) - 2u(x_j, t^n) + u(x_{j-1}, t^n)\big) - f_j^n. \tag{11.15}$$

Inserting Taylor series expansions of $u(x, t)$ about (x_j, t^n) and assuming that $u(x, t)$ is sufficiently smooth, we obtain

$$\begin{aligned} d_j^n &= \frac{\partial u}{\partial t}(x_j, t^n) - \frac{\partial^2 u}{\partial x^2}(x_j, t^n) - f_j^n \\ &\quad + \frac{1}{2}\Delta t \frac{\partial^2 u}{\partial t^2}(x_j, t^n) - \frac{1}{12}\Delta x^2 \frac{\partial^4 u}{\partial x^4}(x_j, t^n) + \mathcal{O}(\Delta t^2) + \mathcal{O}(\Delta x^4) \\ &= \frac{1}{2}\Delta t \frac{\partial^2 u}{\partial t^2}(x_j, t^n) - \frac{1}{12}\Delta x^2 \frac{\partial^4 u}{\partial x^4}(x_j, t^n) + \mathcal{O}(\Delta t^2) + \mathcal{O}(\Delta x^4), \end{aligned} \tag{11.16}$$

which is in agreement with the discretisation errors in (5.29) and (5.31). From (11.16) we conclude that the local discretisation error d_j^n approaches zero for Δt, $\Delta x \to 0$ with $n\Delta t$, $j\Delta x$ fixed; i.e., the scheme is *consistent*; cf. Definition 5.13.

For this scheme there is a simple way to estimate the *global discretisation error* e_j^n (cf. (5.103)), defined by

$$e_j^n := u(x_j, t^n) - u_j^n. \tag{11.17}$$

Subtracting (11.3) from (11.15) and multiplying by Δt, we find the following recurrence relation for e_j^n:

$$e_j^{n+1} = e_j^n + d\big(e_{j+1}^n - 2e_j^n + e_{j-1}^n\big) + \Delta t d_j^n. \tag{11.18}$$

Furthermore, from the boundary conditions (11.7) we see that

$$e_0^{n+1} = e_{M+1}^n = 0. \tag{11.19}$$

Note that the homogeneous part of the recurrence relation (11.18) is exactly equal to the homogeneous part of the numerical scheme (11.4). Consequently, the time step restriction

(11.14) and the expression (11.16) for the local discretisation error d_j^n imply that both the numerical solution u_j^n and the global discretisation error e_j^n remain bounded for increasing n and fixed Δt.

To estimate this global discretisation error e_j^{n+1} we consider its maximum value over all indices j and apply the triangle inequality in (11.18) to find

$$\max_{1\le j\le M} \left|e_j^{n+1}\right| = \left|e_{j^*}^{n+1}\right| \le \left|(1-2d)e_{j^*}^n + d\left(e_{j^*+1}^n + e_{j^*-1}^n\right)\right| + \Delta t\left|d_{j^*}^n\right| \tag{11.20}$$

for some index j^* ($1 \le j^* \le M$). Because of the time step restriction (11.14), all coefficients on the right-hand side of (11.20) are positive and so this inequality can be simplified further to

$$\begin{aligned}\max_{1\le j\le M} \left|e_j^{n+1}\right| &\le (1-2d)\left|e_{j^*}^n\right| + d\left(\left|e_{j^*+1}^n\right| + \left|e_{j^*-1}^n\right|\right) + \Delta t\left|d_{j^*}^n\right| \\ &\le \max_{1\le j\le M} \left|e_j^n\right| + \Delta t \max_{1\le j\le M} \left|d_j^n\right|.\end{aligned} \tag{11.21}$$

The latter inequality can also be written as

$$\|\mathbf{e}^{n+1}\|_\infty \le \|\mathbf{e}^n\|_\infty + \Delta t\|\mathbf{d}^n\|_\infty, \tag{11.22}$$

where the vectors $\mathbf{d}$ and $\mathbf{e}$ are defined by

$$\mathbf{d}^n := \begin{pmatrix} d_1^n \\ d_2^n \\ \vdots \\ d_M^n \end{pmatrix}, \qquad \mathbf{e}^n := \begin{pmatrix} e_1^n \\ e_2^n \\ \vdots \\ e_M^n \end{pmatrix}. \tag{11.23}$$

For the definition of the vector norm $\|\cdot\|_\infty$ the reader is referred to the appendix, Section G. Let us assume that the initial condition (11.6) is exact, so that

$$\|\mathbf{e}^0\|_\infty = 0. \tag{11.24}$$

From the inequality (11.22) and the initial error in (11.24), we can deduce (and show by induction) the following estimate for $\mathbf{e}^n$:

$$\|\mathbf{e}^n\|_\infty \le \Delta t \sum_{l=0}^{n-1} \|\mathbf{d}^l\|_\infty. \tag{11.25}$$

Hence for the special case that $f(x,t) \equiv 0$ and the derivative $\frac{\partial^4}{\partial x^4}u(x,t)$ in the local error is bounded, say

$$\left|\frac{\partial^4 u}{\partial x^4}(x,t)\right| \le C \qquad \text{for all} \qquad (x,t) \in (0,1)\times(0,\infty), \quad C > 0, \tag{11.26}$$

we find from (11.25) and (11.16) the following upper bound for $\|\mathbf{e}^n\|_\infty$:

$$\left\|\mathbf{e}^n\right\|_\infty \le \frac{1}{2}C\left|d - \frac{1}{6}\right| t^n \Delta x^2 + \mathcal{O}(\Delta t^2) + \mathcal{O}(\Delta x^4). \tag{11.27}$$

From the above inequality it is clear that $\|\mathbf{e}^n\|_\infty \to 0$ if $\Delta t, \Delta x \to 0$, while $t^n = n\Delta t \le T$, $T < \infty$ fixed. In that case we deduce that the numerical scheme (11.4) is *convergent*; see Definition 5.17.

11.2 Semidiscretisation

In the previous section we showed how to discretise the heat equation by the simple explicit scheme (11.4). It will turn out to be helpful to use a matrix-vector formulation, both for a more direct and simple description of methods and also for analysis purposes. As was pointed out in Section 5.5.1, we can discretise either the spatial variable or the time first. In this section we consider both types of semidiscretisation.

11.2.1 The Longitudinal Method of Lines

In this subsection we consider the case where the spatial variable is discretised first. This gives rise to the *(longitudinal) method of lines* (MOL). For this we use a finite number of grid points, fixed for all t, where the second order derivative is discretised. So on the grid lines $x = x_j$ $(j = 1, 2, \ldots, M)$ in the interior domain we define vector functions

$$\mathbf{u}(t) := \begin{pmatrix} u(x_1, t) \\ u(x_2, t) \\ \vdots \\ u(x_{M-1}, t) \\ u(x_M, t) \end{pmatrix}. \tag{11.28}$$

If we use a central difference approximation for $\frac{\partial^2}{\partial x^2}u$ in (11.1a), we obtain a set of ODEs (cf. (11.3))

$$\frac{\mathrm{d}u_j}{\mathrm{d}t} = \frac{1}{\Delta x^2}(u_{j+1} - 2u_j + u_{j-1}) + f_j(t) \qquad (j = 1, 2, \ldots, M), \tag{11.29}$$

where $f_j(t) := f(x_j, t)$. The boundary conditions in (11.1c) and (11.1d) plus the source term give a discrete source term $\mathbf{f}$, say. Hence we find the ODE system

$$\frac{\mathrm{d}\mathbf{u}}{\mathrm{d}t} = \mathbf{A}\mathbf{u} + \mathbf{f}, \tag{11.30a}$$

where the *discretisation matrix* $\mathbf{A}$ is defined by

$$\mathbf{A} := \frac{1}{\Delta x^2}\begin{pmatrix} -2 & 1 & & & \\ 1 & -2 & 1 & & \\ & \ddots & \ddots & \ddots & \\ & & 1 & -2 & 1 \\ & & & 1 & -2 \end{pmatrix} \tag{11.30b}$$

and where

$$\mathbf{f}(t) := \begin{pmatrix} f(x_1,t) + \dfrac{1}{\Delta x^2} g_1(t) \\ f(x_2,t) \\ \vdots \\ f(x_{M-1},t) \\ f(x_M,t) + \dfrac{1}{\Delta x^2} g_r(t) \end{pmatrix}. \tag{11.30c}$$

Clearly, the initial condition for (11.30) is given by

$$\mathbf{u}(0) = \begin{pmatrix} v(x_1) \\ v(x_2) \\ \vdots \\ v(x_{M-1}) \\ v(x_M) \end{pmatrix}. \tag{11.30d}$$

In order to analyse the behaviour of the solutions of (11.30a), we can try to find the various possible modes. In this relatively simple case this can be done indeed. This approach is meant to analyse the problem, and not as an option to be carried out when solving problems generally. The point is that the tridiagonal matrix $\mathbf{A}$ can simply be seen to have the eigenvalues λ_k, where

$$\lambda_k := -\frac{4}{\Delta x^2}\sin^2\left(\frac{1}{2}\pi x_k\right) \qquad (k = 1, 2, \ldots, M). \tag{11.31}$$

The corresponding eigenvectors are given by

$$\mathbf{w}_k := \sqrt{\frac{2}{M}} \begin{pmatrix} \sin(\pi x_k) \\ \sin(2\pi x_k) \\ \vdots \\ \sin((M-1)x_k) \\ \sin(M\pi x_k) \end{pmatrix} \qquad (k = 1, 2, \ldots, M). \tag{11.32}$$

The vectors $\mathbf{w}_k$ form a complete basis; i.e., the matrix

$$\mathbf{W} := \left(\mathbf{w}_1, \mathbf{w}_2, \ldots, \mathbf{w}_M\right) \tag{11.33}$$

is nonsingular; in fact, it is even an orthogonal matrix, implying that

$$\mathbf{W}^{-1}\mathbf{A}\mathbf{W} = \mathbf{\Lambda}, \tag{11.34}$$

where

$$\mathbf{\Lambda} := \begin{pmatrix} \lambda_1 & & & & \\ & \lambda_2 & & & \\ & & \ddots & & \\ & & & \lambda_{M-1} & \\ & & & & \lambda_M \end{pmatrix}. \tag{11.35}$$

Using this matrix to transform the variable $\mathbf{u}(t)$ as

$$\tilde{\mathbf{u}}(t) := \mathbf{W}^{-1}\mathbf{u}(t),$$

and likewise $\tilde{\mathbf{f}}(t) := \mathbf{W}^{-1}\mathbf{f}(t)$, we can derive the following decoupled system for (11.30a):

$$\frac{\mathrm{d}\tilde{\mathbf{u}}}{\mathrm{d}t} = \mathbf{\Lambda}\tilde{\mathbf{u}} + \tilde{\mathbf{f}}. \tag{11.36}$$

Hence we can conclude that the stability properties of the system (11.30a) are determined by the scalar equations

$$\frac{\mathrm{d}\tilde{u}_k}{\mathrm{d}t} = \lambda_k \tilde{u}_k + \tilde{f}_k, \qquad (k = 1, 2, \ldots, M). \tag{11.37}$$

For $k = 1$ we find the eigenvalue

$$\lambda_1 = -\frac{4}{\Delta x^2}\sin^2\left(\frac{1}{2}\pi x_1\right) \doteq -\pi^2,$$

whereas for $k = M$ we have

$$\lambda_M = -\frac{4}{\Delta x^2}\sin^2\left(\frac{1}{2}\pi x_M\right) = -4(M+1)^2\sin^2\left(\frac{\pi M}{2(M+1)}\right) \doteq -4(M+1)^2.$$

We thus conclude that (11.37) is *stiff* for larger values of M; see Section 5.4.2. In fact, there is an entire sequence of small time scales, from very small (i.e., $(4(M+1)^2)^{-1}$), gradually increasing to approximately one.

If we want to approximate the solution only moderately accurately by the explicit Euler method, we are faced with time step constraints, i.e., once we have passed the various boundary layers, or, in absence of these, even right from the start. Indeed, explicit Euler would give

$$\mathbf{u}^{n+1} = \big(\mathbf{I} + \Delta t\,\mathbf{A}\big)\mathbf{u}^n + \Delta t\,\mathbf{f}^n, \tag{11.38}$$

where $\mathbf{u}^n$ is the numerical approximation of $\mathbf{u}(t^n)$ and $\mathbf{f}^n = \mathbf{f}(t^n)$. Stability is guaranteed if

$$\big\|\mathbf{I} + \Delta t\,\mathbf{A}\big\| \leq 1.$$

If we take the 2-norm ($\|\cdot\|_2$), then the latter requirement is satisfied if

$$\big\|\mathbf{I} + \Delta t\,\mathbf{\Lambda}\big\|_2 \leq 1.$$

To see this one should use the fact that $\|\mathbf{Q}_1\mathbf{B}\mathbf{Q}_2\|_2 = \|\mathbf{B}\|_2$ for any matrix $\mathbf{B}$ and orthogonal matrices $\mathbf{Q}_1$ and $\mathbf{Q}_2$. Hence (11.38) is stable if

$$\left|1 + \Delta t \lambda_k\right| \leq 1 \qquad (k = 1, 2, \ldots, M).$$

This leads to the same result as obtained from the Fourier mode analysis in the previous section, i.e., $d \leq \frac{1}{2}$.

It should be noted that the generalisation of this analysis to other problems is obvious. In general the resulting matrix is no longer constant or symmetric. This complicates the error analysis, but not in a fundamental way (e.g., one may need a more subtle use of norms). Of course, for nonlinear problems we obtain a nonlinear vector equation for which we cannot directly apply the matrix method as outlined above.

11.2.2 The Transversal MOL

If we discretise the time derivative first, we see that we have a method that differs from the previous type of semidiscretisation only if this discretisation is implicit. In this section we restrict ourselves to the implicit Euler method. So let us denote the numerical approximation of $u(x, t^n)$ by $u^n(x)$, which is defined on the grid lines $t = t^n$ $(n = 0, 1, 2, \ldots)$. Then we obtain from (11.1a) the following equation defined along these transversal lines:

$$u^{n+1} - \Delta t \frac{\mathrm{d}^2}{\mathrm{d}x^2} u^{n+1} = u^n + \Delta t\, f(x, t^{n+1}). \tag{11.39a}$$

Equation (11.39a) is subject to the boundary conditions (cf. (11.1c) and (11.1d))

$$u^{n+1}(0) = g_{\mathrm{l}}(t^{n+1}), \qquad u^{n+1}(1) = g_{\mathrm{r}}(t^{n+1}). \tag{11.39b}$$

Since $u(x, 0)$ is given we can find the solution u of the PDE by successively solving a series of boundary value problems (11.39). The real advantage to discretising the spatial derivative first is the greater flexibility with respect to the choice of the spatial variables at each step. Once we have found u^n on some grid, we can use interpolation to obtain u^n at any other desired point in $(0, 1)$. In particular, for problems that require a finer grid in an area that is moving in time, this can be of great advantage. Indeed, a fixed spatial grid would require a dense grid on a larger subdomain and thus would contain many redundant points at each time level (i.e., that could be omitted without affecting the global accuracy).

Example 11.2 An example of a problem where we have a sharp front moving in the domain $(-1, 1)$ back and forth if time evolves is given by the (slightly contrived but typical) problem

$$\frac{\partial u}{\partial t} = \frac{\partial^2 u}{\partial x^2} + \alpha(2\alpha u - \cos t)\big(1 - u^2\big),$$

which has the solution

$$u(x, t) = \tanh\big(\alpha(x - \sin t)\big)$$

for suitable initial and boundary conditions. In Figure 11.3 we have drawn $u(\cdot, t)$ for $\alpha = 50$ at three different time levels. It is clear that it has a sharp front, requiring a fine spatial grid, but at different positions for different time points. □

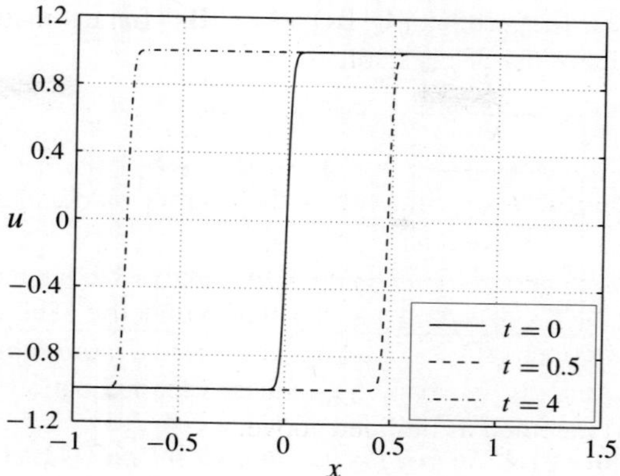

Figure 11.3. *A solution having a sharp moving front of the heat equation with source term at three different time levels.*

At this point we will not go into further detail on how to tackle these types of problems. In a one-dimensional spatial setting one may use any of the existing boundary value methods; see [3]. In Section 11.10 we will encounter this approach when solving a two-phase problem with a moving interface.

11.3 Implicit Schemes

The scheme (11.4) turned out to be first order accurate in the time step Δt and second order accurate in the spatial grid size Δx, as expected. The severe time step restriction (11.14), due to the inherent stiffness (as found in the previous section), is a major drawback indeed. For instance, if we compute the numerical solution of the initial boundary value problem (11.1) on a grid with grid size $\Delta x = 10^{-2}$ at time level say $t^n = 1$, then the number of required time steps will be at least 20,000! In general, the number of time steps is prohibitively large for this scheme. From what we have seen in Chapter 5 we are therefore better to take recourse to implicit schemes.

11.3.1 The Implicit Euler Scheme

Rather than using forward differences for the time discretisation, let us now use the backward difference approximation defined in (5.71) to approximate the derivative $\frac{\partial}{\partial t}u(x_j, t^n)$. If we again use the central difference approximation (5.30) for the space derivative $\frac{\partial^2}{\partial x^2}u(x_j, t^n)$, we get the difference equation

$$\frac{1}{\Delta t}\left(u_j^{n+1} - u_j^n\right) = \frac{1}{\Delta x^2}\left(u_{j+1}^{n+1} - 2u_j^{n+1} + u_{j-1}^{n+1}\right) + f_j^{n+1} \tag{11.40}$$

for the numerical solution u_j^n. This equation can be rewritten as

$$u_j^{n+1} - d\left(u_{j+1}^{n+1} - 2u_j^{n+1} + u_{j-1}^{n+1}\right) = u_j^n + \Delta t f_j^{n+1}. \tag{11.41}$$

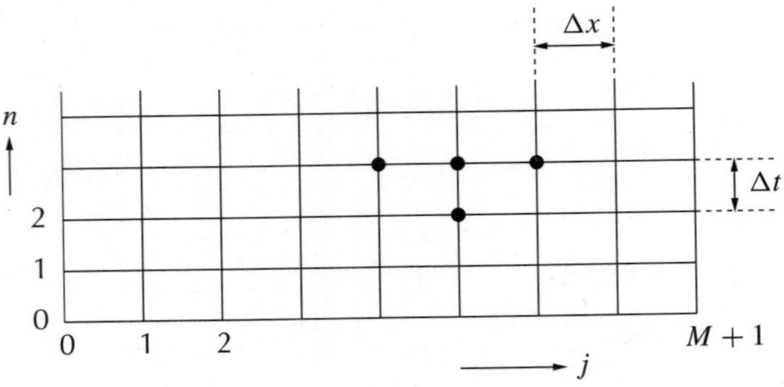

Figure 11.4. *Stencil for the implicit scheme* (11.41).

The stencil of the numerical scheme (11.41), which is called the *implicit Euler scheme*, is given in Figure 11.4. From this stencil it is obvious that the numerical value u_j^{n+1} cannot be computed directly from the numerical values u_j^n at the previous time level t^n. Indeed, writing the difference equations as a matrix-vector recursion like in (11.38), we can formulate the scheme as

$$(\mathbf{I} - \Delta t\mathbf{A})\mathbf{u}^{n+1} = \mathbf{u}^n + \Delta t\mathbf{f}^{n+1}. \tag{11.42}$$

So for each new time level a linear system of equations has to be solved. As before, the boundary conditions are represented in the source term, i.e., $\mathbf{f}^{n+1}$, whereas for the initial value we have $\mathbf{u}^0 = \mathbf{u}(0)$, cf. (11.30d). The coefficient matrix $\mathbf{I} - \Delta t\mathbf{A}$ is a symmetric positive definite tridiagonal matrix and consequently the system (11.42) can be solved very efficiently using LU-decomposition without partial pivoting (in fact, Choleski decomposition; see [52]). In principle the linear system (11.42) has to be solved at each time level. Note that in this (constant coefficient) case the LU-decomposition of the matrix $\mathbf{A}$ has to be done only once; hence only the forward and backward substitutions need to be performed at each time step, making the time stepping very efficient. For nonconstant matrices this advantage is lost, however.

Next we apply Fourier mode analysis to investigate the stability of scheme (11.41). If we substitute the mode (11.10) in the homogeneous scheme (11.41), assuming periodic boundary conditions, and apply the relations (11.11), we find

$$\lambda - d\lambda\left(\mathrm{e}^{\mathrm{i}\varphi} - 2 + \mathrm{e}^{-\mathrm{i}\varphi}\right) = 1,$$

from which we deduce that

$$\lambda(\varphi) = \frac{1}{1 + 4d\sin^2\left(\frac{1}{2}\varphi\right)}. \tag{11.43}$$

It is obvious from (11.43) that the stability condition (11.13) is always satisfied. This means that we do not have to restrict the time step Δt in order to enforce the stability condition (11.13). Therefore the numerical scheme (11.41) is *unconditionally stable*, confirming our expectations.

Due to its similarity with the explicit Euler scheme we expect the accuracy of the implicit scheme (11.40) to be first order as well. We obtain for the local discretisation error d_j^n

$$d_j^n := \frac{1}{\Delta t}\big(u(x_j, t^{n+1}) - u(x_j, t^n)\big) - \frac{1}{\Delta x^2}\big(u(x_{j+1}, t^{n+1}) - 2u(x_j, t^{n+1}) + u(x_{j-1}, t^{n+1})\big) - f_j^{n+1}; \tag{11.44}$$

cf. (11.15). Assuming $u(x, t)$ to be sufficiently smooth, straightforward substitution of Taylor series expansions about (x_j, t^{n+1}) into (11.44) gives

$$d_j^n = -\frac{1}{2}\Delta t \frac{\partial^2 u}{\partial t^2}(x_j, t^{n+1}) - \frac{1}{12}\Delta x^2 \frac{\partial^4 u}{\partial x^4}(x_j, t^{n+1}) + \mathcal{O}(\Delta t^2) + \mathcal{O}(\Delta x^4). \tag{11.45}$$

Hence the implicit scheme (11.40) is consistent.

From this we can find an upper bound for the global discretisation error e_j^{n+1}. From (11.40) and (11.44) we find the following recurrence relation for e_j^n:

$$e_j^{n+1} - d\big(e_{j+1}^{n+1} - 2e_j^{n+1} + e_{j-1}^{n+1}\big) = e_j^n + \Delta t d_j^n. \tag{11.46}$$

Using the triangle inequality, we derive from (11.46)

$$\begin{aligned}(1+2d) \max_{1\le j\le M} \left|e_j^{n+1}\right| &=: (1+2d)\left|e_{j^*}^{n+1}\right| \\ &\le d\big(\left|e_{j^*+1}^{n+1}\right| + \left|e_{j^*-1}^{n+1}\right|\big) + \left|e_{j^*}^n\right| + \Delta t\left|d_{j^*}^n\right| \\ &\le 2d \max_{1\le j\le M}\left|e_j^{n+1}\right| + \max_{1\le j\le M}\left|e_j^n\right| + \Delta t \max_{1\le j\le M}\left|d_j^n\right| \end{aligned} \tag{11.47}$$

for some (possibly not unique) index j^* $(1 \le j^* \le M)$. It is now obvious that (11.47) again leads to the inequality (11.22); however, this time no restriction on Δt is imposed. Following the same line of reasoning as in Section 11.1, we again would obtain the error estimate (11.27) and, consequently, the implicit scheme (11.41) is convergent.

Example 11.3 Continuing Example 11.1, we employ the implicit Euler scheme (11.41) to solve the same initial boundary value problem. We take a grid size $\Delta x = 5 \times 10^{-2}$ and time steps $\Delta t = 1.2 \times 10^{-3}$ and $\Delta t = 1.3 \times 10^{-3}$, respectively. The numerical solutions are shown in Figure 11.5. In this case both solutions are a good approximation of the exact solution. Thus the implicit scheme does not display the erroneous behaviour of the explicit scheme, which was caused by the instability of the scheme. □

11.3.2 The ϑ Scheme

Given the form of the semidiscrete problem (11.30), which is an ODE, it is natural to ask how we can fare with more general time-stepping methods. We shall restrict ourselves here to the class of ϑ methods. These methods are linear combinations of the explicit and implicit Euler schemes, i.e., (11.3) and (11.40). They read

$$\frac{1}{\Delta t}\big(u_j^{n+1} - u_j^n\big) = \frac{1-\vartheta}{\Delta x^2}\big(u_{j+1}^n - 2u_j^n + u_{j-1}^n\big) + \frac{\vartheta}{\Delta x^2}\big(u_{j+1}^{n+1} - 2u_j^{n+1} + u_{j-1}^{n+1}\big) + (1-\vartheta)f_j^n + \vartheta f_j^{n+1} \tag{11.48}$$

for some parameter ϑ with $0 \le \vartheta \le 1$. Obviously, the difference quotient on the left-

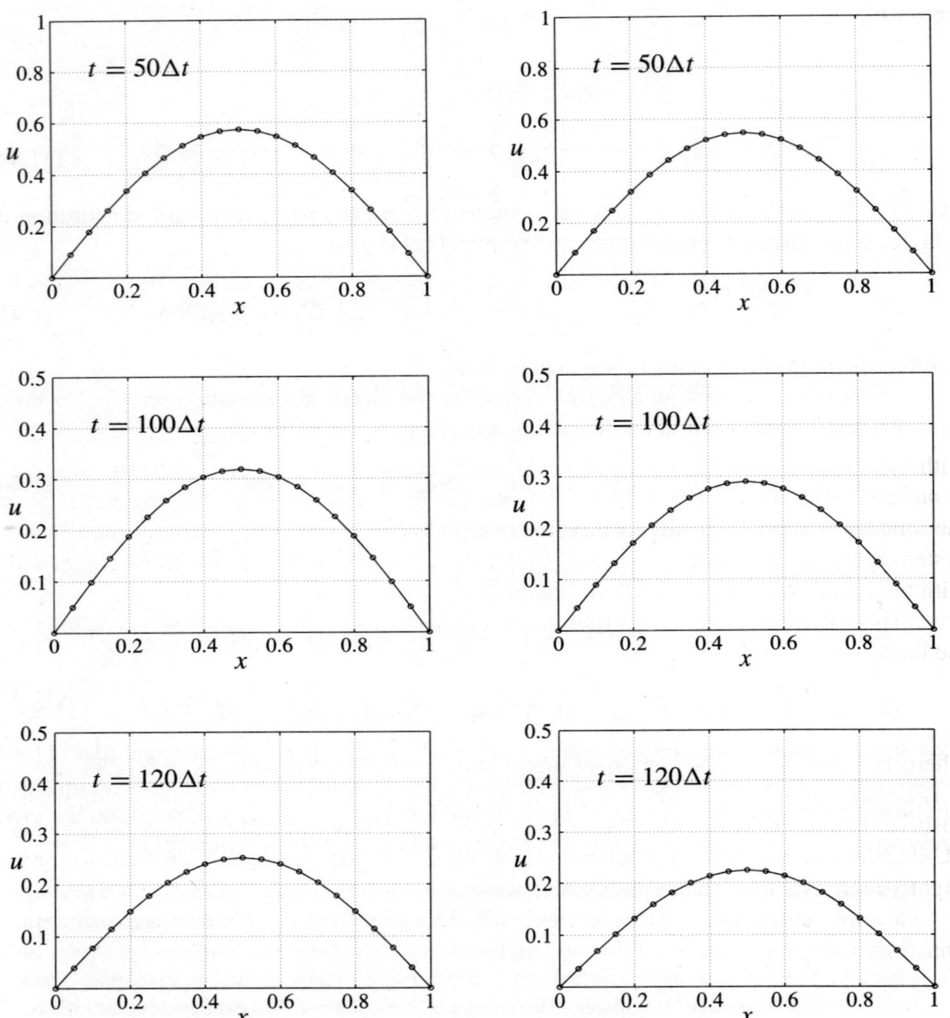

Figure 11.5. *Numerical solution of the heat equation computed with the implicit Euler scheme with time steps $\Delta t = 1.2 \times 10^{-3}$ (left) and $\Delta t = 1.3 \times 10^{-3}$ (right).*

hand side of (11.48) is an approximation of the weighted average time derivative $(1-\vartheta)\frac{\partial}{\partial t}u(x_j, t^n) + \vartheta\frac{\partial}{\partial t}u(x_j, t^{n+1})$. Note that for $\vartheta = 0$ we recover the explicit scheme (11.3) and for $\vartheta = 1$ we have the implicit scheme (11.40). For $\vartheta = \frac{1}{2}$ the scheme (11.48) is called the *Crank–Nicolson scheme*. The parameter ϑ can be used to optimize the accuracy and/or stability of the scheme. Rearranging terms, with all numerical values u_k^{n+1} $(k = j, j \pm 1)$ on the left-hand side and all numerical values u_k^n $(k = j, j \pm 1)$ on the right-hand side, we find

$$\begin{aligned} u_j^{n+1} - \vartheta d\big(u_{j+1}^{n+1} - 2u_j^{n+1} + u_{j-1}^{n+1}\big) &= u_j^n + (1-\vartheta)d\big(u_{j+1}^n - 2u_j^n + u_{j-1}^n\big) \\ &\quad + \Delta t\big((1-\vartheta)f_j^n + \vartheta f_j^{n+1}\big), \end{aligned} \tag{11.49}$$

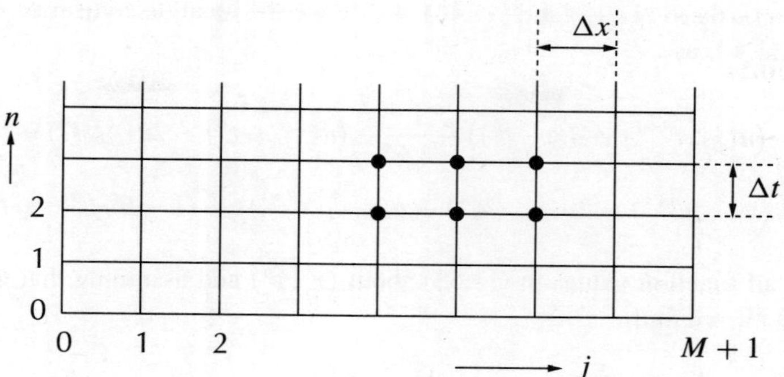

Figure 11.6. *Stencil for the ϑ method* (11.49).

with the diffusion number d defined in (11.5). The numerical scheme (11.49) is called the *ϑ method.* The stencil of (11.49) is shown in Figure 11.6. It is obvious that for $\vartheta \neq 0$ the scheme is implicit. Consequently, to compute the numerical solution requires a linear system to be solved at each time step. Finally, the numerical scheme has to be completed with the initial conditions (11.6) and boundary conditions (11.7).

The difference equations (11.49) together with the boundary conditions (11.7) lead to the linear system

$$(\mathbf{I} - \vartheta \Delta t \mathbf{A})\mathbf{u}^{n+1} = (\mathbf{I} + (1-\vartheta)\Delta t \mathbf{A})\mathbf{u}^n + \Delta t\big((1-\vartheta)\mathbf{f}^n + \vartheta \mathbf{f}^{n+1}\big), \tag{11.50}$$

where the vector $\mathbf{u}^{n+1}$ is the numerical approximation of $\mathbf{u}(t^{n+1})$ and the discretisation matrix $\mathbf{A}$ is defined in (11.30b). The coefficient matrix $\mathbf{B}_\vartheta := \mathbf{I} - \vartheta \Delta t \mathbf{A}$ is again a symmetric positive definite tridiagonal matrix, which makes the system (11.50) easy to solve using LU-decomposition. The computational costs per time step for the ϑ method are only slightly higher than for the implicit Euler scheme.

We investigate the stability of the ϑ scheme (11.49) by the Fourier mode analysis. Thus, substituting the mode (11.10), we find for the amplification factor λ

$$\lambda - \vartheta d \lambda\big(\mathrm{e}^{\mathrm{i}\varphi} - 2 + \mathrm{e}^{-\mathrm{i}\varphi}\big) = 1 + (1-\vartheta)d\big(\mathrm{e}^{\mathrm{i}\varphi} - 2 + \mathrm{e}^{-\mathrm{i}\varphi}\big),$$

from which we conclude that

$$\lambda(\varphi) = \frac{1 - 4(1-\vartheta)d\sin^2(\frac{1}{2}\varphi)}{1 + 4\vartheta d \sin^2(\frac{1}{2}\varphi)}. \tag{11.51}$$

Applying the inequality (11.13) to (11.51) leads to the stability conditions

$$\begin{cases} 0 \le \vartheta < \dfrac{1}{2} \quad \text{and} \quad d \le \dfrac{1}{2(1-2\vartheta)}, \\ \dfrac{1}{2} \le \vartheta \le 1. \end{cases} \tag{11.52}$$

For $\frac{1}{2} \le \vartheta \le 1$ the scheme is unconditionally stable, while for $\vartheta = 0$ the stability condition (11.14) is recovered.

Analogously to (11.15) and (11.44), we define the local discretisation error d_j^n of the ϑ method (11.40) as

$$d_j^n := \frac{1}{\Delta t}\big(u(x_j,t^{n+1}) - u(x_j,t^n)\big) - \frac{1-\vartheta}{\Delta x^2}\big(u(x_{j+1},t^n) - 2u(x_j,t^n) + u(x_{j-1},t^n)\big)$$
$$- \frac{\vartheta}{\Delta x^2}\big(u(x_{j+1},t^{n+1}) - 2u(x_j,t^{n+1}) + u(x_{j-1},t^{n+1})\big) - (1-\vartheta)f_j^n - \vartheta f_j^{n+1}. \quad (11.53)$$

Expanding all function values in (11.53) about (x_j, t^n) and assuming that $u(x,t)$ is sufficiently smooth, we find

$$d_j^n = -\left(\vartheta - \frac{1}{2}\right)\Delta t \frac{\partial^2 u}{\partial t^2}(x_j,t^n) - \frac{1}{12}\Delta x^2 \frac{\partial^4 u}{\partial x^4}(x_j,t^n) + \mathcal{O}(\Delta t^2) + \mathcal{O}(\Delta x^4). \quad (11.54)$$

From this expression we see that the ϑ-method is first order accurate in the time step Δt and second order accurate in the grid size Δx, except for $\vartheta = \frac{1}{2}$, in which case the scheme is also second order accurate in Δt. Apparently, $\vartheta = \frac{1}{2}$ is the best choice, since it gives the smallest discretisation error and unconditional stability; cf. (11.52). This optimality result is well known for the trapezoidal method when solving ODEs (see [97]). However, the Crank–Nicolson scheme does not always produce reliable results, as is demonstrated in the next example.

Example 11.4 In this example we employ the Crank–Nicolson scheme to compute a numerical solution of the initial boundary value problem (11.1) with initial data $v(x) = 0$ and boundary conditions (11.1c) and (11.1d) replaced by $u(0,t) = 0$ and $u(1,t) = 1$. Note that $v(1) \neq u(1,0)$, and we say that the data of the initial boundary value problem (11.1) are discontinuous. We take $\Delta x = \Delta t = 5 \times 10^{-2}$. The numerical solution is shown at time levels $t^n = n\Delta t$ $(n = 1, 2, \ldots, 6)$ in Figure 11.7. Obviously, the numerical solution tends to the steady solution $u_s(x) = x$; however, the oscillatory behaviour in the grid point adjacent to the boundary $x = 1$ is rather peculiar. This erroneous behaviour of the Crank–Nicolson scheme is due to the discontinuity in the data of the initial boundary value problem to be solved, as will be explained in Section 11.5. □

The analysis of the global discretisation error e_j^{n+1} is quite similar to what we have done for the explicit and implicit Euler schemes, and therefore we only give a brief outline. From (11.48) and the expression for the local discretisation error d_j^n in (11.53) we get the following recurrence relation for e_j^{n+1}:

$$e_j^{n+1} - \vartheta d\big(e_{j+1}^{n+1} - 2e_j^{n+1} + e_{j-1}^{n+1}\big) = e_j^n + (1-\vartheta)d\big(e_{j+1}^n - 2e_j^n + e_{j-1}^n\big) + \Delta t d_j^n. \quad (11.55)$$

If $2(1-\vartheta)d \leq 1$, the coefficient of e_j^n in the recurrence relation (11.55) is positive. Proceeding as in the previous sections, we take the maximum value of $|e_j^{n+1}|$ and obtain precisely the inequalities (11.22) and thus (11.25). Combining this with the estimate (11.54), we find the following upper bound for $\|\mathbf{e}^n\|_\infty$:

$$\|\mathbf{e}^n\|_\infty \leq C\left(\left|\frac{1}{2} - \vartheta\right|\Delta t + \frac{1}{12}\Delta x^2\right)t^n + \mathcal{O}(\Delta t^2) + \mathcal{O}(\Delta x^4), \quad (11.56)$$

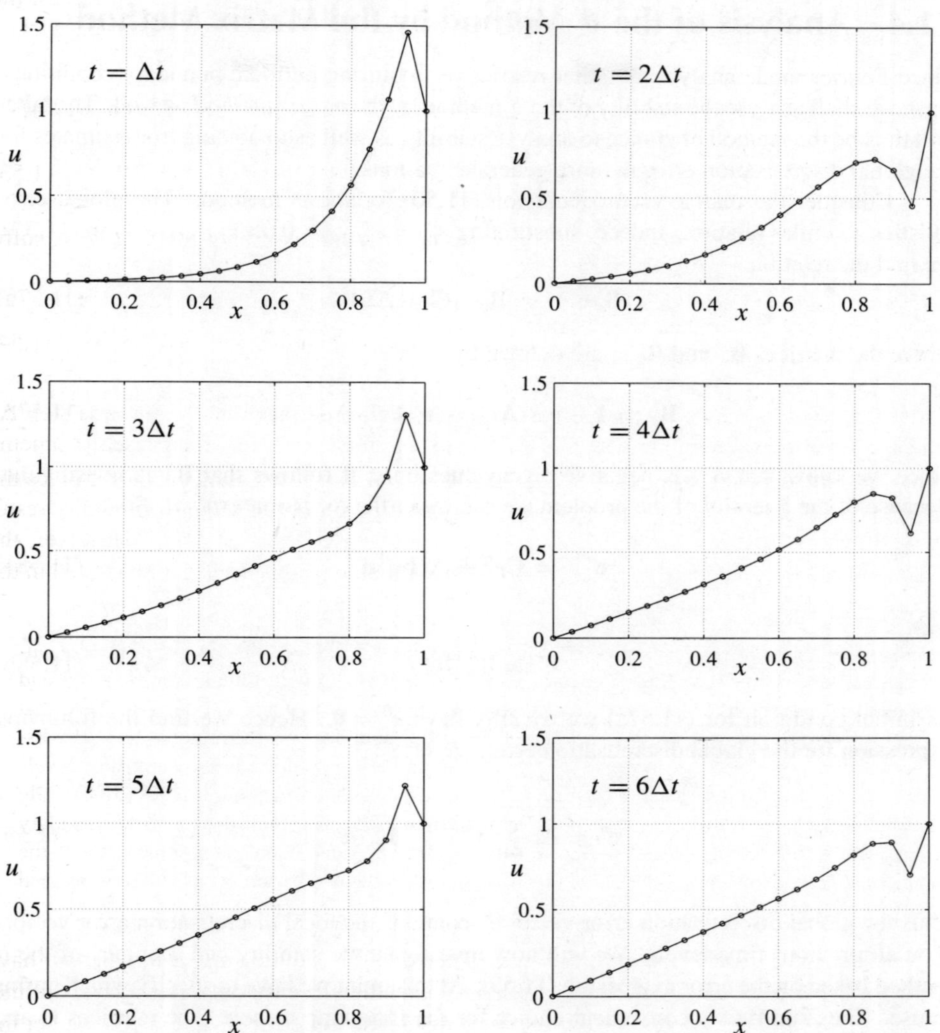

Figure 11.7. *Numerical solution of the heat equation with discontinuous data, computed with the ϑ-scheme.*

with C defined in (11.26). From this inequality we can conclude that the ϑ method is convergent. More precisely, we say that the method is first order convergent in Δt and second order convergent in Δx in the ∞-norm, except when $\vartheta = \frac{1}{2}$, in which case the ϑ method is second order convergent in both Δt and Δx. From a practical point of view a choice $\vartheta = \frac{1}{2} + \mathcal{O}(\Delta t)$ is also advantageous. In order to have unconditional stability, ϑ should be chosen larger than $\frac{1}{2}$.

11.4 Analysis of the ϑ Method by the Matrix Method

Since Fourier mode analysis is rather restrictive (requiring periodic boundary conditions, e.g.), we shall analyse the stability of the ϑ method by the *matrix method* as well. The latter will thus be the method of choice to analyse stability as well as to obtain error estimates for the global discretisation error in more general situations.

Consider the matrix-vector recursion (11.50) for the ϑ method. The global error satisfies a similar relation. Indeed, substituting $e_0^m = e_{M+1}^m = 0\,(m = n, n+1)$ in (11.55), we find the relation

$$\mathbf{B}_\vartheta \mathbf{e}^{n+1} = \mathbf{B}_{\vartheta-1}\mathbf{e}^n + \Delta t\mathbf{d}^n, \tag{11.57a}$$

where the matrices $\mathbf{B}_\vartheta$ and $\mathbf{B}_{\vartheta-1}$ are defined by

$$\mathbf{B}_\nu := \mathbf{I} - \nu\Delta t\mathbf{A} \qquad (\nu = \vartheta,\ \vartheta - 1). \tag{11.57b}$$

Since we know that $\mathbf{A}$ has negative eigenvalues only, it follows that $\mathbf{B}_\vartheta$ is nonsingular. Because of the linearity of the problem we can make the recursion *explicit*. So

$$\mathbf{e}^{n+1} = \mathbf{C}\mathbf{e}^n + \Delta t\mathbf{B}_\vartheta^{-1}\mathbf{d}^n, \tag{11.58a}$$

with

$$\mathbf{C} := \mathbf{B}_\vartheta^{-1}\mathbf{B}_{\vartheta-1}. \tag{11.58b}$$

As initial condition for (11.57a) we trivially have $\mathbf{e}^0 = \mathbf{0}$. Hence we find the following expression for the global discretisation error vector $\mathbf{e}^n$:

$$\mathbf{e}^n = \Delta t \sum_{l=0}^{n-1} \mathbf{C}^{n-1-l}\,\mathbf{B}_\vartheta^{-1}\,\mathbf{d}^l. \tag{11.59}$$

Thus the global discretisation error vector $\mathbf{e}^n$ contains the local discretisation error vectors $\mathbf{d}^l$ at all previous time levels. We will now investigate the stability and accuracy of the ϑ method based on the error expression (11.59). At this point we have to specify which norms to use. The 2-norm is a convenient choice for the following reason. The matrices $\mathbf{C}$ and $\mathbf{B}_\vartheta^{-1}$ can be jointly transformed to diagonal form, for which the 2-norm is simply the spectral radius; see (G.5c). Indeed, using the definitions of $\mathbf{B}_\vartheta$ and $\mathbf{B}_{\vartheta-1}$ in (11.57b), we find

$$\mathbf{W}^{-1}\mathbf{B}_\vartheta^{-1}\mathbf{W} = \big(\mathbf{I} - \vartheta\Delta t\mathbf{\Lambda}\big)^{-1}, \tag{11.60a}$$

$$\mathbf{W}^{-1}\mathbf{C}\mathbf{W} = \big(\mathbf{I} - \vartheta\Delta t\mathbf{\Lambda}\big)^{-1}\big(\mathbf{I} + (1-\vartheta)\Delta t\mathbf{\Lambda}\big), \tag{11.60b}$$

with $\mathbf{W}$ the orthogonal eigenvector matrix and $\mathbf{\Lambda}$ the diagonal matrix defined in (11.33) and (11.35), respectively. The 2-norm of these matrices is given by

$$\big\|\mathbf{W}^{-1}\mathbf{B}_\vartheta^{-1}\mathbf{W}\big\|_2 = \rho\big((\mathbf{I} - \vartheta\Delta t\mathbf{\Lambda})^{-1}\big) = \rho\big(\mathbf{B}_\vartheta^{-1}\big), \tag{11.61a}$$

$$\big\|\mathbf{W}^{-1}\mathbf{C}\mathbf{W}\big\|_2 = \rho\big((\mathbf{I} - \vartheta\Delta t\mathbf{\Lambda})^{-1}(\mathbf{I} + (1-\vartheta)\Delta t\mathbf{\Lambda})\big) = \rho(\mathbf{C}), \tag{11.61b}$$

where $\rho(\cdot)$ denotes the spectral radius of the corresponding matrix. Finally, since the 2-norm is invariant under orthogonal transformations, we have

$$\begin{aligned}\|\mathbf{C}^{n-1-l}\mathbf{B}_\vartheta^{-1}\|_2 &\leq \|\mathbf{C}\|_2^{n-1-l}\|\mathbf{B}_\vartheta^{-1}\|_2\\ &= \|\mathbf{W}^{-1}\mathbf{C}\mathbf{W}\|_2^{n-1-l}\|\mathbf{W}^{-1}\mathbf{B}_\vartheta^{-1}\mathbf{W}\|_2\\ &= \rho(\mathbf{C})^{n-1-l}\rho(\mathbf{B}_\vartheta^{-1}).\end{aligned} \tag{11.62}$$

From (11.59) we obtain the following upper bound for $\|\mathbf{e}^n\|_2$:

$$\|\mathbf{e}^n\|_2 \leq \Delta t \rho(\mathbf{B}_\vartheta^{-1}) \sum_{l=0}^{n-1} \rho(\mathbf{C})^{n-1-l}\|\mathbf{d}^l\|_2. \tag{11.63}$$

If the matrices $\mathbf{C}^{n-1-l}$ are bounded uniformly in l and if $\mathbf{B}_\vartheta$ is nonsingular, we can decide for the *stability* of the ϑ scheme; see Definition 5.15. This is so in particular if

$$\rho(\mathbf{C}) \leq 1, \qquad \rho(\mathbf{B}_\vartheta^{-1}) < \infty. \tag{11.64}$$

In order to compute the spectral radii in (11.64), we employ the eigenvalues λ_k of $\mathbf{A}$; see (11.31). Let the eigenvalues of $\mathbf{B}_\vartheta^{-1}$ and $\mathbf{C}$ be denoted by μ_k and η_k, respectively. Since $\mathbf{B}_\vartheta$ and $\mathbf{C}$ have the same set of eigenvectors as $\mathbf{A}$, it easily follows that

$$\mu_k = (1 - \vartheta\Delta t\lambda_k)^{-1}, \quad \eta_k = \frac{1 + (1-\vartheta)\Delta t\lambda_k}{1 - \vartheta\Delta t\lambda_k} \qquad (k = 1, 2, \ldots, M). \tag{11.65}$$

From (11.31) and (11.65) it is obvious that $0 < \mu_k < 1$ for $k = 1, 2, \ldots, M$, and thus $\rho(\mathbf{B}_\vartheta^{-1}) < 1$. The first inequality in (11.64) is really the interesting one to prove and actually reduces to

$$-1 \leq \frac{1 + (1-\vartheta)\Delta t\lambda_k}{1 - \vartheta\Delta t\lambda_k} \leq 1 \qquad (k = 1, 2, \ldots, M). \tag{11.66}$$

The right inequalities in (11.66) are always satisfied and thus we only have to investigate the left inequalities. Rearranging terms, we find the conditions

$$(2\vartheta - 1)\Delta t\lambda_k \leq 2 \qquad (k = 1, 2, \ldots, M). \tag{11.67}$$

Note that for $\frac{1}{2} \leq \vartheta \leq 1$ (11.67) is simply satisfied. For $0 \leq \vartheta < \frac{1}{2}$ we have to substitute expression (11.31) for the eigenvalues λ_k. This results in the stability conditions

$$\begin{cases} 0 \leq \vartheta < \dfrac{1}{2} \quad \text{and} \quad d \leq \dfrac{1}{2(1-2\vartheta)}, \\ \dfrac{1}{2} \leq \vartheta \leq 1. \end{cases} \tag{11.68}$$

Note that these stability conditions are exactly the same as the ones given in (11.52), which are obtained by Fourier mode analysis.

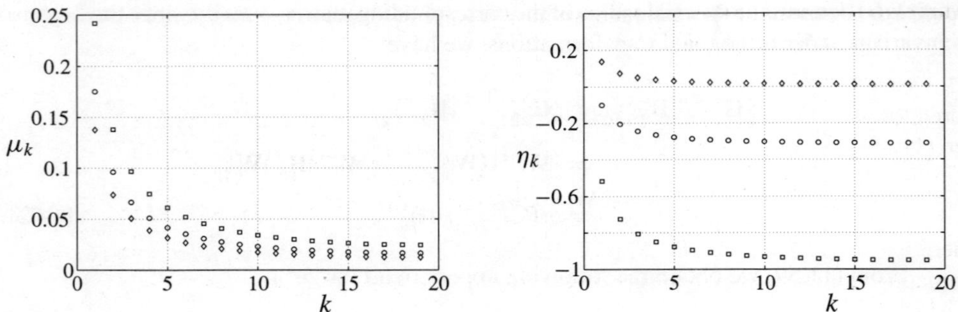

Figure 11.8. *Eigenvalues of the matrices* $\mathbf{B}_\vartheta^{-1}$ *(left) and* $\mathbf{C}$ *(right) for* $\vartheta = 0.5$ (□), 0.75 (○), *and* 1 (◇).

Example 11.5 We have plotted in Figure 11.8 the eigenvalues of $\mathbf{B}_\vartheta^{-1}$ and $\mathbf{C}$ for $\Delta x = \Delta t = 5 \times 10^{-2}$ and $\vartheta = 0.5, 0.75$, and 1. We see that for $\vartheta = 0.5$, we have $\eta_k \approx -1$ when $k \to M$. This means that high-frequency eigenvector components in $\mathbf{e}^n$, corresponding to (in absolute value) large eigenvalues λ_k, are badly damped by the Crank–Nicolson scheme. On the other hand, for $\vartheta = 1$, we have $\eta_k \approx 0$ when $k \to M$, implying that high-frequency components are very well damped for the implicit Euler scheme. □

With the above stability result we can give an upper bound for the global discretisation error. From (11.54) we find the following upper bound for $\|\mathbf{d}^n\|_2$:

$$\begin{aligned} \|\mathbf{d}^n\|_2 &\le \sqrt{M}\|\mathbf{d}^n\|_\infty \\ &\le \frac{1}{\sqrt{\Delta x}}\left(\left(\left|\frac{1}{2} - \vartheta\right|\Delta t + \frac{1}{12}\Delta x^2\right)C + \mathcal{O}(\Delta t^2) + \mathcal{O}(\Delta x^4)\right), \end{aligned} \tag{11.69}$$

where we have assumed that $f(x,t) \equiv 0$ and $u(x,t)$ satisfies (11.26). Now we can determine an upper bound for $\|\mathbf{e}^n\|_2$. Assuming the stability conditions (11.68) to be satisfied, we can deduce from (11.63) that

$$\begin{aligned} \|\mathbf{e}^n\|_2 &\le \Delta t \sum_{l=0}^{n-1}\|\mathbf{d}^l\|_2 \\ &\le \frac{t^n}{\sqrt{\Delta x}}\left(\left(\left|\frac{1}{2} - \vartheta\right|\Delta t + \frac{1}{12}\Delta x^2\right)C + \mathcal{O}(\Delta t^2) + \mathcal{O}(\Delta x^2)\right). \end{aligned} \tag{11.70}$$

This error bound is a bit disappointing in view of the second order accuracy found in (11.56). One way to interpret (11.70) is by considering the *median* of the norm, i.e., $\|\mathbf{e}^n\|_2/\sqrt{M}$. A more technical proof of a sharper ∞-norm for the Crank–Nicolson scheme is given in the following property.

Property 11.6. *If* $\Delta t = \mathcal{O}(\Delta x)$, *then the global error in the* ∞*-norm for the Crank–Nicolson scheme is second order; i.e.,* $\|\mathbf{e}^n\|_\infty = \mathcal{O}(\Delta x^2)$.

Proof. For the proof of this estimate we have to consider the expression

$$\mathbf{C}^{n-1-l}\mathbf{B}_{\vartheta}^{-1}$$

more closely. We note that this matrix is similar to $\mathbf{\Lambda}$. Using the transformation matrix $\mathbf{W}$ (see (11.33)), we obtain

$$\mathbf{W}^{-1}\mathbf{C}^{n-1-l}\mathbf{B}_{\vartheta}^{-1}\mathbf{W} = \operatorname{diag}\left(\eta_k^{n-1-l}\mu_k\right) =: \mathbf{P}_{n-1-l},$$

where η_k and μ_k are given in (11.65). Taking the absolute value of all matrix entries and estimating elementwise, we find

$$|\,\mathbf{C}^{n-1-l}\mathbf{B}_{\vartheta}^{-1}\,| \le \frac{2}{M}\begin{pmatrix} 1 & \cdots & 1 \\ \vdots & & \vdots \\ 1 & \cdots & 1 \end{pmatrix} |\,\mathbf{P}_{n-1-l}\,| \begin{pmatrix} 1 & \cdots & 1 \\ \vdots & & \vdots \\ 1 & \cdots & 1 \end{pmatrix}.$$

After working out the product of the last two matrices, we conclude that

$$\|\mathbf{C}^{n-1-l}\mathbf{B}_{\vartheta}^{-1}\|_{\infty} \le \frac{2}{M} M \sum_{k=1}^{M} |\eta_k|^{n-1-l}\mu_k.$$

Consider the expression for η_k in (11.65). Note that the function $\varphi(\xi) := (1+\frac{1}{2}\xi)/(1-\frac{1}{2}\xi)$ is monotonically increasing. Furthermore, we have $\frac{2}{\pi}\zeta \le \sin\zeta$ for $0 \le \zeta \le \frac{1}{2}\pi$, so $\lambda_k \le -4k^2$. We now distinguish between two possibilities: $\eta_k \ge 0$ (case (i)) and $\eta_k \le 0$ (case (ii)). First, consider case (i). Note that we need the bound for the λ_k here. We obtain

$$\sum_{l=1}^{n-1} |\eta_k|^{n-1-l}\mu_k \le \left(1 - \frac{1+\frac{1}{2}\Delta t\lambda_k}{1-\frac{1}{2}\Delta t\lambda_k}\right)^{-1} \frac{1}{1-\frac{1}{2}\Delta t\lambda_k} \le \frac{1}{4k^2\Delta t}.$$

In case (ii) we find

$$\sum_{l=1}^{n-1} |\eta_k|^{n-1-l}\mu_k \le \left(1 + \frac{\frac{1}{2}\Delta t\lambda_k+1}{1-\frac{1}{2}\Delta t\lambda_k}\right)^{-1} \frac{1}{1-\frac{1}{2}\Delta t\lambda_k} \le \frac{1}{2}.$$

The first case clearly gives an upper bound $\mathcal{O}(\Delta t^{-1})$, while the second gives an upper bound $\mathcal{O}(\Delta x^{-1})$; for the latter one should realize that $\Delta x = \mathcal{O}(M^{-1})$. Combining the results in case (i) and case (ii), we thus conclude that

$$\sum_{l=1}^{n-1}\|\mathbf{C}^{n-1-l}\mathbf{B}_{\vartheta}^{-1}\|_{\infty} \le \sum_{k=1}^{M}\sum_{l=1}^{n-1} |\eta_k|^{n-1-j}\mu_k = \mathcal{O}\left(\frac{1}{\Delta t}\right) + \mathcal{O}\left(\frac{1}{\Delta x}\right).$$

Using this estimate in the expression for the local error in (11.59) yields the required result. □

We remark that one often prefers to have a slightly more dissipative scheme than Crank–Nicolson. In order not to lose the actual second order error, one can take $\vartheta = \frac{1}{2} + \mathcal{O}(\Delta t)$, where of course $\vartheta > \frac{1}{2}$.

11.5 Initial Boundary Value Problems with Discontinuous Data

To show another use of the matrix method we analyse the phenomenon noticed in Example 11.4. Consider the following initial boundary value problem for $u(x,t)$:

$$\frac{\partial u}{\partial t} = \frac{\partial^2 u}{\partial x^2}, \qquad x \in (0,1) \quad t > 0, \tag{11.71a}$$

$$u(x,0) = 0, \qquad x \in (0,1), \tag{11.71b}$$

$$u(0,t) = 0, \quad u(1,t) = 1, \qquad t > 0. \tag{11.71c}$$

Note that at the right boundary $x = 1$ and at time $t = 0$ the solution u abruptly jumps from zero to one; i.e., the initial boundary value problem (11.71) has discontinuous data. If we employ the central difference approximation (5.30) for the spatial derivative and apply the boundary conditions in (11.71c), i.e.,

$$u_0 = 0, \qquad u_{M+1} = 1, \tag{11.72}$$

we obtain the initial value problem (cf. (11.30))

$$\frac{\mathrm{d}\mathbf{u}}{\mathrm{d}t} = \mathbf{A}\mathbf{u} + \mathbf{f}, \tag{11.73a}$$

$$\mathbf{u}(0) = \mathbf{0}, \tag{11.73b}$$

where the vector $\mathbf{f}$ is given by

$$\mathbf{f} := \frac{1}{\Delta x^2}(0, 0, \ldots, 0, 1)^T. \tag{11.74}$$

The exact solution of the initial value problem (11.73) reads

$$\mathbf{u}(t) = \int_0^t \exp\big(\mathbf{A}(t-s)\big)\mathbf{f}\,\mathrm{d}s. \tag{11.75}$$

We can formally compute the matrix exponential $\exp(\mathbf{A}(t-s))$ in (11.75) using the spectral decomposition $\mathbf{A} = \mathbf{W}\boldsymbol{\Lambda}\mathbf{W}^{-1}$; cf. (11.34). We find

$$\exp\big(\mathbf{A}(t-s)\big) = \mathbf{W}\exp\big(\boldsymbol{\Lambda}(t-s)\big)\mathbf{W}^{-1}, \tag{11.76a}$$

$$\exp\big(\boldsymbol{\Lambda}(t-s)\big) = \mathrm{diag}\big(\mathrm{e}^{\lambda_1(t-s)}, \mathrm{e}^{\lambda_2(t-s)}, \ldots, \mathrm{e}^{\lambda_M(t-s)}\big). \tag{11.76b}$$

Substitution of (11.76) in (11.75) and subsequent integration gives

$$\begin{aligned} \mathbf{u}(t) &= \mathbf{W}\boldsymbol{\Lambda}^{-1}\big(\exp(\boldsymbol{\Lambda}t) - \mathbf{I}\big)\mathbf{W}^{-1}\mathbf{f} \\ &= \mathbf{A}^{-1}\big(\exp(\mathbf{A}t) - \mathbf{I}\big)\mathbf{f}. \end{aligned} \tag{11.77}$$

As can be seen from Figure 11.7, we are mainly interested in the transient part of the solution in (11.77), $\mathbf{v}$ say, i.e.,

$$\mathbf{v}(t) = \mathbf{A}^{-1}\exp\big(\mathbf{A}t\big)\mathbf{f}. \tag{11.78}$$

From (11.78) we deduce that the *increment* of $\mathbf{v}(t)$ on a time interval Δt is given by

$$\mathbf{v}(t+\Delta t) = \exp\bigl(\mathbf{A}\Delta t\bigr)\mathbf{v}(t), \tag{11.79}$$

where the increment matrix $\exp(\mathbf{A}\Delta t)$ is given by

$$\exp(\mathbf{A}\Delta t) = \mathbf{W}\,\mathrm{diag}\bigl(\mathrm{e}^{\lambda_1\Delta t}, \mathrm{e}^{\lambda_2\Delta t}, \ldots, \mathrm{e}^{\lambda_M\Delta t}\bigr)\mathbf{W}^{-1}. \tag{11.80}$$

Now suppose we use the ϑ method for the system of differential equations in (11.73). Then in matrix terms we obtain

$$\frac{1}{\Delta t}\bigl(\mathbf{u}^{n+1} - \mathbf{u}^n\bigr) = (1-\vartheta)\bigl(\mathbf{A}\mathbf{u}^n + \mathbf{f}\bigr) + \vartheta\bigl(\mathbf{A}\mathbf{u}^{n+1} + \mathbf{f}\bigr). \tag{11.81}$$

Multiplying by Δt and rearranging terms, we find the recurrence relation

$$\mathbf{u}^{n+1} = \mathbf{C}\mathbf{u}^n + \Delta t\mathbf{B}_\vartheta^{-1}\mathbf{f}, \tag{11.82}$$

where the matrices $\mathbf{B}_\vartheta$ and $\mathbf{C}$ are defined in (11.57b) and (11.58b), respectively. Since one is not an eigenvalue of $\mathbf{C}$, the matrix $\mathbf{C}-\mathbf{I}$ is invertible. Using the initial condition $\mathbf{u}^0 = \mathbf{0}$, we can see that the solution of (11.82) is given by

$$\mathbf{u}^n = \Delta t\left(\sum_{l=0}^{n-1}\mathbf{C}^l\right)\mathbf{B}_\vartheta^{-1}\mathbf{f} = \Delta t(\mathbf{C}-\mathbf{I})^{-1}(\mathbf{C}^n - \mathbf{I})\mathbf{B}_\vartheta^{-1}\mathbf{f}, \tag{11.83}$$

where we used the expression

$$(\mathbf{C}-\mathbf{I})\bigl(\mathbf{I}+\mathbf{C}+\mathbf{C}^2+\cdots+\mathbf{C}^{n-1}\bigr) = \mathbf{C}^n - \mathbf{I}.$$

By further noting that

$$\Delta t(\mathbf{C}-\mathbf{I})^{-1}\mathbf{B}_\vartheta^{-1} = \mathbf{A}^{-1}$$

and that the matrices $(\mathbf{C}-\mathbf{I})^{-1}$ and $\mathbf{C}^n$ commute, we see that the solution of (11.83) is given by

$$\mathbf{u}^n = \mathbf{A}^{-1}\bigl(\mathbf{C}^n - \mathbf{I}\bigr)\mathbf{f}. \tag{11.84}$$

This is a nice (time) discrete analogue of (11.78). In particular, the discrete transient part is given by

$$\mathbf{v}^n := \mathbf{A}^{-1}\mathbf{C}^n\mathbf{f}. \tag{11.85}$$

Clearly, the effect of the initial condition is damped out by the decay of the matrix $\mathbf{C}^n$. Therefore it makes sense to investigate the increment matrix $\mathbf{C}$ as given in (11.82); i.e., we compare this to the increment matrices $\exp(\mathbf{A}\Delta t)$ (the exact decay matrix per step Δt). Using the spectral decomposition of $\mathbf{A}$, we can rewrite the matrix $\mathbf{C}$ as

$$\begin{aligned}\mathbf{C} &= \bigl(\mathbf{I}-\vartheta\Delta t\mathbf{A}\bigr)^{-1}\bigl(\mathbf{I}+(1-\vartheta)\Delta t\mathbf{A}\bigr)\\ &= \mathbf{W}\bigl(\mathbf{I}-\vartheta\Delta t\mathbf{\Lambda}\bigr)^{-1}\bigl(\mathbf{I}+(1-\vartheta)\Delta t\mathbf{\Lambda}\bigr)\mathbf{W}^{-1}\\ &= \mathbf{W}\,\mathrm{diag}\bigl(r(\lambda_1\Delta t), r(\lambda_2\Delta t), \ldots, r(\lambda_M\Delta t)\bigr)\,\mathbf{W}^{-1},\end{aligned} \tag{11.86}$$

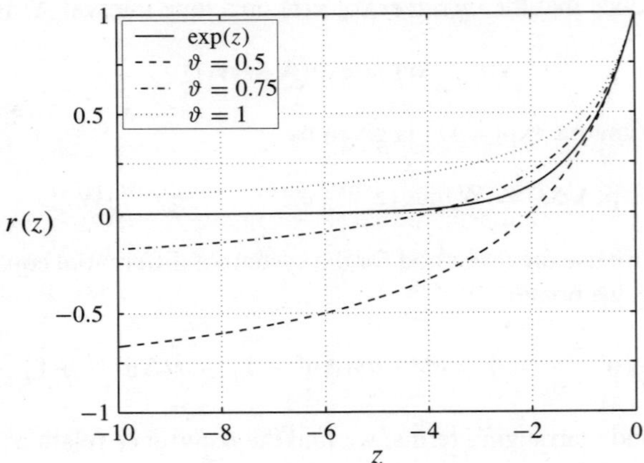

Figure 11.9. *Amplification factor* e^z *compared with the rational function* $r(z)$ *for* $\vartheta = 0.5, 0.75$, *and* 1.

with the rational function $r(z)$ defined by

$$r(z) := \frac{1 + (1 - \vartheta)z}{1 - \vartheta z}. \tag{11.87}$$

Comparing (11.80) and (11.86), we see that the incremental factors $e^{\lambda_k \Delta t}$ in $\exp(\mathbf{A}\Delta t)$ are approximated by the fractions $r(\lambda_k \Delta t)$, which might be close to one in modulus and even become negative for larger negative eigenvalues λ_k; see Figure 11.9. This then introduces oscillatory components in the numerical solution, which explains the curious behaviour in Example 11.4.

11.6 Mixed Boundary Conditions

So far we have only considered Dirichlet boundary conditions. However, in practice the flux of a variable is often prescribed at the boundary of a domain, leading to *mixed boundary conditions*; see Section 10.4. Treating mixed boundary conditions, in particular the fluxes, requires special attention. In this section we investigate how to discretise these and look for sufficient conditions to ensure the stability of the method.

Consider the following initial boundary value problem for $u(x, t)$ (cf. (10.41)):

$$\frac{\partial u}{\partial t} = \frac{\partial^2 u}{\partial x^2}, \qquad x \in (0, 1), \qquad t > 0, \tag{11.88a}$$

$$u(x, 0) = v(x), \qquad x \in (0, 1), \tag{11.88b}$$

$$\frac{\partial u}{\partial x}(0, t) - \alpha\big(u(0, t) - g_\ell(t)\big) = 0, \quad t > 0, \tag{11.88c}$$

$$\frac{\partial u}{\partial x}(1, t) + \beta\big(u(1, t) - g_r(t)\big) = 0, \quad t > 0, \tag{11.88d}$$

with parameters $\alpha, \beta \geq 0$ and $g_\ell(t)$, $g_\mathrm{r}(t)$ some given functions. When $\alpha = \beta = 0$, we obtain the Neumann boundary conditions

$$\frac{\partial u}{\partial x}(0,t) = \frac{\partial u}{\partial x}(1,t) = 0, \qquad t > 0, \tag{11.89}$$

and when $\alpha, \beta \to \infty$ the boundary conditions in (11.88) reduce to the Dirichlet boundary conditions $u(0,t) = g_\ell(t)$ and $u(1,t) = g_\mathrm{r}(t)$.

For the numerical solution of the initial boundary value problem (11.88) we employ central differences for space discretisation and the ϑ method for time integration. This gives the following numerical scheme, which holds at all internal grid points x_j ($j = 1, 2, \ldots, M$):

$$u_j^{n+1} - \vartheta d\big(u_{j+1}^{n+1} - 2u_j^{n+1} + u_{j-1}^{n+1}\big) = u_j^n + (1-\vartheta)d\big(u_{j+1}^n - 2u_j^n + u_{j-1}^n\big). \tag{11.90}$$

To facilitate the notation we take in the following that $g_\ell(t) = g_\mathrm{r}(t) \equiv 0$. For the initial boundary value problem (11.88) the boundary values $u(0,t)$ and $u(1,t)$ are not explicitly known, since they are related to the (unknown) derivatives $\frac{\partial}{\partial x}u(0,t)$ and $\frac{\partial}{\partial x}u(1,t)$, respectively. Therefore the numerical variables u_0^m, u_{M+1}^m ($m = n, n+1$) are unknown and have to be determined somehow. We can achieve this by requiring the PDE to hold on a slightly larger domain, introducing *virtual grid points* $x_{-1} := -\Delta x$ and $x_{M+2} := 1 + \Delta x$, and using the numerical scheme at the boundary points x_0 and x_{M+1}. First consider the left boundary, for which $j = 0$. Substituting $j = 0$ into (11.90), we obtain

$$u_0^{n+1} - \vartheta d\big(u_1^{n+1} - 2u_0^{n+1} + u_{-1}^{n+1}\big) = u_0^n + (1-\vartheta)d\big(u_1^n - 2u_0^n + u_{-1}^n\big). \tag{11.91}$$

This equation contains the numerical values u_{-1}^m ($m = n, n+1$), which can be eliminated from (11.91) by applying a finite difference approximation of the boundary condition at $x = 0$. To discretise the derivative $\frac{\partial}{\partial x}u(0,t)$, we choose the central difference approximation (5.28c), because it is second order accurate like the central difference approximation of $\frac{\partial^2}{\partial x^2}u(x_j,t)$ at all grid points. However, we remark that there is a snag here, which will be addressed at the end of this section. We now obtain

$$\frac{1}{2\Delta x}\big(u_1^m - u_{-1}^m\big) - \alpha u_0^m = 0 \qquad (m = n, n+1), \tag{11.92a}$$

or, equivalently,

$$u_{-1}^m = u_1^m - 2\alpha\Delta x u_0^m \qquad (m = n, n+1). \tag{11.92b}$$

Substituting the discrete boundary condition (11.92b) into (11.91), we get

$$u_0^{n+1} - \vartheta d\big(2u_1^{n+1} - 2(1+\alpha\Delta x)u_0^{n+1}\big) = u_0^n + (1-\vartheta)d\big(2u_1^n - 2(1+\alpha\Delta x)u_0^n\big). \tag{11.93}$$

Likewise, at the right boundary point x_{M+1}, we have the equation

$$u_{M+1}^{n+1} - \vartheta d\big(u_{M+2}^{n+1} - 2u_{M+1}^{n+1} + u_M^{n+1}\big) = u_{M+1}^n + (1-\vartheta)d\big(u_{M+2}^n - 2u_{M+1}^n + u_M^n\big), \tag{11.94}$$

containing the numerical values u_{M+2}^m ($m = n, n+1$) at the virtual grid point $x_{M+2} = 1 + \Delta x$. These values can be eliminated if we combine (11.94) with the central difference approximation of the boundary condition at $x = 1$, i.e.,

$$\frac{1}{2\Delta x}\big(u_{M+2}^m - u_M^m\big) + \beta u_{M+1}^m = 0 \qquad (m = n, n+1). \tag{11.95a}$$

From this equation we immediately conclude that

$$u_{M+2}^m = u_M^m - 2\beta\Delta x u_{M+1}^m \qquad (m = n, n+1). \tag{11.95b}$$

Substituting the latter equation into (11.94), we obtain

$$u_{M+1}^{n+1} - \vartheta d\big(2u_M^{n+1} - 2(1+\beta\Delta x)u_{M+1}^{n+1}\big) = u_{M+1}^n + (1-\vartheta)d\big(2u_M^n - 2(1+\beta\Delta x)u_{M+1}^n\big). \tag{11.96}$$

The numerical scheme for the initial boundary value problem (11.88) consists of the equations (11.90), (11.93), and (11.96) for the unknowns $\mathbf{u}^n := (u_0^n, u_1^n, \ldots, u_{M+1}^n)^T$.

The above numerical scheme can be written as the linear system

$$(\mathbf{I} - \vartheta\Delta t\mathbf{A})\mathbf{u}^{n+1} = (\mathbf{I} + (1-\vartheta)\Delta t\mathbf{A})\mathbf{u}^n, \tag{11.97}$$

with $\mathbf{I}$ the $(M+2)\times(M+2)$ identity matrix and $\mathbf{A}$ the discretisation matrix given by

$$\mathbf{A} := \frac{1}{\Delta x^2}\begin{pmatrix} -2(1+\alpha\Delta x) & 2 & & & & & \\ 1 & -2 & 1 & & & & \\ & 1 & -2 & 1 & & & \\ & & \ddots & \ddots & \ddots & & \\ & & & 1 & -2 & 1 & \\ & & & & 1 & -2 & 1 \\ & & & & & 2 & -2(1+\beta\Delta x) \end{pmatrix}. \tag{11.98}$$

This matrix $\mathbf{A}$ is the discretisation matrix of the differential operator $\frac{\partial^2}{\partial x^2}$ and the boundary conditions together. The first and last rows of $\mathbf{A}$ represent the boundary conditions at $x = 0$ and $x = 1$, respectively. Note that for $\alpha = \beta = 0$, i.e., for Neumann boundary conditions, the matrix $\mathbf{A}$ is singular. In particular, the vector $\mathrm{e} := (1, 1, \ldots, 1)^T$ is then a basis for the null space of $\mathbf{A}$, which in turn implies that the numerical solution computed from (11.97) is determined up to $\gamma\mathrm{e}$, with γ an arbitrary constant.

Next we want to apply the matrix method to investigate the stability of (11.97). Completely analogously to the derivation in Section 11.4, we find that the global discretisation error vector satisfies the recurrence relation (11.57a), but with the discretisation matrix $\mathbf{A}$ now as defined in (11.98). The solution of (11.57a) is given in (11.59), with matrices $\mathbf{B}_\vartheta$, $\mathbf{B}_{\vartheta-1}$, and $\mathbf{C}$ defined like in (11.57b) and (11.58b). The stability conditions are again given in (11.64).

The problem is now that the matrix $\mathbf{A}$ is not symmetric any more as a result of the (discrete) boundary conditions. Consequently, the matrices $\mathbf{B}_\vartheta$, $\mathbf{B}_{\vartheta-1}$, and $\mathbf{C}$ in (11.58b) are also not symmetric and therefore the 2-norm does not seem to be appropriate. However, as we can see from (11.98), the lack of symmetry of $\mathbf{A}$ only occurs at two places and can easily be overcome. To that end we introduce the scaling matrix

$$\mathbf{S} := \mathrm{diag}(\sqrt{2}, 1, 1, \ldots, 1, \sqrt{2}). \tag{11.99}$$

We define a matrix $\hat{\mathbf{A}}$ similar to $\mathbf{A}$ (which implies that $\hat{\mathbf{A}}$ has the same eigenvalues as $\mathbf{A}$) as

$$\hat{\mathbf{A}} := \mathbf{S}^{-1}\mathbf{A}\mathbf{S}. \tag{11.100}$$

Then we find

$$\hat{\mathbf{A}} = \frac{1}{\Delta x^2}\begin{pmatrix} -2(1+\alpha\Delta x) & \sqrt{2} & & & & & \\ \sqrt{2} & -2 & 1 & & & & \\ & 1 & -2 & 1 & & & \\ & & \ddots & \ddots & \ddots & & \\ & & & 1 & -2 & 1 & \\ & & & & 1 & -2 & \sqrt{2} \\ & & & & & \sqrt{2} & -2(1+\beta\Delta x) \end{pmatrix}. \tag{11.101}$$

Clearly, $\hat{\mathbf{A}}$ is symmetric. For ease of notation let us introduce the matrices (cf. (11.57b) and (11.58b))

$$\hat{\mathbf{B}}_\vartheta \;:= \mathbf{S}^{-1}\mathbf{B}_\vartheta\mathbf{S}, \tag{11.102a}$$

$$\hat{\mathbf{B}}_{\vartheta-1} := \mathbf{S}^{-1}\mathbf{B}_{\vartheta-1}\mathbf{S}, \tag{11.102b}$$

$$\hat{\mathbf{C}} \;:= \mathbf{S}^{-1}\mathbf{C}\mathbf{S}. \tag{11.102c}$$

Trivially, these matrices are symmetric, because $\hat{\mathbf{A}}$ is symmetric and the matrices $\hat{\mathbf{B}}_\vartheta$ and $\hat{\mathbf{B}}_{\vartheta-1}$ commute.

Recalling the expression (11.59) for the global discretisation error vector $\mathbf{e}^n$, we find

$$\mathbf{e}^n = \Delta t\sum_{l=0}^{n-1}\mathbf{C}^{n-1-l}\,\mathbf{B}_\vartheta^{-1}\,\mathbf{d}^l = \Delta t\sum_{l=0}^{n-1}\mathbf{S}\,\hat{\mathbf{C}}^{n-1-l}\,\hat{\mathbf{B}}_\vartheta^{-1}\,\mathbf{S}^{-1}\mathbf{d}^l. \tag{11.103}$$

Since all matrices involved are symmetric and $\|\mathbf{S}\|_2=\rho(\mathbf{S})=\sqrt{2}$ and $\|\mathbf{S}^{-1}\|_2=\rho(\mathbf{S}^{-1})=1$, we get the following upper bound for $\|\mathbf{e}^n\|_2$:

$$\begin{aligned}\|\mathbf{e}^n\|_2 &\le \Delta t\sqrt{2}\|\hat{\mathbf{B}}_\vartheta^{-1}\|_2\sum_{l=0}^{n-1}\|\hat{\mathbf{C}}\|_2^{n-1-l}\,\|\mathbf{d}^l\|_2 \\ &= \Delta t\sqrt{2}\rho(\mathbf{B}_\vartheta^{-1})\sum_{l=0}^{n-1}\rho(\mathbf{C})^{n-1-l}\,\|\mathbf{d}^l\|_2,\end{aligned} \tag{11.104}$$

where we used that the matrices $\hat{\mathbf{B}}_\vartheta^{-1}$ and $\hat{\mathbf{C}}$ are similar to $\mathbf{B}_\vartheta^{-1}$ and $\mathbf{C}$, respectively, and consequently have the same eigenvalues.

Next we want to derive sufficient conditions such that the numerical scheme (11.97) is stable. Let the eigenvalues of the matrix $\mathbf{A}$ be denoted by λ_k. Unlike in the derivation in the appendix, Section F, we will not explicitly compute the eigenvalues λ_k. Instead, we apply Gershgorin's theorem to estimate these eigenvalues; see the appendix, Section I. First, note that the matrix $\mathbf{A}$ has real eigenvalues since it is similar to the symmetric matrix $\hat{\mathbf{A}}$. Then by Gershgorin's theorem we conclude that

$$\Delta x^2\lambda_k \in [-4, 0]\cup[-4-2\alpha\Delta x,\, -2\alpha\Delta x]\cup[-4-2\beta\Delta x,\, -2\beta\Delta x]. \tag{11.105}$$

Consequently, we obtain the following inequalities for the eigenvalues λ_k:

$$-\frac{4}{\Delta x^2}-\frac{2}{\Delta x}\max(\alpha,\beta)\leq \lambda_k \leq 0. \tag{11.106}$$

Obviously, the condition $\rho(\mathbf{B}_\vartheta^{-1})<\infty$ is again trivially satisfied. Like we saw in Section 11.4, the requirement $\rho(\mathbf{C})<1$ leads to the inequalities in (11.66). Combining (11.66) and (11.106), we finally obtain that the method is stable if

$$\begin{cases} 0\leq\vartheta<\dfrac{1}{2} \quad \text{and} \quad d\leq\dfrac{1}{1-2\vartheta}\dfrac{1}{2+\Delta x\max(\alpha,\beta)}, \\ \dfrac{1}{2}\leq\vartheta\leq 1. \end{cases} \tag{11.107}$$

We see that for $\alpha,\beta>0$ the stability condition (11.107) is more restrictive than the stability condition given in (11.66).

We would finally like to address the accuracy problem. If we use central differences, we obtain a discretisation error (i.e., the residual) in (11.92a) of order $\mathcal{O}(\Delta x^2)$. As a result, u_{-1}^m has an $\mathcal{O}(\Delta x^3)$ error. After substituting this in (11.93), we get a local error of order $\mathcal{O}(\Delta x)$! So we may expect that this will spoil the second order behaviour of the scheme as a whole. Fortunately, this is not the case. We recall Section 9.3.3, where we saw that the (local) discretisation error in the boundary conditions may be two orders lower than the discretisation error in internal grid points and still not spoil the overall accuracy.

In order to apply the theory developed in Section 9.3.3, we need a property.

Property 11.7. *Let $d\leq 1/(2(1-\vartheta))$. Then the matrix $\mathbf{C}:=(\mathbf{I}-\vartheta\Delta t\mathbf{A})^{-1}(\mathbf{I}+(1-\vartheta)\Delta t\mathbf{A})$ is positive and satisfies*

$$\mathbf{O}<\sum_{k=0}^{n-1}\mathbf{C}^{n-k}\left(\mathbf{I}-\vartheta\Delta t\mathbf{A}\right)^{-1}\leq-\frac{1}{\Delta t}\mathbf{F}\mathbf{A}^{-1}, \tag{11.108}$$

where $\|\mathbf{F}\|_\infty\leq 2$.

Proof. Note that the condition for d implies that $\mathbf{I}+(1-\vartheta)\Delta t\mathbf{A}>\mathbf{O}$; since $\mathbf{I}-\vartheta\Delta t\mathbf{A}$ clearly is an M-matrix, its inverse is positive as well. Also note that one is not an eigenvalue of $\mathbf{C}$, implying that $\mathbf{I}-\mathbf{C}$ is nonsingular. Moreover, we have

$$\left(\mathbf{I}-\mathbf{C}\right)^{-1}=\frac{-1}{\Delta t}\mathbf{A}^{-1}\left(\mathbf{I}-\vartheta\Delta t\mathbf{A}\right).$$

Hence we deduce from (11.83) that

$$\sum_{k=0}^{n-1}\mathbf{C}^k\left(\mathbf{I}-\vartheta\Delta t\mathbf{A}\right)^{-1}=\left(\mathbf{I}-\mathbf{C}^n\right)\left(\mathbf{I}-\mathbf{C}\right)^{-1}\left(\mathbf{I}-\vartheta\Delta t\mathbf{A}\right)^{-1}=-\left(\mathbf{I}-\mathbf{C}^n\right)\frac{1}{\Delta t}\mathbf{A}^{-1}.$$

Let us now define $\mathbf{F}:=\mathbf{I}-\mathbf{C}^n$. It is not complicated to see that

$$\left\|(\mathbf{I}-\vartheta\Delta t\mathbf{A})\mathbf{x}\right\|_\infty\geq\left\|\mathbf{x}\right\|_\infty.$$

Hence $\|(\mathbf{I} - \vartheta \Delta t \mathbf{A})^{-1}\|_\infty \leq 1$. It can also be directly seen that

$$\left\|\mathbf{I} + (1-\vartheta)\Delta t \mathbf{A}\right\|_\infty \leq 1.$$

So we have $\|\mathbf{C}\|_\infty \leq 1$, whence $\|\mathbf{C}^n\|_\infty \leq 1$; i.e., $\|\mathbf{F}\|_\infty \leq 2$. □

Like we did in Section 9.3.3, we will split up the vector of local discretisation errors $\mathbf{d}^k$ into a part arising from discretising the boundary and a part from discretising the interior. Thus let d_{bnd} and d_{int} be estimates such that

$$\left|\mathbf{d}^k\right| \leq d_{\text{bnd}}\, \mathrm{e}_{\text{bnd}} + d_{\text{int}}\, \mathrm{e}_{\text{int}}. \tag{11.109}$$

In order to find the global error, we will again estimate $|\mathbf{A}^{-1}|\mathrm{e}_{\text{bnd}}$ and $|\mathbf{A}^{-1}|\mathrm{e}_{\text{int}}$ separately using Property 11.7. For ease of writing we assume $\alpha = \beta$. We find $-\mathbf{A}\mathrm{e} \geq 2\alpha \mathrm{e}_{\text{bnd}}/\Delta x$, whence

$$\left|\mathbf{A}^{-1}\right|\mathrm{e}_{\text{bnd}} \leq \frac{\Delta x}{2\alpha}\mathrm{e}.$$

Moreover, the vector $\mathbf{v} := (v_0, v_1, \ldots, v_{M+1})^T$ with $v_j := -\frac{1}{2}(x_j^2 - x_j - 1/\alpha)$ yields

$$-\mathbf{A}\mathbf{v} \leq \mathrm{e}.$$

Note that this vector can be found as from the continuous problem, for which the function $u(x) = -\frac{1}{2}x^2 + \frac{1}{2}x + 1/(2\alpha)$ satisfies the boundary conditions and is numerically differentiated exactly. We therefore obtain

$$\left|\mathbf{A}^{-1}\right|\mathrm{e} \leq \mathbf{v} \leq \left(\frac{1}{8} + \frac{1}{2}\alpha\right)\mathrm{e}.$$

Finally, we thus obtain the following estimates for the global discretisation error (cf. (11.59)):

$$\left\|\mathbf{e}^n\right\|_\infty \leq \frac{2}{\Delta t}\frac{\Delta x}{\alpha} d_{\text{bnd}} + \frac{1}{\Delta t}\left(\frac{1}{4} + \alpha\right) d_{\text{int}}.$$

Since $d_{\text{bnd}} = \mathcal{O}(\Delta t \Delta x)$ and $d_{\text{int}} = \mathcal{O}(\Delta t \Delta x^2)$, we have $\|\mathbf{e}^n\|_\infty = \mathcal{O}(\Delta x^2)$ only if $\Delta t \leq \Delta x^2/(2(1-\vartheta))$.

11.7 Problems in Two Space Dimensions

The methods we have encountered so far have obvious counterparts in several space dimensions. In this section we shall restrict ourselves to two dimensions. In order to keep the derivations in this section simple, we assume that the domain Ω is a rectangle in $\mathbb{R}^2$, say $\Omega := \{(x, y) \in \mathbb{R}^2 \mid 0 \leq x \leq a, 0 \leq y \leq b\}$ for some $a, b > 0$. Then we look for the solution $u(x, y, t)$ of the initial boundary value problem

$$\frac{\partial u}{\partial t} = \frac{\partial^2 u}{\partial x^2} + \frac{\partial^2 u}{\partial y^2} + f(x, y, t), \qquad (x, y) \in \Omega, \qquad t > 0, \tag{11.110a}$$

$$u(x, y, 0) = v(x, y), \qquad (x, y) \in \Omega, \tag{11.110b}$$

$$u(x, y, t) = g(x, y, t), \qquad (x, y) \in \partial\Omega, \quad t > 0, \tag{11.110c}$$

where f, v, and g are given functions.

In order to compute a numerical solution, we cover Ω with a rectangular grid with grid sizes Δx in the x direction and Δy in the y direction, with $M+2$ and $N+2$ lines, respectively, i.e., we have $(M+2)^2\,(N+2)^2$ grid points, including the boundary points, which is the same as $M^2\,N^2$ interior points; see Figure 11.10.

Let us indicate the approximation of $u(j\Delta x, k\Delta y, t) =: u(x_j, y_k, t)$ by $u_{jk}(t)$. As remarked in Chapter 9, there is a host of possibilities to order the points in the spatial grid. As before, we shall assume that they are ordered lexicographically; see Figure 11.10. We then define the solution vector as

$$\mathbf{u}(t) := \big(u_{11}(t), \ldots, u_{M1}(t) \,|\, u_{12}(t), \ldots, u_{M2}(t) \,|\, \cdots \,|\, u_{1N}(t), \ldots, u_{MN}(t)\big)^T.$$

This leads via the MOL to the ODE system

$$\frac{\mathrm{d}\mathbf{u}}{\mathrm{d}t} = \mathbf{A}\mathbf{u} + \mathbf{f}(t), \tag{11.111a}$$

where the matrix $\mathbf{A}$ is defined by

$$\mathbf{A} := \left(\begin{array}{ccccc|ccccc|ccccc|ccccc}
\alpha & \beta & & & & \gamma & & & & & & & & & & & & & & \\
\beta & \alpha & \beta & & & & \gamma & & & & & & & & & & & & & \\
 & \ddots & \ddots & \ddots & & & & \ddots & & & & & & & & & & & & \\
 & & \beta & \alpha & \beta & & & & \gamma & & & & & & & & & & & \\
 & & & \beta & \alpha & & & & & \gamma & & & & & & & & & & \\
\hline
\gamma & & & & & \alpha & \beta & & & & \gamma & & & & & & & & & \\
 & \gamma & & & & \beta & \alpha & \beta & & & & \gamma & & & & & & & & \\
 & & \ddots & & & & \ddots & \ddots & \ddots & & & & \ddots & & & & & & & \\
 & & & \gamma & & & & \beta & \alpha & \beta & & & & \gamma & & & & & & \\
 & & & & \gamma & & & & \beta & \alpha & & & & & \gamma & & & & & \\
\hline
 & & & & & \ddots & & & & & \ddots & \ddots & & & & \ddots & & & & \\
 & & & & & & \ddots & & & & \ddots & \ddots & \ddots & & & & \ddots & & & \\
 & & & & & & & \ddots & & & & \ddots & \ddots & \ddots & & & & \ddots & & \\
 & & & & & & & & \ddots & & & & \ddots & \ddots & \ddots & & & & \ddots & \\
 & & & & & & & & & \ddots & & & & \ddots & \ddots & & & & & \ddots \\
\hline
 & & & & & & & & & & \gamma & & & & & \alpha & \beta & & & \\
 & & & & & & & & & & & \gamma & & & & \beta & \alpha & \beta & & \\
 & & & & & & & & & & & & \ddots & & & & \ddots & \ddots & \ddots & \\
 & & & & & & & & & & & & & \gamma & & & & \beta & \alpha & \beta \\
 & & & & & & & & & & & & & & \gamma & & & & \beta & \alpha
\end{array}\right) \tag{11.111b}$$

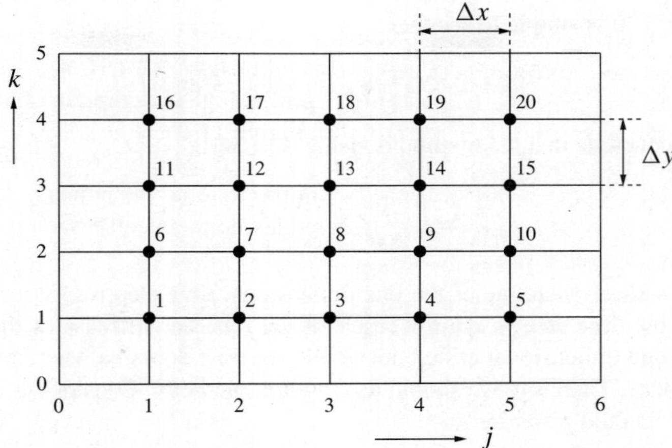

Figure 11.10. *A two-dimensional spatial grid for initial boundary value problem* (11.110) *with $M = 5$ and $N = 4$.*

and $\mathbf{f}(t)$ contains both the source term and contributions from the boundary points. In (11.111b) we have introduced the shorthand notation

$$\alpha := -\frac{2}{\Delta x^2} - \frac{2}{\Delta y^2}, \quad \beta := \frac{1}{\Delta x^2}, \quad \gamma := \frac{1}{\Delta y^2}. \tag{11.111c}$$

As indicated in (11.111b), $\mathbf{A}$ is an $N \times N$ block matrix, each block being an $M \times M$ matrix. Let us have a look at the explicit Euler method first. If u_{jk}^n denotes the numerical approximation of $u(j\Delta x, k\Delta y, n\Delta t) =: u(x_j, y_k, t^n)$ and if we introduce $f_{jk}^n := f(x_j, y_k, t^n)$, we obtain

$$u_{jk}^{n+1} = u_{jk}^n + d_x\big(u_{j+1,k}^n - 2u_{j,k}^n + u_{j-1,k}^n\big) + d_y\big(u_{j,k+1}^n - 2u_{j,k}^n + u_{j,k-1}^n\big) + \Delta t f_{jk}^n, \tag{11.112}$$

where the diffusion numbers d_x and d_y are defined by

$$d_x := \frac{\Delta t}{\Delta x^2}, \quad d_y := \frac{\Delta t}{\Delta y^2}. \tag{11.113}$$

Defining the numerical solution vector as

$$\mathbf{u}^n := \big(u_{1,1}^n, \ldots, u_{M,1}^n \,|\, u_{1,2}^n, \ldots, u_{M,2}^n \,|\cdots|\, u_{1,N}^n, \ldots, u_{M,N}^n\big)^T,$$

we obtain the following matrix formulation for (11.112):

$$\mathbf{u}^{n+1} = \big(\mathbf{I} + \Delta t\mathbf{A}\big)\mathbf{u}^n + \Delta t\mathbf{f}^n, \tag{11.114}$$

with $\mathbf{f}^n := \mathbf{f}(t^n)$. It is simple to see that

$$\left\|\mathbf{I} + \Delta t\mathbf{A}\right\|_\infty = |1 + \alpha\Delta t| + |\alpha\Delta t|.$$

We therefore conclude that the method is stable if $|\alpha\Delta t| \leq \frac{1}{2}$, i.e.,

$$\Delta t\left(\frac{1}{\Delta x^2} + \frac{1}{\Delta y^2}\right) \leq \frac{1}{2}.$$

This result is a sheer analogue of the one-dimensional time step restriction (11.14). This restriction of the time step is again a result of the inherent stiffness of the problem and is, like in the one-dimensional case, undesirable for time domains where the solution u is relatively smooth. Therefore we should use implicit methods. Consider as an example the implicit Euler method given by

$$u_{j,k}^{n+1} - d_x\left(u_{j+1,k}^{n+1} - 2u_{j,k}^{n+1} + u_{j-1,k}^{n+1}\right) - d_y\left(u_{j,k+1}^{n+1} - 2u_{j,k}^{n+1} + u_{j,k-1}^{n+1}\right) = u_{j,k}^n + \Delta t f_{jk}^{n+1}. \tag{11.115}$$

Written as a system this is

$$\left(\mathbf{I} - \Delta t\mathbf{A}\right)\mathbf{u}^{n+1} = \mathbf{u}^n + \Delta t\mathbf{f}^{n+1}. \tag{11.116}$$

This system is less trivial to solve than the one-dimensional case. The sparsity structure, i.e., five nonzero (co)diagonals, will be lost if an LU-decomposition is carried out; by this we mean that the lower triangular and the upper triangular matrices will have *fill-in*, i.e., have potential nonzeros on the first M codiagonals. Moreover, the *complexity*, i.e., the number of elementary operations, is significantly larger due to this fill-in. A further discussion of this linear algebra topic is outside the scope of this book. We refer to [52]. In the next section we shall consider an alternative that has low complexity and good stability properties.

11.8 Splitting Methods

In some situations it will turn out to be useful to make a distinction between implicit and explicit discretisations on the operator level; i.e., if one part of the operator is discretised on the new time level and another part on the old time level, some efficiency gains can be achieved. While such a "splitting" may lead to lower complexity of the computations, it may also interfere with the stability of the overall method. In Section 11.8.1 we treat methods that split the elliptic operator as such and in Section 11.8.2 we discuss methods that deal with advection and diffusion operators separately.

11.8.1 The ADI Method

In order to lower the complexity needed for solving (11.116), we may try to take only the x derivative implicitly in (11.115). This would give

$$u_{j,k}^{n+1} - d_x\left(u_{j+1,k}^{n+1} - 2u_{j,k}^{n+1} + u_{j-1,k}^{n+1}\right) = u_{j,k}^{n} + d_y\left(u_{j,k+1}^{n} - 2u_{j,k}^{n} + u_{j,k-1}^{n}\right) + \Delta t f_{jk}^{*}. \tag{11.117}$$

The source term f_{jk}^{*} may be evaluated at some combination of the time points t^n and t^{n+1}. Introducing the splitting $\mathbf{A} = \mathbf{A}_x + \mathbf{A}_y$, where the matrices $\mathbf{A}_x$ and $\mathbf{A}_y$ are defined by

$$\mathbf{A}_x := \frac{1}{\Delta x^2}\left(\begin{array}{c|c|c|c}
\begin{matrix} -2 & 1 & & & \\ 1 & -2 & 1 & & \\ & \ddots & \ddots & \ddots & \\ & & 1 & -2 & 1 \\ & & & 1 & -2 \end{matrix} & & & \\ \hline
 & \begin{matrix} -2 & 1 & & & \\ 1 & -2 & 1 & & \\ & \ddots & \ddots & \ddots & \\ & & 1 & -2 & 1 \\ & & & 1 & -2 \end{matrix} & & \\ \hline
 & & \begin{matrix} \ddots & \ddots & & & \\ \ddots & \ddots & \ddots & & \\ & \ddots & \ddots & \ddots & \\ & & \ddots & \ddots & \ddots \\ & & & \ddots & \ddots \end{matrix} & \\ \hline
 & & & \begin{matrix} -2 & 1 & & & \\ 1 & -2 & 1 & & \\ & \ddots & \ddots & \ddots & \\ & & 1 & -2 & 1 \\ & & & 1 & -2 \end{matrix}
\end{array}\right), \tag{11.118a}$$

$$
\mathbf{A}_y := \frac{1}{\Delta y^2}\left(\begin{array}{ccccc|ccccc|ccccc|ccccc}
-2 & & & & & 1 & & & & & & & & & & & & & & \\
& -2 & & & & & 1 & & & & & & & & & & & & & \\
& & \ddots & & & & & \ddots & & & & & & & & & & & & \\
& & & -2 & & & & & 1 & & & & & & & & & & & \\
& & & & -2 & & & & & 1 & & & & & & & & & & \\
\hline
1 & & & & & -2 & & & & & 1 & & & & & & & & & \\
& 1 & & & & & -2 & & & & & 1 & & & & & & & & \\
& & \ddots & & & & & \ddots & & & & & \ddots & & & & & & & \\
& & & 1 & & & & & -2 & & & & & 1 & & & & & & \\
& & & & 1 & & & & & -2 & & & & & 1 & & & & & \\
\hline
& & & & & \ddots & & & & & \ddots & & & & & \ddots & & & & \\
& & & & & & \ddots & & & & & \ddots & & & & & \ddots & & & \\
& & & & & & & \ddots & & & & & \ddots & & & & & \ddots & & \\
& & & & & & & & \ddots & & & & & \ddots & & & & & \ddots & \\
& & & & & & & & & \ddots & & & & & \ddots & & & & & \ddots \\
\hline
& & & & & & & & & & 1 & & & & & -2 & & & & \\
& & & & & & & & & & & 1 & & & & & -2 & & & \\
& & & & & & & & & & & & \ddots & & & & & \ddots & & \\
& & & & & & & & & & & & & 1 & & & & & -2 & \\
& & & & & & & & & & & & & & 1 & & & & & -2
\end{array}\right), \tag{11.118b}
$$

we can write (11.117) as

$$
\big(\mathbf{I} - \Delta t \mathbf{A}_x\big)\mathbf{u}^{n+1} = \big(\mathbf{I} + \Delta t \mathbf{A}_y\big)\mathbf{u}^n + \Delta t \mathbf{f}_x^*. \tag{11.119}
$$

Note that the coefficient matrix $\mathbf{I} - \Delta t\mathbf{A}_x$ is tridiagonal again and hence solving the system (11.119) is inexpensive (cf. the one-dimensional case).

Let us now investigate the stability of this system, i.e., $\|(\mathbf{I} - \Delta t\mathbf{A}_x)^{-1}(\mathbf{I} + \Delta t\mathbf{A}_y)\|$. We choose the 2-norm again. Recall the transformation matrix $\mathbf{W}$ in (11.33) that brings the one-dimensional discretisation matrix in (11.30b) into diagonal form (11.35). We now build a larger matrix, composed of such transformation matrices, $\tilde{\mathbf{W}}$ say, defined by

$$
\tilde{\mathbf{W}} := \begin{pmatrix} \mathbf{W} & & & \\ & \mathbf{W} & & \\ & & \ddots & \\ & & & \mathbf{W} \end{pmatrix}, \tag{11.120}
$$

and thus transform $\mathbf{I} - \Delta t \mathbf{A}_x$ as follows:

$$\tilde{\mathbf{W}}^{-1}(\mathbf{I} - \Delta t \mathbf{A}_x)\tilde{\mathbf{W}} = \begin{pmatrix} \mathbf{\Lambda}_x & & & \\ & \mathbf{\Lambda}_x & & \\ & & \ddots & \\ & & & \mathbf{\Lambda}_x \end{pmatrix}, \tag{11.121}$$

where $\mathbf{\Lambda}_x$ is defined by

$$\mathbf{\Lambda}_x := \mathbf{I} - \Delta t \mathbf{\Lambda}; \tag{11.122}$$

cf. (11.35). We now obtain

$$\tilde{\mathbf{W}}^{-1}(\mathbf{I} - \Delta t \mathbf{A}_x)^{-1}\tilde{\mathbf{W}}\tilde{\mathbf{W}}^{-1}(\mathbf{I} + \Delta t \mathbf{A}_y)\tilde{\mathbf{W}} = \begin{pmatrix} \mathbf{\Lambda}_x^{-1} & & & \\ & \mathbf{\Lambda}_x^{-1} & & \\ & & \ddots & \\ & & & \mathbf{\Lambda}_x^{-1} \end{pmatrix} (\mathbf{I} + \Delta t \mathbf{A}_y), \tag{11.123}$$

which follows from the simple structure of the matrix $\mathbf{A}_y$. Next we can bring the split matrix $\mathbf{A}_y$ into tridiagonal form, like $\mathbf{A}_x$, if we renumber the unknowns along vertical grid lines. This renumbering corresponds to a permutation of the rows and columns of $\mathbf{A}_y$, and we can formally define a tridiagonal matrix $\mathbf{B}_y$ by

$$\mathbf{B}_y := \tilde{\mathbf{P}}^{-1}\mathbf{A}_y\tilde{\mathbf{P}},$$

where $\tilde{\mathbf{P}}$ is the corresponding permutation matrix. Applying the same permutations to the matrix in (11.123), we obtain

$$\begin{aligned} &\tilde{\mathbf{P}}^{-1}\tilde{\mathbf{W}}^{-1}(\mathbf{I} - \Delta t \mathbf{A}_x)^{-1}\tilde{\mathbf{W}}\tilde{\mathbf{P}}\tilde{\mathbf{P}}^{-1}\tilde{\mathbf{W}}^{-1}(\mathbf{I} + \Delta t \mathbf{A}_y)\tilde{\mathbf{W}}\tilde{\mathbf{P}} \\ &\qquad = \tilde{\mathbf{P}}^{-1} \begin{pmatrix} \mathbf{\Lambda}_x^{-1} & & & \\ & \mathbf{\Lambda}_x^{-1} & & \\ & & \ddots & \\ & & & \mathbf{\Lambda}_x^{-1} \end{pmatrix} \tilde{\mathbf{P}}(\mathbf{I} + \Delta t \mathbf{B}_y). \end{aligned} \tag{11.124}$$

It is simple to see then that $\tilde{\mathbf{P}}$ transforms the matrix with the $\mathbf{\Lambda}_x^{-1}$ blocks into another diagonal matrix:

$$\begin{pmatrix} (1 - \Delta t \lambda_1)\mathbf{I} & & & \\ & (1 - \Delta t \lambda_2)\mathbf{I} & & \\ & & \ddots & \\ & & & (1 - \Delta t \lambda_M)\mathbf{I} \end{pmatrix}.$$

Since the 2-norm of a matrix does not change when pre- or postmultiplied by orthogonal (permutation) matrices, it is easy to see that

$$\left\| (\mathbf{I} - \Delta t \mathbf{A}_x)^{-1}(\mathbf{I} + \Delta t \mathbf{A}_y) \right\|_2 = \max_{\substack{1 \le j \le M \\ 1 \le k \le N}} \left| \frac{1 - 4d_y \sin^2(\frac{\pi}{2} y_k)}{1 + 4d_x \sin^2(\frac{\pi}{2} x_j)} \right|. \tag{11.125}$$

We conclude that we end up with a similar stability restriction as before and that at this stage not much is gained. However, we now do a second step, thereby alternating the implicit direction, which gave the method its name *ADI*. In matrix terms the method goes as follows:

$$(\mathbf{I} - \Delta t \mathbf{A}_x)\mathbf{u}^{n+1} = (\mathbf{I} + \Delta t \mathbf{A}_y)\mathbf{u}^n + \Delta t \mathbf{f}_x^*, \tag{11.126a}$$

$$(\mathbf{I} - \Delta t \mathbf{A}_y)\mathbf{u}^{n+2} = (\mathbf{I} + \Delta t \mathbf{A}_x)\mathbf{u}^{n+1} + \Delta t \mathbf{f}_y^*, \tag{11.126b}$$

with $\mathbf{f}_x^*$ and $\mathbf{f}_y^*$ the vectors of the source term and boundary values. For stability we now investigate the matrix $\mathbf{K}$ and its similar form $\mathbf{L}$, defined by

$$\mathbf{K} := (\mathbf{I} - \Delta t \mathbf{A}_y)^{-1}(\mathbf{I} + \Delta t \mathbf{A}_x)(\mathbf{I} - \Delta t \mathbf{A}_x)^{-1}(\mathbf{I} + \Delta t \mathbf{A}_y), \tag{11.127a}$$

$$\mathbf{L} := (\mathbf{I} - \Delta t \mathbf{A}_y)\mathbf{K}(\mathbf{I} - \Delta t \mathbf{A}_y)^{-1}. \tag{11.127b}$$

Actually, we need to transform $\mathbf{L}$ first, like we did above in (11.124), and work with the matrix $\tilde{\mathbf{L}}$, say. Then we easily obtain

$$\|\mathbf{L}\|_2 = \|\tilde{\mathbf{L}}\|_2 = \max_{\substack{1 \le j \le M \\ 1 \le k \le N}} \left| \frac{1 - 4d_y \sin^2(\frac{\pi}{2} y_k)}{1 + 4d_x \sin^2(\frac{\pi}{2} x_j)} \frac{1 - 4d_x \sin^2(\frac{\pi}{2} x_j)}{1 + 4d_y \sin^2(\frac{\pi}{2} y_k)} \right|, \tag{11.128}$$

which is obviously bounded by one, independent of the grid sizes. Although $\mathbf{L}$ may be a skew transformation of $\mathbf{K}$, this does not affect the stability as such (one can give a sharper estimate of the increment matrices, which we shall omit here).

There is another way to look at this process. Consider the matrix ODE (11.111a). As discussed for the one-dimensional case, the homogeneous solution of this problem has an increment matrix

$$\exp(2\Delta t \mathbf{A})$$

relating the solution at time levels t^n and t^{n+2}. Since $\mathbf{A} = \mathbf{A}_x + \mathbf{A}_y$ and the matrices $\mathbf{A}_x$ and $\mathbf{A}_y$ commute, i.e., $\mathbf{A}_x\mathbf{A}_y = \mathbf{A}_y\mathbf{A}_x$, we can split the increment matrix as follows:

$$\begin{aligned}
\exp(2\Delta t \mathbf{A}) &= \exp(2\Delta t \mathbf{A}_x + 2\Delta t \mathbf{A}_y) \\
&= \exp(\Delta t \mathbf{A}_y + \Delta t \mathbf{A}_x + \Delta t \mathbf{A}_x + \Delta t \mathbf{A}_y) \\
&= \exp(\Delta t \mathbf{A}_y)\exp(\Delta t \mathbf{A}_x)\exp(\Delta t \mathbf{A}_x)\exp(\Delta t \mathbf{A}_y).
\end{aligned} \tag{11.129}$$

ADI can thus be interpreted as taking an implicit Euler approximation for the first and third factors in this product and an explicit Euler approximation for the second and fourth factors. There are many other splitting methods, such as local one-dimensional methods; see, e.g., [96].

In (9.67) we have already encountered ADI. For the solution of the linear system $\mathbf{A}\mathbf{u} + \mathbf{f} = \mathbf{0}$ it reads

$$\begin{aligned}
(\mathbf{I} - \tau \mathbf{A}_y)\mathbf{u}^{l+1} &= (\mathbf{I} + \tau \mathbf{A}_x)\,\mathbf{u}^l + \tau \mathbf{f}, \\
(\mathbf{I} - \tau \mathbf{A}_x)\mathbf{u}^{l+2} &= (\mathbf{I} + \tau \mathbf{A}_y)\mathbf{u}^{l+1} + \tau \mathbf{f},
\end{aligned}$$

where τ is a time-step-like parameter that still needs to be determined. These recursions have the same form as (11.126) if we take $\Delta t = \tau$. In order to estimate the convergence rate of ADI as an iterative method for solving Poisson problems, we use the bound for the increment matrix (11.128) and work this out a bit. First we note that if

$$\alpha \leq d\lambda_k \leq \beta \qquad (k = 1, 2, \ldots, M),$$

an upper bound for (11.128) (with $M = N$ and $d_x = d_y = d$) is given by

$$\max_{\alpha,\beta} \left(\left| \frac{1 - \tau\alpha}{1 + \tau\alpha} \right|, \left| \frac{1 - \tau\beta}{1 + \tau\beta} \right| \right).$$

Clearly this is minimal if both factors are equal. Hence τ should satisfy

$$\tau^2 - \frac{1}{\alpha\beta} = 0.$$

So we obtain the estimate

$$\tau = \frac{1}{\sqrt{\alpha\beta}} \doteq \frac{1}{2}\Delta x. \tag{11.130}$$

Using this value for $d = \tau/\Delta x^2$ in (11.128), we find that the convergence factor is approximately equal to $(1 - \sin(\pi\Delta x))/(1 + \sin(\pi\Delta x)) \doteq 1 - 2\pi\Delta x$.

11.8.2 Mixed Explicit/Implicit Discretisation of the Advection-Diffusion Equation

The stability requirement, forcing the time step to be taken very small or else forcing the use of an implicit method, is one of the most cumbersome problems in solving evolutionary equations. There are various ways to tackle this problem for equations including an advection term, i.e., first order spatial derivatives. So consider first the linear *advection-diffusion equation*

$$\frac{\partial u}{\partial t} + b\frac{\partial u}{\partial x} = D\frac{\partial^2 u}{\partial x^2}, \tag{11.131}$$

where b and $D > 0$ are the advection velocity and diffusion coefficient, respectively. Applying the MOL to this equation, we obtain an ODE system of the form

$$\frac{\mathrm{d}\mathbf{u}}{\mathrm{d}t} = \mathbf{A}_{\mathrm{d}}\mathbf{u} - \mathbf{A}_{\mathrm{a}}\mathbf{u} + \mathbf{f}, \tag{11.132}$$

where $\mathbf{A}_{\mathrm{d}}\mathbf{u}$ originates from discretising the diffusion term $D\frac{\partial^2}{\partial x^2}u$ and $\mathbf{A}_{\mathrm{a}}\mathbf{u}$ comes from discretising the advection term $b\frac{\partial}{\partial x}u$. For the diffusion term we usually employ central differences, whereas for the advection term we can choose either central or one-sided differences. In Chapter 13 we will discuss this matter in detail.

Since the discrete diffusion operator $\mathbf{A}_{\mathrm{d}}\mathbf{u}$ causes the problem to be very stiff on a fine spatial grid, we decide to use the implicit Euler method for time integration. This way, we obtain the linear system

$$\big(\mathbf{I} + \Delta t\big(\mathbf{A}_{\mathrm{a}} - \mathbf{A}_{\mathrm{d}}\big)\big)\mathbf{u}^{n+1} = \mathbf{u}^n + \Delta t\mathbf{f}^{n+1}. \tag{11.133}$$

The coefficient matrix $\mathbf{I} + \Delta t(\mathbf{A}_\mathrm{a} - \mathbf{A}_\mathrm{d})$ is tridiagonal so that (11.133) can be very efficiently solved at each time step. However, the symmetry is lost due to the term $\mathbf{A}_\mathrm{a}$. This is not a problem for equations in one space dimension. On the other hand, for two- or three-dimensional problems we have to solve (11.133) iteratively, which is often considerably more expensive than solving a symmetric system by an iterative procedure. Based on these considerations, we might take the advection term explicitly, leading to the system

$$\big(\mathbf{I} - \Delta t\mathbf{A}_\mathrm{d}\big)\mathbf{u}^{n+1} = \big(\mathbf{I} - \Delta t\mathbf{A}_\mathrm{a}\big)\mathbf{u}^{n} + \Delta t\mathbf{f}^{\,n+1}. \tag{11.134}$$

Concerning stability, we will show in Chapter 13 that the explicit advection term leads to a favorable time step restriction of the form $\Delta t/\Delta x \leq C$ for some (small) constant $C > 0$. Other combinations of implicit discretisations of the diffusion term and explicit discretisations of the advection term are of course possible. We leave this as an exercise.

The need to take the advection term explicitly becomes more urgent if the advection term is nonlinear. We will demonstrate this for the nonlinear advection-diffusion equation

$$\frac{\partial u}{\partial t} + b(u)\frac{\partial u}{\partial x} = D\frac{\partial^2 u}{\partial x^2}, \tag{11.135}$$

where the advection velocity $b(u)$ is now a (nonlinear) function of u. After space discretisation, we obtain the ODE system

$$\frac{\mathrm{d}\mathbf{u}}{\mathrm{d}t} = \mathbf{A}_\mathrm{d}\mathbf{u} - \mathbf{A}_\mathrm{a}(\mathbf{u})\mathbf{u} + \mathbf{f}, \tag{11.136}$$

where the discretisation matrix $\mathbf{A}_\mathrm{a}(\mathbf{u})$ for the advection term depends on $\mathbf{u}$. Once more applying the implicit Euler method, we find the nonlinear system

$$\big(\mathbf{I} - \Delta t\mathbf{A}_\mathrm{d}\big)\mathbf{u}^{n+1} + \Delta t\mathbf{A}_\mathrm{a}(\mathbf{u}^{n+1})\mathbf{u}^{n+1} = \mathbf{u}^{n} + \Delta t\mathbf{f}^{\,n+1}, \tag{11.137}$$

which we have to solve using a Newton-type method. We can circumvent this nonlinear system by using $\mathbf{A}_\mathrm{a}(\mathbf{u}^n)$ instead of $\mathbf{A}_\mathrm{a}(\mathbf{u}^{n+1})$. This way, we obtain the linear system

$$\big(\mathbf{I} + \Delta t\big(\mathbf{A}_\mathrm{a}(\mathbf{u}^n) - \mathbf{A}_\mathrm{d}\big)\big)\mathbf{u}^{n+1} = \mathbf{u}^{n} + \Delta t\mathbf{f}^{\,n+1}. \tag{11.138}$$

This method has comparable stability properties to (11.137), whereas it only requires the solution of a linear system at each time step.

11.9 Solving Nonlinear Problems

Nonlinear parabolic equations are in principle not much different from linear ones. In order to solve them, one can often employ the same discretisations as in the linear case. The main difficulty is the solution of the nonlinear algebraic system resulting from an implicit time-stepping method. Yet, or rather because of this, one may take recourse to special methods that alleviate the burden of nonlinearity.

To start with let us investigate stability for a problem with nonlinear diffusion. So consider the equation

$$\frac{\partial u}{\partial t} = \frac{\partial}{\partial x}\left(D(u)\frac{\partial u}{\partial x}\right) + f(x,t), \qquad x \in (0,1), \quad t > 0, \tag{11.139}$$

where the diffusion coefficient $D(u) > 0$ is a function of u. We can find a difference scheme that has a symmetric form for the spatial part of the equation; see also (5.32b), where this was derived for the linear case. If we define

$$D_j^n := D\big(u(x_j, t^n)\big), \qquad D_{j+1/2}^n := \frac{1}{2}(D_j^n + D_{j+1}^n),$$

we have, e.g., with the explicit Euler method, that

$$u_j^{n+1} = u_j^n + d\big(D_{j+1/2}^n(u_{j+1}^n - u_j^n) - D_{j-1/2}^n(u_j^n - u_{j-1}^n)\big) + \Delta t f_j^n. \tag{11.140}$$

By "freezing" the coefficients, we can investigate stability for a particular value of D. Clearly, we have $\tilde{d} := D\Delta t/\Delta x^2 \leq \frac{1}{2}$. By requiring this to hold for all values of D that are relevant for the problem, we can find an intuitively reasonable bound for the time step.

Example 11.8 If the value of D in (11.139) is given by $D(u) := 1 + \mathrm{e}^{-u}$, then $1 < D < 2$ for $u > 0$. In this case stability requires that $\Delta t \leq 1/(4\Delta x^2)$. □

From the foregoing it again follows that one may need a fairly small time step. However, if we want to overcome this by using an implicit method, there is a big problem since the resulting system is nonlinear. For moderate dependencies of D on u we can simply use the values of u found at previous time levels. This gives the following implicit alternative for (11.140):

$$u_j^{n+1} = u_j^n + d\Big(D_{j+1/2}^n(u_{j+1}^{n+1} - u_j^{n+1}) - D_{j-1/2}^n(u_j^{n+1} - u_{j-1}^{n+1})\Big) + \Delta t f_j^n. \tag{11.141}$$

Finally, let us consider problems with a nonlinear source term, i.e.,

$$\frac{\partial u}{\partial t} = \frac{\partial^2 u}{\partial x^2} + f(u) \qquad x \in (0, 1), \quad t > 0. \tag{11.142}$$

In Section 10.6.2 we encountered Fisher's equation as an example of this type. The treatment of this term follows the same principles as before. If f is not strongly dependent on u (i.e., $|f'(u)|$ is not large), then we can again simply take the values at the previous time step, so

$$u_j^{n+1} = u_j^n + \frac{\Delta t}{\Delta x^2}(u_{j+1}^n - 2u_j^n + u_{j-1}^n) + \Delta t f\big(u_j^n\big). \tag{11.143}$$

However, if $|f'(u)|$ is large, this may require very small time steps again, as demonstrated in the next example.

Example 11.9 Consider the problem

$$\frac{\partial u}{\partial t} = \frac{\partial^2 u}{\partial x^2} + \exp\left(-\frac{1}{\varepsilon}u\right), \qquad x \in (0, 1), \quad t > 0,$$

where ε is a small positive number. Such equations typically occur in chemical kinetics (Arrhenius law), where ε is something like the reaction temperature. Clearly, $f'(u) = -f(u)/\varepsilon$. For (11.143) to be stable an intuitively reasonable guess is

$$\Delta t\left(\frac{1}{\Delta x^2} + \frac{1}{\varepsilon}\max_u \exp\left(-\frac{1}{\varepsilon}u\right)\right) \leq 1,$$

where the maximum should be taken over the actually attained values. If ε is much smaller than Δx^2, this is obviously a strong time step restriction indeed. □

For very stiff problems, as in Example 11.9, we have to use an implicit method for the source term. In fact, if we use an explicit time-stepping method for the diffusion term, we have to solve the following nonlinear system for $\mathbf{y} := \mathbf{u}^{n+1}$:

$$\mathbf{F}(\mathbf{y}) := \mathbf{y} - \Delta t \mathbf{f}(\mathbf{y}) - \mathbf{u}^n - \Delta t \big(\mathbf{A}\mathbf{u}^n + \mathbf{b}^n\big) = \mathbf{0}, \tag{11.144}$$

where $\mathbf{b}^n$ is a term coming from the boundary conditions and $\mathbf{A}$ is the matrix in (11.30b). In order to find the solution of (11.144), we can use Newton's method, which produces a sequence $\mathbf{y}^{(1)}, \mathbf{y}^{(2)}, \ldots$, with $\mathbf{y}^{(l)} \to \mathbf{u}^{n+1}$, satisfying

$$\mathbf{y}^{(l+1)} = \mathbf{y}^{(l)} - \mathbf{J}^{-1}\big(\mathbf{y}^{(l)}\big)\mathbf{F}\big(\mathbf{y}^{(l)}\big). \tag{11.145a}$$

Here $\mathbf{J}(\mathbf{y})$ is the Jacobi matrix given by

$$\mathbf{J}(\mathbf{y}) = \mathbf{I} - \Delta t \,\mathrm{diag}(f'(y_j)). \tag{11.145b}$$

A suitable initial guess for (11.145) is $\mathbf{y}^{(0)} = \mathbf{u}^n$, provided that the time step Δt is small enough.

11.10 Stefan Problems

In Section 10.5 we encountered parabolic problems with moving boundaries or interfaces. In this section we will discuss some numerical issues, though it should be stated that we cannot cover this vast area in any detail in this book. Let us start with the following simple one-phase problem (cf. (10.43) to (10.45)):

$$\frac{\partial u}{\partial t} = \frac{\partial^2 u}{\partial x^2}, \qquad x \in (0, S(t)), \quad t > 0, \tag{11.146a}$$

$$u(0,t) - \beta \frac{\partial u}{\partial x}(0,t) = 1, \qquad t > 0, \tag{11.146b}$$

$$u(S(t),t) = 0, \qquad t > 0, \tag{11.146c}$$

$$\frac{\partial u}{\partial x}(S(t),t) = -\alpha \frac{\mathrm{d}S}{\mathrm{d}t}, \qquad S(0) = 0, \qquad t > 0. \tag{11.146d}$$

In the following we will employ a fixed spatial grid. The points to the right of the interface $x = S(t)$ are redundant and may be discarded. During the computation new grid points are added if the boundary passes by one or a few grid points. Typically, we have a grid as sketched in Figure 11.11. Here the boundary point $x = S(t^n)$ lies between the grid points $x_{\mathrm{i}(n)}$ and $x_{\mathrm{i}(n)+1}$. Let S^n be the numerical approximation of $S(t^n)$. The distance between S^n and $x_{\mathrm{i}(n)}$ is written as

$$S^n - x_{\mathrm{i}(n)} =: r^n \Delta x, \qquad 0 < r^n < 1.$$

We want to apply the explicit Euler time integration method to (11.146). Applying the explicit Euler central difference scheme (11.4) to (11.146a) in any grid point not neighbouring the boundary S^n is trivial. However, for the grid point $(x_{\mathrm{i}(n)}, t^n)$ the scheme

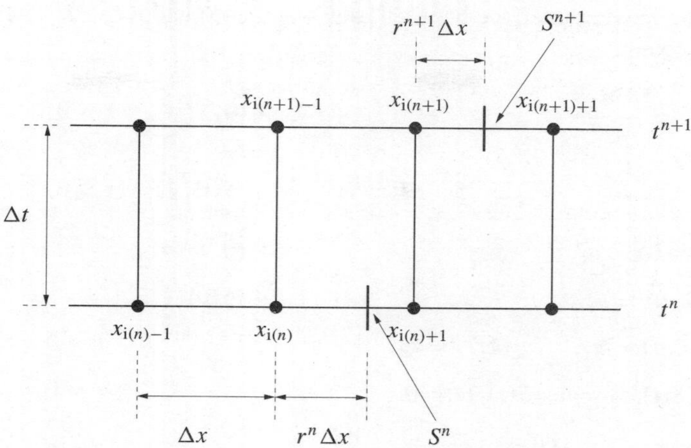

Figure 11.11. *Patch of the grid with a moving interface* $x = S(t)$.

is not standard. Discretising the spatial derivative in (11.146a) by the central difference approximation at $(x_{i(n)-1}, t^n)$, $(x_{i)(n)}, t^n)$, and (S^n, t^n), we obtain (cf. (5.30))

$$\frac{\partial^2 u}{\partial x^2}(S^n, t^n) \doteq \frac{2}{\Delta x^2}\left(\frac{1}{1+r^n}u\big(x_{i(n)-1}, t^n\big) - \frac{1}{r^n}u\big(x_{i(n)}, t^n\big)\right), \tag{11.147}$$

where we have used the condition $u(S^n, t^n) = 0$ because of (11.146c). The explicit Euler scheme in $(x_{i(n)}, t^n)$ then reads

$$u_{i(n)}^{n+1} = u_{i(n)}^n + 2d\left(\frac{1}{1+r^n}u_{i(n)-1}^n - \frac{1}{r^n}u_{i(n)}^n\right). \tag{11.148}$$

The next step is to use condition (11.146d). In order to find an approximation of $\frac{\partial}{\partial x}u(S(t^n), t^n)$, we employ the derivative of the second order Lagrangian interpolation polynomial through the grid points $(x_{i(n)-1}, t^n)$, $(x_{i(n)}, t^n)$, and (S^n, t^n); see Example 5.2. Given the values of u in these points, we find

$$\frac{\partial u}{\partial x}(S(t^n), t^n) \doteq \frac{1}{\Delta x}\left(\frac{r^n}{1+r^n}u\big(x_{i(n)-1}, t^n\big) - \frac{1+r^n}{r^n}u\big(x_{i(n)}, t^n\big)\right). \tag{11.149}$$

Hence the new interface location is given by

$$S^{n+1} = S^n - \frac{\Delta t}{\alpha\Delta x}\left(\frac{r^n}{1+r^n}u_{i(n)-1}^n - \frac{1+r^n}{r^n}u_{i(n)}^n\right). \tag{11.150}$$

Finally, note that (11.146) has mixed boundary conditions. However, this problem has already been addressed in Section 11.6.

We have to use implicit methods to overcome the inherent stiffness resulting from discretising the term $\frac{\partial^2}{\partial x^2}u$. We now demonstrate how to do this for a two-phase problem, describing the moving interface between, say, water (wa) and ice (ic). Thus consider the following problem:

$$\frac{\partial u_{\rm wa}}{\partial t} = \frac{\partial^2 u_{\rm wa}}{\partial x^2}, \qquad x \in (0, S(t)), \quad t > 0, \tag{11.151a}$$

$$\frac{\partial u_{\rm ic}}{\partial t} = \frac{\partial^2 u_{\rm ic}}{\partial x^2}, \qquad x \in (S(t), 1), \quad t > 0, \tag{11.151b}$$

$$u_{\rm wa}(x, 0) = u_\ell, \qquad x \in (0, S(0)), \tag{11.151c}$$

$$u_{\rm ic}(x, 0) = u_{\rm r}, \qquad x \in (S(0), 1), \tag{11.151d}$$

$$u_{\rm wa}(0, t) = u_\ell, \quad u_{\rm ic}(1, t) = u_{\rm r}, \qquad t > 0, \tag{11.151e}$$

$$u_{\rm wa}(S(t), t) = u_{\rm ic}(S(t), t) = 0, \qquad t > 0, \tag{11.151f}$$

$$\frac{\partial u_{\rm ic}}{\partial x}(S(t), t) - \frac{\partial u_{\rm wa}}{\partial x}(S(t), t) = \alpha \frac{{\rm d}S}{{\rm d}t}, \qquad t > 0, \tag{11.151g}$$

with $u_\ell > 0 > u_{\rm r}$. Since we have to deal with a moving interface, the transversal MOL is a suitable approach; i.e., we discretise all time derivatives first. Suppose we employ the implicit Euler method. This way, for (11.151a) and (11.151b), we have

$$u_{\rm wa}^{n+1} - \Delta t \frac{{\rm d}^2}{{\rm d}x^2} u_{\rm wa}^{n+1} = u_{\rm wa}^n, \quad x \in (0, S^{n+1}), \tag{11.152a}$$

$$u_{\rm ic}^{n+1} - \Delta t \frac{{\rm d}^2}{{\rm d}x^2} u_{\rm ic}^{n+1} = u_{\rm ic}^n, \quad x \in (S^{n+1}, 1), \tag{11.152b}$$

and, for the Stefan condition (11.151g), we have

$$\frac{\rm d}{{\rm d}x} u_{\rm ic}^{n+1}(S^{n+1}) - \frac{\rm d}{{\rm d}x} u_{\rm wa}^{n+1}(S^{n+1}) = \frac{\alpha}{\Delta t}\big(S^{n+1} - S^n\big). \tag{11.153}$$

Note that (11.152) and (11.153) are coupled, since the numerical interface position S^{n+1} is not known.

Next we have to choose the proper space discretisation. For the second derivatives in (11.152) we take the central difference approximation. For grid points other than $(x_{{\rm i}(n+1)}, t^{n+1})$ and $(x_{{\rm i}(n+1)+1}, t^{n+1})$ this is trivial. However, for these two grid points, neighbouring the interface at $x = S^{n+1}$, we have to apply the central difference approximation on a nonuniform grid. As an example, consider the grid point $(x_{{\rm i}(n+1)+1}, t^{n+1})$ just left of the interface. In this point we have the approximation

$$\frac{{\rm d}^2}{{\rm d}x^2} u_{\rm ic}(x_{{\rm i}(n+1)+1}) \doteq \frac{2}{\Delta x^2}\left(\frac{-1}{1 - r^{n+1}} u_{\rm ic}(x_{{\rm i}(n+1)+1}) + \frac{1}{2 - r^{n+1}} u_{\rm ic}(x_{{\rm i}(n+1)+2})\right), \tag{11.154}$$

which we subsequently have to combine with the implicit Euler time integration method. In the derivation of (11.154) we have assumed that $u_{\rm ic}(S^{n+1}) = 0$; cf. (11.151f). Proceeding this way, we obtain the linear systems

$$(\mathbf{I} - \Delta t \mathbf{A}_{\rm wa})\mathbf{u}_{\rm wa}^{n+1} = \mathbf{u}_{\rm wa}^n + \Delta t\, \mathbf{f}_{\rm wa}^{n+1}, \tag{11.155a}$$

$$(\mathbf{I} - \Delta t \mathbf{A}_{\rm ic})\mathbf{u}_{\rm ic}^{n+1} = \mathbf{u}_{\rm ic}^n + \Delta t\, \mathbf{f}_{\rm ic}^{n+1}, \tag{11.155b}$$

where the discretisation matrices $\mathbf{A}_{\text{wa}}$ and $\mathbf{A}_{\text{ic}}$ are given by

$$\mathbf{A}_{\text{wa}} := \frac{1}{\Delta x^2}\begin{pmatrix} -2 & 1 & & & \\ 1 & -2 & 1 & & \\ & \ddots & \ddots & \ddots & \\ & & 1 & -2 & 1 \\ & & & \dfrac{2}{1+r^{n+1}} & \dfrac{-2}{r^{n+1}} \end{pmatrix},$$

$$\mathbf{A}_{\text{ic}} := \frac{1}{\Delta x^2}\begin{pmatrix} \dfrac{-2}{1-r^{n+1}} & \dfrac{2}{2-r^{n+1}} & & & \\ 1 & -2 & 1 & & \\ & \ddots & \ddots & \ddots & \\ & & 1 & -2 & 1 \\ & & & 1 & -2 \end{pmatrix}. \tag{11.156}$$

Note that $\mathbf{A}_{\text{wa}}$ is an $(\mathrm{i}(n+1)-1) \times (\mathrm{i}(n+1)-1)$ matrix, so its dimension is still unknown. Likewise, $\mathbf{A}_{\text{ic}}$ is an $(M-1-\mathrm{i}(n+1)) \times (M-1-\mathrm{i}(n+1))$ matrix with M the total number of spatial grid points. For the two derivatives in (11.153) we use one-sided differences based on the second order Lagrangian interpolation polynomial through (S^{n+1}, t^{n+1}) and the two closest grid points on either side of the interface. For the first derivative we have, e.g.,

$$\frac{\mathrm{d}}{\mathrm{d}x}u_{\text{ic}}(S^{n+1}) \doteq \frac{1}{\Delta x}\left(\frac{2-r^{n+1}}{1-r^{n+1}}u(x_{\mathrm{i}(n+1)+1}, t^{n+1}) - \frac{1-r^{n+1}}{2-r^{n+1}}u(x_{\mathrm{i}(n+1)+2}, t^{n+1})\right), \tag{11.157}$$

where we have once more used the condition $u_{\text{ic}}(S^{n+1}) = 0$. We can obtain a similar expression for $\frac{\mathrm{d}}{\mathrm{d}x}u_{\text{wa}}(S^{n+1})$. The new location of the interface then follows from (11.153) and is given by

$$\begin{aligned} S^{n+1} = S^n - \frac{\Delta t}{\alpha \Delta x}\bigg(&-\frac{2-r^{n+1}}{1-r^{n+1}}u^{n+1}_{\text{ic},\mathrm{i}(n+1)+1} + \frac{1-r^{n+1}}{2-r^{n+1}}u^{n+1}_{\text{ic},\mathrm{i}(n+1)+2} \\ &+ \frac{r^{n+1}}{1+r^{n+1}}u^{n+1}_{\text{wa},\mathrm{i}(n+1)-1} - \frac{1+r^{n+1}}{r^{n+1}}u^{n+1}_{\text{wa},\mathrm{i}(n+1)}\bigg). \end{aligned} \tag{11.158}$$

Summarizing, we have the linear systems (11.155) together with (11.158) for the unknowns $\mathbf{p} := \mathbf{u}^{n+1}_{\text{wa}}$, $\mathbf{q} := \mathbf{u}^{n+1}_{\text{ic}}$, and $\sigma := S^{n+1}$. Furthermore, to facilitate the discussion, we introduce the shorthand notation $\mathrm{I} := \mathrm{i}(n+1)$ and $R = r^{n+1}$. An iterative solution procedure for (11.155) and (11.158) is the following:

- *initialisation*, $l = 0$:
  - Choose the initial guess $\sigma^{(l)} := S^n$ and set $\mathrm{I}^{(l)} = \mathrm{i}(n)$ and $R^{(l)} = r^n$. Compute the coefficient matrices in (11.155).

- *iteration*, $l = 1, 2, \ldots$:
  - Compute $\mathbf{p}^{(l)}$ from (11.155a).
  - Compute $\mathbf{q}^{(l)}$ from (11.155b).
  - Compute $\sigma^{(l)}$ from (11.158), determine $\mathrm{I}^{(l)}$ and $R^{(l)}$, and update the coefficient matrices in (11.155).

Note that when $\mathrm{I}^{(l)}$ changes, then the dimensions of the linear systems in (11.155) also change. We decide to stop the iteration when $|\sigma^{(l+1)} - \sigma^{(l)}| < \text{tol}$ for some tolerance tol. Typically, one or two iteration steps are sufficient to get a convergent solution, depending on the time step.

Example 11.10 We have actually computed a numerical solution of (11.151) with $u_\ell = 4$, $u_r = -1$, $\alpha = 1$, and $S(0) = 1/3$. For the grid size and time step we have chosen $\Delta x = 10^{-2}$ and $\Delta t = 2 \times 10^{-3}$. The numerical solution after 1, 5, 10, and 50 time steps is shown in Figure 11.12. Clearly, the solution tends to a steady state where derivatives $\frac{\partial}{\partial x} u_{\text{ic}}$ and $\frac{\partial}{\partial x} u_{\text{wa}}$ balance at the interface. □

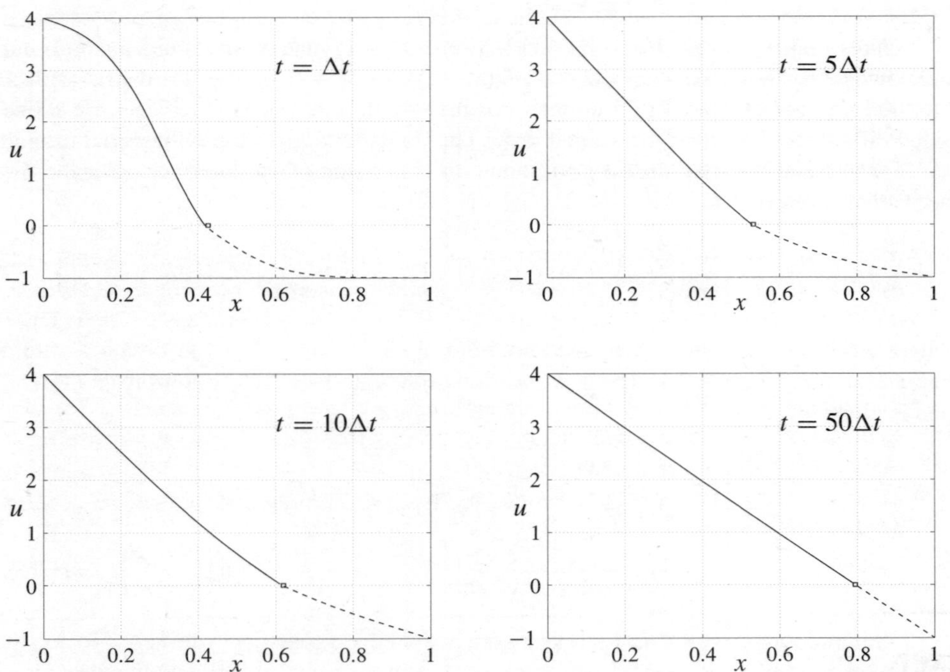

Figure 11.12. *Numerical solution of the Stefan problem* (11.151).

Another possible approach for solving (11.151) is to use a coordinate transformation. In particular, if we introduce the normalized coordinates ξ and η defined by

$$\xi(x, t) := x/S(t), \qquad \eta(x, t) := (1 - x)/(1 - S(t)), \tag{11.159}$$

we obtain the following equations instead of (11.151a) and (11.151b).

$$\frac{\partial u_{\mathrm{wa}}}{\partial t} = \frac{1}{S^2}\frac{\partial^2 u_{\mathrm{wa}}}{\partial \xi^2} + \frac{\xi}{S}\frac{\mathrm{d}S}{\mathrm{d}t}\frac{\partial u_{\mathrm{wa}}}{\partial \xi}, \qquad \xi \in (0,1), t > 0 \qquad (11.160a)$$

$$\frac{\partial u_{\mathrm{ic}}}{\partial t} = \frac{1}{(S-1)^2}\frac{\partial^2 u_{\mathrm{ic}}}{\partial \eta^2} + \frac{\eta}{S-1}\frac{\mathrm{d}S}{\mathrm{d}t}\frac{\partial u_{\mathrm{ic}}}{\partial \eta}, \qquad \xi \in (0,1), t > 0. \qquad (11.160b)$$

Note that both equations are defined on the interval (0, 1). Furthermore, for the interface condition we have

$$\frac{1}{S-1}\frac{\partial u_{\mathrm{ic}}}{\partial \eta}(1,t) - \frac{1}{S}\frac{\partial u_{\mathrm{wa}}}{\partial \xi}(1,t) = \alpha\frac{\mathrm{d}S}{\mathrm{d}t}. \qquad (11.161)$$

Analogously to the above, we can apply the implicit Euler central difference scheme to the system (11.160) and (11.161). We will not go into further detail.

11.11 Discussion

- As appears from the text, the stability of a scheme is of paramount importance when approximating the solution of a parabolic equation. Generally speaking, explicit methods require a very small time step for stability, making them prohibitively expensive. Therefore we seem to have to use implicit schemes, which do not suffer from a time step restriction. However, the effort in solving the resulting equations is often so large that explicit methods may still be competitive. This is even more the case when there is high temporal activity.

- Problems often exhibit convection as well as diffusion. Technically, problems with a large convection term are still parabolic. However, these convection-dominated problems "behave" more like hyperbolic problems, which is why we defer their discussion to Chapter 13; see Section 13.3.

- In Chapter 16 we discuss a few problems of a parabolic nature. In Section 16.1 we encounter a problem that deals with the effect of heat in the pressing of (resin) plates. A thermal explosion of materials can be found in Section 16.3, the melting of material (a free boundary problem) in laser drilling in Section 16.12, and a pulse-tube cooler in Section 16.14.

Exercises

11.1. Consider the heat equation (11.1a) defined on the domain $\Omega := [0,1] \times [0,\infty)$. Suppose that we discretise both $\frac{\partial}{\partial t}u$ and $\frac{\partial^2}{\partial x^2}u$ by central differences, so that we have the scheme

$$\frac{1}{2\Delta t}\left(u_j^{n+1} - u_j^{n-1}\right) = \frac{1}{\Delta x^2}\left(u_{j+1}^n - 2u_j^n + u_{j-1}^n\right).$$

Show that this scheme is unconditionally unstable.

11.2. Suppose we apply the second order backward difference formula (BDF2) to the ODE system (11.30a).

(a) Give the resulting scheme.

(b) Compute the local discretisation error and show that it is second order in Δt and Δx.

(c) Investigate the stability of the scheme.

11.3. Consider the problem of Section 11.6 with mixed boundary conditions.

(a) Show that on a finite time interval the requirement $\rho(\mathbf{C}) \leq 1 + \mathcal{O}(\Delta t)$ is sufficient for stability.

(b) How is the stability restriction (11.107) changed by this?

11.4. Rather than using virtual grid points to solve a problem with mixed boundary conditions, we can employ one-sided differences.

(a) Find the scheme that would result from discretising (11.88c) with the forward difference approximation and (11.88d) with the backward difference approximation.

(b) Show that the method found in part (a) is stable.

(c) If we used the backward difference approximation for (11.88c) and the forward difference approximation for (11.88d), we would find an unstable method. Prove this.

11.5. Consider the initial boundary value problem

$$\frac{\partial u}{\partial t} + b\frac{\partial u}{\partial x} = D\frac{\partial^2 u}{\partial x^2}, \qquad x \in (0,1), \quad t > 0,$$

$$u(x,0) = 0, \qquad x \in (0,1),$$

$$u(0,t) = u(1,t) = \sin t, \qquad t > 0.$$

We want to solve this problem with the explicit Euler central difference scheme. Give sufficient conditions for Δt, Δx to have a stable scheme. Hint: Use Gershgorin's theorem; see the appendix, Section I.

11.6. Consider problem (11.88) with $\alpha = 0$ in (11.88c); i.e., we have a Neumann boundary condition at $x = 0$.

(a) Show that the discretisation matrix $\mathbf{A}$ in (11.98) is nonsingular.

(b) Investigate the stability and accuracy of the ϑ method.

11.7. Consider the viscous Burgers' equation

$$\frac{\partial u}{\partial t} + \frac{\partial}{\partial x}\left(\frac{1}{2}u^2\right) = \frac{\partial^2 u}{\partial x^2}, \qquad x \in (-1,1), \quad t > 0.$$

(a) Show that a particular solution is given by

$$u(x,t) = 1 - 2\alpha \tanh\big(\alpha(x-t)\big), \qquad \alpha \in \mathbb{R}.$$

(b) Plot $u(x,t)$, taking α rather large.

(c) Compute a numerical solution, choosing the initial and boundary conditions corresponding to the given solution. Give a proper motivation for why you chose a particular method.

11.8. Consider scheme (11.141). Compute the local discretisation error and show that it is second order in Δx and first order in Δt.

11.9. The *DuFort–Frankel scheme* for the heat equation reads

$$u_j^{n+1} = u_j^{n-1} + 2d\bigl(u_{j+1}^n - \bigl(u_j^{n+1} + u_j^{n-1}\bigr) + u_{j-1}^n\bigr).$$

(a) Obviously this is a two-step method (three time levels involved). What is the stencil of this method?

(b) Show that this method has a local discretisation error estimated by $\mathcal{O}(\Delta t^2) + \mathcal{O}(\Delta x^2) + \mathcal{O}((\Delta x/\Delta t)^2)$.

(c) Assume we have homogeneous Dirichlet boundary conditions. Show that the method in matrix-vector form can be written as

$$(1+2d)\mathbf{u}^{n+1} = (1-2d)\mathbf{u}^{n-1} + 2d\mathbf{B}\mathbf{u}^n,$$

where the matrix $\mathbf{B}$ is given by

$$\mathbf{B} := \begin{pmatrix} 0 & 1 & & & \\ 1 & 0 & 1 & & \\ & \ddots & \ddots & \ddots & \\ & & 1 & 0 & 1 \\ & & & 1 & 0 \end{pmatrix}.$$

(d) Show that the matrix $\frac{1}{\Delta x^2}\mathbf{B}$ has eigenvalues $\lambda_k + \frac{2}{\Delta x^2}$ $(k = 1, 2, \ldots, M)$, where λ_k are the eigenvalues of the discretisation matrix $\mathbf{A}$ in (11.30b).

(e) Use the matrix $\mathbf{W}$ in (11.33) to transform the matrix recursion to diagonal form and show that it is unconditionally stable.

11.10. The *leapfrog scheme* is based on using central differences for both the space and the time derivatives. Applied to the heat equation it reads

$$u_j^{n+1} = u_j^{n-1} + 2d\bigl(u_{j+1}^n - 2u_j^n + u_{j-1}^n\bigr).$$

(a) Give the stencil of this scheme.

(b) Find a matrix-vector formulation of this method, like in Exercise 11.9.

(c) Bring the recursion in part (b) to diagonal form and show that it is unstable.

11.11. Consider the following spherically symmetric problem:

$$\begin{aligned}
&\frac{\partial u}{\partial t} = \frac{1}{r^2}\frac{\partial}{\partial r}\left(r^2\frac{\partial u}{\partial r}\right), && r \in (0,1), \quad t > 0,\\
&u(r,0) = 1, && r \in (0,1),\\
&\frac{\partial u}{\partial r}(0,t) = 0, \quad u(1,t) = 0, && t > 0.
\end{aligned}$$

For space discretisation we use the finite volume method.

(a) Give the implicit Euler scheme.

(b) Find the matrix-vector formulation (11.42).

11.12. Consider the problem

$$\frac{\partial u}{\partial t} = \frac{\partial^2 u}{\partial x^2} + \frac{\partial^2 u}{\partial y^2}, \qquad (x, y) \in \Omega, \quad t > 0,$$

with domain $\Omega := \{(x, y) \in \mathbb{R}^2 \mid x^2/4 + y^2 \leq 1\}$. This domain is covered by a uniform grid with grid size h. This means that we have to use different approximations for grid points near the boundary of Ω, which are first order in h. Show that the resulting (large) local discretisation error does not influence the global error.

11.13. Consider the advection-diffusion equation (11.131). For space discretisation we use central differences. To integrate the resulting ODE system (11.132) we use the second order *Adams–Bashforth scheme* for the advection term and the Crank–Nicolson scheme for the diffusion term; i.e.,

$$\frac{1}{\Delta t}\left(\mathbf{u}^{n+1} - \mathbf{u}^n\right) = \frac{1}{2}\left(\mathbf{A}_\mathrm{d}\mathbf{u}^n + \mathbf{A}_\mathrm{d}\mathbf{u}^{n+1}\right) - \frac{1}{2}\left(3\mathbf{A}_\mathrm{a}\mathbf{u}^n - \mathbf{A}_\mathrm{a}\mathbf{u}^{n-1}\right) + \mathbf{f}.$$

(a) Give the scheme.

(b) Compute the local discretisation error and show that it is second order in Δt and Δx.

(c) Investigate the stability of the scheme.

11.14. Consider Fisher's equation, i.e.,

$$\frac{\partial u}{\partial t} = \frac{\partial^2 u}{\partial x^2} + u(1 - u).$$

(a) Give the implicit Euler scheme for this equation.

(b) Give the nonlinear system (11.144) for $\mathbf{y} := \mathbf{u}^{n+1}$.

Chapter 12

Analysis of Hyperbolic Equations

This chapter is devoted to hyperbolic problems. We mainly restrict ourselves to equations with one space variable only. In Section 12.1 we start with first order scalar equations and describe solutions in terms of characteristics. Difficulties are encountered for nonlinear equations, for which characteristics may intersect or fan out. We introduce the corresponding solutions, i.e., shock waves and rarefaction waves. It turns out that a classical approach, with smooth solutions, is not suited to treating these phenomena. For that reason, we introduce weak solutions in Section 12.2. As an important example, we investigate the solution of the Riemann problem, which is an initial value problem with piecewise constant initial condition. We extend the above to systems in the next two sections. First, in Section 12.3, we introduce the definition of hyperbolicity. A system is hyperbolic if its Jacobi matrix has real eigenvalues and linearly independent eigenvectors. This then means that the system can be diagonalised. Subsequently, we can apply the scalar theory to each of the resulting equations. Furthermore, for a linear system of two unknowns we give a concise description of the method of characteristics. Next, in Section 12.4, we consider weak solutions of the Riemann problem. In particular, we introduce the elementary solutions, i.e., shock waves, rarefaction waves, and contact discontinuities. In Section 12.5 we apply this theory to the shallow-water equations. In Section 12.6 the wave equation is treated. We derive the solution in one space dimension first and then study higher-dimensional problems. The last section deals with choosing the proper boundary conditions. Since the characteristics determine the propagation directions of the solution, these determine what boundary conditions can be prescribed.

12.1 First Order Scalar Equations

In Chapter 2 we introduced first order scalar equations. In this section we shall have a closer look at these equations, starting with the linear case. After that, we will deal with the nonlinear case in more detail. A crucial element in the construction of solutions will turn out to be the behaviour of the characteristics.

12.1.1 Semilinear Equations

Consider the first order scalar equation (cf. (2.1))

$$a\frac{\partial u}{\partial t} + b\frac{\partial u}{\partial x} = c \qquad (a, b \neq 0). \tag{12.1}$$

Equation (12.1) is a scalar *hyperbolic* equation. In this subsection we allow the coefficients a, b, and c to depend on x and t but not on u, so that we have a linear equation in u. We showed in Chapter 2 that the PDE (12.1) leads to the set of ODEs

$$\frac{\mathrm{d}x}{\mathrm{d}t} = \frac{b}{a}, \tag{12.2a}$$

$$\frac{\mathrm{d}u}{\mathrm{d}t} = \frac{c}{a}. \tag{12.2b}$$

Equation (12.2a) defines the location of the *(base) characteristics* $\mathcal{C}$ in the (x, t) plane and (12.2b) defines the solution u, as a function of t, along these base characteristics. In the following we will omit the adjective "base," and use the term "characteristics" to denote solution curves of (12.2a) in the (x, t) plane. In order to find solutions of (12.1), it will be sufficient to study the ODEs (12.2). Let us first consider the homogeneous problem, i.e., $c(x, t) = 0$. If u is given at a single point on some characteristic $\mathcal{C}$, it is completely determined along $\mathcal{C}$ by virtue of (12.2b). Consequently, we may not prescribe a solution $u(x, t)$ on a characteristic unless it satisfies (12.2b). The solution would then exist on this single characteristic only. Thus consider a curve $\mathcal{J}$ that is not a characteristic and that intersects each characteristic at most once. So for some interval $I \subset \mathbb{R}$ let

$$\mathcal{J} := \{(x(\sigma), t(\sigma)) \mid \sigma \in I\}. \tag{12.3}$$

Then we may prescribe u on $\mathcal{J}$, say

$$u(x(\sigma), t(\sigma)) = v(\sigma), \qquad \sigma \in I, \tag{12.4}$$

where v is some given function. We can now compute u from the differential equations (12.2) and the condition (12.4).

Next we assume that a and b are constant. Special cases are when $\mathcal{J}$ coincides with either the x axis or the t axis. In the first case we have

$$\mathcal{J} = \{(x(\sigma), 0)^T \mid \sigma \in \mathbb{R}\},$$

and we may identify σ with x. The characteristic intersecting $\mathcal{J}$ at a point $(x_0, 0)^T$, say, is given by

$$x - \frac{b}{a}t = x_0.$$

The solution along this characteristic reads

$$u(x, t) = v(x_0) = v\left(x - \frac{b}{a}t\right). \tag{12.5}$$

In a similar vein, if we have

$$\mathcal{J} = \{(0, t(\sigma))^T \mid \sigma \geq 0\},$$

we may identify σ with t. At the characteristic intersecting $\mathcal{J}$ at a point $(0, t_0)^T$ we then have

$$t - \frac{a}{b}x = t_0.$$

Consequently, for the solution along this characteristic we find

$$u(x, t) = v(t_0) = v\left(t - \frac{a}{b}x\right). \tag{12.6}$$

Note that b/a can be interpreted as a velocity, implying that the solution in (12.5) is just the initial solution $v(x)$ propagated over a distance bt/a. A similar interpretation holds for (12.6).

Example 12.1 Consider the PDE

$$\frac{\partial u}{\partial t} + \frac{\partial u}{\partial x} = 0.$$

The characteristics of this equation are given by $x - t = C$ ($C \in \mathbb{R}$). If we prescribe $u(x, t) = v(t)$ on $\mathcal{C} = \{(x, t) \mid x = t\}$, i.e., the characteristic through the origin, we apparently need $v(t)$ to be constant. The solution then only exists on $\mathcal{C}$. Prescribing $u(x, t)$ along any line $\mathcal{J}$ that is not parallel to this curve $\mathcal{C}$ will then be meaningful. Suppose we take for $\mathcal{J}$ the x axis and choose

$$u(x, 0) = \sin x.$$

Clearly, we then have

$$u(x, t) = \sin(x - t),$$

implying that $u(x, t)$ is indeed constant along the characteristics. □

If a and b are not constant and $c \neq 0$, we can still find an integral representation of the solution along characteristics from (12.2b). Suppose a characteristic $\mathcal{C}$ intersects the "initial" line $\mathcal{J}$ at the point $(x_0, t_0)^T = (x(s_0), t(s_0)^T)$. We can express the value $u(x, t)$, for $(x, t)^T \in \mathcal{C}$, in terms of the initial value $u(x(s_0), t(s_0))$ as follows:

$$u(x, t) = u(x(s_0), t(s_0)) + \int_{t_0}^{t} \frac{c(x(\tau), \tau))}{a(x(\tau), \tau)}\, \mathrm{d}\tau. \tag{12.7}$$

In particular, if u is given along the x-axis, i.e.,

$$u(x, 0) = v(x),$$

and if the coefficients a and b are constant, we find from (12.7) that

$$\begin{aligned} u(x, t) &= v(x_0) + \frac{1}{a}\int_0^t c\big(x(\tau), \tau\big)\, \mathrm{d}s \\ &= v\left(x - \frac{b}{a}t\right) + \frac{1}{a}\int_0^t c\left(x - \frac{b}{a}(t - \tau), \tau\right) \mathrm{d}\tau, \end{aligned} \tag{12.8}$$

where we have used that $x - \frac{b}{a}t = x_0 = x(\tau) - \frac{b}{a}\tau$ along characteristics. Also, for more general c, i.e., $c = c(x, t, u)$, in which case the PDE is sometimes called *semilinear*, it is fairly simple to find a solution.

Example 12.2 Consider the initial value problem

$$\frac{\partial u}{\partial t} + \frac{\partial u}{\partial x} = -u, \qquad x \in \mathbb{R}, \quad t > 0,$$
$$u(x,0) = \sin x, \qquad x \in \mathbb{R}.$$

From (12.2) we conclude that $u(x(t),t) = C\mathrm{e}^{-t}$ ($C \in \mathbb{R}$) along the characteristic with $x(t) - t = x_0$. Applying the initial condition, we obtain $u(x_0,0) = \sin x_0 = C$. So the solution is given by

$$u(x,t) = \mathrm{e}^{-t}\sin(x-t).$$ □

If a and b are not constant, the characteristics are not straight lines in general. Yet the preceding outline for the construction of a solution is still valid, as is shown by the following example.

Example 12.3 Consider the initial value problem

$$\frac{\partial u}{\partial t} + x\frac{\partial u}{\partial x} = 0, \qquad x \in \mathbb{R}, \quad t > 0,$$
$$u(x,0) = \sin x, \qquad x \in \mathbb{R}.$$

From (12.2) we see that the characteristic through the initial point $(x_0,0)^T$ is given by $x(t) = x_0\,\mathrm{e}^t$, along which $u(x(t),t) = C$. From the initial condition we find $u(x_0,0) = \sin x_0 = C$ and, consequently, the solution is given by

$$u(x,t) = \sin\left(x\,\mathrm{e}^{-t}\right).$$ □

It is important to realise that the initial condition function v (the function of initial values) does not need to be smooth, or even continuous. Indeed, it is just a representation of a collection of initial values for the solution defined on the various characteristics. If v has a discontinuity, then the solution is not smooth either. In fact, one may wonder whether a solution can still satisfy (12.1). We shall see that it can, in a so-called weak sense, in Section 12.2.

12.1.2 Quasi-linear Equations

If the coefficients a, b, and c depend on u as well, we call the PDE *quasi-linear*; cf. Example 1.2. Since we have assumed that $a \neq 0$, it is not restrictive to take $a(x,t,u) \equiv 1$. So consider the equation

$$\frac{\partial u}{\partial t} + b(x,t,u)\frac{\partial u}{\partial x} = c(x,t,u). \tag{12.9}$$

Moreover, if we assume that $b = b(u)$, i.e., b does not explicitly depend on x or t, we can define a *flux function* $f(u)$ as

$$f(u) := \int_{u_0}^{u} b(v)\,\mathrm{d}v, \tag{12.10}$$

where u_0 is some reference value for u. We can then rewrite equation (12.9) in *conservation form* as

$$\frac{\partial u}{\partial t} + \frac{\partial f(u)}{\partial x} = c(x,t,u). \tag{12.11}$$

Equation (12.11) has the form of a *conservation equation*. It is also known as a *transport equation*.

Below we consider the case $c(x, t, u) \equiv 0$. In order to determine a solution of (12.11), or equivalently (12.9), we proceed like before. So the governing equations for the solution along a characteristic C read

$$\frac{\mathrm{d}x}{\mathrm{d}t} = b(u) = f'(u), \tag{12.12a}$$

$$\frac{\mathrm{d}u}{\mathrm{d}t} = 0. \tag{12.12b}$$

From (12.12a) we see that the location of characteristics depends on u, in contrast to the linear case where we could compute the characteristics independently from the solution u. We can, in principle, compute u from these equations if u is given on some initial curve $\mathcal{J}$ intersecting the characteristics at most once. Let $\mathcal{J}$ be the x axis. Then

$$u(x, 0) = v(x) \tag{12.13}$$

for some given function v. From (12.12b) we conclude that u, and therefore also $b(u)$, is constant along a characteristic. Integration of the differential equations in (12.12) is then trivial, and we find the solution

$$x - x_0 = b(\, v(x_0)\,)t, \tag{12.14a}$$

$$u(x, t) = v(x_0) = v(\,x - b(v(x_0))t\,), \tag{12.14b}$$

which holds on the characteristic through the point $(x_0, 0)^T$; see also Example 2.2. Equation (12.14a) implicitly defines x_0 as a function of x and t; i.e., $x_0 = x_0(x, t)$. By substituting the latter relation into (12.14b), we can find the solution $u(x, t)$.

A well-known example of (12.9) is the (inviscid) Burgers' equation. (In Chapter 12 and 13 we simply write "Burgers' equation" to denote the inviscid Burgers' equation.)

Example 12.4 Consider the Burgers' equation

$$\frac{\partial u}{\partial t} + u\frac{\partial u}{\partial x} = 0.$$

Clearly $b(u) = u$ and we have $u(x, t) = v(x - v(x_0)t)$ along the characteristics given by $x - x_0 = v(x_0)t$. □

As before, we note that the initial condition is propagated along characteristics, now with a speed that depends on the location. This induces a problem, certainly from a mathematical point of view.

Example 12.5 Consider the Burgers' equation again. Let the initial condition $v(x)$ be piecewise constant, say

$$v(x) := \begin{cases} \alpha & \text{if} \quad x < 0, \\ \beta & \text{if} \quad x \geq 0. \end{cases}$$

The characteristic through a point $(x_0, 0)^T$ is given by

$$x = x_0 + v(x_0)t, \qquad x_0 \in \mathbb{R},$$

and has a slope $\frac{dt}{dx} = 1/v(x_0)$. Therefore we can encounter two typical situations: $\alpha > \beta$ and $\alpha \leq \beta$. In the first case the slope of characteristics emanating from the negative x-axis is smaller than the slope of characteristics emanating from the positive x axis; see Figure 12.1. This would imply a multivalued solution where characteristics intersect. In the second case the characteristics on the left have a larger slope than the characteristics on the right, leading to a wedge-shaped region in the (x, t) plane where the solution is not defined. □

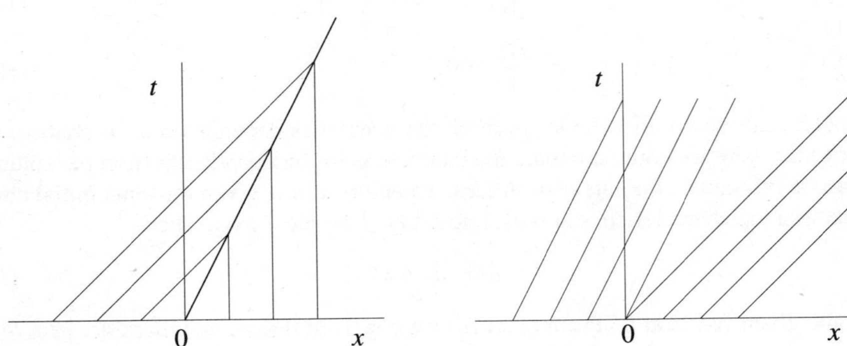

Figure 12.1. *Characteristics of the Burgers' equation that intersect (left) or fan out (right).*

Apparently, the nonlinearity is causing the problems met in Example 12.5. We emphasize that this is not necessarily a consequence of discontinuous initial data. To see this we shall analyse the influence of the initial condition in more detail. For simplicity we restrict ourselves to the Burgers' equation. Consider as an example the initial condition $v(x) = \sin \pi x$ $(0 \leq x \leq 1)$ shown in Figure 12.2; $v(x)$ is monotonically increasing on $(0, \frac{1}{2})$ and monotonically decreasing on $(\frac{1}{2}, 1)$. From (12.14) we conclude that the initial condition $v(x)$ is propagated along characteristics with velocity $b(v(x)) = v(x)$. This means that the characteristics emanating from $(0, \frac{1}{2})$ fan out and consequently the initial solution on this interval expands. On the other hand, characteristics originating from $(\frac{1}{2}, 1)$ are approaching each other, leading to a compression of the initial solution on this interval. This means that the left part of the solution overtakes the right part, leading to an increasingly steep profile, as shown in Figure 12.2. The solution will eventually break down when $\frac{\partial u}{\partial x}$ tends to infinity at some point $(x^*, t^*)^T$, say, where a discontinuity starts. We can compute t^* as follows. Consider the characteristic through $(x_0, 0)^T$ with $x_0 \in (\frac{1}{2}, 1)$ where the initial solution is monotonically decreasing. The location of this characteristic is given by the relation (12.14a), from which we can deduce

$$\left(1 + v'(x_0)t\right)\frac{\partial x_0}{\partial x} = 1. \tag{12.15}$$

Furthermore, the solution u along this characteristic is implicitly given by (12.14b). From this relation and (12.15) we can compute $\frac{\partial u}{\partial x}$ to find

$$\frac{\partial u}{\partial x} = v'(x_0)\frac{\partial x_0}{\partial x} = v'(x_0)\left(1 + v'(x_0)t\right)^{-1}. \tag{12.16}$$

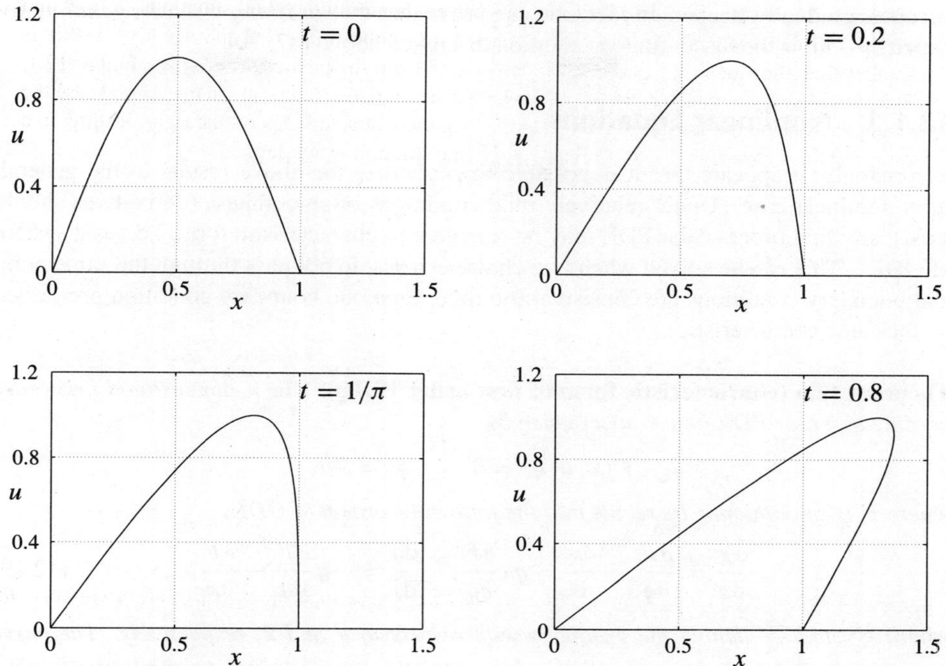

Figure 12.2. *The solution of* (12.12) *at* $t = 0$ *and* 0.2 *and* $t^* = 1/\pi$ *and* 0.8 *for the initial condition* $v(x) = \sin \pi x$.

Obviously $\frac{\partial u}{\partial x} \to -\infty$ when $1 + v'(x_0)t^* = 0$ for some $t^* > 0$. Note that this condition also implies that x_0 cannot be determined any more from relation (12.14a). The time t^* when a discontinuity first emerges is thus given by

$$t^* = -1/\min_{0\le x\le 1} v'(x) \tag{12.17}$$

and only holds when $v'(x) < 0$ somewhere.

If we computed the solution from (12.14) beyond the critical time t^*, we would obtain the physically incorrect triple-valued function shown in Figure 12.2. To determine the correct physical behaviour one should realise that the true physical model is an equation of the form

$$\frac{\partial u}{\partial t} + \frac{\partial f(u)}{\partial x} = \varepsilon \frac{\partial^2 u}{\partial x^2}, \tag{12.18}$$

including the viscous term $\varepsilon \frac{\partial^2 u}{\partial x^2}$ with $0 < \varepsilon \ll 1$, instead of (12.11) with $c(x, t, u) = 0$. Equation (12.11) is an appropriate model only if ε is small and the solution is smooth. In this case the viscous term $\varepsilon \frac{\partial^2 u}{\partial x^2}$ is negligible. However, when a discontinuity starts to develop, (12.11) loses its validity and we must return to (12.18). In the vicinity of the emerging discontinuity the term $\varepsilon \frac{\partial^2 u}{\partial x^2}$ becomes gradually larger, thus balancing the left-hand side in (12.18) and preventing breakdown of the solution. For decreasing ε the solution becomes gradually steeper. In fact, one can prove that the *vanishing viscosity solution*, for which $\varepsilon \to 0$, is the discontinuous solution discussed above [87, 70].

12.1.3 Nonlinear Equations

Surprisingly, it appears that it is possible to generalize the above results to the general, fully nonlinear case. Under relatively mild conditions of smoothness (derivatives should exist), any first order scalar PDE may be rewritten in characteristic form, i.e., as a system of ODEs. This can be solved when two characteristics do not pass through the same point and boundary conditions are consistent (no more than one boundary condition prescribed on the same characteristic).

Theorem 12.6 (characteristic form of first order PDEs). *The n-dimensional first order nonlinear scalar PDE in $u = u(\boldsymbol{x})$ given by*

$$F(\boldsymbol{x}, u, \boldsymbol{q}) = 0, \qquad \boldsymbol{q} := \nabla u,$$

where F is smooth, may be recast into the following system of ODEs

$$\frac{\mathrm{d}\boldsymbol{\chi}}{\mathrm{d}s} = \frac{\partial F}{\partial \boldsymbol{q}}, \quad \frac{\mathrm{d}u}{\mathrm{d}s} = \boldsymbol{q} \cdot \frac{\partial F}{\partial \boldsymbol{q}}, \quad \frac{\mathrm{d}\boldsymbol{q}}{\mathrm{d}s} = -\boldsymbol{q}\frac{\partial F}{\partial u} - \frac{\partial F}{\partial \boldsymbol{x}}, \tag{12.19}$$

where $\frac{\partial F}{\partial \boldsymbol{q}}$ and $\frac{\partial F}{\partial \boldsymbol{x}}$ denote the gradients with respect to $\boldsymbol{q}$ and $\boldsymbol{x}$, respectively. The curve $\boldsymbol{x} = \boldsymbol{\chi}(s)$, with parameter s, is called a characteristic. Since s is only an auxiliary variable, other equivalent forms exist. They may be constructed by varying the defining equation $F(\boldsymbol{x}, u, \boldsymbol{q}) = 0$.

Proof. We have the problem $F(x_1, \ldots, x_n, u, q_1, \ldots, q_n) = 0$, where $q_i = \frac{\partial u}{\partial x_i}$ $(i = 1, 2, \ldots, n)$. We look for characteristics, given by $x_i = \chi_i(s)$, along which the PDE can be written as an ODE. If we compare the expressions

$$\frac{\mathrm{d}F}{\mathrm{d}x_i} = \sum_{j=1}^{n} \frac{\partial q_j}{\partial x_i}\frac{\partial F}{\partial q_j} + \frac{\partial u}{\partial x_i}\frac{\partial F}{\partial u} + \frac{\partial F}{\partial x_i} = 0,$$

$$\frac{\mathrm{d}q_i}{\mathrm{d}s} = \sum_{j=1}^{n} \frac{\partial q_i}{\partial x_j}\frac{\mathrm{d}\chi_j}{\mathrm{d}s} = \sum_{j=1}^{n} \frac{\partial^2 u}{\partial x_j \partial x_i}\frac{\mathrm{d}\chi_j}{\mathrm{d}s} = \sum_{j=1}^{n} \frac{\partial q_j}{\partial x_i}\frac{\mathrm{d}\chi_j}{\mathrm{d}s},$$

we observe that if we take

$$\frac{\mathrm{d}\chi_j}{\mathrm{d}s} = \frac{\partial F}{\partial q_j},$$

we obtain the relations

$$\frac{\mathrm{d}q_i}{\mathrm{d}s} = \sum_{j=1}^{n} \frac{\partial q_j}{\partial x_i}\frac{\partial F}{\partial q_j} = -q_i\frac{\partial F}{\partial u} - \frac{\partial F}{\partial x_i},$$

$$\frac{\mathrm{d}u}{\mathrm{d}s} = \sum_{j=1}^{n} \frac{\partial u}{\partial x_j}\frac{\mathrm{d}\chi_j}{\mathrm{d}s} = \sum_{j=1}^{n} q_j\frac{\partial F}{\partial q_j}$$

and the problem is indeed rewritten in characteristic form. □

Example 12.7 A famous example is the *eikonal equation*

$$F(\boldsymbol{x}, u, \boldsymbol{q}) = \boldsymbol{q}\cdot\boldsymbol{q} - \frac{1}{c^2(\boldsymbol{x})} = 0, \qquad \boldsymbol{q} := \nabla u,$$

which becomes, along the characteristics $\boldsymbol{x} = \boldsymbol{\chi}(s)$,

$$\frac{\mathrm{d}\boldsymbol{\chi}}{\mathrm{d}s} = 2\boldsymbol{q}, \quad \frac{\mathrm{d}u}{\mathrm{d}s} = 2\boldsymbol{q}\cdot\boldsymbol{q} = \frac{2}{c^2(\boldsymbol{x})}, \quad \frac{\mathrm{d}\boldsymbol{q}}{\mathrm{d}s} = -\frac{2}{c^3(\boldsymbol{x})}\nabla c.$$

A much neater form, with the same characteristics but a different parametrization, is found if we take

$$F(\boldsymbol{x}, u, \boldsymbol{q}) = \frac{1}{2}\big(c^2(\boldsymbol{x})\,\boldsymbol{q}\cdot\boldsymbol{q} - 1\big) = 0, \qquad \boldsymbol{q} := \nabla u.$$

Then we have

$$\frac{\mathrm{d}\boldsymbol{\chi}}{\mathrm{d}s} = c^2(\boldsymbol{x})\,\boldsymbol{q}, \quad \frac{\mathrm{d}u}{\mathrm{d}s} = c^2(\boldsymbol{x})\,\boldsymbol{q}\cdot\boldsymbol{q} = 1, \quad \frac{\mathrm{d}\boldsymbol{q}}{\mathrm{d}s} = -\frac{1}{c}\nabla c.$$

Obviously, $u = s + u(0)$, implying that the parametrization corresponds now to the level surfaces of u. □

A geometrical interpretation is possible for the special case of two independent variables, say x and y; see also Section 2.1. Therefore consider the equation

$$F(x, y, u, p, q) = 0, \qquad p := \frac{\partial u}{\partial x}, \quad q := \frac{\partial u}{\partial y}, \tag{12.20}$$

for $u = u(x, y)$. The solution(s) of (12.20) can be represented as vectors $(x, y, u)^T$ lying on an integral surface $\mathcal{S}$ in $\mathbb{R}^3$. All possible integral surfaces of (12.20) at a given point $\mathrm{P}_0\,(x_0, y_0, u_0)$ have normal vectors $\boldsymbol{n} = (p, q, -1)^T$, where p and q satisfy the nonlinear equation $F(x_0, y_0, u_0, p, q) = 0$. This means that all possible normal vectors at P_0 sweep out a curved surface with apex at P_0, referred to as the *cone of normals*. This nomenclature is based on the eikonal equation, for which the cone of normals is actually a right circular cone. Consequently, all possible tangent vectors at P_0, each of which is perpendicular to the cone of normals, generate a curved surface with apex at P_0 called the *Monge cone*. Hence all possible integral surfaces at P_0 are tangent to the Monge cone.

If (12.20) is quasi-linear, the relation $F(x_0, y_0, u_0, p, q) = 0$ is linear in p and q, defining only one direction vector for the normal at P_0. In that case the Monge cone reduces to a single characteristic direction.

Suppose that the solution is given along some initial value curve $\mathcal{J} = \{(x(\sigma), y(\sigma), u(\sigma)) \mid \sigma \in I \subset \mathbb{R}\}$. We want to construct the integral surface through $\mathcal{J}$. In the quasi-linear case this is relatively straightforward—compute all characteristics emanating from $\mathcal{J}$. In the fully nonlinear case this approach does not give a unique solution since there is more than one integral surface that is tangent to all Monge cones on $\mathcal{J}$. To determine the *generator* of the cone that is tangent to the integral surface we proceed as follows. Suppose we parametrize the solution of $F(x_0, y_0, u_0, p, q) = 0$; i.e., $p = p(\lambda)$ and $q = q(\lambda)$ for some λ. Then we have

$$F(x_0, y_0, u_0, p(\lambda), q(\lambda)) = 0. \tag{12.21}$$

Differentiating with respect to λ, we obtain

$$\frac{\partial F}{\partial p}p'(\lambda) + \frac{\partial F}{\partial q}q'(\lambda) = 0, \tag{12.22}$$

where the derivatives of F should be taken at $(x_0, y_0, u_0, p(\lambda), q(\lambda))$. A tangent plane at P_0 on the integral surface is given by

$$u - u_0 = (x - x_0)p(\lambda) + (y - y_0)q(\lambda). \tag{12.23}$$

The Monge cone is the *envelope* of all these tangent planes and can be obtained if we differentiate (12.23) with respect to λ; i.e.,

$$0 = (x - x_0)p'(\lambda) + (y - y_0)q'(\lambda). \tag{12.24}$$

This relation should hold for the generator of the Monge cone. Comparing (12.22) and (12.24), we conclude that

$$\frac{\mathrm{d}x}{F_p} = \frac{\mathrm{d}y}{F_q} = \mathrm{d}s, \tag{12.25a}$$

with s the parameter along the generator. Obviously, these relations can be rewritten as

$$\frac{\mathrm{d}x}{\mathrm{d}s} = \frac{\partial F}{\partial p}, \qquad \frac{\mathrm{d}y}{\mathrm{d}s} = \frac{\partial F}{\partial q}. \tag{12.25b}$$

Along the generator the solution u satisfies

$$\frac{\mathrm{d}u}{\mathrm{d}s} = \frac{\partial u}{\partial x}\frac{\mathrm{d}x}{\mathrm{d}s} + \frac{\partial u}{\partial y}\frac{\mathrm{d}y}{\mathrm{d}s} = p\frac{\partial F}{\partial p} + q\frac{\partial F}{\partial q}. \tag{12.26}$$

Since (12.25b) and (12.26) contain p and q, we need differential equations for these variables as well. Differentiating (12.20) with respect to x and y, respectively, we obtain

$$\frac{\partial F}{\partial x} + \frac{\partial F}{\partial u}p + \frac{\partial F}{\partial p}\frac{\partial p}{\partial x} + \frac{\partial F}{\partial q}\frac{\partial q}{\partial x} = 0, \tag{12.27a}$$

$$\frac{\partial F}{\partial y} + \frac{\partial F}{\partial u}q + \frac{\partial F}{\partial p}\frac{\partial p}{\partial y} + \frac{\partial F}{\partial q}\frac{\partial q}{\partial y} = 0. \tag{12.27b}$$

Using (12.25b), (12.27), and the consistency relation $p_y = q_x$, we obtain the following ODEs for p and q:

$$\frac{\mathrm{d}p}{\mathrm{d}s} = \frac{\partial p}{\partial x}\frac{\mathrm{d}x}{\mathrm{d}s} + \frac{\partial p}{\partial y}\frac{\mathrm{d}y}{\mathrm{d}s} = \frac{\partial F}{\partial p}\frac{\partial p}{\partial x} + \frac{\partial F}{\partial q}\frac{\partial q}{\partial x} = -\left(\frac{\partial F}{\partial x} + \frac{\partial F}{\partial u}p\right), \tag{12.28a}$$

$$\frac{\mathrm{d}q}{\mathrm{d}s} = \frac{\partial q}{\partial x}\frac{\mathrm{d}x}{\mathrm{d}s} + \frac{\partial q}{\partial y}\frac{\mathrm{d}y}{\mathrm{d}s} = \frac{\partial F}{\partial p}\frac{\partial p}{\partial y} + \frac{\partial F}{\partial q}\frac{\partial q}{\partial y} = -\left(\frac{\partial F}{\partial y} + \frac{\partial F}{\partial u}q\right). \tag{12.28b}$$

To summarize, we have the ODE system (12.25b), (12.26), and (12.28). These equations are called the *characteristic strip equations* or *Charpit's equations*. They are a special case of the ODE system in Theorem 12.6.

Initial conditions for x, y, and u are determined by the initial curve $\mathcal{J}$; i.e., we have

$$x(0,\sigma) = x_0(\sigma), \quad y(0,\sigma) = y_0(\sigma), \quad u(0,\sigma) = u_0(\sigma). \tag{12.29a}$$

The remaining initial conditions for p and q follow from the relations

$$F\big(x_0(\sigma), y_0(\sigma), u_0(\sigma), p_0(\sigma), q_0(\sigma)\big) = 0, \tag{12.29b}$$

$$u_0'(\sigma) = p_0(\sigma)x_0'(\sigma) + q_0(\sigma)y_0'(\sigma), \tag{12.29c}$$

where the prime ($'$) denotes differentiation with respect to σ. The last relation allows us to move characteristics along the initial curve $\mathcal{J}$ to generate the integral surface $\mathcal{S}$. It follows from the condition that $\mathcal{J} \subset \mathcal{S}$.

Using (12.25b), (12.26), and (12.27), we can show that

$$\frac{\mathrm{d}}{\mathrm{d}s} F(x(s), y(s), u(s), p(s), q(s)) = 0.$$

This means that if $F(x_0, y_0, u_0, p_0, q_0) = 0$ holds, then the curve determined by the characteristic strip equations and the initial conditions (12.29) satisfies the differential equation $F(x, y, u, p, q) = 0$.

The formal solution procedure is as follows. We solve the characteristic strip equations subject to the initial conditions (12.29) to find $x = x(s, \sigma)$, $y = y(s, \sigma)$, $u = u(s, \sigma)$, $p = p(s, \sigma)$, and $q = q(s, \sigma)$. Inverting the first two relations, which is only possible if $x_s y_\sigma - x_\sigma y_s \neq 0$, we obtain $s = s(x, y)$ and $\sigma = \sigma(x, y)$, which we then substitute in the expression for u. In this way we find the representation $u = u(s(x, y), \sigma(x, u)) =: u^*(x, y)$.

12.2 Weak Formulation of First Order Scalar Equations

As we saw in the previous section, a solution of (12.11) that is computed from the ODE system (12.12) is not necessarily continuous, let alone differentiable. This means that we have to reconsider our concept of a solution of (12.11). For this we need distribution theory; cf. Chapter 4.

12.2.1 Weak Solutions

Let us start by observing that hyperbolic conservation equations are often derived in integral form rather than as a differential equation. As an example, think of gas flowing in a tube of constant cross section. Let x denote the coordinate along the tube and $\rho(x, t)$ and $v(x, t)$ denote the mass density and flow velocity, respectively, at position x and time t. Then conservation of mass in an arbitrary segment (x_1, x_2) is given by the relation

$$\frac{\mathrm{d}}{\mathrm{d}t} \int_{x_1}^{x_2} \rho(x, t)\,\mathrm{d}x = (\rho v)(x_1, t) - (\rho v)(x_2, t), \tag{12.30}$$

stating that the increase of mass in (x_1, x_2) is balanced by the net influx of mass. If we replace ρ by a variable u and the mass flux ρv by a generic flux $f(u)$, (12.30) generalises to

$$\frac{\mathrm{d}}{\mathrm{d}t} \int_{x_1}^{x_2} u(x, t)\,\mathrm{d}x = f(u(x_1, t)) - f(u(x_2, t)). \tag{12.31}$$

By integrating this equation over an arbitrary time interval $[t_1, t_2]$, we find

$$\int_{x_1}^{x_2} \big(u(x,t_2) - u(x,t_1)\big)\,\mathrm{d}x = \int_{t_1}^{t_2} \big(f(u(x_1,t)) - f(u(x_2,t))\big)\,\mathrm{d}t. \tag{12.32}$$

If u and $f(u)$ are continuously differentiable, this equation is equivalent to

$$\int_{t_1}^{t_2}\int_{x_1}^{x_2} \left(\frac{\partial u}{\partial t} + \frac{\partial f(u)}{\partial x}\right) \mathrm{d}x\mathrm{d}t = 0. \tag{12.33}$$

Since this equation should hold for arbitrary x_1, x_2, t_1, and t_2, the integrand has to be zero necessarily; i.e.,

$$\frac{\partial u}{\partial t} + \frac{\partial f(u)}{\partial x} = 0. \tag{12.34}$$

A function u is called a *weak solution* of (12.34) if it satisfies (12.32) for arbitrary x_1, x_2, t_1, and t_2. Note that a solution of (12.34) is always a solution of (12.32); the converse need not be true.

Since the verification of (12.32) for arbitrary x_1, x_2, t_1, and t_2 is rather cumbersome, we prefer another definition of weak solution, which is based on distribution theory; see Chapter 4. Here we define the space of *test functions* $\mathcal{D}$ as follows:

$$\mathcal{D} = C_0^1(\mathbb{R}\times[0,\infty)) := \{\varphi \in C^1(\mathbb{R}\times[0,\infty)) \mid \varphi \text{ has compact support for any } t\}. \tag{12.35}$$

The basic idea is then to multiply (12.34) by such a test function $\varphi(x,t)$, integrate over $\mathbb{R}\times[0,\infty)$, and subsequently apply integration by parts. Using the fact that $\varphi(x,t)$ vanishes for $|x| + t \to \infty$, we obtain

$$\int_0^{\infty}\int_{-\infty}^{\infty} \left(u\frac{\partial \varphi}{\partial t} + f(u)\frac{\partial \varphi}{\partial x}\right) \mathrm{d}x\mathrm{d}t = -\int_{-\infty}^{\infty} u(x,0)\varphi(x,0)\,\mathrm{d}x. \tag{12.36}$$

This relation then gives rise to the following definition.

Definition 12.8. *A function $u(x,t)$ is called a weak solution of conservation law* (12.34) *if* (12.36) *holds for all test functions $\varphi \in C_0^1(\mathbb{R}\times[0,\infty))$.*

Obviously, when $u(x,t)$ satisfies (12.34), it is a weak solution. The converse is only true when $u(x,t)$ is continuously differentiable. In the following, when we speak of a solution of (12.34), we mean a weak solution in the sense of this definition.

Note that relation (12.36) allows for discontinuous solutions. However, not every discontinuous function can be a solution of (12.34), as is shown in the following theorem.

Theorem 12.9. *Let u be a piecewise smooth solution of* (12.34) *that has a discontinuity across a curve $\mathcal{E} : x = x(t)$. Then u satisfies the condition*

$$[f(u)]_-^+ = s[u]_-^+, \tag{12.37}$$

where $[v]_-^+ := v(x(t)+,t) - v(x(t)-,t)$ $(v = u, f(u))$ is the jump of v across $\mathcal{E}$ and s is the speed of $\mathcal{E}$.

Proof. Assume that $\mathcal{E}$ separates a domain $\Omega \supset \text{supp}(\varphi)$ into a left part Ω_ℓ and a right part Ω_r; see Figure 12.3. The solution is smooth both in Ω_ℓ and in Ω_r. Since (12.34) holds in Ω_ℓ, we have

$$\iint_{\Omega_\ell} \left(\frac{\partial u}{\partial t} + \frac{\partial f(u)}{\partial x} \right) \varphi \, \mathrm{d}x\mathrm{d}t = 0$$

for every test function $\varphi \in \mathcal{D}$. We can rewrite this equation as

$$\iint_{\Omega_\ell} \left(\frac{\partial (u\varphi)}{\partial t} + \frac{\partial (f(u)\varphi)}{\partial x} \right) \mathrm{d}x\mathrm{d}t = \iint_{\Omega_\ell} \left(u\frac{\partial \varphi}{\partial t} + f(u)\frac{\partial \varphi}{\partial x} \right) \mathrm{d}x\mathrm{d}t.$$

If we apply the two-dimensional Gauss's theorem to the integral on the left-hand side, we find

$$-\oint_{\partial\Omega_\ell} \varphi\big(u\mathrm{d}x - f(u)\mathrm{d}t\big) = \iint_{\Omega_\ell} \left(u\frac{\partial \varphi}{\partial t} + f(u)\frac{\partial \varphi}{\partial x} \right) \mathrm{d}x\mathrm{d}t,$$

with $\partial\Omega_\ell$ the boundary of Ω_ℓ. Next, using that $\varphi(x,t) = 0$ for $(x,t) \in \partial\Omega \cap \{t > 0\}$, with $\partial\Omega$ the boundary of Ω, we obtain

$$-\int_a^{x_s} u(x,0)\varphi(x,0)\,\mathrm{d}x - \int_{\mathcal{E}} \varphi\big(u_\ell \mathrm{d}x - f(u_\ell)\mathrm{d}t\big) = \iint_{\Omega_\ell} \left(u\frac{\partial \varphi}{\partial t} + f(u)\frac{\partial \varphi}{\partial x} \right) \mathrm{d}x\mathrm{d}t, \tag{$*$}$$

where x_s is the intersection of $\mathcal{E}$ with the x axis and $u_\ell := u(x(t)-, t)$ is the limit value of u just left of the discontinuity. Carrying out a similar procedure for Ω_r, we find

$$-\int_{x_s}^{b} u(x,0)\varphi(x,0)\,\mathrm{d}x + \int_{\mathcal{E}} \varphi\big(u_r \mathrm{d}x - f(u_r)\mathrm{d}t\big) = \iint_{\Omega_r} \left(u\frac{\partial \varphi}{\partial t} + f(u)\frac{\partial \varphi}{\partial x} \right) \mathrm{d}x\mathrm{d}t, \tag{$**$}$$

with $u_r := u(x(t)+, t)$ the limit value of u from the right of $\mathcal{E}$. The integral over $\mathcal{E}$ in ($**$) is evaluated in the same direction as in ($*$) (see Figure 12.3) and therefore has a + sign in

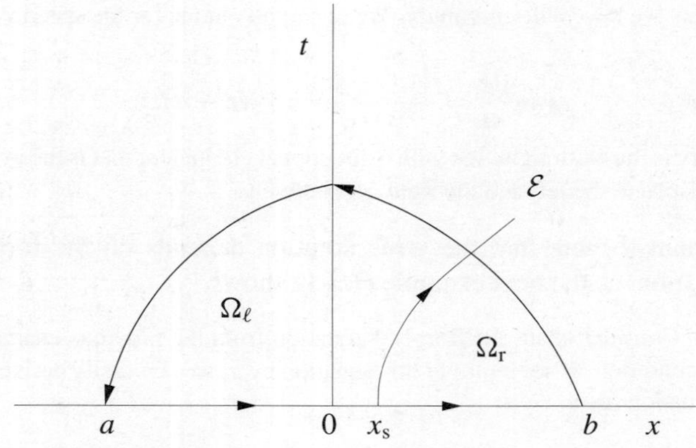

Figure 12.3. *Weak solution, discontinuous across $\mathcal{E}$.*

front. Adding (∗) and (∗∗), we obtain

$$-\int_a^b u(x,0)\varphi(x,0)\,\mathrm{d}x + \int_{\mathcal{E}} \varphi\big([u]_-^+\mathrm{d}x - [f(u)]_-^+\mathrm{d}t\big) = \iint_\Omega \left(u\frac{\partial\varphi}{\partial t} + f(u)\frac{\partial\varphi}{\partial x}\right)\mathrm{d}x\mathrm{d}t.$$

Combining this relation with (12.36) and taking into account that $\varphi(x,t)$ vanishes outside Ω, we find

$$\int_{\mathcal{E}} \varphi\big([u]_-^+\mathrm{d}x - [f(u)]_-^+\big)\mathrm{d}t = 0.$$

This relation holds for arbitrary test functions $\varphi \in \mathcal{D}$, so that $[u]_-^+\mathrm{d}x - [f(u)]_-^+\mathrm{d}t = 0$. Finally, since $s = \frac{\mathrm{d}x}{\mathrm{d}t}$, this implies relation (12.37). □

The propagation speed s of a discontinuity is thus given by

$$f(u_\mathrm{r}) - f(u_\ell) = s(u_\mathrm{r} - u_\ell), \tag{12.38}$$

with $u_\mathrm{r} := u(x(t)+,t)$ and $u_\ell := u(x(t)-,t)$ the limit values of $u(x,t)$ just right and left of the discontinuity, respectively. Relation (12.38) is called the *Rankine–Hugoniot jump condition*. Inserting (12.10) into (12.38), we find the following alternative expression for s:

$$s = \frac{1}{u_\mathrm{r} - u_\ell}\int_{u_\ell}^{u_\mathrm{r}} b(v)\,\mathrm{d}v; \tag{12.39}$$

i.e. s is the average advection velocity $b(v)$ over the interval $\mathrm{int}(u_\ell, u_\mathrm{r})$.

Example 12.10 Consider the Burgers' equation, written in conservation form,

$$\frac{\partial u}{\partial t} + \frac{\partial}{\partial x}\left(\frac{1}{2}u^2\right) = 0,$$

subject to the piecewise constant initial condition

$$u(x,0) = \begin{cases} \alpha & \text{if } x < 0, \\ \beta & \text{if } x > 0. \end{cases}$$

Let $\alpha > \beta$, so that we have a discontinuity. We apparently have, for the speed s of the discontinuity,

$$s = \frac{\mathrm{d}x}{\mathrm{d}t} = \frac{\frac{1}{2}(\beta^2 - \alpha^2)}{\beta - \alpha} = \frac{1}{2}(\beta + \alpha).$$

The discontinuity is thus a straight line with a directional coefficient that is the average of those of the characteristics to the left and the right, respectively. □

It is important to note that the weak solution depends on the formulation of the conservation equation, as the next example clearly shows.

Example 12.11 Consider again the Burgers' equation from the previous example, subject to the same initial condition. If we multiply this equation by u, we can easily derive the following conservation equation for $w := u^2$:

$$\frac{\partial w}{\partial t} + \frac{\partial}{\partial x}\left(\frac{2}{3}w^{3/2}\right) = 0. \tag{$*$}$$

If we apply the Rankine–Hugoniot jump condition (12.38) to (∗), we find, for the propagation speed s of a discontinuity,

$$s = \frac{2}{3}\frac{w_r^{3/2} - w_\ell^{3/2}}{w_r - w_\ell} = \frac{2}{3}\frac{\beta^3 - \alpha^3}{\beta^2 - \alpha^2} = \frac{2}{3}\frac{\beta^2 + \beta\alpha + \alpha^2}{\beta + \alpha}.$$

This is generally not equal to the propagation speed $\frac{1}{2}(\beta + \alpha)$ found in Example 12.10. So, although the Burgers' equation and (∗) are equivalent for smooth solutions, they have different weak solutions. □

What we learn from this example is that we cannot manipulate the strong formulation of a problem when dealing with discontinuous solutions. In fact, the Rankine–Hugoniot jump condition is the reformulation of the correct physical conservation law across a discontinuity. Stated otherwise, the Rankine–Hugoniot jump condition is an extra condition that should be imposed for discontinuous solutions along with the corresponding PDE.

It is instructive to look back now to the problem depicted in Figure 12.2. The breaking at $t = t^*$ starts a shock, which then propagates at speed s given by (12.38). From the shock speed we can determine the location of the shock. Now consider the initial profile in Figure 12.2, which is zero outside $(0, 1)$. Then if (a, b) is a sufficiently large interval, we have from the integral form of the conservation equation (12.34) that

$$\frac{\mathrm{d}}{\mathrm{d}t}\int_a^b u(x,t)\,\mathrm{d}x = f(u(a,t)) - f(u(b,t)) = 0. \tag{12.40}$$

Hence $\int_a^b u(x,t)\,\mathrm{d}x$ is constant. At a time point $t > t^*$ a classical solution would have a positive and a negative integral part, as shown in Figure 12.4. Of course, the shape as depicted does not make sense practically as we would have a multivalued solution, yet we may formally do the integration. The total effect would be the same as when we integrated up to the point x_s (starting from a); see Figure 12.4. Conservation means that our weak solution should also be conserved, and hence we may identify the point x_s with the propagated breaking point on the shock line. Actually, one may reverse the argument and determine x_s graphically: choose the point x_s such that the shaded regions on either side of the line $x = x_s$ have equal areas.

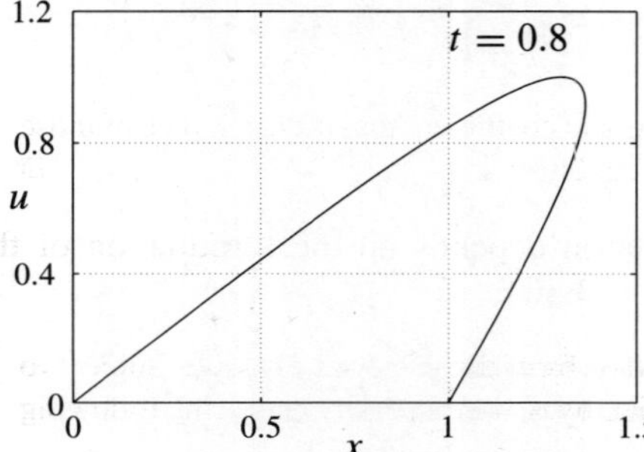

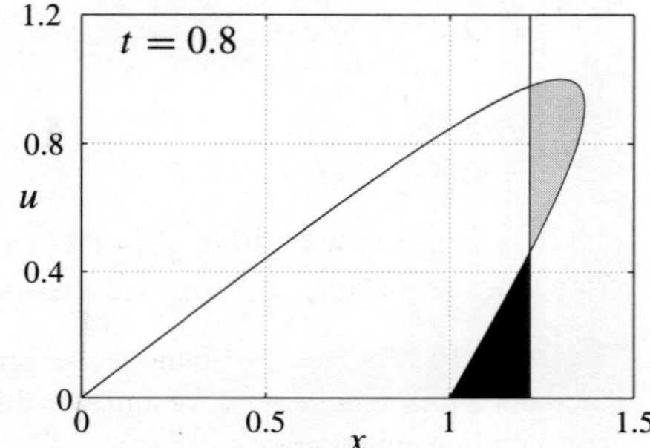

Figure 12.4. *Illustration of the equal-area rule: The multivalued "solution" on the left should be replaced by the shock on the right.*

12.2.2 The Riemann Problem

The initial value problem for a conservation equation with piecewise constant initial condition and a single discontinuity is known as the *Riemann problem*. The canonical form of the Riemann problem for (12.34) reads

$$\frac{\partial u}{\partial t} + \frac{\partial f(u)}{\partial x} = 0, \qquad x \in \mathbb{R}, \quad t > 0, \tag{12.41a}$$

$$u(x,0) = \begin{cases} u_\ell & \text{if} \quad x < 0, \\ u_\mathrm{r} & \text{if} \quad x > 0, \end{cases} \tag{12.41b}$$

Note that if $u(x,t)$ is a solution of (12.41), then so is $u(\alpha x, \alpha t)$ for arbitrary $\alpha > 0$, implying that the solution of (12.41) is a similarity solution of the form $u(x,t) = \hat{u}(x/t)$; cf. Section 2.5. In Section 12.1 we showed that solutions of an initial value problem like (12.41) are propagated along characteristics with velocity $b(u) = f'(u)$. Based on this observation we can distinguish two cases, $b(u_\ell) > b(u_\mathrm{r})$ and $b(u_\ell) < b(u_\mathrm{r})$, to be discussed below separately.

Case 1. $b(u_\ell) > b(u_\mathrm{r})$.

The characteristics emanating from the negative x axis have a slope smaller than the slope of the characteristics coming from the positive x axis. As a consequence, characteristics intersect, which would lead to multivalued solutions. Instead, we have a discontinuous solution. We can easily verify by substitution into (12.36) that the solution of the Riemann problem (12.41) is indeed given by

$$u(x,t) = \begin{cases} u_\ell & \text{if} \quad x/t < s, \\ u_\mathrm{r} & \text{if} \quad x/t > s, \end{cases} \tag{12.42}$$

where s is defined in (12.38). The solution in (12.42) represents a discontinuity traveling with speed s and is called a *shock wave*; s is called the *shock speed*. A typical shock wave and the corresponding characteristics are shown in Figure 12.5. Note that the characteristics move into the shock for increasing t.

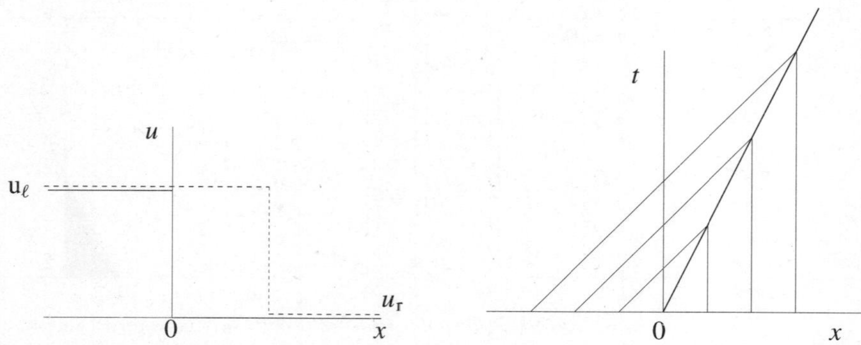

Figure 12.5. *Shock wave and corresponding characteristics.*

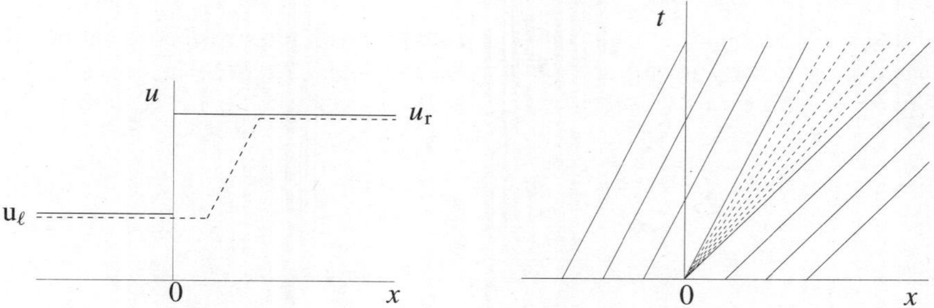

Figure 12.6. *Rarefaction wave and corresponding characteristics.*

Case 2. $b(u_\ell) < b(u_r)$.

In this case the characteristics emanating from the negative x axis have a larger slope than those emanating from the positive x axis. So we have separating characteristics and we do not expect a discontinuous solution. The solution of (12.41) is now given by

$$u(x,t) = \begin{cases} u_\ell & \text{if} \quad x/t < b(u_\ell), \\ w(x/t) & \text{if} \quad b(u_\ell) < x/t < b(u_r), \\ u_r & \text{if} \quad x/t > b(u_r), \end{cases} \tag{12.43}$$

where $w(\eta)$ is the solution of the relation $b(w(\eta)) = \eta$. This solution is called a *rarefaction wave*. It is continuous, despite the fact that the initial condition is discontinuous, and consists of the constant states $u(x,t) = u_\ell$ and $u(x,t) = u_r$, connected by the intermediate solution $u(x,t) = w(x/t)$. The latter solution is a similarity solution of (12.34), as we can easily verify by direct substitution. The constant states are an immediate consequence of (12.12). A typical rarefaction wave and the corresponding characteristics are depicted in Figure 12.6.

Example 12.12 Consider once more the Burgers' equation; cf. Example 12.10. Since the convection velocity $b(u) = u$, we have for the solution either a shock wave when $u_\ell > u_r$ or a rarefaction wave when $u_\ell < u_r$. The shock wave solution reads

$$u(x,t) = \begin{cases} u_\ell & \text{if} \quad x/t < s, \\ u_r & \text{if} \quad x/t > s, \end{cases}$$

with shock speed $s = \frac{1}{2}(u_\ell + u_r)$. The rarefaction wave is given by

$$u(x,t) = \begin{cases} u_\ell & \text{if} \quad x/t < u_\ell, \\ x/t & \text{if} \quad u_\ell < x/t < u_r, \\ u_r & \text{if} \quad x/t > u_r. \end{cases}$$ □

We can show that (12.42) is a solution of the Riemann problem (12.41) by substitution into (12.36), regardless of whether $b(u_\ell) > b(u_r)$ or $b(u_\ell) < b(u_r)$. In the latter case this solution is called an *expansion shock*. In Figure 12.7 we have sketched the corresponding characteristics. Note that the characteristics move out of the shock for increasing t. This

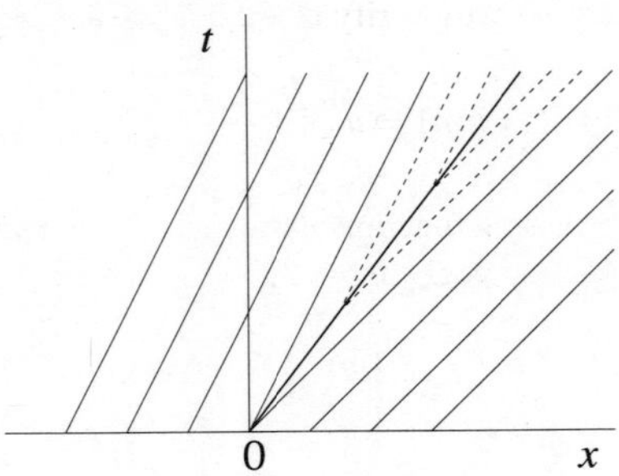

Figure 12.7. *Characteristics corresponding to an expansion shock of the Burgers' equation.*

poses a problem, since for $b(u_\ell) < b(u_\mathrm{r})$ we have at least two solutions of the Riemann problem (12.41). The expansion shock is physically not correct and should be discarded. One of the reasons is that this solution is not stable under small perturbations in the initial data, as is demonstrated in the next example.

Example 12.13 Consider the Burgers' equation subject to the piecewise linear initial condition

$$u(x,0) = \begin{cases} 0 & \text{if} \quad x < 0, \\ x/\delta & \text{if} \quad 0 < x < \delta, \\ 1 & \text{if} \quad x > \delta, \end{cases}$$

with $0 < \delta \ll 1$. Note that for $\delta \to 0$ we obtain the standard Riemann problem. The characteristics in this initial value problem fan out and we can compute its solution from (12.14). This way, we find

$$u(x,t) = \begin{cases} 0 & \text{if} \quad x < 0, \\ x/(t+\delta) & \text{if} \quad 0 < x < t+\delta, \\ 1 & \text{if} \quad x > t+\delta. \end{cases}$$

Clearly, for $\delta \to 0$ this solution changes into the rarefaction wave (12.43), which is completely different from an expansion shock. □

The next example concerns the modeling of traffic flow; cf. Example 1.2.

Example 12.14 A simple model for *traffic flow* on a highway is given by the following conservation equation:

$$\frac{\partial n}{\partial t} + \frac{\partial f(n)}{\partial x} = 0, \qquad f(n) := u_\mathrm{m} n(1 - n/n_\mathrm{m}), \tag{$*$}$$

where x is the coordinate along the highway, $n(x,t)$ is the density of cars, and n_m is the maximum density and u_m is the maximum speed of vehicles. Obviously, $n_\mathrm{m}, u_\mathrm{m} > 0$. Consider the corresponding Riemann problem with initial condition

$$n(x,0) = \begin{cases} n_\ell & \text{if} \quad x < 0, \\ n_\mathrm{r} & \text{if} \quad x > 0, \end{cases}$$

where $0 \leq n_\ell, n_r \leq n_m$. We can easily verify that the advection velocity $b(n)$ is given by

$$b(n) = u_m \left(1 - \frac{2n}{n_m}\right),$$

which is a monotonically decreasing function of n. This implies that a shock wave occurs when $n_\ell < n_r$. The shock wave solution is given by

$$n(x,t) = \begin{cases} n_\ell & \text{if} \quad x/t < s, \\ n_r & \text{if} \quad x/t > s, \end{cases}$$

with shock speed $s = u_m\big(1 - (n_\ell + n_r)/n_m\big)$. Note that the shock speed can be either positive or negative, depending on the values of n_ℓ and n_r. This corresponds, e.g., to the situation when cars approach a red traffic light. On the other hand, when $n_\ell > n_r$, we have a rarefaction wave given by

$$n(x,t) = \begin{cases} n_\ell & \text{if} \quad x/t < b(n_\ell), \\ \dfrac{1}{2} n_m \left(1 - \dfrac{x}{u_m t}\right) & \text{if} \quad b(n_\ell) < x/t < b(n_r), \\ n_r & \text{if} \quad x/t > b(n_r). \end{cases}$$

This solution describes, e.g., the situation when cars speed up after the traffic light has turned green. □

We have seen that not every discontinuous solution of (12.41) is physically correct. Therefore we would like to have a simple criterion to determine whether a discontinuous solution is admissible. In fact, the physically relevant solution is the solution of (12.18) for $\varepsilon \to 0$. One can show that this vanishing viscosity solution for the Burgers' equation reduces to a shock wave when $u_\ell > u_r$ and to a rarefaction wave when $u_\ell < u_r$; see, e.g., [70]. However, the computation of the solution of (12.18) is often very tedious and not very practical to work with. A simple criterion is suggested by the requirement that characteristics move into a shock for increasing t, as shown in Figure 12.5. This then gives rise to the following definition.

Definition 12.15 (Lax entropy condition). *A discontinuous solution of* (12.34) *that has a convex flux function satisfies the entropy condition if*

$$b(u_\ell) > s > b(u_r), \tag{12.44}$$

with s the propagation speed of the discontinuity, given by (12.38).

One can show that this *entropy solution* is the unique physically correct solution; cf. [142]. When this condition is generalized to the Euler equations for compressible gas flow, one can prove that the entropy of the flow increases across the discontinuity, in agreement with the second law of thermodynamics, and therefore (12.44) is referred to as the *entropy condition*.

Example 12.16 The flux function for the Burgers' equation is convex, and therefore the entropy condition (12.44) simply reduces to $u_\ell > u_r$. On the other hand, for the traffic flow problem we have a concave flux function, leading to the criterion $n_\ell < n_r$. □

Integrating the ODE system (12.12) along the characteristic that goes into the shock from the left gives a relation between u_ℓ and the initial data. Likewise, the characteristic going into the shock from the right provides a relation for u_r. Together with the jump condition (12.38), these relations suffice to compute the three unknowns u_ℓ, u_r, and s.

A more general definition, which is also applicable when $f(u)$ is neither convex nor concave, is the following; cf. [105].

Definition 12.17 (Oleinik's entropy condition). *A weak solution of* (12.34) *is the entropy solution if all discontinuities, which propagate at speed s given by* (12.38), *satisfy*

$$\frac{f(u)-f(u_\ell)}{u-u_\ell} \geq s \geq \frac{f(u)-f(u_r)}{u-u_r} \tag{12.45}$$

for all u between u_ℓ and u_r.

We have seen that for convex or concave flux functions the Riemann problem (12.41) has either a shock or a rarefaction wave as solution. For general flux functions the entropy solution might involve both, as demonstrated by the next example.

Example 12.18 A model equation for two-phase flow is the *Buckley–Leverett equation*, given by (12.34) with flux function

$$f(u) := \frac{u^2}{u^2 + a(1-u)^2}; \tag{$*$}$$

see Figure 12.8. Here u typically represents the saturation of water, so that $0 \leq u \leq 1$ and a is a constant. It is applied in oil reservoir simulation; for more details see, e.g., [61, 87]. Consider the Riemann problem for $(*)$ with initial data

$$u(x,0) = \begin{cases} 1 & \text{if} \quad x < 0, \\ 0 & \text{if} \quad x > 0; \end{cases}$$

The characteristic velocity $b(u)$ is given by

$$b(u) = \frac{2au(1-u)}{\left(u^2 + a(1-u)^2\right)^2}.$$

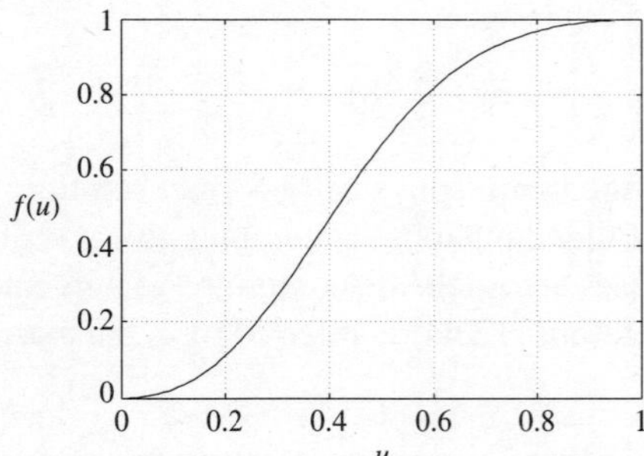

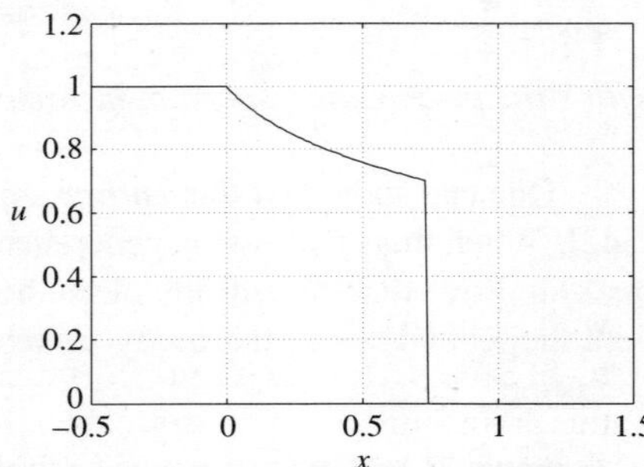

Figure 12.8. *Flux function and solution of the Buckley–Leverett equation.*

Note that $b(1) = b(0) = 0$, so that all characteristics emanate vertically from the x axis. For all intermediate values u $(0 < u < 1)$ we have $b(u) > 0$. As a consequence, the two constant states $u_r = 0$ and $u_\ell = 1$ have to be connected by a shock followed by a rarefaction wave. The solution is thus given by

$$u(x,t) = \begin{cases} 1 & \text{if } \; x/t < 0, \\ w(x/t) & \text{if } \; 0 < x/t < s, \\ 0 & \text{if } \; x/t > s, \end{cases}$$

where $w = w(\eta)$ is a similarity solution satisfying $b(w(\eta)) = \eta$ and where s is the shock speed; see Figure 12.8. Suppose the shock connects the right state $u_r = 0$ with an intermediate state u_s $(0 < u_s < 1)$. Then according to (12.38), we have

$$s = f(u_s)/u_s.$$

To determine the value of u_s we have to invoke the entropy condition (12.45), which in this case reduces to

$$\frac{f(u) - f(u_s)}{u - u_s} \geq s \geq \frac{f(u)}{u}. \tag{**}$$

It will turn out that u_s satisfies the relation

$$f(u_s)/u_s = f'(u_s),$$

which means that the straight line through $(0, 0)$ and $(u_s, f(u_s))$ is tangent to $y = f(u)$ in u_s. Moreover, the shock moving with speed $s = f'(u_s)$ is then parallel to the characteristic just left of it. Suppose the shock were connected to a state $u^* < u_s$. Then the shock speed $f(u^*)/u^*$ would be smaller than $f'(u^*)$, leading to a triple-valued function. On the other hand, if the shock were connected to a state $u^* > u_s$, then the entropy condition $(**)$ would be violated.□

12.3 First Order Systems

In this section we shall deal with first order systems of hyperbolic equations, starting with linear systems in Section 12.3.1, followed by nonlinear ones in Section 12.3.2. Finally, in Section 12.3.3, we briefly describe the method of characteristics for a system of two equations, which gives a nice illustration of the role of characteristics for hyperbolic equations.

12.3.1 Linear Systems

Consider the first order system of equations

$$\boldsymbol{A}\,\frac{\partial \boldsymbol{u}}{\partial t} + \boldsymbol{B}\,\frac{\partial \boldsymbol{u}}{\partial x} = \boldsymbol{c}, \tag{12.46}$$

with $\boldsymbol{A}$ and $\boldsymbol{B}$ constant $m \times m$ matrices and $\boldsymbol{c}$ an m vector, possibly depending on x, t, and $\boldsymbol{u}$. In Section 2.2 we considered the case where $\boldsymbol{A}$ was nonsingular. We can easily generalise this procedure for singular matrices. For this we have to consider the *generalised eigenvalue problem*: Find *left eigenvectors* $\boldsymbol{t}^T$ and corresponding eigenvalues λ such that

$$\lambda\, \boldsymbol{t}^T \boldsymbol{A} = \boldsymbol{t}^T \boldsymbol{B}. \tag{12.47}$$

Obviously, (12.47) only holds for nonzero $\boldsymbol{t}^T$ if

$$\det\big(\lambda \boldsymbol{A} - \boldsymbol{B}\big) = 0. \tag{12.48}$$

The linear combination

$$\lambda \boldsymbol{A} - \boldsymbol{B}, \qquad \lambda \in \mathbb{C}, \tag{12.49}$$

is called a *matrix pencil* [49]. This matrix pencil is called *singular* if $\det(\lambda \boldsymbol{A} - \boldsymbol{B}) = 0$ for all $\lambda \in \mathbb{C}$; otherwise it is called *regular*. We now assume the matrix pencil (12.49) to be regular. In this case there exist m eigenvalues $\lambda_1, \lambda_2, \ldots, \lambda_m$ such that (12.48) holds. Furthermore, we assume that these eigenvalues are real and that the matrix is not defect; i.e., there exists a complete set of (left) eigenvectors $\boldsymbol{t}_1^T, \boldsymbol{t}_2^T, \ldots, \boldsymbol{t}_m^T$. We can take together all (generalised) eigenvalue problems, giving the relation

$$\boldsymbol{T}\boldsymbol{B} = \boldsymbol{\Lambda}\boldsymbol{T}\boldsymbol{A}, \tag{12.50}$$

where the matrices $\boldsymbol{\Lambda}$ and $\boldsymbol{T}$ are defined by

$$\boldsymbol{\Lambda} := \operatorname{diag}(\lambda_1, \lambda_2, \ldots, \lambda_m), \qquad \boldsymbol{T} := \begin{pmatrix} \boldsymbol{t}_1^T \\ \boldsymbol{t}_2^T \\ \vdots \\ \boldsymbol{t}_m^T \end{pmatrix}. \tag{12.51}$$

Note that $\boldsymbol{T}$ is nonsingular since its rows $\boldsymbol{t}_k^T$ are linearly independent.

Left multiplication of (12.46) by $\boldsymbol{T}$ results in

$$\boldsymbol{T}\boldsymbol{A}\frac{\partial \boldsymbol{u}}{\partial t} + \boldsymbol{\Lambda}\boldsymbol{T}\boldsymbol{A}\frac{\partial \boldsymbol{u}}{\partial x} = \boldsymbol{T}\boldsymbol{c}. \tag{12.52}$$

In analogy of what we did in Section 2.2, we define the characteristic variable $\tilde{\boldsymbol{u}}$ by

$$\tilde{\boldsymbol{u}} := \boldsymbol{T}\boldsymbol{A}\boldsymbol{u}\ , \tag{12.53}$$

and since $\boldsymbol{T}$ is a constant matrix, we obtain the decoupled system (cf. (2.13a))

$$\frac{\partial \tilde{\boldsymbol{u}}}{\partial t} + \boldsymbol{\Lambda}\frac{\partial \tilde{\boldsymbol{u}}}{\partial x} = \tilde{\boldsymbol{c}} := \boldsymbol{T}\boldsymbol{c}. \tag{12.54a}$$

Written componentwise, this is

$$\frac{\partial \tilde{u}_k}{\partial t} + \lambda_k \frac{\partial \tilde{u}_k}{\partial x} = \tilde{c}_k \qquad (k = 1, 2, \ldots, m). \tag{12.54b}$$

When this decoupling is possible, we call the system (12.46) *hyperbolic*. We thus have the following formal definition.

Definition 12.19. *The system* (12.46) *is called hyperbolic if there exists a real diagonal matrix* $\boldsymbol{\Lambda}$ *and a nonsingular matrix* $\boldsymbol{T}$ *such that* (12.50) *holds.*

Example 12.20 Consider *tidal waves* travelling along a straight canal of uniform depth h. Let x denote the coordinate along the canal. For small-amplitude waves the water elevation $\eta(x, t)$ above the still-water level satisfies the standard wave equation [116]

$$\frac{\partial^2 \eta}{\partial t^2} = a^2 \frac{\partial^2 \eta}{\partial x^2},$$

with $a := \sqrt{gh}$ and g the gravitational acceleration. Introducing the auxiliary variables

$$u_1 := \frac{1}{a}\frac{\partial \eta}{\partial t}, \quad u_2 := \frac{\partial \eta}{\partial x}, \quad \boldsymbol{u} := \begin{pmatrix} u_1 \\ u_2 \end{pmatrix},$$

we can, e.g., reformulate the wave equation for η as the 2×2 linear system

$$\frac{\partial \boldsymbol{u}}{\partial t} + \boldsymbol{B}\frac{\partial \boldsymbol{u}}{\partial x} = \boldsymbol{0}, \tag{$*$}$$

with the coefficient matrix $\boldsymbol{B}$ given by

$$\boldsymbol{B} := -\begin{pmatrix} 0 & a \\ a & 0 \end{pmatrix}.$$

The eigenvalues λ_1, λ_2 and corresponding eigenvectors $\boldsymbol{t}_1^T, \boldsymbol{t}_2^T$ are given by

$$\lambda_1 = -a < 0, \quad \lambda_2 = a > 0, \quad \boldsymbol{t}_1^T = (1 \ \ 1), \quad \boldsymbol{t}_2^T = (1 \ \ -1),$$

and, consequently, the linear system $(*)$ is hyperbolic. □

Since there is no bias with respect to $\boldsymbol{A}$ or $\boldsymbol{B}$ we may consider pairs (λ, μ) of solutions of

$$\det(\lambda \boldsymbol{A} - \mu \boldsymbol{B}) = 0. \tag{12.55}$$

Note that for each pair (λ, μ) satisfying (12.55) any nonzero multiple of it also satisfies this equation. Analogously to the above, the matrix pencil

$$\lambda \boldsymbol{A} - \mu \boldsymbol{B}, \qquad \lambda, \mu \in \mathbb{C}, \tag{12.56}$$

is called singular if (12.55) holds for all $\lambda, \mu \in \mathbb{C}$; otherwise it is called regular. We shall again assume that the matrix pencil is regular. In this case there exist m *eigenvalues pairs* $(\lambda_1, \mu_1), (\lambda_2, \mu_2), \ldots, (\lambda_m, \mu_m)$, apart from a multiplicative constant. Note that for each nontrivial eigenvalue pair (λ_k, μ_k), at least one eigenvalue is nonzero. The appropriate generalisation of (12.50) then reads as follows: There exist nonsingular matrices $\boldsymbol{T}, \boldsymbol{S}$ and real diagonal matrices $\boldsymbol{\Lambda}_A, \boldsymbol{\Lambda}_B$ such that

$$\boldsymbol{T}\boldsymbol{A}\boldsymbol{S}^{-1} = \boldsymbol{\Lambda}_A, \qquad \boldsymbol{T}\boldsymbol{B}\boldsymbol{S}^{-1} = \boldsymbol{\Lambda}_B \tag{12.57}$$

i.e., the matrices $\boldsymbol{A}$ and $\boldsymbol{B}$ can be diagonalised simultaneously. In this case (12.46) is called *hyperbolic*; note that both $\boldsymbol{A}$ and $\boldsymbol{B}$ may be singular.

Premultiplying (12.46) by $\boldsymbol{T}$ and substituting

$$\tilde{\boldsymbol{u}} := \boldsymbol{S}\boldsymbol{u} \tag{12.58}$$

yields the decoupled system

$$\boldsymbol{\Lambda}_A \frac{\partial \tilde{\boldsymbol{u}}}{\partial t} + \boldsymbol{\Lambda}_B \frac{\partial \tilde{\boldsymbol{u}}}{\partial x} = \tilde{\boldsymbol{c}} := \boldsymbol{T}\boldsymbol{c} \,. \tag{12.59}$$

If a diagonal element of $\boldsymbol{\Lambda}_A$, say $\lambda_{A,k}$, is zero, then for the corresponding characteristic $\mathcal{C}_k$ we have

$$\frac{\mathrm{d}t}{\mathrm{d}x} = 0;$$

i.e., $\mathcal{C}_k$ is parallel to the x axis. This implies that the information is propagating with infinite velocity along this characteristic. We shall explicitly exclude such cases here and in the rest of this book. As a consequence, we may assume $\boldsymbol{A}$ to be nonsingular.

12.3.2 Quasi-linear Systems

We now consider the more general case where the matrices $\boldsymbol{A}$ and $\boldsymbol{B}$ may depend on x, t, and $\boldsymbol{u}$:

$$\boldsymbol{A}(x,t,\boldsymbol{u}) \frac{\partial \boldsymbol{u}}{\partial t} + \boldsymbol{B}(x,t,\boldsymbol{u}) \frac{\partial \boldsymbol{u}}{\partial x} = \boldsymbol{c}(x,t,\boldsymbol{u}). \tag{12.60}$$

Assuming now that $\boldsymbol{A}$ is nonsingular (cf. what we said in the previous section), we may as well take it equal to $\boldsymbol{I}$; i.e., we consider the system

$$\frac{\partial \boldsymbol{u}}{\partial t} + \boldsymbol{B}(x,t,\boldsymbol{u}) \frac{\partial \boldsymbol{u}}{\partial x} = \boldsymbol{c}(x,t,\boldsymbol{u}). \tag{12.61}$$

As a straightforward generalisation of Definition 12.19, we can define hyperbolicity of the system (12.61).

Definition 12.21. *The system* (12.61) *is called hyperbolic in* $(x,t,\boldsymbol{u})$ *if there exists a real diagonal matrix* $\boldsymbol{\Lambda}(x,t,\boldsymbol{u})$ *and a nonsingular matrix* $\boldsymbol{T}(x,t,\boldsymbol{u})$ *such that*

$$\boldsymbol{T}(x,t,\boldsymbol{u})\boldsymbol{B}(x,t,\boldsymbol{u}) = \boldsymbol{\Lambda}(x,t,\boldsymbol{u})\boldsymbol{T}(x,t,\boldsymbol{u}). \tag{12.62}$$

In this definition $\boldsymbol{\Lambda}(x,t,\boldsymbol{u}) = \mathrm{diag}(\lambda_1(x,t,\boldsymbol{u}), \lambda_2(x,t,\boldsymbol{u}), \ldots, \lambda_m(x,t,\boldsymbol{u}))$ and $\boldsymbol{T}(x,t,\boldsymbol{u})$ is the matrix of corresponding left eigenvectors; cf. (12.51). Note that the hyperbolicity of (12.61) depends on x, t, and $\boldsymbol{u}$. In the following we will suppress this dependency and simply write $\boldsymbol{\Lambda}$ instead of $\boldsymbol{\Lambda}(x,t,\boldsymbol{u})$, etc.

Left multiplication of (12.61) by the matrix $\boldsymbol{T}$ gives

$$\boldsymbol{T}\frac{\partial \boldsymbol{u}}{\partial t} + \boldsymbol{\Lambda}\boldsymbol{T}\frac{\partial \boldsymbol{u}}{\partial x} = \boldsymbol{T}\boldsymbol{c} =: \tilde{\boldsymbol{c}}. \tag{12.63}$$

Next we introduce the *characteristic variables* $\tilde{\boldsymbol{u}}$ by the relation

$$\mathrm{d}\tilde{\boldsymbol{u}} := \boldsymbol{T}\mathrm{d}\boldsymbol{u}, \tag{12.64}$$

where d is an arbitrary differential. Equation (12.64) is a *Pfaffian differential equation* and expresses the differentials of $\tilde{\boldsymbol{u}}$ as a linear combination of the differentials of $\boldsymbol{u}$. Conditions for the solvability of (12.64) can be found in, e.g., [173]. Note that for constant $\boldsymbol{T}$ relation

(12.64) is equivalent to (12.53) in the case $\boldsymbol{A} = \boldsymbol{I}$. Furthermore, for many nonlinear systems of conservation equations the eigenvectors $\boldsymbol{t}_k^T(\boldsymbol{u})$ can be scaled such that (12.64) is integrable. Assuming (12.64) has a solution, system (12.63) reduces to

$$\frac{\partial \tilde{\boldsymbol{u}}}{\partial t} + \boldsymbol{\Lambda} \frac{\partial \tilde{\boldsymbol{u}}}{\partial x} = \tilde{\boldsymbol{c}}. \tag{12.65}$$

Thus, like in the linear case, system (12.61) can be diagonalised. Note, however, that system (12.65) is still coupled through the eigenvalue matrix $\boldsymbol{\Lambda}$, which in general depends on $\boldsymbol{u}$.

Alternatively, we can premultiply (12.61) by an arbitrary left eigenvector $\boldsymbol{t}_k^T$ of $\boldsymbol{B}$, giving

$$\boldsymbol{t}_k^T \frac{\partial \boldsymbol{u}}{\partial t} + \boldsymbol{t}_k^T \boldsymbol{B} \frac{\partial \boldsymbol{u}}{\partial x} = \boldsymbol{t}_k^T \left(\frac{\partial \boldsymbol{u}}{\partial t} + \lambda_k \frac{\partial \boldsymbol{u}}{\partial x} \right) = \boldsymbol{t}_k^T \boldsymbol{c} =: \tilde{c}_k. \tag{12.66}$$

This is in fact a linear combination of the equations of (12.61). We would like (12.66) to be equivalent to an ODE of the form

$$\boldsymbol{t}_k^T \frac{\mathrm{d}\boldsymbol{u}}{\mathrm{d}s} = \tilde{c}_k, \tag{12.67}$$

which should hold on some curve $\mathcal{K} := \{(x(s), t(s)) \mid s \in I \subset \mathbb{R}\}$. Since we have

$$\frac{\mathrm{d}\boldsymbol{u}}{\mathrm{d}s} = \frac{\partial \boldsymbol{u}}{\partial t}\frac{\mathrm{d}t}{\mathrm{d}s} + \frac{\partial \boldsymbol{u}}{\partial x}\frac{\mathrm{d}x}{\mathrm{d}s}, \tag{12.68}$$

we thus find by comparing (12.66) and (12.67) and using relation (12.64) that

$$\frac{\mathrm{d}t}{\mathrm{d}s} = 1, \qquad \frac{\mathrm{d}x}{\mathrm{d}s} = \lambda_k, \qquad \frac{\mathrm{d}\tilde{u}_k}{\mathrm{d}t} = \tilde{c}_k. \tag{12.69}$$

The curve $\mathcal{K}$ is apparently the characteristic $\mathcal{C}_k$ corresponding to the kth eigenvalue λ_k. If system (12.61) is hyperbolic, there exist m such characteristics. Equation (12.67) is said to be in *normal form* or *characteristic form*. The variables $\tilde{\boldsymbol{u}}$ are also called *Riemann variables*. If $\boldsymbol{c} = \boldsymbol{0}$, the variables $\tilde{u}_k$ are constant along the corresponding characteristic $\mathcal{C}_k$ and are therefore often referred to as *Riemann invariants*.

In many practical applications the coefficient matrix $\boldsymbol{B}$ and the right-hand side vector $\boldsymbol{c}$ depend only on $\boldsymbol{u}$. In the rest of this book, we will restrict ourselves to this case. Then system (12.61) can be rewritten as

$$\frac{\partial \boldsymbol{u}}{\partial t} + \frac{\partial \boldsymbol{f}(\boldsymbol{u})}{\partial x} = \boldsymbol{c}(\boldsymbol{u}), \tag{12.70}$$

where the *flux vector* $\boldsymbol{f}(\boldsymbol{u})$ is related to $\boldsymbol{B}(\boldsymbol{u})$ through

$$\boldsymbol{B}(\boldsymbol{u}) = \frac{\partial \boldsymbol{f}(\boldsymbol{u})}{\partial \boldsymbol{u}} = \left(\frac{\partial f_i(\boldsymbol{u})}{\partial u_j} \right). \tag{12.71}$$

So $\boldsymbol{B}(\boldsymbol{u})$ is the Jacobi matrix of $\boldsymbol{f}(\boldsymbol{u})$. The formulation in (12.70) is again called the *conservation form*. According to Definition 12.21, the system (12.70) is hyperbolic if the Jacobi matrix $\boldsymbol{B}(\boldsymbol{u})$ is diagonalisable through its left eigenvectors. Alternatively, $\boldsymbol{B}(\boldsymbol{u})$ can be brought into diagonal form by its right eigenvectors. Indeed, introducing the matrix

$$\boldsymbol{S} = (\boldsymbol{s}_1, \boldsymbol{s}_2, \ldots, \boldsymbol{s}_m) := \boldsymbol{T}^{-1}, \tag{12.72}$$

we readily see from (12.62) that

$$\boldsymbol{B}\boldsymbol{S} = \boldsymbol{S}\boldsymbol{\Lambda}; \tag{12.73}$$

i.e., the kth column $\boldsymbol{s}_k$ of $\boldsymbol{S}$ is the right eigenvector of $\boldsymbol{B}(\boldsymbol{u})$ corresponding to the eigenvalue λ_k. Note that the right eigenvectors are linearly independent since $\boldsymbol{S}$ is nonsingular. Changing to the characteristic variables $\tilde{\boldsymbol{u}}$, which are now defined by $\mathrm{d}\tilde{\boldsymbol{u}} = \boldsymbol{S}^{-1}\mathrm{d}\boldsymbol{u}$, we obtain in a similar way as before the diagonalised system (12.65).

Example 12.22 Referring to Chapter 6, we note that the Euler equations for isentropic gas flow can be written in the standard form (12.70), with $\boldsymbol{u}$, $\boldsymbol{f}(\boldsymbol{u})$, and $\boldsymbol{c}(\boldsymbol{u})$ given by

$$\boldsymbol{u} = \begin{pmatrix} \rho \\ \rho u \end{pmatrix}, \quad \boldsymbol{f}(\boldsymbol{u}) = \begin{pmatrix} \rho u \\ \rho u^2 + p(\rho) \end{pmatrix}, \quad \boldsymbol{c}(\boldsymbol{u}) = \boldsymbol{0},$$

where ρ, u, and p are the density, velocity, and pressure, respectively, of the gas flow. For isentropic flow, the pressure is given by the relation

$$p(\rho) = p_0\rho^{\gamma}, \tag{$*$}$$

with $\gamma = C_P/C_V$ the specific heat ratio and p_0 a reference pressure. The Jacobi matrix is given by

$$\boldsymbol{B}(\boldsymbol{u}) = \begin{pmatrix} 0 & 1 \\ -u^2 + p'(\rho) & 2u \end{pmatrix},$$

and its eigenvalues $\lambda_k(\boldsymbol{u})$ and (left) eigenvectors $\boldsymbol{t}_k^T(\boldsymbol{u})$ are given by

$$\lambda_1(\boldsymbol{u}) = u - c, \quad \lambda_2(\boldsymbol{u}) = u + c, \qquad c := \sqrt{p'(\rho)} = \sqrt{\gamma p/\rho},$$

$$\boldsymbol{t}_1^T(\boldsymbol{u}) = (-u - c, 1)^T, \qquad \boldsymbol{t}_2^T(\boldsymbol{u}) = (-u + c, 1).$$

The variable c is the speed of sound. Note that the eigenvectors are determined up to a multiplicative constant. Clearly, the isentropic Euler equations are hyperbolic. To decouple these equations we introduce the characteristic variables $\tilde{\boldsymbol{u}}$ through the relation (12.64). This way, we obtain the system

$$\mathrm{d}\tilde{u}_1 = -(u + c)\mathrm{d}\rho + \mathrm{d}(\rho u) = \rho \mathrm{d}u - c\mathrm{d}\rho,$$

$$\mathrm{d}\tilde{u}_2 = (-u + c)\mathrm{d}\rho + \mathrm{d}(\rho u) = \rho \mathrm{d}u + c\mathrm{d}\rho.$$

Unfortunately, these equations are not integrable. However, scaling the eigenvectors by a factor $1/\rho$, we find the relations

$$\mathrm{d}\tilde{u}_1 = \mathrm{d}u - \frac{c}{\rho}\mathrm{d}\rho, \qquad \mathrm{d}\tilde{u}_2 = \mathrm{d}u + \frac{c}{\rho}\mathrm{d}\rho.$$

Using relation ($*$), these equations are easy to integrate, and we find

$$\tilde{u}_1 = u - \frac{2c}{\gamma - 1}, \qquad \tilde{u}_2 = u + \frac{2c}{\gamma - 1}.$$

Finally, we obtain the decoupled system

$$\frac{\partial}{\partial t}\left(u - \frac{2c}{\gamma - 1}\right) + (u - c)\frac{\partial}{\partial x}\left(u - \frac{2c}{\gamma - 1}\right) = 0,$$

$$\frac{\partial}{\partial t}\left(u + \frac{2c}{\gamma - 1}\right) + (u + c)\frac{\partial}{\partial x}\left(u + \frac{2c}{\gamma - 1}\right) = 0,$$

implying that the Riemann variables $u - \frac{2c}{\gamma-1}$ and $u + \frac{2c}{\gamma-1}$ are constant along the $\mathcal{C}_1$ and $\mathcal{C}_2$ characteristics, respectively. □

12.3.3 Method of Characteristics

For systems of two equations the normal form (12.67) together with the equations for the characteristics (12.69) lend themselves to a (theoretically) simple solution method. We shall work this out below for the general linear case (12.46), which can be written as

$$\begin{aligned} a_{11}\frac{\partial u_1}{\partial t} + a_{12}\frac{\partial u_2}{\partial t} + b_{11}\frac{\partial u_1}{\partial x} + b_{12}\frac{\partial u_2}{\partial x} &= c_1, \\ a_{21}\frac{\partial u_1}{\partial t} + a_{22}\frac{\partial u_2}{\partial t} + b_{21}\frac{\partial u_1}{\partial x} + b_{22}\frac{\partial u_2}{\partial x} &= c_2. \end{aligned} \tag{12.74}$$

Now let $\boldsymbol{v}^T := (v_1, v_2)$ be a left eigenvector corresponding to the eigenvalue $\lambda_{\boldsymbol{v}}$ of (12.47); i.e.,

$$\boldsymbol{v}^T \boldsymbol{B} = \lambda_{\boldsymbol{v}} \boldsymbol{v}^T \boldsymbol{A}. \tag{12.75}$$

If we write

$$\boldsymbol{v}^T \boldsymbol{A} =: (\tilde{v}_1, \tilde{v}_2), \qquad \boldsymbol{v}^T \boldsymbol{c} =: \tilde{c}_1 \tag{12.76}$$

and let ξ be the independent variable along the characteristic defined by

$$\frac{\mathrm{d}t}{\mathrm{d}\xi} = 1, \qquad \frac{\mathrm{d}x}{\mathrm{d}\xi} = \lambda_{\boldsymbol{v}}, \tag{12.77}$$

then we have

$$\tilde{v}_1 \frac{\partial u_1}{\partial \xi} + \tilde{v}_2 \frac{\partial u_2}{\partial \xi} = \tilde{c}_1 . \tag{12.78}$$

There also exists a second left eigenvector $\boldsymbol{w}$ corresponding to the other eigenvalue $\lambda_{\boldsymbol{w}}$, say. Define

$$\boldsymbol{w}^T \boldsymbol{A} =: (\tilde{w}_1, \tilde{w}_2), \qquad \boldsymbol{w}^T \boldsymbol{c} = \tilde{c}_2, \tag{12.79}$$

and, finally, let η be the independent variable along the corresponding characteristic, with

$$\frac{\mathrm{d}t}{\mathrm{d}\eta} = 1, \qquad \frac{\mathrm{d}x}{\mathrm{d}\eta} = \lambda_{\boldsymbol{w}}. \tag{12.80}$$

Then

$$\tilde{w}_1 \frac{\partial u_1}{\partial \eta} + \tilde{w}_2 \frac{\partial u_2}{\partial \eta} = \tilde{c}_2 . \tag{12.81}$$

The equations (12.77), (12.78), (12.80), and (12.81) form a complete system that determines both the characteristics and the solutions along them. This system lends itself to a numerical method in an obvious way. Using, e.g., forward differences leads to the *method of Massau*. In Figure 12.9 we have sketched the idea. We denote by t_{P} the value of the variable t at the point P, etc.

First we discretise (12.77) and (12.80), relating them to step sizes $\Delta\xi$ and $\Delta\eta$, respectively, which can be chosen to be constant during the process, i.e.,

$$t_{\mathrm{Q}} - t_{\mathrm{P}} \doteq \Delta\xi, \qquad x_{\mathrm{Q}} - x_{\mathrm{P}} \doteq \lambda_{\boldsymbol{v}} \Delta\xi, \tag{12.82a}$$

$$t_{\mathrm{Q}} - t_{\mathrm{R}} \doteq \Delta\eta, \qquad x_{\mathrm{Q}} - x_{\mathrm{R}} \doteq \lambda_{\boldsymbol{w}} \Delta\eta. \tag{12.82b}$$

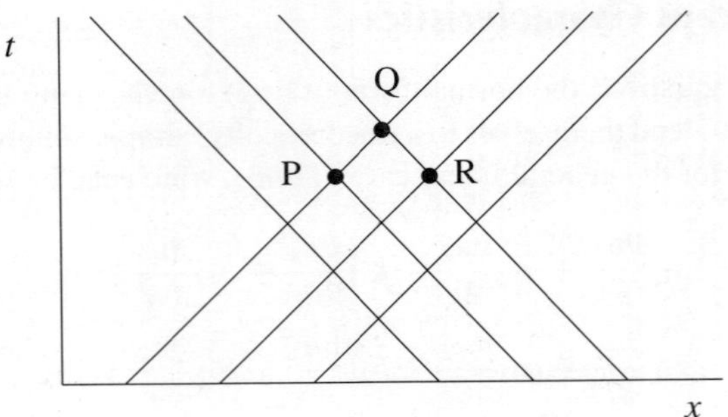

Figure 12.9. *Illustration of the method of Massau.*

These then are used in the discretised form of (12.78) and (12.81) to give

$$\tilde{v}_{1,\mathrm{P}}\big(u_{1,\mathrm{Q}} - u_{1,\mathrm{P}}\big) + \tilde{v}_{2,\mathrm{P}}\big(u_{2,\mathrm{Q}} - u_{2,\mathrm{P}}\big) = \tilde{c}_1\big(t_\mathrm{Q} - t_\mathrm{P}\big), \tag{12.83a}$$

$$\tilde{w}_{1,\mathrm{R}}\big(u_{1,\mathrm{Q}} - u_{1,\mathrm{R}}\big) + \tilde{w}_{2,\mathrm{R}}\big(u_{2,\mathrm{Q}} - u_{2,\mathrm{R}}\big) = \tilde{c}_2\big(t_\mathrm{Q} - t_\mathrm{R}\big), \tag{12.83b}$$

respectively. Suppose we are dealing with a *Cauchy problem*, so the data are given at $t = 0$. Then typically (12.82) determines the point Q, while from (12.83) we can find $u_{1,\mathrm{Q}}$ and $u_{2,\mathrm{Q}}$. From the previous approximation method it immediately follows that we will face complications if $m > 2$.

12.4 Weak Formulation of First Order Systems

In this section we consider the nonlinear system (12.70). An example of such a system is the shallow-water equations, which we will discuss in detail in the next section. In many applications nonlinear systems have discontinuous solutions, which is why we need to take recourse to weak solutions. In Section 12.4.1 we generalise the concept of weak solutions for hyperbolic systems and in Section 12.4.2 we investigate (weak) solutions of the Riemann problem.

12.4.1 Weak Solutions

In this section we give the mathematical definition of a weak solution, which is a straightforward generalization of Definition 12.8. The discussion is rather concise, since it is very similar to the scalar case.

Let $\boldsymbol{\varphi} \in \mathcal{D}^m$ be the space of test functions, with $\mathcal{D}$ defined in (12.35). If we take the inner product of (12.70) with $\boldsymbol{\varphi}$, integrate over $\mathbb{R} \times [0, \infty)$, and subsequently apply Green's theorem, we get the relation

$$\int_0^\infty \int_{-\infty}^\infty \left(\boldsymbol{u}\cdot\frac{\partial\boldsymbol{\varphi}}{\partial t} + \boldsymbol{f}(\boldsymbol{u})\cdot\frac{\partial\boldsymbol{\varphi}}{\partial x}\right) \mathrm{d}x\mathrm{d}t = -\int_{-\infty}^\infty \boldsymbol{u}(x,0)\cdot\boldsymbol{\varphi}(x,0)\,\mathrm{d}x - \int_0^\infty \int_{-\infty}^\infty \boldsymbol{c}\cdot\boldsymbol{\varphi}\,\mathrm{d}x\mathrm{d}t. \tag{12.84}$$

In the derivation of (12.84) we have used that $\varphi(x,t)$ vanishes for $|x|+t \to \infty$. We then have the following definition.

Definition 12.23. *A function $\boldsymbol{u}(x,t)$ is called a weak solution of system* (12.70) *if relation* (12.84) *holds for all test functions $\varphi \in (C_0^1(\mathbb{R}\times[0,\infty)))^m$.*

A weak solution that often occurs is a piecewise smooth solution, where the smooth parts are connected by discontinuities. These discontinuities cannot be of arbitrary size, as is apparent from the following theorem.

Theorem 12.24. *Let $\boldsymbol{u}$ be a piecewise smooth solution of* (12.70) *that has a discontinuity across a curve $\mathcal{E} : x = x(t)$. Then $\boldsymbol{u}$ satisfies the condition*

$$[\boldsymbol{f}(\boldsymbol{u})]_-^+ = s[\boldsymbol{u}]_-^+, \tag{12.85}$$

where $[\boldsymbol{v}]_-^+ := \boldsymbol{v}(x(t)+,t) - \boldsymbol{v}(x(t)-,t)$ ($\boldsymbol{v} = \boldsymbol{u}, \boldsymbol{f}(\boldsymbol{u})$) is the jump of $\boldsymbol{v}$ across $\mathcal{E}$ and s is the speed of $\mathcal{E}$.

Proof. The curve $\mathcal{E}$ separates a domain $\Omega \supset \mathrm{supp}(\varphi)$ into a left part Ω_ℓ and a right part Ω_r. The solution $\boldsymbol{u}$ is smooth in both subdomains. Since (12.70) holds in Ω_ℓ, we have

$$\iint_{\Omega_\ell} \left(\frac{\partial \boldsymbol{u}}{\partial t} + \frac{\partial \boldsymbol{f}(\boldsymbol{u})}{\partial x}\right)\cdot\varphi \,\mathrm{d}x\mathrm{d}t = \iint_{\Omega_\ell} \boldsymbol{c}\cdot\varphi\,\mathrm{d}x\mathrm{d}t$$

for every test function $\varphi \in \mathcal{D}^m$. Using the product rule of differentiation, we can rewrite this relation as

$$\iint_{\Omega_\ell} \left(\frac{\partial}{\partial t}(\boldsymbol{u}\cdot\varphi) + \frac{\partial}{\partial x}(\boldsymbol{f}(\boldsymbol{u})\cdot\varphi)\right)\mathrm{d}x\mathrm{d}t$$
$$= \iint_{\Omega_\ell} \left(\boldsymbol{u}\cdot\frac{\partial\varphi}{\partial t} + \boldsymbol{f}(\boldsymbol{u})\cdot\frac{\partial\varphi}{\partial x}\right)\mathrm{d}x\mathrm{d}t + \iint_{\Omega_\ell} \boldsymbol{c}\cdot\varphi\,\mathrm{d}x\mathrm{d}t.$$

Next we apply the two-dimensional Gauss theorem to the integral on the left-hand side to find

$$\iint_{\Omega_\ell} \left(\frac{\partial}{\partial t}(\boldsymbol{u}\cdot\varphi) + \frac{\partial}{\partial x}(\boldsymbol{f}(\boldsymbol{u})\cdot\varphi)\right)\mathrm{d}x\mathrm{d}t = -\oint_{\partial\Omega_\ell} \varphi\cdot(\boldsymbol{u}\,\mathrm{d}x - \boldsymbol{f}(\boldsymbol{u})\,\mathrm{d}t)$$
$$= -\int_a^{x_s} \boldsymbol{u}(x,0)\cdot\varphi(x,0)\,\mathrm{d}x - \int_{\mathcal{E}} \varphi\cdot\big(\boldsymbol{u}_\ell\,\mathrm{d}x - \boldsymbol{f}(\boldsymbol{u}_\ell)\,\mathrm{d}t\big),$$

with $\partial\Omega_\ell$ the boundary of Ω_ℓ and $\boldsymbol{u}_\ell := \boldsymbol{u}(x(t)-,t)$ the limit value of $\boldsymbol{u}$ just left of $\mathcal{E}$; see Figure 12.3. In the derivation of this relation we have used that $\varphi(x,t) = \boldsymbol{0}$ for $(x,t) \in \partial\Omega \cap \{t > 0\}$, with $\partial\Omega$ the boundary of Ω. Combining these relations, we obtain

$$-\int_a^{x_s} \boldsymbol{u}(x,0)\cdot\varphi(x,0)\,\mathrm{d}x - \int_{\mathcal{E}} \varphi\cdot\big(\boldsymbol{u}_\ell\mathrm{d}x - \boldsymbol{f}(\boldsymbol{u}_\ell)\mathrm{d}t\big)$$
$$= \iint_{\Omega_\ell} \left(\boldsymbol{u}\cdot\frac{\partial\varphi}{\partial t} + \boldsymbol{f}(\boldsymbol{u})\cdot\frac{\partial\varphi}{\partial x}\right)\mathrm{d}x\mathrm{d}t + \iint_{\Omega_\ell} \boldsymbol{c}\cdot\varphi\,\mathrm{d}x\mathrm{d}t. \quad (*)$$

In the same fashion we find for the right subdomain Ω_{r} that

$$-\int_{x_s}^{b} \boldsymbol{u}(x,0)\cdot\boldsymbol{\varphi}(x,0)\,\mathrm{d}x + \int_{\mathcal{E}} \boldsymbol{\varphi}\cdot(\boldsymbol{u}_{\mathrm{r}}\mathrm{d}x - \boldsymbol{f}(\boldsymbol{u}_{\mathrm{r}})\mathrm{d}t)$$
$$= \iint_{\Omega_{\mathrm{r}}} \left(\boldsymbol{u}\cdot\frac{\partial\boldsymbol{\varphi}}{\partial t} + \boldsymbol{f}(\boldsymbol{u})\cdot\frac{\partial\boldsymbol{\varphi}}{\partial x}\right)\mathrm{d}x\mathrm{d}t + \iint_{\Omega_{\mathrm{r}}} \boldsymbol{c}\cdot\boldsymbol{\varphi}\,\mathrm{d}x\mathrm{d}t, \quad (**)$$

with $\boldsymbol{u}_{\mathrm{r}} := \boldsymbol{u}(x(t)+, t)$ the limit value of $\boldsymbol{u}$ just right of $\mathcal{E}$. The integral over $\mathcal{E}$ in $(**)$ is evaluated in the same direction as in $(*)$ (see Figure 12.3) and therefore has a plus sign in front of it. Adding the relations $(*)$ and $(**)$, we find

$$-\int_{a}^{b} \boldsymbol{u}(x,0)\cdot\boldsymbol{\varphi}(x,0)\,\mathrm{d}x + \int_{\mathcal{E}} \boldsymbol{\varphi}\cdot\left([\boldsymbol{u}]_-^+\mathrm{d}x - [\boldsymbol{f}(\boldsymbol{u})]_-^+\mathrm{d}t\right)$$
$$= \iint_{\Omega} \left(\boldsymbol{u}\cdot\frac{\partial\boldsymbol{\varphi}}{\partial t} + \boldsymbol{f}(\boldsymbol{u})\cdot\frac{\partial\boldsymbol{\varphi}}{\partial x}\right)\mathrm{d}x\mathrm{d}t + \iint_{\Omega} \boldsymbol{c}\cdot\boldsymbol{\varphi}\,\mathrm{d}x\mathrm{d}t.$$

Combining this last relation with (12.84) and taking into account that $\boldsymbol{\varphi}(x,t)$ vanishes outside Ω, we have

$$\int_{\mathcal{E}} \boldsymbol{\varphi}\cdot\left([\boldsymbol{u}]_-^+\mathrm{d}x - [\boldsymbol{f}(\boldsymbol{u})]_-^+\right)\mathrm{d}t = \mathbf{0}.$$

Since this relation holds for arbitrary test functions $\boldsymbol{\varphi} \in \mathcal{D}^m$, we conclude that

$$[\boldsymbol{u}]_-^+\mathrm{d}x - [\boldsymbol{f}(\boldsymbol{u})]_-^+\mathrm{d}t = \mathbf{0},$$

and since $s = \frac{\mathrm{d}x}{\mathrm{d}t}$ it is obvious that (12.85) holds. □

Written in full, the jump relation (12.85) reads

$$\boldsymbol{f}(\boldsymbol{u}_{\mathrm{r}}) - \boldsymbol{f}(\boldsymbol{u}_{\ell}) = s(\boldsymbol{u}_{\mathrm{r}} - \boldsymbol{u}_{\ell}), \tag{12.86}$$

with $\boldsymbol{u}_{\mathrm{r}} := \boldsymbol{u}(x(t)+, t)$ and $\boldsymbol{u}_{\ell} := \boldsymbol{u}(x(t)-, t)$, and is also called the *Rankine–Hugoniot jump condition*. Relation (12.86) provides m equations for the $2m+1$ variables $\boldsymbol{u}_{\ell}$, $\boldsymbol{u}_{\mathrm{r}}$, and s. This result is needed to construct the shock wave solutions; see next section.

12.4.2 The Riemann Problem

In this section we will derive the elementary solutions of the Riemann problem for (12.70). We assume the system to be homogeneous; i.e., $\boldsymbol{c} = \mathbf{0}$. The canonical form of the Riemann problem for (12.70) reads

$$\frac{\partial \boldsymbol{u}}{\partial t} + \frac{\partial \boldsymbol{f}(\boldsymbol{u})}{\partial x} = \mathbf{0}, \qquad x \in \mathbb{R}, \quad t > 0, \tag{12.87a}$$

$$\boldsymbol{u}(x,0) = \begin{cases} \boldsymbol{u}_{\ell} & \text{if} \quad x < 0, \\ \boldsymbol{u}_{\mathrm{r}} & \text{if} \quad x > 0. \end{cases} \tag{12.87b}$$

Like in the scalar case, any solution of (12.87) is a similarity solution of the form $\boldsymbol{u}(x,t) = \hat{\boldsymbol{u}}(x/t)$.

We first consider the linear case; i.e., $\boldsymbol{f}(\boldsymbol{u}) = \boldsymbol{B}\boldsymbol{u}$ with $\boldsymbol{B}$ a constant matrix. Since the eigenvectors $\boldsymbol{s}_1, \boldsymbol{s}_2, \ldots, \boldsymbol{s}_m$ of the coefficient matrix $\boldsymbol{B}$ are linearly independent, we can decompose the initial state vectors $\boldsymbol{u}_\ell$ and $\boldsymbol{u}_\mathrm{r}$ as

$$\boldsymbol{u}_\ell = \sum_{k=1}^{m} \alpha_k \boldsymbol{s}_k, \qquad \boldsymbol{u}_\mathrm{r} = \sum_{k=1}^{m} \beta_k \boldsymbol{s}_k. \tag{12.88}$$

Alternatively, the initial condition can be written as

$$\boldsymbol{u}(x,0) = \boldsymbol{S}\tilde{\boldsymbol{u}}(x,0) = \sum_{k=1}^{m} \tilde{u}_k(x,0)\boldsymbol{s}_k. \tag{12.89}$$

Comparing (12.87b) and (12.89) and using that the eigenvectors $\boldsymbol{s}_k$ are linearly independent, we see that

$$\tilde{u}_k(x,0) = \begin{cases} \alpha_k & \text{if } x < 0, \\ \beta_k & \text{if } x > 0. \end{cases} \tag{12.90}$$

Since the eigenvalues λ_k are constant, the variables $\tilde{u}_k$ can be readily computed from the initial value problem (12.54b) and (12.90), and we find

$$\tilde{u}_k(x,t) = \tilde{u}_k(x - \lambda_k t, 0) = \begin{cases} \alpha_k & \text{if } x/t < \lambda_k, \\ \beta_k & \text{if } x/t > \lambda_k. \end{cases} \tag{12.91}$$

Inserting this in the relation $\boldsymbol{u} = \boldsymbol{S}\tilde{\boldsymbol{u}}$, we obtain the solution

$$\boldsymbol{u}(x,t) = \sum_{k=1}^{m} \tilde{u}_k(x,t)\boldsymbol{s}_k = \sum_{x/t<\lambda_k} \alpha_k \boldsymbol{s}_k + \sum_{x/t>\lambda_k} \beta_k \boldsymbol{s}_k. \tag{12.92}$$

The solution $\boldsymbol{u}$ is thus piecewise constant because the initial discontinuity at $x = 0$ propagates along all characteristics. The patches of constant $\boldsymbol{u}$ in the (x,t) plane are separated by the characteristics. As an example, we show in Figure 12.10 the solution for a 3×3 system with $\lambda_1 < 0$ and $\lambda_2, \lambda_3 > 0$.

Next we consider the quasi-linear case. Note that since $\boldsymbol{B}$ depends on $\boldsymbol{u}$, all eigenvalues and eigenvectors depend on $\boldsymbol{u}$ as well. The general solution of a Riemann problem is hard to obtain. Instead, we will derive specific elementary wave solutions corresponding to an eigenvalue, i.e., a simple wave, a contact discontinuity, and a shock. In the next section we will solve the Riemann problem for the shallow-water equations in full detail. First, we introduce the following definitions.

Definition 12.25. *An eigenvector $\boldsymbol{s}_k$ is called genuinely nonlinear if*

$$\left(\frac{\partial \lambda_k}{\partial \boldsymbol{u}}(\boldsymbol{u}), \boldsymbol{s}_k(\boldsymbol{u})\right) \neq 0 \tag{12.93a}$$

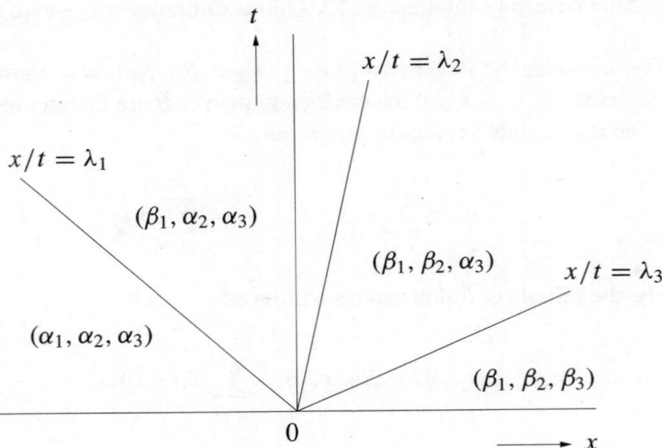

Figure 12.10. *Similarity solution of the Riemann problem for a* 3×3 *linear system. The triple* $(\beta_1, \alpha_2, \alpha_3)$ *denotes the solution* $\boldsymbol{u} = \beta_1 \boldsymbol{s}_1 + \alpha_2 \boldsymbol{s}_2 + \alpha_3 \boldsymbol{s}_3$, *etc.*

for all $\boldsymbol{u}$, *where* $\frac{\partial}{\partial \boldsymbol{u}} := (\partial/\partial u_1, \partial/\partial u_2, \ldots, \partial/\partial u_m)$. *Likewise, an eigenvector* $\boldsymbol{s}_k$ *is called linearly degenerate if*

$$\left(\frac{\partial \lambda_k}{\partial \boldsymbol{u}}(\boldsymbol{u}), \boldsymbol{s}_k(\boldsymbol{u}) \right) = 0 \tag{12.93b}$$

for all $\boldsymbol{u}$.

Case 1. Simple wave solution.

Assume that $\lambda_k(\boldsymbol{u}_\ell) < \lambda_k(\boldsymbol{u}_\mathrm{r})$ and that $\boldsymbol{s}_k$ is genuinely nonlinear. In this case we can normalize $\boldsymbol{s}_k$ such that

$$\left(\frac{\partial \lambda_k}{\partial \boldsymbol{u}}(\boldsymbol{u}), \boldsymbol{s}_k(\boldsymbol{u}) \right) = 1 \tag{12.94}$$

for all $\boldsymbol{u}$. For an arbitrary left state vector $\boldsymbol{u}_\ell$ we consider the initial value problem

$$\frac{\mathrm{d}\hat{\boldsymbol{u}}(\eta)}{\mathrm{d}\eta} = \boldsymbol{s}_k(\hat{\boldsymbol{u}}(\eta)), \qquad \eta > 0, \tag{12.95a}$$

$$\hat{\boldsymbol{u}}(0) = \boldsymbol{u}_\ell, \tag{12.95b}$$

which defines an integral curve in phase space that is everywhere tangent to $\boldsymbol{s}_k$. Let $\boldsymbol{u}_\mathrm{r} = \hat{\boldsymbol{u}}(\eta_\mathrm{r})$ for some $\eta_\mathrm{r} > 0$. Since

$$\frac{\mathrm{d}}{\mathrm{d}\eta}\big(\lambda_k(\hat{\boldsymbol{u}}(\eta))\big) = \left(\frac{\partial \lambda_k}{\partial \boldsymbol{u}}(\hat{\boldsymbol{u}}(\eta)), \boldsymbol{s}_k(\hat{\boldsymbol{u}}(\eta)) \right) = 1,$$

we have

$$\lambda_k(\hat{\boldsymbol{u}}(\eta)) = \eta + \lambda_k(\boldsymbol{u}_\ell), \qquad \lambda_k(\boldsymbol{u}_\mathrm{r}) = \eta_\mathrm{r} + \lambda_k(\boldsymbol{u}_\ell) > \lambda_k(\boldsymbol{u}_\ell).$$

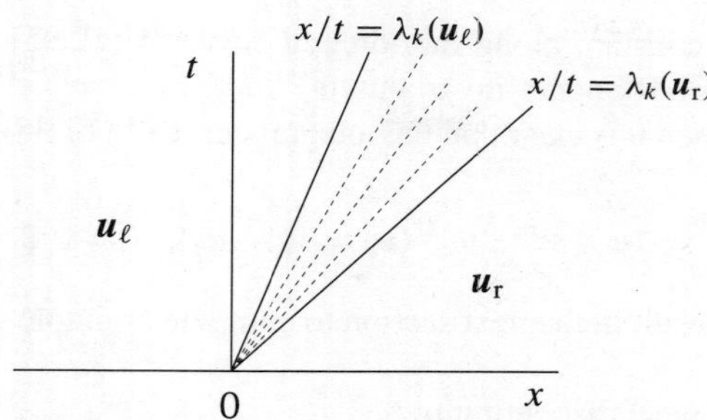

Figure 12.11. *Wave pattern of a k-simple wave.*

Next we will show that the solution of the Riemann problem (12.87) is given by

$$\boldsymbol{u}(x,t) = \begin{cases} \boldsymbol{u}_\ell & \text{if } x/t < \lambda_k(\boldsymbol{u}_\ell), \\ \hat{\boldsymbol{u}}(x/t - \lambda_k(\boldsymbol{u}_\ell)) & \text{if } \quad \lambda_k(\boldsymbol{u}_\ell) < x/t < \lambda_k(\boldsymbol{u}_\mathrm{r}), \\ \boldsymbol{u}_\mathrm{r} & \text{if } \quad \lambda_k(\boldsymbol{u}_\mathrm{r}) < x/t. \end{cases} \tag{12.96}$$

This solution is called a *k-simple wave* or a *k-rarefaction wave*. We restrict ourselves to the nontrivial case $\lambda_k(\boldsymbol{u}_\ell) < x/t < \lambda_k(\boldsymbol{u}_\mathrm{r})$. We readily see that

$$\lambda_k(\boldsymbol{u}(x,t)) = \lambda_k(\hat{\boldsymbol{u}}(x/t - \lambda_k(\boldsymbol{u}_\ell))) = x/t - \lambda_k(\boldsymbol{u}_\ell) + \lambda_k(\boldsymbol{u}_\ell) = x/t.$$

Therefore, using (12.95a) and the previous equation, we have

$$\begin{aligned} \frac{\partial \boldsymbol{u}}{\partial t} + \frac{\partial \boldsymbol{f}(\boldsymbol{u})}{\partial x} &= \frac{\partial \boldsymbol{u}}{\partial t} + \boldsymbol{B}(\boldsymbol{u})\frac{\partial \boldsymbol{u}}{\partial x} \\ &= \frac{-x}{t^2}\boldsymbol{s}_k(\hat{\boldsymbol{u}}(\eta)) + \frac{1}{t}\boldsymbol{B}(\hat{\boldsymbol{u}}(\eta))\boldsymbol{s}_k(\hat{\boldsymbol{u}}(\eta)) \\ &= \frac{1}{t}\left(-\frac{x}{t} + \lambda_k(\hat{\boldsymbol{u}}(\eta))\right)\boldsymbol{s}_k(\hat{\boldsymbol{u}}(\eta)) = \boldsymbol{0}. \end{aligned}$$

So $\boldsymbol{u}(x,t)$ defined in (12.96) is indeed the solution of the Riemann problem (12.87). An illustration of this solution is given in Figure 12.11.

For the computation of k-simple waves from the initial value problem (12.95) the *k-Riemann invariants* are useful. They are defined as follows.

Definition 12.26. *A k-Riemann invariant of* (12.87a) *is a continuously differentiable function* $w_k : \mathbb{R}^m \to \mathbb{R}$ *such that*

$$\left(\frac{\partial w_k}{\partial \boldsymbol{u}}(\boldsymbol{u}), \boldsymbol{s}_k(\boldsymbol{u})\right) = 0 \tag{12.97}$$

for all $\boldsymbol{u}$.

Relation (12.97) is a first order equation that can usually be integrated exactly. Let $\hat{\boldsymbol{u}}(\eta)$ be a solution of (12.95). Then we have

$$\frac{\mathrm{d}}{\mathrm{d}\eta}w_k(\hat{\boldsymbol{u}}(\eta)) = \left(\frac{\partial w_k}{\partial \boldsymbol{u}}(\hat{\boldsymbol{u}}(\eta)), \boldsymbol{s}_k(\hat{\boldsymbol{u}}(\eta))\right) = 0; \tag{12.98}$$

i.e., $w_k(\hat{\boldsymbol{u}}(\eta))$ is constant along the integral curve of (12.95). One can prove that there exist $m-1$ such k-Riemann invariants $w_k^{(1)}, w_k^{(2)}, \dots, w_k^{(m-1)}$ with linearly independent gradients [50]. Then it is clear that the integral curve of (12.95) is part of the curve $\mathcal{K}$ given by

$$\mathcal{K} := \{\boldsymbol{u} \in \mathbb{R}^m \mid w_k^{(j)}(\boldsymbol{u}) = w_k^{(j)}(\boldsymbol{u}_\ell),\ j = 1, 2, \dots, m-1\}. \tag{12.99}$$

We will use this result in the next section to compute k-simple waves for the shallow-water equations.

Case 2. Contact discontinuity.

Assume that $\boldsymbol{s}_k$ is linearly degenerate. Let $\hat{\boldsymbol{u}}(\eta)$ be the solution of (12.95) with $\hat{\boldsymbol{u}}(\eta_r) = \boldsymbol{u}_r$. Then we readily see that

$$\frac{\mathrm{d}}{\mathrm{d}\eta}(\lambda_k(\hat{\boldsymbol{u}}(\eta))) = \left(\frac{\partial \lambda_k}{\partial \boldsymbol{u}}(\hat{\boldsymbol{u}}(\eta)), \boldsymbol{s}_k(\hat{\boldsymbol{u}}(\eta))\right) = 0,$$

implying that $\lambda_k(\hat{\boldsymbol{u}}(\eta)) = \lambda_k(\boldsymbol{u}_\ell) = \lambda_k(\boldsymbol{u}_r)$ for all $\eta \in [0, \eta_r]$. Thus $\lambda_k(\hat{\boldsymbol{u}}(\eta))$ is constant along integral curves in the phase space that are tangent to $\boldsymbol{s}_k$. We will show that the solution of the Riemann problem is given by

$$\boldsymbol{u}(x,t) = \begin{cases} \boldsymbol{u}_\ell & \text{if } x/t < \lambda_k(\boldsymbol{u}_\ell), \\ \boldsymbol{u}_r & \text{if } \qquad \lambda_k(\boldsymbol{u}_\ell) < x/t. \end{cases} \tag{12.100}$$

This solution is called a *contact discontinuity*. To this purpose, we have to show that (12.100) satisfies the Rankine–Hugoniot condition (12.86). Since $\lambda_k(\hat{\boldsymbol{u}}(\eta))$ is constant, we have

$$\begin{aligned}\frac{\mathrm{d}}{\mathrm{d}\eta}\big(\boldsymbol{f}(\hat{\boldsymbol{u}}(\eta)) - \lambda_k(\hat{\boldsymbol{u}}(\eta))\hat{\boldsymbol{u}}(\eta)\big) &= \boldsymbol{B}(\hat{\boldsymbol{u}}(\eta))\frac{\mathrm{d}\hat{\boldsymbol{u}}}{\mathrm{d}\eta} - \lambda_k(\hat{\boldsymbol{u}}(\eta))\frac{\mathrm{d}\hat{\boldsymbol{u}}}{\mathrm{d}\eta} \\ &= (\boldsymbol{B}(\hat{\boldsymbol{u}}(\eta)) - \lambda_k(\hat{\boldsymbol{u}}(\eta))\boldsymbol{I})\boldsymbol{s}_k(\hat{\boldsymbol{u}}(\eta)) = \boldsymbol{0},\end{aligned}$$

and, consequently, $\boldsymbol{f}(\hat{\boldsymbol{u}}(\eta)) - \lambda_k(\hat{\boldsymbol{u}}(\eta))\hat{\boldsymbol{u}}(\eta) = \boldsymbol{C}$ ($\boldsymbol{C} \in \mathbb{R}^m$). In particular, we have $\boldsymbol{f}(\boldsymbol{u}_r) - \lambda_k(\boldsymbol{u}_r)\boldsymbol{u}_r = \boldsymbol{f}(\boldsymbol{u}_\ell) - \lambda_k(\boldsymbol{u}_\ell)\boldsymbol{u}_\ell$, so that the Rankine–Hugoniot condition is satisfied with $s = \lambda_k(\boldsymbol{u}_\ell) = \lambda_k(\boldsymbol{u}_r)$. An illustration is given in Figure 12.12.

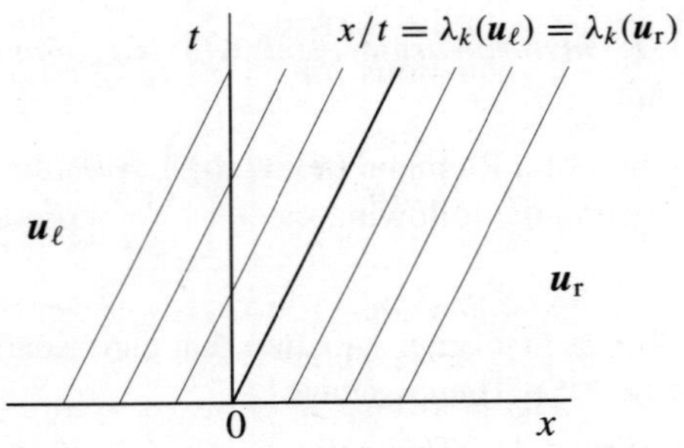

Figure 12.12. *Wave pattern of a contact discontinuity.*

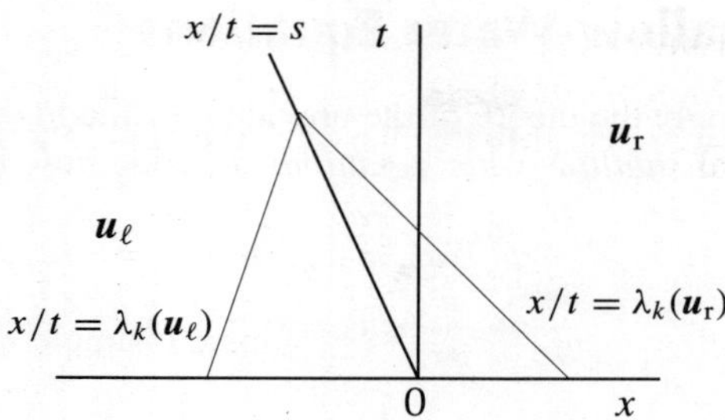

Figure 12.13. *Wave pattern of a k-shock.*

Case 3. Shock wave.

Assume that $\lambda_k(\boldsymbol{u}_\ell) > \lambda_k(\boldsymbol{u}_\mathrm{r})$ and that $\boldsymbol{s}_k$ is genuinely nonlinear. In this case the solution of the Riemann problem (12.87) is given by the *k-shock wave*

$$\boldsymbol{u}(x,t) = \begin{cases} \boldsymbol{u}_\ell & \text{if } x/t < s, \\ \boldsymbol{u}_\mathrm{r} & \text{if } \qquad s < x/t, \end{cases} \tag{12.101}$$

where the shock speed s has to satisfy the Rankine–Hugoniot jump condition (12.86). An illustration of a k-shock wave is given in Figure 12.13.

Like in the scalar case, we need a simple criterion to determine whether a shock wave is physically correct. This is given by the following definition.

Definition 12.27 (Lax entropy condition). *The k-shock wave* (12.101) *satisfies the entropy condition if the inequalities*

$$\lambda_{k-1}(\boldsymbol{u}_\ell) < s < \lambda_k(\boldsymbol{u}_\ell), \tag{12.102a}$$

$$\lambda_k(\boldsymbol{u}_\mathrm{r}) < s < \lambda_{k+1}(\boldsymbol{u}_\mathrm{r}) \tag{12.102b}$$

hold with s the shock speed given by (12.86).

These inequalities imply that $m-k+1$ characteristics move into the shock from the left and k characteristics move into the shock from the right. Integrating the ODE system (12.69) along these characteristics, we find $m-k+1$ relations between $\boldsymbol{u}_\ell$ and the initial condition left of the shock and k relations between $\boldsymbol{u}_\mathrm{r}$ and the initial condition right of the shock. Together with the Rankine–Hugoniot jump condition these constitute $m-k+1+k+1 = 2m+1$ equations for the same number of unknowns, i.e., $\boldsymbol{u}_\ell$, $\boldsymbol{u}_\mathrm{r}$, and s.

The general solution of a Riemann problem involves the elementary solutions introduced above, as described in the following theorem; for a proof see, e.g., [142].

Theorem 12.28. *Suppose that the system* (12.87a) *is hyperbolic and that each eigenvector of the Jacobi matrix of* $\boldsymbol{f}(\boldsymbol{u})$ *is either genuinely nonlinear or linearly degenerate. Then for any* $\boldsymbol{u}_\ell \in \mathbb{R}^m$ *there exists a neighbourhood* $\mathcal{N}$ *of* $\boldsymbol{u}_\ell$ *such that the Riemann problem* (12.87) *has a unique solution if* $\boldsymbol{u}_\mathrm{r} \in \mathcal{N}$*. This solution consists of at most* $m+1$ *constant states separated by shocks, simple waves, or contact discontinuities.*

12.5 The Shallow-Water Equations

In this section we apply the theory of the previous section to the shallow-water equations. The one-dimensional *shallow-water equations* describe flow in a straight canal and read [159]

$$\frac{\partial \varphi}{\partial t} + \frac{\partial}{\partial x}(\varphi u) = 0, \tag{12.103a}$$

$$\frac{\partial}{\partial t}(\varphi u) + \frac{\partial}{\partial x}\left(\varphi u^2 + \frac{1}{2}\varphi^2\right) = 0, \tag{12.103b}$$

where x is the coordinate along the canal, u is the flow velocity, and $\varphi := gh$ is the *geopotential*, with $h > 0$ the depth of the canal and g the gravitational acceleration. The first equation in (12.103) describes conservation of mass and the second describes conservation of momentum. An alternative formulation is presented in Chapter 15. These equations can be written in the standard form (12.70) with $\boldsymbol{c}(\boldsymbol{u}) = \boldsymbol{0}$ and $\boldsymbol{u}$ and $\boldsymbol{f}(\boldsymbol{u})$ defined by

$$\boldsymbol{u} := \begin{pmatrix} \varphi \\ \varphi u \end{pmatrix}, \qquad \boldsymbol{f}(\boldsymbol{u}) := \begin{pmatrix} \varphi u \\ \varphi u^2 + \frac{1}{2}\varphi^2 \end{pmatrix}. \tag{12.104}$$

The Jacobi matrix $\boldsymbol{B}(\boldsymbol{u})$ of the flux is

$$\boldsymbol{B}(\boldsymbol{u}) = \begin{pmatrix} 0 & 1 \\ \varphi - u^2 & 2u \end{pmatrix}, \tag{12.105}$$

and its eigenvalues $\lambda_k(\boldsymbol{u})$ and eigenvectors $\boldsymbol{s}_k(\boldsymbol{u})$ $(k = 1, 2)$ are given by

$$\lambda_1(\boldsymbol{u}) = u - c, \quad \lambda_2(\boldsymbol{u}) = u + c, \qquad c := \sqrt{\varphi}, \tag{12.106a}$$

$$\boldsymbol{s}_1(\boldsymbol{u}) = \begin{pmatrix} 1 \\ u - c \end{pmatrix}, \qquad \boldsymbol{s}_2(\boldsymbol{u}) = \begin{pmatrix} 1 \\ u + c \end{pmatrix}. \tag{12.106b}$$

The shallow-water equations are thus a 2×2 hyperbolic system of equations. Moreover, by direct substitution in (12.97), we see that $w_1(\boldsymbol{u}) = u + 2c$ and $w_2(\boldsymbol{u}) = u - 2c$ are the 1- and 2-Riemann invariants of the shallow-water equations, respectively.

Consider the Riemann problem for these equations, i.e.,

$$\frac{\partial \boldsymbol{u}}{\partial t} + \frac{\partial \boldsymbol{f}(\boldsymbol{u})}{\partial x} = \boldsymbol{0}, \qquad x \in \mathbb{R}, \quad t > 0, \tag{12.107a}$$

$$\boldsymbol{u}(x, 0) = \begin{cases} \boldsymbol{u}_\ell & \text{if } x < 0, \\ \boldsymbol{u}_\mathrm{r} & \text{if } x > 0, \end{cases} \tag{12.107b}$$

with $\boldsymbol{u}$ and $\boldsymbol{f}(\boldsymbol{u})$ as defined in (12.104). Since the eigenvectors $\boldsymbol{s}_k(\boldsymbol{u})$ are genuinely nonlinear, the solution of (12.107) consists of (at most) three constant states, separated by shocks and/or rarefaction waves. The possible wave patterns for the Riemann problem (12.107) are shown in Figure 12.14.

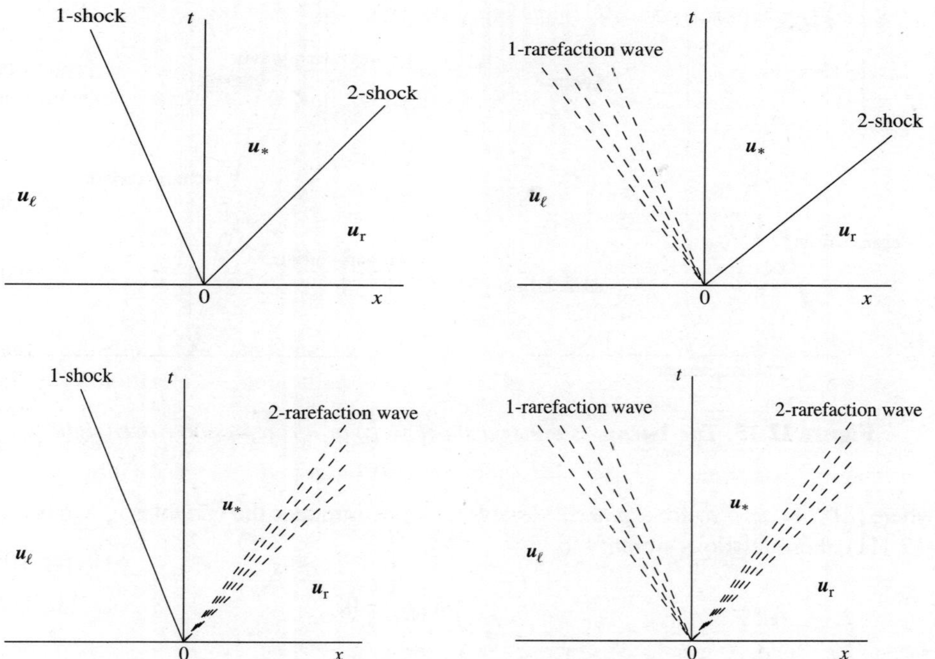

Figure 12.14. *Possible wave patterns of the Riemann problem for the shallow-water equations.*

Suppose that the constant states $\boldsymbol{u}_\ell$ and $\boldsymbol{u}_*$ are separated by a shock, referred to as the 1-shock. We want to establish a relation between $\boldsymbol{u}_\ell$ and $\boldsymbol{u}_*$ using the entropy condition and the Rankine–Hugoniot jump condition. Let s_1 denote the speed of the shock. For ease of presentation we introduce the variables

$$v_1 := u - s_1, \qquad m_1 := \varphi v_1; \tag{12.108}$$

i.e., v_1 is the flow velocity relative to the 1-shock. The entropy condition for the 1-shock reads

$$s_1 < \lambda_1(\boldsymbol{u}_\ell), \tag{12.109a}$$

$$\lambda_1(\boldsymbol{u}_*) < s_1 < \lambda_2(\boldsymbol{u}_*), \tag{12.109b}$$

which means that three characteristics go into the shock, as shown in Figure 12.15. From (12.108) and the entropy condition (12.109) we derive the inequalities

$$v_{1,\ell} > c_\ell, \qquad |v_{1,*}| < c_*. \tag{12.110}$$

The 1-shock also satisfies the jump conditions

$$s_1[\varphi]_-^+ = [\varphi u]_-^+, \tag{12.111a}$$

$$s_1[\varphi u]_-^+ = \left[\varphi u^2 + \frac{1}{2}\varphi^2\right]_-^+, \tag{12.111b}$$

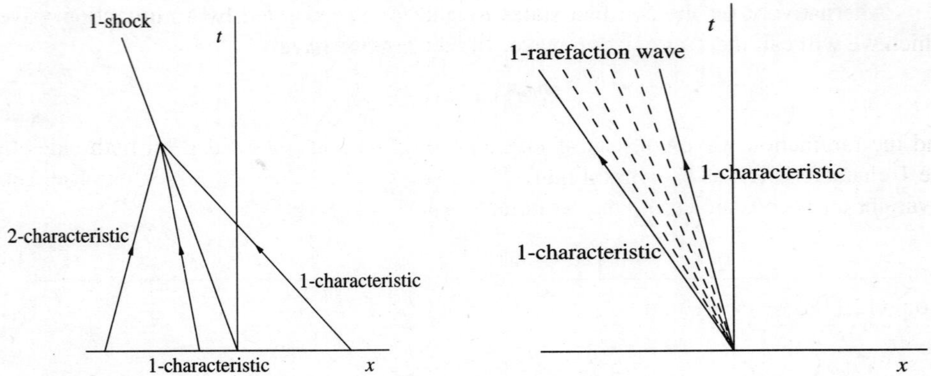

Figure 12.15. *The* 1*-wave is either a shock (left) or a rarefaction wave (right).*

where $[a]_-^+ := a_* - a_\ell$ for a generic variable a. If we introduce the variables v_1 and m_1 into (12.111), these relations simplify to

$$[m_1]_-^+ = 0, \tag{12.112a}$$

$$\left[m_1 v_1 + \frac{1}{2}\varphi^2\right]_-^+ = 0. \tag{12.112b}$$

Combining both jump conditions in (12.112) and using that $m_1 = \varphi_\ell v_{1,\ell} > 0$, we find, for m_1,

$$m_1 = \sqrt{\frac{1}{2}(\varphi_\ell + \varphi_*)\varphi_\ell \varphi_*}. \tag{12.113}$$

Furthermore, from the second jump condition (12.112b) we obtain the relation

$$u_* - u_\ell = -\frac{1}{2}\frac{\varphi_*^2 - \varphi_\ell^2}{m_1}. \tag{12.114}$$

Let the variable z_1 be defined by

$$z_1 := \frac{\varphi_*}{\varphi_\ell}. \tag{12.115}$$

From the first jump condition (12.112a) and the inequalities in (12.110) we can conclude that $z_1 > 1$; i.e., the geopotential increases when crossing the 1-shock from left to right. Substituting (12.113) into (12.114), we get the relation

$$\frac{u_* - u_\ell}{c_\ell} = \frac{1}{2}\sqrt{2}(1 - z_1)\sqrt{1 + \frac{1}{z_1}}, \tag{12.116}$$

from which we conclude that the flow velocity decreases when passing the 1-shock from left to right.

Alternatively, let the constant states $\boldsymbol{u}_\ell$ and $\boldsymbol{u}_*$ be separated by a rarefaction wave, which we will call the 1-rarefaction wave. In this case we have

$$\lambda_1(\boldsymbol{u}_*) > \lambda_1(\boldsymbol{u}_\ell), \tag{12.117}$$

and the rarefaction wave consists of rays $x/t = \lambda_1(\boldsymbol{u})$ and is bounded on both sides by the 1-characteristics, as shown in Figure 12.15. As shown in Section 12.4.2, the Riemann invariant w_1 is constant across the 1-rarefaction wave, implying that

$$u + 2c = u_\ell + 2c_\ell = u_* + 2c_*. \tag{12.118}$$

From (12.118) we easily find

$$\frac{u_* - u_\ell}{c_\ell} = 2(1 - \sqrt{z_1}), \tag{12.119}$$

with z_1 defined in (12.115). From (12.117) and (12.118) we see that $0 < z_1 < 1$; i.e., the geopotential decreases and the flow velocity increases when passing the 1-rarefaction wave from left to right

Summarizing, we have the following relation for the 1-wave connecting the states $\boldsymbol{u}_\ell$ and $\boldsymbol{u}_*$:

$$\frac{u_* - u_\ell}{c_\ell} = g(z_1), \tag{12.120}$$

with z_1 defined in (12.115) and the function $g(z)$ defined by

$$g(z) := \begin{cases} 2(1 - \sqrt{z}) & \text{if} \quad 0 < z \leq 1, \\ \frac{1}{2}\sqrt{2}(1 - z)\sqrt{1 + \frac{1}{z}} & \text{if} \quad z > 1; \end{cases} \tag{12.121}$$

see Figure 12.16. The case $0 < z_1 \leq 1$ corresponds to a 1-rarefaction wave, and for $z_1 > 1$ we have a 1-shock.

Now we consider the 2-wave connecting the constant states $\boldsymbol{u}_*$ and $\boldsymbol{u}_r$. First, consider the case of a 2-shock. Let s_2 denote the speed of the shock. Analogously to (12.108), we introduce the variables

$$v_2 := u - s_2, \qquad m_2 := \varphi v_2. \tag{12.122}$$

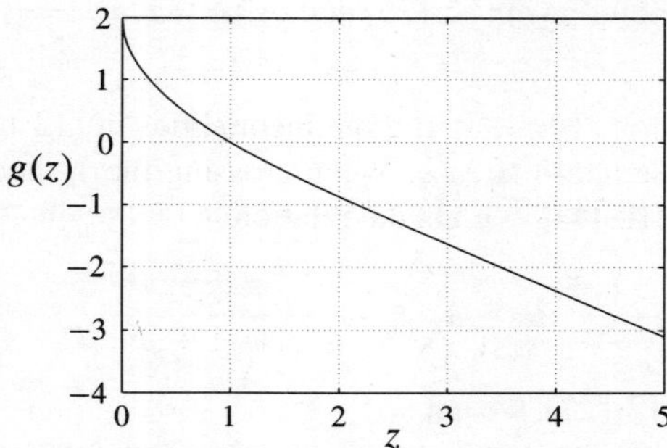

Figure 12.16. *The function* $g(z)$.

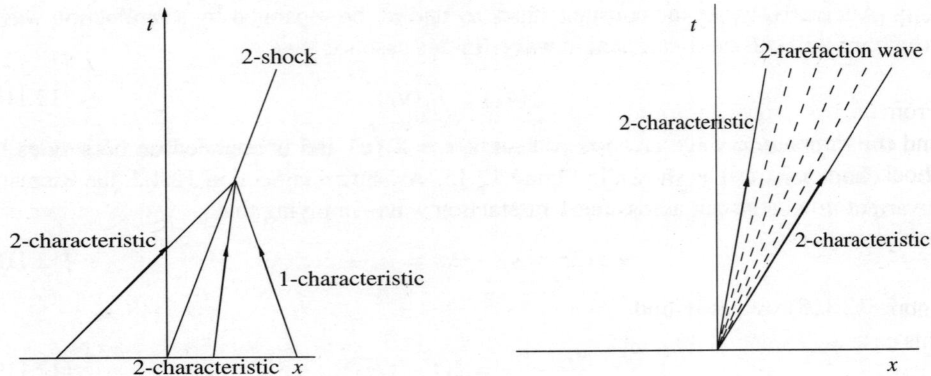

Figure 12.17. *The 2-wave is either a shock (left) or a rarefaction wave (right).*

The entropy condition for the 2-shock reads

$$\lambda_2(\boldsymbol{u}_{\mathrm{r}}) < s_2, \tag{12.123a}$$

$$\lambda_1(\boldsymbol{u}_*) < s_2 < \lambda_2(\boldsymbol{u}_*), \tag{12.123b}$$

and in this case three characteristics go into the shock, as shown in Figure 12.17. From (12.122) and the entropy condition (12.123) we find the inequalities

$$v_{2,\mathrm{r}} < -c_{\mathrm{r}}, \qquad |v_{2,*}| < c_*. \tag{12.124}$$

The jump conditions for the 2-shock are identical to (12.111), with s_1 replaced by s_2 and where now $[a]_-^+ := a_{\mathrm{r}} - a_*$ for a generic variable a. In a similar way as for the 1-shock we then obtain the jump conditions

$$[m_2]_-^+ = 0, \tag{12.125a}$$

$$\left[m_2 v_2 + \frac{1}{2}\varphi^2\right]_-^+ = 0. \tag{12.125b}$$

Combining both jump conditions in (12.125) and using that $m_2 = \varphi_{\mathrm{r}} v_{2,\mathrm{r}} < 0$, we find

$$m_2 = -\sqrt{\frac{1}{2}(\varphi_{\mathrm{r}} + \varphi_*)\varphi_{\mathrm{r}}\varphi_*}. \tag{12.126}$$

Also, from the second jump condition (12.125b) we can derive the relation

$$u_{\mathrm{r}} - u_* = -\frac{1}{2}\frac{\varphi_{\mathrm{r}}^2 - \varphi_*^2}{m_2}. \tag{12.127}$$

Substituting (12.126) in (12.127), we get

$$\frac{u_* - u_{\mathrm{r}}}{c_{\mathrm{r}}} = \frac{1}{2}\sqrt{2}(z_2 - 1)\sqrt{1 + \frac{1}{z_2}}, \tag{12.128}$$

with the variable z_2 defined by

$$z_2 := \frac{\varphi_*}{\varphi_\mathrm{r}}. \tag{12.129}$$

From the first jump condition in (12.125) and the inequalities in (12.124) we see that $z_2 > 1$, and consequently both the geopotential and the flow velocity decrease when passing the 2-shock from left to right.

Alternatively, when the constant states $\boldsymbol{u}_*$ and $\boldsymbol{u}_\mathrm{r}$ are connected by a 2-rarefaction wave, we have

$$\lambda_2(\boldsymbol{u}_\mathrm{r}) > \lambda_2(\boldsymbol{u}_*), \tag{12.130}$$

and rays $x/t = \lambda_2(\boldsymbol{u})$ are bounded by the 2-characteristics, as shown in Figure 12.17. In this case the Riemann invariant w_2 is constant across the rarefaction wave, leading to

$$u - 2c = u_* - 2c_* = u_\mathrm{r} - 2c_\mathrm{r}. \tag{12.131}$$

Rearranging terms, we obtain

$$\frac{u_* - u_\mathrm{r}}{c_\mathrm{r}} = 2(\sqrt{z_2} - 1). \tag{12.132}$$

We conclude from (12.130) and (12.131) that $0 < z_2 < 1$ and consequently both the geopotential and the flow velocity increase when passing the 2-rarefaction wave from left to right.

Summarizing, for the 2-wave connecting the states $\boldsymbol{u}_*$ and $\boldsymbol{u}_\mathrm{r}$, the following relation holds:

$$\frac{u_* - u_\mathrm{r}}{c_\mathrm{r}} = -g(z_2), \tag{12.133}$$

with the variable z_2 and the function $g(z)$ defined in (12.129) and (12.121), respectively. For $0 < z_2 < 1$ the 2-wave is a rarefaction wave and for $z_2 > 1$ it is a shock.

By eliminating of the intermediate values φ_* and u_* from the definitions in (12.115) and (12.129) and the equations (12.120) and (12.133), we obtain the equations

$$z_1\varphi_\ell = z_2\varphi_\mathrm{r}, \tag{12.134a}$$

$$u_\ell + c_\ell g(z_1) = u_\mathrm{r} - c_\mathrm{r} g(z_2). \tag{12.134b}$$

Introducing the auxiliary variables

$$A := \frac{\varphi_\mathrm{r}}{\varphi_\ell}, \qquad B := \frac{u_\mathrm{r} - u_\ell}{c_\ell} \tag{12.135}$$

and eliminating $z_2 = z_1/A$, we find the following nonlinear equation for z_1:

$$G(z_1) := g(z_1) + \sqrt{A}\, g(z_1/A) - B = 0. \tag{12.136}$$

We can verify that the function $G(z_1)$, defined in (12.136), has the properties $G(0) = 2(1 + \sqrt{A}) - B$, $G'(z_1) < 0$, and $G''(z_1) > 0$. These conditions imply that the nonlinear equation (12.136) has a unique solution provided that $G(0) > 0$. In terms of the variables $\boldsymbol{u}_\ell$ and $\boldsymbol{u}_\mathrm{r}$, this latter inequality boils down to

$$u_\mathrm{r} - u_\ell < 2(c_\ell + c_\mathrm{r}). \tag{12.137}$$

The Riemann problem (12.107) has a unique solution if the inequality (12.137) holds.

To summarize, the Riemann problem (12.107) can be solved as follows:

1. Compute A and B from (12.135).
2. Solve (12.136) for z_1.
3. Compute $z_2 = z_1/A$.
4. Determine φ_* and u_* for the intermediate state, from, e.g., (12.115) and (12.120).

In the last step we have computed the intermediate state $\boldsymbol{u}_*$ from the left state $\boldsymbol{u}_\ell$. We could have equally well computed $\boldsymbol{u}_*$ from $\boldsymbol{u}_\mathrm{r}$ using (12.129) and (12.133). One should note that step 2 above can conveniently be done numerically. Indeed, since $G(z_1) > 0$ and $G''(z_1) > 0$ on the interval $[0, z_1]$, Newton iteration for the numerical solution of (12.136) is bound to converge for an initial guess in $[0, z_1]$; cf. [149]. We still have to determine which wave pattern from Figure 12.14 is the solution of the Riemann problem. We have seen that the 1-wave is a shock if $z_1 > 1$. This condition is equivalent to $G(1) > 0$, or, stated in terms of the variables A and B, $B < \sqrt{A}g(1/A)$. Otherwise, if $B \geq \sqrt{A}g(1/A)$, the 1-wave is a rarefaction wave. Likewise, the 2-wave is a shock if $z_2 > 1$, or, equivalently, $G(A) > 0$. This latter inequality gives the condition $B < g(A)$. On the other hand, if $B \geq g(A)$, the 2-wave is a rarefaction wave.

Putting everything together, we have the following similarity solution $\boldsymbol{u}(x,t) = \hat{\boldsymbol{u}}(x/t; \boldsymbol{u}_\ell, \boldsymbol{u}_\mathrm{r})$ of the Riemann problem (12.107):

1-shock if $B < \sqrt{A}g(1/A)$:

$$\boldsymbol{u}(x,t) = \begin{cases} \boldsymbol{u}_\ell & \text{if} \quad x/t < s_1, \\ \boldsymbol{u}_* & \text{if} \quad x/t > s_1, \end{cases} \tag{12.138a}$$

with shock speed s_1 given by

$$s_1 = u_\ell - c_\ell \sqrt{\frac{1}{2}(1+z_1)z_1} \tag{12.138b}$$

1-rarefaction wave if $B \geq \sqrt{A}g(1/A)$:

$$\begin{aligned} &\boldsymbol{u}(x,t) = \boldsymbol{u}_\ell && \text{if} \quad x/t < u_\ell - c_\ell \\ &\begin{cases} u + 2c = u_\ell + 2c_\ell \\ u - \ c = x/t \end{cases} && \text{if} \quad u_\ell - c_\ell < x/t < u_* - c_*, \\ &\boldsymbol{u}(x,t) = \boldsymbol{u}_* && \text{if} \quad x/t > u_* - c_*. \end{aligned} \tag{12.139}$$

2-shock if $B < g(A)$:

$$\boldsymbol{u}(x,t) = \begin{cases} \boldsymbol{u}_* & \text{if} \quad x/t < s_2, \\ \boldsymbol{u}_\mathrm{r} & \text{if} \quad x/t > s_2, \end{cases} \tag{12.140a}$$

with shock speed s_2 given by

$$s_2 = u_\mathrm{r} + c_\mathrm{r} \sqrt{\frac{1}{2}(1+z_2)z_2} \tag{12.140b}$$

2-rarefaction wave if $B \geq g(A)$:

$$
\begin{aligned}
&\boldsymbol{u}(x,t) = \boldsymbol{u}_* && \text{if} \quad x/t < u_* + c_*, \\
&\begin{cases} u - 2c = u_\mathrm{r} - 2c_\mathrm{r} \\ u + \ \ c = x/t \end{cases} && \text{if} \quad u_* + c_* < x/t < u_\mathrm{r} + c_\mathrm{r}, \\
&\boldsymbol{u}(x,t) = \boldsymbol{u}_\mathrm{r} && \text{if} \quad x/t > u_\mathrm{r} + c_\mathrm{r}.
\end{aligned}
\tag{12.141}
$$

Expression (12.138b) for the shock speed s_1 follows readily from (12.108) and (12.113), and, likewise, expression (12.140b) for s_2 can be derived from (12.122) and (12.126).

12.6 The Wave Equation

A special kind of hyperbolic equation is given by second order problems, as discussed in Section 2.3. In particular, the *wave equation*

$$\frac{\partial^2 u}{\partial t^2} = a^2 \frac{\partial^2 u}{\partial x^2} \tag{12.142}$$

(given here in a one-dimensional medium) occurs in the modeling of many phenomena. We shall first consider solutions of this equation in one space dimension in Section 12.6.1. In Section 12.6.2 we discuss solutions in more space dimensions.

12.6.1 One-Dimensional Problems

We can easily derive a first order system associated with (12.142). Introducing the variables

$$p := \frac{\partial u}{\partial t}, \qquad q := \frac{\partial u}{\partial x},$$

we obtain

$$\frac{\partial}{\partial t}\begin{pmatrix} p \\ q \end{pmatrix} - \begin{pmatrix} 0 & a^2 \\ 1 & 0 \end{pmatrix} \frac{\partial}{\partial x}\begin{pmatrix} p \\ q \end{pmatrix} = \mathbf{0}. \tag{12.143}$$

Hence we immediately see that the system matrix has two eigenvalues $-a$ and a corresponding to the two characteristics $\mathcal{C}_1$ and $\mathcal{C}_2$ given by

$$\mathcal{C}_1 := \{(x,t) \mid x + at = \text{constant}\}, \tag{12.144a}$$

$$\mathcal{C}_2 := \{(x,t) \mid x - at = \text{constant}\}. \tag{12.144b}$$

These characteristics imply that we should describe two initial boundary conditions. In the simple case of a Cauchy problem, i.e., data given on the line $t = 0$, we can, e.g., prescribe u and $\frac{\partial u}{\partial t}$. So let

$$u(x,0) = v(x), \tag{12.145a}$$

$$\frac{\partial u}{\partial t}(x,0) = w(x) \tag{12.145b}$$

for some given functions $v(x)$ and $w(x)$. There is a nice way to construct a solution, named after d'Alembert. Since (12.143) has constant coefficients, we first note that we can find the normal form (cf. (12.69))

$$\frac{\mathrm{d}\tilde{u}_k}{\mathrm{d}s} = 0 \quad \text{on} \quad C_k \quad (k = 1, 2). \tag{12.146}$$

Because of linearity we have

$$u(x, t) = \tilde{u}_1(x + at) + \tilde{u}_2(x - at). \tag{12.147}$$

Note that $\tilde{u}_1$ and $\tilde{u}_2$ are unique but for a multiplicative constant. From (12.145a) we derive

$$\tilde{u}_1(x) + \tilde{u}_2(x) = v(x), \tag{12.148}$$

and from (12.145b) we derive

$$a\left(\tilde{u}_1'(x) - \tilde{u}_2'(x)\right) = w(x),$$

where the prime (′) denotes differentiation with respect to x. Integration of the latter equation results in

$$\tilde{u}_1(x) - \tilde{u}_2(x) = \frac{1}{a}\int_0^x w(\xi)\,\mathrm{d}\xi + C, \qquad C \in \mathbb{R}. \tag{12.149}$$

From (12.148) and (12.149) we then deduce

$$u(x, t) = \frac{1}{2}\left(v(x + at) + v(x - at)\right) + \frac{1}{2a}\int_{x-at}^{x+at} w(\xi)\,\mathrm{d}\xi. \tag{12.150}$$

Formula (12.150) is the *d'Alembert solution* of (12.142) and (12.145) and holds on the whole real line.

Wave phenomena are often defined on a semi-infinite or finite interval. One may, e.g., think of a (simplified) model for a string attached at one end or two ends. A natural way to solve such problems is by employing *reflections*. Consider first (12.142), subject to the boundary and initial conditions

$$u(0, t) = 0, \qquad\qquad t \geq 0 \tag{12.151a}$$

$$u(x, 0) = v(x), \quad \frac{\partial u}{\partial t}(x, 0) = w(x), \qquad x > 0. \tag{12.151b}$$

We then define a problem on $(-\infty, \infty)$ by continuing v and w as odd functions for negative values of the argument; i.e., the points in the graph are reflected with respect to the origin. So we have functions $\bar{v}$ and $\bar{w}$ with

$$\bar{v}(x) = -v(-x), \quad \bar{w}(x) = -w(-x), \qquad x < 0. \tag{12.152}$$

We shall omit the bar below again and consider the condition (12.151b) as defined on $(-\infty, \infty)$ now. The solution is then formally given by (12.150). At $x = 0$ we thus find

$$u(0, t) = \frac{1}{2}\left(v(at) + v(-at)\right) + \frac{1}{2a}\int_{-at}^{at} w(\xi)\,\mathrm{d}\xi = 0, \tag{12.153}$$

as is required by (12.151a). The complete solution of (12.142) and (12.151) is given by

$$u(x,t) = \begin{cases} \frac{1}{2}\big(v(x+at) + v(x-at)\big) + \frac{1}{2a}\int_{x-at}^{x+at} w(\xi)\,\mathrm{d}\xi & \text{if} \quad x > at, \\ \frac{1}{2}\big(v(at+x) - v(at-x)\big) + \frac{1}{2a}\int_{at-x}^{at+x} w(\xi)\,\mathrm{d}\xi & \text{if} \quad x < at. \end{cases} \tag{12.154}$$

We assume $w(x) = 0$. Then the initial profile $v(x)$ is split into two parts, one travelling to the right and one to the left. The left-travelling part is reflected and "inverted" at $x = 0$. This part of the solution can be interpreted as the inverted profile originating from $-x$. For a finite interval the procedure is similar, taking a reflection on the right boundary as well.

12.6.2 Solutions in Several Dimensions

We now turn to the wave equation in three space variables. So consider

$$\frac{\partial^2 u}{\partial t^2} = a^2\,\nabla^2 u \tag{12.155}$$

subject to the initial conditions

$$u(\boldsymbol{x},0) = v(\boldsymbol{x}), \qquad \frac{\partial u}{\partial t}(\boldsymbol{x},0) = w(\boldsymbol{x}). \tag{12.156}$$

One can find a d'Alembert-type solution to this problem using *averaging*. Define for $\boldsymbol{\xi} \in \mathbb{R}^3$ the average of u on a sphere $\mathcal{S}(\boldsymbol{\xi}; r)$ with centre $\boldsymbol{\xi}$ and radius r; i.e.,

$$\bar{u}(r,t;\boldsymbol{\xi}) := \frac{1}{4\pi r^2}\int_{\mathcal{S}(\boldsymbol{\xi};r)} u(\boldsymbol{x},t)\,\mathrm{d}S, \qquad r \neq 0. \tag{12.157}$$

Introducing the variable $\tilde{\boldsymbol{x}} := (\boldsymbol{x} - \boldsymbol{\xi})/r = \boldsymbol{n}$, with $\boldsymbol{n}$ the outward unit normal on $\mathcal{S}(\boldsymbol{\xi}; r)$, we may as well take the average over the unit sphere; i.e.,

$$\bar{u}(r,t;\boldsymbol{\xi}) := \frac{1}{4\pi}\int_{\mathcal{S}(\boldsymbol{0};1)} u(\boldsymbol{\xi} + r\boldsymbol{n},t)\,\mathrm{d}\tilde{S}, \tag{12.158}$$

with $\mathrm{d}\tilde{S} = \mathrm{d}S/r^2$. One easily verifies that u satisfies

$$u(\boldsymbol{\xi},t) = \bar{u}(0,t;\boldsymbol{\xi}).$$

For $\bar{u}$ we have the following property.

Property 12.29. *The variable $r\bar{u}$, with $\bar{u}$ defined in* (12.158)*, satisfies the one-dimensional equation*

$$\frac{\partial^2}{\partial t^2}(r\bar{u}) = a^2\frac{\partial^2}{\partial r^2}(r\bar{u}). \tag{12.159}$$

Proof. From (12.158) we find

$$\frac{\partial^2 \bar{u}}{\partial t^2} = \frac{1}{4\pi} \int_{\mathcal{S}(\mathbf{0};1)} \frac{\partial^2 u}{\partial t^2}(\boldsymbol{\xi} + r\boldsymbol{n}, t)\,\mathrm{d}\tilde{S} = \frac{a^2}{4\pi} \int_{\mathcal{S}(\mathbf{0};1)} \nabla^2 u(\boldsymbol{\xi} + r\boldsymbol{n}, t)\,\mathrm{d}\tilde{S}. \qquad (*)$$

On the other hand, we find from Gauss's theorem (cf. the appendix, Section J) that

$$\begin{aligned} r^2 \frac{\partial \bar{u}}{\partial r} &= \frac{r^2}{4\pi} \int_{\mathcal{S}(\mathbf{0};1)} \nabla u(\boldsymbol{\xi} + r\boldsymbol{n}, t)\cdot\boldsymbol{n}\,\mathrm{d}\tilde{S} \\ &= \frac{1}{4\pi} \int_{\mathcal{S}(\boldsymbol{\xi};r)} \nabla u(\boldsymbol{x}, t)\cdot\boldsymbol{n}\,\mathrm{d}S \\ &= \frac{1}{4\pi} \int_{\mathcal{B}(\boldsymbol{\xi};r)} \nabla^2 u(\boldsymbol{x}, t)\,\mathrm{d}V, \end{aligned}$$

where $\mathcal{B}(\boldsymbol{\xi}; r)$ is the ball of the sphere $S(\boldsymbol{\xi}; r)$. The latter integral can be rewritten as

$$\frac{1}{4\pi} \int_0^r \mathrm{d}r' \int_{\mathcal{S}(\boldsymbol{\xi};r')} \nabla^2 u(\boldsymbol{x}, t)\,\mathrm{d}S'.$$

Hence we have

$$\frac{1}{r^2}\frac{\partial}{\partial r}\left(r^2 \frac{\partial \bar{u}}{\partial r}\right) = \frac{1}{4\pi r^2} \int_{\mathcal{S}(\boldsymbol{\xi};r)} \nabla^2 u(\boldsymbol{x}, t)\,\mathrm{d}S = \frac{1}{4\pi} \int_{\mathcal{S}(\mathbf{0};1)} \nabla^2 u(\boldsymbol{\xi} + r\boldsymbol{n}, t)\,\mathrm{d}\tilde{S}. \qquad (**)$$

From $(*)$ and $(**)$ we conclude that

$$\frac{\partial^2 \bar{u}}{\partial t^2} = \frac{a^2}{r^2}\frac{\partial}{\partial r}\left(r^2 \frac{\partial \bar{u}}{\partial r}\right).$$

By straightforward manipulation this can then be rewritten as (12.159). □

From (12.159) we can obtain an expression for the solution $r\bar{u}$ as a function of r and t; i.e.,

$$r\bar{u}(r, t; \boldsymbol{\xi}) = \tilde{v}_1(r + at) + \tilde{v}_2(r - at) \qquad (12.160)$$

for some $\tilde{v}_1$, $\tilde{v}_2$. By substituting $r = 0$, we immediately see that $\tilde{v}_1(at) = -\tilde{v}_2(-at)$, so that (12.160) turns into

$$r\bar{u}(r, t; \boldsymbol{\xi}) = \tilde{v}_1(r + at) - \tilde{v}_1(at - r). \qquad (12.161)$$

We can obtain a simpler relation now for $u(\boldsymbol{\xi}, t) = \bar{u}(0, t; \boldsymbol{\xi})$ by differentiating both sides in (12.161) with respect to r and setting $r = 0$ (the prime ($'$) denotes differentiation with respect to the argument of $\tilde{v}_1$); i.e.,

$$u(\boldsymbol{\xi}, t) = 2\tilde{v}_1'(at). \qquad (12.162)$$

We now determine $\tilde{v}_1$. First we differentiate (12.161) with respect to r and set $t = 0$, resulting in

$$\frac{\partial}{\partial r}(r\bar{u})(r, 0; \boldsymbol{\xi}) = \tilde{v}_1'(r) + \tilde{v}_1'(-r). \qquad (12.163)$$

Next we differentiate (12.161) with respect to t and set $t = 0$ to find

$$\frac{\partial}{\partial t}(r\bar{u})(r, 0; \boldsymbol{\xi}) = a\big(\tilde{v}_1'(r) - \tilde{v}_1'(-r)\big). \tag{12.164}$$

In (12.163) and (12.164) we can eliminate $\tilde{v}_1'(-r)$ to find an expression for $2\tilde{v}_1(r)$ and use this to determine $u(\boldsymbol{\xi}, t)$. We obtain

$$\begin{aligned} 2\tilde{v}_1'(r) &= \frac{\partial}{\partial r}(r\bar{u})(r, 0; \boldsymbol{\xi}) + \frac{r}{a}\frac{\partial \bar{u}}{\partial t}(r, 0; \boldsymbol{\xi}) \\ &= \frac{\partial}{\partial r}\left(\frac{r}{4\pi}\int_{\mathcal{S}(\mathbf{0};1)} u(\boldsymbol{\xi} + r\boldsymbol{n}; 0)\,\mathrm{d}\tilde{S}\right) + \frac{r}{4\pi a}\frac{\partial}{\partial t}\int_{\mathcal{S}(\mathbf{0};1)} u(\boldsymbol{\xi} + r\boldsymbol{n}; 0)\,\mathrm{d}\tilde{S} \\ &= \frac{\partial}{\partial r}\left(\frac{r}{4\pi}\int_{\mathcal{S}(\mathbf{0};1)} v(\boldsymbol{\xi} + r\boldsymbol{n})\,\mathrm{d}\tilde{S}\right) + \frac{r}{4\pi a}\int_{\mathcal{S}(\mathbf{0};1)} w(\boldsymbol{\xi} + r\boldsymbol{n})\,\mathrm{d}\tilde{S}. \end{aligned} \tag{12.165}$$

The anticipated expression for $u(\boldsymbol{\xi}, t)$ is now obtained from (12.165) by substituting $r = at$:

$$\begin{aligned} u(\boldsymbol{\xi}, t) &= \frac{\partial}{\partial t}\left(\frac{t}{4\pi}\int_{\mathcal{S}(\mathbf{0};1)} v(\boldsymbol{\xi} + at\boldsymbol{n})\,\mathrm{d}\tilde{S}\right) + \frac{t}{4\pi}\int_{\mathcal{S}(\mathbf{0};1)} w(\boldsymbol{\xi} + at\boldsymbol{n})\,\mathrm{d}\tilde{S} \\ &= \frac{\partial}{\partial t}\left(\frac{1}{4\pi a^2 t}\int_{\mathcal{S}(\boldsymbol{\xi};at)} v(\boldsymbol{x})\,\mathrm{d}S\right) + \frac{1}{4\pi a^2 t}\int_{\mathcal{S}(\boldsymbol{\xi};at)} w(\boldsymbol{x})\,\mathrm{d}S \\ &= \frac{\partial}{\partial t}\big(t\bar{v}(at; \boldsymbol{\xi})\big) + t\bar{w}(at; \boldsymbol{\xi}), \end{aligned} \tag{12.166}$$

where $\bar{v}$ and $\bar{w}$ are defined analogously to (12.157). The expression in (12.166) lends itself to obtaining the solution in $\mathbb{R}^2$ by employing the fact that v and w depend on two variables only. We leave this as an exercise.

We can also find a solution of an inhomogeneous problem employing the Duhamel principle. Consider the problem

$$\frac{\partial^2 u}{\partial t^2} = a^2\nabla^2 u + s(\boldsymbol{x}, t), \tag{12.167}$$

subject to homogeneous initial conditions (12.156); i.e., $v(x) = w(x) = 0$. We recall from Section 4.6 that we can solve this initial value problem if we can find a solution $\bar{u}(\boldsymbol{x}, t; \tau)$ such that

$$\frac{\partial^2 \bar{u}}{\partial t^2} = a^2\nabla^2 \bar{u}, \qquad \boldsymbol{x} \in \mathbb{R}^3, \quad t > \tau, \tag{12.168a}$$

$$\bar{u}(\boldsymbol{x}, \tau; \tau) = 0, \qquad \boldsymbol{x} \in \mathbb{R}^3, \tag{12.168b}$$

$$\frac{\partial \bar{u}}{\partial t}(\boldsymbol{x}, \tau; \tau) = s(\boldsymbol{x}, \tau), \qquad \boldsymbol{x} \in \mathbb{R}^3. \tag{12.168c}$$

Clearly, $\bar{u}$ follows from (12.166) and is given by

$$\bar{u}(\boldsymbol{\xi}, t; \tau) = \frac{1}{4\pi a^2(t - \tau)}\int_{\mathcal{S}(\boldsymbol{\xi};a(t-\tau))} s(\boldsymbol{x}, \tau)\,\mathrm{d}S. \tag{12.169}$$

Hence we find the following representation for the solution of (12.167):

$$u(\boldsymbol{\xi}, t) = \frac{1}{4\pi a^2} \int_0^t \frac{\mathrm{d}\tau}{t-\tau} \int_{\mathcal{S}(\boldsymbol{\xi}; a(t-\tau))} s(\boldsymbol{x}, \tau)\, \mathrm{d}S. \tag{12.170}$$

Example 12.30 Consider the one-dimensional case. We can directly apply Duhamel's principle. We then find the representation

$$u(\xi, t) = \frac{1}{2a} \int_0^t \mathrm{d}\tau \int_{\xi - a(t-\tau)}^{\xi + a(t-\tau)} s(x, \tau)\, \mathrm{d}x. \qquad \square$$

Example 12.31 An important application of solution (12.160) is the following initial value problem in $\mathbb{R}^3$ of the field generated by a spherical source at $r = R$, where $r = |\boldsymbol{x}|$:

$$\begin{aligned}
&\frac{\partial^2 u}{\partial t^2} = a^2 \nabla^2 u, && r > R, \quad t > 0 \\
&u = U(t), && r = R, \quad t > 0 \\
&u \equiv 0, && r > R, \quad t < 0.
\end{aligned}$$

The general solution is thus

$$u = \frac{1}{r} F\left(t - \frac{r}{a}\right) + \frac{1}{r} G\left(t + \frac{r}{a}\right),$$

where F corresponds apparently to outgoing waves and G to incoming waves. From the causality condition $u \equiv 0$ for $t < 0$ it follows that $G = 0$. From the boundary condition we obtain $U(t) = F(t - R/a)/R$, and so we have

$$u = \frac{R}{r} U\left(t - \frac{r-R}{a}\right) \qquad \text{for} \quad r \geq R, \quad t > 0. \qquad \square$$

12.7 Boundary Conditions

Proper initial and boundary conditions are crucial for having unique solutions of PDEs. As we have seen, for hyperbolic equations these conditions are propagated along characteristics. In order to find out whether the problem is well posed, it is therefore important to know where the characteristics emanate. As we shall see, the number of conditions we can impose at a boundary, often referred to as *physical boundary conditions*, depends on (the sign of) the eigenvalues of the system.

We first discuss the linear case. Consider the following m-dimensional linear initial boundary value problem for $\boldsymbol{u}(x, t)$:

$$\frac{\partial \boldsymbol{u}}{\partial t} + \boldsymbol{B} \frac{\partial \boldsymbol{u}}{\partial x} = \boldsymbol{0}, \qquad x \in (0, 1), \quad t > 0, \tag{12.171a}$$

$$\boldsymbol{u}(x, 0) = \boldsymbol{v}(x), \qquad x \in (0, 1), \tag{12.171b}$$

$$\boldsymbol{C}_\ell \boldsymbol{u}(0, t) = \boldsymbol{g}_\ell(t), \quad \boldsymbol{C}_\mathrm{r} \boldsymbol{u}(1, t) = \boldsymbol{g}_\mathrm{r}(t), \qquad t > 0. \tag{12.171c}$$

Since the system in (12.171) is hyperbolic, the coefficient matrix $\boldsymbol{B}$ has m real eigenvalues $\lambda_1, \lambda_2, \ldots, \lambda_m$, of which p, say, are positive and $m - p$ are negative. Without restriction we may assume that they are ordered as

$$\lambda_1 \leq \lambda_2 \leq \cdots \leq \lambda_{m-p} < 0 < \lambda_{m-p+1} \leq \cdots \leq \lambda_m. \tag{12.172}$$

Furthermore, let $\boldsymbol{C}_\ell$ be an $m_\ell \times m$ matrix and $\boldsymbol{g}_\ell$ an m_ℓ vector, which means that we impose m_ℓ boundary conditions at the left boundary $x = 0$. Likewise, let $\boldsymbol{C}_\mathrm{r}$ be an $m_\mathrm{r} \times m$ matrix and $\boldsymbol{g}_\mathrm{r}$ be an m_r vector, implying that we have m_r boundary conditions at $x = 1$. Obviously, $0 \leq m_\ell, m_\mathrm{r} \leq m$.

To investigate the boundary conditions in (12.171) we need to decouple the system and write it in terms of the characteristic variables $\tilde{\boldsymbol{u}}$. In Section 12.3.2 we showed that the characteristic variable $\tilde{u}_k$ corresponding to the kth eigenvalue and its characteristic satisfy the differential equations

$$\frac{\mathrm{d}\tilde{u}_k}{\mathrm{d}t} = 0, \qquad \frac{\mathrm{d}x}{\mathrm{d}t} = \lambda_k, \tag{12.173}$$

implying that $\tilde{u}_k(x,t) = C_k$ $(C_k \in \mathbb{R})$ along the characteristic $\mathcal{C}_k : x - \lambda_k t = C$ $(C \in \mathbb{R})$. Consider the case $\lambda_k < 0$ first. From Figure 12.18 it is clear that through every point on the left boundary $x = 0$ a characteristic passes emanating from a point on either the initial line $t = 0$ or the right boundary $x = 1$. This means that we have to prescribe $\tilde{u}_k$ at $t = 0$ and $x = 1$; then $\tilde{u}_k$ is completely determined at the boundary $x = 0$, where the characteristics leave the domain. Consequently, we may not even impose any boundary condition for $\tilde{u}_k$ at $x = 0$. The case $\lambda_k > 0$ is similar. Characteristics $\mathcal{C}_k$ enter the domain at $t = 0$ and $x = 0$, and $\tilde{u}_k$ has to be specified there. On the other hand, characteristics $\mathcal{C}_k$ leave the domain at $x = 1$, so that no boundary conditions for $\tilde{u}_k$ may be given there. Summarizing, the characteristic variable $\tilde{u}_k$ must be given at the boundary from which characteristics $\mathcal{C}_k$ emanate.

We introduce the following partitioning of the characteristic variables $\tilde{\boldsymbol{u}}$:

$$\tilde{\boldsymbol{u}} = \begin{pmatrix} \tilde{\boldsymbol{u}}^- \\ \tilde{\boldsymbol{u}}^+ \end{pmatrix} \quad \text{with} \quad \tilde{\boldsymbol{u}}^- := \begin{pmatrix} \tilde{u}_1 \\ \vdots \\ \tilde{u}_{m-p} \end{pmatrix}, \quad \tilde{\boldsymbol{u}}^+ := \begin{pmatrix} \tilde{u}_{m-p+1} \\ \vdots \\ \tilde{u}_m \end{pmatrix}; \tag{12.174}$$

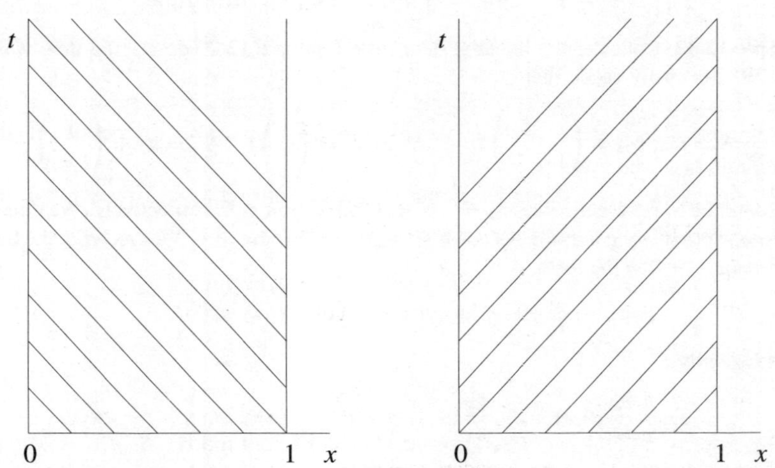

Figure 12.18. *Characteristics of the hyperbolic system* (12.171a) *corresponding to* $\lambda_k < 0$ *(left) and* $\lambda_k > 0$ *(right).*

i.e., $\tilde{\boldsymbol{u}}^-$ and $\tilde{\boldsymbol{u}}^+$ contain the characteristic variables corresponding to negative and positive eigenvalues, respectively. Likewise, we split the right eigenvector matrix $\boldsymbol{S} := \boldsymbol{T}^{-1}$ as follows:

$$\boldsymbol{S} = \left(\boldsymbol{S}^- \mid \boldsymbol{S}^+\right) \quad \text{with} \quad \boldsymbol{S}^- := \left(\boldsymbol{s}_1, \ldots, \boldsymbol{s}_{m-p}\right), \quad \boldsymbol{S}^+ := \left(\boldsymbol{s}_{m-p+1}, \ldots, \boldsymbol{s}_m\right). \tag{12.175}$$

The matrices $\boldsymbol{S}^-$ and $\boldsymbol{S}^+$ contain the eigenvectors corresponding to negative and positive eigenvalues, respectively. Now consider the boundary condition at $x = 0$ for the variable $\boldsymbol{u}$. In terms of the characteristic variable $\tilde{\boldsymbol{u}}$, it can be written as

$$\boldsymbol{C}_\ell \boldsymbol{S} \tilde{\boldsymbol{u}}(0, t) = \boldsymbol{g}_\ell(t). \tag{12.176}$$

Substituting the partitionings of (12.174) and (12.175) in (12.176), we find

$$\boldsymbol{C}_\ell \boldsymbol{S}^- \tilde{\boldsymbol{u}}^-(0, t) + \boldsymbol{C}_\ell \boldsymbol{S}^+ \tilde{\boldsymbol{u}}^+(0, t) = \boldsymbol{g}_\ell(t). \tag{12.177}$$

From the preceding discussion we conclude that $\tilde{\boldsymbol{u}}^-(0, t)$ cannot be prescribed. On the other hand, $\tilde{\boldsymbol{u}}^+(0, t)$ has to be specified. This means that the $m_\ell \times p$ matrix $\boldsymbol{C}_\ell \boldsymbol{S}^+$ in (12.177) has to be invertible. Then at least $m_\ell = p$ should hold; i.e., the number of boundary conditions at $x = 0$ is equal to the number of positive eigenvalues. Analogously to (12.177), we obtain, for the boundary condition at $x = 1$, that

$$\boldsymbol{C}_\mathrm{r} \boldsymbol{S}^- \tilde{\boldsymbol{u}}^-(1, t) + \boldsymbol{C}_\mathrm{r} \boldsymbol{S}^+ \tilde{\boldsymbol{u}}^+(1, t) = \boldsymbol{g}_\mathrm{r}(t). \tag{12.178}$$

In this case $\tilde{\boldsymbol{u}}^-(1, t)$ has to be given, which means that the $m_\mathrm{r} \times (m - p)$ matrix $\boldsymbol{C}_\mathrm{r}\boldsymbol{S}^-$ in (12.178) has to be invertible. This in turn implies that we must have $m_\mathrm{r} = m - p$, so the number of boundary conditions at $x = 1$ is equal to the number of negative eigenvalues. Finally, the boundary conditions in (12.171) have to satisfy the conditions

$$x = 0: \quad m_\ell = p, \qquad \boldsymbol{C}_\ell \boldsymbol{S}^+ \text{ invertible}, \tag{12.179a}$$

$$x = 1: \quad m_\mathrm{r} = m - p, \qquad \boldsymbol{C}_\mathrm{r} \boldsymbol{S}^- \text{ invertible}. \tag{12.179b}$$

Example 12.32 Consider the linear system from Example 12.20 describing tidal waves in a canal. We can easily verify that

$$\boldsymbol{S} = (\boldsymbol{s}_1, \boldsymbol{s}_2) = \begin{pmatrix} 1 & 1 \\ 1 & -1 \end{pmatrix}, \qquad \boldsymbol{S}^- = \boldsymbol{s}_1 = \begin{pmatrix} 1 \\ 1 \end{pmatrix}, \quad \boldsymbol{S}^+ = \boldsymbol{s}_2 = \begin{pmatrix} 1 \\ -1 \end{pmatrix},$$

corresponding to the eigenvalues $\lambda_1 = -a < 0$ and $\lambda_2 = a > 0$, respectively. We thus have to impose one boundary condition at each boundary $x = 0$ and $x = 1$. We can write the boundary condition at $x = 0$ in the form

$$\boldsymbol{c}_\ell^T \boldsymbol{u}(0, t) = g_\ell(t) \qquad \text{with} \quad \boldsymbol{c}_\ell^T \boldsymbol{s}_2 \neq 0,$$

or, equivalently,

$$c_{\ell,1} \frac{1}{a} \frac{\partial \eta}{\partial t}(0, t) + c_{\ell,2} \frac{\partial \eta}{\partial x}(0, t) = g_\ell(t) \qquad \text{with} \quad c_{\ell,1} - c_{\ell,2} \neq 0.$$

Likewise, we have at $x = 1$ the boundary condition

$$c_{\mathrm{r},1} \frac{1}{a} \frac{\partial \eta}{\partial t}(1, t) + c_{\mathrm{r},2} \frac{\partial \eta}{\partial x}(1, t) = g_\mathrm{r}(t) \qquad \text{with} \quad c_{\mathrm{r},1} + c_{\mathrm{r},2} \neq 0.$$

□

Next consider the following nonlinear initial boundary value problem for $\boldsymbol{u}(x,t)$:

$$\frac{\partial \boldsymbol{u}}{\partial t} + \frac{\partial \boldsymbol{f}(\boldsymbol{u})}{\partial x} = \boldsymbol{0}, \qquad x \in (0,1),\ t > 0, \tag{12.180a}$$

$$\boldsymbol{u}(x,0) = \boldsymbol{v}(x), \qquad x \in (0,1), \tag{12.180b}$$

$$\boldsymbol{C}_\ell(\boldsymbol{u})(0,t) = \boldsymbol{g}_\ell(t),\ \boldsymbol{C}_\mathrm{r}(\boldsymbol{u})(1,t) = \boldsymbol{g}_\mathrm{r}(t), \qquad t > 0. \tag{12.180c}$$

The nonlinear system (12.180a) is hyperbolic, which means that the Jacobi matrix $\boldsymbol{B}(\boldsymbol{u})$ has m real eigenvalues $\lambda_k(\boldsymbol{u})$. The ordering is the same as in (12.172). Since the eigenvalues depend on the solution $\boldsymbol{u}$, so does the number of positive eigenvalues p. In particular, p can be different on the boundaries $x = 0$ and $x = 1$. Therefore, we will use the notation $p(\xi, t)$ ($\xi = 0, 1$) to denote the number of positive eigenvalues at $x = \xi$. In (12.180c) $\boldsymbol{C}_\ell(\boldsymbol{u})$ is a vector function mapping the m-dimensional space onto the m_ℓ-dimensional space and $\boldsymbol{g}_\ell$ is an m_ℓ vector. We thus impose m_ℓ boundary conditions at $x = 0$. At $x = 1$ we have m_r boundary conditions, where $\boldsymbol{C}_\mathrm{r}(\boldsymbol{u})$ is a vector function mapping the m-dimensional space onto the m_r-dimensional space and $\boldsymbol{g}_\mathrm{r}$ is an m_r vector.

To investigate the boundary conditions for (12.180) we introduce the characteristic variables $\tilde{\boldsymbol{u}}$ by the Pfaffian differential equation

$$\boldsymbol{T}\mathrm{d}\boldsymbol{u} = \mathrm{d}\tilde{\boldsymbol{u}}. \tag{12.181}$$

Assuming that (12.181) is integrable, we can determine $\tilde{\boldsymbol{u}}$ as a function of $\boldsymbol{u}$. Formally inverting this function, we may write

$$\boldsymbol{u} = \boldsymbol{\mathcal{C}}(\tilde{\boldsymbol{u}}). \tag{12.182}$$

In Section 12.3.2, we showed that the characteristic variables $\tilde{u}_k$ satisfy the differential equations (12.69) and, consequently, $\tilde{u}_k$ is constant along the characteristic $\mathcal{C}_k$. Repeating the previous discussion on boundary conditions for $\tilde{u}_k$, we conclude that $\tilde{u}_k$ must be specified at the boundary at which the corresponding characteristic $\mathcal{C}_k$ enters the domain. This means that $\tilde{u}_k(0,t)$ and $\tilde{u}_k(1,t)$ must be specified if $\lambda_k(\boldsymbol{u})(0,t) > 0$ and $\lambda_k(\boldsymbol{u})(1,t) < 0$, respectively. Alternatively, in terms of the partitioning (12.174), this means that $\tilde{\boldsymbol{u}}^+(0,t)$ and $\tilde{\boldsymbol{u}}^-(1,t)$ must be given.

Now consider the boundary conditions for $\boldsymbol{u}$. Substituting the relation (12.182), we can write the boundary condition at $x = 0$ as

$$\big(\boldsymbol{C}_\ell \circ \boldsymbol{\mathcal{C}}\big)(\tilde{\boldsymbol{u}}^-, \tilde{\boldsymbol{u}}^+)(0,t) - \boldsymbol{g}_\ell(t) = 0, \tag{12.183}$$

where $\boldsymbol{C}_\ell \circ \boldsymbol{\mathcal{C}}$ is the composition of the mappings $\boldsymbol{C}_\ell$ and $\boldsymbol{\mathcal{C}}$, i.e., $(\boldsymbol{C}_\ell \circ \boldsymbol{\mathcal{C}})(\tilde{\boldsymbol{u}}) := \boldsymbol{C}_\ell(\boldsymbol{\mathcal{C}}(\tilde{\boldsymbol{u}}))$. In (12.183) we explicitly distinguish the variables $\tilde{\boldsymbol{u}}^-$ and $\tilde{\boldsymbol{u}}^+$. From this relation $\tilde{\boldsymbol{u}}^+(0,t)$ has to be determined, and according to the implicit function theorem, this can be done if the $m_\ell \times p(0,t)$ matrix $\frac{\partial}{\partial \tilde{\boldsymbol{u}}^+}\big(\boldsymbol{C}_\ell \circ \boldsymbol{\mathcal{C}}\big)(0,t)$ is invertible. Therefore we must at least have that $m_\ell = p(0,t)$; i.e., the number of boundary conditions at $x = 0$ equals the number of positive eigenvalues at $x = 0$. Analogously to (12.183), the boundary condition at $x = 1$ can be written as

$$\big(\boldsymbol{C}_\mathrm{r} \circ \boldsymbol{\mathcal{C}}\big)(\tilde{\boldsymbol{u}}^-, \tilde{\boldsymbol{u}}^+)(1,t) - \boldsymbol{g}_\mathrm{r}(t) = 0, \tag{12.184}$$

from which $\tilde{\boldsymbol{u}}^-(1,t)$ has to be determined. This leads to the requirement that the $m_r \times (m - p(1,t))$ matrix $\frac{\partial}{\partial \tilde{\boldsymbol{u}}^-}(\boldsymbol{C}_r \circ \boldsymbol{\mathcal{C}})(1,t)$ be invertible. This in turn implies that at least $m_r = m - p(1,t)$; i.e., the number of boundary conditions at $x = 1$ is equal to the number of negative eigenvalues at $x = 1$. To conclude, we have the following requirements on the boundary conditions in (12.180):

$$x = 0: \qquad m_\ell = p, \qquad \frac{\partial}{\partial \tilde{\boldsymbol{u}}^+}\big(\boldsymbol{C}_\ell \circ \boldsymbol{\mathcal{C}}\big) \text{ invertible}, \tag{12.185a}$$

$$x = 1: \qquad m_r = n - p, \qquad \frac{\partial}{\partial \tilde{\boldsymbol{u}}^-}\big(\boldsymbol{C}_r \circ \boldsymbol{\mathcal{C}}\big) \text{ invertible}. \tag{12.185b}$$

Example 12.33 Recall from Section 12.5 that the eigenvalues $\lambda_k(\boldsymbol{u})$ and eigenvectors $\boldsymbol{s}_k(\boldsymbol{u})$ of the shallow-water equations are given by

$$\lambda_1(\boldsymbol{u}) = u - c, \quad \lambda_2(\boldsymbol{u}) = u + c, \qquad c := \sqrt{\varphi},$$
$$\boldsymbol{s}_1(\boldsymbol{u}) = \begin{pmatrix} 1 \\ u - c \end{pmatrix}, \qquad \boldsymbol{s}_2(\boldsymbol{u}) = \begin{pmatrix} 1 \\ u + c \end{pmatrix}.$$

Next we have to scale the eigenvectors such that the Pfaffian differential equation $\mathrm{d}\tilde{\boldsymbol{u}} = \boldsymbol{T}\mathrm{d}\boldsymbol{u}$ with $\boldsymbol{T} = \boldsymbol{S}^{-1}$ is integrable. It turns out that we have to choose

$$\boldsymbol{T} = \frac{1}{c^2}\begin{pmatrix} -(u+c) & 1 \\ -u + c & 1 \end{pmatrix}.$$

Inserting this matrix into (12.181), we obtain the relations

$$\mathrm{d}\tilde{u}_1 = \left(\mathrm{d}u - \frac{\mathrm{d}\varphi}{\sqrt{\varphi}}\right), \qquad \mathrm{d}\tilde{u}_2 = \left(\mathrm{d}u + \frac{\mathrm{d}\varphi}{\sqrt{\varphi}}\right). \tag{$*$}$$

These equations can be trivially solved, and we find

$$\tilde{u}_1 = u - 2c, \qquad \tilde{u}_2 = u + 2c.$$

Subsequently, we can give the explicit inverse

$$\boldsymbol{u} = \boldsymbol{\mathcal{C}}(\tilde{\boldsymbol{u}}) := \frac{1}{16}(\tilde{u}_2 - \tilde{u}_1)^2\begin{pmatrix} 1 \\ \frac{1}{2}(\tilde{u}_1 + \tilde{u}_2) \end{pmatrix}. \tag{$**$}$$

The number of boundary conditions for this problem is summarized in Table 12.1. We can distinguish three cases at each boundary. We only discuss the boundary conditions at $x = 0$; the situation at $x = 1$ is similar. First, when $u \geq c$, i.e., $\lambda_1(\boldsymbol{u}) \geq 0$, we have *supercritical* inflow and both characteristic variables $\tilde{u}_1$ and $\tilde{u}_2$ have to be specified; in other words, $\tilde{\boldsymbol{u}}^+ = \tilde{\boldsymbol{u}}$. From

	$u \geq c$	$\lvert u \rvert < c$	$u \leq -c$
$x = 0$	2	$1, \tilde{u}_2$	0
$x = 1$	0	$1, \tilde{u}_1$	2

Table 12.1. *Number of boundary conditions for the shallow-water equations.*

(12.185) we deduce that the matrix $\frac{\partial}{\partial \tilde{\boldsymbol{u}}}(\boldsymbol{C}_\ell \circ \boldsymbol{C}) = \frac{\partial}{\partial \boldsymbol{u}}\boldsymbol{C}_1 \frac{\partial}{\partial \tilde{\boldsymbol{u}}}\boldsymbol{C}$ has to be invertible. Using the mappings ($*$) and ($**$), we can see that this is equivalent to the requirement

$$\det\left(\frac{\partial}{\partial \tilde{\boldsymbol{u}}}(\boldsymbol{C}_\ell \circ \boldsymbol{C})\right) = -\frac{1}{2}c^3 \det\left(\frac{\partial \boldsymbol{C}_\ell}{\partial \boldsymbol{u}}\right) \neq 0.$$

Second, when $|u| < c$ or, equivalently $\lambda_1(\boldsymbol{u}) < 0 < \lambda_2(\boldsymbol{u})$, the inflow or outflow is *subcritical*. The characteristics C_2 enter the domain and consequently $\tilde{u}_2$ has to be imposed; i.e., $\tilde{\boldsymbol{u}}^+ = \tilde{u}_2$. From (12.185) and ($**$) we can easily see that $c_\ell(\varphi, u)$ should satisfy the condition

$$\frac{\partial c_\ell}{\partial \varphi} + (u + c)\frac{\partial c_\ell}{\partial(\varphi u)} \neq 0.$$

Third, when $u \leq -c$, i.e., $\lambda_2(\boldsymbol{u}) \leq 0$, we have supercritical outflow and no boundary conditions are required. □

12.8 Discussion

- Hyperbolic equations describe propagation phenomena such as the evolution of surfaces or the propagation of waves. Examples of the latter are water waves [90, 32, 150, 169], sound waves [114, 101], and electromagnetic waves [67, 94].
- Hyperbolic problems in more than one spatial dimension are more complicated than those in one dimension. For one thing the characteristics generalise to Monge cones and are already much more complex from a geometrical point of view. A large problem area is that of fluid dynamics. We refer the interested reader to [95].
- In Chapter 16 a number of problems deal with hyperbolic equations. In Section 16.2 the etching effect of high-speed powder blasting on glass is investigated. In Sections 16.9 and 16.10 the effect of atmospheric conditions on noise and the flow along curved surfaces (as in bowls or gutters) for thin layers, respectively, are investigated.

Exercises

12.1. Determine the solution of the problem

$$x\frac{\partial u}{\partial t} - t\frac{\partial u}{\partial x} = 0, \qquad x \in (0, \infty), \quad t > 0,$$

$$u(x, 0) = x^2, \qquad x \in (0, \infty).$$

12.2. Consider the equation

$$\frac{\partial u}{\partial t} + \frac{\partial u}{\partial x} = u^2$$

subject to the condition

$$u(x, t) = t \qquad \text{for} \quad (x, t) \in \mathcal{J} := \{(x, t) \in \mathbb{R}^2 \mid x + t = 0\}.$$

Determine the solution on the relevant domain.

12.3. Determine the solution of the Cauchy problem

$$\frac{\partial u}{\partial t} + u\frac{\partial u}{\partial x} = 1, \qquad x \in \mathbb{R}, \quad t > 0,$$
$$u(x,0) = x, \qquad x \in \mathbb{R}.$$

12.4. Find the weak solution of the Burgers' equation if the initial condition is given by

$$u(0,x) = \begin{cases} 1 & \text{if} \quad x \in (-\infty, -1), \\ 0 & \text{if} \quad x \in [-1, 1], \\ 1 & \text{if} \quad x \in (1, \infty). \end{cases}$$

12.5. Verify that the shock wave (12.42) is a weak solution of (12.34).

12.6. Verify that the rarefaction wave (12.43) is a weak solution of (12.34).

12.7. Consider the system of equations

$$\frac{\partial u_1}{\partial t} + \frac{\partial u_1}{\partial x} + \frac{\partial u_2}{\partial x} = 0,$$
$$\frac{\partial u_2}{\partial t} + \frac{\partial u_1}{\partial x} + 2\frac{\partial u_2}{\partial x} + \frac{\partial u_3}{\partial x} = 0,$$
$$\frac{\partial u_3}{\partial t} - \frac{\partial u_1}{\partial x} + 2\frac{\partial u_2}{\partial x} = 0.$$

(a) Show that this system is hyperbolic.
(b) Determine the Riemann invariants.
(c) Discuss possible boundary conditions on an interval $(0, L)$.

12.8. A one-dimensional model problem from linear acoustics reads

$$\frac{\partial p}{\partial t} + K_0\frac{\partial u}{\partial x} = 0,$$
$$\rho_0\frac{\partial u}{\partial t} + \frac{\partial p}{\partial x} = 0,$$

where p and u are small perturbations to the pressure and velocity of the ambient fluid. K_0 and ρ_0 are constants. Repeat the questions from Exercise 12.7.

12.9. A model equation for one-dimensional electromagnetic waves propagating in the x direction reads

$$\frac{\partial E}{\partial t} + \frac{1}{\epsilon_0\mu_0}\frac{\partial B}{\partial x} = 0,$$
$$\frac{\partial B}{\partial t} + \frac{\partial E}{\partial x} = 0,$$

where E and B are the electric field and magnetic inductance, respectively. Repeat the questions from Exercise 12.7.

12.10. A model problem from gas dynamics is the p system [50], given by

$$\frac{\partial v}{\partial t} - \frac{\partial u}{\partial x} = 0,$$
$$\frac{\partial u}{\partial t} + \frac{\partial p(v)}{\partial x} = 0,$$

where v, u, and $p(v)$ are the specific volume, velocity, and pressure of the gas, respectively.

(a) Show that this system is hyperbolic if $p'(v) < 0$ for all $v > 0$.

(b) Investigate the Riemann problem for this system.

(c) Discuss possible boundary conditions on an interval $(0, L)$.

12.11. Show that the Euler equations for a perfect gas are hyperbolic. Compute the Riemann invariants and give the decoupled system.

12.12. Consider the wave equation with a source term, i.e.,

$$\frac{\partial^2 u}{\partial t^2} = \frac{\partial^2 u}{\partial x^2} + 2 - 6x, \quad x \in (0, 1), \quad t > 0,$$

subject to the boundary and initial conditions

$$u(0, t) = 0, \quad u(1, t) = 1, \qquad t > 0,$$
$$u(x, 0) = x^3, \quad \frac{\partial u}{\partial t}(x, 0) = 2x, \qquad x \in (0, 1).$$

Determine the solution.

12.13. The expression (12.166) lends itself to obtaining the solution in $\mathbb{R}^2$ by employing the fact that v and w depend on x and y only.

(a) Show, using spherical coordinates, that the average $\bar{v}(r, t; \boldsymbol{\xi})$ with $\boldsymbol{\xi} = (\xi, \eta, 0)$ reduces to

$$\bar{v}(r, t; \boldsymbol{\xi}) = \frac{1}{2\pi a t} \int_{\mathcal{D}(\boldsymbol{\xi}; at)} \frac{v(\boldsymbol{x})}{\sqrt{a^2 t^2 - \|\boldsymbol{x} - \boldsymbol{\xi}\|_2^2}} \, \mathrm{d}x\mathrm{d}y,$$

with $\mathcal{D}(\boldsymbol{\xi}; at)$ the disc with centre $\boldsymbol{\xi}$ and radius at, i.e., $\mathcal{D}(\boldsymbol{\xi}; at) := \{\boldsymbol{x} \in \mathbb{R}^2 \mid \|\boldsymbol{x} - \boldsymbol{\xi}\|_2 \leq at\}$.

(b) Show then that the solution of the two-dimensional wave equation, subject to the initial conditions (12.156), is given by

$$u(\boldsymbol{\xi}, t) = \frac{\partial}{\partial t} \left(\frac{1}{2\pi a} \int_{\mathcal{D}(\boldsymbol{\xi}; at)} \frac{v(\boldsymbol{x})}{\sqrt{a^2 t^2 - \|\boldsymbol{x} - \boldsymbol{\xi}\|_2^2}} \, \mathrm{d}x\mathrm{d}y \right)$$
$$+ \frac{1}{2\pi a} \int_{\mathcal{D}(\boldsymbol{\xi}; at)} \frac{w(\boldsymbol{x})}{\sqrt{a^2 t^2 - \|\boldsymbol{x} - \boldsymbol{\xi}\|_2^2}} \, \mathrm{d}x\mathrm{d}y.$$

12.14. Consider the damped wave equation

$$\frac{\partial u}{\partial t} + \frac{\partial^2 u}{\partial t^2} = a^2 \frac{\partial^2 u}{\partial x^2}.$$

(a) Write this equation as a linear first order system.

(b) Show that this system is hyperbolic.

(c) Give the decoupled system.

Chapter 13

Numerical Methods for Scalar Hyperbolic Equations

In this chapter we discuss a few numerical methods for scalar hyperbolic equations in one space dimension. We pay special attention to the advection equation as a model problem to introduce and analyse numerical methods. These numerical methods can then be generalized to nonlinear equations and even systems of equations. Therefore we start in Section 13.1 with the upwind and Lax–Wendroff schemes for the advection equation. These two schemes are the building blocks for many other numerical methods that we consider. Next, in Section 13.2, we investigate dissipation and dispersion of a numerical scheme. An alternative description of many numerical methods is based on the advection-diffusion equation. This will be discussed in Section 13.3. A special class of schemes are the nondissipative schemes, which are considered in Section 13.4. The Godunov method is a generalization of the upwind scheme to nonlinear equations and is presented in Section 13.5. From the analysis in Section 13.2 we can conclude that the upwind scheme is too dissipative and that the Lax–Wendroff scheme causes spurious oscillations in the vicinity of discontinuities. These shortcomings motivate us to introduce high-resolution schemes in Section 13.6. Two special examples of high-resolution schemes are the flux limiter method and slope limiter method. These are discussed in Sections 13.7 and 13.8, respectively. Finally, in Section 13.9 we give a concise description of numerical boundary conditions needed to carry out a computation on a finite domain.

13.1 Explicit One-Step Schemes for the Advection Equation

In Section 12.1 we introduced the advection equation, which will be our model equation here, with $a = 1$ and $c = 0$. So we shall study the equation

$$\frac{\partial u}{\partial t} + b\frac{\partial u}{\partial x} = 0, \qquad x \in \mathbb{R}, \quad t > 0, \tag{13.1a}$$

where the advection velocity b is assumed constant. The methods we will discuss will, in principle, be applied to a Cauchy problem: i.e., for u the initial condition

$$u(x, 0) = v(x), \qquad x \in \mathbb{R}, \tag{13.1b}$$

is given, where v may be a discontinuous function. We recall that the exact solution of the initial value problem (13.1) is given by

$$u(x,t) = v(x - bt). \tag{13.2}$$

Clearly, we can use this problem as a benchmark to derive and analyse numerical schemes; these schemes are easily generalised for more complex problems (albeit these are often not as easy to analyse in such situations). In this section we will consider two typical explicit one-step methods: the upwind and Lax–Wendroff schemes. There exist a few more that are used in practice, but their derivation is similar to the two types discussed below.

13.1.1 The Upwind Scheme

In order to compute a numerical solution of the initial value problem (13.1), we use a grid with points (x_j, t^n) defined by

$$x_j := j\Delta x \quad (j = 0, \pm 1, \pm 2, \ldots), \qquad t^n := n\Delta t \quad (n = 0, 1, 2, \ldots), \tag{13.3}$$

for some spatial grid size Δx and time step Δt. As we saw in Chapter 12, the characteristics play an important role and so we make a distinction between the cases $b > 0$ and $b < 0$. First, we take $b > 0$; i.e., (13.2) is a "wave" propagating in the positive x direction. For the time derivative $\frac{\partial u}{\partial t}(x_j, t^n)$ we use the forward difference approximation (5.70). In order to make the proper choice for the spatial derivative $\frac{\partial u}{\partial x}(x_j, t^n)$, we note that $u(x,t)$ is constant along characteristics $x - bt = C$ ($C \in \mathbb{R}$); see (12.5). In particular, we have $u(x_j, t^{n+1}) = u(x_j - b\Delta t, t^n)$ and, since $b > 0$, this equation suggests the backward difference approximation (5.28b) for the spatial derivative $\frac{\partial u}{\partial x}(x_j, t^n)$. Denoting by u_j^n the numerical approximation of $u(x_j, t^n)$, we thus obtain the difference scheme

$$\frac{1}{\Delta t}\left(u_j^{n+1} - u_j^n\right) + \frac{b}{\Delta x}\left(u_j^n - u_{j-1}^n\right) = 0. \tag{13.4a}$$

This scheme is called the *upwind scheme* because it uses the numerical solution at the grid point (x_{j-1}, t^n), which lies upwind of (x_j, t^{n+1}). The dependence of the numerical value u_j^{n+1} at the new time level t^{n+1} on the numerical values u_k^n ($k = j, j-1$) at the old time level t^n is displayed in the stencil in Figure 13.1. The scheme (13.4a) can be rewritten as

$$u_j^{n+1} = u_j^n - c\left(u_j^n - u_{j-1}^n\right), \tag{13.4b}$$

where c is the *Courant number* defined as

$$c := \frac{b\Delta t}{\Delta x}. \tag{13.5}$$

An alternative derivation of the upwind scheme is the following. Assume that $b\Delta t \le \Delta x$, so that $x_j - b\Delta t \in [x_{j-1}, x_j)$. Then $u(x_j - b\Delta t, t^n)$, and thus also $u(x_j, t^{n+1})$, can easily be approximated by linear interpolation, resulting in the scheme (13.4b); see Figure 13.1.

To analyse the stability of the upwind scheme we look for a planar wave solution of the form (cf. (3.33))

$$u_j^n = \mathrm{e}^{\mathrm{i}\kappa x_j - \mathrm{i}\omega t^n}, \tag{13.6}$$

where κ is the wave number and ω is the frequency. This planar wave is a solution of (13.4b), disregarding any boundary conditions, if the following *discrete dispersion relation*

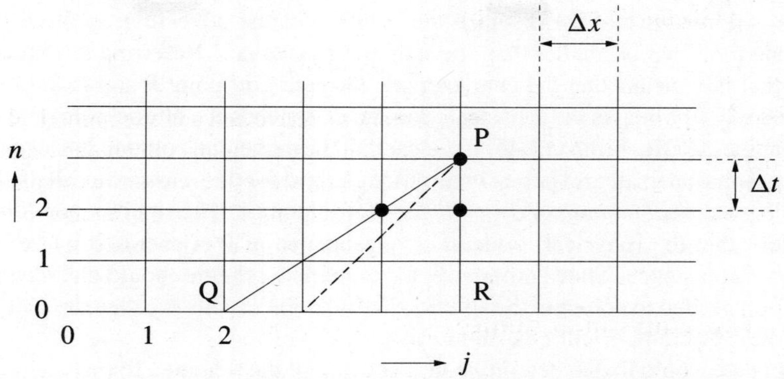

Figure 13.1. *Stencil for the upwind scheme. The dotted line is the characteristic through the point (x_j, t^{n+1}) facing back to the initial line $t = 0$.*

holds (see Section 3.4):

$$e^{-i\omega\Delta t} = 1 - c\big(1 - e^{-i\kappa\Delta x}\big). \tag{13.7}$$

This relation is rather difficult to handle and therefore we introduce, dropping the dependencies on frequency and wave number, two new variables, the *amplification factor* λ and *phase angle* φ, i.e.,

$$\lambda := e^{-i\omega\Delta t}, \qquad \varphi := \kappa\Delta x. \tag{13.8}$$

Note that we thus have

$$u_j^n = \lambda^n e^{ij\varphi}. \tag{13.9}$$

Combining (13.7) and (13.8) results in

$$\lambda(\varphi) = 1 - c\big(1 - e^{-i\varphi}\big) = 1 - 2c\sin^2\left(\frac{1}{2}\varphi\right) - ic\sin\varphi. \tag{13.10}$$

The Fourier modes (13.9) that can be represented on a fixed spatial grid have a phase φ ranging from zero, corresponding to the constant mode $u_j^n = \lambda^n$, to π, corresponding to the high-frequency mode $u_j^n = \lambda^n(-1)^j$ of wavelength $2\pi/\kappa = 2\Delta x$. Therefore we have to investigate $\lambda(\varphi)$ for $0 < \varphi < \pi$. Taking the modulus of (13.10), we find

$$\begin{aligned} |\lambda(\varphi)|^2 &= \left(1 - 2c\sin^2\left(\frac{1}{2}\varphi\right)\right)^2 + c^2\sin^2\varphi \\ &= 1 - 4c(1-c)\sin^2\left(\frac{1}{2}\varphi\right). \end{aligned} \tag{13.11}$$

Stability requires that $|\lambda(\varphi)| \leq 1$ and, obviously, this requirement is fulfilled if

$$c = \frac{b\Delta t}{\Delta x} \leq 1. \tag{13.12}$$

Therefore (13.12) yields the *stability condition* of the upwind scheme. Since it restricts the choice of the time step, the upwind scheme is only *conditionally stable*.

An interpretation of the stability condition (13.12) is the following. The inequality $b\Delta t/\Delta x \leq 1$ means that the distance the solution of the advection equation propagates during one time step is smaller than the spatial grid size Δx. Referring to Figure 13.1, we can see that this means that the characteristic through grid point P and facing back to the initial line $t = 0$, which is the (physical) domain of dependence of grid point P, is contained in the triangle $\triangle$PQR. From (13.4b) it is clear that the numerical solution in P depends on the numerical solution in all grid points in this triangle, and for this reason we call this set of grid points the *numerical domain of dependence* of grid point P. The stability condition (13.12) thus means that the (physical) domain of dependence of P is enclosed by the numerical domain of dependence. Stated otherwise, the numerical scheme should include all physical information needed to compute the solution in grid point P. This is called the *CFL condition* (named after Courant, Friedrichs, and Lewy).

Next we would like to determine the accuracy of the scheme. To that end we compute the *local discretisation error* d_j^n defined as

$$d_j^n := \frac{1}{\Delta t}\big(u(x_j, t^{n+1}) - u(x_j, t^n)\big) + \frac{b}{\Delta x}\big(u(x_j, t^n) - u(x_{j-1}, t^n)\big). \tag{13.13}$$

The local discretisation error d_j^n is thus the residual of the difference scheme (13.4a) after substituting the exact solution of (13.1a). Assuming that $u(x, t)$ is sufficiently smooth and using Taylor series expansions of $u(x, t)$ about (x_j, t^n), we find

$$\begin{aligned} d_j^n &= \frac{1}{2}\Delta t \frac{\partial^2 u}{\partial t^2}(x_j, t^n) - \frac{1}{2} b\Delta x \frac{\partial^2 u}{\partial x^2}(x_j, t^n) + \mathcal{O}(\Delta t^2) + \mathcal{O}(\Delta x^2) \\ &= \frac{1}{2}\Delta t\left(1 - \frac{1}{c}\right)\frac{\partial^2 u}{\partial t^2}(x_j, t^n) + \mathcal{O}(\Delta t^2). \end{aligned} \tag{13.14}$$

The second line in (13.14) follows from the relation $\frac{\partial^2 u}{\partial t^2} = b^2 \frac{\partial^2 u}{\partial x^2}$, which readily follows from (13.1a). Consequently, the upwind scheme is *first order consistent* in the time step Δt and the spatial grid size Δx, which was to be expected, since we only used one-sided differences for both $\frac{\partial u}{\partial t}$ and $\frac{\partial u}{\partial x}$.

For the other situation, where $b < 0$, the derivation is similar. We then take the forward difference approximation (5.28a) for the spatial derivative $\frac{\partial u}{\partial x}(x_j, t^n)$ to obtain

$$\frac{1}{\Delta t}\left(u_j^{n+1} - u_j^n\right) + \frac{b}{\Delta x}\left(u_{j+1}^n - u_j^n\right) = 0. \tag{13.15}$$

The Courant number is negative and the stability condition now reads

$$\frac{b\Delta t}{\Delta x} \geq -1. \tag{13.16}$$

This scheme is again first order consistent in Δt and Δx, since we used one-sided differences for the derivatives $\frac{\partial u}{\partial t}$ and $\frac{\partial u}{\partial x}$.

We can combine the difference schemes (13.4a) and (13.15) for the general case, i.e., when $\operatorname{sign}(b)$ is not known beforehand. To do this we introduce a splitting of the velocity b as follows:

$$b = b^+ + b^-, \qquad b^+ := \max(b, 0) \geq 0, \quad b^- := \min(b, 0) \leq 0. \tag{13.17}$$

A straightforward generalisation of both upwind schemes then reads

$$\frac{1}{\Delta t}(u_j^{n+1} - u_j^n) + \frac{b^+}{\Delta x}(u_j^n - u_{j-1}^n) + \frac{b^-}{\Delta x}(u_{j+1}^n - u_j^n) = 0. \tag{13.18}$$

The b^+ term in (13.18) is the backward difference approximation of $b^+\frac{\partial u}{\partial x}(x_j, t^n)$ and corresponds to a wave propagating in the positive x direction. Likewise, the b^- term in (13.18) is the forward difference approximation of $b^-\frac{\partial u}{\partial x}(x_j, t^n)$ and corresponds to a wave propagating in the negative x direction. Thus the sign of $b^\pm$ determines the upwind direction, i.e., the direction of the one-sided approximation of $\frac{\partial u}{\partial x}(x_j, t^n)$. In Section 14.7.1 we will generalize this approach for hyperbolic systems. It is obvious that this version of the upwind scheme is first order consistent in Δt and Δx.

Finally, we want to assess the stability of the scheme (13.18). Substituting the Fourier mode (13.9) in (13.18), we find (in a similar fashion as done for (13.11)) the following expression for the amplification factor:

$$|\lambda(\varphi)|^2 = 1 - 4|c|(1 - |c|)\sin^2\left(\frac{1}{2}\varphi\right). \tag{13.19}$$

From (13.19) we deduce that the upwind scheme (13.18) is stable under the condition

$$\frac{|b|\Delta t}{\Delta x} \leq 1. \tag{13.20}$$

This condition can again be interpreted as a CFL condition; i.e., the domain of dependence of a grid point is entirely enclosed by its numerical domain of dependence.

Example 13.1 In this example we computed the upwind numerical solution of (13.1a) subject to the initial condition $v(x) = \sin 8\pi x$. We have chosen the (numerical) parameters $b = 1$, $\Delta x = 2.5 \times 10^{-2}$, and $\Delta t = 2 \times 10^{-2}$, resulting in a stable scheme for which $c = 0.8$. Comparing the numerical solution after 50 time steps with the exact solution (dotted line), we conclude that the upwind scheme suffers from severe damping (see Figure 13.2). □

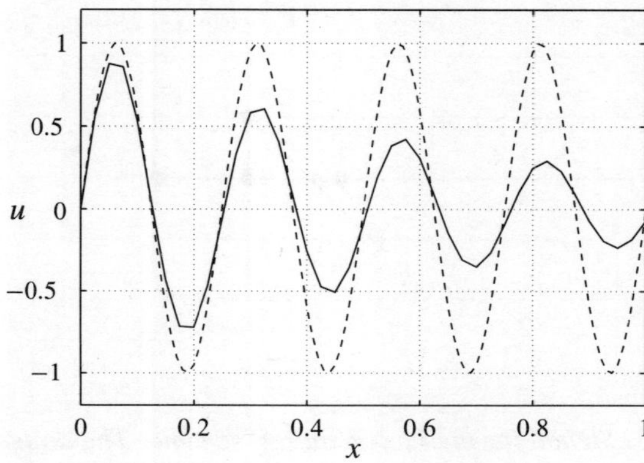

Figure 13.2. *Numerical approximation of a sine wave computed with the upwind scheme.*

13.1.2 The Lax–Wendroff Scheme

The upwind scheme in the previous section is only first order accurate. For practical problems, however, first order accuracy is often not enough. There are various schemes that give higher-order results, the best known being the Lax–Wendroff scheme; it is second order accurate in both Δt and Δx. The derivation is somewhat unconventional. In Section 13.3 we address the idea behind this. For now let us consider the Taylor series expansion

$$u(x_j, t^{n+1}) = u(x_j, t^n) + \Delta t \frac{\partial u}{\partial t}(x_j, t^n) + \frac{1}{2}\Delta t^2 \frac{\partial^2 u}{\partial t^2}(x_j, t^n) + \mathcal{O}(\Delta t^3), \tag{13.21}$$

which includes terms up to and including $\frac{1}{2}\Delta t^2 \frac{\partial^2 u}{\partial t^2}$ in order to achieve second order accuracy in Δt. Next we replace the time derivatives in (13.21) by space derivatives. We use the relation $\frac{\partial^2 u}{\partial t^2} = b^2 \frac{\partial^2 u}{\partial x^2}$ (cf. (13.1a)) and substitute this in (13.21) to obtain the expansion

$$u(x_j, t^{n+1}) = u(x_j, t^n) - b\Delta t \frac{\partial u}{\partial x}(x_j, t^n) + \frac{1}{2} b^2 \Delta t^2 \frac{\partial^2 u}{\partial x^2}(x_j, t^n) + \mathcal{O}(\Delta t^3). \tag{13.22}$$

In order to get a scheme that is also second order accurate in Δx, we approximate the space derivatives in (13.22) by central differences to find the *Lax–Wendroff scheme*

$$u_j^{n+1} = u_j^n - \frac{1}{2}c\big(u_{j+1}^n - u_{j-1}^n\big) + \frac{1}{2}c^2\big(u_{j+1}^n - 2u_j^n + u_{j-1}^n\big). \tag{13.23}$$

The stencil of this scheme is shown in Figure 13.3. An alternative derivation of (13.23) is to approximate $u(x_j, t^{n+1}) = u(x_j - b\Delta t, t^n)$ by quadratic interpolation from the values u_k^n $(k = j, j \pm 1)$ at time level t^n.

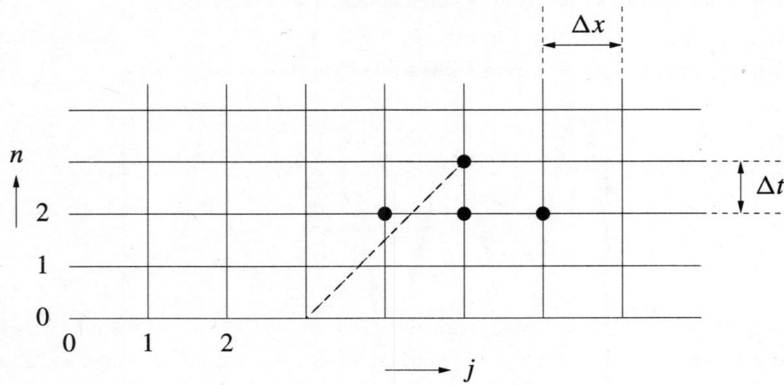

Figure 13.3. *Stencil for the Lax–Wendroff scheme. The dotted line is the characteristic through the point (x_j, t^{n+1}) facing back to the initial line $t = 0$.*

As before, we investigate the stability of (13.23) by substituting the Fourier mode (13.9). We find the following expression for the amplification factor:

$$\lambda(\varphi) = 1 - 2c^2 \sin^2\left(\frac{1}{2}\varphi\right) - ic\sin\varphi. \tag{13.24}$$

Taking the absolute value of $\lambda(\varphi)$, we find

$$|\lambda(\varphi)|^2 = 1 - 4c^2(1-c^2)\sin^4\left(\frac{1}{2}\varphi\right), \tag{13.25}$$

from which we see that the Lax–Wendroff scheme is stable if

$$|c| = \frac{|b|\Delta t}{\Delta x} \le 1. \tag{13.26}$$

In order to find the local discretisation error, we have to scale (13.23) so that it is compatible with the advection equation (13.1a), i.e.,

$$\frac{1}{\Delta t}(u_j^{n+1} - u_j^n) + \frac{b}{2\Delta x}(u_{j+1}^n - u_{j-1}^n) - \frac{b^2\Delta t}{2\Delta x^2}(u_{j+1}^n - 2u_j^n + u_{j-1}^n) = 0. \tag{13.27}$$

The local discretisation error d_j^n of the Lax–Wendroff scheme is then found as the residual of (13.27) after substituting the exact solution. So we have

$$\begin{aligned} d_j^n :=& \frac{1}{\Delta t}\big(u(x_j,t^{n+1}) - u(x_j,t^n)\big) + \frac{b}{2\Delta x}\big(u(x_{j+1},t^n) - u(x_{j-1},t^n)\big) \\ &- \frac{b^2\Delta t}{2\Delta x^2}\big(u(x_{j+1},t^n) - 2u(x_j,t^n) + u(x_{j-1},t^n)\big). \end{aligned} \tag{13.28}$$

Assuming $u(x,t)$ is sufficiently smooth, we can compute d_j^n by substituting Taylor expansions of $u(x,t)$ around (x_j,t^n) in (13.28) to find

$$\begin{aligned} d_j^n &= \frac{1}{6}\Delta t^2 \frac{\partial^3 u}{\partial t^3}(x_j,t^n) + \frac{1}{6}b\Delta x^2 \frac{\partial^3 u}{\partial x^3}(x_j,t^n) + \mathcal{O}(\Delta t^3) + \mathcal{O}(\Delta x^3) \\ &= \frac{1}{6}\Delta t^2\left(1 - \frac{1}{c^2}\right)\frac{\partial^3 u}{\partial t^3}(x_j,t^n) + \mathcal{O}(\Delta t^3). \end{aligned} \tag{13.29}$$

The second line in (13.29) follows from the relation $\frac{\partial^3 u}{\partial t^3} = -b^3\frac{\partial^3 u}{\partial x^3}$; see (13.1a). Thus, as was to be expected, the Lax–Wendroff scheme is second order consistent in Δt and Δx.

Example 13.2 We repeat the computation of Example 13.1 using the Lax–Wendroff scheme. This time the damping of the numerical solution is less severe; however, now the numerical solution lags the exact solution (see Figure 13.4). □

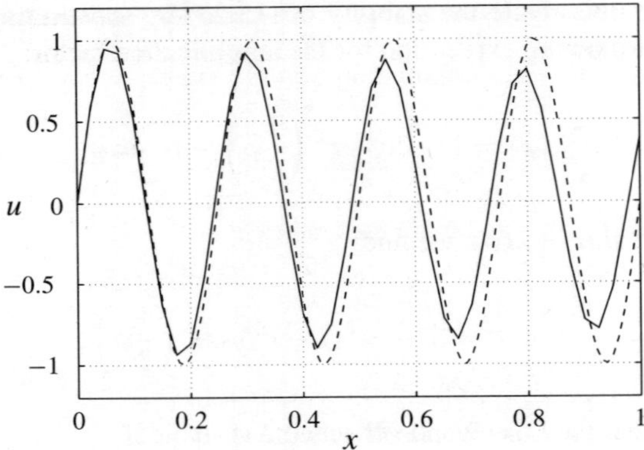

Figure 13.4. *Numerical approximation of a sine wave computed with the Lax–Wendroff scheme.*

13.2 Dissipation and Dispersion of Numerical Schemes

In this section we will give a Fourier analysis for both the upwind scheme and the Lax–Wendroff scheme. As it will turn out, the numerical behaviour of the solution may differ from that of the exact solution not only in growth character (cf. stability criteria for the former) but also in what is called the *phase*. We will investigate this for the advection equation.

Consider the equation (13.1a). Like in Section 3.4, we look for a planar wave solution of the form

$$u(x,t) = \mathrm{e}^{\mathrm{i}(\kappa x - \omega t)}, \tag{13.30}$$

where κ is the wave number and ω is the frequency. Substituting (13.30) in (13.1a), we find the dispersion relation

$$\omega(\kappa) = b\kappa. \tag{13.31}$$

This means that every mode of the form (13.30) propagates with *phase velocity* $b = \omega/\kappa$. We thus could also write

$$u(x,t) = \mathrm{e}^{\mathrm{i}\kappa(x-bt)}. \tag{13.32}$$

From (13.30) and (13.31) we conclude that the exact planar wave amplifies by a factor $\mathrm{e}^{-\mathrm{i}b\kappa\Delta t} = \mathrm{e}^{-\mathrm{i}c\varphi}$ per time step Δt, so we have

$$u(x_j, t^{n+1}) = \mathrm{e}^{-\mathrm{i}b\kappa\Delta t}\, u(x_j, t^n) = \mathrm{e}^{-\mathrm{i}c\varphi}\, u(x_j, t^n). \tag{13.33}$$

We compare this amplification to that of the numerical solution. Using the numerical dispersion relation, which we preferred to formulate in terms of the amplification factor and the phase angle, we obtain a relation $\lambda = \lambda(\varphi)$. Obviously, the amplification factor $\lambda(\varphi)$ of the numerical solution is an approximation of the amplification factor $\mathrm{e}^{-\mathrm{i}c\varphi}$ of the exact planar wave. So it makes sense to write λ in polar form:

$$\lambda = |\lambda|\, \mathrm{e}^{-\mathrm{i}\psi}, \tag{13.34}$$

where $|\lambda|$ is called the *amplitude* and ψ is the *phase* of the amplification factor. The argument ψ is apparently an approximation of $c\varphi = b\kappa\Delta t$. Hence we define a *numerical phase velocity* β as

$$\beta := \frac{\psi}{\kappa \Delta t}. \tag{13.35}$$

Inserting (13.34) and (13.35), we obtain the following representation of the numerical mode (13.9):

$$u_j^n = |\lambda|^n \, \mathrm{e}^{\mathrm{i}\kappa(x_j - \beta t^n)}. \tag{13.36}$$

Thus the mode propagates with speed β indeed.

In order to compare (13.36) with (13.32), it turns out to be convenient to define an *amplitude error* ϵ_{a} and a *relative phase error* ϵ_{f}:

$$\epsilon_{\mathrm{a}} := 1 - |\lambda|, \qquad \epsilon_{\mathrm{f}} := \frac{c\varphi - \psi}{c\varphi} = 1 - \frac{\beta}{b}. \tag{13.37}$$

In order to understand the definition of the amplitude error ϵ_{a}, one should realise that the exact solution has an amplitude $|\,\mathrm{e}^{-\mathrm{i}c\varphi}\,| = 1$. Clearly, only if $0 \le \epsilon_{\mathrm{a}} < 1$ is the scheme stable. Furthermore, we conclude that for $\epsilon_{\mathrm{f}} > 0$ the numerical solution lags behind the exact solution, while for $\epsilon_{\mathrm{f}} < 0$ it propagates faster. Combining the definitions of ϵ_{a} and ϵ_{f} with the representation (13.36), we find the following relation between the exact and numerical modes:

$$u_j^n = ((1 - \epsilon_{\mathrm{a}}) \exp(\mathrm{i}\epsilon_{\mathrm{f}} c\varphi))^n u(x_j, t^n), \tag{13.38}$$

which confirms our previous conclusions.

Let us now apply this to the schemes we have seen so far, to begin with to the upwind scheme (13.4b). Combining the polar representation (13.34) with the expression in (13.10), we find the following relations for the amplitude $|\lambda|$ and phase ψ:

$$|\lambda| = \sqrt{1 - 4c(1-c)\sin^2\left(\frac{1}{2}\varphi\right)}, \qquad \tan\psi = \frac{c \sin\varphi}{1 - 2c\sin^2(\frac{1}{2}\varphi)}. \tag{13.39}$$

From the first relation we can trivially determine the amplitude error. In order to derive an expression for the phase error, we have to invert the second relation in (13.39), and since $0 \le \psi \le \pi$, we have

$$\psi = \tan^{-1}\left(\frac{c\sin\varphi}{1 - 2c\sin^2(\frac{1}{2}\varphi)}\right), \tag{13.40}$$

where $\tan^{-1} : \mathbb{R} \to [0, \pi)$ is given by

$$\tan^{-1}(q) := \begin{cases} \arctan q & \text{if } \; q \ge 0, \\ \arctan q + \pi & \text{if } \; q < 0. \end{cases} \tag{13.41}$$

In Figure 13.5 we have drawn ϵ_{a} and ϵ_{f} as functions of φ. We see from these pictures that low-frequency components, characterized by a small wave number κ and a small phase angle φ, indeed have amplitude and phase errors close to zero. This means that these components are hardly damped and propagate with the correct velocity. On the other hand,

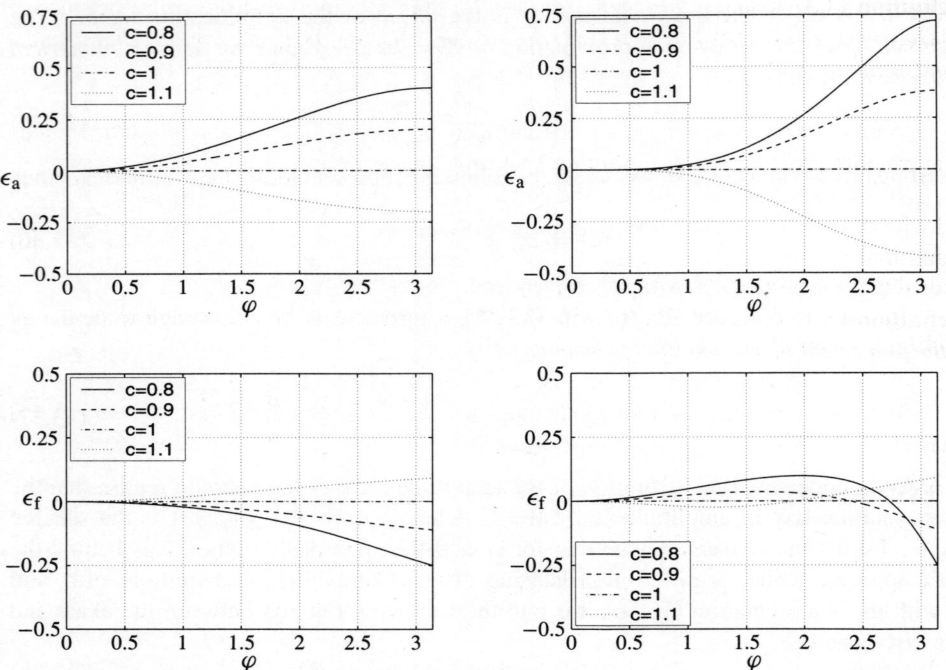

Figure 13.5. *Amplitude and phase errors of the upwind scheme (left) and the Lax–Wendroff scheme (right).*

high-frequency components, for which $\kappa \approx \pi/\Delta x$ and $\varphi \approx \pi$, have a significant amplitude and phase errors. These components are badly damped and propagate at the wrong speed.

Next consider the Lax–Wendroff scheme (13.23). The amplification factor $\lambda(\varphi)$ is given in (13.24). For the amplitude $|\lambda|$ and the phase ψ we find

$$|\lambda| = \sqrt{1 - 4c^2(1-c^2)\sin^4\left(\frac{1}{2}\varphi\right)}, \qquad \tan\psi = \frac{c\sin\varphi}{1 - 2c^2\sin^2(\frac{1}{2}\varphi)}. \tag{13.42}$$

Combining (13.37) and (13.42), we trivially obtain the amplitude and phase errors of the Lax–Wendroff scheme, which are shown in Figure 13.5. Essentially the same conclusions hold as for the upwind scheme. Moreover, comparing the upwind scheme with the Lax–Wendroff scheme, we see that the amplitude error of the latter is much closer to zero for low-frequency modes. This means that these Fourier modes are not so badly damped by the Lax–Wendroff scheme as by the upwind scheme.

The growth/decay of the numerical mode (13.36) is determined by the amplitude $|\lambda(\varphi)|$. Ideally, we would like to have $|\lambda(\varphi)| = 1$, as is true for the exact solution. In practice we have to be content with $|\lambda(\varphi)| \leq 1$; i.e., the scheme has to be stable. Based on this observation we introduce the following definition.

Definition 13.3. *A numerical scheme for the advection equation* (13.1a) *is called* dissipative *if* $|\lambda(\varphi)| \leq 1$ *for all* φ $(0 \leq \varphi \leq \pi)$ *and the inequality holds for at least one* φ. *On the other hand, a numerical scheme is called nondissipative if* $|\lambda(\varphi)| = 1$ *for all* φ $(0 \leq \varphi \leq \pi)$.

A nondissipative scheme does not damp the numerical solution and thus corresponds to an amplitude error $\epsilon_a = 0$. In Section 13.4 we will introduce two nondissipative schemes.

It is more difficult to recover the correct phase velocity. In fact, the numerical phase velocity β of any scheme depends on the wave number κ. This leads to the following definition.

Definition 13.4. *A numerical scheme for the advection equation* (13.1a) *is called* dispersive *if the numerical phase velocity* β *depends on the wave number* κ, *i.e.*, $\mathrm{d}\beta/\mathrm{d}\kappa \neq 0$.

In a dispersive scheme each mode in a solution will propagate with its own numerical phase velocity β depending on the wave number κ, so that the solution will disperse.

Example 13.5 Let us compute numerical solutions of the initial value problem (13.1) using the upwind and the Lax–Wendroff schemes for the initial function

$$v(x) = \begin{cases} 1 & \text{if} \quad 0 \leq x \leq 0.2, \\ 0 & \text{if} \quad x < 0, \quad x > 0.2. \end{cases}$$

We choose $\Delta x = 10^{-2}$, $\Delta t = 0.8 \times 10^{-2}$, and $b = 1$, and consequently $c = 0.8$. The results are shown in Figure 13.6. From this figure we see that the upwind solution is damped too much, whereas the Lax–Wendroff solution contains oscillations. These oscillations are due to the large phase errors for high-frequency components: each component in the Fourier series of the initial function $v(x)$ has a different numerical propagation velocity $\beta = (1 - \epsilon_f)b$. In Section 13.7 we will combine the upwind and Lax–Wendroff schemes into a high-resolution scheme, which has an amplitude error comparable to the Lax–Wendroff scheme but does not produce spurious oscillations. □

13.3 The Advection-Diffusion Equation

In practice we often encounter an advection-diffusion equation which has both an advective term and a (possibly small) diffusion term. Although this equation is of parabolic type, we have to use schemes from this chapter to discretise the advective term. Furthermore, it turns out that many numerical schemes for the advection equation (13.1a) can be described in terms of an advection-diffusion equation. Therefore we will introduce in Section 13.3.1 an explicit scheme for the advection-diffusion equation and investigate its stability. Next, in Section 13.3.2, we will give an alternative formulation of the upwind and Lax–Wendroff schemes.

13.3.1 An Explicit Scheme for the Advection-Diffusion Equation

In this section we consider the *advection-diffusion equation*

$$\frac{\partial u}{\partial t} + b\frac{\partial u}{\partial x} = D\frac{\partial^2 u}{\partial x^2}, \tag{13.43}$$

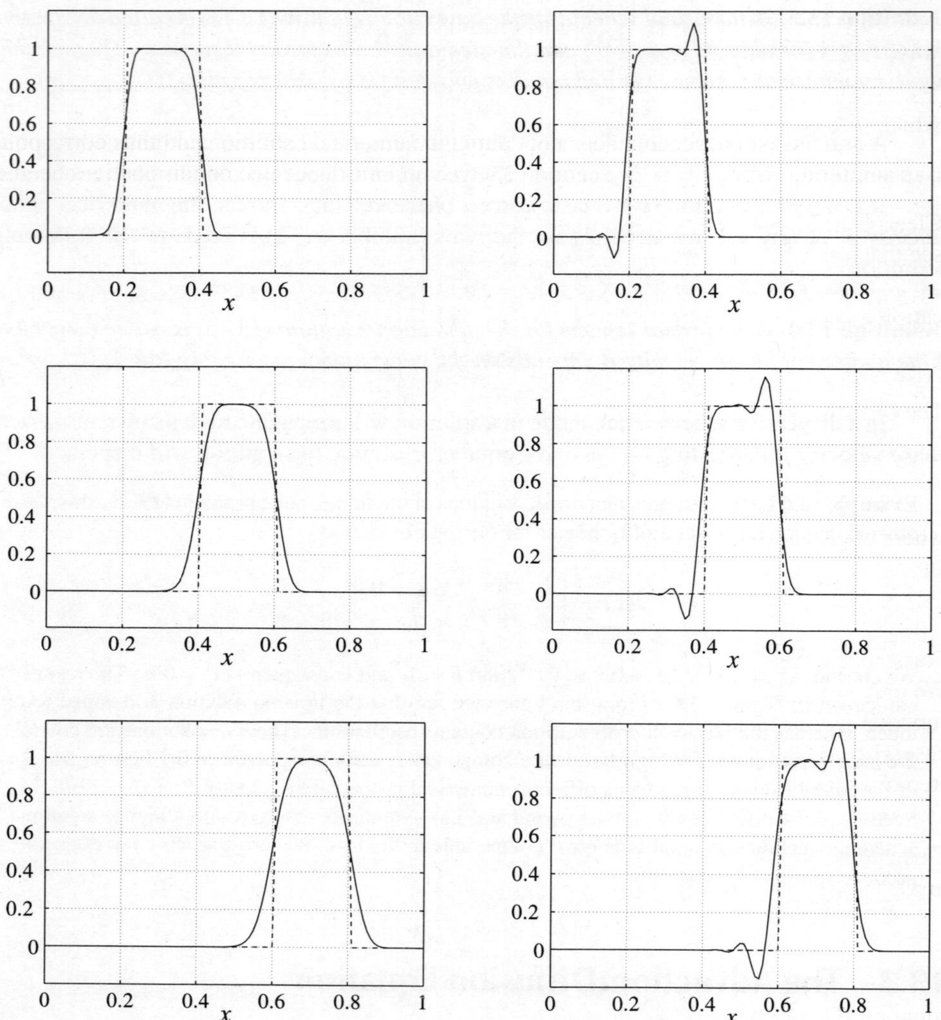

Figure 13.6. *Numerical solution computed with the upwind scheme (left) and the Lax–Wendroff scheme (right), after* 25, 50, *and* 75 *time steps, respectively.*

where $D > 0$ is the diffusion coefficient. If we discretise the spatial derivatives by central differences and use the explicit Euler method for time integration, we obtain the scheme

$$u_j^{n+1} = u_j^n - \frac{1}{2}c\big(u_{j+1}^n - u_{j-1}^n\big) + d\big(u_{j+1}^n - 2u_j^n + u_{j-1}^n\big), \tag{13.44}$$

where $d := D\Delta t/\Delta x^2$ is the diffusion number; cf. (11.5). Obviously, this scheme is first order accurate in Δt and second order in Δx.

It is instructive to derive the dispersion relation for (13.43). Substituting the planar wave (13.30), we find

$$\omega(\kappa) = b\kappa - \mathrm{i}D\kappa^2, \tag{13.45}$$

which then results in the damped mode

$$u(x,t) = \mathrm{e}^{-D\kappa^2 t}\,\mathrm{e}^{\mathrm{i}\kappa(x-bt)}\,. \tag{13.46}$$

Clearly, the addition of a diffusion term leads to damping of the planar wave mode. We will see that a similar conclusion holds for numerical schemes.

To analyse the stability of (13.44) we substitute the mode (13.9) to find the amplification factor

$$\lambda(\varphi) = 1 - 4d\sin^2\left(\frac{1}{2}\varphi\right) - \mathrm{i}c\sin\varphi. \tag{13.47}$$

The modulus of $\lambda(\varphi)$ is given by

$$|\lambda(\varphi)|^2 = 1 - 4g(\varphi)\sin^2\left(\frac{1}{2}\varphi\right), \tag{13.48a}$$

where the function $g(\varphi)$ is defined as

$$g(\varphi) := 2d - 4d^2\sin^2\left(\frac{1}{2}\varphi\right) - c^2\cos^2\left(\frac{1}{2}\varphi\right). \tag{13.48b}$$

For stability we need $|\lambda(\varphi)|^2 \leq 1$ and this requirement is met if $g(\varphi) \geq 0$ for $0 \leq \varphi \leq \pi$. We can easily verify that $g'(\varphi) = (-2d^2 + \frac{1}{2}c^2)\sin\varphi$, implying that $g(\varphi)$ is a monotone function, either increasing or decreasing, depending on the sign of $-2d^2 + \frac{1}{2}c^2$. In either case $g(\varphi) \geq 0$ only if $g(0) \geq 0$ and $g(\pi) \geq 0$. The inequality $g(0) \geq 0$ leads to the time step restriction

$$\Delta t \leq 2D/b^2. \tag{13.49a}$$

This condition is rather peculiar since it does not depend on Δx. It is sometimes referred to as the *advection-diffusion barrier*. The inequality $g(\pi) \geq 0$ results in

$$\Delta t \leq \Delta x^2/(2D), \tag{13.49b}$$

which is the usual time step restriction for the explicit Euler scheme applied to the heat equation. Combining both results, we have

$$\Delta t \leq \min\left(2D/b^2, \Delta x^2/(2D)\right). \tag{13.49c}$$

From (13.49c) we conclude that the scheme (13.44) becomes unconditionally unstable if $D = 0$, or, stated otherwise, a diffusion term is needed to keep the scheme stable.

There is another issue that needs further analysis, as is apparent from the following example.

Example 13.6 We compute a numerical solution of the advection-diffusion equation (13.43) at $t = 1$ with $b = 1$ and $D = 2.5 \times 10^{-3}$. This means that the advective term is highly dominant, and, consequently, the exact solution, indicated by the dotted line in Figure 13.7, has a very thin boundary layer near $x = 1$. We have applied scheme (13.44) with $\Delta x = 2.5 \times 10^{-2}$ and $\Delta t = 4 \times 10^{-3}$. We can easily check that the stability condition (13.49c) is satisfied. However, the condition $c/d \leq 2$ is clearly violated, leading to the numerical oscillations in the vicinity of $x = 1$. □

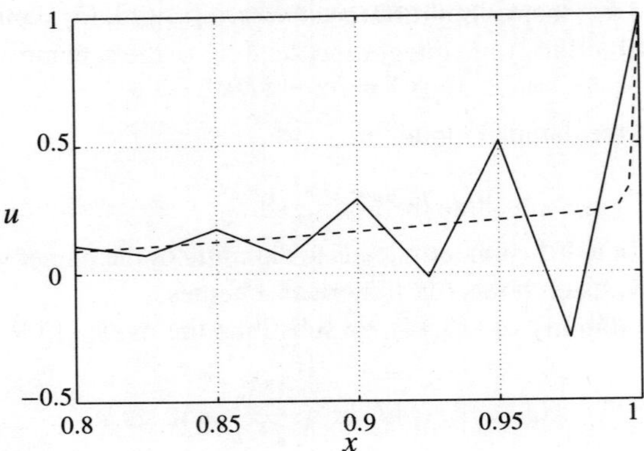

Figure 13.7. *Numerical solution of the advection-diffusion equation computed with the central difference scheme.*

From the example above it is clear that, although the scheme (13.44) is stable, it can produce unrealistic numerical solutions. To investigate the oscillatory behaviour of the numerical solution we write (13.44) in the alternative form

$$u_j^{n+1} = u_j^n - d\left(1 + \frac{1}{2}P\right)\left(u_j^n - u_{j-1}^n\right) + d\left(1 - \frac{1}{2}P\right)\left(u_{j+1}^n - u_j^n\right), \tag{13.50}$$

where $Pe := c/d = b\Delta x/D$ is the *Péclet number*. Replacing index j by $j+1$ and subtracting (13.50) from the resulting equation, we find the relation

$$\begin{aligned} u_{j+1}^{n+1} - u_j^{n+1} &= (1-2d)\left(u_{j+1}^n - u_j^n\right) + d\left(1 + \frac{1}{2}Pe\right)\left(u_j^n - u_{j-1}^n\right) \\ &\quad + \left(1 - \frac{1}{2}Pe\right)\left(u_{j+2}^n - u_{j+1}^n\right). \end{aligned} \tag{13.51}$$

If all coefficients on the right-hand side of (13.51) are positive, i.e., if $d \leq \frac{1}{2}$ and $|Pe| \leq 2$, we see that

$$\sum_j \left|u_{j+1}^{n+1} - u_j^{n+1}\right| \leq \sum_j \left|u_{j+1}^n - u_j^n\right|, \tag{13.52}$$

where the summation is over all possible values of j. This inequality suggests that an initially monotone solution cannot be corrupted by oscillations as in the example above. Indeed, we can prove that unphysical oscillations cannot occur if (13.52) holds; for a more rigorous analysis the reader is referred to Section 13.6. To conclude, scheme (13.44) should satisfy the stability condition (13.49c) and in addition the inequality $|Pe| \leq 2$ in order to prevent unphysical oscillations.

13.3.2 Numerical and Artificial Diffusion

In this section we consider again the advection equation (13.1a). As was noted in Section 13.1.2, the derivation of the Lax–Wendroff scheme was somewhat unconventional. Instead, let us (naively) try to use the central difference approximation for the space derivative and the forward Euler method for time integration, leading to the scheme

$$u_j^{n+1} = u_j^n - \frac{1}{2}c\big(u_{j+1}^n - u_{j-1}^n\big). \tag{13.53}$$

To investigate stability we substitute the Fourier mode (13.9), resulting in the amplification factor

$$\lambda(\varphi) = 1 - \mathrm{i}c\sin\varphi. \tag{13.54}$$

Hence we find

$$|\lambda(\varphi)|^2 = 1 + c^2\sin^2\varphi \geq 1. \tag{13.55}$$

We therefore conclude that this scheme is unconditionally unstable, which is in agreement with the analysis from the previous section. With the derivation of the Lax–Wendroff scheme in mind, we try to add a diffusion term to the scheme such that $\mathrm{Re}(\lambda) < 1$, resulting in a (conditionally) stable scheme. Comparing (13.23) and (13.53), we see that the Lax–Wendroff scheme is precisely the central difference scheme (13.53) with a diffusion term added. Indeed, introducing the coefficient $D := \frac{1}{2}b^2\Delta t$, we can rewrite the Lax–Wendroff scheme in the form

$$u_j^{n+1} = u_j^n - \frac{1}{2}c\big(u_{j+1}^n - u_{j-1}^n\big) + D\frac{\Delta t}{\Delta x^2}\big(u_{j+1}^n - 2u_j^n + u_{j-1}^n\big). \tag{13.56}$$

Viewing D as a constant, (13.56) is the central difference approximation in space and the forward Euler approximation in time of the advection-diffusion equation (13.43). Consequently, the Lax–Wendroff scheme introduces the discretisation of the term $D\frac{\partial^2 u}{\partial x^2}$, which can be interpreted as a diffusion. Therefore the central difference approximation of $D\frac{\partial^2 u}{\partial x^2}$ is often called the *numerical diffusion*. We can now easily derive a stability condition from the theory of the previous section. For the particular choice $D = \frac{1}{2}b^2\Delta t$, the inequality (13.49a) is trivially satisfied, whereas (13.49b) is equivalent to $|c| \leq 1$; i.e., the stability condition (13.26) of the Lax–Wendroff scheme is recovered. The main conclusion of the analysis above is that the Lax–Wendroff scheme stabilizes the central difference scheme by adding numerical diffusion.

For the upwind scheme a similar situation applies, though it is less obvious. To see how we rewrite the upwind scheme (13.4b) for $b > 0$ as the central difference scheme (13.53) with a correction term added:

$$\begin{aligned} u_j^{n+1} &= u_j^n - \frac{1}{2}c\big(u_{j+1}^n - u_{j-1}^n\big) - c\big(u_j^n - u_{j-1}^n\big) + \frac{1}{2}c\big(u_{j+1}^n - u_{j-1}^n\big) \\ &= u_j^n - \frac{1}{2}c\big(u_{j+1}^n - u_{j-1}^n\big) + \frac{1}{2}c\big(u_{j+1}^n - 2u_j^n + u_{j-1}^n\big). \end{aligned} \tag{13.57}$$

Defining $D := \frac{1}{2}b\Delta x$ and taking it constant, we see that (13.57) is the forward Euler approximation in time and the central difference approximation in space of (13.43). Since $D > 0$, the upwind scheme can thus be interpreted as the central difference scheme stabilized by adding numerical diffusion. Note that for $0 < c < 1$ the upwind scheme has more numerical diffusion than the Lax–Wendroff scheme. This then explains the damping character of the upwind scheme. Of course, similar reasoning holds for $b < 0$.

Since it appears to be the diffusion (or in some physical applications the viscosity) that makes a scheme stable, one can also try to add some *artificial diffusion* before discretisation. Hence, instead of a typical hyperbolic problem, say

$$\frac{\partial u}{\partial t} + \mathcal{L}[u] = 0, \tag{13.58a}$$

where $\mathcal{L}$ is some quasi-linear first order operator, one then solves

$$\frac{\partial u}{\partial t} + \mathcal{L}[u] = D\frac{\partial^2 u}{\partial x^2}. \tag{13.58b}$$

The parameter D is then chosen to produce a stable solution.

It may be clear that stability obtained by either numerical or artificial diffusion will have an effect on the accuracy of the numerical solution. This is demonstrated by the next example.

Example 13.7 We once more compute a numerical solution of the advection-diffusion equation (13.43), this time using the upwind discretisation for the advective term $b\frac{\partial u}{\partial x}$. All parameters are the same as in Example 13.6. The result is shown in Figure 13.8. Clearly, the numerical oscillations are gone; however, the boundary layer at $x = 1$ is only inaccurately resolved. □

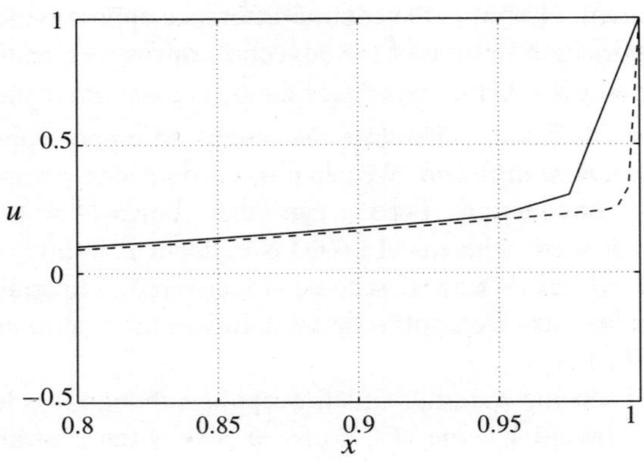

Figure 13.8. *Numerical solution of the advection-diffusion equation computed with the upwind scheme.*

13.4 Nondissipative Schemes

The numerical schemes we have encountered so far are dissipative, implying that the numerical solution is damped. In this section we introduce two nondissipative schemes: the box scheme and the leapfrog scheme. These schemes are very useful for the computing smooth solutions of (13.1a). On the other hand, for nonsmooth solutions these methods are not suitable due to the phase error and the lack of damping.

13.4.1 The Box Scheme

Apart from the explicit schemes studied so far, implicit schemes are also used. Consider, e.g., the scheme we obtain by employing the trapezoidal rule for time integration; cf. (5.73). Besides being second order accurate, this method is interesting for another reason. To see this note that the time-dependent behaviour of the planar wave (13.30) is characterized by the ODE

$$\frac{\mathrm{d}u}{\mathrm{d}t} = -\mathrm{i}\omega u, \quad \omega \in \mathbb{R}. \tag{13.59}$$

A typical solution of (13.59) is given by $u(x,t) = C\,\mathrm{e}^{-\mathrm{i}\omega t}$, which has the property that $|u(t)| = |C|$ for all t. Applying the trapezoidal rule to (13.59) gives

$$u^{n+1} = \frac{1 - \frac{1}{2}\mathrm{i}\omega\Delta t}{1 + \frac{1}{2}\mathrm{i}\omega\Delta t} u^n, \tag{13.60}$$

with u^n the numerical approximation of $u(t^n)$. Clearly, $|u^{n+1}| = |u^n|$ which is a "non-growth/nondecay" property it shares with the exact solution. Therefore we call the time integration scheme (13.60) *nondissipative*.

Next we return to the advection equation (13.1a). In the following we assume that (13.1a) is defined on a finite interval, let's say $x \in (0,1)$. Thus, apart from an initial condition, we need a boundary condition as well. If $b > 0$, we have to prescribe u at the left boundary $x = 0$, and if $b < 0$, we have to prescribe it at $x = 1$; see Chapter 12 for a discussion of boundary conditions. In order to compute a numerical solution, we cover the domain $\Omega = [0,1] \times [0,\infty)$ with a grid containing points (x_j, t^n) defined as

$$x_j := j\Delta x \quad (j = 0, 1, 2, \ldots, M), \qquad t^n := n\Delta t \quad (n = 0, 1, 2, \ldots), \tag{13.61}$$

with $\Delta x = 1/M$ the spatial grid size and $\Delta t > 0$ the time step. For time integration we use the trapezoidal rule because we expect it to be nondissipative. In order to match the second order behaviour of the trapezoidal rule, we use the central difference approximation for the space derivative. However, in order to keep the stencil compact, we apply the central difference scheme at intermediate points $x_{j+1/2}$ given by

$$x_{j+1/2} := \frac{1}{2}(x_j + x_{j+1}), \qquad (j = 0, 1, \ldots, M-1).$$

This way, we obtain the scheme

$$\frac{1}{\Delta t}\left(u_{j+1/2}^{n+1} - u_{j+1/2}^{n}\right) + \frac{1}{2}\frac{b}{\Delta x}\left(\left(u_{j+1}^{n} - u_j^n\right) + \left(u_{j+1}^{n+1} - u_j^{n+1}\right)\right) = 0. \tag{13.62}$$

Finally, we approximate the intermediate values $u(x_{j+1/2}, t^m)$ $(m = n, n+1)$ by straightforward linear interpolation, which is second order accurate as well, to find the *box scheme*

$$u_j^{n+1} + u_{j+1}^{n+1} - \left(u_j^n + u_{j+1}^n\right) + c\left(u_{j+1}^n - u_j^n + u_{j+1}^{n+1} - u_j^{n+1}\right) = 0. \tag{13.63}$$

The stencil of this scheme is displayed in Figure 13.9.

This scheme is implicit with two unknown variables at the new time level t^{n+1}, thus resulting in a bidiagonal linear system. Such a system can be very efficiently solved without

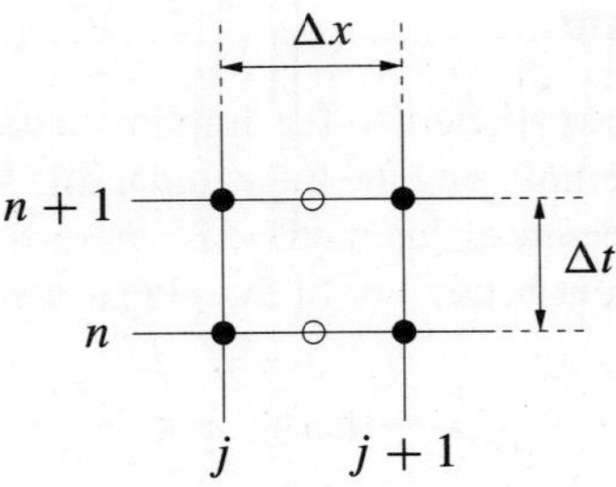

Figure 13.9. *Stencil for the box scheme. The open circles indicate the intermediate grid points.*

the need to compute an LU-decomposition; one sweep through the spatial grid is sufficient. Indeed, if $b > 0$, then u_0^{n+1} is given and we can compute all other values u_j^{n+1} in increasing order of j from

$$u_{j+1}^{n+1} = u_j^n + \frac{1-c}{1+c}\left(u_{j+1}^n - u_j^{n+1}\right), \qquad (j = 0, 1, \ldots, M-1). \tag{13.64a}$$

Likewise, if $b < 0$, the boundary value u_M^{n+1} is given and we can compute u_j^{n+1} in decreasing order of j from

$$u_j^{n+1} = u_{j+1}^n + \frac{1+c}{1-c}\left(u_j^n - u_{j+1}^{n+1}\right), \qquad (j = M-1, M-2, \ldots, 0). \tag{13.64b}$$

Although the box scheme is implicit, the computational costs per time step are comparable to those of an explicit scheme!

Let us now investigate the numerical dispersion relation. Substituting the mode (13.9) in (13.64a), we obtain the relation

$$\lambda\, \mathrm{e}^{\mathrm{i}\varphi} = 1 + \frac{1-c}{1+c}\left(\mathrm{e}^{\mathrm{i}\varphi} - \lambda\right), \tag{13.65}$$

where $\varphi = \kappa \Delta x$. Hence we find

$$\lambda(\varphi) = \frac{(1+c)\,\mathrm{e}^{-\mathrm{i}\varphi/2} + (1-c)\,\mathrm{e}^{\mathrm{i}\varphi/2}}{(1+c)\,\mathrm{e}^{\mathrm{i}\varphi/2} + (1-c)\,\mathrm{e}^{-\mathrm{i}\varphi/2}} = \frac{\cos\frac{1}{2}\varphi - \mathrm{i}c\sin\frac{1}{2}\varphi}{\cos\frac{1}{2}\varphi + \mathrm{i}c\sin\frac{1}{2}\varphi}. \tag{13.66}$$

We conclude that $|\lambda(\varphi)| = 1$, as we hoped for. Although this scheme looks attractive, there is a snag since it suffers from dispersion. Comparing (13.66) with the polar representation (13.34) and using elementary trigonometric formulas, we find

$$\cos\frac{1}{2}\varphi\,\cos\psi + c\sin\frac{1}{2}\varphi\,\sin\psi = \cos\frac{1}{2}\varphi, \tag{13.67a}$$

$$-c\sin\frac{1}{2}\varphi\,\cos\psi + \cos\frac{1}{2}\varphi\,\sin\psi = c\sin\frac{1}{2}\varphi. \tag{13.67b}$$

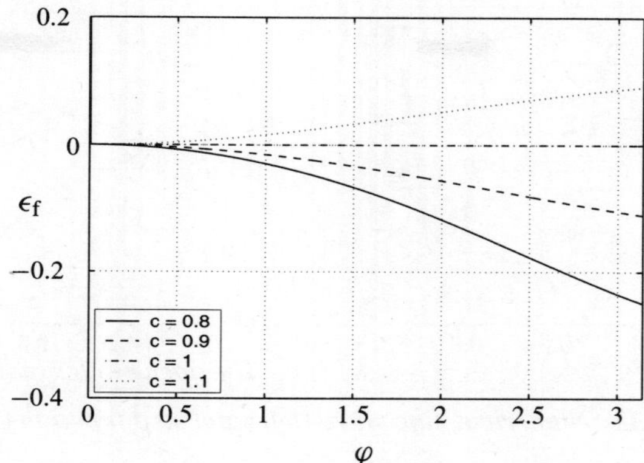

Figure 13.10. *Phase error of the box scheme.*

From this system we obtain for the phase angle ψ that

$$\tan\frac{1}{2}\psi = c\tan\frac{1}{2}\varphi. \tag{13.68}$$

The phase error of the box scheme is then given by

$$\epsilon_f = 1 - \frac{2}{c\varphi}\tan^{-1}\left(c\tan\frac{1}{2}\varphi\right), \tag{13.69}$$

with $\tan^{-1}$ defined in (13.41); see Figure 13.10.

Example 13.8 Let us compute two numerical solutions of the advection equation (13.1a) in the interval (0, 1) using the box scheme (13.64a). The initial condition is either $v(x) = \sin 8\pi x$ or the step function

$$v(x) = \begin{cases} 1 & \text{if } x \leq 0.2, \\ 0 & \text{if } x > 0.2. \end{cases}$$

Furthermore, we choose the parameter values $b = 0.2$, $\Delta x = 2.5 \times 10^{-2}$, and $\Delta t = 5 \times 10^{-2}$. The numerical solution after 40 time steps is displayed in Figure 13.11. The sine wave is well resolved and only a small phase shift is visible. On the other hand, the step function is completely corrupted by oscillations caused by the large phase error of the high-frequency components in the solution. □

13.4.2 The Leapfrog Scheme

It is also possible to use explicit methods and still have nondissipation. The method we shall discuss in this section is, however, a two-step method; i.e., it uses the numerical solution at two previous time levels.

Like in the previous section, we consider the advection equation on the interval (0, 1) and for its numerical solution we employ the grid defined in (13.61). The numerical scheme

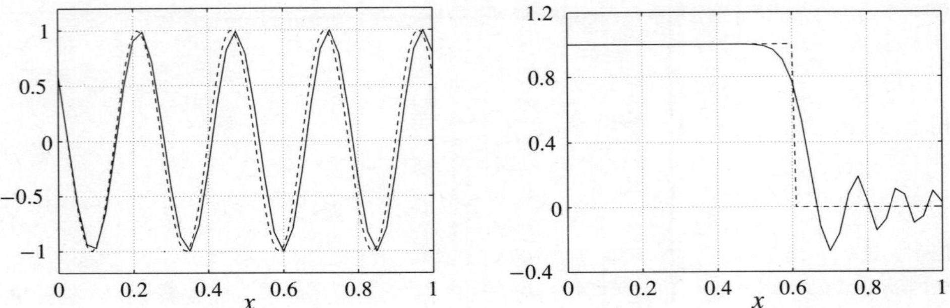

Figure 13.11. *Numerical sine wave (left) and step function (right) computed with the box scheme.*

is the midpoint rule for time integration in combination with the central difference scheme for the space derivative. Both are second order accurate. The resulting scheme reads

$$\frac{1}{2\Delta t}\left(u_j^{n+1} - u_j^{n-1}\right) + \frac{b}{2\Delta x}\left(u_{j+1}^n - u_{j-1}^n\right) = 0. \tag{13.70a}$$

Rearranging terms, we find the *leapfrog scheme*

$$u_j^{n+1} = u_j^{n-1} - c\left(u_{j+1}^n - u_{j-1}^n\right); \tag{13.70b}$$

see Figure 13.12 for its stencil. Clearly, we have to find a second "initial" condition at $t^1 = \Delta t$ in order to be able to start the computation. A possibility is to use a one-step method like the box scheme. For the second order wave equation, as will be treated in Chapter 14, a special gridding takes care of this problem more directly.

As before, we can obtain a numerical dispersion relation:

$$\lambda = \frac{1}{\lambda} - c\left(\mathrm{e}^{\mathrm{i}\varphi} - \mathrm{e}^{-\mathrm{i}\varphi}\right). \tag{13.71}$$

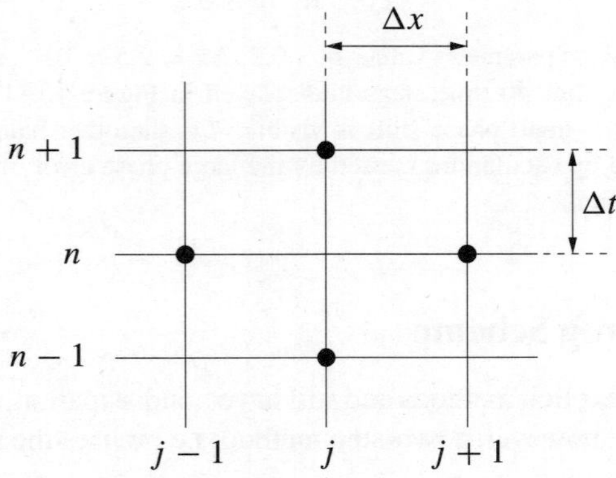

Figure 13.12. *Stencil for the leapfrog scheme.*

Hence the amplification factor satisfies the quadratic equation

$$\lambda^2 + 2ic\lambda \sin\varphi - 1 = 0, \tag{13.72}$$

which has the two solutions

$$\lambda(\varphi) = \pm\sqrt{1 - c^2 \sin^2\varphi} - ic\sin\varphi. \tag{13.73}$$

Clearly, if $|c| \le 1$, then $|\lambda(\varphi)| = 1$. Furthermore, the product of the roots is -1; hence they are truly complex, i.e., have a nonzero imaginary part. As a consequence, both roots are unimodular, so that the scheme is nondissipative indeed. On the other hand, if $|c| > 1$, then $|\lambda(\varphi)| > 1$ for some φ, resulting in an unstable scheme.

A particular feature of the leapfrog scheme, and in fact of any multistep method, is the fact that it introduces additional solutions, as is evidenced by the fact that more than one initial condition is needed to compute the numerical solution. In order to investigate this, we apply the midpoint rule to the ODE model problem (13.59). The result is

$$u^{n+1} = u^{n-1} - 2i\omega\Delta t\, u^n,$$

with u^n the approximation of $u(t^n)$. A typical solution $u^n = \mu^n$ exists if μ satisfies the *characteristic equation*

$$\mu^2 + 2i\omega\Delta t\,\mu - 1 = 0.$$

For $\Delta t \to 0$ we obtain the two expansions for the two roots

$$\mu_1 \doteq 1 - i\omega\Delta t - \frac{1}{2}\omega^2\Delta t^2, \qquad \mu_2 \doteq -1 - i\omega\Delta t + \frac{1}{2}\omega^2\Delta t^2$$

as approximations of the exact amplification factor $e^{-i\omega\Delta t}$. Obviously, only the first *essential root* makes sense, giving rise to an *essential solution*, i.e., a numerical solution that is related to the actual solution of (13.59). The other root is often referred to as *spurious*; see, e.g., [97]. This root generates a spurious solution that does not have a physical meaning at all. The spurious solution is often generated by initial or boundary conditions, and since it is not damped, it will manifest itself by oscillatory components in the solution.

For the essential root we can deduce from (13.73) and its polar representation (13.34) the relation

$$\sin\psi = c\sin\varphi, \qquad \cos\psi \ge 0 \tag{13.74}$$

for the phase ψ. Inverting this relation, we obtain for the phase error

$$\epsilon_f = 1 - \frac{1}{c\varphi}\arcsin(c\sin\varphi); \tag{13.75}$$

see Figure 13.13. From this figure we conclude that the phase error of the leapfrog scheme is significant, and, consequently, the scheme is highly dispersive. The performance of the leapfrog scheme is comparable to that of the box scheme: smooth solutions are well resolved on a fine enough grid and discontinuous solutions are completely destroyed by spurious oscillations.

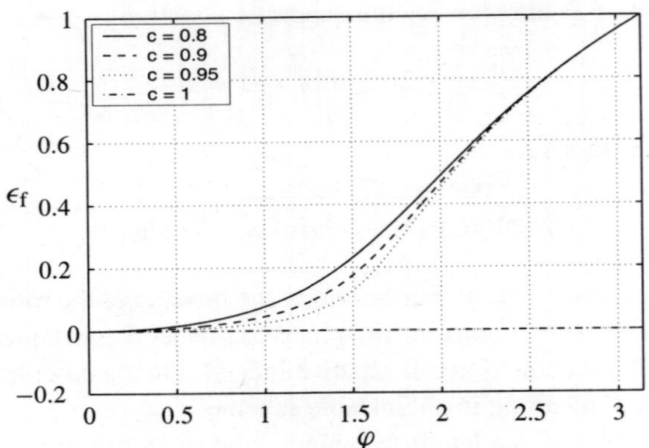

Figure 13.13. *Relative phase error of the leapfrog scheme.*

13.4.3 Propagation of Wave Packets

In Section 3.4 we encountered wave packets, i.e., superpositions of planar waves all having nearly the same wave number and phase velocity. It is interesting to see what this implies for a numerical scheme. Recall the wave packet at time $t = 0$, as was defined in (3.36a); i.e., for a particular wave number κ_0 and a parameter ε we have

$$u(x, 0) = \mathrm{e}^{\mathrm{i}\kappa_0 x} f(\varepsilon x). \tag{13.76}$$

Suppose that the constituent modes of the wave packet satisfy a dispersion relation $\omega = \omega(\kappa)$. In Section 3.4 we showed that the wave packet corresponding to (13.76) is then given by

$$u(x, t) = \mathrm{e}^{\mathrm{i}\kappa_0 x - \mathrm{i}\omega_0 t} f\big(\varepsilon(x - \omega_0' t)\big), \tag{13.77}$$

with $\omega_0 = \omega(\kappa_0)$ and $\omega_0' = \omega'(\kappa_0)$. This means that all wave crests propagate with phase velocity ω_0/κ_0, whereas the envelope of the packet propagates with the group velocity ω_0'. For the special case of the advection equation we have $\omega(\kappa) = b\kappa$, and, consequently, $\omega_0/\kappa_0 = \omega_0' = b$.

Modes propagate at different phase velocities with numerical approximations of wave packets, since any numerical scheme is dispersive. In order to investigate dispersion of a nondissipative scheme, it is convenient to formulate the discrete dispersion relation in terms of the (real) *numerical frequency* Ω as

$$\Omega(\kappa) := \beta\kappa = \psi/\Delta t, \tag{13.78}$$

with β the numerical phase velocity; cf. (13.35). In analogy with the continuous case, we define the *numerical group velocity* v_{ng} by

$$v_{\mathrm{ng}} := \frac{\mathrm{d}\Omega}{\mathrm{d}\kappa} = \frac{1}{\Delta t}\frac{\mathrm{d}\psi}{\mathrm{d}\kappa}. \tag{13.79}$$

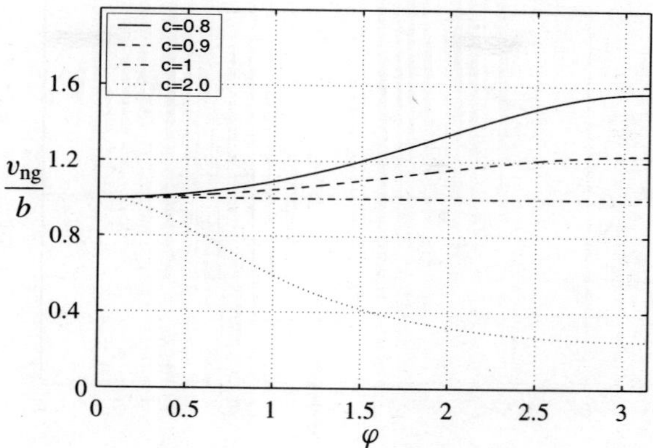

Figure 13.14. *Numerical group velocity of the box scheme.*

As an illustration we compute the numerical group velocity of the box scheme (13.63). Combining relation (13.68) with the definition of v_{ng}, we obtain the expression

$$\frac{v_{ng}}{b} = \frac{\cos^2 \frac{1}{2}\psi}{\cos^2 \frac{1}{2}\varphi} = \left(\cos^2\left(\frac{1}{2}\varphi\right) + c^2 \sin^2\left(\frac{1}{2}\varphi\right)\right)^{-1}$$

for the numerical group velocity. In Figure 13.14 we have drawn graphs of this function for several values of c.

Example 13.9 Let us illustrate the effect of the numerical group velocity on a wave packet. To that purpose, we compute a numerical solution of the advection equation (13.1a) on the interval (0, 4), subject to the initial condition

$$v(x) = \mathrm{e}^{-\varepsilon(x-1)^2} \sin(\alpha\pi x).$$

We choose the parameter values $b = 1$, $\varepsilon = 10$, and $\alpha = 16$. We apply the box scheme with $\Delta x = 1.25 \times 10^{-2}$ and $\Delta t = 2.5 \times 10^{-2}$, so that $c = 2$. The situation at $t = 2$ is given in Figure 13.15; the solid line denotes the numerical wave packet and the dotted line denotes the envelope of the exact solution. In this case the numerical solution has a phase $\varphi = \alpha\pi\,\Delta x = \pi/5$ and, consequently, $v_{ng} \approx 0.78$; see Figure 13.14. The numerical and the exact wave packets have travelled distances 1.55 and 2, respectively, at $t = 2$, as can be seen from Figure 13.15 when comparing the maximums of both solutions. □

13.5 The Godunov Scheme for Nonlinear Conservation Laws

In the preceding sections we analysed schemes based on finite difference methods. Moreover, we only dealt with the advection equation. In this section we will consider finite volumes instead (see Section 5.3). At the same time we will view them in the more general

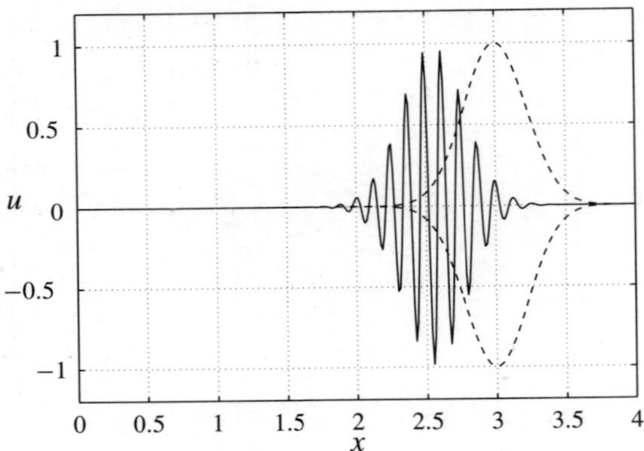

Figure 13.15. *The effect of the numerical group velocity on a wave packet computed with the box scheme.*

context of nonlinear conservation laws. So consider the initial value problem

$$\frac{\partial u}{\partial t} + \frac{\partial f(u)}{\partial x} = 0, \qquad x \in \mathbb{R}, \quad t > 0, \tag{13.80a}$$

$$u(x,0) = v(x), \qquad x \in \mathbb{R}, \tag{13.80b}$$

for $u(x,t)$, where the flux $f(u)$ is generally a nonlinear function of u. Recall that for the advection equation we have $f(u) = bu$. For the most part we will assume that $f(u)$ is a convex function; i.e., $f''(u) > 0$ for all u. Nonlinear problems are essentially more difficult than linear ones. In particular, nonlinear equations allow for discontinuous solutions even if the initial condition is smooth; see Chapter 12. As we will see, the correct numerical computation of a discontinuous solution to a nonlinear equation puts extra requirements on a scheme.

The finite volume method we will discuss can be seen as a generalisation of the upwind scheme in the sense that both schemes use the exact local solution of the conservation law. Let us begin by introducing control volumes or cells V_j as follows:

$$V_j := [x_{j-\frac{1}{2}}, x_{j+\frac{1}{2}}), \qquad x_{j+\frac{1}{2}} := \frac{1}{2}(x_j + x_{j+1}), \quad j = 0, \pm 1, \pm 2, \ldots, \tag{13.81}$$

with x_j defined in (13.3); see Figure 13.16. The numerical solution u_j^n has to be interpreted as an approximation of the average value of $u(x,t)$ over control volume V_j at time level t^n. Associated with u_j^n is the function $\bar{u}(x,t)$, defined as the solution of the following initial value problem with piecewise constant initial condition at $t = t^n$:

$$\frac{\partial \bar{u}}{\partial t} + \frac{\partial f(\bar{u})}{\partial x} = 0, \qquad x \in \mathbb{R}, \quad t > t^n, \tag{13.82a}$$

$$\bar{u}(x,t^n) = u_j^n, \qquad x \in V_j \quad (j = 0, \pm 1, \pm 2, \ldots). \tag{13.82b}$$

As shown in Section 5.3 for a stationary conservation law, the finite volume method is based on the integral formulation of the corresponding equation. We modify this derivation for the

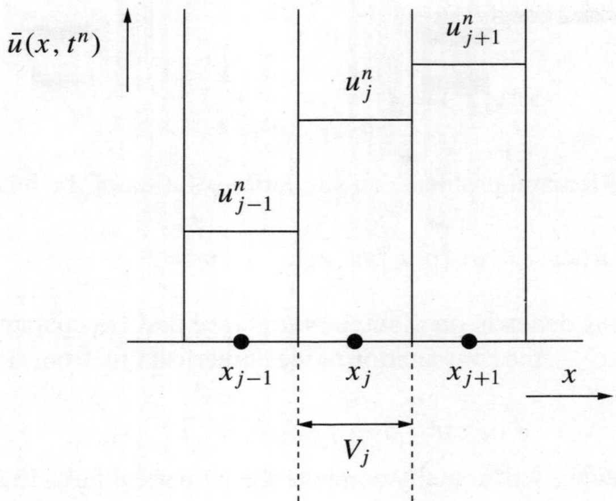

Figure 13.16. *Piecewise constant numerical solution at time level t^n.*

initial value problem (13.82). Integration of the conservation law (13.82a) over V_j results in

$$\frac{\mathrm{d}}{\mathrm{d}t}\int_{V_j} \bar{u}(x,t)\,\mathrm{d}x + f(\bar{u}(x_{j+\frac{1}{2}},t)) - f(\bar{u}(x_{j-\frac{1}{2}},t)) = 0. \tag{13.83}$$

Subsequent integration over the time interval $[t^n, t^{n+1}]$ gives

$$\int_{V_j} \bar{u}(x,t^{n+1})\,\mathrm{d}x - \int_{V_j} \bar{u}(x,t^n)\,\mathrm{d}x \\ + \int_{t^n}^{t^{n+1}} f(\bar{u}(x_{j+\frac{1}{2}},t))\,\mathrm{d}t - \int_{t^n}^{t^{n+1}} f(\bar{u}(x_{j-\frac{1}{2}},t))\,\mathrm{d}t = 0. \tag{13.84}$$

Introducing the *numerical flux* $F(u^n_j, u^n_{j+1})$ as the time average flux at the control volume edge $x_{j+\frac{1}{2}}$, i.e.,

$$F(u^n_j, u^n_{j+1}) := \frac{1}{\Delta t}\int_{t^n}^{t^{n+1}} f(\bar{u}(x_{j+\frac{1}{2}},t))\,\mathrm{d}t, \tag{13.85}$$

we can write the "integral balance" (13.84) in the form

$$\int_{V_j} \bar{u}(x,t^{n+1})\,\mathrm{d}x - \int_{V_j} \bar{u}(x,t^n)\,\mathrm{d}x = -\Delta t\big(F(u^n_j, u^n_{j+1}) - F(u^n_{j-1}, u^n_j)\big). \tag{13.86}$$

In (13.85) we have explicitly written the numerical flux at the cell boundary $x_{j+\frac{1}{2}}$ as a function of the numerical values u^n_j and u^n_{j+1} in the adjacent cells. The next step is to compute the numerical fluxes.

To compute the numerical flux $F(u^n_j, u^n_{j+1})$ we solve a *local* Riemann problem (see Section 12.2.2) at the cell edge $x_{j+\frac{1}{2}}$ consisting of the conservation law (13.82a) and the piecewise constant initial condition

$$\bar{u}(x, t^n) = \begin{cases} u_j^n & \text{if} \quad x < x_{j+\frac{1}{2}}, \\ u_{j+1}^n & \text{if} \quad x > x_{j+\frac{1}{2}}. \end{cases} \tag{13.87}$$

The solution of this Riemann problem is a similarity solution of the form

$$\bar{u}(x, t) = u_{\mathrm{R}}\big(\eta; u_j^n, u_{j+1}^n\big), \qquad \eta := \frac{x - x_{j+\frac{1}{2}}}{t - t^n}, \tag{13.88}$$

where the function u_{R} depends on a single variable η and two parameters u_j^n and u_{j+1}^n. Since $\eta = 0$ for $x = x_{j+\frac{1}{2}}$, the computation of the numerical flux from (13.85) is particularly simple, and we find

$$F(u_j^n, u_{j+1}^n) = f\big(u_{\mathrm{R}}(0; u_j^n, u_{j+1}^n)\big). \tag{13.89}$$

In the examples below we will actually compute the numerical flux (13.89) for two specific cases: the advection equation and the Burgers' equation. In general the procedure may be rather involved; see, e.g., [87, 158].

As a final step, the numerical value u_j^{n+1} is then found as the average value of $\bar{u}(x, t^{n+1})$ over the control volume V_j:

$$u_j^{n+1} := \frac{1}{\Delta x} \int_{V_j} \bar{u}(x, t^{n+1})\,\mathrm{d}x. \tag{13.90}$$

Substituting (13.90) in (13.86) leads to the *Godunov scheme*

$$u_j^{n+1} = u_j^n - \frac{\Delta t}{\Delta x}\big(F(u_j^n, u_{j+1}^n) - F(u_{j-1}^n, u_j^n)\big), \tag{13.91}$$

with the numerical flux $F(u_j^n, u_{j+1}^n)$ as defined in (13.89).

Example 13.10 Consider the advection equation (13.1a). The Riemann problem associated with this equation reads

$$\frac{\partial \bar{u}}{\partial t} + b\frac{\partial \bar{u}}{\partial x} = 0, \qquad x \in \mathbb{R}, \quad t > t^n,$$

$$\bar{u}(x, t^n) = \begin{cases} u_j^n & \text{if} \quad x < x_{j+\frac{1}{2}}, \\ u_{j+1}^n & \text{if} \quad x > x_{j+\frac{1}{2}}. \end{cases}$$

We can easily verify that the solution of this problem is given by $\bar{u}(x, t) = \bar{u}(x - b(t - t^n), t^n)$. Consequently, the similarity solution $u_{\mathrm{R}}(\eta; u_j^n, u_{j+1}^n)$ is given by

$$u_{\mathrm{R}}(\eta; u_j^n, u_{j+1}^n) = \begin{cases} u_j^n & \text{if} \quad \eta < b, \\ u_{j+1}^n & \text{if} \quad \eta > b. \end{cases}$$

For the numerical flux $F(u_j^n, u_{j+1}^n) = b\,u_{\mathrm{R}}(0; u_j^n, u_{j+1}^n)$ we then find

$$F(u_j^n, u_{j+1}^n) = \begin{cases} b\,u_j^n & \text{if} \quad b > 0, \\ b\,u_{j+1}^n & \text{if} \quad b < 0. \end{cases}$$

Substituting this flux into the Godunov scheme (13.91), we recover the upwind scheme (13.18). □

Example 13.11 Consider the Burgers' equation, for which $f(u) = \frac{1}{2}u^2$. The Riemann problem is then given by

$$\frac{\partial \bar{u}}{\partial t} + \frac{\partial}{\partial x}\left(\frac{1}{2}\bar{u}^2\right) = 0, \qquad x \in \mathbb{R}, \quad t > t^n,$$

$$\bar{u}(x, t^n) = \begin{cases} u_j^n & \text{if} \quad x < x_{j+\frac{1}{2}}, \\ u_{j+1}^n & \text{if} \quad x > x_{j+\frac{1}{2}}. \end{cases}$$

For the similarity solution of this problem we distinguish two cases: a shock wave and a rarefaction wave; see Example 12.13. If $u_j^n > u_{j+1}^n$, the solution of the Riemann problem is a shock wave propagating with speed $s_j^n := \frac{1}{2}(u_j^n + u_{j+1}^n)$. The similarity solution $u_R(\eta; u_j^n, u_{j+1}^n)$ is then given by

$$u_R(\eta; u_j^n, u_{j+1}^n) = \begin{cases} u_j^n & \text{if} \quad \eta < s_j^n, \\ u_{j+1}^n & \text{if} \quad \eta > s_j^n. \end{cases}$$

Alternatively, if $u_j^n < u_{j+1}^n$, the solution of the Riemann problem is the rarefaction wave

$$u_R(\eta; u_j^n, u_{j+1}^n) = \begin{cases} u_j^n & \text{if} \quad \eta < u_j^n, \\ \eta & \text{if} \quad u_j^n < \eta < u_{j+1}^n, \\ u_{j+1}^n & \text{if} \quad \eta > u_{j+1}^n. \end{cases}$$

For the numerical flux $F(u_j^n, u_{j+1}^n) = \frac{1}{2}\, u_R^2(0; u_j^n, u_{j+1}^n)$ we obtain the following result:

- If $u_j^n > u_{j+1}^n$, then

$$F(u_j^n, u_{j+1}^n) = \begin{cases} \frac{1}{2}\left(u_j^n\right)^2 & \text{if} \quad s_j^n > 0, \\ \frac{1}{2}\left(u_{j+1}^n\right)^2 & \text{if} \quad s_j^n < 0, \end{cases}$$

  with $s_j^n := \frac{1}{2}(u_j^n + u_{j+1}^n)$.
- If $u_j^n < u_{j+1}^n$, then

$$F(u_j^n, u_{j+1}^n) = \begin{cases} \frac{1}{2}\left(u_j^n\right)^2 & \text{if} \quad u_j^n > 0, \\ 0 & \text{if} \quad u_j^n < 0 < u_{j+1}^n, \\ \frac{1}{2}\left(u_{j+1}^n\right)^2 & \text{if} \quad u_{j+1}^n < 0. \end{cases}$$

The Godunov scheme for the Burgers' equation is thus given by (13.91) with numerical flux $F(u_j^n, u_{j+1}^n)$ given above. □

Next we will briefly comment on the stability and accuracy of the Godunov scheme. In the derivation of the Godunov scheme we have tacitly assumed that $\bar{u}(x, t)$ at the cell edges is not influenced by adjacent Riemann problems, so that $\bar{u}(x_{j+1/2}, t) = u_R(0; u_j^n, u_{j+1}^n) = C$ $(C \in \mathbb{R})$ for $t^n \le t \le t^{n+1}$. This condition is satisfied if the time step restriction

$$\frac{\Delta t}{\Delta x}|f'(u_j^n)| \le 1 \qquad (j = 0, \pm 1, \pm 2, \ldots), \tag{13.92}$$

holds, which means that the fastest of all waves emanating from the cell boundaries traverses at most one control volume during a time step; for more details see, e.g., [158, 168, 88]. The time step restriction (13.92) is a nonlinear variant of the CFL condition (13.20). Furthermore, it can be proved that the Godunov scheme is only first order accurate; see exercises. This is to be expected since it reduces to the upwind scheme for the linear advection equation.

The Godunov scheme is an example of a *conservative scheme*, which has the general form

$$u_j^{n+1} = u_j^n - \frac{\Delta t}{\Delta x}\big(F_{j+\frac{1}{2}}^n - F_{j-\frac{1}{2}}^n\big), \tag{13.93}$$

with $F_{j+\frac{1}{2}}^n$ the numerical flux at the cell interface $x_{j+\frac{1}{2}}$. Conservative schemes have the following nice property. Summing the relation (13.93) over any number of adjacent control volumes, ranging from $j = J_1$ to $j = J_2$, and rearranging terms, we find

$$\Delta x \sum_{j=J_1}^{J_2} \big(u_j^{n+1} - u_j^n\big) = -\Delta t\big(F_{J_2+\frac{1}{2}}^n - F_{J_1-\frac{1}{2}}^n\big), \tag{13.94}$$

since at each interface numerical fluxes cancel, except at the utmost left and right ones at $x = x_{J_1-\frac{1}{2}}$ and $x_{J_2+\frac{1}{2}}$, respectively. This relation can be seen as a discrete version of the integral balance over the interval $(a, b) := (x_{J_1-\frac{1}{2}}, x_{J_2+\frac{1}{2}})$, i.e.,

$$\int_a^b \big(u(x, t^{n+1}) - u(x, t^n)\big)\,\mathrm{d}x = -\int_{t^n}^{t^{n+1}} \big(f(u(b, t)) - f(u(a, t))\big)\,\mathrm{d}t, \tag{13.95}$$

which is the basic conservation law; see Section 12.2. Consequently, (discrete) conservation is guaranteed for any union of adjacent control volumes. For this reason a scheme of the form (13.93) is called conservative. The conservative form of a scheme is sometimes advantageous, as is evident from the following example.

Example 13.12 We would like to compute the shock wave solution of the Burgers' equation corresponding to the initial condition

$$u(x, 0) = \begin{cases} 2 & \text{if } x < 0, \\ 1 & \text{if } x > 0. \end{cases}$$

According to (12.38), this shock wave is propagating at speed $s = 1.5$. In this case the Godunov scheme reduces to

$$u_j^{n+1} = u_j^n - \frac{\Delta t}{\Delta x}\left(\frac{1}{2}\big(u_j^n\big)^2 - \frac{1}{2}\big(u_{j-1}^n\big)^2\right).$$

Alternatively, we can write the Burgers' equation in the equivalent form

$$\frac{\partial u}{\partial t} + u\frac{\partial u}{\partial x} = 0$$

and apply the upwind scheme to this equation. Since $u > 0$, we then find the nonconservative formulation

$$u_j^{n+1} = u_j^n - \frac{\Delta t}{\Delta x}u_j^n\big(u_j^n - u_{j-1}^n\big).$$

For both schemes we choose $\Delta x = 2.5 \times 10^{-2}$ and $\Delta t = 1.25 \times 10^{-2}$ and compute the solution at $t = 1$. The result is shown in Figure 13.17. Clearly, the Godunov scheme gives the correct location of the shock, although it is not very sharp. On the other hand, the upwind numerical solution propagates too slowly, resulting in the wrong position of the shock. □

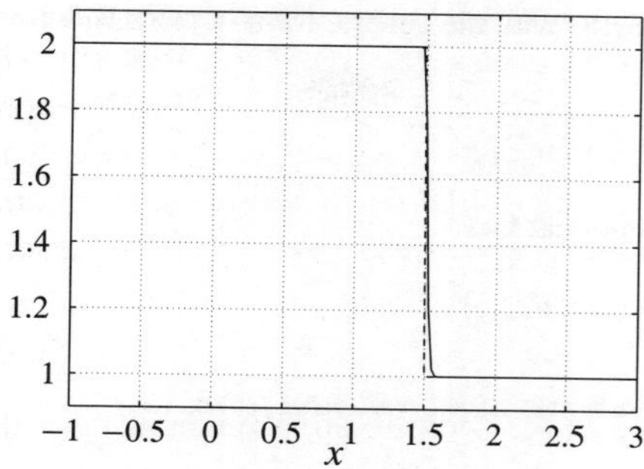

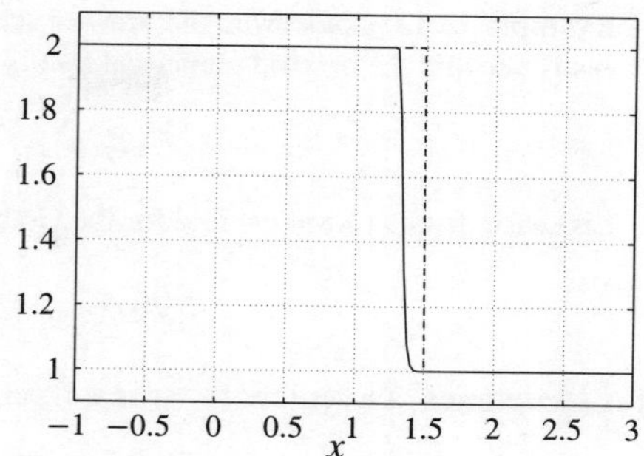

Figure 13.17. *A shock wave solution of the Burgers equation computed with Godunov's scheme (left) and the upwind scheme (right).*

From this example we conclude that nonconservative schemes are not suitable for computing discontinuous solutions of a nonlinear conservation law, since the exact location of the discontinuity is not correctly captured. The reason for this is that the discrete conservation property (13.94) is violated. On the other hand, conservative schemes do produce numerical solutions with the discontinuity at the correct location. Moreover, it can be shown that under certain (mild) conditions, the solution of a conservative scheme converges to the correct, entropy-satisfying, weak solution of the conservation law (13.80a); for more details see, e.g., [88].

13.6 High-Resolution Schemes

In Section 13.2 we found that the upwind scheme is very dissipative; i.e., the numerical solution is damped very much. On the other hand, the Lax–Wendroff scheme is less dissipative, so that numerical dispersion may become annoying. The latter is visible as spurious oscillations in the vicinity of discontinuities. In this and the following sections we will construct *high-resolution schemes*, which satisfy the following two requirements:

1. They are (at least) second order accurate for smooth solutions.

2. They give an accurate, nonoscillatory resolution of discontinuities.

Many numerical schemes for the scalar conservation law (13.80a) can be written in the conservative form (13.93), where the numerical flux $F^n_{j+\frac{1}{2}}$ is an approximation of the exact flux $f(u(x_{j+\frac{1}{2}}, t^n))$. The numerical flux $F^n_{j+\frac{1}{2}}$ depends on the neighbouring values u^n_j and u^n_{j+1}, and we write

$$F^n_{j+\frac{1}{2}} = F(u^n_j, u^n_{j+1}), \tag{13.96}$$

where $F = F(u, v)$ is called the *numerical flux function*. The dependence of the numerical flux function F on two values of the numerical solution leads, of course, to a three-point scheme.

Example 13.13 Comparing the upwind scheme (13.18) with the generic form in (13.93), we easily see that the upwind numerical flux is given by

$$F(u_j^n, u_{j+1}^n) = b^+ u_j^n + b^- u_{j+1}^n.$$

Likewise, from (13.89) we find for the Godunov numerical flux

$$F(u_j^n, u_{j+1}^n) = f\big(u_{\mathrm{R}}(0; u_j^n, u_{j+1}^n)\big).$$

□

It is sometimes convenient to write a three-point scheme in *incremental form*, i.e.,

$$u_j^{n+1} = u_j^n - C_{j-\frac{1}{2}}^n \big(u_j^n - u_{j-1}^n\big) + D_{j+\frac{1}{2}}^n \big(u_{j+1}^n - u_j^n\big), \tag{13.97}$$

with coefficients $C_{j-\frac{1}{2}}^n$ and $D_{j+\frac{1}{2}}^n$ depending on the numerical values u_k^n $(k = j, j \pm 1)$.

We first address the second requirement for high-resolution schemes mentioned above; i.e., the scheme should not produce spurious oscillations in the vicinity of discontinuities. This means that we require a scheme to be *monotonicity preserving*, according to the following definition.

Definition 13.14. *A numerical scheme for conservation law* (13.80a) *is called monotonicity preserving if the following holds: If the set* $\{u_j^n\}$ *is monotonically increasing, then* $\{u_j^{n+1}\}$ *is also monotonically increasing, or, if* $\{u_j^n\}$ *is monotonically decreasing, so is* $\{u_j^{n+1}\}$.

In order to quantify oscillations in a numerical solution, we define the *total variation* *TV* of a numerical solution $\mathbf{u}^n = (u_j^n)$ as

$$TV(\mathbf{u}^n) := \sum_{j=-\infty}^{\infty} |u_{j+1}^n - u_j^n|. \tag{13.98}$$

Hence we also have the following definition.

Definition 13.15. *A numerical scheme for conservation law* (13.80a) *is called total variation diminishing (TVD) if its solution satisfies* $TV(\mathbf{u}^{n+1}) \le TV(\mathbf{u}^n)$ *for any set of data* $\mathbf{u}^n$ *and for all* $n \ge 0$.

It is clear that the total variation of the Lax–Wendroff numerical solution in Figure 13.6 is larger than the total variation of the initial solution due to the presence of oscillations. It is intuitively clear that a numerical scheme is monotonicity preserving when it is TVD, as is confirmed by the following theorem.

Theorem 13.16. *A TVD scheme for conservation law* (13.80a) *is monotonicity preserving.*

For a proof one may consult [59, 87].

Consequently, we have to look for schemes that are TVD. Conditions under which a numerical scheme in incremental form (13.97) is TVD are given in the following theorem.

Theorem 13.17 (Harten). *A numerical scheme for conservation law* (13.80a)*, written in incremental form* (13.97)*, is TVD if the following inequalities hold:*

$$C^n_{j+\frac{1}{2}} \geq 0, \quad D^n_{j+\frac{1}{2}} \geq 0, \quad 0 \leq C^n_{j+\frac{1}{2}} + D^n_{j+\frac{1}{2}} \leq 1. \tag{13.99}$$

Proof. Replacing index j by $j+1$ in difference scheme (13.97) yields

$$u^{n+1}_{j+1} = u^n_{j+1} - C^n_{j+\frac{1}{2}}\big(u^n_{j+1} - u^n_j\big) + D^n_{j+\frac{3}{2}}\big(u^n_{j+2} - u^n_{j+1}\big).$$

Subtracting (13.97) from the latter equation, we get

$$u^{n+1}_{j+1} - u^{n+1}_j = \big(1 - C^n_{j+\frac{1}{2}} - D^n_{j+\frac{1}{2}}\big)(u^n_{j+1} - u^n_j) + D^n_{j+\frac{3}{2}}(u^n_{j+2} - u^n_{j+1}) + C^n_{j-\frac{1}{2}}(u^n_j - u^n_{j-1}).$$

Taking the absolute value of $u^{n+1}_{j+1} - u^{n+1}_j$, summing over all indices j, and using the triangle inequality results in

$$\sum_{j=-\infty}^{\infty} |u^{n+1}_{j+1} - u^{n+1}_j| \leq \sum_{j=-\infty}^{\infty} \big(1 - C^n_{j+\frac{1}{2}} - D^n_{j+\frac{1}{2}}\big)|u^n_{j+1} - u^n_j|$$

$$+ \sum_{j=-\infty}^{\infty} D^n_{j+\frac{3}{2}}|u^n_{j+2} - u^n_{j+1}| + \sum_{j=-\infty}^{\infty} C^n_{j-\frac{1}{2}}|u^n_j - u^n_{j-1}|.$$

The coefficients in front of the difference terms are all positive by virtue of the inequalities in (13.99). If we apply a shift in index in the second ($j' = j+1$) and in the third ($j' = j-1$) sums on the right-hand side of the inequality above, it is clear that these terms cancel against the corresponding terms in the first sum on the right-hand side. Consequently, this inequality reduces to

$$\sum_{j=-\infty}^{\infty} |u^{n+1}_{j+1} - u^{n+1}_j| \leq \sum_{j=-\infty}^{\infty} |u^n_{j+1} - u^n_j|,$$

or, in other words, $TV(\mathbf{u}^{n+1}) \leq TV(\mathbf{u}^n)$, proving the statement. □

Example 13.18 If we compare the upwind scheme (13.18) to the incremental form (13.97), we see that

$$C^n_{j-\frac{1}{2}} = c^+ := b^+\Delta t/\Delta x, \qquad -D^n_{j+\frac{1}{2}} = c^- := b^-\Delta t/\Delta x.$$

Hence the inequalities $C^n_{j-\frac{1}{2}} \geq 0$ and $D^n_{j+\frac{1}{2}} \geq 0$ are trivially satisfied. The third inequality in Theorem 13.17, $C^n_{j+\frac{1}{2}} + D^n_{j+\frac{1}{2}} \leq 1$, is satisfied if the stability condition (13.20) holds. Thus the upwind scheme is TVD if the CFL condition holds. □

Example 13.19 The Lax–Wendroff scheme (13.23) can be written in incremental form as follows:

$$\begin{aligned} u^{n+1}_j &= u^n_j - \frac{1}{2}c\big((u^n_{j+1} - u^n_j) + (u^n_j - u^n_{j-1})\big) + \frac{1}{2}c^2\big((u^n_{j+1} - u^n_j) - (u^n_j - u^n_{j-1})\big) \\ &= u^n_j - \frac{1}{2}c(c+1)(u^n_j - u^n_{j-1}) + \frac{1}{2}c(c-1)(u^n_{j+1} - u^n_j). \end{aligned}$$

Using (13.97), we can conclude that $C^n_{j-\frac{1}{2}} = \frac{1}{2}c(c+1)$ and $D^n_{j+\frac{1}{2}} = \frac{1}{2}c(c-1)$. If $0 < c \leq 1$, we have the inequalities $C^n_{j-\frac{1}{2}} \geq 0$, $D^n_{j+\frac{1}{2}} \leq 0$, and, alternatively, if $-1 \leq c < 0$, then $C^n_{j-\frac{1}{2}} \leq 0$, $D^n_{j+\frac{1}{2}} \geq 0$, and, consequently, the Lax–Wendroff scheme can never be TVD. □

Now we have to address the first requirement for high-resolution schemes, i.e., the second order accuracy. So let us determine the local discretisation error d^n_j of the numerical scheme (13.93) with the numerical flux given in (13.96). The local discretisation error for this scheme is defined by

$$d^n_j := \frac{1}{\Delta t}\big(u(x_j, t^{n+1}) - u(x_j, t^n)\big) + \frac{1}{\Delta x}\big(F(u(x_j, t^n), u(x_{j+1}, t^n)) - F(u(x_{j-1}, t^n), u(x_j, t^n))\big), \quad (13.100)$$

with $u(x, t)$ the solution of the conservation law (13.80a). A general expression for the local discretisation error is given in the following property.

Property 13.20. *If the numerical flux function $F(u, v)$ satisfies $F(u, u) = f(u)$ and if it is twice continuously differentiable, then the local discretisation error is given by*

$$d^n_j = \frac{1}{2}\Delta t \frac{\partial^2 u}{\partial t^2}(x_j, t^n) + \frac{1}{2}\Delta x \left(\frac{\partial F}{\partial v} - \frac{\partial F}{\partial u}\right)\frac{\partial^2 u}{\partial x^2}(x_j, t^n) + \frac{1}{2}\Delta x \left(\frac{\partial^2 F}{\partial v^2} - \frac{\partial^2 F}{\partial u^2}\right)\left(\frac{\partial u}{\partial x}\right)^2 (x_j, t^n) + \mathcal{O}(\Delta t^2) + \mathcal{O}(\Delta x^2), \quad (13.101)$$

where all derivatives of F have to be evaluated at $(u(x_j, t^n), u(x_j, t^n))$.

Proof. Straightforward Taylor series expansion of $u(x_j, t^{n+1})$ gives, for the first term on the right-hand side of (13.100),

$$\frac{1}{\Delta t}\big(u(x_j, t^{n+1}) - u(x_j, t^n)\big) = \frac{\partial u}{\partial t}(x_j, t^n) + \frac{1}{2}\Delta t \frac{\partial^2 u}{\partial t^2}(x_j, t^n) + \mathcal{O}(\Delta t^2). \quad (*)$$

Let the auxiliary variables A and B be defined by

$$A := u(x_{j+1}, t^n) - u(x_j, t^n), \quad B := u(x_{j-1}, t^n) - u(x_j, t^n).$$

Again, straightforward Taylor series expansion of $u(x_{j\pm1}, t^n)$ gives

$$A = \Delta x \frac{\partial u}{\partial x}(x_j, t^n) + \frac{1}{2}\Delta x^2 \frac{\partial^2 u}{\partial x^2}(x_j, t^n) + \mathcal{O}(\Delta x^3),$$

$$B = -\Delta x \frac{\partial u}{\partial x}(x_j, t^n) + \frac{1}{2}\Delta x^2 \frac{\partial^2 u}{\partial x^2}(x_j, t^n) + \mathcal{O}(\Delta x^3).$$

Expanding the flux terms on the right-hand side of (13.100), we get

$$F(u(x_j, t^n), u(x_{j+1}, t^n)) = F + A\frac{\partial F}{\partial v} + \frac{1}{2}A^2\frac{\partial^2 F}{\partial v^2} + \mathcal{O}(A^3),$$

$$F(u(x_{j-1}, t^n), u(x_j, t^n)) = F + B\frac{\partial F}{\partial u} + \frac{1}{2}B^2\frac{\partial^2 F}{\partial u^2} + \mathcal{O}(B^3),$$

where F and all its derivatives on the right-hand sides have to be evaluated at $(u(x_j, t^n), u(x_j, t^n))$. Substituting the expressions for A and B into the fluxes above, we find

$$F(u(x_j, t^n), u(x_{j+1}, t^n)) - F(u(x_{j-1}, t^n), u(x_j, t^n))$$

$$= \Delta x \left(\frac{\partial F}{\partial u} + \frac{\partial F}{\partial v}\right) \frac{\partial u}{\partial x}(x_j, t^n) + \frac{1}{2}\Delta x^2 \left(\frac{\partial F}{\partial v} - \frac{\partial F}{\partial u}\right) \frac{\partial^2 u}{\partial x^2}(x_j, t^n)$$

$$+ \frac{1}{2}\Delta x^2 \left(\frac{\partial^2 F}{\partial v^2} - \frac{\partial^2 F}{\partial u^2}\right) \left(\frac{\partial u}{\partial x}\right)^2 (x_j, t^n) + \mathcal{O}(\Delta x^3). \quad (**)$$

The first term on the right-hand side of this equation equals

$$\Delta x \frac{\partial}{\partial x} F(u, u) = \Delta x \frac{\partial}{\partial x} f(u),$$

since $F(u, u) = f(u)$. Substitution of $(*)$ and $(**)$ into (13.100) finally gives the expression in (13.101). □

Consider, as an example, the advection equation (13.1a) with $b > 0$. In this case $F(u, v) = bu$ and, substituting this numerical flux function in (13.101), we precisely find (13.14).

In the remainder of this section we restrict ourselves to the advection equation (13.1a) with $b > 0$. A high-resolution scheme for this equation should be TVD and second order accurate. These two requirements are not easily realized, as shown by the following theorem.

Theorem 13.21. *A linear, three-point scheme for the advection equation* (13.1a) *that is TVD is at most first order accurate.*

Proof. Since we are dealing with a linear equation, the numerical flux function $F(u, v)$ can generally be written as

$$F(u, v) = bu + \gamma b(v - u)$$

for some constant γ. Note that this flux function satisfies $F(u, u) = bu$ and, consequently, the resulting scheme is consistent. Substituting this flux function in (13.93) gives

$$u_j^{n+1} = u_j^n - \frac{\Delta t}{\Delta x}\big(bu_j^n + \gamma b(u_{j+1}^n - u_j^n) - bu_{j-1}^n - \gamma b(u_j^n - u_{j-1}^n)\big)$$

$$= u_j^n - c\big((1 - \gamma)(u_j^n - u_{j-1}^n) + \gamma(u_{j+1}^n - u_j^n)\big). \quad (*)$$

Using (13.101), we obtain for the local discretisation error

$$d_j^n = \frac{1}{2}\Delta t \frac{\partial^2 u}{\partial t^2}(x_j, t^n) + \frac{1}{2}\Delta x(2\gamma - 1)b\frac{\partial^2 u}{\partial x^2}(x_j, t^n) + \mathcal{O}(\Delta t^2) + \mathcal{O}(\Delta x^2).$$

Since $\frac{\partial^2 u}{\partial t^2} = b^2 \frac{\partial^2 u}{\partial x^2}$, we thus find

$$d_j^n = \frac{1}{2} b\Delta x(c + 2\gamma - 1)\frac{\partial^2 u}{\partial x^2}(x_j, t^n) + \mathcal{O}(\Delta t^2) + \mathcal{O}(\Delta x^2).$$

Let the scheme be second order accurate. From the expression for d_j^n we see that this second order accuracy can only be obtained if $\gamma = (1-c)/2$. Consider now the following numerical solution at time level t^n:

$$u_j^n = \begin{cases} 0 & \text{if} \quad j < 0, \\ 1 & \text{if} \quad j \geq 0. \end{cases}$$

It is obvious that $TV(\mathbf{u}^n) = 1$. We want to compute $TV(\mathbf{u}^{n+1})$. Replacing index j by $j+1$ in scheme $(*)$, we find

$$u_{j+1}^{n+1} = u_{j+1}^n - c\big((1-\gamma)(u_{j+1}^n - u_j^n) + \gamma(u_{j+2}^n - u_{j+1}^n)\big). \qquad (**)$$

Subtracting $(*)$ from $(**)$, we obtain

$$u_{j+1}^{n+1} - u_j^{n+1} = (1 + c(2\gamma - 1))(u_{j+1}^n - u_j^n) - c\gamma(u_{j+2}^n - u_{j+1}^n) + c(1-\gamma)(u_j^n - u_{j-1}^n).$$

This then makes it possible to compute $TV(\mathbf{u}^{n+1})$. For the specific choice of u_j^n only three terms contribute to $TV(\mathbf{u}^{n+1})$, i.e., the terms corresponding to $j = -2, -1, 0$ in $(**)$. Thus we find

$$TV(\mathbf{u}^{n+1}) = c(|\gamma| + |1-\gamma|) + |1 + c(2\gamma - 1)|.$$

For the choice $\gamma = (1-c)/2$, which is necessary to have second order accuracy, this expression reduces to

$$TV(\mathbf{u}^{n+1}) = \frac{1}{2}c\big(|1-c| + |1+c|\big) + |1 - c^2|.$$

We only have to consider the case $0 < c < 1$, because for $c > 1$ the scheme is unstable. For $0 < c < 1$ we have $TV(\mathbf{u}^{n+1}) = 1 + c(1-c) > 1 = TV(\mathbf{u}^n)$, implying that the scheme is not TVD. This is a contradiction, and therefore our assumption in the beginning that the scheme was second order accurate is false. □

13.7 A Flux Limiter Scheme for the Advection Equation

In the previous section we introduced the notion of high-resolution schemes for the scalar conservation law (13.80a). For the special case of the advection equation (13.1a) we proved that the two requirements for high-resolution schemes, second order accuracy and nonoscillatory representation of discontinuities, cannot be combined into a linear scheme. Consequently, we have to look for nonlinear high-resolution schemes. In this section we will construct a high-resolution scheme for the advection equation (13.1a) by combining the upwind and Lax–Wendroff schemes in a suitable manner. More specifically, we will combine the numerical fluxes of both schemes and apply a *flux limiter* in order to make the resulting scheme second order accurate and TVD.

Recall the general form of a three-point conservative scheme for (13.1a):

$$u_j^{n+1} = u_j^n - \frac{\Delta t}{\Delta x}\big(F(u_j^n, u_{j+1}^n) - F(u_{j-1}^n, u_j^n)\big). \qquad (13.102)$$

We now take for the numerical flux $F(u_j^n, u_{j+1}^n)$ a combination of the upwind flux $F_{\text{uw}}(u_j^n, u_{j+1}^n)$ and the Lax–Wendroff flux $F_{\text{LW}}(u_j^n u_{j+1}^n)$ as follows:

$$F(u_j^n, u_{j+1}^n) := F_{\text{uw}}(u_j^n, u_{j+1}^n) + \Phi_{j+\frac{1}{2}}^n\big(F_{\text{LW}}(u_j^n, u_{j+1}^n) - F_{\text{uw}}(u_j^n, u_{j+1}^n)\big), \tag{13.103}$$

where $\Phi_{j+\frac{1}{2}}^n$ is the flux limiter, which depends on the numerical solution at time level t^n. In regions where the solution is smooth we would like to have $\Phi_{j+\frac{1}{2}}^n \approx 1$, so that (13.102) reduces to the (second order) Lax–Wendroff scheme. Near discontinuities we have to be satisfied with first order accuracy. Moreover, near extrema we will choose $\Phi_{j+\frac{1}{2}}^n = 0$, resulting in the upwind scheme. The limiter $\Phi_{j+\frac{1}{2}}^n$ is given by

$$\Phi_{j+\frac{1}{2}}^n := \Phi\big(r_{j+\frac{1}{2}}^n\big), \tag{13.104}$$

with $\Phi(r) \geq 0$ the *limiter function* and the variable $r_{j+\frac{1}{2}}^n$ a measure of the smoothness of the numerical solution near the cell boundary $x_{j+\frac{1}{2}}$. There are many possible choices for $r_{j+\frac{1}{2}}^n$. Usually, it is defined by

$$r_{j+\frac{1}{2}}^n := \begin{cases} \dfrac{u_j^n - u_{j-1}^n}{u_{j+1}^n - u_j^n} & \text{if} \quad b > 0, \\[2ex] \dfrac{u_{j+2}^n - u_{j+1}^n}{u_{j+1}^n - u_j^n} & \text{if} \quad b < 0; \end{cases} \tag{13.105}$$

i.e., $r_{j+\frac{1}{2}}^n$ is the ratio of the upwind difference and the central difference of the numerical solution at the cell boundary $x_{j+\frac{1}{2}}$. Several limiter functions are used in practice; we will consider a few later on.

Let us first elaborate the numerical fluxes of the upwind and Lax–Wendroff schemes. For the upwind scheme (13.18) we find

$$F_{\text{uw}}(u_j^n, u_{j+1}^n) = b^+ u_j^n + b^- u_{j+1}^n, \tag{13.106}$$

with b^+ and b^- as defined in (13.17). The determination of the Lax–Wendroff numerical flux is somewhat more involved. We actually need the flux difference $F_{\text{LW}}(u_j^n, u_{j+1}^n) - F_{\text{uw}}(u_j^n, u_{j+1}^n)$, and for that reason we write the Lax–Wendroff scheme (13.23) as the upwind scheme plus a correction term. If $b > 0$, we have

$$\begin{aligned} u_j^{n+1} &= u_j^n - c\big(u_j^n - u_{j-1}^n\big) - \frac{1}{2}c\big(u_{j+1}^n - 2u_j^n + u_{j-1}^n\big) + \frac{1}{2}c^2\big(u_{j+1}^n - 2u_j^n + u_{j-1}^n\big) \\ &= u_j^n - c\big(u_j^n - u_{j-1}^n\big) - \frac{1}{2}c(1-c)\big(u_{j+1}^n - 2u_j^n + u_{j-1}^n\big). \end{aligned} \tag{13.107}$$

Since stability requires $0 < c < 1$, we can interpret the Lax–Wendroff scheme as the upwind scheme with the central difference approximation of the *antidiffusion* term $-\frac{1}{2}b(1-c)\Delta x \frac{\partial^2 u}{\partial x^2}(x_j, t^n)$ added. Using the relation $u_{j+1}^n - 2u_j^n + u_{j-1}^n = (u_{j+1}^n - u_j^n) - (u_j^n - u_{j-1}^n)$, we can rewrite (13.107) in the generic form (13.102), with numerical flux

$$F_{\text{LW}}(u_j^n, u_{j+1}^n) = bu_j^n + \frac{1}{2}b(1-c)(u_{j+1}^n - u_j^n). \tag{13.108}$$

Likewise, if $b < 0$, the Lax–Wendroff scheme can be written as

$$\begin{aligned} u_j^{n+1} &= u_j^n - c\big(u_{j+1}^n - u_j^n\big) + \frac{1}{2}c\big(u_{j+1}^n - 2u_j^n + u_{j-1}^n\big) + \frac{1}{2}c^2\big(u_{j+1}^n - 2u_j^n + u_{j-1}^n\big) \\ &= u_j^n - c\big(u_{j+1}^n - u_j^n\big) + \frac{1}{2}c(1+c)\big(u_{j+1}^n - 2u_j^n + u_{j-1}^n\big). \end{aligned} \tag{13.109}$$

Since we now need $-1 < c < 0$ to have stability, we thus find that

$$F_{\mathrm{LW}}(u_j^n, u_{j+1}^n) = bu_{j+1}^n - \frac{1}{2}b(1+c)\big(u_{j+1}^n - u_j^n\big). \tag{13.110}$$

Combining (13.108) and (13.110) for the Lax–Wendroff flux into a single expression yields

$$F_{\mathrm{LW}}(u_j^n, u_{j+1}^n) = F_{\mathrm{uw}}(u_j^n, u_{j+1}^n) + \frac{1}{2}|b|(1-|c|)\big(u_{j+1}^n - u_j^n\big). \tag{13.111}$$

The correction term in (13.111) is an antidiffusion flux that has the effect of steepening the upwind approximation of a discontinuity. Substituting (13.106) and (13.111) into (13.103), we obtain for the *high-resolution flux*

$$F(u_j^n, u_{j+1}^n) = b^+u_j^n + b^-u_{j+1}^n + \frac{1}{2}|b|(1-|c|)\Phi_{j+\frac{1}{2}}^n\big(u_{j+1}^n - u_j^n\big). \tag{13.112}$$

In what follows we will derive constraints on the limiter function $\Phi(r)$ such that the scheme (13.102) with numerical flux (13.112) is both TVD and second order accurate.

We now investigate when the scheme (13.102) is TVD. Substituting (13.112) into (13.102), we obtain the *high-resolution scheme*

$$\begin{aligned} u_j^{n+1} &= u_j^n - c^+\big(u_j^n - u_{j-1}^n\big) - c^-\big(u_{j+1}^n - u_j^n\big) \\ &\quad - \frac{1}{2}|c|(1-|c|)\Big(\Phi_{j+\frac{1}{2}}^n\big(u_{j+1}^n - u_j^n\big) - \Phi_{j-\frac{1}{2}}^n\big(u_j^n - u_{j-1}^n\big)\Big), \end{aligned} \tag{13.113}$$

where $c := b\Delta t/\Delta x$, $c^+ := b^+\Delta t/\Delta x$, and $c^- := b^-\Delta t/\Delta x$. In order to determine under what conditions (13.113) is TVD, we have to rewrite it in the incremental form (13.97). An obvious choice for the coefficients $C_{j-\frac{1}{2}}^n$ and $D_{j+\frac{1}{2}}^n$ is

$$C_{j-\frac{1}{2}}^n = c^+ - \frac{1}{2}|c|(1-|c|)\Phi_{j-\frac{1}{2}}^n, \qquad D_{j+\frac{1}{2}}^n = -c^- - \frac{1}{2}|c|(1-|c|)\Phi_{j+\frac{1}{2}}^n.$$

Unfortunately, $D_{j+\frac{1}{2}}^n < 0$ if $b > 0$ and $C_{j-\frac{1}{2}}^n < 0$ if $b < 0$, so that we cannot apply Theorem 13.17 to prove that (13.113) is TVD. However, other choices for the coefficients $C_{j-\frac{1}{2}}^n$ and $D_{j+\frac{1}{2}}^n$ are possible. Let $b > 0$. Using the definition of $r_{j+\frac{1}{2}}^n$, we can rewrite (13.113) as a modification of the upwind scheme in the following way:

$$u_j^{n+1} = u_j^n - c\left(1 + \frac{1}{2}(1-c)\left(\frac{\Phi_{j+\frac{1}{2}}^n}{r_{j+\frac{1}{2}}^n} - \Phi_{j-\frac{1}{2}}^n\right)\right)\big(u_j^n - u_{j-1}^n\big). \tag{13.114}$$

This then corresponds to the choice

$$C^n_{j-\frac{1}{2}} = c\left(1 + \frac{1}{2}(1-c)\left(\frac{\Phi^n_{j+\frac{1}{2}}}{r^n_{j+\frac{1}{2}}} - \Phi^n_{j-\frac{1}{2}}\right)\right), \qquad D^n_{j+\frac{1}{2}} = 0. \tag{13.115}$$

From (13.99) we see that (13.114) is TVD if $0 \leq C^n_{j-\frac{1}{2}} \leq 1$. For $b < 0$ we can rewrite (13.113) as

$$u^{n+1}_j = u^n_j - c\left(1 - \frac{1}{2}(1+c)\left(\Phi^n_{j+\frac{1}{2}} - \frac{\Phi^n_{j-\frac{1}{2}}}{r^n_{j-\frac{1}{2}}}\right)\right)\left(u^n_{j+1} - u^n_j\right). \tag{13.116}$$

Obviously, the coefficients $C^n_{j-\frac{1}{2}}$ and $D^n_{j+\frac{1}{2}}$ are now given by

$$C^n_{j-\frac{1}{2}} = 0, \qquad D^n_{j+\frac{1}{2}} = -c\left(1 - \frac{1}{2}(1+c)\left(\Phi^n_{j+\frac{1}{2}} - \frac{\Phi^n_{j-\frac{1}{2}}}{r^n_{j-\frac{1}{2}}}\right)\right). \tag{13.117}$$

From (13.99) we conclude that scheme (13.116) is TVD if $0 \leq D^n_{j+\frac{1}{2}} \leq 1$.

We have to elaborate the inequalities for $C^n_{j-\frac{1}{2}}$ and $D^n_{j+\frac{1}{2}}$ mentioned above. For $b > 0$ and $0 < c \leq 1$ the inequalities $0 \leq C^n_{j-\frac{1}{2}} \leq 1$ are equivalent to

$$\frac{-2}{1-c} \leq \frac{\Phi^n_{j+\frac{1}{2}}}{r^n_{j+\frac{1}{2}}} - \Phi^n_{j-\frac{1}{2}} \leq \frac{2}{c}. \tag{13.118}$$

Likewise, for $b < 0$ and $-1 \leq c < 0$ the inequalities $0 \leq D^n_{j+\frac{1}{2}} \leq 1$ can be reformulated as

$$\frac{2}{c} \leq \Phi^n_{j+\frac{1}{2}} - \frac{\Phi^n_{j-\frac{1}{2}}}{r^n_{j-\frac{1}{2}}} \leq \frac{2}{1+c}. \tag{13.119}$$

Maximising the lower bounds and minimising the upper bounds, we can derive sufficient conditions for the inequalities in (13.118) and (13.119) to hold. More precisely, the upper bound $2/c$ in (13.118) attains its minimum value for $c = 1$ and the lower bound $2/c$ in (13.119) has its maximum value for $c = -1$. Likewise, the choice $c = 0$ gives the maximum value for the lower bound $-2/(1-c)$ in (13.118) and the minimum value for the upper bound $2/(1+c)$ in (13.119). This way, we find the sufficient condition

$$\left|\frac{\Phi(r_1)}{r_1} - \Phi(r_2)\right| \leq 2, \qquad \text{for all} \quad r_1, r_2 \tag{13.120}$$

for the inequalities in (13.118) and (13.119). For $r_1, r_2 > 0$ the inequality in (13.120) holds if $\Phi(r)$ satisfies

$$0 \leq \Phi(r) \leq \min(2, 2r) \qquad \text{for all} \quad r > 0. \tag{13.121}$$

The region in the (r, Φ) plane satisfying (13.121) is referred to as the *TVD region* and is shown in Figure 13.18. Recall that near extrema we apply the upwind scheme, which

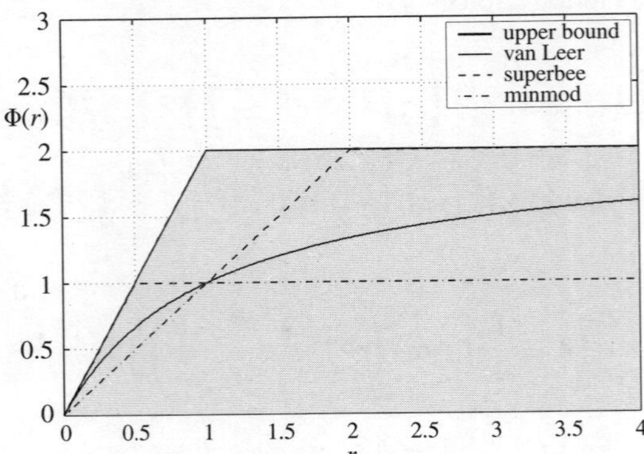

Figure 13.18. *The van Leer, superbee, and minmod limiters in the TVD region.*

means that $\Phi(r) = 0$ for $r \leq 0$. Thus near extrema the generation of spurious oscillations is prevented at the cost of a lower accuracy.

It remains to be investigated whether (13.113) is second order accurate. From the foregoing it is clear that (13.113) can only be second order accurate away from extrema. This is stated in the following theorem.

Theorem 13.22. *If the limiter function $\Phi(r)$ satisfies $\Phi(1) = 1$ and is Lipschitz continuous near $r = 1$, then the high-resolution scheme* (13.113) *with $\Phi^n_{j+\frac{1}{2}}$ given by* (13.104) *and* (13.105) *is second order consistent on smooth regions where $\frac{\partial u}{\partial x}$ is bounded away from zero.*

Proof. We have to prove that the local discretisation error is second order in Δt and Δx. The high-resolution scheme (13.113) can be written in the form

$$\frac{1}{\Delta t}\left(u_j^{n+1} - u_j^n\right) + \frac{b^+}{\Delta x}\left(u_j^n - u_{j-1}^n\right) + \frac{b^-}{\Delta x}\left(u_{j+1}^n - u_j^n\right)$$
$$+ \frac{1}{2}|b|(1-|c|)\left[\Phi^n_{j+\frac{1}{2}}\frac{1}{\Delta x}\left(u_{j+1}^n - u_j^n\right) - \Phi^n_{j-\frac{1}{2}}\frac{1}{\Delta x}\left(u_j^n - u_{j-1}^n\right)\right] = 0.$$

Substituting the exact solution of the advection equation (13.1a) into this scheme gives the following expression for the local discretisation error d_j^n:

$$d_j^n = \frac{1}{\Delta t}\left(u(x_j, t^{n+1}) - u(x_j, t^n)\right) + \frac{b^+}{\Delta x}\left(u(x_j, t^n) - u(x_{j-1}, t^n)\right)$$
$$+ \frac{b^-}{\Delta x}\left(u(x_{j+1}, t^n) - u(x_j, t^n)\right) + \frac{1}{2}|b|(1-|c|)\left[\widetilde{\Phi}^n_{j+\frac{1}{2}}\frac{1}{\Delta x}\left(u(x_{j+1}, t^n) - u(x_j, t^n)\right)\right.$$
$$\left. - \widetilde{\Phi}^n_{j-\frac{1}{2}}\frac{1}{\Delta x}\left(u(x_j, t^n) - u(x_{j-1}, t^n)\right)\right],$$

with $\widetilde{\Phi}^n_{j+\frac{1}{2}} = \Phi(\widetilde{r}^n_{j+\frac{1}{2}})$ and $\widetilde{r}^n_{j+\frac{1}{2}}$ given by

$$\widetilde{r}^n_{j+\frac{1}{2}} = \begin{cases} \dfrac{u(x_j, t^n) - u(x_{j-1}, t^n)}{u(x_{j+1}, t^n) - u(x_j, t^n)} & \text{if} \quad b > 0, \\[2ex] \dfrac{u(x_{j+2}, t^n) - u(x_{j+1}, t^n)}{u(x_{j+1}, t^n) - u(x_j, t^n)} & \text{if} \quad b < 0. \end{cases}$$

Introducing the auxiliary variables

$$A := u(x_{j+1}, t^n) - u(x_j, t^n), \qquad B := u(x_j, t^n) - u(x_{j-1}, t^n),$$

we find the following condensed expression for d^n_j:

$$\begin{aligned} d^n_j = \frac{1}{\Delta t}\big(u(x_j, t^{n+1}) - u(x_j, t^n)\big) &+ \frac{1}{\Delta x}\left(b^- + \frac{1}{2}|b|\big(1 - |c|\big)\widetilde{\Phi}^n_{j+\frac{1}{2}}\right) A \\ &+ \frac{1}{\Delta x}\left(b^+ - \frac{1}{2}|b|\big(1 - |c|\big)\widetilde{\Phi}^n_{j-\frac{1}{2}}\right) B. \quad (*) \end{aligned}$$

Straightforward Taylor series expansion of $u(x_j, t^{n+1})$ gives, for the first term on the right-hand side of $(*)$,

$$\frac{1}{\Delta t}\big(u(x_j, t^{n+1}) - u(x_j, t^n)\big) = \frac{\partial u}{\partial t}(x_j, t^n) + \frac{1}{2}\Delta t \frac{\partial^2 u}{\partial t^2}(x_j, t^n) + \mathcal{O}(\Delta t^2).$$

Likewise, Taylor series expansion of $u(x_{j\pm1}, t^n)$ gives

$$A = \Delta x \frac{\partial u}{\partial x}(x_j, t^n) + \frac{1}{2}\Delta x^2 \frac{\partial^2 u}{\partial x^2}(x_j, t^n) + \mathcal{O}(\Delta x^3),$$

$$B = \Delta x \frac{\partial u}{\partial x}(x_j, t^n) - \frac{1}{2}\Delta x^2 \frac{\partial^2 u}{\partial x^2}(x_j, t^n) + \mathcal{O}(\Delta x^3).$$

Substituting these Taylor series expansions into expression $(*)$ for the local discretisation error, we obtain

$$\begin{aligned} d^n_j = {} & \frac{\partial u}{\partial t}(x_j, t^n) + \frac{1}{2}\Delta t \frac{\partial^2 u}{\partial t^2}(x_j, t^n) + \mathcal{O}(\Delta t^2) \\ & + \left(b^- + \frac{1}{2}|b|\big(1 - |c|\big)\widetilde{\Phi}^n_{j+\frac{1}{2}}\right)\left(\frac{\partial u}{\partial x}(x_j, t^n) + \frac{1}{2}\Delta x \frac{\partial^2 u}{\partial x^2}(x_j, t^n) + \mathcal{O}(\Delta x^2)\right) \\ & + \left(b^+ - \frac{1}{2}|b|\big(1 - |c|\big)\widetilde{\Phi}^n_{j-\frac{1}{2}}\right)\left(\frac{\partial u}{\partial x}(x_j, t^n) - \frac{1}{2}\Delta x \frac{\partial^2 u}{\partial x^2}(x_j, t^n) + \mathcal{O}(\Delta x^2)\right). \end{aligned}$$

Replacing in this expression the time derivatives by the appropriate space derivatives, i.e., $\frac{\partial u}{\partial t} = -b\frac{\partial u}{\partial x}$ and $\frac{\partial^2 u}{\partial t^2} = b^2\frac{\partial^2 u}{\partial x^2}$, and rearranging terms, we find

$$\begin{aligned} d^n_j = {} & \frac{1}{2}|b|\big(1 - |c|\big)\Big(\widetilde{\Phi}^n_{j+\frac{1}{2}} - \widetilde{\Phi}^n_{j-\frac{1}{2}}\Big)\frac{\partial u}{\partial x}(x_j, t^n) \\ & + \frac{1}{4}\Delta x|b|\big(1 - |c|\big)\Big(\widetilde{\Phi}^n_{j+\frac{1}{2}} + \widetilde{\Phi}^n_{j-\frac{1}{2}} - 2\Big)\frac{\partial^2 u}{\partial x^2}(x_j, t^n) + \mathcal{O}(\Delta t^2) + \mathcal{O}(\Delta x^2). \end{aligned}$$

Applying the triangle inequality and assuming that $u(x,t)$ is sufficiently smooth, we find the following upper bound for d_j^n:

$$\begin{aligned} |d_j^n| &\leq \frac{1}{2}|b|(1-|c|)\left|\widetilde{\Phi}^n_{j+\frac{1}{2}} - \widetilde{\Phi}^n_{j-\frac{1}{2}}\right|C_1 \\ &+ \frac{1}{4}\Delta x|b|(1-|c|)\left|\widetilde{\Phi}^n_{j+\frac{1}{2}} + \widetilde{\Phi}^n_{j-\frac{1}{2}} - 2\right|C_2 + \mathcal{O}(\Delta t^2) + \mathcal{O}(\Delta x^2) \quad (**) \end{aligned}$$

for some constants $C_1, C_2 > 0$. Next we have to estimate the terms involving the flux limiters $\widetilde{\Phi}^n_{j\pm\frac{1}{2}}$. Since the limiter function $\Phi(r)$ is Lipschitz continuous near $r = 1$, with Lipschitz constant L, say, we have the following estimates:

$$\begin{aligned} \left|\widetilde{\Phi}^n_{j+\frac{1}{2}} - \widetilde{\Phi}^n_{j-\frac{1}{2}}\right| &\leq L\left|\widetilde{r}^n_{j+\frac{1}{2}} - \widetilde{r}^n_{j-\frac{1}{2}}\right|, \\ \left|\widetilde{\Phi}^n_{j+\frac{1}{2}} + \widetilde{\Phi}^n_{j-\frac{1}{2}} - 2\right| &= \left|\Phi(\widetilde{r}^n_{j+\frac{1}{2}}) - \Phi(1) + \Phi(\widetilde{r}^n_{j-\frac{1}{2}}) - \Phi(1)\right| \\ &\leq L\left(\left|\widetilde{r}^n_{j+\frac{1}{2}} - 1\right| + \left|\widetilde{r}^n_{j-\frac{1}{2}} - 1\right|\right). \quad (***) \end{aligned}$$

In the derivation of the second inequality in $(***)$ we have used that $\Phi(1) = 1$. Finally, we have to determine the ratios $\widetilde{r}^n_{j\pm\frac{1}{2}}$. To that purpose, we assume that $b > 0$. The case $b < 0$ is completely similar. Expanding $u(x_{j\pm1}, t^n)$ in a Taylor series, we find

$$\widetilde{r}^n_{j+\frac{1}{2}} = 1 - \Delta x \frac{\partial^2 u}{\partial x^2}(x_j, t^n) \Big/ \frac{\partial u}{\partial x}(x_j, t^n) + \mathcal{O}(\Delta x^2), \quad (****)$$

provided that $\frac{\partial u}{\partial x}(x_j, t^n) \neq 0$. A similar expression holds for $\widetilde{r}^n_{j-\frac{1}{2}}$. Combining $(**)$, $(***)$, and $(****)$, we find that $d_j^n = \mathcal{O}(\Delta t^2) + \mathcal{O}(\Delta x^2)$ in regions where $\frac{\partial u}{\partial x}$ is bounded away from zero. □

We conclude this section by giving two limiter functions. A smooth *limiter function* is due to van Leer [85] and is given for $r > 0$ by

$$\Phi(r) = \frac{2r}{1+r}. \tag{13.122a}$$

Another limiter function, which is especially suitable for representing discontinuities, is the *superbee limiter* function of Roe [133]. It is defined for $r > 0$ by

$$\Phi(r) = \max(\min(1, 2r), \min(r, 2)). \tag{13.122b}$$

Both limiter functions are shown in Figure 13.18.

13.8 Slope Limiter Methods

In Section 13.7 we constructed a high-resolution scheme for the advection equation (13.1a) by combining the upwind and Lax–Wendroff numerical fluxes and applying a flux limiter to make the scheme TVD. In this section we will present an alternative formulation of high-resolution schemes that is easier to generalise to nonlinear equations. This formulation is based on a piecewise linear representation of the numerical solution. Applying this to the advection equation gives the Lax–Wendroff scheme, which is not TVD. In order to make the scheme TVD, we need *slope limiters*. As we will show, the resulting slope limiter method is identical to the flux limiter method derived in the previous section. Finally, we will briefly describe the extension to nonlinear equations.

13.8.1 A Slope Limiter Method for the Advection Equation

As we did for the derivation of the Godunov scheme discussed in Section 13.5, we introduce a function $\bar{u}(x,t)$ that satisfies the initial value problem

$$\frac{\partial \bar{u}}{\partial t} + b\frac{\partial \bar{u}}{\partial x} = 0, \qquad x \in \mathbb{R}, \quad t > t^n, \tag{13.123a}$$

$$\bar{u}(x,t^n) = u_j^n + s_j^n(x - x_j), \qquad x \in V_j \quad (j = 0, \pm 1, \pm 2, \ldots), \tag{13.123b}$$

where u_j^n and s_j^n are the average value and slope, respectively, of $\bar{u}(x,t^n)$ in the control volume V_j; see Figure 13.19. The initial value $\bar{u}(x,t^n)$ is thus piecewise linear in each control volume V_j. The slope s_j^n is still unknown and has to be determined such that the resulting scheme is TVD. Note that for $s_j^n = 0$ we recover the upwind scheme. The slope limiter method is based on the integral formulation of the initial value problem (13.123).

Integrating the advection equation (13.123a) over control volume V_j gives

$$\frac{\mathrm{d}}{\mathrm{d}t}\int_{V_j} \bar{u}(x,t)\,\mathrm{d}x + b\left(\bar{u}(x_{j+\frac{1}{2}},t) - \bar{u}(x_{j-\frac{1}{2}},t)\right) = 0, \tag{13.124}$$

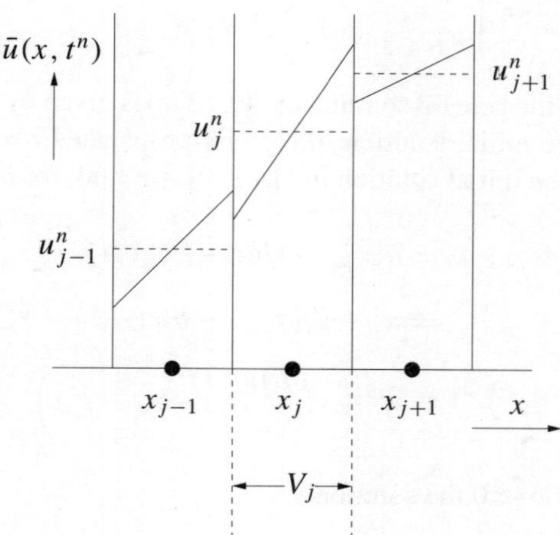

Figure 13.19. *Piecewise linear numerical solution at time level t^n.*

and subsequent integration over the time interval $[t^n, t^{n+1}]$ results in

$$\int_{V_j} \bar{u}(x, t^{n+1})\,\mathrm{d}x - \int_{V_j} \bar{u}(x, t^n)\,\mathrm{d}x + b\left[\int_{t^n}^{t^{n+1}} \bar{u}(x_{j+\frac{1}{2}}, t)\,\mathrm{d}t - \int_{t^n}^{t^{n+1}} \bar{u}(x_{j-\frac{1}{2}}, t)\,\mathrm{d}t\right] = 0. \tag{13.125}$$

We can write the integral formulation (13.125) in the condensed form

$$\int_{V_j} \bar{u}(x, t^{n+1})\,\mathrm{d}x - \int_{V_j} \bar{u}(x, t^n)\,\mathrm{d}x = -\Delta t\big(F(u_j^n, u_{j+1}^n) - F(u_{j-1}^n, u_j^n)\big), \tag{13.126}$$

where the numerical flux $F(u_j^n, u_{j+1}^n)$ is now given by

$$F(u_j^n, u_{j+1}^n) = \frac{b}{\Delta t}\int_{t^n}^{t^{n+1}} \bar{u}(x_{j+\frac{1}{2}}, t)\,\mathrm{d}t; \tag{13.127}$$

cf. (13.85). In order to complete the derivation of this version of the Godunov scheme, we assume that $\bar{u}(x, t^{n+1})$ also has a piecewise linear distribution analogous to (13.123b). The integrals in (13.126) are then trivial and we find

$$u_j^{n+1} = u_j^n - \frac{\Delta t}{\Delta x}\Big(F(u_j^n, u_{j+1}^n) - F(u_{j-1}^n, u_j^n)\Big). \tag{13.128}$$

The next step in the derivation of the slope limiter scheme is to compute the numerical fluxes from (13.127).

In order to compute the numerical flux $F(u_j^n, u_{j+1}^n)$, we have to compute $\bar{u}(x_{j+\frac{1}{2}}, t)$. Locally, i.e., in the vicinity of the cell boundary $x_{j+\frac{1}{2}}$, the function $\bar{u}(x, t)$ is the solution of the *generalized Riemann problem* consisting of (13.123a) and the piecewise linear initial condition

$$\bar{u}(x, t^n) = \begin{cases} u_j^n + s_j^n(x - x_j) & \text{if} \quad x < x_{j+\frac{1}{2}}, \\ u_{j+1}^n + s_{j+1}^n(x - x_{j+1}) & \text{if} \quad x > x_{j+\frac{1}{2}}. \end{cases} \tag{13.129}$$

As we have seen before, the general solution of (13.123a) is given by $\bar{u}(x, t) = \bar{u}(x - b(t - t^n), t^n)$ $(t > t^n)$; i.e., the initial solution $\bar{u}(x, t^n)$ is propagated over a distance $b(t - t^n)$. Applying this result to the initial solution in (13.129), we find, for $b > 0$, that

$$\begin{aligned} \bar{u}(x_{j+\frac{1}{2}}, t) &= \bar{u}(x_{j+\frac{1}{2}} - b(t - t^n), t^n) \\ &= u_j^n + s_j^n\big(x_{j+\frac{1}{2}} - b(t - t^n) - x_j\big) \\ &= u_j^n + s_j^n\left(\frac{1}{2}\Delta x - b(t - t^n)\right). \end{aligned} \tag{13.130}$$

Likewise, we obtain for $b < 0$ the solution

$$\bar{u}(x_{j+\frac{1}{2}}, t) = u_{j+1}^n - s_{j+1}^n\left(\frac{1}{2}\Delta x + b(t - t^n)\right). \tag{13.131}$$

We can now compute the numerical flux by substituting (13.130) and (13.131) into (13.127). First, we consider the case $b > 0$. Then we find for the numerical flux

$$\begin{aligned} F(u_j^n, u_{j+1}^n) &= \frac{b}{\Delta t} \int_{t^n}^{t^{n+1}} \left(u_j^n + s_j^n \left(\frac{1}{2}\Delta x - b(t - t^n) \right) \right) \mathrm{d}t \\ &= \frac{b}{\Delta t} \left(u_j^n \Delta t + s_j^n \left(\frac{1}{2}\Delta x \Delta t - b\frac{1}{2}\Delta t^2 \right) \right) \\ &= bu_j^n + \frac{1}{2}b(1 - c)\Delta x s_j^n. \end{aligned} \tag{13.132}$$

The computation of the numerical flux $F(u_j^n, u_{j+1}^n)$ for $b < 0$ is completely similar to what we found in (13.132), and we obtain

$$F(u_j^n, u_{j+1}^n) = bu_{j+1}^n - \frac{1}{2}b(1 + c)\Delta x s_{j+1}^n. \tag{13.133}$$

For either sign of b we can combine the expressions (13.132) and (13.133) to find for the numerical flux

$$F(u_j^n, u_{j+1}^n) = F_{\mathrm{uw}}(u_j^n, u_{j+1}^n) + \frac{1}{2}|b|(1 - |c|)\Delta x s_{j_\mathrm{u}}^n, \tag{13.134}$$

where the index j_u is defined by

$$j_\mathrm{u} := \begin{cases} j & \text{if} \quad b > 0, \\ j+1 & \text{if} \quad b < 0. \end{cases} \tag{13.135}$$

The numerical flux in (13.134) can thus be interpreted as the upwind flux plus a correction term depending on the slope in the cell upwind of the boundary $x_{j+\frac{1}{2}}$.

Finally, we have to specify the slope s_j^n. Note that the numerical flux (13.134) is identical to the Lax–Wendroff flux (13.111) for the choice $s_{j_\mathrm{u}} = (u_{j+1}^n - u_j^n)/\Delta x$, i.e., when the slope $s_{j_\mathrm{u}}^n$ in the upwind cell is given by the central difference approximation of $\frac{\partial u}{\partial x}(x_{j+\frac{1}{2}}, t^n)$. As seen in the previous section, the Lax–Wendroff scheme is not TVD, and therefore this choice is not allowed. However, if we take

$$s_{j_\mathrm{u}} = \Phi_{j+\frac{1}{2}}^n \frac{1}{\Delta x}(u_{j+1}^n - u_j^n), \tag{13.136}$$

where $\Phi_{j+\frac{1}{2}}^n$ is a limiter on the slope $(u_{j+1}^n - u_j^n)/\Delta x$, the numerical flux (13.134) is precisely the high-resolution flux (13.112). $\Phi_{j+\frac{1}{2}}^n$ is now referred to as the *slope limiter* and is defined by (13.104) and (13.105). If we choose a limiter function $\Phi(r)$ in the TVD region shown in Figure 13.18, the resulting scheme (13.128) is TVD. The numerical scheme (13.128) with numerical flux defined by (13.134) to (13.136) is also referred to as the *slope limiter scheme*.

Example 13.23 We repeat the computations of Example 13.5, now using the slope limiter method (13.128) with numerical flux given by (13.134) to (13.136). We have used the van

Leer limiter function in (13.122a). The numerical parameters are once more $\Delta x = 10^{-2}$, $\Delta t = 0.8 \times 10^{-2}$, and $b = 1$. The results are shown in Figure 13.20. From this figure we can conclude that the slope limiter method gives a sharp and nonoscillatory approximation of the discontinuities in the solution. Apparently, this method is much better suited than either the upwind or Lax–Wendroff scheme for the numerical simulation of problems involving discontinuities. □

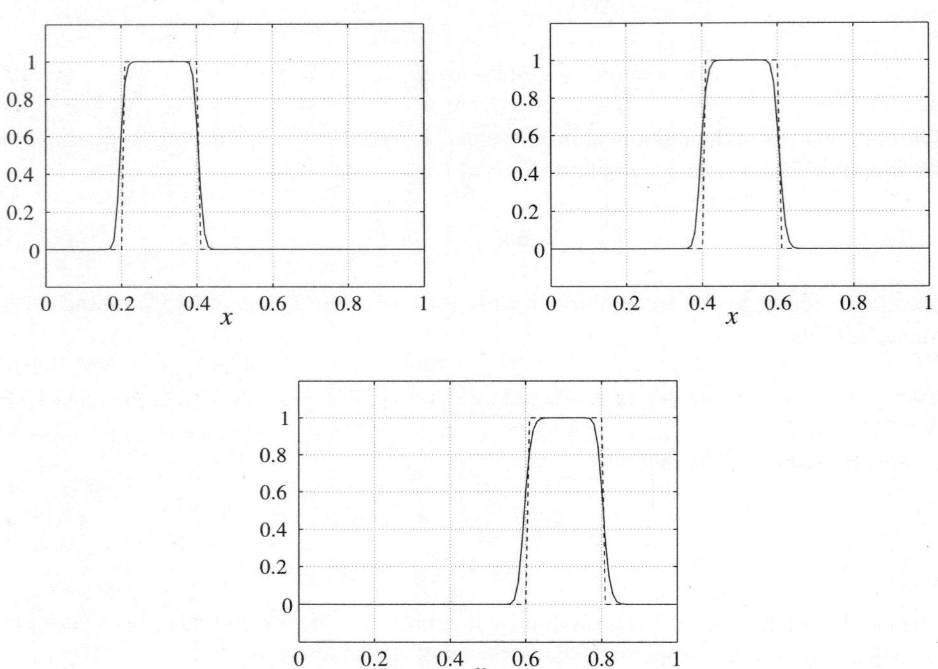

Figure 13.20. *Numerical solution computed with the slope limiter method, after 25, 50, and 75 time steps, respectively.*

13.8.2 A Slope Limiter Method for Nonlinear Conservation Laws

For nonlinear conservation laws one can also use a slope limiter method, which we outline below. As with the advection equation, the slope limiter method is based on the Godunov scheme in combination with a piecewise linear numerical solution. In particular, we need an expression for the numerical flux $F(u_j^n, u_{j+1}^n)$. For the sake of simplicity we restrict ourselves to the case $b(u) := f'(u) > 0$. Other situations can be dealt with in a similar way; cf. [53, 156].

The generic form of a slope limiter method for the conservation law (13.80a) is given by (13.102). In order to compute the numerical flux $F(u_j^n, u_{j+1}^n)$, we have to solve the generalized Riemann problem consisting of the conservation law (13.80a) and the piecewise linear initial condition (13.123b). In general this is a difficult problem. Since we are approximating the solution anyway it makes sense to overcome this problem by replacing the flux $f(u)$ by its linear interpolant $\widetilde{f}(u)$, i.e.,

$$\widetilde{f}(u) := f(u_j^n) + \widetilde{b}_{j+\frac{1}{2}}^n (u - u_j^n), \qquad u \in \text{int}(u_j^n, u_{j+1}^n). \tag{13.137}$$

Here $\widetilde{b}_{j+\frac{1}{2}}^n$ is the finite difference approximation of $b(u(x_{j+\frac{1}{2}}, t^n))$ given by

$$\widetilde{b}_{j+\frac{1}{2}}^n := \frac{f(u_{j+1}^n) - f(u_j^n)}{u_{j+1}^n - u_j^n} \quad \text{provided that} \quad u_j^n \neq u_{j+1}^n. \tag{13.138}$$

Since the interpolated flux $\widetilde{f}(u)$ is piecewise linear, we obtain linear Riemann problems similar to (13.123), from which we can compute the numerical flux. Hence the generalised Riemann problem reduces to the linear initial value problem

$$\frac{\partial \bar{u}}{\partial t} + \widetilde{b}_{j+\frac{1}{2}}^n \frac{\partial \bar{u}}{\partial x} = 0, \qquad x \in \mathbb{R}, \quad t > t^n, \tag{13.139a}$$

$$\bar{u}(x, t^n) = u_j^n + s_j^n (x - x_j), \quad x \in V_j \quad (j = 0, \pm 1, \pm 2, \ldots), \tag{13.139b}$$

where $\widetilde{b}_{j+\frac{1}{2}}^n$ is defined in (13.138) and where u_j^n and s_j^n are the average value and slope, respectively, of $\bar{u}(x, t^n)$ in the control volume V_j; cf. (13.123). Like in (13.85) the numerical flux $F(u_j^n, u_{j+1}^n)$ is the time average of $\widetilde{f}(\bar{u}(x_{j+\frac{1}{2}}, t))$ over $[t^n, t^{n+1}]$; i.e.,

$$F(u_j^n, u_{j+1}^n) := \frac{1}{\Delta t} \int_{t^n}^{t^{n+1}} \widetilde{f}(\bar{u}(x_{j+\frac{1}{2}}, t))\, \mathrm{d}t. \tag{13.140}$$

The computation of $\bar{u}(x_{j+\frac{1}{2}}, t)$ from (13.139) is straightforward, and we obtain

$$\bar{u}(x_{j+\frac{1}{2}}, t) = u_j^n + s_j^n \left(\frac{1}{2}\Delta x - \widetilde{b}_{j+\frac{1}{2}}^n (t - t^n) \right), \tag{13.141}$$

and, subsequently, substituting this expression into (13.140), we find for the numerical flux

$$F(u_j^n, u_{j+1}^n) = f(u_j^n) + \frac{1}{2}\widetilde{b}_{j+\frac{1}{2}}^n \left(1 - \frac{\Delta t}{\Delta x} \widetilde{b}_{j+\frac{1}{2}}^n \right) \Delta x s_j^n. \tag{13.142}$$

Note that for $f(u) = bu$ $(b > 0)$ we recover (13.132).

As a final step we have to determine the slopes s_j^n such that the resulting scheme is TVD. A possible choice for s_j^n is the *minmod slope* [88]

$$s_j^n = \frac{1}{\Delta x} \text{minmod}(u_{j+1}^n - u_j^n, u_j^n - u_{j-1}^n), \tag{13.143}$$

where the minmod function is defined as

$$\text{minmod}(y, z) := \begin{cases} y & \text{if} \quad |y| < |z| \quad \text{and} \quad yz > 0, \\ z & \text{if} \quad |z| < |y| \quad \text{and} \quad yz > 0, \\ 0 & \text{if} \quad yz \leq 0; \end{cases} \tag{13.144}$$

see also Figure 13.18. We can rewrite the minmod slope (13.143) in the standard form (13.136) if we set $\Phi(r) = \max(0, \min(1, r))$. The minmod limiter thus chooses from the upwind slope $(u_j^n - u_{j-1}^n)/\Delta x$ and the downwind slope $(u_{j+1}^n - u_j^n)/\Delta x$ the one that is smallest in absolute value if at least both have the same sign. If the two slopes have opposite signs, then s_j^n is set to zero.

Example 13.24 We apply the slope limiter scheme to compute the rarefaction wave of the Burgers' equation with $u_\ell = 1$ and $u_r = 2$; cf. (12.43). We choose $\Delta x = 2.5 \times 10^{-2}$ and $\Delta t = 10^{-2}$, resulting in a stable scheme. The result after 40 time steps is shown in Figure 13.21. The slope limiter scheme gives an accurate numerical approximation, with the largest errors occurring near the head and tail of the rarefaction wave. □

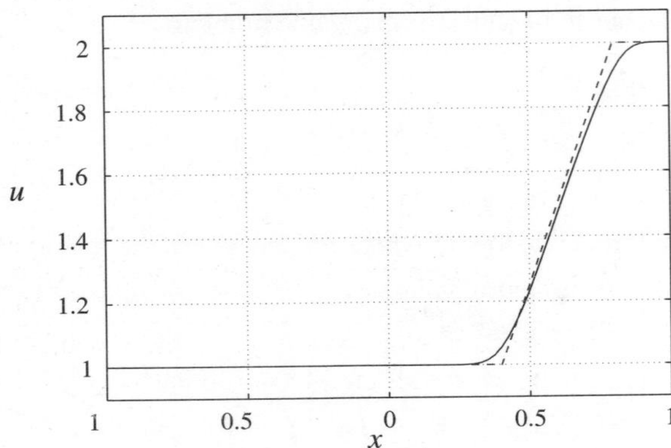

Figure 13.21. *Rarefaction wave computed with the slope limiter method.*

13.9 Numerical Boundary Conditions

For most schemes one needs to provide additional boundary conditions in order to carry out the computations. In general this may be a complicated and nontrivial matter. We will not discuss the issue in full detail, so we restrict ourselves to the advection equation (13.1a) defined on the domain $[0, 1] \times [0, \infty)$. Furthermore, we assume that $b > 0$, and consequently we can prescribe the solution at the left boundary $x = 0$. However, some numerical boundary condition is required at the right boundary $x = 1$. We will elaborate this for two schemes. In particular, we will investigate the effect of the numerical boundary condition on stability.

We must prescribe numerical boundary conditions very carefully, as is demonstrated in the next example.

Example 13.25 We apply the leapfrog scheme to compute the numerical solution of the advection equation subject to the initial and boundary conditions $u(x, 0) = \sin(8\pi x)$ and $u(0, t) = -\sin(8\pi bt)$, respectively. Furthermore, we apply the erroneous boundary condition $u(1, t) = 0$! We choose the (numerical) parameters $b = 1$, $t_{\max} = 0.5$, $\Delta x = 1.25 \times 10^{-2}$, and $\Delta t = 6.25 \times 10^{-3}$, and consequently $c = 0.5$. Hence the scheme is stable. The result is shown in Figure 13.22. From this figure we seen that the error propagates undamped to the left over a distance $bt_{\max} = 0.5$ and thereby completely corrupts the solution. The left part of the solution only suffers from slight dispersion. □

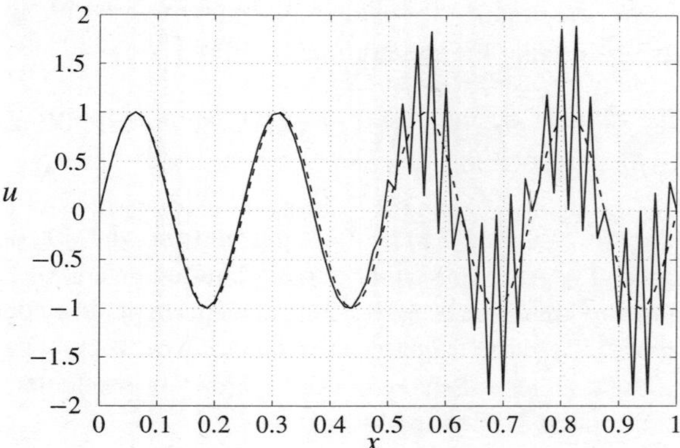

Figure 13.22. *Numerical solution of the leapfrog scheme, corrupted by the erroneous boundary condition* $u(1,t) = 0$.

First we would like to apply the Lax–Wendroff scheme (13.23) to the advection equation. Assume that the domain is covered with the grid defined in (13.61). The numerical value u_M^{n+1} at the right boundary has to be approximated somehow. For this we cannot use the Lax–Wendroff scheme, since it has a three-point coupling in the x direction, resulting in a virtual grid point outside the domain. Note, however, that the upwind scheme can be applied without difficulty. Thus we have for all interior grid points x_j $(j = 1, 2, \dots, M-1)$ the scheme

$$u_j^{n+1} = u_j^n - \frac{1}{2}c\big(u_{j+1}^n - u_{j-1}^n\big) + \frac{1}{2}c^2\big(u_{j+1}^n - 2u_j^n + u_{j-1}^n\big), \tag{13.145a}$$

whereas at the right boundary we have the *numerical boundary condition*

$$u_M^{n+1} = u_M^n - c\big(u_M^n - u_{M-1}^n\big). \tag{13.145b}$$

Clearly (13.145) gives a method of computing the numerical solution at all grid points x_j $(j = 1, 2, \dots, M)$. We will now prove that this method is stable as long as the CFL condition $(c \leq 1)$ is satisfied. Since boundary conditions are involved, we have to employ the matrix method for stability; cf. Chapter 11. Introducing the numerical solution vector at time t^n, i.e., $\mathbf{u}^n := (u_1^n, u_2^n, \dots, u_M^n)^T$, we can write (13.145) in the form

$$\mathbf{u}^{n+1} = \big(\mathbf{I} + \mathbf{A}\big)\mathbf{u}^n + \Delta t\mathbf{f}^n, \tag{13.146a}$$

where the matrix $\mathbf{A}$ is defined as

$$\mathbf{A} := \begin{pmatrix} -c^2 & \frac{1}{2}c(c-1) & & & \\ \frac{1}{2}c(c+1) & -c^2 & \frac{1}{2}c(c-1) & & \\ & \ddots & \ddots & \ddots & \\ & & \frac{1}{2}c(c+1) & -c^2 & \frac{1}{2}c(c-1) \\ & & & c & -c \end{pmatrix}. \tag{13.146b}$$

The right-hand side vector $\mathbf{f}^n$ contains the boundary data at $x = 0$. We will investigate the stability of (13.146) by estimating the spectral radius $\rho(\mathbf{I}+\mathbf{A})$.

Property 13.26. *If $0 < c \leq 1$, then scheme* (13.145), *i.e., the Lax–Wendroff scheme with upwind numerical boundary condition, is stable.*

Proof. Let us consider the case $c = 1$ first. We then find that the diagonal and the upper codiagonal of the matrix $\mathbf{I}+\mathbf{A}$ contain only zeros. The lower codiagonal contains only ones. Clearly, the norm of this matrix (a 1-, 2-, or ∞-norm) is bounded by one. So let us now assume $0 < c < 1$. We first transform the matrix $\mathbf{A}$ in such a way that it becomes skew symmetric, i.e., the codiagonals have opposite signs. In particular, we introduce the diagonal transformation matrix $\mathbf{D}$, defined by

$$\mathbf{D} := \mathrm{diag}(d_1, d_2, \ldots, d_M),$$

$$d_l := \left(\frac{1+c}{1-c}\right)^{(l-1)/2} \quad (l = 1, 2, \ldots, M-1), \qquad d_M := \sqrt{\frac{2}{1-c}}\, d_{M-1}.$$

This indeed transforms the matrix $\mathbf{A}$ into the required form:

$$\widetilde{\mathbf{A}} := \mathbf{D}^{-1}\mathbf{A}\mathbf{D} = \begin{pmatrix} -c^2 & -\gamma & & & \\ \gamma & -c^2 & -\gamma & & \\ & \ddots & \ddots & \ddots & \\ & & \gamma & -c^2 & -\delta \\ & & & \delta & -c \end{pmatrix},$$

where $\gamma := \frac{1}{2}c\sqrt{1-c^2}$ and $\delta := c\sqrt{(1-c)/2}$. We proceed as follows: First, we introduce a complex inner product, as we have to deal with complex eigenvalues. For vectors with elements consisting of complex numbers the inner product reads $(\mathbf{x}, \mathbf{y}) := \bar{\mathbf{y}}^T\mathbf{x}$, where the overbar denotes the complex conjugate. For any eigenvalue λ of $\mathbf{I}+\mathbf{A}$ and corresponding eigenvector $\mathbf{x}$ we have

$$\big((\mathbf{I}+\mathbf{A})\mathbf{x}, \mathbf{x}\big) = \lambda\big(\mathbf{x}, \mathbf{x}\big).$$

Since $\mathbf{A}$ and $\widetilde{\mathbf{A}}$ have the same set of eigenvectors, we find for the inner product on the left-hand side

$$\big((\mathbf{I}+\mathbf{A})\mathbf{x}, \mathbf{x}\big) = \big((\mathbf{I}+\widetilde{\mathbf{A}})\mathbf{x}, \mathbf{x}\big) = \gamma \sum_{k=1}^{M-2} \big(x_k\,\bar{x}_{k+1} - x_{k+1}\bar{x}_k\big)$$
$$+ \,\delta\big(x_{M-1}\bar{x}_M - x_M\bar{x}_{M-1}\big) + (1-c^2)\sum_{k=1}^{M-1} x_k\bar{x}_k + (1-c)x_M\bar{x}_M.$$

On the other hand, we find, splitting λ into its real and imaginary parts, i.e., $\lambda =: \alpha + \mathrm{i}\beta$, that $\lambda\,(\mathbf{x}, \mathbf{x}) = (\alpha + \mathrm{i}\beta)\sum_{k=1}^{M} x_k\bar{x}_k$. We can take $(\mathbf{x}, \mathbf{x}) = 1$ without restriction. Writing $\rho := |x_M|^2 = x_M\bar{x}_M$, we thus find, from comparing the real parts, that

$$\alpha = (1-c^2)(1-\rho) + (1-c)\rho = 1 - c^2 + \rho(c^2 - c),$$

and, from the imaginary parts, we find

$$\beta = 2\gamma \,\mathrm{Im}\left(\sum_{k=1}^{M-2} x_k\bar{x}_{k+1}\right) + 2\delta\,\mathrm{Im}(x_{M-1}\bar{x}_M).$$

Note that $|\,\mathrm{Im}(\sum_{k=1}^{M-2} x_k\bar{x}_{k+1})| \le \sum_{k=1}^{M-2}|x_k\bar{x}_{k+1}|$. Also note that $2\gamma = c\sqrt{1-c^2} \le c\sqrt{2-2c} = 2\delta$. We thus obtain

$$|\beta| \le c\sqrt{2-2c}\sum_{k=1}^{M-1}|x_k\bar{x}_{k+1}|.$$

We can estimate the latter by applying Cauchy's inequality (cf. (G.3)), giving

$$|\beta| \le c\sqrt{2-2c}\sum_{k=1}^{M-1}|x_k\bar{x}_{k+1}| \;\le\; c\sqrt{2-2c}\sum_{k=1}^{M-1}|x_k\bar{x}_k|\sum_{k=2}^{M}|x_k\bar{x}_k| \;\le\; c\sqrt{2-2c}.$$

Here we have used that $\sum_{k=1}^{M-1}|x_k\bar{x}_k|$ and $\sum_{k=2}^{M}|x_k\bar{x}_k|$ are not greater than 1. Hence we find

$$\alpha^2 + \beta^2 < (1-c^2)^2 + 2c^2 - 2c^3 = 1 - 2c^3 + c^4 \le 1.$$

This indeed proves that $|\lambda| = \sqrt{\alpha^2+\beta^2} \le 1$. □

Another scheme that needs a numerical boundary condition at $x = 1$ is the leapfrog scheme. Like for the Lax–Wendroff scheme, we apply the upwind scheme at grid point x_M, leading to the vector recursion

$$\mathbf{u}^{n+1} = \mathbf{A}\mathbf{u}^n + \mathbf{D}\mathbf{u}^{n-1} + \Delta t\mathbf{f}^n, \tag{13.147a}$$

where

$$\mathbf{A} := \begin{pmatrix} 0 & -c & & & \\ c & 0 & -c & & \\ & \ddots & \ddots & \ddots & \\ & & c & 0 & -c \\ & & & c & 1-c \end{pmatrix}, \qquad \mathbf{D} := \mathrm{diag}(1, 1, \ldots, 1, 0). \tag{13.147b}$$

Property 13.27. *If* $0 < c \le 1$, *then scheme* (13.147), *i.e., the leapfrog scheme with upwind numerical boundary condition, is stable.*

Proof. We leave it as an exercise to show that the case $c = 1$ is correct and hence can assume that $0 < c < 1$. Let us rewrite (13.147a) as a first order recursion whose homogeneous part reads

$$\mathbf{v}^{n+1} = \widehat{\mathbf{A}}\,\mathbf{v}^n, \qquad \text{where } \widehat{\mathbf{A}} := \begin{pmatrix} \mathbf{A} & \mathbf{D} \\ \mathbf{I} & \mathbf{O} \end{pmatrix}, \quad \mathbf{v}^n := \begin{pmatrix} \mathbf{u}^n \\ \mathbf{u}^{n-1} \end{pmatrix}.$$

For stability we investigate the spectral radius $\rho(\widehat{\mathbf{A}})$. Note that zero is always an eigenvalue of $\widehat{A}$. In the following we only consider eigenvalues $\lambda \neq 0$. Let $\mathbf{x}$ be the corresponding eigenvector, which we partition into two parts $\mathbf{y}$ and $\mathbf{z}$, analogous to the partitioning of $\mathbf{v}^n$, i.e., $\mathbf{x}^T = (\mathbf{y}^T, \mathbf{z}^T)$. Writing out the relation $\widehat{\mathbf{A}}\mathbf{x} = \lambda\mathbf{x}$ gives

$$\mathbf{A}\mathbf{y} + \mathbf{D}\mathbf{z} = \lambda\mathbf{y}, \qquad \mathbf{y} = \lambda\mathbf{z}.$$

This then implies the relation

$$(\mathbf{A}\mathbf{z}, \mathbf{z}) = \lambda(\mathbf{z}, \mathbf{z}) - \frac{1}{\lambda}(\mathbf{D}\mathbf{z}, \mathbf{z}). \tag{$*$}$$

Writing $\lambda =: \alpha + \mathrm{i}\beta$, we note that $1/\lambda = (\alpha - \mathrm{i}\beta)/(\alpha^2 + \beta^2)$. Collecting the real parts of $(*)$ gives

$$(1-c) z_M \bar{z}_M = \alpha \sum_{k=1}^{M} z_k \bar{z}_k - \frac{\alpha}{\alpha^2 + \beta^2} \sum_{k=1}^{M-1} z_k \bar{z}_k,$$

while for the imaginary parts we have

$$2c \,\mathrm{Im}\left(\sum_{k=1}^{M-1} z_k \, \bar{z}_{k+1} \right) = \beta \sum_{k=1}^{M} z_k \, \bar{z}_k + \frac{\beta}{\alpha^2 + \beta^2} \sum_{k=1}^{M-1} z_k \bar{z}_k.$$

As before it is not restrictive to take $\sum_{k=1}^{M} z_k \, \bar{z}_k = 1$. Defining $\rho := z_M \bar{z}_M$, we thus have the relations

$$(1-c)\rho = \alpha \left(1 - \frac{1-\rho}{\alpha^2 + \beta^2}\right), \qquad 2c\sqrt{1-\rho} \geq \left|\beta\left(1 + \frac{1-\rho}{\alpha^2+\beta^2}\right)\right|, \tag{$**$}$$

where we have used the Cauchy–Schwarz inequality. Now suppose that we have no stability; i.e., $\alpha^2 + \beta^2 > 1$. Then we find $|\alpha| < (1-c)\rho/(1-(1-\rho)) = 1-c$. As a consequence, we must have that $\beta^2 > 1 - (1-c)^2 = 2c - c^2$. This would then give a lower bound for $|\beta|$. Comparing this lower bound with the upper bound in $(**)$ results in

$$\sqrt{2c - c^2} < \frac{2c|\lambda|^2\sqrt{1-\rho}}{|\lambda|^2 + 1 - \rho}.$$

After some rearrangement, this leads to

$$\left(2c\sqrt{\frac{1-\rho}{2c-c^2}} - 1\right) |\lambda|^2 > 1 - \rho.$$

Note that the factor in brackets must be positive in order to let this inequality make sense. Moreover, due to our assumption, we have $|\lambda|^2 = \alpha^2 + \beta^2 > 1$; hence the inequality is still correct if we require it to hold without the factor $|\lambda|^2$, i.e., $2c\sqrt{(1-\rho)/(2c-c^2)} > 2 - \rho$. After squaring and some simple rewriting, we thus find that we should have

$$g(\rho) := (2-c)\rho^2 + 8(c-1)\rho + 8(1-c) < 0.$$

It is easy to see that $g(\rho)$ is positive definite. So this inequality cannot be fulfilled for any choice of ρ. We conclude that assuming $|\lambda| > 1$ leads to a false statement. Thus we have shown that $|\lambda| \leq 1$. $\square$

Example 13.28 For the leapfrog scheme with upwind numerical boundary condition we have drawn in Figure 13.23 the (complex) eigenvalues of the matrix $\widehat{A}$ for $\Delta x = 5 \times 10^{-2}$ and $c = 0.99$. Clearly, all eigenvalues are smaller than one in modulus, and some are even significantly smaller than one. This then leads to the conclusion that the nondissipation property of the leapfrog scheme is annihilated. □

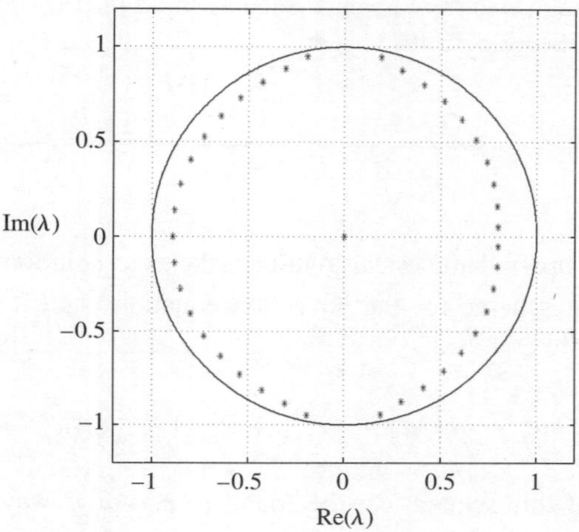

Figure 13.23. *Eigenvalues of the leapfrog scheme with upwind numerical boundary condition.*

13.10 Discussion

- Hyperbolic problems very rarely appear in practice without some small diffusive (or viscous) term. In this sense they are an idealisation. However, as can be seen from the analysis in this chapter, methods for advection-dominated problems are best studied by analysing methods for advection problems proper. If higher-order terms are present, one may be forced to resort to implicit methods for the latter. As outlined in Chapter 11, the advection term can then often be discretised by an explicit method using operator splitting.

- Most of the methods discussed in this chapter have higher-dimensional analogues. However, as was already indicated in Chapter 12, hyperbolic equations in several spatial dimensions may have solutions with quite complicated behaviour. This has its impact on the numerical treatment of such problems, for instance shock capturing. These complications are reflected in the literature, which is much less extensive for the latter class of problems. For some further reading see, e.g., [42, 88, 158].

- A wealth of methods and applications can be found in fluid dynamics. Indeed, one of the most advanced areas of numerical simulation of problems arising in practice is computational fluid dynamics (CFD). Here all sorts of problems in areas such as aerospace or management of waterways occur. CFD is often also the main ingredient in complex physical problems like combustion (flow of gas mixtures and chemistry) and plasma physics (flow of gas and electromagnetism). Some books on CFD are, e.g., [63, 64, 45, 168].

- In Chapter 16 some typical problems modeled by the advection-diffusion equation, with dominant advection, are the following: the determination of chemical species by separation through electrophoresis in Section 16.13 and the cooling device employing a pulse tube in Section 16.14.

Exercises

13.1. Show that the upwind numerical solution is the exact solution of (13.1a) when $c = 1$.

13.2. A well-known scheme for the advection equation (13.1a) is the *Beam–Warming scheme* given by

$$u_j^{n+1} = u_j^n - \frac{1}{2}c\big(3u_j^n - 4u_{j-1}^n + u_{j-2}^n\big) + \frac{1}{2}c^2\big(u_j^n - 2u_{j-1}^n + u_{j-2}^n\big).$$

(a) Show that this scheme can be found in a similar way as the Lax–Wendroff scheme, but using backward difference approximations for the spatial derivatives in (13.22). Draw the stencil.

(b) Show that the scheme is second order.

(c) Find a sufficient condition for stability.

13.3. Compute the amplitude error and the relative phase error of the Beam–Warming scheme in Exercise 13.2. Investigate the dissipation and dispersion of the scheme.

13.4. Sometimes it is useful to use higher-order differences for the space discretisation. To this end consider the scheme applied to the advection equation (13.1a):

$$u_j^{n+1} = u_j^n - \frac{1}{12}c\big(-u_{j+2}^n + 8u_{j+1}^n - 8u_{j-1}^n + u_{j-2}^n\big).$$

(a) Give the stencil of this scheme.

(b) Show that this method is third order in space and first order in time.

(c) Investigate the stability of this method.

13.5. Consider the following scheme for the advection equation (13.1a):

$$u_j^{n+1} = \frac{1}{2}\big(u_{j+1}^n + u_{j-1}^n\big) - \frac{1}{2}c\big(u_{j+1}^n - u_{j-1}^n\big).$$

(a) Draw the stencil of this *Lax–Friedrichs scheme*.

(b) Show that this scheme is stable if $|c| \leq 1$.

(c) Show that the local discretisation error of the method is $\mathcal{O}(\Delta t + \frac{\Delta x^2}{\Delta t} + \Delta x^2)$.

(d) Comment on the result in part (c).

13.6. Consider the advection equation (13.1a).

(a) Show what scheme you get if you use the standard central difference approximation to discretise the spatial derivative and the implicit Euler method for time integration.

(b) Is this scheme stable?

13.7. As is well known, backward difference formula (BDF) methods are dissipative; see, e.g., Section 5.4.2 and also [97]. In fact, the Euler backward method is the lowest-order member of this family (BDF-1). If we use BDF-2 for time integration and the standard central difference approximation to discretise the spatial derivative, we obtain the following scheme for (13.1a):

$$3u_j^{n+1} = 4u_j^n - u_j^{n-1} - \frac{1}{2}c\left(u_{j+1}^{n+1} - u_{j-1}^{n+1}\right).$$

(a) Give the stencil.

(b) Show that the scheme is second order in Δx and Δt.

(c) Show that the scheme is unconditionally stable.

13.8. Compute the local discretisation error of scheme (13.44) and show that it is second order in Δx and first order in Δt.

13.9. Prove that scheme (13.44) for the advection-diffusion equation is TVD provided that $d \le \frac{1}{2}$ and $|Pe| \le 2$.

13.10. Instead of (13.44), we could use the upwind approximation for the advective term in (13.43).

(a) Give the resulting scheme.

(b) Compute the local discretisation error and show that it is first order in Δx and Δt.

(c) Investigate the stability of the scheme.

(d) Prove that the scheme is TVD.

13.11. Compute the local discretisation error of the box scheme (13.63) and show that it is second order in Δx and Δt.

13.12. Compute the local discretisation error of the leapfrog scheme (13.70b) and show that it is second order in Δx and Δt.

13.13. Compute the numerical group velocity of the leapfrog scheme.

13.14. Compute the Godunov numerical flux for the traffic flow equation from Example 12.16.

13.15. Compute the local discretisation error of the Godunov scheme applied to the Burgers' equation and show that it is first order in Δx and Δt.

13.16. Consider the nonlinear conservation law (13.80a) with convex flux function $f(u)$. Derive a general expression for the Godunov numerical flux in terms of $f(u)$.

13.17. Apply Property 13.20 to derive a general expression for the local discretisation error of the Godunov scheme.

13.18. Give the slope limiter method for the Burgers' equation when $u > 0$.

13.19. Consider the advection equation (13.1a) with $b < 0$ defined on the domain $[0, 1] \times [0, \infty)$. Show that the Lax–Wendroff scheme with upwind numerical boundary condition at $x = 0$ is stable if $-1 \leq c < 0$.

13.20. Repeat Exercise 13.19 for the leapfrog scheme.

Chapter 14

Numerical Methods for Hyperbolic Systems

In this chapter we extend the numerical methods in the previous chapter to systems. We start in Section 14.1 with a description of the upwind scheme for linear systems. In Section 14.2 we present the Godunov scheme. For linear systems the method reduces to the upwind scheme, like in the scalar case. For nonlinear systems the computational costs of the Godunov scheme are high since it requires the solution of local Riemann problems. Approximate Riemann solvers are often a cheaper alternative. The Roe scheme is a well-known example of these and is presented in Section 14.3. Next, in Section 14.4, we construct high-resolution schemes based on slope limiters. As an example, we present in Section 14.5 a detailed elaboration of both the Godunov and Roe schemes applied to the shallow-water equations. The wave equation can be reformulated as a 2×2 hyperbolic system and for that reason it is covered in this chapter. In Section 14.6 we present a few numerical methods for this equation, based on either the scalar form or the equivalent system formulation. Finally, in Section 14.7, we address the important topic of boundary conditions.

14.1 The Upwind Scheme

The simplest scheme we encountered in Chapter 13 was the upwind scheme, introduced in Section 13.1.1. To generalise this to linear hyperbolic systems consider the following initial value problem for $\boldsymbol{u}(x,t) \in \mathbb{R}^m$:

$$\frac{\partial \boldsymbol{u}}{\partial t} + \boldsymbol{B}\frac{\partial \boldsymbol{u}}{\partial x} = \boldsymbol{0}, \qquad x \in \mathbb{R}, \quad t > 0, \tag{14.1a}$$

$$\boldsymbol{u}(x,0) = \boldsymbol{v}(x), \qquad x \in \mathbb{R}, \tag{14.1b}$$

with $\boldsymbol{B} \in \mathbb{R}^{m\times m}$ a constant matrix. The linear system (14.1a) is a model problem for advective processes; e.g., the linearised equations of gas dynamics can be written in this form [158]. Moreover, the one-dimensional wave equation can be reformulated as in (14.1a); see Example 12.20. We assume that the linear system (14.1a) is hyperbolic. Then, according to Definition 12.19, the matrix $\boldsymbol{B}$ has m real eigenvalues $\lambda_1, \lambda_2, \ldots, \lambda_m$ and m linearly independent left eigenvectors $\boldsymbol{t}_1^T, \boldsymbol{t}_2^T, \ldots, \boldsymbol{t}_m^T$ that make up the matrix $\boldsymbol{T}$; see (12.51). Alternatively, $\boldsymbol{B}$ has a complete set of (right) eigenvectors $\boldsymbol{s}_1, \boldsymbol{s}_2, \ldots, \boldsymbol{s}_m$ such that

$$\boldsymbol{B} = \boldsymbol{S}\boldsymbol{\Lambda}\boldsymbol{S}^{-1}, \tag{14.2}$$

where the matrices $\boldsymbol{\Lambda}$ and $\boldsymbol{S}$ are defined as

$$\boldsymbol{\Lambda} := \mathrm{diag}(\lambda_1, \lambda_2, \ldots, \lambda_m), \qquad \boldsymbol{S} := (\boldsymbol{s}_1, \boldsymbol{s}_2, \ldots, \boldsymbol{s}_m) = \boldsymbol{T}^{-1}. \tag{14.3}$$

In Section 12.3 we showed that the characteristic variable $\tilde{\boldsymbol{u}}$, defined by

$$\tilde{\boldsymbol{u}} := \boldsymbol{T}\boldsymbol{u} = \boldsymbol{S}^{-1}\boldsymbol{u}, \tag{14.4}$$

satisfies the decoupled system

$$\frac{\partial \tilde{\boldsymbol{u}}}{\partial t} + \boldsymbol{\Lambda}\frac{\partial \tilde{\boldsymbol{u}}}{\partial x} = \boldsymbol{0}. \tag{14.5}$$

Written componentwise, this is

$$\frac{\partial \tilde{u}_k}{\partial t} + \lambda_k \frac{\partial \tilde{u}_k}{\partial x} = 0 \qquad (k = 1, 2, \ldots, m). \tag{14.6a}$$

From (14.1b) we can derive the following initial condition for $\tilde{\boldsymbol{u}}$:

$$\tilde{\boldsymbol{u}}(x, 0) = \boldsymbol{S}^{-1}\boldsymbol{v}(x) =: \left(\tilde{v}_1(x), \tilde{v}_2(x), \ldots, \tilde{v}_m(x)\right)^T. \tag{14.6b}$$

Each of the equations in (14.6a) is an advection equation, and we can verify that the solutions of these equations, together with the initial conditions in (14.6b), are given by

$$\tilde{u}_k(x, t) = \tilde{v}_k(x - \lambda_k t). \tag{14.7}$$

The solution of (14.1) is then given by

$$\boldsymbol{u}(x, t) = \boldsymbol{S}\tilde{\boldsymbol{u}}(x, t) = \sum_{k=1}^{m} \tilde{v}_k(x - \lambda_k t)\boldsymbol{s}_k. \tag{14.8}$$

Thus the solution $\boldsymbol{u}(x, t)$ consists of m components $\tilde{v}_k(x - \lambda_k t)\boldsymbol{s}_k$, each propagating with advection velocity λ_k along the x axis.

In order to compute a numerical solution of (14.1), we cover the domain $\Omega := \mathbb{R} \times [0, \infty)$ with grid points (x_j, t^n) defined by

$$x_j := j\Delta x \quad (j = 0, \pm 1, \pm 2, \ldots), \qquad t^n := n\Delta t \quad (n = 0, 1, 2, \ldots), \tag{14.9}$$

with Δx the spatial grid size and Δt the time step. We first introduce the upwind scheme for the linear advection equations in (14.6a) and subsequently give the scheme for $\boldsymbol{u}$. By virtue of (14.8), each component $\tilde{v}_k(x - \lambda_k t)\boldsymbol{s}_k$ of the solution is constant along characteristics $\mathcal{C}_k : x - \lambda_k t = C$ ($C \in \mathbb{R}$) and therefore has an upwind direction determined by the sign of the corresponding eigenvalue λ_k. Introducing the splitting of the eigenvalues λ_k as

$$\lambda_k = \lambda_k^+ + \lambda_k^-, \qquad \lambda_k^+ := \max(\lambda_k, 0), \quad \lambda_k^- := \min(\lambda_k, 0), \tag{14.10}$$

we can write the upwind schemes for the advection equations in (14.6a) as

$$\frac{1}{\Delta t}\left(\tilde{u}_{k,j}^{n+1} - \tilde{u}_{k,j}^{n}\right) + \frac{\lambda_k^+}{\Delta x}\left(\tilde{u}_{k,j}^{n} - \tilde{u}_{k,j-1}^{n}\right) + \frac{\lambda_k^-}{\Delta x}\left(\tilde{u}_{k,j+1}^{n} - \tilde{u}_{k,j}^{n}\right) = 0 \qquad (k = 1, 2, \ldots, m), \quad (14.11)$$

where $\tilde{u}_{k,j}^n$ denotes the numerical approximation of $\tilde{u}_k(x_j, t^n)$, etc.; see Section 13.1. The upwind schemes for all characteristic variables $\tilde{u}_k$ can be combined in vector form as

$$\frac{1}{\Delta t}\left(\tilde{\boldsymbol{u}}_j^{n+1} - \tilde{\boldsymbol{u}}_j^{n}\right) + \frac{1}{\Delta x}\boldsymbol{\Lambda}^+\left(\tilde{\boldsymbol{u}}_j^{n} - \tilde{\boldsymbol{u}}_{j-1}^{n}\right) + \frac{1}{\Delta x}\boldsymbol{\Lambda}^-\left(\tilde{\boldsymbol{u}}_{j+1}^{n} - \tilde{\boldsymbol{u}}_j^{n}\right) = \boldsymbol{0}, \quad (14.12)$$

with

$$\boldsymbol{\Lambda}^+ := \mathrm{diag}(\lambda_1^+, \lambda_2^+, \ldots, \lambda_m^+), \qquad \boldsymbol{\Lambda}^- := \mathrm{diag}(\lambda_1^-, \lambda_2^-, \ldots, \lambda_m^-). \quad (14.13)$$

We obtain the *upwind scheme* in terms of the original variable $\boldsymbol{u}$ by substituting (14.4) in (14.12) and multiplying the resulting scheme by the matrix $\boldsymbol{S}$, giving

$$\frac{1}{\Delta t}\left(\boldsymbol{u}_j^{n+1} - \boldsymbol{u}_j^{n}\right) + \frac{1}{\Delta x}\boldsymbol{B}^+\left(\boldsymbol{u}_j^{n} - \boldsymbol{u}_{j-1}^{n}\right) + \frac{1}{\Delta x}\boldsymbol{B}^-\left(\boldsymbol{u}_{j+1}^{n} - \boldsymbol{u}_j^{n}\right) = \boldsymbol{0}, \quad (14.14)$$

where $\boldsymbol{u}_j^n$ is obviously the numerical approximation of $\boldsymbol{u}(x_j, t^n)$ and the matrices $\boldsymbol{B}^+$ and $\boldsymbol{B}^-$ are defined by

$$\boldsymbol{B}^+ := \boldsymbol{S}\boldsymbol{\Lambda}^+\boldsymbol{S}^{-1}, \qquad \boldsymbol{B}^- := \boldsymbol{S}\boldsymbol{\Lambda}^-\boldsymbol{S}^{-1}. \quad (14.15)$$

We see that the coefficient matrix $\boldsymbol{B}$ is split as $\boldsymbol{B} = \boldsymbol{B}^+ + \boldsymbol{B}^-$, with $\boldsymbol{B}^+$ having the nonnegative eigenvalues λ_k^+ and $\boldsymbol{B}^-$ the nonpositive eigenvalues λ_k^-. Thus the $\boldsymbol{B}^+$ term in (14.14) is the backward difference approximation of $\boldsymbol{B}^+\frac{\partial \boldsymbol{u}}{\partial x}$ and corresponds to components propagating in the positive x direction. Likewise, the $\boldsymbol{B}^-$ term in (14.14) is the forward difference approximation of $\boldsymbol{B}^-\frac{\partial \boldsymbol{u}}{\partial x}$ and corresponds to components propagating in the negative x direction.

Stability of the upwind scheme (14.14) is ensured if the corresponding scalar upwind schemes for all variables $\tilde{u}_k$ are stable. This latter requirement gives the Courant, Friedrichs, and Lewy (CFL) stability condition

$$\max_k |\lambda_k| \frac{\Delta t}{\Delta x} \le 1; \quad (14.16)$$

cf. (13.20). This requirement means that all characteristics through the point (x_j, t^{n+1}) intersect the grid line $t = t^n$ at points in the interval $[x_{j-1}, x_{j+1}]$.

The upwind scheme (14.12) is first order accurate in Δt and Δx. Consequently, this also holds for scheme (14.14). We leave it as an exercise to show this.

Example 14.1 Consider the hyperbolic system in Example 12.20, which describes small-amplitude waves in a canal of uniform depth h:

$$\frac{\partial \boldsymbol{u}}{\partial t} + \boldsymbol{B}\frac{\partial \boldsymbol{u}}{\partial x} = \boldsymbol{0}, \quad (*)$$

where $\boldsymbol{u}$ and $\boldsymbol{B}$ are given by

$$\boldsymbol{u} = \begin{pmatrix} \frac{1}{a}\frac{\partial}{\partial t}\eta \\ \frac{\partial}{\partial x}\eta \end{pmatrix}, \qquad \boldsymbol{B} = -\begin{pmatrix} 0 & a \\ a & 0 \end{pmatrix}, \qquad a = \sqrt{gh},$$

with $\eta(x,t)$ the water elevation. The eigenvalues λ_1, λ_2 and corresponding eigenvectors $\boldsymbol{s}_1$, $\boldsymbol{s}_2$ are given by

$$\lambda_1 = -a < 0, \quad \lambda_2 = a > 0, \qquad \boldsymbol{s}_1 = \begin{pmatrix} 1 \\ 1 \end{pmatrix}, \quad \boldsymbol{s}_2 = \begin{pmatrix} 1 \\ -1 \end{pmatrix}.$$

The splitting of the eigenvalues according to (14.10) is particularly simple, and we find

$$\lambda_1^- = -a, \quad \lambda_1^+ = 0, \quad \lambda_2^- = 0, \quad \lambda_2^+ = a.$$

From (14.13) and (14.15) we can compute the matrices $\boldsymbol{B}^+$ and $\boldsymbol{B}^-$ to obtain

$$\boldsymbol{B}^+ = \frac{1}{2}\begin{pmatrix} a & -a \\ -a & a \end{pmatrix}, \qquad \boldsymbol{B}^- = -\frac{1}{2}\begin{pmatrix} a & a \\ a & a \end{pmatrix}.$$

We can verify that the matrix $\boldsymbol{B}^+$ has the eigenvalues zero and a, whereas $\boldsymbol{B}^-$ has the eigenvalues $-a$ and zero. Substituting these matrices in (14.14), we obtain the upwind scheme for system (∗):

$$u_{1,j}^{n+1} = u_{1,j}^n + \frac{1}{2}c\left(u_{1,j+1}^n - 2u_{1,j}^n + u_{1,j-1}^n + u_{2,j+1}^n - u_{2,j-1}^n\right),$$

$$u_{2,j}^{n+1} = u_{2,j}^n + \frac{1}{2}c\left(u_{1,j+1}^n - u_{1,j-1}^n + u_{2,j+1}^n - 2u_{2,j}^n + u_{2,j-1}^n\right),$$

with $c := a\Delta t/\Delta x$ the Courant number. Finally, the water elevation η has to be computed numerically from either $\frac{\partial}{\partial x}\eta = u_2$ or $\frac{\partial}{\partial t}\eta = au_1$. □

As we have seen in Section 13.3.2, the scalar upwind scheme can be interpreted as the central difference scheme stabilized by adding some numerical diffusion. The upwind scheme in the example above also contains terms reminiscent of numerical diffusion. Indeed, we can rewrite the upwind scheme (14.14) as the central difference scheme with some numerical diffusion added, as follows:

$$\begin{aligned}
\boldsymbol{u}_j^{n+1} &= \boldsymbol{u}_j^n - \frac{\Delta t}{\Delta x}\left(\boldsymbol{B}^+\left(\boldsymbol{u}_j^n - \boldsymbol{u}_{j-1}^n\right) + \boldsymbol{B}^-\left(\boldsymbol{u}_{j+1}^n - \boldsymbol{u}_j^n\right)\right) \\
&= \boldsymbol{u}_j^n - \frac{\Delta t}{2\Delta x}\boldsymbol{B}\left(\boldsymbol{u}_{j+1}^n - \boldsymbol{u}_j^n + \boldsymbol{u}_j^n - \boldsymbol{u}_{j-1}^n\right) \\
&\quad + \frac{\Delta t}{2\Delta x}\left(\left(\boldsymbol{B} - 2\boldsymbol{B}^+\right)\left(\boldsymbol{u}_j^n - \boldsymbol{u}_{j-1}^n\right) + \left(\boldsymbol{B} - 2\boldsymbol{B}^-\right)\left(\boldsymbol{u}_{j+1}^n - \boldsymbol{u}_j^n\right)\right) \\
&= \boldsymbol{u}_j^n - \frac{\Delta t}{2\Delta x}\boldsymbol{B}\left(\boldsymbol{u}_{j+1}^n - \boldsymbol{u}_{j-1}^n\right) + \frac{\Delta t}{2\Delta x}\boldsymbol{B}^*\left(\boldsymbol{u}_{j+1}^n - 2\boldsymbol{u}_j^n + \boldsymbol{u}_{j-1}^n\right),
\end{aligned} \tag{14.17}$$

where the matrix $\boldsymbol{B}^*$ is defined as

$$\boldsymbol{B}^* := \boldsymbol{B}^+ - \boldsymbol{B}^-. \tag{14.18}$$

Clearly, $\boldsymbol{B}^*$ is a positive definite matrix since we have

$$\boldsymbol{B}^* = \boldsymbol{S}\boldsymbol{\Lambda}^*\boldsymbol{S}^{-1}, \tag{14.19a}$$

$$\boldsymbol{\Lambda}^* := \boldsymbol{\Lambda}^+ - \boldsymbol{\Lambda}^- = |\boldsymbol{\Lambda}| = \mathrm{diag}(|\lambda_1|, |\lambda_2|, \ldots, |\lambda_m|). \tag{14.19b}$$

The formulation in (14.17) can be interpreted as the central difference approximation of the *modified equation*

$$\frac{\partial \boldsymbol{u}}{\partial t} + \boldsymbol{B}\frac{\partial \boldsymbol{u}}{\partial x} = \frac{1}{2}\Delta x \boldsymbol{B}^* \frac{\partial^2 \boldsymbol{u}}{\partial x^2}, \tag{14.20}$$

and since the coefficient matrix on the right-hand side is positive definite, it is of advection-diffusion type. For this reason the upwind scheme produces overly damped numerical solutions.

Example 14.2 We continue with the previous example. Substituting $\boldsymbol{B}^* = a\boldsymbol{I}$ in (14.17), we obtain the scheme

$$u_{1,j}^{n+1} = u_{1,j}^n + \frac{1}{2}c\big(u_{2,j+1}^n - u_{2,j-1}^n\big) + \frac{1}{2}c\big(u_{1,j+1}^n - 2u_{1,j}^n + u_{1,j-1}^n\big),$$

$$u_{2,j}^{n+1} = u_{2,j}^n + \frac{1}{2}c\big(u_{1,j+1}^n - u_{1,j-1}^n\big) + \frac{1}{2}c\big(u_{2,j+1}^n - 2u_{2,j}^n + u_{2,j-1}^n\big),$$

which is of course equivalent to the scheme in Example 14.1. □

14.2 The Godunov Scheme

The Godunov scheme that was discussed in the previous chapter can also be generalised to systems. Below we shall first derive the general scheme in Section 14.2.1, after which we will work this scheme out for the linear case in Section 14.2.2. The application of the Godunov scheme to the shallow-water equations, as an example of a nonlinear system, is deferred to Section 14.5.1.

14.2.1 Derivation of the Scheme

Consider the following initial value problem for $\boldsymbol{u}(x,t) \in \mathbb{R}^m$:

$$\frac{\partial \boldsymbol{u}}{\partial t} + \frac{\partial \boldsymbol{f}(\boldsymbol{u})}{\partial x} = \boldsymbol{0}, \qquad x \in \mathbb{R}, \quad t > 0, \tag{14.21a}$$

$$\boldsymbol{u}(x,0) = \boldsymbol{v}(x), \qquad x \in \mathbb{R}, \tag{14.21b}$$

where the flux $\boldsymbol{f}(\boldsymbol{u})$ is generally a nonlinear function of $\boldsymbol{u}$. Note that $\boldsymbol{f}(\boldsymbol{u}) = \boldsymbol{B}\boldsymbol{u}$ for the linear system (14.1a). An example of such a system is the shallow-water equations, introduced in Section 12.5. Let us assume that the system (14.21a) is hyperbolic; i.e., the Jacobi matrix $\boldsymbol{B}(\boldsymbol{u}) := \frac{\partial}{\partial \boldsymbol{u}}\boldsymbol{f}(\boldsymbol{u})$ has m real eigenvalues, assumed to be ordered as $\lambda_1(\boldsymbol{u}) \le \lambda_2(\boldsymbol{u}) \le \cdots \le \lambda_m(\boldsymbol{u})$, and m corresponding linearly independent right eigenvectors $\boldsymbol{s}_1(\boldsymbol{u}), \boldsymbol{s}_2(\boldsymbol{u}), \ldots, \boldsymbol{s}_m(\boldsymbol{u})$; see Section 12.3.2.

The Godunov scheme for the hyperbolic system (14.21a) that we derive below is a natural generalisation of the scalar method and therefore our presentation can be brief. Like in Section 13.5, we introduce control volumes or cells V_j as follows:

$$V_j := [x_{j-\frac{1}{2}}, x_{j+\frac{1}{2}}), \qquad x_{j+\frac{1}{2}} = \frac{1}{2}(x_j + x_{j+1}) \quad (j = 0, \pm 1, \pm 2, \ldots), \tag{14.22}$$

with x_j defined in (14.9). The numerical solution $\boldsymbol{u}_j^n$ has to be interpreted as an approximation of the average value vector of $\boldsymbol{u}(x,t)$ over the control volume V_j at time level t^n. Associated with $\boldsymbol{u}_j^n$ is defined the vector function $\bar{\boldsymbol{u}}(x,t)$:

$$\frac{\partial \bar{\boldsymbol{u}}}{\partial t} + \frac{\partial \boldsymbol{f}(\bar{\boldsymbol{u}})}{\partial x} = \boldsymbol{0}, \qquad x \in \mathbb{R}, \quad t > t^n, \tag{14.23a}$$

$$\bar{\boldsymbol{u}}(x, t^n) = \boldsymbol{u}_j^n, \qquad x \in V_j \quad (j = 0, \pm 1, \pm 2, \ldots). \tag{14.23b}$$

The Godunov scheme is based on the integral formulation of the initial value problem (14.23).

Integrating the conservation law (14.23a) over V_j, we find

$$\frac{\mathrm{d}}{\mathrm{d}t}\int_{V_j} \bar{\boldsymbol{u}}(x,t)\,\mathrm{d}x + \boldsymbol{f}(\bar{\boldsymbol{u}}(x_{j+\frac{1}{2}}, t)) - \boldsymbol{f}(\bar{\boldsymbol{u}}(x_{j-\frac{1}{2}}, t)) = \boldsymbol{0}. \tag{14.24}$$

Subsequently integrating (14.24) over the time interval $[t^n, t^{n+1}]$, we obtain

$$\begin{aligned} &\int_{V_j} \bar{\boldsymbol{u}}(x, t^{n+1})\,\mathrm{d}x - \int_{V_j} \bar{\boldsymbol{u}}(x, t^n)\,\mathrm{d}x \\ &\quad + \int_{t^n}^{t^{n+1}} \boldsymbol{f}(\bar{\boldsymbol{u}}(x_{j+\frac{1}{2}}, t))\,\mathrm{d}t - \int_{t^n}^{t^{n+1}} \boldsymbol{f}(\bar{\boldsymbol{u}}(x_{j-\frac{1}{2}}, t))\,\mathrm{d}t = \boldsymbol{0}. \end{aligned} \tag{14.25}$$

Introducing the numerical flux $\boldsymbol{F}(\boldsymbol{u}_j^n, \boldsymbol{u}_{j+1}^n)$ as the time average flux at the control volume edge $x_{j+\frac{1}{2}}$, i.e.,

$$\boldsymbol{F}(\boldsymbol{u}_j^n, \boldsymbol{u}_{j+1}^n) := \frac{1}{\Delta t}\int_{t^n}^{t^{n+1}} \boldsymbol{f}(\bar{\boldsymbol{u}}(x_{j+\frac{1}{2}}, t))\,\mathrm{d}t, \tag{14.26}$$

we can write the integral balance (14.25) in the form

$$\int_{V_j} \bar{\boldsymbol{u}}(x, t^{n+1})\,\mathrm{d}x - \int_{V_j} \bar{\boldsymbol{u}}(x, t^n)\,\mathrm{d}x = -\Delta t\big(\boldsymbol{F}(\boldsymbol{u}_j^n, \boldsymbol{u}_{j+1}^n) - \boldsymbol{F}(\boldsymbol{u}_{j-1}^n, \boldsymbol{u}_j^n)\big). \tag{14.27}$$

In (14.26) we have explicitly written the numerical flux at the cell boundary $x_{j+\frac{1}{2}}$ as a function of the numerical vectors $\boldsymbol{u}_j^n$ and $\boldsymbol{u}_{j+1}^n$ in the adjacent cells. The next step is to compute the numerical fluxes.

To compute the numerical flux $\boldsymbol{F}(\boldsymbol{u}_j^n, \boldsymbol{u}_{j+1}^n)$ we solve the local Riemann problem at the cell edge $x_{j+\frac{1}{2}}$, consisting of the conservation law (14.23a) and the piecewise constant initial condition

$$\bar{\boldsymbol{u}}(x, t^n) = \begin{cases} \boldsymbol{u}_j^n & \text{if } \; x < x_{j+\frac{1}{2}}, \\ \boldsymbol{u}_{j+1}^n & \text{if } \; x > x_{j+\frac{1}{2}}. \end{cases} \tag{14.28}$$

The solution of this Riemann problem is a similarity solution of the form

$$\bar{\boldsymbol{u}}(x,t) = \boldsymbol{u}_{\mathrm{R}}\big(\eta; \boldsymbol{u}_j^n, \boldsymbol{u}_{j+1}^n\big), \qquad \eta := \frac{x - x_{j+\frac{1}{2}}}{t - t^n}, \tag{14.29}$$

where the function $\boldsymbol{u}_{\mathrm{R}}$ depends on a single variable η and the parameter vectors $\boldsymbol{u}_j^n$ and $\boldsymbol{u}_{j+1}^n$. Since $\eta = 0$ for $x = x_{j+\frac{1}{2}}$, the numerical flux from (14.26) is given by

$$\boldsymbol{F}(\boldsymbol{u}_j^n, \boldsymbol{u}_{j+1}^n) = \boldsymbol{f}\big(\boldsymbol{u}_{\mathrm{R}}(0; \boldsymbol{u}_j^n, \boldsymbol{u}_{j+1}^n)\big). \tag{14.30}$$

Thus we need the solution of local Riemann problems. For general nonlinear problems, i.e., for general flux vectors $\boldsymbol{f}(\boldsymbol{u})$, this can be quite complicated.

The numerical vector $\boldsymbol{u}_j^{n+1}$ is defined as the average value vector of $\bar{\boldsymbol{u}}(x, t^{n+1})$ over the control volume V_j; i.e.,

$$\boldsymbol{u}_j^{n+1} := \frac{1}{\Delta x}\int_{V_j} \bar{\boldsymbol{u}}(x, t^{n+1})\,\mathrm{d}x. \tag{14.31}$$

If we finally substitute (14.31) in (14.27), we obtain the *Godunov scheme*, i.e.,

$$\boldsymbol{u}_j^{n+1} = \boldsymbol{u}_j^n - \frac{\Delta t}{\Delta x}\big(\boldsymbol{F}(\boldsymbol{u}_j^n, \boldsymbol{u}_{j+1}^n) - \boldsymbol{F}(\boldsymbol{u}_{j-1}^n, \boldsymbol{u}_j^n)\big), \tag{14.32}$$

with the numerical flux $\boldsymbol{F}(\boldsymbol{u}_j^n, \boldsymbol{u}_{j+1}^n)$ defined in (14.30).

As for stability, we note that in the derivation of the Godunov scheme it is tacitly assumed that $\bar{\boldsymbol{u}}(x,t)$ at the cell boundaries is not influenced by neighbouring Riemann problems, implying that $\bar{\boldsymbol{u}}(x_{j+\frac{1}{2}}, t) = \boldsymbol{u}_{\mathrm{R}}(0; \boldsymbol{u}_j^n, \boldsymbol{u}_{j+1}^n) = C$ ($C \in \mathbb{R}$) for $t^n \le t \le t^{n+1}$. This condition is satisfied if the time step restriction

$$\frac{\Delta t}{\Delta x} S_{\max}^n \le 1 \tag{14.33}$$

holds, with $S_{\max}^n$ the maximum wave speed (in absolute value) occurring in the Riemann problems involved at time level t^n; see [158] for more details.

14.2.2 The Linear Case

If the flux is linear, we can derive an explicit form of the Godunov flux and hence for the method as such. Consider the hyperbolic system (14.1a) with flux $\boldsymbol{f}(\boldsymbol{u}) := \boldsymbol{B}\boldsymbol{u}$. In order to compute the Godunov numerical flux $\boldsymbol{F}(\boldsymbol{u}_\ell, \boldsymbol{u}_{\mathrm{r}}) = \boldsymbol{f}(\boldsymbol{u}_{\mathrm{R}}(0; \boldsymbol{u}_\ell, \boldsymbol{u}_{\mathrm{r}}))$ for generic parameter vectors $\boldsymbol{u}_\ell$ and $\boldsymbol{u}_{\mathrm{r}}$, we consider the Riemann problem

$$\frac{\partial \boldsymbol{u}}{\partial t} + \boldsymbol{B}\frac{\partial \boldsymbol{u}}{\partial x} = \boldsymbol{0}, \qquad x \in \mathbb{R}, \quad t > 0, \tag{14.34a}$$

$$\boldsymbol{u}(x,0) = \begin{cases} \boldsymbol{u}_\ell & \text{if } x < 0, \\ \boldsymbol{u}_{\mathrm{r}} & \text{if } x > 0. \end{cases} \tag{14.34b}$$

Since the right eigenvectors $\boldsymbol{s}_k$ of $\boldsymbol{B}$ are linearly independent, we can decompose $\boldsymbol{u}_\ell$ and $\boldsymbol{u}_\mathrm{r}$ as

$$\boldsymbol{u}_\ell = \sum_{k=1}^{m} \alpha_k \boldsymbol{s}_k, \qquad \boldsymbol{u}_\mathrm{r} = \sum_{k=1}^{m} \beta_k \boldsymbol{s}_k \tag{14.35}$$

for some α_k and β_k. Note that the coefficients α_k and β_k depend on the parameter vectors $\boldsymbol{u}_\ell$ and $\boldsymbol{u}_\mathrm{r}$; i.e., $\alpha_k = \alpha_k(\boldsymbol{u}_\ell, \boldsymbol{u}_\mathrm{r})$ and $\beta_k = \beta_k(\boldsymbol{u}_\ell, \boldsymbol{u}_\mathrm{r})$. In the following we often suppress this dependency. In Section 12.4.2 we showed that the similarity solution of the Riemann problem (14.34) is given by

$$\boldsymbol{u}_\mathrm{R}(x/t; \boldsymbol{u}_\ell, \boldsymbol{u}_\mathrm{r}) = \sum_{\lambda_k < x/t} \beta_k \boldsymbol{s}_k + \sum_{\lambda_k > x/t} \alpha_k \boldsymbol{s}_k. \tag{14.36}$$

The solution $\boldsymbol{u}_\mathrm{R}$ is thus piecewise constant because the initial discontinuity at $x = 0$ propagates along all characteristics. The patches of constant-value vectors in the (x, t) plane are separated by the characteristics. Along the centre line $x/t = 0$ the similarity solution is given by

$$\boldsymbol{u}_\mathrm{R}(0; \boldsymbol{u}_\ell, \boldsymbol{u}_\mathrm{r}) = \sum_{\lambda_k < 0} \beta_k \boldsymbol{s}_k + \sum_{\lambda_k > 0} \alpha_k \boldsymbol{s}_k; \tag{14.37}$$

i.e., $\boldsymbol{u}_\mathrm{R}(0; \boldsymbol{u}_\ell, \boldsymbol{u}_\mathrm{r})$ is a superposition of left-going (first term) and right-going (second term) waves. Rearranging terms in (14.37), we can derive the alternative expressions

$$\begin{aligned} \boldsymbol{u}_\mathrm{R}(0; \boldsymbol{u}_\ell, \boldsymbol{u}_\mathrm{r}) &= \boldsymbol{u}_\ell + \sum_{\lambda_k < 0} (\beta_k - \alpha_k) \boldsymbol{s}_k \\ &= \boldsymbol{u}_\mathrm{r} - \sum_{\lambda_k > 0} (\beta_k - \alpha_k) \boldsymbol{s}_k. \end{aligned} \tag{14.38}$$

Using both expressions for $\boldsymbol{u}_\mathrm{R}(0; \boldsymbol{u}_\ell, \boldsymbol{u}_\mathrm{r})$ in (14.38), we find for the Godunov numerical flux $\boldsymbol{F}(\boldsymbol{u}_\ell, \boldsymbol{u}_\mathrm{r}) = \boldsymbol{B}\boldsymbol{u}_\mathrm{R}(0; \boldsymbol{u}_\ell, \boldsymbol{u}_\mathrm{r}))$ the two representations

$$\begin{aligned} \boldsymbol{F}(\boldsymbol{u}_\ell, \boldsymbol{u}_\mathrm{r}) &= \boldsymbol{B}\boldsymbol{u}_\ell + \sum_{\lambda_k < 0} (\beta_k - \alpha_k) \lambda_k \boldsymbol{s}_k \\ &= \boldsymbol{B}\boldsymbol{u}_\mathrm{r} - \sum_{\lambda_k > 0} (\beta_k - \alpha_k) \lambda_k \boldsymbol{s}_k. \end{aligned} \tag{14.39}$$

We can find still another expression for the numerical flux as follows. With λ_k^- as defined in (14.10), we can simplify the sum in the first expression in (14.39) as

$$\sum_{\lambda_k < 0} (\beta_k - \alpha_k) \lambda_k \boldsymbol{s}_k = \sum_{k=1}^{m} (\beta_k - \alpha_k) \lambda_k^- \boldsymbol{s}_k = \boldsymbol{B}^-(\boldsymbol{u}_\mathrm{r} - \boldsymbol{u}_\ell) \tag{14.40}$$

to find for the numerical flux

$$\boldsymbol{F}(\boldsymbol{u}_\ell, \boldsymbol{u}_\mathrm{r}) = \boldsymbol{B}\boldsymbol{u}_\ell + \boldsymbol{B}^-(\boldsymbol{u}_\mathrm{r} - \boldsymbol{u}_\ell) = \boldsymbol{B}^+\boldsymbol{u}_\ell + \boldsymbol{B}^-\boldsymbol{u}_\mathrm{r}. \tag{14.41}$$

The expression (14.41) can also be found using the second representation for the numerical flux in (14.39).

All three alternative expressions for the numerical flux can be combined with the Godunov scheme. If we insert, for example, the expression (14.41) in the Godunov scheme (14.32), we obtain

$$\begin{aligned} \boldsymbol{u}_j^{n+1} &= \boldsymbol{u}_j^n - \frac{\Delta t}{\Delta x}\Big(\big(\boldsymbol{B}^+\boldsymbol{u}_j^n + \boldsymbol{B}^-\boldsymbol{u}_{j+1}^n\big) - \big(\boldsymbol{B}^+\boldsymbol{u}_{j-1}^n + \boldsymbol{B}^-\boldsymbol{u}_j^n\big)\Big) \\ &= \boldsymbol{u}_j^n - \frac{\Delta t}{\Delta x}\Big(\boldsymbol{B}^+\big(\boldsymbol{u}_j^n - \boldsymbol{u}_{j-1}^n\big) + \boldsymbol{B}^-\big(\boldsymbol{u}_{j+1}^n - \boldsymbol{u}_j^n\big)\Big). \end{aligned} \tag{14.42}$$

Like in the scalar case, the Godunov scheme reduces to the upwind scheme for linear hyperbolic systems.

We can derive an alternative formulation of the Godunov scheme if we use the first expression in (14.39) for the numerical flux $\boldsymbol{F}(\boldsymbol{u}_j^n, \boldsymbol{u}_{j+1}^n)$ at the right cell interface $x_{j+1/2}$ and the second expression in (14.39) for the flux $\boldsymbol{F}(\boldsymbol{u}_{j-1}^n, \boldsymbol{u}_j^n)$ at the left interface $x_{j-1/2}$. Then the common term $\boldsymbol{B}\boldsymbol{u}_j^n$ cancels and we find the scheme

$$\boldsymbol{u}_j^{n+1} = \boldsymbol{u}_j^n - \frac{\Delta t}{\Delta x}\left(\sum_{\lambda_k<0} \gamma_{k,j+1/2}^n \lambda_k \boldsymbol{s}_k + \sum_{\lambda_k>0} \gamma_{k,j-1/2}^n \lambda_k \boldsymbol{s}_k\right), \tag{14.43}$$

with $\gamma_{k,j+1/2}^n := \beta_k(\boldsymbol{u}_j^n, \boldsymbol{u}_{j+1}^n) - \alpha_k(\boldsymbol{u}_j^n, \boldsymbol{u}_{j+1}^n)$, etc. Thus the cell average in cell V_j is only affected by left-going waves from $x_{j+1/2}$ and right-going waves from $x_{j-1/2}$. This form of Godunov's scheme is sometimes referred to as the *wave propagation form*.

Example 14.3 Consider once more the linear system from Example 14.1. Taking into account that $\lambda_1 = -a < 0$ and $\lambda_2 = a > 0$, we obtain the following wave propagation form of the Godunov scheme:

$$\begin{aligned} \boldsymbol{u}_j^{n+1} &= \boldsymbol{u}_j^n - \frac{\Delta t}{\Delta x}\big(\gamma_{1,j+1/2}^n \lambda_1 \boldsymbol{s}_1 + \gamma_{2,j-1/2}^n \lambda_2 \boldsymbol{s}_2\big) \\ &= \boldsymbol{u}_j^n + c\big(\gamma_{1,j+1/2}^n \boldsymbol{s}_1 - \gamma_{2,j-1/2}^n \boldsymbol{s}_2\big). \end{aligned}$$

We have to compute the coefficients $\gamma_{k,j+1/2}^n$ $(k = 1, 2)$ from the relation

$$\boldsymbol{u}_{j+1}^n - \boldsymbol{u}_j^n = \gamma_{1,j+1/2}^n \boldsymbol{s}_1 + \gamma_{2,j+1/2}^n \boldsymbol{s}_2.$$

The solution is given by

$$\gamma_{1,j+1/2}^n = \frac{1}{2}\big(u_{1,j+1}^n - u_{1,j}^n + u_{2,j+1}^n - u_{2,j}^n\big),$$

$$\gamma_{2,j+1/2}^n = \frac{1}{2}\big(u_{1,j+1}^n - u_{1,j}^n - u_{2,j+1}^n + u_{2,j}^n\big).$$

Substituting this result in the wave propagation form above, we recover the upwind scheme from Example 14.1. □

14.3 Roe's Approximate Riemann Solver

The Godunov scheme requires the solution of local Riemann problems at control volume interfaces for the computation of the numerical flux; cf. (14.30). For nonlinear systems this is a difficult and expensive computation and, moreover, it is not necessary since we only use the average value of the state variable over the control volumes. Therefore one often employs schemes based on *approximate Riemann solvers*, where the numerical flux is computed from an approximate and easier to solve Riemann problem. A straightforward choice is to replace the nonlinear Riemann problem by a linear one. In this section we consider a method named after Roe [134].

Like we did for the Godunov scheme (14.32), we will write the *Roe scheme* as

$$\boldsymbol{u}_j^{n+1} = \boldsymbol{u}_j^n - \frac{\Delta t}{\Delta x}\big(\boldsymbol{F}_{\mathrm{R}}(\boldsymbol{u}_j^n, \boldsymbol{u}_{j+1}^n) - \boldsymbol{F}_{\mathrm{R}}(\boldsymbol{u}_{j-1}^n, \boldsymbol{u}_j^n)\big), \tag{14.44}$$

where the *Roe numerical flux* $\boldsymbol{F}_{\mathrm{R}}(\boldsymbol{u}_j^n, \boldsymbol{u}_{j+1}^n)$ is an approximation of the Godunov flux (14.30). The problem thus is to find $\boldsymbol{F}_{\mathrm{R}}(\boldsymbol{u}_\ell, \boldsymbol{u}_{\mathrm{r}})$ for generic state vectors $\boldsymbol{u}_\ell$ and $\boldsymbol{u}_{\mathrm{r}}$. The derivation goes briefly as follows. First, we derive integral expressions for the Godunov numerical flux of the original Riemann problem. Subsequently, we replace this Riemann problem by a linearisation and again determine integral expressions for the corresponding Godunov numerical flux. Finally, combining these Godunov fluxes, we can determine the Roe numerical flux.

Consider the general Riemann problem

$$\frac{\partial \boldsymbol{u}}{\partial t} + \frac{\partial \boldsymbol{f}(\boldsymbol{u})}{\partial x} = \boldsymbol{0}, \qquad x \in \mathbb{R}, \quad t > 0, \tag{14.45a}$$

$$\boldsymbol{u}(x,0) = \begin{cases} \boldsymbol{u}_\ell & \text{if} \quad x < 0, \\ \boldsymbol{u}_{\mathrm{r}} & \text{if} \quad x > 0. \end{cases} \tag{14.45b}$$

A schematic picture of the similarity solution $\boldsymbol{u}_{\mathrm{R}}(x/t; \boldsymbol{u}_\ell, \boldsymbol{u}_{\mathrm{r}})$ in the (x, t) plane is given in Figure 14.1. Here s_ℓ and s_{r} are the minimum and maximum propagation speeds, respectively, involved in the Riemann problem. Thus the similarity solution holds in the wedge-shaped region in the centre and is bounded by the constant states $\boldsymbol{u}_\ell$ and $\boldsymbol{u}_{\mathrm{r}}$. Integrating (14.45a) over the left control volume $[x_\ell, 0] \times [0, \Delta t]$ for some time step $\Delta t > 0$ and taking into account that the solution is constant along the t axis and outside the wedge-shaped region, we find

$$\boldsymbol{F}(\boldsymbol{u}_\ell, \boldsymbol{u}_{\mathrm{r}}) = \boldsymbol{f}(\boldsymbol{u}_\ell) - \frac{1}{\Delta t}\bigg(\int_{x_\ell}^{0} \boldsymbol{u}(x, \Delta t)\,\mathrm{d}x + x_\ell \boldsymbol{u}_\ell\bigg), \tag{14.46}$$

with $\boldsymbol{F}(\boldsymbol{u}_\ell, \boldsymbol{u}_{\mathrm{r}})$ the Godunov flux corresponding to (14.45a), i.e.,

$$\boldsymbol{F}(\boldsymbol{u}_\ell, \boldsymbol{u}_{\mathrm{r}}) = \boldsymbol{f}\big(\boldsymbol{u}_{\mathrm{R}}(0; \boldsymbol{u}_\ell, \boldsymbol{u}_{\mathrm{r}})\big). \tag{14.47}$$

Likewise, integration of (14.45a) over the right control volume $[0, x_{\mathrm{r}}] \times [0, \Delta t]$ gives

$$\boldsymbol{F}(\boldsymbol{u}_\ell, \boldsymbol{u}_{\mathrm{r}}) = \boldsymbol{f}(\boldsymbol{u}_{\mathrm{r}}) + \frac{1}{\Delta t}\bigg(\int_{0}^{x_{\mathrm{r}}} \boldsymbol{u}(x, \Delta t)\,\mathrm{d}x - x_{\mathrm{r}} \boldsymbol{u}_{\mathrm{r}}\bigg). \tag{14.48}$$

Equality of the right-hand sides of (14.46) and (14.48) can be established by integration of (14.45a) over $[x_\ell, x_{\mathrm{r}}] \times [0, \Delta t]$.

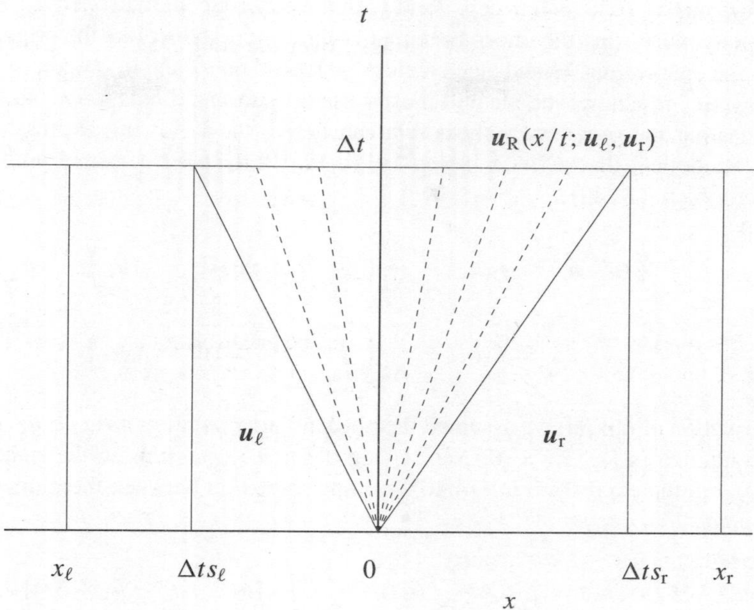

Figure 14.1. *Schematic representation of a similarity solution in the* (x, t) *plane.*

The computation of the Roe numerical flux is based on the approximate Riemann problem

$$\frac{\partial \widetilde{\boldsymbol{u}}}{\partial t} + \widetilde{\boldsymbol{B}}(\boldsymbol{u}_\ell, \boldsymbol{u}_r)\frac{\partial \widetilde{\boldsymbol{u}}}{\partial x} = \boldsymbol{0}, \qquad x \in \mathbb{R}, \quad t > 0, \tag{14.49a}$$

$$\widetilde{\boldsymbol{u}}(x, 0) = \begin{cases} \boldsymbol{u}_\ell & \text{if } x < 0, \\ \boldsymbol{u}_r & \text{if } x > 0, \end{cases} \tag{14.49b}$$

with $\widetilde{\boldsymbol{B}}(\boldsymbol{u}_\ell, \boldsymbol{u}_r)$ a constant matrix depending only on the initial states $\boldsymbol{u}_\ell$ and $\boldsymbol{u}_r$. Associated with (14.49) is the linear flux $\widetilde{\boldsymbol{f}}(\boldsymbol{u}) := \widetilde{\boldsymbol{B}}(\boldsymbol{u}_\ell, \boldsymbol{u}_r)\boldsymbol{u}$. The flux $\widetilde{\boldsymbol{f}}(\boldsymbol{u})$ and the *Roe matrix* $\widetilde{\boldsymbol{B}}(\boldsymbol{u}_\ell, \boldsymbol{u}_r)$ should be an approximation of the flux $\boldsymbol{f}(\boldsymbol{u})$ in (14.45a) and the corresponding Jacobi matrix $\boldsymbol{B}(\boldsymbol{u}) := \frac{\partial}{\partial \boldsymbol{u}}\boldsymbol{f}(\boldsymbol{u})$, respectively. The following three requirements are imposed on the Roe matrix $\widetilde{\boldsymbol{B}}(\boldsymbol{u}_\ell, \boldsymbol{u}_r)$:

1. $\boldsymbol{f}(\boldsymbol{u}_r) - \boldsymbol{f}(\boldsymbol{u}_\ell) = \widetilde{\boldsymbol{B}}(\boldsymbol{u}_\ell, \boldsymbol{u}_r)(\boldsymbol{u}_r - \boldsymbol{u}_\ell)$.
2. $\widetilde{\boldsymbol{B}}(\boldsymbol{u}, \boldsymbol{u}) = \boldsymbol{B}(\boldsymbol{u})$.
3. $\widetilde{\boldsymbol{B}}(\boldsymbol{u}_\ell, \boldsymbol{u}_r)$ has m real eigenvalues $\widetilde{\lambda}_1, \widetilde{\lambda}_2, \dots, \widetilde{\lambda}_m$ and m corresponding linearly independent eigenvectors $\widetilde{\boldsymbol{s}}_1, \widetilde{\boldsymbol{s}}_2, \dots, \widetilde{\boldsymbol{s}}_m$.

The first property implies that an isolated shock or contact discontinuity connecting the constant states $\boldsymbol{u}_\ell$ and $\boldsymbol{u}_r$ in the true Riemann solution is approximated by a single discontinuity by virtue of the Rankine–Hugoniot jump relations; cf. (12.86). The second property means that the Roe matrix $\widetilde{\boldsymbol{B}}(\boldsymbol{u}_\ell, \boldsymbol{u}_\mathrm{r})$ is consistent with the exact Jacobi matrix $\boldsymbol{B}(\boldsymbol{u})$. Finally, the third property means that the linear system (14.49a) is hyperbolic, like the original system. Note that the eigenvalues $\widetilde{\lambda}_k$ and eigenvectors $\widetilde{\boldsymbol{s}}_k$ depend on $\boldsymbol{u}_\ell$ and $\boldsymbol{u}_\mathrm{r}$, i.e., $\widetilde{\lambda}_k = \widetilde{\lambda}_k(\boldsymbol{u}_\ell, \boldsymbol{u}_\mathrm{r})$ and $\widetilde{\boldsymbol{s}}_k = \widetilde{\boldsymbol{s}}_k(\boldsymbol{u}_\ell, \boldsymbol{u}_\mathrm{r})$; however, we will mostly use the shorthand notation $\widetilde{\lambda}_k$ and $\widetilde{\boldsymbol{s}}_k$.

Repeating the previous analysis, we can derive the following expressions for the Godunov numerical flux $\widetilde{\boldsymbol{F}}(\boldsymbol{u}_\ell, \boldsymbol{u}_\mathrm{r}) := \widetilde{\boldsymbol{B}}(\boldsymbol{u}_\ell, \boldsymbol{u}_\mathrm{r})\boldsymbol{u}_\mathrm{R}(0; \boldsymbol{u}_\ell, \boldsymbol{u}_\mathrm{r})$ associated with the linear Riemann problem (14.49):

$$\begin{aligned}\widetilde{\boldsymbol{F}}(\boldsymbol{u}_\ell, \boldsymbol{u}_\mathrm{r}) &= \widetilde{\boldsymbol{f}}(\boldsymbol{u}_\ell) - \frac{1}{\Delta t}\left(\int_{x_\ell}^{0} \widetilde{\boldsymbol{u}}(x, \Delta t)\,\mathrm{d}x + x_\ell \boldsymbol{u}_\ell\right)\\ &= \widetilde{\boldsymbol{f}}(\boldsymbol{u}_\mathrm{r}) + \frac{1}{\Delta t}\left(\int_{0}^{x_\mathrm{r}} \widetilde{\boldsymbol{u}}(x, \Delta t)\,\mathrm{d}x - x_\mathrm{r} \boldsymbol{u}_\mathrm{r}\right).\end{aligned} \tag{14.50}$$

In the derivation of (14.50) we assumed that the similarity solution $\widetilde{\boldsymbol{u}}_\mathrm{R}(x/t; \boldsymbol{u}_\ell, \boldsymbol{u}_\mathrm{r})$ is completely contained in $[x_\ell, x_\mathrm{r}] \times [0, \Delta t]$. Using the first expression on the right-hand side in (14.50) and the expression in (14.46), we find a relation between the numerical fluxes $\boldsymbol{F}(\boldsymbol{u}_\ell, \boldsymbol{u}_\mathrm{r})$ and $\widetilde{\boldsymbol{F}}(\boldsymbol{u}_\ell, \boldsymbol{u}_\mathrm{r})$:

$$\boldsymbol{F}(\boldsymbol{u}_\ell, \boldsymbol{u}_\mathrm{r}) - \widetilde{\boldsymbol{F}}(\boldsymbol{u}_\ell, \boldsymbol{u}_\mathrm{r}) = \boldsymbol{f}(\boldsymbol{u}_\ell) - \widetilde{\boldsymbol{f}}(\boldsymbol{u}_\ell) - \frac{1}{\Delta t}\int_{x_\ell}^{0} \big(\boldsymbol{u}(x, \Delta t) - \widetilde{\boldsymbol{u}}(x, \Delta t)\big)\,\mathrm{d}x. \tag{14.51}$$

Omitting the last, presumably small, term in (14.51), we obtain the following expression for the Roe numerical flux $\boldsymbol{F}_\mathrm{R}(\boldsymbol{u}_\ell, \boldsymbol{u}_\mathrm{r})$:

$$\boldsymbol{F}_\mathrm{R}(\boldsymbol{u}_\ell, \boldsymbol{u}_\mathrm{r}) = \boldsymbol{f}(\boldsymbol{u}_\ell) + \widetilde{\boldsymbol{F}}(\boldsymbol{u}_\ell, \boldsymbol{u}_\mathrm{r}) - \widetilde{\boldsymbol{f}}(\boldsymbol{u}_\ell). \tag{14.52}$$

Likewise, using (14.48) and the second expression in (14.50) and omitting the resulting integral, we find the alternative expression

$$\boldsymbol{F}_\mathrm{R}(\boldsymbol{u}_\ell, \boldsymbol{u}_\mathrm{r}) = \boldsymbol{f}(\boldsymbol{u}_\mathrm{r}) + \widetilde{\boldsymbol{F}}(\boldsymbol{u}_\ell, \boldsymbol{u}_\mathrm{r}) - \widetilde{\boldsymbol{f}}(\boldsymbol{u}_\mathrm{r}). \tag{14.53}$$

The equality of the right-hand sides of (14.52) and (14.53) is an immediate consequence of the first property of the Roe matrix.

Hence we have to find suitable expressions for the numerical flux $\widetilde{\boldsymbol{F}}(\boldsymbol{u}_\ell, \boldsymbol{u}_\mathrm{r})$. To do this let us introduce the decomposition

$$\boldsymbol{u}_\mathrm{r} - \boldsymbol{u}_\ell = \sum_{k=1}^{m} \gamma_k \widetilde{\boldsymbol{s}}_k, \tag{14.54}$$

where the coefficients $\gamma_k = \gamma_k(\boldsymbol{u}_\ell, \boldsymbol{u}_\mathrm{r})$ are referred to as the *wave strengths*. Then the exact solution of the linear Riemann problem (14.49) at $x/t = 0$ can be written as (see (14.38)),

$$\begin{aligned}\widetilde{\boldsymbol{u}}_\mathrm{R}\big(0; \boldsymbol{u}_\ell, \boldsymbol{u}_\mathrm{r}\big) &= \boldsymbol{u}_\ell + \sum_{\widetilde{\lambda}_k<0} \gamma_k \widetilde{\boldsymbol{s}}_k\\ &= \boldsymbol{u}_\mathrm{r} - \sum_{\widetilde{\lambda}_k>0} \gamma_k \widetilde{\boldsymbol{s}}_k.\end{aligned} \tag{14.55}$$

We can now compute $\widetilde{\boldsymbol{F}}(\boldsymbol{u}_\ell, \boldsymbol{u}_\mathrm{r})$ to find

$$\begin{aligned}\widetilde{\boldsymbol{F}}(\boldsymbol{u}_\ell, \boldsymbol{u}_\mathrm{r}) &= \widetilde{\boldsymbol{f}}(\boldsymbol{u}_\ell) + \sum_{\widetilde{\lambda}_k<0} \gamma_k \widetilde{\lambda}_k \widetilde{\boldsymbol{s}}_k \\ &= \widetilde{\boldsymbol{f}}(\boldsymbol{u}_\mathrm{r}) - \sum_{\widetilde{\lambda}_k>0} \gamma_k \widetilde{\lambda}_k \widetilde{\boldsymbol{s}}_k.\end{aligned} \tag{14.56}$$

Combining (14.52), (14.53), and (14.56), we thus obtain for the Roe numerical flux

$$\begin{aligned}\boldsymbol{F}_\mathrm{R}(\boldsymbol{u}_\ell, \boldsymbol{u}_\mathrm{r}) &= \boldsymbol{f}(\boldsymbol{u}_\ell) + \sum_{\widetilde{\lambda}_k<0} \gamma_k \widetilde{\lambda}_k \widetilde{\boldsymbol{s}}_k \\ &= \boldsymbol{f}(\boldsymbol{u}_\mathrm{r}) - \sum_{\widetilde{\lambda}_k>0} \gamma_k \widetilde{\lambda}_k \widetilde{\boldsymbol{s}}_k.\end{aligned} \tag{14.57}$$

From the first expression in (14.57) we conclude that $\boldsymbol{F}_\mathrm{R}(\boldsymbol{u}_\ell, \boldsymbol{u}_\mathrm{r}) = \boldsymbol{f}(\boldsymbol{u}_\ell)$ if all eigenvalues $\widetilde{\lambda}_k$ are positive; i.e., all waves in the Riemann problem (14.49) propagate in the positive x direction. On the other hand, $\widetilde{\boldsymbol{F}}(\boldsymbol{u}_\ell, \boldsymbol{u}_\mathrm{r}) = \boldsymbol{f}(\boldsymbol{u}_\mathrm{r})$ if all $\widetilde{\lambda}_k$ are negative. Taking the average value of both expressions in (14.57), we finally obtain the alternative expression

$$\boldsymbol{F}_\mathrm{R}(\boldsymbol{u}_\ell, \boldsymbol{u}_\mathrm{r}) = \frac{1}{2}\big(\boldsymbol{f}(\boldsymbol{u}_\ell) + \boldsymbol{f}(\boldsymbol{u}_\mathrm{r})\big) - \frac{1}{2}\sum_{k=1}^{m} \gamma_k\, |\widetilde{\lambda}_k|\, \widetilde{\boldsymbol{s}}_k. \tag{14.58}$$

The computation of the Roe numerical flux $\boldsymbol{F}_\mathrm{R}(\boldsymbol{u}_\ell, \boldsymbol{u}_\mathrm{r})$ thus requires the following steps: Compute

- the Roe matrix $\widetilde{\boldsymbol{B}}(\boldsymbol{u}_\ell, \boldsymbol{u}_\mathrm{r})$ from $\boldsymbol{f}(\boldsymbol{u}_\mathrm{r}) - \boldsymbol{f}(\boldsymbol{u}_\ell) = \widetilde{\boldsymbol{B}}(\boldsymbol{u}_\ell, \boldsymbol{u}_\mathrm{r})(\boldsymbol{u}_\mathrm{r} - \boldsymbol{u}_\ell)$,
- its eigenvalues $\widetilde{\lambda}_k$ and eigenvectors $\widetilde{\boldsymbol{s}}_k$,
- the wave strengths γ_k from (14.54), and
- the flux from either expression in (14.57).

The crucial step is the computation of the Roe matrix $\widetilde{\boldsymbol{B}}(\boldsymbol{u}_\ell, \boldsymbol{u}_\mathrm{r})$. This is generally a difficult and expensive task, and we will not discuss it here. However, for many relevant conservation laws there do exist efficient methods to compute the Roe matrix. As an example, we will compute in Section 14.5.2 the Roe matrix for the shallow-water equations.

A drawback of the Roe scheme is that it cannot represent rarefaction waves because the underlying linear Riemann problem (14.49) does not allow such solutions; cf. Chapter 12. A way to overcome this disadvantage is the *Harten–Hyman entropy fix*. We will discuss this as well in Section 14.5.2.

An alternative derivation of the Roe scheme is based on the wave propagation form (14.43) of the Godunov scheme. Consider the Godunov scheme (14.32) rewritten in the form

$$\begin{aligned}\boldsymbol{u}_j^{n+1} = \boldsymbol{u}_j^n - \frac{\Delta t}{\Delta x}\Big(&\big(\boldsymbol{f}(\boldsymbol{u}_\mathrm{R}(0; \boldsymbol{u}_j^n, \boldsymbol{u}_{j+1}^n)) - \boldsymbol{f}(\boldsymbol{u}_j^n)\big) \\ &+ \big(\boldsymbol{f}(\boldsymbol{u}_j^n) - \boldsymbol{f}(\boldsymbol{u}_\mathrm{R}(0; \boldsymbol{u}_{j-1}^n, \boldsymbol{u}_j^n))\big)\Big).\end{aligned} \tag{14.59}$$

The flux differences $\boldsymbol{f}(\boldsymbol{u}_{\mathrm{R}}(0; \boldsymbol{u}_j^n, \boldsymbol{u}_{j+1}^n)) - \boldsymbol{f}(\boldsymbol{u}_j^n)$ and $\boldsymbol{f}(\boldsymbol{u}_j^n) - \boldsymbol{f}(\boldsymbol{u}_{\mathrm{R}}(0; \boldsymbol{u}_{j-1}^n, \boldsymbol{u}_j^n))$ include the waves emanating from the interfaces $x = x_{j+1/2}$ and $x = x_{j-1/2}$, respectively, and for this reason (14.59) is referred to as the wave propagation form. Using (14.39), we see that for the special case $\boldsymbol{f}(\boldsymbol{u}) = \boldsymbol{B}\boldsymbol{u}$ the scheme (14.59) simply reduces to the linear wave propagation form (14.43). The basic idea of the Roe scheme is to approximate the flux differences in (14.59) from local linear Riemann problems of the form (14.49). Obviously, we take the Roe matrices $\widetilde{\boldsymbol{B}}(\boldsymbol{u}_j^n, \boldsymbol{u}_{j+1}^n)$ and $\widetilde{\boldsymbol{B}}(\boldsymbol{u}_{j-1}^n, \boldsymbol{u}_j^n)$ to approximate $\boldsymbol{f}(\boldsymbol{u}_{\mathrm{R}}(0; \boldsymbol{u}_j^n, \boldsymbol{u}_{j+1}^n)) - \boldsymbol{f}(\boldsymbol{u}_j^n)$ and $\boldsymbol{f}(\boldsymbol{u}_j^n) - \boldsymbol{f}(\boldsymbol{u}_{\mathrm{R}}(0; \boldsymbol{u}_{j-1}^n, \boldsymbol{u}_j^n))$, respectively. Using the appropriate expressions for the Roe flux in (14.57), we obtain the wave propagation form

$$\boldsymbol{u}_j^{n+1} = \boldsymbol{u}_j^n - \frac{\Delta t}{\Delta x}\left(\sum_{\widetilde{\lambda}_{k,j+1/2}^n < 0} \gamma_{k,j+1/2}^n \widetilde{\lambda}_{k,j+1/2}^n \widetilde{\boldsymbol{s}}_{k,j+1/2}^n + \sum_{\widetilde{\lambda}_{k,j-1/2}^n > 0} \gamma_{k,j-1/2}^n \widetilde{\lambda}_{k,j-1/2}^n \widetilde{\boldsymbol{s}}_{k,j-1/2}^n\right), \tag{14.60}$$

where $\widetilde{\lambda}_{k,j+1/2}^n = \widetilde{\lambda}_k(\boldsymbol{u}_j^n, \boldsymbol{u}_{j+1}^n)$, $\widetilde{\boldsymbol{s}}_{k,j+1/2}^n = \widetilde{\boldsymbol{s}}_k(\boldsymbol{u}_j^n, \boldsymbol{u}_{j+1}^n)$, etc., and where the wave strengths $\gamma_{k,j+1/2}^n$ follow from the decomposition (14.54) with $\boldsymbol{u}_\ell = \boldsymbol{u}_j^n$ and $\boldsymbol{u}_{\mathrm{r}} = \boldsymbol{u}_{j+1}^n$.

14.4 Slope Limiter Methods

For the linear system (14.1a) we have derived the upwind scheme (14.14), which is only first order accurate. This is not very satisfactory, and therefore we look for a high-resolution scheme, i.e., a scheme that is second order accurate in regions where the solution is smooth and suppresses spurious oscillations in the vicinity of discontinuities. There are two possible approaches to achieve this: the flux limiter method and the slope limiter method. We only treat the slope limiter method, because this approach, which is considered in Section 14.4.1 for the linear case, can easily be generalised to nonlinear systems, as shown in Section 14.4.2.

14.4.1 A Slope Limiter Method for Linear Systems

In the derivation of the slope limiter method for (14.1a), we first decouple the system and then apply the basic slope limiter method to the advection equation for each characteristic variable $\tilde{u}_k$. Thus consider the decoupled system of equations (14.6a). The basic slope limiter method for these equations is based on the Godunov scheme in combination with a piecewise linear distribution for the numerical solution in each control volume. Applying the basic slope limiter method from Section 13.8 to (14.6a), we obtain the following scheme for the characteristic variables $\tilde{u}_k$:

$$\tilde{u}_{k,j}^{n+1} = \tilde{u}_{k,j}^n - \frac{\Delta t}{\Delta x}\Bigg(\lambda_k^+\left(\tilde{u}_{k,j}^n - \tilde{u}_{k,j-1}^n\right) + \lambda_k^-\left(\tilde{u}_{k,j+1}^n - \tilde{u}_{k,j}^n\right) + \frac{1}{2}|\lambda_k|\left(1 - \frac{\Delta t}{\Delta x}|\lambda_k|\right)\Delta x\left(\tilde{\sigma}_{k,j_{\mathrm{u}}(k)}^n - \tilde{\sigma}_{k,j_{\mathrm{u}}(k)-1}^n\right)\Bigg), \tag{14.61}$$

with $\tilde{\sigma}^n_{k,j_{\mathrm{u}}(k)}$ the slope of the numerical solution in the control volume labeled $j_{\mathrm{u}}(k)$, where the index $j_{\mathrm{u}}(k)$ is defined by

$$j_{\mathrm{u}}(k) := \begin{cases} j & \text{if} \quad \lambda_k > 0, \\ j+1 & \text{if} \quad \lambda_k < 0; \end{cases} \tag{14.62}$$

cf. (13.134) and (13.135). The slopes $\tilde{\sigma}^n_{k,j_{\mathrm{u}}(k)}$ are chosen according to

$$\tilde{\sigma}^n_{k,j_{\mathrm{u}}(k)} = \Phi^n_{k,j+1/2} \frac{1}{\Delta x}\big(\tilde{u}^n_{k,j+1} - \tilde{u}^n_{k,j}\big), \tag{14.63}$$

with $\Phi^n_{k,j+1/2}$ the slope limiter given by

$$\Phi^n_{k,j+1/2} = \Phi(r^n_{k,j+1/2}), \qquad r^n_{k,j+1/2} := \begin{cases} \dfrac{\tilde{u}^n_{k,j} - \tilde{u}^n_{k,j-1}}{\tilde{u}^n_{k,j+1} - \tilde{u}^n_{k,j}} & \text{if} \quad \lambda_k > 0, \\[2ex] \dfrac{\tilde{u}^n_{k,j+2} - \tilde{u}^n_{k,j+1}}{\tilde{u}^n_{k,j+1} - \tilde{u}^n_{k,j}} & \text{if} \quad \lambda_k < 0; \end{cases} \tag{14.64}$$

cf. (13.104) and (13.105). The limiter function $\Phi(r)$ should lie in the total variation diminishing (TVD) region of Figure 13.18 in order that scheme (14.61) is TVD.

It is most convenient to formulate the slope limiter method for (14.1a) in terms of the variable $\boldsymbol{u}$. Therefore we multiply (14.61) by the eigenvector $\boldsymbol{s}_k$ and sum over all indices k. Taking into account that $\boldsymbol{u} = \boldsymbol{S}\tilde{\boldsymbol{u}} = \sum_{k=1}^m \tilde{u}_k \boldsymbol{s}_k$, we obtain

$$\begin{aligned} \boldsymbol{u}^{n+1}_j = \boldsymbol{u}^n_j - \frac{\Delta t}{\Delta x}\Bigg(& \sum_{k=1}^m \big(\tilde{u}^n_{k,j} - \tilde{u}^n_{k,j-1}\big)\lambda_k^+ \boldsymbol{s}_k + \sum_{k=1}^m \big(\tilde{u}^n_{k,j+1} - \tilde{u}^n_{k,j}\big)\lambda_k^- \boldsymbol{s}_k + \\ & + \frac{1}{2}\sum_{k=1}^m |\lambda_k| \left(1 - \frac{\Delta t}{\Delta x}|\lambda_k|\right) \Delta x \big(\tilde{\sigma}^n_{k,j_{\mathrm{u}}(k)} - \tilde{\sigma}^n_{k,j_{\mathrm{u}}(k)-1}\big)\boldsymbol{s}_k \Bigg). \end{aligned} \tag{14.65}$$

Since λ_k^+ and λ_k^- are the eigenvalues of the split matrices $\boldsymbol{B}^+$ and $\boldsymbol{B}^-$, respectively, the first and second sums on the right-hand side of (14.65) reduce to

$$\sum_{k=1}^m \big(\tilde{u}^n_{k,j} - \tilde{u}^n_{k,j-1}\big)\lambda_k^+ \boldsymbol{s}_k = \boldsymbol{B}^+\big(\boldsymbol{u}^n_j - \boldsymbol{u}^n_{j-1}\big), \tag{14.66a}$$

$$\sum_{k=1}^m \big(\tilde{u}^n_{k,j+1} - \tilde{u}^n_{k,j}\big)\lambda_k^- \boldsymbol{s}_k = \boldsymbol{B}^-\big(\boldsymbol{u}^n_{j+1} - \boldsymbol{u}^n_j\big). \tag{14.66b}$$

Substitution of (14.66) in (14.65) gives the slope limiter scheme

$$\begin{aligned} \boldsymbol{u}^{n+1}_j = \boldsymbol{u}^n_j - \frac{\Delta t}{\Delta x}\Bigg(& \boldsymbol{B}^+\big(\boldsymbol{u}^n_j - \boldsymbol{u}^n_{j-1}\big) + \boldsymbol{B}^-\big(\boldsymbol{u}^n_{j+1} - \boldsymbol{u}^n_j\big) \\ & + \frac{1}{2}\sum_{k=1}^m |\lambda_k| \left(1 - \frac{\Delta t}{\Delta x}|\lambda_k|\right) \Delta x \big(\tilde{\sigma}^n_{k,j_{\mathrm{u}}(k)} - \tilde{\sigma}^n_{k,j_{\mathrm{u}}(k)-1}\big)\boldsymbol{s}_k \Bigg). \end{aligned} \tag{14.67}$$

The numerical flux $\boldsymbol{F}(\boldsymbol{u}_j^n, \boldsymbol{u}_{j+1}^n)$ for (14.67) is obviously given by

$$\boldsymbol{F}(\boldsymbol{u}_j^n, \boldsymbol{u}_{j+1}^n) = \boldsymbol{B}^+\boldsymbol{u}_j^n + \boldsymbol{B}^-\boldsymbol{u}_{j+1}^n + \frac{1}{2}\sum_{k=1}^{m}|\lambda_k|\left(1 - \frac{\Delta t}{\Delta x}|\lambda_k|\right)\Delta x\, \tilde{\sigma}_{k,j_u(k)}^n \boldsymbol{s}_k \tag{14.68}$$

and is a straightforward generalisation of the slope limiter flux in (13.134) for the linear advection equation. The numerical flux $\boldsymbol{F}(\boldsymbol{u}_j^n, \boldsymbol{u}_{j+1}^n)$ is thus equivalent to the upwind flux plus a correction term containing the slopes $\tilde{\sigma}_{k,j_u(k)}^n$ of the numerical approximations of the variables $\tilde{u}_k$, taken in the control volume that lies upwind (with respect to $\tilde{u}_k$) of the cell boundary $x_{j+1/2}$.

The coefficients $\gamma_{k,j+1/2}^n := \tilde{u}_{k,j+1}^n - \tilde{u}_{k,j}^n$ in (14.63) for the slopes $\tilde{\sigma}_{k,j_u(k)}^n$ can be computed from the decomposition

$$\boldsymbol{u}_{j+1}^n - \boldsymbol{u}_j^n = \sum_{k=1}^{m}\gamma_{k,j+1/2}^n \boldsymbol{s}_k. \tag{14.69}$$

Combining (14.62), (14.63), and (14.64), we find the following expressions for the slope $\tilde{\sigma}_{k,j}^n$:

$$\tilde{\sigma}_{k,j}^n = \begin{cases} \Phi\left(\dfrac{\gamma_{k,j-1/2}^n}{\gamma_{k,j+1/2}^n}\right)\dfrac{1}{\Delta x}\gamma_{k,j+1/2}^n & \text{if} \quad \lambda_k > 0, \\[2ex] \Phi\left(\dfrac{\gamma_{k,j+1/2}^n}{\gamma_{k,j-1/2}^n}\right)\dfrac{1}{\Delta x}\gamma_{k,j-1/2}^n & \text{if} \quad \lambda_k < 0. \end{cases} \tag{14.70}$$

Example 14.4 Consider the linear system from Example 14.1. For this system expression (14.68) for the numerical flux $\boldsymbol{F}(\boldsymbol{u}_j^n, \boldsymbol{u}_{j+1}^n)$ reduces to

$$\boldsymbol{F}(\boldsymbol{u}_j^n, \boldsymbol{u}_{j+1}^n) = \boldsymbol{B}^+\boldsymbol{u}_j^n + \boldsymbol{B}^-\boldsymbol{u}_{j+1}^n + \frac{1}{2}a(1-c)\Delta x\big(\tilde{\sigma}_{1,j+1}^n\boldsymbol{s}_1 + \tilde{\sigma}_{2,j}^n\boldsymbol{s}_2\big),$$

with $c = a\Delta t/\Delta x$. The matrices $\boldsymbol{B}^\pm$ and the eigenvectors $\boldsymbol{s}_k$ $(k = 1, 2)$ are given in Example 14.1. Since $\lambda_1 < 0$ and $\lambda_2 > 0$, we have to select the second expression in (14.70) for $\tilde{\sigma}_{1,k}^n$ and the first one for $\tilde{\sigma}_{2,k}^n$. The coefficients $\gamma_{k,j+1/2}^n$ can be determined from (14.69), and we find

$$\gamma_{1,j+1/2}^n = \frac{1}{2}\big(u_{1,j+1}^n - u_{1,j}^n + u_{2,j+1}^n - u_{2,j}^n\big),$$

$$\gamma_{2,j+1/2}^n = \frac{1}{2}\big(u_{1,j+1}^n - u_{1,j}^n - u_{2,j+1}^n + u_{2,j}^n\big). \qquad \square$$

14.4.2 A Slope Limiter Method for Nonlinear Systems

In this section we briefly discuss the slope limiter method for the nonlinear system (14.21a). In particular, we give a generalization of the numerical flux $\boldsymbol{F}(\boldsymbol{u}_j^n, \boldsymbol{u}_{j+1}^n)$ in (14.68). The basic idea is to replace the nonlinear system (14.21a) by a suitable linearization in the neighbourhood of each cell interface $x_{j+1/2}$ and apply the slope limiter method of the previous section.

We compute the numerical flux $\boldsymbol{F}(\boldsymbol{u}_j^n, \boldsymbol{u}_{j+1}^n)$ from the local linearization

$$\frac{\partial \widetilde{\boldsymbol{u}}}{\partial t} + \widetilde{\boldsymbol{B}}_{j+1/2}^n \frac{\partial \widetilde{\boldsymbol{u}}}{\partial x} = \boldsymbol{0}, \tag{14.71}$$

which holds in the vicinity of the cell interface $x_{j+1/2}$. We take $\widetilde{\boldsymbol{B}}_{j+1/2}^n = \widetilde{\boldsymbol{B}}(\boldsymbol{u}_j^n, \boldsymbol{u}_{j+1}^n)$, where $\widetilde{\boldsymbol{B}}(\boldsymbol{u}_\ell, \boldsymbol{u}_\mathrm{r})$ is the Roe matrix satisfying the three conditions in Section 14.3. Let the eigenvalues and eigenvectors of $\widetilde{\boldsymbol{B}}_{j+1/2}^n$ be denoted by $\widetilde{\lambda}_{k,j+1/2}^n$ and $\widetilde{\boldsymbol{s}}_{k,j+1/2}^n$, respectively. Then the Roe numerical flux $\boldsymbol{F}_\mathrm{R}(\boldsymbol{u}_j^n, \boldsymbol{u}_{j+1}^n)$ is given by

$$\begin{aligned} \boldsymbol{F}_\mathrm{R}(\boldsymbol{u}_j^n, \boldsymbol{u}_{j+1}^n) &= \boldsymbol{f}(\boldsymbol{u}_j^n) + \sum_{\widetilde{\lambda}_{k,j+\frac{1}{2}}^n < 0} \gamma_{k,j+\frac{1}{2}}^n \widetilde{\lambda}_{k,j+\frac{1}{2}}^n \widetilde{\boldsymbol{s}}_{k,j+\frac{1}{2}}^n \\ &= \boldsymbol{f}(\boldsymbol{u}_{j+1}^n) - \sum_{\widetilde{\lambda}_{k,j+\frac{1}{2}}^n > 0} \gamma_{k,j+\frac{1}{2}}^n \widetilde{\lambda}_{k,j+\frac{1}{2}}^n \widetilde{\boldsymbol{s}}_{k,j+\frac{1}{2}}^n, \end{aligned} \tag{14.72}$$

where the coefficients $\gamma_{k,j+1/2}^n$ have to be computed from the decomposition

$$\boldsymbol{u}_{j+1}^n - \boldsymbol{u}_j^n = \sum_{k=1}^m \gamma_{k,j+1/2}^n \widetilde{\boldsymbol{s}}_{k,j+1/2}^n. \tag{14.73}$$

Recall that the decomposition (14.73) is always possible, since the eigenvectors $\widetilde{\boldsymbol{s}}_{k,j+1/2}^n$ are linearly independent. For the linear system (14.1a) we have $\boldsymbol{f}(\boldsymbol{u}) = \boldsymbol{B}\boldsymbol{u}$, and the expressions for the Roe numerical flux in (14.72) reduce to

$$\boldsymbol{F}_\mathrm{R}(\boldsymbol{u}_j^n, \boldsymbol{u}_{j+1}^n) = \boldsymbol{B}^+ \boldsymbol{u}_j^n + \boldsymbol{B}^- \boldsymbol{u}_{j+1}^n, \tag{14.74}$$

which is the first part of the numerical flux in (14.68). In the derivation of (14.74) we have used that $\lambda_k^\pm$ are the eigenvalues of $\boldsymbol{B}^\pm$; see Section 14.1.

A natural generalization of the numerical flux (14.68) to the nonlinear system (14.21a) is

$$\boldsymbol{F}(\boldsymbol{u}_j^n, \boldsymbol{u}_{j+1}^n) = \boldsymbol{F}_\mathrm{R}(\boldsymbol{u}_j^n, \boldsymbol{u}_{j+1}^n) + \frac{1}{2}\sum_{k=1}^m |\widetilde{\lambda}_{k,j+1/2}^n| \left(1 - \frac{\Delta t}{\Delta x}|\widetilde{\lambda}_{k,j+1/2}^n|\right) \Delta x\, \widetilde{\sigma}_{k,j_\mathrm{u}(k)}^n \widetilde{\boldsymbol{s}}_{k,j+1/2}^n, \tag{14.75}$$

where $\boldsymbol{F}_\mathrm{R}(\boldsymbol{u}_j^n, \boldsymbol{u}_{j+1}^n)$ is the Roe numerical flux given in (14.72) and $\widetilde{\sigma}_{k,j_\mathrm{u}(k)}^n$ is the limited slope in control volume $V_{j_\mathrm{u}(k)}$ of the characteristic variable $\tilde{u}_{k,j+1/2}^n$ associated with the linear system (14.71). The index $j_\mathrm{u}(k)$ defined in (14.62), with λ_k replaced by $\widetilde{\lambda}_{k,j+1/2}^n$, denotes the control volume that lies upwind (with respect to $\tilde{u}_{k,j+1/2}^n$) of the interface $x_{j+1/2}$. Analogously to (14.70), the slopes $\widetilde{\sigma}_{k,j}^n$ are given by

$$\widetilde{\sigma}_{k,j}^n = \begin{cases} \Phi\left(\dfrac{\gamma_{k,j-1/2}^n}{\gamma_{k,j+1/2}^n}\right) \dfrac{1}{\Delta x}\gamma_{k,j+1/2}^n & \text{if } \widetilde{\lambda}_{k,j+1/2}^n > 0, \\[2ex] \Phi\left(\dfrac{\gamma_{k,j+1/2}^n}{\gamma_{k,j-1/2}^n}\right) \dfrac{1}{\Delta x}\gamma_{k,j-1/2}^n & \text{if } \widetilde{\lambda}_{k,j-1/2}^n < 0, \end{cases} \tag{14.76}$$

where the limiter function Φ has to be chosen such that the resulting numerical method is TVD; see Section 13.7.

14.5 Numerical Solution of the Shallow-Water Equations

The shallow-water equations were introduced in Section 12.5 as an example of a nonlinear hyperbolic system. In this section we will give a detailed elaboration of the Godunov and Roe schemes for this system.

14.5.1 Numerical Solution by Godunov's Method

In this section we will determine the Godunov numerical flux $\boldsymbol{F}(\boldsymbol{u}_\ell, \boldsymbol{u}_\mathrm{r})$, which is defined in (14.30), for the shallow-water equations. Obviously, this flux depends on the solution of the corresponding Riemann problem, which is presented in Section 12.5. The four possible wave patterns of the Riemann problem are displayed in Figure 12.14. In all four cases we have the constant states $\boldsymbol{u}_\ell$, $\boldsymbol{u}_*$, and $\boldsymbol{u}_\mathrm{r}$ connected by shocks and/or rarefaction waves. We compute the intermediate state $\boldsymbol{u}_*$ from the following algorithm:

1. Compute $A = \varphi_\mathrm{r}/\varphi_\ell$ and $B = (u_\mathrm{r} - u_\ell)/c_\ell$.
2. Solve (12.136) for z_1.
3. Compute $z_2 = z_1/A$.
4. Compute $\varphi_* = z_1\varphi_\ell$ and $u_* = u_\ell + g(z_1)c_\ell$.

Recall that $c = \sqrt{\varphi}$. The function $g = g(z)$ in this algorithm is defined in (12.121). The conditions for which shocks and/or rarefaction waves occur are given in Table 14.1. The speeds s_k $(k = 1, 2)$ of possible shocks are given by

$$s_1 = u_\ell - c_\ell\sqrt{\frac{1}{2}(1+z_1)z_1}, \qquad s_2 = u_\mathrm{r} + c_\mathrm{r}\sqrt{\frac{1}{2}(1+z_2)z_2}. \tag{14.77}$$

The four elementary waves of the Riemann problem are detailed in (12.138) to (12.141). Combining these, we can determine the four possible wave patterns in Figure 12.14. However, for the Godunov flux we only need the similarity solution along the t axis, i.e., $\boldsymbol{u}_\mathrm{R}(0; \boldsymbol{u}_\ell, \boldsymbol{u}_\mathrm{r})$, for which we can distinguish the following four cases:

Case 1. 1- and 2-shocks.

$$\boldsymbol{u}_\mathrm{R}(0; \boldsymbol{u}_\ell, \boldsymbol{u}_\mathrm{r}) = \begin{cases} \boldsymbol{u}_\ell & \text{if} \quad s_1 > 0, \\ \boldsymbol{u}_* & \text{if} \quad s_1 < 0 < s_2, \\ \boldsymbol{u}_\mathrm{r} & \text{if} \quad s_2 < 0. \end{cases} \tag{14.78}$$

k-wave	shock	rarefaction wave
1	$B < \sqrt{A}g(1/A)$	$B \geq \sqrt{A}g(1/A)$
2	$B < g(A)$	$B \geq g(A)$

Table 14.1. *Conditions for shocks and rarefaction waves.*

Case 2. 1-rarefaction wave and 2-shock.

$$\boldsymbol{u}_{\mathrm{R}}(0;\boldsymbol{u}_\ell,\boldsymbol{u}_{\mathrm{r}}) = \begin{cases} \boldsymbol{u}_\ell & \text{if} \quad u_\ell - c_\ell > 0, \\ \varphi_1 \begin{pmatrix} 1 \\ u_1 \end{pmatrix} & \text{if} \quad u_\ell - c_\ell < 0 < u_* - c_*, \\ \boldsymbol{u}_* & \text{if} \quad u_* - c_* < 0 < s_2, \\ \boldsymbol{u}_{\mathrm{r}} & \text{if} \quad s_2 < 0, \end{cases} \tag{14.79, 14.80}$$

with $\varphi_1 := \frac{1}{9}\big(u_\ell + 2c_\ell\big)^2, \qquad u_1 := \frac{1}{3}\big(u_\ell + 2c_\ell\big).$

Case 3. 1-shock and 2-rarefaction wave.

$$\boldsymbol{u}_{\mathrm{R}}(0;\boldsymbol{u}_\ell,\boldsymbol{u}_{\mathrm{r}}) = \begin{cases} \boldsymbol{u}_\ell & \text{if} \quad s_1 > 0, \\ \boldsymbol{u}_* & \text{if} \quad s_1 < 0 < u_* + c_*, \\ \varphi_2 \begin{pmatrix} 1 \\ u_2 \end{pmatrix} & \text{if} \quad u_* + c_* < 0 < u_{\mathrm{r}} + c_{\mathrm{r}}, \\ \boldsymbol{u}_{\mathrm{r}} & \text{if} \quad u_{\mathrm{r}} + c_{\mathrm{r}} < 0, \end{cases} \tag{14.81, 14.82}$$

with $\varphi_2 := \frac{1}{9}\big(-u_{\mathrm{r}} + 2c_{\mathrm{r}}\big)^2, \qquad u_2 := \frac{1}{3}\big(u_{\mathrm{r}} - 2c_{\mathrm{r}}\big).$

Case 4. 1- and 2-rarefaction waves.

$$\boldsymbol{u}_{\mathrm{R}}(0;\boldsymbol{u}_\ell,\boldsymbol{u}_{\mathrm{r}}) = \begin{cases} \boldsymbol{u}_\ell & \text{if} \quad u_\ell - c_\ell > 0, \\ \varphi_1 \begin{pmatrix} 1 \\ u_1 \end{pmatrix} & \text{if} \quad u_\ell - c_\ell < 0 < u_* - c_*, \\ \boldsymbol{u}_* & \text{if} \quad u_* - c_* < 0 < u_* + c_*, \\ \varphi_2 \begin{pmatrix} 1 \\ u_2 \end{pmatrix} & \text{if} \quad u_* + c_* < 0 < u_{\mathrm{r}} + c_{\mathrm{r}}, \\ \boldsymbol{u}_{\mathrm{r}} & \text{if} \quad u_{\mathrm{r}} + c_{\mathrm{r}} < 0. \end{cases} \tag{14.83}$$

We can now simply compute the Godunov flux by substituting the various expressions for $\boldsymbol{u}_{\mathrm{R}}(0;\boldsymbol{u}_\ell,\boldsymbol{u}_{\mathrm{r}})$ given in (14.78) to (14.83) in the flux vector $\boldsymbol{f}(\boldsymbol{u})$ defined in (12.104).

Example 14.5 We apply the Godunov scheme to compute the numerical solution of the *dam break problem*. This is a special case of the Riemann problem for the shallow-water equations with $u_\ell = u_{\mathrm{r}} = 0$ and $h_\ell \neq h_{\mathrm{r}}$, and it describes the situation where a dam situated at $x = 0$ separates two water levels. At $t = 0$ the dam is removed and waves start to propagate from $x = 0$. In this example we choose $\varphi_\ell = 100g$ and $\varphi_{\mathrm{r}} = 50g$ with g the gravitational acceleration, so that $A = 0.5$ and $B = 0$. We can easily verify the inequalities $\sqrt{A}g(1/A) \leq B < g(A)$, implying that the solution consists of a 1-rarefaction wave and a 2-shock. This is indeed confirmed by the numerical results presented in Figure 14.2, for which we take $\Delta x = 0.8$ and $\Delta t = 10^{-2}$. We see that the 2-shock is slightly smeared due to the first order accuracy of the Godunov scheme. □

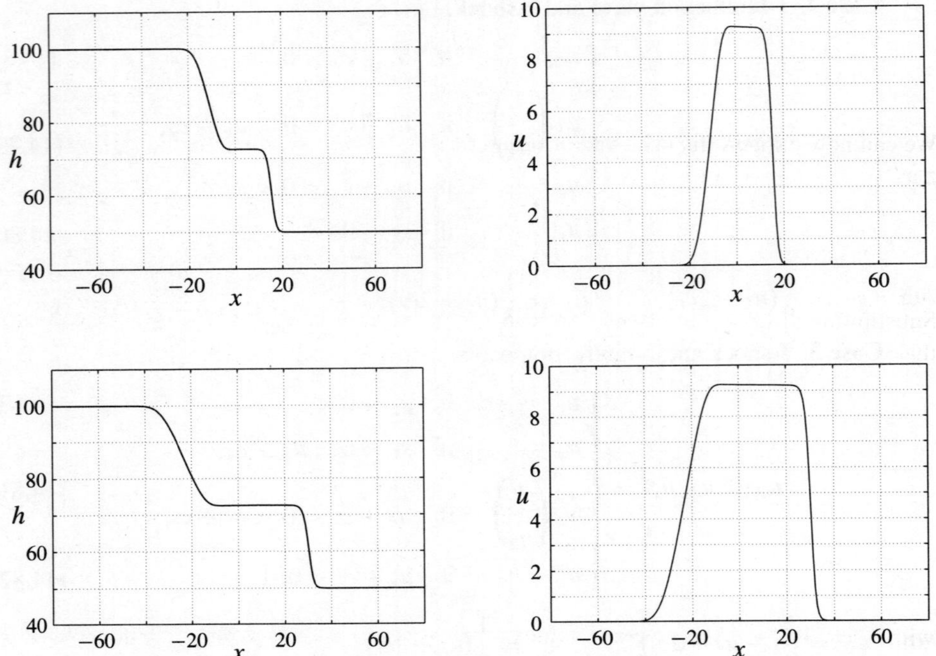

Figure 14.2. *Numerical approximation of the dam break problem at time levels* 0.5 *(top) and* 1.0 *(bottom).*

14.5.2 Numerical Solution by Roe's Method

We can also apply the Roe scheme (14.44), with numerical flux given in (14.57), to the shallow-water equations. In particular, we focus on the computation of the Roe matrix $\widetilde{\boldsymbol{B}}(\boldsymbol{u}_\ell, \boldsymbol{u}_\mathrm{r})$. We will also discuss the *entropy fix*, as mentioned above.

Consider the shallow-water equations (12.103). In order to compute the numerical flux $\boldsymbol{F}_\mathrm{R}(\boldsymbol{u}_\ell, \boldsymbol{u}_\mathrm{r})$, we first have to compute the Roe matrix $\widetilde{\boldsymbol{B}}(\boldsymbol{u}_\ell, \boldsymbol{u}_\mathrm{r})$ from the relation

$$\boldsymbol{f}(\boldsymbol{u}_\mathrm{r}) - \boldsymbol{f}(\boldsymbol{u}_\ell) = \widetilde{\boldsymbol{B}}(\boldsymbol{u}_\ell, \boldsymbol{u}_\mathrm{r})(\boldsymbol{u}_\mathrm{r} - \boldsymbol{u}_\ell). \tag{14.84}$$

The basic idea is to introduce a parameter vector $\boldsymbol{z}$ such that the differences $\boldsymbol{u}_\mathrm{r} - \boldsymbol{u}_\ell$ and $\boldsymbol{f}(\boldsymbol{u}_\mathrm{r}) - \boldsymbol{f}(\boldsymbol{u}_\ell)$ in (14.84) can be expressed in terms of the difference $\boldsymbol{z}_\mathrm{r} - \boldsymbol{z}_\ell$. In particular, we try to find relations of the form

$$\boldsymbol{u}_\mathrm{r} - \boldsymbol{u}_\ell = \widetilde{\boldsymbol{P}}(\boldsymbol{u}_\ell, \boldsymbol{u}_\mathrm{r})(\boldsymbol{z}_\mathrm{r} - \boldsymbol{z}_\ell), \tag{14.85a}$$

$$\boldsymbol{f}(\boldsymbol{u}_\mathrm{r}) - \boldsymbol{f}(\boldsymbol{u}_\ell) = \widetilde{\boldsymbol{Q}}(\boldsymbol{u}_\ell, \boldsymbol{u}_\mathrm{r})(\boldsymbol{z}_\mathrm{r} - \boldsymbol{z}_\ell), \tag{14.85b}$$

where $\widetilde{\boldsymbol{P}}(\boldsymbol{u}_\ell, \boldsymbol{u}_\mathrm{r})$ and $\widetilde{\boldsymbol{Q}}(\boldsymbol{u}_\ell, \boldsymbol{u}_\mathrm{r})$ are 2×2 matrices depending on the state vectors $\boldsymbol{u}_\ell$ and $\boldsymbol{u}_\mathrm{r}$. Combining (14.84) and (14.85) and assuming that $\widetilde{\boldsymbol{P}}(\boldsymbol{u}_\ell, \boldsymbol{u}_\mathrm{r})$ is nonsingular, we readily see that the Roe matrix is given by

$$\widetilde{\boldsymbol{B}}(\boldsymbol{u}_\ell, \boldsymbol{u}_\mathrm{r}) = \widetilde{\boldsymbol{Q}}(\boldsymbol{u}_\ell, \boldsymbol{u}_\mathrm{r})\widetilde{\boldsymbol{P}}^{-1}(\boldsymbol{u}_\ell, \boldsymbol{u}_\mathrm{r}). \tag{14.86}$$

In the following we will suppress the arguments $\boldsymbol{u}_\ell$ and $\boldsymbol{u}_\mathrm{r}$.

A suitable choice for the parameter vector z appears to be

$$z := \frac{1}{\sqrt{\varphi}}\boldsymbol{u} \quad \text{or} \quad \begin{pmatrix} z_1 \\ z_2 \end{pmatrix} = \begin{pmatrix} \sqrt{\varphi} \\ \sqrt{\varphi}u \end{pmatrix}. \tag{14.87}$$

We can now express the state vector $\boldsymbol{u}$ and the flux vector $\boldsymbol{f}(\boldsymbol{u})$ from (12.104) in terms of z as

$$\boldsymbol{u} = \begin{pmatrix} z_1^2 \\ z_1 z_2 \end{pmatrix}, \qquad \boldsymbol{f}(\boldsymbol{u}) = \begin{pmatrix} z_1 z_2 \\ z_2^2 + \dfrac{1}{2} z_1^4 \end{pmatrix}. \tag{14.88}$$

Substituting (14.88) into (14.85), we can determine the matrices $\widetilde{\boldsymbol{P}}$ and $\widetilde{\boldsymbol{Q}}$. We can verify that

$$\varphi_{\mathrm{r}} - \varphi_\ell = 2\bar{z}_1(z_{1,\mathrm{r}} - z_{1,\ell}), \tag{14.89a}$$

$$\varphi_{\mathrm{r}} u_{\mathrm{r}} - \varphi_\ell u_\ell = \bar{z}_2(z_{1,\mathrm{r}} - z_{1,\ell}) + \bar{z}_1(z_{2,\mathrm{r}} - z_{2,\ell}), \tag{14.89b}$$

$$\varphi_{\mathrm{r}} u_{\mathrm{r}}^2 - \varphi_\ell u_\ell^2 = 2\bar{z}_2(z_{2,\mathrm{r}} - z_{2,\ell}), \tag{14.89c}$$

$$\frac{1}{2}(\varphi_{\mathrm{r}}^2 - \varphi_\ell^2) = 2\,\bar{\varphi}\,\bar{z}_1(z_{1,\mathrm{r}} - z_{1,\ell}), \tag{14.89d}$$

with $\bar{v} := \frac{1}{2}(v_\ell + v_{\mathrm{r}})$ the average value of the variable v, and from these relations we obtain

$$\widetilde{\boldsymbol{P}} = \begin{pmatrix} 2\bar{z}_1 & 0 \\ \bar{z}_2 & \bar{z}_1 \end{pmatrix}, \qquad \widetilde{\boldsymbol{Q}} = \begin{pmatrix} \bar{z}_2 & \bar{z}_1 \\ 2\bar{\varphi}\bar{z}_1 & 2\bar{z}_2 \end{pmatrix}. \tag{14.90}$$

The matrix $\widetilde{\boldsymbol{P}}$ is obviously nonsingular since $\det(\widetilde{\boldsymbol{P}}) = 2\bar{z}_1^2 \neq 0$. From (14.86) and (14.90) we can determine the Roe matrix in terms of the average $\bar{z}$ of the parameter vector:

$$\widetilde{\boldsymbol{B}} = \begin{pmatrix} 0 & 1 \\ \bar{\varphi} - \left(\dfrac{\bar{z}_2}{\bar{z}_1}\right)^2 & \dfrac{2\bar{z}_2}{\bar{z}_1} \end{pmatrix}. \tag{14.91a}$$

In terms of the state vector $\boldsymbol{u}$, the Roe matrix reads

$$\widetilde{\boldsymbol{B}} = \begin{pmatrix} 0 & 1 \\ \bar{\varphi} - \tilde{u}^2 & 2\tilde{u} \end{pmatrix}, \tag{14.91b}$$

where we have introduced the weighted average velocity $\tilde{u}$ defined by

$$\tilde{u} := \frac{\left(u\sqrt{\varphi}\right)_\ell + \left(u\sqrt{\varphi}\right)_{\mathrm{r}}}{\sqrt{\varphi_\ell} + \sqrt{\varphi_{\mathrm{r}}}} = \frac{\bar{z}_2}{\bar{z}_1}. \tag{14.92}$$

Note that the Roe matrix $\widetilde{\boldsymbol{B}}$ in (14.92) is precisely the Jacobi matrix $\boldsymbol{B}$ in (12.105), with the geopotential φ replaced by its average value $\bar{\varphi}$ and the flow velocity u by the weighted average $\tilde{u}$ defined in (14.92). Consequently, its eigenvalues $\widetilde{\lambda}_k$ and eigenvectors $\widetilde{\boldsymbol{s}}_k$ $(k = 1, 2)$

are given by

$$\widetilde{\lambda}_1(\boldsymbol{u}_\ell, \boldsymbol{u}_\mathrm{r}) = \tilde{u} - \tilde{c}, \quad \widetilde{\lambda}_2(\boldsymbol{u}_\ell, \boldsymbol{u}_\mathrm{r}) = \tilde{u} + \tilde{c}, \qquad \tilde{c} := \sqrt{\bar{\varphi}}, \tag{14.93a}$$

$$\widetilde{\boldsymbol{s}}_1(\boldsymbol{u}_\ell, \boldsymbol{u}_\mathrm{r}) = \begin{pmatrix} 1 \\ \tilde{u} - \tilde{c} \end{pmatrix}, \qquad \widetilde{\boldsymbol{s}}_2(\boldsymbol{u}_\ell, \boldsymbol{u}_\mathrm{r}) = \begin{pmatrix} 1 \\ \tilde{u} + \tilde{c} \end{pmatrix}, \tag{14.93b}$$

and thus the third property of the Roe matrix is automatically satisfied. Moreover, if $\boldsymbol{u}_\ell = \boldsymbol{u}_\mathrm{r} = \boldsymbol{u}$, then $\bar{\varphi} = \varphi$ and $\tilde{u} = u$, and, consequently, the second property of the Roe matrix is also trivially fulfilled.

We now compute the Roe numerical flux $\boldsymbol{F}_\mathrm{R}(\boldsymbol{u}_\ell, \boldsymbol{u}_\mathrm{r})$ as follows. From (14.54) and (14.93) we can compute the coefficients γ_k $(k = 1, 2)$:

$$\gamma_1 = \frac{1}{2\tilde{c}}\big((\tilde{u} + \tilde{c})(\varphi_\mathrm{r} - \varphi_\ell) - (\varphi_\mathrm{r} u_\mathrm{r} - \varphi_\ell u_\ell)\big), \tag{14.94a}$$

$$\gamma_2 = \frac{1}{2\tilde{c}}\big(- (\tilde{u} - \tilde{c})(\varphi_\mathrm{r} - \varphi_\ell) + (\varphi_\mathrm{r} u_\mathrm{r} - \varphi_\ell u_\ell)\big). \tag{14.94b}$$

In the evaluation of the expressions for the numerical flux in (14.57) we have to distinguish three cases: $\widetilde{\lambda}_1 > 0$, $\widetilde{\lambda}_1 \le 0 \le \widetilde{\lambda}_2$, and $\widetilde{\lambda}_2 < 0$. In the case $\widetilde{\lambda}_1 > 0$ the first expression in (14.57) is most appropriate, since the sum in it is void. We then have $\boldsymbol{F}_\mathrm{R}(\boldsymbol{u}_\ell, \boldsymbol{u}_\mathrm{r}) = \boldsymbol{f}(\boldsymbol{u}_\ell)$. Similarly, when $\widetilde{\lambda}_2 < 0$, we have to choose the second expression in (14.57), and we obtain $\boldsymbol{F}_\mathrm{R}(\boldsymbol{u}_\ell, \boldsymbol{u}_\mathrm{r}) = \boldsymbol{f}(\boldsymbol{u}_\mathrm{r})$. Most interesting is the second case, for which we can use either expression in (14.57); we choose the first one. Summarizing, we obtain for the Roe numerical flux

$$\boldsymbol{F}_\mathrm{R}(\boldsymbol{u}_\ell, \boldsymbol{u}_\mathrm{r}) = \begin{cases} \boldsymbol{f}(\boldsymbol{u}_\ell) & \text{if } \ \tilde{u} > \tilde{c}, \\ \boldsymbol{f}(\boldsymbol{u}_\ell) + \gamma_1 \widetilde{\lambda}_1 \widetilde{\boldsymbol{s}}_1 & \text{if } \ |\tilde{u}| \le \tilde{c}, \\ \boldsymbol{f}(\boldsymbol{u}_\mathrm{r}) & \text{if } \ \tilde{u} < -\tilde{c}, \end{cases} \tag{14.95}$$

with γ_1, $\widetilde{\lambda}_1$, and $\widetilde{\boldsymbol{s}}_1$ defined in (14.94) and (14.93).

We conclude this section with an outline of the Harten–Hyman entropy fix for (12.103); for more details see, e.g., [60, 87, 158]. Consider the particular solution of the Riemann problem (14.45) depicted in Figure 14.3. In this case the solution consists of three constant states $\boldsymbol{u}_\ell$, $\boldsymbol{u}_*$, and $\boldsymbol{u}_\mathrm{r}$ connected by a 1-rarefaction wave and a 2-shock. Moreover, the 1-rarefaction wave is *critical*, which means that the corresponding eigenvalue $\lambda_1 = u - c$ changes sign across the wave. In particular, we have

$$\lambda_{1,\ell} < 0 < \lambda_{1,*}, \tag{14.96}$$

where $\lambda_{1,\ell}$ and $\lambda_{1,*}$ denote the value of λ_1 at the left and right ends of the 1-rarefaction wave, respectively. Thus $\lambda_{1,\ell}$ is the speed of the head of the rarefaction wave and $\lambda_{1,*}$ is the speed of the tail.

The computation of the Roe numerical flux $\boldsymbol{F}_\mathrm{R}(\boldsymbol{u}_\ell, \boldsymbol{u}_\mathrm{r})$ is based on the linearized Riemann problem (14.49). Its solution consists of the constant states $\boldsymbol{u}_\ell$, $\boldsymbol{u}_*$, and $\boldsymbol{u}_\mathrm{r}$, now connected by two shocks. The 1-rarefaction wave is thus approximated by a 1-shock propagating with speed $\widetilde{\lambda}_1$, which is obviously not correct. Suppose we want to compute the Roe numerical flux using the first expression in (14.57). The sum in this expression

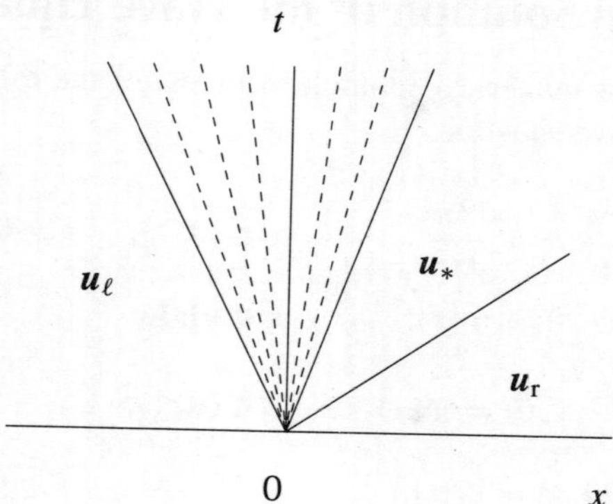

Figure 14.3. *Schematic representation of a critical* 1*-rarefaction wave for the shallow-water equations.*

accounts for the effect of jumps in the solution of (14.49) propagating with negative speeds $\widetilde{\lambda}_k$. Straightforward application of this formula would lead to an entropy-violating shock as an approximation of the 1-rarefaction wave [158]. In order to prevent this, we apply the Harten–Hyman entropy fix. The idea is to split the jump $\boldsymbol{u}_* - \boldsymbol{u}_\ell$ travelling at speed $\widetilde{\lambda}_1$ into two jumps $\boldsymbol{u}_\mathrm{c} - \boldsymbol{u}_\ell$ and $\boldsymbol{u}_* - \boldsymbol{u}_\mathrm{c}$ travelling at speeds $\lambda_{1,\ell} < 0$ and $\lambda_{1,*} > 0$, respectively. The intermediate state $\boldsymbol{u}_\mathrm{c}$ is referred to as the *critical state* and is as yet unknown. Note that the single jump $\boldsymbol{u}_* - \boldsymbol{u}_\ell$ can be written as

$$\boldsymbol{u}_* - \boldsymbol{u}_\ell = \gamma_1 \widetilde{\boldsymbol{s}}_1. \tag{14.97}$$

Integrating the linear conservation law (14.49a) over a sufficiently large interval, we can derive the following relation between the three jumps:

$$\widetilde{\lambda}_1(\boldsymbol{u}_* - \boldsymbol{u}_\ell) = \lambda_{1,\ell}(\boldsymbol{u}_\mathrm{c} - \boldsymbol{u}_\ell) + \lambda_{1,*}(\boldsymbol{u}_* - \boldsymbol{u}_\mathrm{c}). \tag{14.98}$$

From (14.97) and (14.98) we obtain

$$\boldsymbol{u}_\mathrm{c} - \boldsymbol{u}_\ell = \frac{\lambda_{1,*} - \widetilde{\lambda}_1}{\lambda_{1,*} - \lambda_{1,\ell}} \gamma_1 \widetilde{\boldsymbol{s}}_1. \tag{14.99}$$

In the first expression for the Roe numerical flux in (14.57) we only take into account the wave due to the jump $\boldsymbol{u}_\mathrm{c} - \boldsymbol{u}_\ell$ propagating with negative speed $\lambda_{1,\ell}$. Consequently, the Roe flux reduces to

$$\boldsymbol{F}_\mathrm{R}(\boldsymbol{u}_\ell, \boldsymbol{u}_\mathrm{r}) = \boldsymbol{f}(\boldsymbol{u}_\ell) + \frac{\lambda_{1,*} - \widetilde{\lambda}_1}{\lambda_{1,*} - \lambda_{1,\ell}} \gamma_1 \lambda_{1,\ell} \widetilde{\boldsymbol{s}}_1 \tag{14.100a}$$

or, in a more condensed form,

$$\boldsymbol{F}_\mathrm{R}(\boldsymbol{u}_\ell, \boldsymbol{u}_\mathrm{r}) = \boldsymbol{f}(\boldsymbol{u}_\ell) + \gamma_1 \bar{\lambda}_1 \widetilde{\boldsymbol{s}}_1, \quad \bar{\lambda}_1 := \frac{\lambda_{1,*} - \widetilde{\lambda}_1}{\lambda_{1,*} - \lambda_{1,\ell}} \lambda_{1,\ell}. \tag{14.100b}$$

To conclude we need an estimate for the intermediate state $\boldsymbol{u}_*$ in order to compute $\lambda_{1,*}$. One possibility is to compute $\boldsymbol{u}_*$ from (14.97). Other possibilities are to use approximate Riemann solvers; see, e.g., [159] for more details. The case of a sonic 2-rarefaction wave can be handled in a similar manner; we will not discuss this here.

14.6 Numerical Solution of the Wave Equation

In this section we discuss numerical solution methods for the following initial boundary value problem for the wave equation:

$$\frac{\partial^2 u}{\partial t^2} = a^2 \frac{\partial^2 u}{\partial x^2}, \qquad x \in (0,1), \quad t > 0, \tag{14.101a}$$

$$u(x,0) = v(x), \qquad x \in (0,1), \tag{14.101b}$$

$$\frac{\partial u}{\partial t}(x,0) = w(x), \qquad x \in (0,1), \tag{14.101c}$$

$$u(0,t) = g_\ell(t), \qquad t \geq 0, \tag{14.101d}$$

$$u(1,t) = g_r(t), \qquad t \geq 0. \tag{14.101e}$$

As we have seen in Section 12.6, we can reduce (14.101a) to a first order system. As such it can be dealt with by the methods we have encountered so far in this chapter. However, we will consider two alternative approaches here. The first one is based on direct discretisation of the wave equation by central differences and is discussed in Section 14.6.1. The second alternative employs the leapfrog scheme for the equivalent first order system, but on a staggered grid, and is the topic of Section 14.6.2.

14.6.1 Difference Methods Based on the Scalar Form

The first method we discuss is based on using central differences for the second order time and space derivatives. Hence let us assume that we have an equispaced grid covering $\Omega := [0,1] \times [0,\infty)$ as follows:

$$x_j := j\Delta x \quad (j = 0, 1, 2, \ldots, M+1), \qquad t^n := n\Delta t \quad (n = 0, 1, 2, \ldots),$$

with $\Delta x = 1/(M+1)$ the (spatial) grid size and $\Delta t > 0$ the time step. Let u_j^n denote the numerical approximation of $u(x_j, t^n)$. An obvious way to combine the central difference approximations of both derivatives is to centre them around the grid point (x_j, t^n), leading to the scheme

$$\frac{1}{\Delta t^2}\left(u_j^{n+1} - 2u_j^n + u_j^{n-1}\right) = \frac{a^2}{\Delta x^2}\left(u_{j+1}^n - 2u_j^n + u_{j-1}^n\right) \tag{14.102a}$$

or, equivalently,

$$u_j^{n+1} = 2(1-c^2)u_j^n + c^2\left(u_{j+1}^n + u_{j-1}^n\right) - u_j^{n-1}, \tag{14.102b}$$

where the parameter c is defined by

$$c := \frac{a\Delta t}{\Delta x} > 0. \tag{14.103}$$

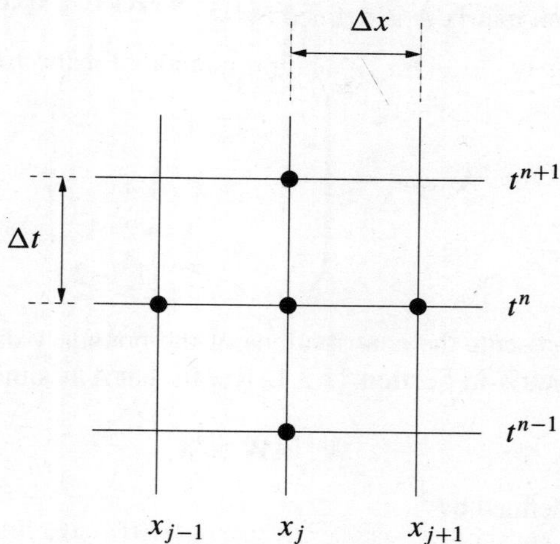

Figure 14.4. *Stencil for the centred scheme* (14.102b).

The stencil of scheme (14.102b) is shown in Figure 14.4. In order to compute the numerical values u_j^n $(n = 2, 3, \ldots)$, we need a second "initial" value u_j^1, besides u_j^0, for which we use (14.101c). To that end we expand $u(x_j, t^1)$ in a Taylor series to obtain

$$u(x_j, t^1) = u(x_j, 0) + \Delta t \frac{\partial u}{\partial t}(x_j, 0) + \frac{1}{2}\Delta t^2 \frac{\partial^2 u}{\partial t^2}(x_j, 0) + \mathcal{O}(\Delta t^3). \tag{14.104}$$

Substituting $\frac{\partial^2}{\partial t^2} u = a^2 \frac{\partial^2}{\partial x^2} u$, approximating the latter derivative by central differences, and applying the initial conditions, we obtain

$$\begin{aligned} u(x_j, t^1) &= v(x_j) + \Delta t w(x_j) + \frac{1}{2}(a\Delta t)^2 v''(x_j) + \mathcal{O}(\Delta t^3) \\ &= v(x_j) + \Delta t w(x_j) + \frac{1}{2}c^2\big(v(x_{j+1}) - 2v(x_j) + v(x_{j-1})\big) \\ &\qquad + \mathcal{O}(\Delta t^3) + \mathcal{O}((\Delta t \Delta x)^2). \end{aligned} \tag{14.105}$$

From (14.105) we easily deduce the scheme

$$u_j^1 = (1 - c^2)v_j + \frac{1}{2}c^2\big(v_{j+1} + v_{j-1}\big) + \Delta t w_j \tag{14.106}$$

for u_j^1, where we have introduced the obvious notation $v_j = v(x_j)$, etc.

Writing the unknown values u_j^n $(j = 1, 2, \ldots, M)$ as a vector $\mathbf{u}^n$, we obtain from (14.102b) the (*three-term*) recursion

$$\mathbf{u}^{n+1} - (2\mathbf{I} + (a\Delta t)^2\mathbf{A})\mathbf{u}^n + \mathbf{u}^{n-1} = (a\Delta t)^2\mathbf{f}^n, \tag{14.107}$$

where the discretisation matrix $\mathbf{A}$ is defined by

$$\mathbf{A} := \frac{1}{\Delta x^2}\begin{pmatrix} -2 & 1 & & & \\ 1 & -2 & 1 & & \\ & \ddots & \ddots & \ddots & \\ & & 1 & -2 & 1 \\ & & & 1 & -2 \end{pmatrix}$$

and the vector $\mathbf{f}^n$ represents the contributions of the boundary data. We encountered the same discretisation matrix in Section 11.2.1. Recall that $\mathbf{A}$ is similar to a diagonal matrix; i.e.,

$$\mathbf{W}^{-1}\mathbf{A}\mathbf{W} = \mathbf{\Lambda}, \tag{14.108a}$$

where $\mathbf{\Lambda}$ and $\mathbf{W}$ are defined by

$$\mathbf{\Lambda} := \text{diag}(\lambda_1, \lambda_2, \ldots, \lambda_M), \qquad \mathbf{W} := (\mathbf{w}_1, \mathbf{w}_2, \ldots, \mathbf{w}_M), \tag{14.108b}$$

with λ_k and $\mathbf{w}_k$ the eigenvalues and corresponding eigenvectors of $\mathbf{A}$. The eigenvalues λ_k are given by

$$\lambda_k = \frac{-4}{\Delta x^2}\sin^2\left(\frac{\pi}{2}x_k\right) \qquad (k = 1, 2, \ldots, M) \tag{14.109}$$

and will be needed in the following stability analysis.

To investigate stability we perform the variable transformation

$$\tilde{\mathbf{u}}^n := \mathbf{W}^{-1}\mathbf{u}^n. \tag{14.110}$$

Inserting (14.110) into (14.107), we obtain the decoupled system

$$\tilde{\mathbf{u}}^{n+1} - \left(2\mathbf{I} + (a\Delta t)^2\mathbf{\Lambda}\right)\tilde{\mathbf{u}}^n + \tilde{\mathbf{u}}^{n-1} = (a\Delta t)^2\tilde{\mathbf{f}}^n, \tag{14.111}$$

with $\tilde{\mathbf{f}}^n := \mathbf{W}^{-1}\mathbf{f}^n$. The recursion in (14.111) actually consists of M decoupled recursions; i.e.,

$$\tilde{u}_k^{n+1} - \left(2 + (a\Delta t)^2\lambda_k\right)\tilde{u}_k^n + \tilde{u}_k^{n-1} = (a\Delta t)^2\tilde{f}_k^n \qquad (k = 1, 2, \ldots, M). \tag{14.112}$$

The homogeneous recursion determines the stability of the scheme (14.102a) and has a solution of the form

$$\tilde{u}_k^n = A_k\mu_{k,1}^n + B_k\mu_{k,2}^n, \qquad A_k, B_k \in \mathbb{R}, \tag{14.113a}$$

where $\mu_{k,1}$ and $\mu_{k,2}$ are the roots of the characteristic polynomial

$$p_k(\mu) := \mu^2 - \left(2 + (a\Delta t)^2\lambda_k\right)\mu + 1. \tag{14.113b}$$

Clearly the method is stable if $|\mu_{k,1}|,\ |\mu_{k,2}| \le 1$. The stability condition is given in the next property.

Property 14.6. *The centred scheme* (14.102a) *is stable if* $c \le 1$.

Proof. We note that the product $\mu_{k,1}\mu_{k,2}$ must be equal to one. For ease of notation we introduce the variable $q_k := 2 + (a\Delta t)^2\lambda_k = 2 - 4c^2\sin^2(\frac{\pi}{2}x_k)$. Obviously, the solutions of (14.113b) are given by

$$\mu_{k,l} = \frac{1}{2}q_k \pm \sqrt{\left(\frac{1}{2}q_k\right)^2 - 1} \qquad (l = 1, 2).$$

Hence if $|q_k| \leq 2$, we have two complex conjugate roots with a product of one. This implies that each root has modulus one. If, on the other hand, $|q_k| > 2$, we always have a real root of modulus larger than one. It is simple to check that $c \leq 1$ implies that $|q_k| \leq 2$. □

Note that the result of Property 14.6 also follows from the CFL condition.

As usual, the local discretisation error is the residual after substituting the exact solution of (14.101a) in (14.102a), i.e.,

$$\begin{aligned} d_j^n :=& \frac{1}{\Delta t^2}\big(u(x_j, t^{n+1}) - 2u(x_j, t^n) + u(x_j, t^{n-1})\big) \\ &- \frac{a^2}{\Delta x^2}\big(u(x_{j+1}, t^n) - 2u(x_j, t^n) + u(x_{j-1}, t^n)\big). \end{aligned} \tag{14.114}$$

Expanding all function values in a Taylor series around (x_j, t^n) and assuming that $u(x, t)$ is sufficiently smooth, we find

$$\begin{aligned} d_j^n &= \frac{1}{12}\left(\Delta t^2\frac{\partial^4 u}{\partial t^4}(x_j, t^n) - a^2\Delta x^2\frac{\partial^4 u}{\partial x^4}(x_j, t^n)\right) + \mathcal{O}(\Delta t^4) + \mathcal{O}(\Delta x^4) \\ &= -\frac{1}{12}a^2(1 - c^2)\Delta x^2\frac{\partial^4 u}{\partial x^4}(x_j, t^n) + \mathcal{O}(\Delta x^4), \end{aligned} \tag{14.115}$$

where we have used that $\frac{\partial^4}{\partial t^4}u = a^4\frac{\partial^4}{\partial x^4}u$. Clearly, scheme (14.102a) is second order accurate in Δt and Δx.

Assuming that the boundary conditions are exact, we can derive the following recursion for the global discretisation error vector $\mathbf{e}^n$:

$$\mathbf{e}^{n+1} - \big(2\mathbf{I} + (a\Delta t)^2\mathbf{A}\big)\mathbf{e}^n + \mathbf{e}^{n-1} = \Delta t^2\mathbf{d}^n. \tag{14.116}$$

Analogously to the derivation of (14.112), we find the componentwise error recursion (in an obvious notation)

$$\tilde{e}_k^{n+1} - \big(2 + (a\Delta t)^2\lambda_k\big)\tilde{e}_k^n + \tilde{e}_k^{n-1} = \Delta t^2\tilde{d}_k^n \qquad (k = 1, 2, \ldots, M). \tag{14.117}$$

Using (E.7), we find

$$|\tilde{e}_k^n| \leq \Delta t^2\frac{\sum_{l=1}^{n-1}|\mu_{k,2}^{n-l} - \mu_{k,1}^{n-l}|}{|\mu_{k,2} - \mu_{k,1}|}\max_{1\leq l\leq n-1}|\tilde{d}_k^l|. \tag{14.118}$$

Let us now assume that $0 \leq t \leq T$ for some $T > 0$. Since $|\mu_{k,1}| = |\mu_{k,2}| = 1$, we find for $n\Delta t \leq T$ that

$$\sum_{l=1}^{n-1}|\mu_{k,2}^{n-l} - \mu_{k,1}^{n-l}| \leq 2(n-1) \leq 2T/\Delta t.$$

For the denominator in (14.118) we have

$$|\mu_{k,2} - \mu_{k,1}| = 4c \sin\left(\frac{\pi}{2} x_k\right) \sqrt{1 - c^2 \sin^2\left(\frac{\pi}{2} x_k\right)}.$$

It is simple to check that its minimum value is $\mathcal{O}(\Delta t)$. Hence we obtain

$$|\tilde{e}_k^n| \leq C \max_{1 \leq l \leq n-1} |\tilde{d}_k^l|$$

for some $C > 0$. From this inequality and the relation $\Delta t = \mathcal{O}(\Delta x)$ we eventually obtain the infinity norm estimate

$$\|\mathbf{e}\|_\infty = \mathcal{O}(\Delta x^{3/2}). \tag{14.119}$$

Of course, one can use other methods to compute solutions of the wave equation. In particular, in view of the linearity, implicit methods may be attractive to avoid time step constraints. As an example, consider the central difference spatial discretisation at three time levels (a kind of generalisation of Crank–Nicolson). This leads to the scheme

$$\begin{aligned} \frac{1}{\Delta t^2}\left(u_j^{n+1} - 2u_j^n + u_j^{n-1}\right) &= \frac{a^2}{4}\left(\frac{1}{\Delta x^2}\left(u_{j+1}^{n-1} - 2u_j^{n-1} + u_{j-1}^{n-1}\right)\right. \\ &\quad \left. + \frac{2}{\Delta x^2}\left(u_{j+1}^n - 2u_j^n + u_{j-1}^n\right) + \frac{1}{\Delta x^2}\left(u_{j+1}^{n+1} - 2u_j^{n+1} + u_{j-1}^{n+1}\right)\right). \end{aligned} \tag{14.120}$$

This results in the matrix recursion

$$\left(\mathbf{I} - \frac{1}{4}(a\Delta t)^2\mathbf{A}\right)\mathbf{u}^{n+1} - 2\left(\mathbf{I} + \frac{1}{4}(a\Delta t)^2\mathbf{A}\right)\mathbf{u}^n + \left(\mathbf{I} - \frac{1}{4}(a\Delta t)^2\mathbf{A}\right)\mathbf{u}^{n-1} = (a\Delta t)^2\mathbf{f}^*, \tag{14.121}$$

for $\mathbf{u}$, where $\mathbf{f}^*$ is the vector containing boundary data at all three time levels involved. After transforming this to diagonal form, as we did in (14.111), we find out that in this case the basis solutions of the thus found decoupled recursions are determined by the roots of the characteristic polynomial

$$p_k(\mu) := \mu^2 - 2\frac{1 + \frac{1}{4}(a\Delta t)^2\lambda_k}{1 - \frac{1}{4}(a\Delta t)^2\lambda_k} + 1. \tag{14.122}$$

Clearly, we have two complex conjugate roots $\mu_{k,1}$ and $\mu_{k,2}$ with a product of one. Hence $|\mu_{k,1}| = |\mu_{k,2}| = 1$. Since $\mu_{k,1} \neq \mu_{k,2}$, i.e., the eigenvalues are simple, stability is guaranteed.

14.6.2 The Staggered Leapfrog Scheme

We can rewrite equation (14.101a) as a system using the new variables

$$p := \frac{1}{a}\frac{\partial u}{\partial t}, \qquad q := \frac{\partial u}{\partial x};$$

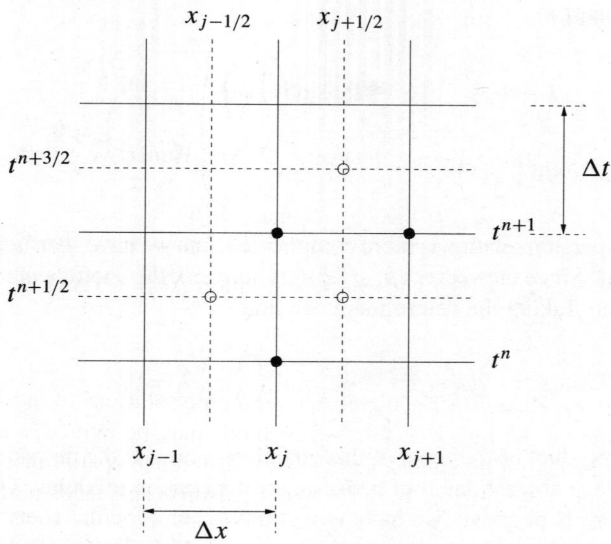

Figure 14.5. *Patch of the staggered grid and stencil for the leapfrog scheme* (14.124).

cf. Section 12.6. This way, we obtain the first order system

$$\frac{\partial p}{\partial t} - a\frac{\partial q}{\partial x} = 0, \tag{14.123a}$$

$$\frac{\partial q}{\partial t} - a\frac{\partial p}{\partial x} = 0. \tag{14.123b}$$

As we see from (14.123), we have two coupled equations of identical form, where the time derivative of one variable is coupled to the space derivative of the other. This makes the leapfrog scheme in combination with a staggered grid an attractive choice. Note that we introduced this type of grid in Section 9.7. A patch of such a grid is shown in Figure 14.5. Typically, we take the p values at the grid points (x_j, t^n), whereas we take the q values at the shifted points $(x_{j+1/2}, t^{n+1/2})$. The resulting leapfrog scheme then reads

$$p_j^{n+1} = p_j^n + c\big(q_{j+1/2}^{n+1/2} - q_{j-1/2}^{n+1/2}\big), \tag{14.124a}$$

$$q_{j+1/2}^{n+3/2} = q_{j+1/2}^{n+1/2} + c\big(p_{j+1}^{n+1} - p_j^{n+1}\big); \tag{14.124b}$$

see (13.70b). We can interpret this scheme as a one-step method for the vector $(p_j^n, q_{j+1/2}^{n+1/2})^T$.

Next we will investigate the stability of (14.124). Let us now assume that we have periodic boundary conditions. This enables us to carry out a Fourier mode analysis. We find, upon substituting the Fourier modes

$$p_j^n = \hat{p}\lambda^n \,\mathrm{e}^{\mathrm{i}j\varphi}, \qquad q_{j+1/2}^{n+1/2} = \hat{q}\lambda^{n+1/2}\,\mathrm{e}^{\mathrm{i}(j+1/2)\varphi},$$

the dispersion relation

$$\begin{pmatrix} \lambda - 1 & -2ic \sin\left(\frac{1}{2}\varphi\right) \\ -2ic \sin\left(\frac{1}{2}\varphi\right)\lambda & \lambda - 1 \end{pmatrix} \begin{pmatrix} \hat{p} \\ \sqrt{\lambda}\hat{q} \end{pmatrix} = \mathbf{0}. \tag{14.125}$$

Note that this dispersion relation is more complicated than we have seen before. It is implicit but also vectorial. Since the vector $(\hat{p}, \sqrt{\lambda}\hat{q})^T$ is nonzero, the coefficient matrix in (14.125) has to be singular. Taking the determinant, we find

$$(\lambda - 1)^2 + 4c^2 \sin^2\left(\frac{1}{2}\varphi\right)\lambda = 0. \tag{14.126}$$

Apparently, the product of the roots of this equation is one. If the discriminant of (14.126) is positive, we have at least one root being larger than one in modulus. On the other hand, if the discriminant is negative, we have two complex, unimodular roots. It is easy to see that this is the case if $c \leq 1$; cf. the proof of Property 14.6. For a stability analysis in the case of nonperiodic boundaries we refer to the next section.

Once we have determined $p = \frac{1}{a}\frac{\partial u}{\partial t}$ and $q = \frac{\partial u}{\partial x}$ numerically, we can formally integrate either one to obtain u. There is, however, a more direct way to find u. Indeed, let r be an auxiliary variable such that

$$\frac{\partial u}{\partial t} - a\frac{\partial r}{\partial x} = 0, \tag{14.127a}$$

$$\frac{\partial r}{\partial t} - a\frac{\partial u}{\partial x} = 0. \tag{14.127b}$$

One can simply check that u satisfies the wave equation (14.101a). In other words, the leapfrog method can also be applied to system (14.127) directly. As for the initial conditions, we directly find from (14.101b) the value of u_j^0. In order to find the initial value $r_{j+1/2}^{1/2}$, we have to use system (14.127) and the initial conditions (14.101b) and (14.101c). First, integrating (14.127a) and applying initial condition (14.101c), we find

$$r(x_{j+1/2}, 0) = r(0,0) + \frac{1}{a}\int_0^{x_{j+1/2}} w(x)\mathrm{d}x. \tag{14.128}$$

If $r(x,t)$ is a solution of (14.127), then so is $r(x,t) + C$ for any $C \in \mathbb{R}$, so we can choose $r(0,0) = 0$. Second, from (14.127b) and (14.101b) we obtain

$$\frac{\partial r}{\partial t}(x_{j+1/2}, 0) = av'(x_{j+1/2}) \doteq \frac{a}{\Delta x}\big(v(x_{j+1}) - v(x_j)\big). \tag{14.129}$$

Finally, combining the Taylor expansion

$$r\left(x_{j+1/2}, \frac{1}{2}\Delta t\right) = r(x_{j+1/2}, 0) + \frac{1}{2}\Delta t\frac{\partial r}{\partial t}(x_{j+1/2}, 0) + \mathcal{O}(\Delta t^2) \tag{14.130}$$

with (14.128) and (14.129), we find the following approximation of $r_{j+1/2}^{1/2}$:

$$r_{j+1/2}^{1/2} = \frac{1}{a} Q(w)(x_{j+1/2}) + \frac{c}{2}\big(v(x_{j+1}) - v(x_j)\big), \tag{14.131}$$

where $Q(w)(x_{j+1/2})$ is some quadrature rule for the integral in (14.128).

There is an interesting observation to be made from the alternative formulation (14.127) of the wave equation. If we apply the leapfrog scheme (14.124) to (14.127) and take a one-sided backward numerical time derivative of (14.124a), we obtain

$$u_j^{n+1} - 2u_j^n + u_j^{n-1} = c\big(r_{j+1/2}^{n+1/2} - r_{j-1/2}^{n+1/2} - r_{j+1/2}^{n-1/2} + r_{j-1/2}^{n-1/2}\big). \tag{14.132}$$

Likewise, if we take a one-sided backward numerical space derivative of (14.124b) and replace n by $n-1$, we find

$$r_{j+1/2}^{n+1/2} - r_{j+1/2}^{n-1/2} - r_{j-1/2}^{n+1/2} + r_{j-1/2}^{n-1/2} = c\big(u_{j+1}^n - 2u_j^n + u_{j-1}^n\big). \tag{14.133}$$

Combining (14.132) and (14.133), we retrieve the central difference scheme (14.102b)!

14.7 Numerical Boundary Conditions

An important question is how to deal with boundary conditions. We will not cover this subject in full detail; instead we will discuss a few selected topics. First, we introduce nonreflecting and numerical boundary conditions for two model problems: the system formulation of the wave equation and the shallow-water equations. Next we investigate the impact of boundary conditions on the behaviour of a numerical solution. In particular, we discuss parasitic solutions of a multistep scheme and we conclude with a stability analysis of the leapfrog scheme.

Hyperbolic systems are often defined on a very large or even infinite spatial domain. In order to make a numerical computation feasible, we have to cut off this domain at some point. At this point we want to impose an artificial boundary condition such that the numerical solution on the smaller domain agrees well with the numerical solution on the larger domain, assuming that there are no incoming waves at this point. Moreover, we almost never have enough physical boundary conditions to carry out a numerical computation. Therefore we have to derive additional numerical boundary conditions. We will illustrate these two issues for an initial boundary value problem of the form

$$\frac{\partial \boldsymbol{u}}{\partial t} + \boldsymbol{B}(\boldsymbol{u})\frac{\partial \boldsymbol{u}}{\partial x} = \boldsymbol{0}, \qquad x \in (0,\infty), \quad t > 0, \tag{14.134a}$$

$$\boldsymbol{u}(x,0) = \boldsymbol{0}, \qquad x \in (0,\infty), \tag{14.134b}$$

$$\boldsymbol{C}_\ell \boldsymbol{u}(0,t) = \boldsymbol{g}_\ell(t), \qquad t > 0. \tag{14.134c}$$

Note that the initial condition $\boldsymbol{u}(x,0) = \boldsymbol{0}$ implies the *radiation condition* of no incoming waves from infinity. In the following we take for the hyperbolic system (14.134a) either the linear system from Example 14.1 or the shallow-water equations. In both cases the boundary conditions depend on the characteristic variable $\tilde{\boldsymbol{u}}$. The solution of (14.134) is a right-travelling wave generated by the boundary condition (14.134c) at $x = 0$.

First, consider the linear system from Example 14.1. The characteristic variable $\tilde{\boldsymbol{u}} := \boldsymbol{S}^{-1}\boldsymbol{u}$, with $\boldsymbol{S}$ the (right) eigenvector matrix, is given by

$$\tilde{u}_1 = \frac{1}{2}(u_1 + u_2), \qquad \tilde{u}_2 = \frac{1}{2}(u_1 - u_2). \tag{14.135}$$

These variables satisfy the scalar equations

$$\frac{\partial \tilde{u}_1}{\partial t} - a\frac{\partial \tilde{u}_1}{\partial x} = 0, \qquad \frac{\partial \tilde{u}_2}{\partial t} + a\frac{\partial \tilde{u}_2}{\partial x} = 0, \tag{14.136}$$

implying that $\tilde{u}_1$ is a left-going wave and $\tilde{u}_2$ is a right-going wave. Suppose we cut off the spatial domain at $x = L$ for some $L > 0$. At $x = L$ we impose the condition $\tilde{u}_1(L, t) = 0$; i.e., there is no incoming wave at $x = L$. This gives the *absorbing* or *nonreflecting boundary condition*

$$u_1(L, t) + u_2(L, t) = 0, \qquad t > 0. \tag{14.137a}$$

According to the theory in Section 12.7, this is an admissible boundary condition. The numerical approximation of (14.137a) at time level t^{n+1} is straightforward and is given by

$$u^{n+1}_{1,M+1} + u^{n+1}_{2,M+1} = 0, \tag{14.137b}$$

assuming $[0, L]$ is covered with a uniform (spatial) grid as follows:

$$x_j = j\Delta x \qquad (j = 0, 1, 2, \ldots, M+1), \quad \Delta x := L/(M+1).$$

Most schemes for (14.134a) require two conditions at the boundary point $x_{M+1} = L$. Therefore we have to complete the boundary condition (14.137b) with a numerical boundary condition. One way to determine such a numerical boundary condition is to use that $\tilde{u}_2$ is constant along the $\mathcal{C}_2$ characteristic through the grid point (x_{M+1}, t^{n+1}); see Figure 14.6. If the stability condition $c := a\Delta t/\Delta x \le 1$ is satisfied, this characteristic intersects the grid

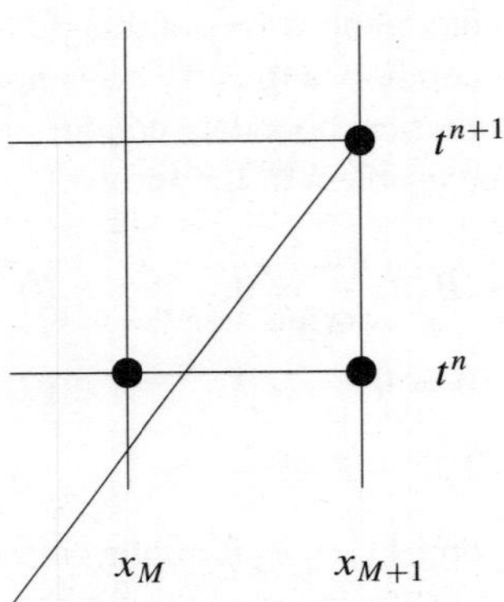

Figure 14.6. *The $\mathcal{C}_2$ characteristic through the boundary grid point (x_{M+1}, t^{n+1}).*

line $t = t^n$ between the grid points (x_M, t^n) and (x_{M+1}, t^n). Applying linear interpolation, we find

$$\tilde{u}_{2,M+1}^{n+1} = c\tilde{u}_{2,M}^n + (1-c)\tilde{u}_{2,M+1}^n,$$

which is just the upwind scheme applied to the advection equation for the characteristic variable $\tilde{u}_2$ in (14.136) at grid point (x_{M+1}, t^n). In terms of the original variables we find the *numerical boundary condition*

$$u_{1,M+1}^{n+1} - u_{2,M+1}^{n+1} = c\big(u_{1,M}^n - u_{2,M}^n\big) + (1-c)\big(u_{1,M+1}^n - u_{2,M+1}^n\big), \tag{14.138}$$

and together with (14.137b) this is sufficient to determine both u_1 and u_2 at the boundary $x = L$.

Next suppose system (14.134a) represents the shallow-water equations. In this case the characteristic variables are defined by the Pfaffian differential equation $\mathrm{d}\,\tilde{\boldsymbol{u}} = \boldsymbol{S}^{-1}\mathrm{d}\,\boldsymbol{u}$. In Example 12.33 we saw that

$$\tilde{u}_1 = u - 2c, \quad \tilde{u}_2 = u + 2c, \qquad c := \sqrt{\varphi}, \tag{14.139}$$

and these variables satisfy the diagonalised system

$$\frac{\partial \tilde{u}_1}{\partial t} + (u-c)\frac{\partial \tilde{u}_1}{\partial x} = 0, \qquad \frac{\partial \tilde{u}_2}{\partial t} + (u+c)\frac{\partial \tilde{u}_2}{\partial x} = 0. \tag{14.140}$$

Assume the in/outflow at the artificial boundary $x = L$ is subcritical, i.e., $|u| < c$. Then we have $\lambda_1(\boldsymbol{u}) < 0 < \lambda_2(\boldsymbol{u})$, implying that we have to prescribe one physical boundary condition and derive one numerical boundary condition at $x = L$; see Section 12.7. For the physical boundary condition we take again the nonreflecting boundary condition. In this case it reads: the amplitude of the incoming wave $\tilde{u}_1$ is constant in time [146]. This means that the incoming wave $\tilde{u}_1$ vanishes, since it is generated by changes in amplitude. Thus we have

$$\frac{\partial u}{\partial t}(L,t) - 2\frac{\partial c}{\partial t}(L,t) = 0, \qquad t > 0. \tag{14.141a}$$

The numerical approximation is straightforward, and we find

$$u_{M+1}^{n+1} - 2c_{M+1}^{n+1} = u_{M+1}^n - 2c_{M+1}^n. \tag{14.141b}$$

The derivation of the numerical boundary condition is again based on the observation that $\tilde{u}_2$ is constant along the $\mathcal{C}_2$ characteristic through (x_{M+1}, t^{n+1}). One approach is then to apply linear interpolation in order to express $\tilde{u}_{2,M+1}^{n+1}$ in terms of the values $\tilde{u}_{2,M}^n$ and $\tilde{u}_{2,M+1}^n$. Alternatively, we can apply the upwind scheme to the scalar equation for $\tilde{u}_2$ in (14.140). This way we obtain

$$\begin{aligned} &u_{M+1}^{n+1} + 2c_{M+1}^{n+1} \\ &\quad = u_{M+1}^n + 2c_{M+1}^n - \frac{\Delta t}{\Delta x}\big(u_{M+1}^n + c_{M+1}^n\big)\big(u_{M+1}^n + 2c_{M+1}^n - u_M^n - 2c_M^n\big). \end{aligned} \tag{14.142}$$

From (14.141b) and (14.142) we can determine u_{M+1}^{n+1} and c_{M+1}^{n+1} and subsequently we can compute φ_{M+1}^{n+1}.

Example 14.7 We compute tidal waves in a canal, which we model by the shallow-water equations. We assume that at $x = 0$ the mouth of the canal enters the sea, where we prescribe the periodic boundary condition

$$\varphi(0, t) = g(h_s + H \sin \omega t),$$

with $H = 0.1 h_s = 0.2$ and $\omega = 8$. At $x = 1$ we put an artificial boundary where we prescribe the nonreflecting and numerical boundary conditions (14.141b) and (14.142), respectively. We have applied the Roe scheme to compute a numerical solution. Numerical parameters are $\Delta x = 2.5 \times 10^{-2}$ and $\Delta t = 4 \times 10^{-3}$. The results are shown in Figure 14.7. Clearly, the wave generated by the boundary condition at $x = 0$ leaves the domain at the artificial boundary $x = 1$

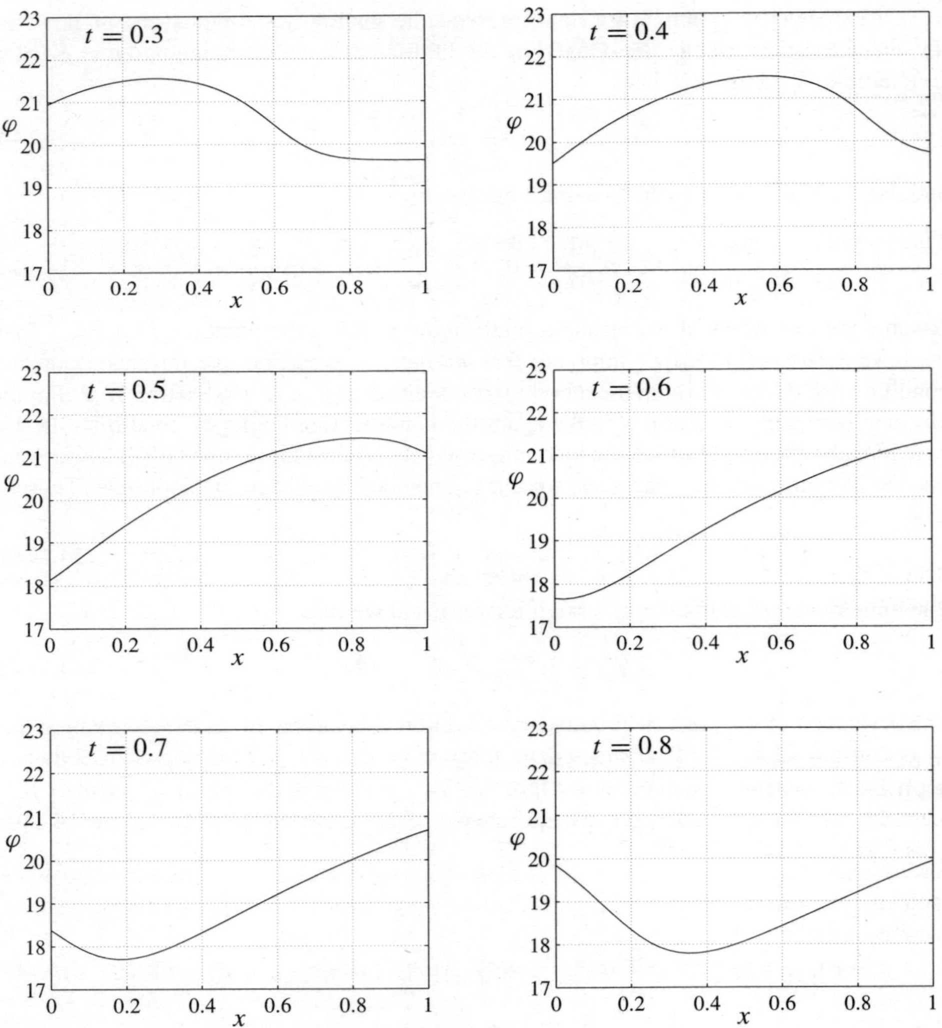

Figure 14.7. *Numerical approximation of the geopotential $\varphi = gh$ at time levels* $0.3, 0.4, \ldots, 0.8$.

undisturbed and does not produce spurious reflections. Moreover, we notice that the amplitude of the solution is slightly damped due to the dissipative character of the Roe scheme. □

A related question is the occurrence of parasitic solutions of a multistep scheme for (14.134a). These parasitic solutions should be properly controlled by appropriate boundary conditions. We will work this out for the leapfrog scheme, which is a two-step difference scheme; cf. Section 13.4.2. Applying the leapfrog scheme, we obtain

$$\boldsymbol{u}_j^{n+1} = -\frac{\Delta t}{\Delta x}\boldsymbol{B}\big(\boldsymbol{u}_{j+1}^n - \boldsymbol{u}_{j-1}^n\big) + \boldsymbol{u}_j^{n-1}. \tag{14.143}$$

Inserting the planar wave solution

$$\boldsymbol{u}_j^n = \boldsymbol{s}\,\mathrm{e}^{\mathrm{i}(\kappa x_j - \omega t^n)} = \boldsymbol{s}\,\lambda^n \mu^j, \qquad \lambda := \mathrm{e}^{-\mathrm{i}\omega\Delta t}, \quad \mu := \mathrm{e}^{\mathrm{i}\kappa\Delta x}, \tag{14.144}$$

for some vector $\boldsymbol{s} \neq \boldsymbol{0}$, we find the relation

$$\left(\lambda - \frac{1}{\lambda}\right)\boldsymbol{s} = -\frac{\Delta t}{\Delta x}\left(\mu - \frac{1}{\mu}\right)\boldsymbol{B}\boldsymbol{s}.$$

This relation is actually an eigenvalue problem. Indeed, assuming $\mu^2 \neq 1$, we have

$$\boldsymbol{B}\boldsymbol{s} = -\frac{\Delta x}{\Delta t}\frac{\mu(\lambda^2-1)}{\lambda(\mu^2-1)}\boldsymbol{s}. \tag{14.145}$$

Using the eigenvalues of $\boldsymbol{B}$, we can compute λ as a function of μ or vice versa.

To make the subsequent analysis more transparent, we take for $\boldsymbol{B}$ the matrix

$$\boldsymbol{B} = -\begin{pmatrix} 0 & a \\ a & 0 \end{pmatrix}$$

occurring in Example 14.1 on small-amplitude waves. Clearly its eigenvalues[1] ν_k and eigenvectors $\boldsymbol{s}_k$ are given by

$$\nu_1 = -a < 0, \quad \nu_2 = a > 0, \qquad \boldsymbol{s}_1 = \begin{pmatrix} 1 \\ 1 \end{pmatrix}, \quad \boldsymbol{s}_2 = \begin{pmatrix} 1 \\ -1 \end{pmatrix}.$$

In this case the solution of (14.134a) consists of left- and right-travelling waves and is of the form

$$\boldsymbol{u}(x,t) = \hat{w}_1(x+at)\boldsymbol{s}_1 + \hat{w}_2(x-at)\boldsymbol{s}_2$$

for arbitrary continuously differentiable functions $\hat{w}_1(\xi)$ and $\hat{w}_2(\xi)$. The first component typically has planar wave modes with dispersion relation $\omega(\kappa) = -a\kappa$, whereas for the second component we have $\omega(\kappa) = a\kappa$.

From the eigenvalue problem (14.145) we conclude that

$$\mu(\lambda^2-1) = \pm c\lambda(\mu^2-1), \tag{14.146}$$

[1] Here we use ν_k to denote the eigenvalues of $\boldsymbol{B}$ instead of λ_k in order to avoid confusion with the amplification factor $\lambda = \mathrm{e}^{-\mathrm{i}\omega\Delta t}$.

with $c := a\Delta t/\Delta x$. For given wave number κ, or equivalently for given μ, we can determine the amplification factor λ. In particular, for the plus sign in (14.146), corresponding to $\nu_1 = -a$, we obtain the quadratic equation

$$\lambda^2 - \alpha\lambda - 1 = 0, \qquad \alpha := c\frac{\mu^2 - 1}{\mu}, \tag{14.147a}$$

assuming $\mu \neq 0$. This equation has the solutions

$$\lambda_{+,1} = \frac{1}{2}\alpha + \sqrt{1 + \frac{1}{4}\alpha^2}, \qquad \lambda_{+,2} = \frac{1}{2}\alpha - \sqrt{1 + \frac{1}{4}\alpha^2}. \tag{14.147b}$$

For small $\omega\Delta t$ and $\kappa\Delta x$ we find that $\alpha \doteq 2\mathrm{i}a\kappa\Delta t$. Clearly, $\lambda_{+,1}$ is the essential root and we have

$$\lambda_{+,1} =: \mathrm{e}^{-\mathrm{i}\omega\Delta t} \doteq 1 + \mathrm{i}a\kappa\Delta t \doteq \mathrm{e}^{\mathrm{i}a\kappa\Delta t}, \tag{14.148}$$

and the corresponding dispersion relation reads $\omega(\kappa) = -a\kappa$ for a left-going wave. The other root $\lambda_{+,2}$ does not belong to a physical mode; it is parasitic (see Section 5.4). Indeed, we find

$$\mathrm{e}^{-\mathrm{i}\omega\Delta t} \doteq -1 + \mathrm{i}a\kappa\Delta t \doteq -\mathrm{e}^{-\mathrm{i}a\kappa\Delta t}; \tag{14.149}$$

i.e., $\omega(\kappa) = a\kappa - \pi/\Delta t$. This would correspond to a mode that changes sign each time step and travels in the wrong direction. In the same way, when we take the minus sign in (14.146), corresponding to $\nu_2 = a$, we obtain the relation

$$\lambda^2 + \alpha\lambda - 1 = 0. \tag{14.150a}$$

It roots are

$$\lambda_{-,1} = -\frac{1}{2}\alpha + \sqrt{1 + \frac{1}{4}\alpha^2}, \qquad \lambda_{-,2} = -\frac{1}{2}\alpha - \sqrt{1 + \frac{1}{4}\alpha^2}. \tag{14.150b}$$

In this case $\lambda_{-,1}$ is the essential root corresponding to a right-going mode, while $\lambda_{-,2}$ corresponds to a parasitic mode. Observe that $\lambda_{+,1}\lambda_{-,1} = 1$. Thus we have four linearly independent solutions of the form (14.144). The general solution of (14.134a) is a linear combination of these and reads

$$\boldsymbol{u}_j^n = \Big(C_1\lambda_{+,1}^n\boldsymbol{s}_1 + C_2\lambda_{+,2}^n\boldsymbol{s}_1 + C_3\lambda_{-,1}\boldsymbol{s}_2 + C_4\lambda_{-,2}\boldsymbol{s}_2\Big)\mu^j \tag{14.151}$$

for arbitrary constants C_l $(l = 1, 2, 3, 4)$. The second and fourth terms in (14.151) are parasitic modes.

Because of symmetry with respect to μ and λ, we may as well express μ in terms of λ. Completely analogously to the previous derivation, we obtain for μ the quadratic equations

$$\mu^2 \mp \beta\mu - 1 = 0, \qquad \beta := \frac{\lambda^2 - 1}{c\lambda}, \tag{14.152}$$

the roots of which are given by

$$\mu_{+,1} = \frac{1}{2}\beta + \sqrt{1 + \frac{1}{4}\beta^2}, \qquad \mu_{+,2} = \frac{1}{2}\beta - \sqrt{1 + \frac{1}{4}\beta^2}, \tag{14.153a}$$

$$\mu_{-,1} = -\frac{1}{2}\beta + \sqrt{1 + \frac{1}{4}\beta^2}, \qquad \mu_{-,2} = -\frac{1}{2}\beta - \sqrt{1 + \frac{1}{4}\beta^2}. \tag{14.153b}$$

The essential roots are $\mu_{+,1}$ and $\mu_{-,1}$, and, furthermore, $\mu_{+,1}\mu_{-,1} = 1$. The general solution of (14.134a) can thus be rewritten as

$$\boldsymbol{u}_j^n = \lambda^n \left(D_1 \mu_{+,1}^j \boldsymbol{s}_1 + D_2 \mu_{+,2}^j \boldsymbol{s}_1 + D_3 \mu_{-,1}^j \boldsymbol{s}_2 + D_4 \mu_{-,2}^j \boldsymbol{s}_2 \right) \tag{14.154}$$

for arbitrary coefficients D_l ($l = 1, 2, 3, 4$). Also, in this representation, the second and fourth terms are parasitic. Suppose we prescribe a physical boundary condition at $x = L$ of the form

$$\boldsymbol{C}_\text{r}\mathbf{u}(L, t) = 0, \qquad t > 0, \tag{14.155}$$

with $\boldsymbol{C}_\text{r}$ a 1×2 matrix; cf. Section 12.7. Inserting (14.154) into (14.155), we obtain the relation

$$D_1 \mu_{+,1}^{M+1} \boldsymbol{C}_\text{r}\boldsymbol{s}_1 + D_2 \mu_{+,2}^{M+1} \boldsymbol{C}_\text{r}\boldsymbol{s}_1 + D_3 \mu_{-,1}^{M+1} \boldsymbol{C}_\text{r}\boldsymbol{s}_2 + D_4 \mu_{-,2}^{M+1} \boldsymbol{C}_\text{r}\boldsymbol{s}_2 = 0. \tag{14.156}$$

In either case we see that this boundary condition may annihilate either the left- or right-going parasitic wave; i.e., we can choose $\boldsymbol{C}_\text{r}$ such that either $\boldsymbol{C}_\text{r}\boldsymbol{s}_1 = 0$ or $\boldsymbol{C}_r\boldsymbol{s}_2 = 0$. Thus one parasitic mode is persistent; i.e., it cannot be removed in this way. However, there is a variety of ways to remove these parasitic waves. One is to choose boundary conditions that include derivatives. Below we give an example.

Example 14.8 Consider the general solution (14.154) and assume that the left-going parasitic mode is absent; i.e., $D_4 = 0$. In this case we have

$$\boldsymbol{u}_j^n = \begin{pmatrix} u_{1,j}^n \\ u_{2,j}^n \end{pmatrix} = \lambda^n \left[D_1 \mu_{+,1}^j \begin{pmatrix} 1 \\ 1 \end{pmatrix} + D_2 \mu_{+,2}^j \begin{pmatrix} 1 \\ 1 \end{pmatrix} + D_3 \mu_{-,1}^j \begin{pmatrix} 1 \\ -1 \end{pmatrix} \right].$$

We want to prescribe boundary conditions at $x_{M+1} = L$ such that the remaining parasitic mode is eliminated; i.e., the boundary conditions should imply $D_2 = 0$. A successful choice for this is

$$u_{1,M+1}^n - u_{1,M}^n = 0, \qquad u_{2,M+1}^n + u_{2,M}^n = 0.$$

The latter condition can be interpreted as a "smoothing" approximation of $u_2(L, t) = 0$. Applying these boundary conditions, we obtain the linear system

$$(\mu_{+,1} - 1)\hat{D}_1 + (\mu_{+,2} - 1)\hat{D}_2 = 1 - \mu_{-,1},$$
$$(\mu_{+,1} + 1)\hat{D}_1 + (\mu_{+,2} + 1)\hat{D}_2 = 1 + \mu_{-,1}$$

for $\hat{D}_1 := (\mu_{+,1}/\mu_{-,1})^M D_1/D_3$ and $\hat{D}_2 := (\mu_{+,2}/\mu_{-,1})^M D_2/D_3$. Since $\mu_{+,1}\mu_{-,1} = 1$ and $\mu_{+,1}\mu_{+,2} = -1$, this system can be simplified to

$$\left(\frac{1}{\eta} - 1\right)\hat{D}_1 - (\eta + 1)\hat{D}_2 = 1 - \eta,$$
$$\left(\frac{1}{\eta} + 1\right)\hat{D}_1 + (-\eta + 1)\hat{D}_2 = 1 + \eta,$$

with $\eta := \mu_{-,1}$. It is simple to see that the solution is given by $\hat{D}_1 = \eta$ and $\hat{D}_2 = 0$, and so $D_2 = 0$. The latter result is the desired one, showing that no parasitic solutions are being generated. □

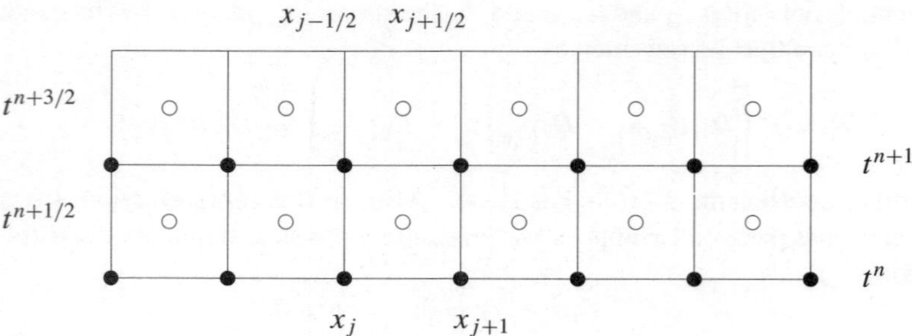

Figure 14.8. *Two time levels in the staggered grid for the leapfrog scheme. The variables u and r are defined at points marked by • and ∘, respectively.*

Finally, we want to investigate the impact of boundary conditions on the stability of a scheme. We consider in particular the staggered leapfrog scheme (14.124) applied to system (14.127). We assume that u satisfies the boundary conditions (14.101d) and (14.101e). Since we know the value of u at the boundaries $x = 0$ and $x = 1$, it is tempting to choose the grid in Figure 14.8, where grid points for u are located on the boundaries. Note that there is one more unknown value of r than of u at every time level, because the u values at the boundary are known. Let the number of unknown u values at each line $t = t^n$ be M. Introducing the vectors

$$\mathbf{u}^n := \begin{pmatrix} u_1^n \\ u_2^n \\ \vdots \\ u_M^n \end{pmatrix}, \qquad \mathbf{r}^{n+1/2} := \begin{pmatrix} r_{1/2}^{n+1/2} \\ r_{3/2}^{n+1/2} \\ \vdots \\ r_{M+1/2}^{n+1/2} \end{pmatrix}$$

and assuming homogeneous boundary conditions, we see that the scheme (14.124) applied to (14.127) can be written as the matrix-vector recursion

$$\begin{pmatrix} \mathbf{I} & \mathbf{O} \\ \mathbf{C}^T & \mathbf{I} \end{pmatrix} \begin{pmatrix} \mathbf{u}^{n+1} \\ \mathbf{r}^{n+3/2} \end{pmatrix} = \begin{pmatrix} \mathbf{I} & \mathbf{C} \\ \mathbf{O} & \mathbf{I} \end{pmatrix} \begin{pmatrix} \mathbf{u}^{n} \\ \mathbf{r}^{n+1/2} \end{pmatrix}, \tag{14.157}$$

where the $M \times (M+1)$ matrix C is defined by

$$\mathbf{C} := c \begin{pmatrix} -1 & 1 & & & & \\ & -1 & 1 & & & \\ & & \ddots & \ddots & & \\ & & & -1 & 1 & \\ & & & & -1 & 1 \end{pmatrix}. \tag{14.158}$$

In order to analyse the stability of (14.157), we investigate the eigenvalues of the generalised eigenvalue problem

$$\begin{pmatrix} \mathbf{I} & \mathbf{O} \\ \mathbf{C}^T & \mathbf{I} \end{pmatrix} \begin{pmatrix} \mathbf{u} \\ \mathbf{r} \end{pmatrix} = \lambda \begin{pmatrix} \mathbf{I} & \mathbf{C} \\ \mathbf{O} & \mathbf{I} \end{pmatrix} \begin{pmatrix} \mathbf{u} \\ \mathbf{r} \end{pmatrix}. \tag{14.159}$$

We leave it as an exercise to show that $\lambda \neq 0$. We find from (14.159) that

$$(1-\lambda)\mathbf{u} = \lambda \mathbf{C}\mathbf{r}, \tag{14.160a}$$

$$\mathbf{C}^T\mathbf{u} = (\lambda - 1)\mathbf{r}. \tag{14.160b}$$

Multiplying (14.160a) by $\mathbf{C}^T$ and substituting $\mathbf{C}^T\mathbf{u}$ from (14.160b) yields the following eigenvalue problem for $\mathbf{r}$:

$$\mathbf{C}^T\mathbf{C}\mathbf{r} = -\frac{(\lambda-1)^2}{\lambda}\mathbf{r}. \tag{14.161}$$

Next we readily verify that $\mathbf{C}^T\mathbf{C}$ is given by

$$\mathbf{C}^T\mathbf{C} = c^2 \begin{pmatrix} 1 & -1 & & & \\ -1 & 2 & -1 & & \\ & \ddots & \ddots & \ddots & \\ & & -1 & 2 & -1 \\ & & & -1 & 1 \end{pmatrix}. \tag{14.162}$$

Let us write $\mu := -\lambda^{-1}(\lambda-1)^2$. Applying Gershgorin's theorem (cf. Theorem I.7) to $\mathbf{C}^T\mathbf{C}$ (see (14.161)), we conclude that $0 \le \mu \le 4c^2 \le 4$. The last inequality is an immediate consequence of the stability condition $c \le 1$. Moreover, $\mathbf{C}^T\mathbf{C}$ has a simple eigenvalue $\mu = 0$, with eigenvector $\mathrm{e} := (1, 1, \ldots, 1)^T$. Clearly, λ satisfies

$$\lambda^2 + (\mu - 2)\lambda + 1 = 0,$$

the roots of which are complex conjugates and have modulus one. However, the eigenvalue $\mu = 0$ implies a double root $\lambda = 1$, which turns out to have only one eigenvector $(\mathbf{0}^T, \mathrm{e}^T)^T$. Hence (14.159), and consequently (14.157), cannot be stable; cf. Definition I.5. We therefore conclude that one cannot use the u values at both boundaries directly.

Alternatively, if we assume that r is given at the right boundary $x = 0$, we obtain a similar system as (14.157), but now with an $M \times M$ matrix $\mathbf{C}$ given by

$$\mathbf{C} := c \begin{pmatrix} 1 & & & & \\ -1 & 1 & & & \\ & \ddots & \ddots & & \\ & & -1 & 1 & \\ & & & -1 & 1 \end{pmatrix}. \tag{14.163}$$

If we carry out a similar analysis, now with this matrix, it will be clear that $\mathbf{C}^T\mathbf{C}$ is nonsingular and hence all eigenvalues λ in (14.159) are complex conjugate pairs with nonzero imaginary parts. We leave it as an exercise to show that there are no double eigenvalues. Thus the scheme is stable.

14.8 Discussion

- We have only discussed problems in one space dimension. All methods in this chapter can be extended to several space dimensions by dimensional splitting, i.e., the one-dimensional scheme is applied in each direction; see, e.g., [159, 88].
- Hyperbolic systems often include a source term, describing, e.g., chemical reactions. One approach to solving these systems is to use an operator-splitting method. In such a method the system is split into a "convection part" and a "chemistry part," which are alternately solved at each time step. For a detailed discussion see, e.g., [88].
- Boundary conditions provide a real source of concern. We have only dealt with this problem briefly. In particular, the treatment of reflective waves has been, and still is, a subject of much research. A good survey paper dealing with radiation boundary conditions (Robin type) is [56]. Another good source for the study of numerical boundary conditions is [72].

Exercises

14.1. Consider the upwind scheme (14.14).

(a) Verify the stability condition (14.16). Explain that it is just the CFL condition applied to (14.14).

(b) Show that the scheme is first order accurate in Δt and Δx.

14.2. Consider the model problem from Exercise 12.9, describing one-dimensional electromagnetic waves.

(a) Give the upwind scheme (14.14).

(b) Give the wave propagation form (14.43).

(c) Elaborate the slope limiter method (14.67).

14.3. Compute the Godunov numerical flux (14.30) for the p system in Exercise 12.10. Then elaborate the Godunov scheme (14.32) and the stability condition (14.33).

14.4. Show that an alternative expression for the Roe numerical flux is given by

$$\boldsymbol{F}_{\mathrm{R}}(\boldsymbol{u}_\ell, \boldsymbol{u}_{\mathrm{r}}) = \frac{1}{2}\big(\boldsymbol{f}(\boldsymbol{u}_\ell) + \boldsymbol{f}(\boldsymbol{u}_{\mathrm{r}})\big) - \frac{1}{2}\widetilde{\boldsymbol{B}}^*(\boldsymbol{u}_\ell, \boldsymbol{u}_{\mathrm{r}})(\boldsymbol{u}_{\mathrm{r}} - \boldsymbol{u}_\ell),$$

where $\widetilde{\boldsymbol{B}}^*(\boldsymbol{u}_\ell, \boldsymbol{u}_{\mathrm{r}})$ is defined analogously to (14.18); i.e., the Roe numerical flux can be interpreted as the central difference flux $\frac{1}{2}(\boldsymbol{f}(\boldsymbol{u}_\ell) + \boldsymbol{f}(\boldsymbol{u}_{\mathrm{r}}))$ stabilized by the diffusive flux $-\frac{1}{2}\widetilde{\boldsymbol{B}}^*(\boldsymbol{u}_\ell, \boldsymbol{u}_{\mathrm{r}})(\boldsymbol{u}_{\mathrm{r}} - \boldsymbol{u}_\ell)$.

14.5. Consider the p system from Exercise 12.10.

(a) Show that the Roe matrix $\widetilde{\boldsymbol{B}}(\boldsymbol{u}_\ell, \boldsymbol{u}_{\mathrm{r}})$ is given by

$$\widetilde{\boldsymbol{B}}(\boldsymbol{u}_\ell, \boldsymbol{u}_{\mathrm{r}}) = \begin{pmatrix} 0 & -1 \\ \dfrac{p(v_{\mathrm{r}}) - p(v_\ell)}{v_{\mathrm{r}} - v_\ell} & 0 \end{pmatrix}.$$

(b) Compute the Roe numerical flux from either expression in (14.57). Then elaborate the Roe scheme (14.44).

(c) Give the slope limiter flux (14.75).

14.6. Discuss the Harten–Hyman entropy fix for a sonic 2-rarefaction wave of the shallow-water equations.

14.7. Consider the *damped wave equation*

$$\frac{\partial u}{\partial t} + \frac{\partial^2 u}{\partial t^2} = a^2 \frac{\partial^2 u}{\partial x^2}, \qquad x \in (0, 1), \quad t > 0, \quad a > 0.$$

(a) Use central differences to discretise this equation.

(b) Investigate for which values of $c = a\Delta t/\Delta x$ the scheme is stable.

14.8. For the wave equation (14.101a) we can use a variant of the implicit Euler method by discretising $\frac{\partial^2 u}{\partial x^2}$ at the time point t^{n+1}.

(a) Write down the resulting scheme.

(b) Show that this scheme is unconditionally stable.

(c) Find the dispersion relation and show that this method is dissipative.

14.9. Consider the nonlinear initial value problem

$$\begin{aligned} u^2 \frac{\partial^2 u}{\partial t^2} &= \frac{\partial^2 u}{\partial x^2}, && x \in \mathbb{R}, \quad t > 0, \\ u(x, 0) &= 1 + x^2, && x \in \mathbb{R}, \\ \frac{\partial u}{\partial t}(x, 0) &= 0, && x \in \mathbb{R}. \end{aligned}$$

Show how to construct an explicit scheme.

14.10. Show that we can associate with the wave equation $\frac{\partial^2 u}{\partial t^2} = a^2 \frac{\partial^2 u}{\partial x^2}$ the system

$$\begin{aligned} \frac{\partial u}{\partial t} \pm a \frac{\partial v}{\partial x} &= 0, \\ \frac{\partial v}{\partial t} \pm a \frac{\partial u}{\partial x} &= 0, \end{aligned}$$

where either the plus or the minus sign is assumed in both equations.

14.11. Show the stability of the recursion (14.107) by writing it as a one-step recursion and investigating the eigenvalues of the matrix involved in the latter.

14.12. Prove that the matrix $\mathbf{C}^T \mathbf{C}$ with $\mathbf{C}$ defined in (14.163) has eigenvalues on the unit circle but no double ones.

14.13. Consider the wave equation for $x, t > 0$ in the form

$$\begin{aligned} \frac{\partial u}{\partial t} + a \frac{\partial v}{\partial x} &= 0, \\ \frac{\partial v}{\partial t} + a \frac{\partial u}{\partial x} &= 0. \end{aligned}$$

(a) Derive the Lax–Wendroff scheme for this system.

(b) Assume we use a special boundary condition similar to the one in Example 14.8, i.e.,

$$\frac{\partial u}{\partial x}(0, t) = 0, \quad v(0, t) = 0, \qquad t > 0.$$

Show that there is no parasitic solution arising from this boundary.

Chapter 15

Perturbation Methods

In this chapter we give a concise introduction to a number of important analytical techniques that use available small parameters in a problem. These methods are called perturbation or asymptotic methods, and follow naturally from the ideas of systematic modeling set forth in Chapter 7. After an introduction, explaining why small parameters occur frequently in practice, we start in Section 15.2 by stating some principles of asymptotic approximations and expansions. Then we continue in Section 15.3 with two methods for solving regular perturbation problems: the method of slow variation and the Lindstedt–Poincaré method. Two of the most important methods for singular perturbation problems are considered in Section 15.4: the method of matched asymptotic expansions (MAEs) and some versions of the method of multiple scales.

15.1 Introduction

We saw in Chapter 7 that a real-world problem can be described by a hierarchy of models such that a higher-level model is more comprehensive and more accurate than one from a lower level. Now suppose that we have a fairly good model describing the dominating phenomena in good order of magnitude. And suppose that we are interested in improving this model by adding some previously ignored aspects or effects. In general, this implies a very abrupt change in our model. The equations are more complex and more difficult to solve. As an illustration, consider the simple model $x^2 = a^2$ and the more complete model $x^2 + \varepsilon x^5 = a^2$. The first one can easily be solved analytically, the second one, with much more effort, only numerically. So it seems that the relation between solution and model is not continuous in the problem parameters. However small ε we take, from a transparent and exact solution of the simple model at $\varepsilon = 0$, we abruptly face a far more complicated solution of a model that is just a little bit better. This is a pity, because certain types of useful information (parametric dependencies, trends) become increasingly difficult to dig out of the more complicated solution of the complex model. This discontinuity of models in the parameter ε may therefore be an argument for retaining the simpler model.

The (complexity of the) model is, however, only discontinuous if we are merely interested in exact or numerically "exact" solutions (e.g., for reasons of benchmarking or validation of solution methods). This is not always the case. As far as our modeling objectives are concerned, we have to keep in mind that the improved model is only the next step in the modeling hierarchy and is not exact in any absolute sense. So there is no reason to require the solution to be more exact than the corresponding model, as an exact solution of an approximate model is no better than an approximate solution of an exact model. Moreover, the *type* of information that analytical solutions may provide (functional relationships, etc.) is sometimes so important that it may be worth sacrificing numerical accuracy.

Let us go back to our "fairly good" improved model. The effects we added are relatively small. Otherwise, the previous lower-level model was not fairly good, as we assumed, but just completely wrong. Usually, this smallness is quantified by small dimensionless parameters occurring in the equations and (or) boundary conditions. This is the generic situation. The transition from a lower-level to a higher-level theory is characterized by the appearance of one or more modeling parameters, which are (when made dimensionless) small or large, and yield in the limit a simpler description. Examples are infinitely large or small geometries with circular or spherical symmetry that reduce the number of spatial dimensions, small amplitudes allowing linearization, low velocities and long time scales in flow problems allowing incompressible description, and small relative viscosity allowing inviscid models. In fact, in any practical problem it is really the rule rather than the exception that dimensionless numbers are either small or large (cf. [76]).

If we accept approximate solutions, where the approximation is based on the inherently small or large modeling parameters, it is possibile to gradually increase the complexity of a model and study small but significant effects in the most efficient way. The methods utilizing this approach systematically are called "perturbation methods." The approximation constructed is almost always asymptotic, i.e., where the error decreases with the small or large parameter.

Usually, a distinction is made between *regular* and *singular* perturbations. A (loose definition of a) regular perturbation problem is where the approximate problem is everywhere close to the unperturbed problem. This, however, depends of course on the domain of interest and, as we will see, on the choice of coordinates. If a problem is regular without any need for other than trivial reformulations, the construction of an asymptotic solution is straightforward. In fact, this forms the usual strategy in modeling when terms are linearised or effects are neglected. The more interesting perturbation problems are those where this straightforward approach fails.

We will consider here four methods relevant in the presented modeling problems. The first two are examples of regular perturbation methods, but only after a suitable coordinate transformation. The first is called the method of slow variation, where the typical axial length scale is much greater than the transverse length scale. The second is the Lindstedt–Poincaré method or the method of strained coordinates for periodic processes. Here the intrinsic time scale ($\sim$ the period of the solution) is unknown and has to be found. The other two methods are of singular perturbation type, because there is no coordinate transformation possible that renders the problem into one of regular type. The third one is the method of Matched Asymptotic Expansions. To render the problem into one of regular type, different scalings are necessary in spatially distinct regions (boundary layers). The fourth singular perturbation method considered here is the method of multiple scales and may be considered as a combination of the method of slow variation and the method of strained coordinates, as now several (long, short, shorter) length scales occur in parallel. This cannot be repaired by a single coordinate transformation. Therefore the problem is temporarily reformulated into a higher-dimensional problem by taking the various length scales apart. Then the problem is regular again, and can be solved. A refinement of this method is the Wentzel, Kramers, and Brillouin (WKB) method, where the coordinate transformation of the fast variable becomes itself slowly varying.

15.2 Asymptotic Approximations and Expansions

Before we can introduce the methods, we have to define our terminology for asymptotic approximations and asymptotic expansions.

15.2.1 Asymptotic Approximations

In order to give a qualitative description of the behaviour of a function f with parameter ε near a point of interest, say $\varepsilon = 0$ (equivalent to any other value by a simple translation), we have the order symbols $\mathcal{O}$, o, and $\mathcal{O}_s$; see the appendix, Section A. Often $\varepsilon = 0$ is the lower limit of a parameter range, and we have the tacit assumption that $\varepsilon \downarrow 0$.

Definition 15.1. *$\varphi(\varepsilon)$ is an asymptotic approximation to $f(\varepsilon)$ as $\varepsilon \to 0$ if*

$$f(\varepsilon) = \varphi(\varepsilon) + o(\varphi(\varepsilon)) \qquad \textit{as} \quad \varepsilon \to 0.$$

This is sometimes more compactly denoted by $f \sim \varphi$.

If f and φ depend on $\boldsymbol{x}$, this definition remains valid pointwise, i.e., for $\boldsymbol{x}$ fixed. It is, however, useful to extend the definition to uniformly valid approximations.

Definition 15.2. *Let $f(\boldsymbol{x};\varepsilon)$ and $\varphi(\boldsymbol{x};\varepsilon)$ be continuous functions for $\boldsymbol{x} \in \mathcal{D}$ and $0 < \varepsilon < a$. We call $\varphi(\boldsymbol{x};\varepsilon)$ a uniform asymptotic approximation to $f(\boldsymbol{x};\varepsilon)$ for $\boldsymbol{x} \in \mathcal{D}$ as $\varepsilon \to 0$ if for any positive number δ there is an ε_1 (independent of $\boldsymbol{x}$ and ε) such that*

$$|f(\boldsymbol{x};\varepsilon) - \varphi(\boldsymbol{x};\varepsilon)| \le \delta|\varphi(\boldsymbol{x};\varepsilon)| \qquad \textit{for} \quad \boldsymbol{x} \in \mathcal{D} \quad \textit{and} \quad 0 < \varepsilon < \varepsilon_1.$$

We write $f(\boldsymbol{x};\varepsilon) = \varphi(\boldsymbol{x};\varepsilon) + o(\varphi(\boldsymbol{x};\varepsilon))$ uniformly in $\boldsymbol{x} \in \mathcal{D}$ as $\varepsilon \to 0$. Note that $\mathcal{D}$ may depend on ε.

Example 15.3 Let $\mathcal{D} = [0, 1]$ and $0 < \varepsilon < 1$. Then we have $\cos(\varepsilon x) = 1 + o(1)$ uniformly in $\mathcal{D}$ as $\varepsilon \to 0$, since for any given δ we can choose $\varepsilon_1 = \sqrt{\delta}$ such that $|\cos(\varepsilon x) - 1| \le \varepsilon^2 x^2 \le \varepsilon_1^2 = \delta$. □

Example 15.4 Although $\cos(x/\varepsilon) = \mathcal{O}(1)$ uniformly in $x \in [0, 1]$ for $\varepsilon \to 0$, there is no constant K such that $\cos(x/\varepsilon) = K + o(1)$. □

Example 15.5 $x + \sin(\varepsilon x) + \mathrm{e}^{-x/\varepsilon} = x + \varepsilon x + \mathcal{O}(\varepsilon^3)$ as $\varepsilon \to 0$ for all $x \neq 0$, but this is not the case uniformly in $x \in [0, 1]$. More precisely, for $x \in [\delta(\varepsilon), 1]$, $x + \sin(\varepsilon x) + e^{-x/\varepsilon} = x + \varepsilon x + o(\varepsilon^3)$ uniformly if $\varepsilon = o(\delta)$, but it is not uniformly equal for any $\delta = \mathcal{O}(\varepsilon)$. If $x = \mathcal{O}(\varepsilon)$, the otherwise exponentially small term is not small any more. This is illustrated in Figure 15.1. The difference between the original function and its nonuniform asymptotic approximation is

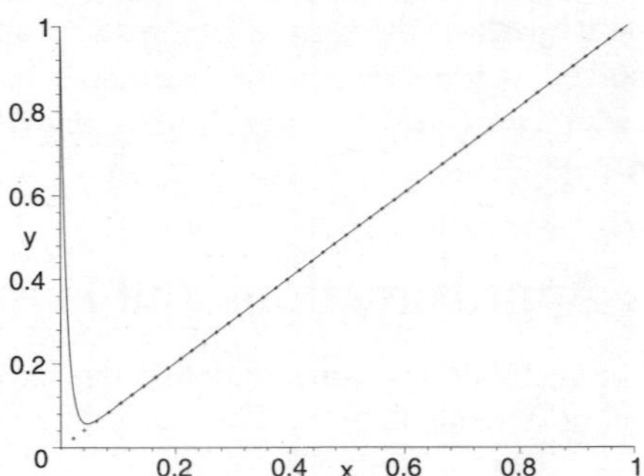

Figure 15.1. *A plot of* $x + \sin(\varepsilon x) + \mathrm{e}^{-x/\varepsilon}$ *and its nonuniform asymptotic approximation* $x + \varepsilon x$ *for* $\varepsilon = 0.01$.

typically large in a neighbourhood of $x = 0$, while the size of this neighbourhood is $x = \mathcal{O}(\varepsilon)$. This neighbourhood is an example of a boundary layer. The occurrence and behaviour of boundary layers will be discussed in more detail in Section 15.4.1. □

15.2.2 Asymptotic Expansions

Asymptotic approximations are usually structured in the form of a series expansion that helps us to construct an approximation systematically.

Definition 15.6. *The sequence* $\{\mu_n(\varepsilon)\}_{n=0}^{\infty}$ *is called an asymptotic sequence if* $\mu_{n+1}(\varepsilon) = o(\mu_n(\varepsilon))$ *as* $\varepsilon \to 0$ *for each* $n = 0, 1, 2, \ldots$.

Example 15.7 Examples of asymptotic sequences (as $\varepsilon \to 0$) are

$$\mu_n(\varepsilon) = \varepsilon^n; \quad \mu_n(\varepsilon) = \varepsilon^{\frac{1}{2}n}; \quad \mu_n(\varepsilon) = \tan^n(\varepsilon); \quad \mu_n(\varepsilon) = \ln(\varepsilon)^{-n};$$

$$\mu_n(\varepsilon) = \varepsilon^p \ln(\varepsilon)^q, \qquad \text{where} \quad p = 0, 1, 2\ldots,$$

$$q = 0, \ldots, p, \quad \text{and} \quad n = \frac{1}{2}p(p+3) - q. \qquad \square$$

Definition 15.8. *If* $\{\mu_n(\varepsilon)\}_{n=0}^{\infty}$ *is an asymptotic sequence, then* $f(\varepsilon)$ *has an asymptotic expansion of* N *terms with respect to this sequence denoted by*

$$f(\varepsilon) \sim \sum_{n=0}^{N-1} a_n \mu_n(\varepsilon),$$

where the coefficients a_n *are independent of* ε, *if*

$$f(\varepsilon) - \sum_{n=0}^{M} a_n \mu_n(\varepsilon) = o(\mu_M(\varepsilon)) \quad as \ \varepsilon \to 0$$

for each $M = 0, \dots, N-1$. $\mu_n(\varepsilon)$ *is called a* gauge function. *If* $\mu_n(\varepsilon) = \varepsilon^n$, *we call the expansion an* asymptotic power series.

Definition 15.9. *Two functions* f *and* g *are asymptotically equal up to* N *terms with respect to the asymptotic sequence* $\{\mu_n\}$ *if* $f - g = o(\mu_N)$. *If the remaining error is clear from the context, this is sometimes denoted as* $f \sim g$.

Asymptotic expansions based on the same gauge functions may be added. They may be multiplied if the products of the gauge functions can be asymptotically ordered.

In contrast to ordinary series expansions, defined for an infinite number of terms, in asymptotic expansions only a finite (N) number of terms are considered. For $N \to \infty$ the series may either converge or diverge, but this is irrelevant for the asymptotic behaviour. In addition it may be worthwhile to note that it is not necessary for a convergent asymptotic expansion to converge to the expanded function.

For given $\{\mu_n(\varepsilon)\}_{n=0}^{\infty}$ the coefficients a_n can be determined uniquely by the following recursive procedure (provided μ_n are nonzero for ε near zero and each of the limits below exists):

$$a_0 = \lim_{\varepsilon\to 0} \frac{f(\varepsilon)}{\mu_0(\varepsilon)}, \quad a_1 = \lim_{\varepsilon\to 0} \frac{f(\varepsilon) - a_0\mu_0(\varepsilon)}{\mu_1(\varepsilon)}, \quad \dots, \quad a_{N-1} = \lim_{\varepsilon\to 0} \frac{f(\varepsilon) - \sum_{n=0}^{N-2} a_n\mu_n(\varepsilon)}{\mu_{N-1}(\varepsilon)}.$$

Example 15.10 A function may have different asymptotic expansions:

$$\begin{aligned}\tan(\varepsilon) &= \varepsilon + \frac{1}{3}\varepsilon^3 + \frac{2}{15}\varepsilon^5 + \mathcal{O}(\varepsilon^7) \\ &= \sin\varepsilon + \frac{1}{2}(\sin\varepsilon)^3 + \frac{3}{8}(\sin\varepsilon)^5 + \mathcal{O}((\sin\varepsilon)^7) \\ &= \varepsilon\cos\varepsilon + \frac{5}{6}(\varepsilon\cos\varepsilon)^3 + \frac{161}{120}(\varepsilon\cos\varepsilon)^5 + \mathcal{O}((\varepsilon\cos\varepsilon)^7).\end{aligned}$$ □

Example 15.11 The asymptotic expansion

$$\varepsilon^{-1}\,\mathrm{e}^{-1/\varepsilon}\,Ei(1/\varepsilon) = \sum_{n=0}^{N} n!\,\varepsilon^n + o(\varepsilon^N), \qquad \text{where} \quad Ei(x) = \fint_{-\infty}^{x} \frac{\mathrm{e}^t}{t}\,\mathrm{d}t,$$

which is related to the exponential integral Ei, diverges as $N \to \infty$ if $\varepsilon \neq 0$. The accuracy increases with N, but on a smaller interval [1]. □

Example 15.12 Different functions may have the same asymptotic expansion:

$$\cos(\varepsilon) = 1 - \frac{1}{2}\varepsilon^2 + \frac{1}{24}\varepsilon^4 + \mathcal{O}(\varepsilon^6),$$

$$\cos(\varepsilon) + \mathrm{e}^{-1/\varepsilon} = 1 - \frac{1}{2}\varepsilon^2 + \frac{1}{24}\varepsilon^4 + \mathcal{O}(\varepsilon^6).$$

Note that both asymptotic expansions, considered as regular power series in ε, converge to $\cos(\varepsilon)$ rather than $\cos(\varepsilon) + \mathrm{e}^{-1/\varepsilon}$. □

Theorem 15.13. *An asymptotic expansion vanishes only if the coefficients vanish; i.e.,*

$$\{a_0\mu_0(\varepsilon) + a_1\mu_1(\varepsilon) + a_2\mu_2(\varepsilon) + \cdots = 0 \quad (\varepsilon \to 0)\} \quad \Leftrightarrow \quad \{a_0 = a_1 = a_2 = \cdots = 0\}.$$

Proof. The sequence $\{\mu_n\}$ is asymptotically ordered, so both $\mu_0 a_0 = -\mu_1 a_1 - \cdots = \mathcal{O}(\mu_1)$ and $\mu_1 = o(\mu_0)$. So there is a positive constant K such that for any positive δ there is an ε interval where $|a_0\mu_0| < \delta K|\mu_0|$, which is only possible if $a_0 = 0$. This may now be repeated for a_1, etc. This proves $\Rightarrow$. The proof of $\Leftarrow$ is trivial. □

15.2.3 Perturbation Problems

The assumed existence of an asymptotic expansion yields a class of methods to solve problems that depend on a parameter that is typically small in the range of interest. Such methods are called *perturbation methods*. The importance of these methods is twofold. They provide analytic solutions to otherwise intractable problems, and the asymptotic structure of the solution provides instant insight into the dominating qualities.

If $a(\varepsilon)$ is implicitly given as the solution of an algebraic equation

$$\mathcal{F}(a; \varepsilon) = 0 \tag{15.1}$$

and both $a(\varepsilon)$ and $\mathcal{F}(a; \varepsilon)$ have an asymptotic series expansion with the same gauge functions, $a(\varepsilon)$ may be determined asymptotically by the following perturbation method. We expand a, substitute this expansion in $\mathcal{F}$, and expand $\mathcal{F}$ to obtain

$$a(\varepsilon) = a_0\mu_0(\varepsilon) + a_1\mu_1(\varepsilon) + \cdots, \tag{15.2a}$$

$$\mathcal{F}(a; \varepsilon) = \mathcal{F}_0(a_0)\mu_0(\varepsilon) + \mathcal{F}_1(a_1, a_0)\mu_1(\varepsilon) + \mathcal{F}_2(a_2, a_1, a_0)\mu_2(\varepsilon) + \cdots = 0. \tag{15.2b}$$

From Theorem 15.13 it follows that each term $\mathcal{F}_n$ vanishes, and the sequence of coefficients (a_n) can be determined by induction:

$$\mathcal{F}_0(a_0) = 0, \quad \mathcal{F}_1(a_1, a_0) = 0, \quad \mathcal{F}_2(a_2, a_1, a_0) = 0, \quad \text{etc.} \tag{15.3}$$

It should be noted that finding the sequence of gauge functions (μ_n) is of particular importance. This is done iteratively. First, the order of magnitude of a should be determined by seeking the asymptotic scaling $a(\varepsilon) = \gamma(\varepsilon)A(\varepsilon)$ that yields a meaningful $A = \mathcal{O}_s(1)$ in the limit $\varepsilon \to 0$. This is called a *distinguished limit*, while the reduced equation for $A(0)$, i.e., $\mathcal{F}_0(A) = 0$, is called a *significant degeneration* (there may be more than one). The first gauge function that occurs is now $\mu_0(\varepsilon) = \gamma(\varepsilon)$, while $a_0 = A(0)$. The procedure may be repeated for the new unknown $a(\varepsilon) - \mu_0(\varepsilon)a_0$, and so on. Often the rest of the sequence (μ_n) can be guessed from the structure of the defining equation $\mathcal{F} = 0$.

We illustrate this procedure by the following example.

Example 15.14 Consider the roots for $\varepsilon \to 0$ of the equation

$$x^3 - \varepsilon x^2 + 2\varepsilon^3 x + 2\varepsilon^6 = 0.$$

From the structure of the equation it seems reasonable to assume that the solutions $x^{(1)}, x^{(2)}, x^{(3)}$ have an asymptotic expansion in powers of ε. However, the order of magnitude of the leading-order term is not immediately clear:

$$x(\varepsilon) = \varepsilon^n\big(X_0 + \varepsilon X_1 + \varepsilon^2 X_2 + \mathcal{O}(\varepsilon^3)\big).$$

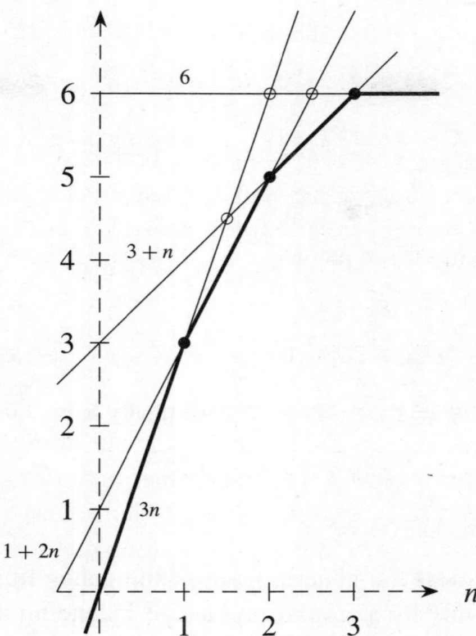

Figure 15.2. *Analysis of distinguished limits.*

Therefore we have to determine the exponent n first. This is done by balancing terms. We scale

$$x = \varepsilon^n X(\varepsilon), \qquad X = \mathcal{O}(1),$$

and we seek an n that produces a nontrivial limit as $\varepsilon \to 0$. We compare asymptotically the coefficients in the equation that remain after scaling:

$$\varepsilon^{3n} X^3 - \varepsilon^{1+2n} X^2 + 2\varepsilon^{3+n} X + 2\varepsilon^6 = 0.$$

In order to have a meaningful (or "significant") degenerate solution $X(0) = \mathcal{O}_s(1)$, at least two terms of the equation should be asymptotically equivalent and at the same time of leading order when $\varepsilon \to 0$. So this leaves us with the task of comparing the exponents $3n$, $1 + 2n$, $3 + n$, and 6 as a function of n. Consider Figure 15.2. The solid lines denote the exponents of the powers of ε that occur in the coefficients of the equation considered. At the intersections of these lines, denoted by the open and solid circles, we find the candidates of distinguished limits, i.e., the points where at least two coefficients are asymptotically equivalent. Finally, only the solid circles are the distinguished limits, because these are located along the lower envelope (thick solid line) and therefore correspond to leading-order terms when $\varepsilon \to 0$. We now have three cases.

Case 1. $n = 1$

$$\varepsilon^3 X^3 - \varepsilon^3 X^2 + 2\varepsilon^4 X + 2\varepsilon^6 = 0 \quad \text{or} \quad X^3 - X^2 + 2\varepsilon X + 2\varepsilon^3 = 0.$$

If we assume the expansion $X = X_0 + \varepsilon X_1 + \cdots$, we finally have

$$X_0^3 - X_0^2 = 0, \quad 3X_0^2 X_1 - 2X_0 X_1 + 2X_0 = 0, \quad \text{etc.},$$

and so $X_0 = 1$ and $X_1 = -2$, etc., leading to $x(\varepsilon) = \varepsilon - 2\varepsilon^2 + \cdots$. Note that the solution $X_0 = 0$ is excluded because that would change the order of the scaling!

Case 2. $n = 2$

$$\varepsilon^6 X^3 - \varepsilon^5 X^2 + 2\varepsilon^5 X + 2\varepsilon^6 = 0 \quad \text{or} \quad \varepsilon X^3 - X^2 + 2X + 2\varepsilon = 0.$$

If we assume the expansion $X = X_0 + \varepsilon X_1 + \cdots$, we finally have

$$-X_0^2 + 2X_0 = 0, \qquad \text{etc.,}$$

and so $X_0 = 2$, etc., leading to $x(\varepsilon) = 2\varepsilon^2 + \cdots$.

Case 3. $n = 3$

$$\varepsilon^9 X^3 - \varepsilon^7 X^2 + 2\varepsilon^6 X + 2\varepsilon^6 = 0 \quad \text{or} \quad \varepsilon^3 X^3 - \varepsilon X^2 + 2X + 2 = 0.$$

If we assume the expansion $X = X_0 + \varepsilon X_1 + \cdots$, we finally have

$$2X_0 + 2 = 0, \qquad \text{etc.,}$$

and so $X_0 = -1$, etc., leading to $x(\varepsilon) = -\varepsilon^3 + \cdots$, □

It is not always so easy to guess the general form of the gauge functions. Then all terms have to be estimated iteratively by a similar process of balancing as for the leading-order term. See Exercise 15.4.

15.2.4 Asymptotic Expansions of Poincaré Type

An asymptotic expansion of a function of variable $\boldsymbol{x}$ and small parameter ε of the form

$$f(\boldsymbol{x};\varepsilon) = \sum_{n=0}^{N-1} a_n(\boldsymbol{x};\varepsilon)\mu_n(\varepsilon) + \mathcal{O}(\mu_N), \qquad a_n = \mathcal{O}(1),$$

appears to be too general to be of practical use. The restriction that the coefficients a_n depend on $\boldsymbol{x}$ only appears to be fruitful. This is called a *Poincaré expansion*, defined more precisely here.

Definition 15.15. *If* $\{\mu_n(\varepsilon)\}_{n=0}^{\infty}$ *is an asymptotic sequence and* $f(\boldsymbol{x};\varepsilon)$ *has an asymptotic expansion of N terms with respect to this sequence given by*

$$f(\boldsymbol{x};\varepsilon) \sim \sum_{n=0}^{N-1} a_n(\boldsymbol{x})\mu_n(\varepsilon),$$

where the shape functions $a_n(\boldsymbol{x})$ *are independent of* ε*, this expansion is called a* Poincaré expansion. *The domain of* $\boldsymbol{x}$ *may depend on* ε.

Example 15.16 For $x > 0$, but not for $x \in (-\varepsilon, 0]$, we have the Poincaré expansion

$$\ln(x+\varepsilon) = \ln x + \frac{\varepsilon}{x} - \frac{\varepsilon^2}{2x^2} + \frac{\varepsilon^3}{3x^3} + \mathcal{O}(\varepsilon^4).$$ □

Example 15.17 $\sin\left(\frac{x}{\varepsilon}\right)$ has no Poincaré expansion for $x \neq 0$. □

Example 15.18 The Poincaré expansion of $e^{-x/\varepsilon}$ with respect to $\{\varepsilon^n\}$ is equal to zero for $x \in [A, \infty)$, is equal to one for $x = 0$, and does not exist for $x \in (-\infty, -A]$. (A is fixed and positive.) □

Definition 15.19. *If a Poincaré expansion is uniform in* $\boldsymbol{x}$ *on a given domain* $\mathcal{D}$ *(Definition* 15.2*), this expansion is called a* regular expansion. *Otherwise, the expansion is called a* singular expansion.

(Note that there is no uniformity in the literature on the definition of regular and singular expansions.) Regular expansions may be differentiated with respect to the independent variable $\boldsymbol{x}$.

Example 15.20 The Poincaré expansion

$$\cos(x+\varepsilon) = \cos(x) - \varepsilon\sin(x) - \frac{1}{2}\varepsilon^2\cos(x) + \frac{1}{6}\varepsilon^3\sin(x) + \mathcal{O}(\varepsilon^4),$$

with respect to the gauge functions $\mu_n(\varepsilon) = \varepsilon^n$ and with domain $\mathcal{D} = \mathbb{R}$, is uniform since $\cos(x)$ and $\sin(x)$ are bounded for all $x \in \mathbb{R}$. It follows that it is a regular expansion. □

Example 15.21 The expansion

$$\cos(x+\varepsilon)\,e^{-x/\varepsilon} + \sin(x+\varepsilon) = \sin x + \varepsilon\cos x + \mathcal{O}(\varepsilon^2)$$

is a uniform, and therefore regular, expansion on any interval $[A, \infty)$, where $A > 0$. However, it is a nonuniform, and therefore singular, expansion on $[0, \infty)$. In fact, on any interval $[A\varepsilon^\alpha, \infty)$ it is regular if $\alpha < 1$ and singular if $\alpha \geq 1$. □

It is important to appreciate the central role of the *choice of the independent variable* $\boldsymbol{x}$ in a Poincaré expansion. By suitable linear coordinate transformations of the type $\boldsymbol{x} = \lambda(\varepsilon) + \delta(\varepsilon)\boldsymbol{\xi}$ we can change and optimize the domain of uniformity. This filters out specific behaviour that belongs to one asymptotic length scale.

Example 15.22

- $\sin(x + \varepsilon x + \varepsilon^2 x) = \sin(x) + \varepsilon x\cos(x) + \varepsilon^2(x\cos x - \frac{1}{2}x^2\sin x) + \mathcal{O}(\varepsilon^3)$, is only uniform on an interval $[0, A]$, but if we introduce $\xi = (1+\varepsilon)x$, we have $\sin(\xi + \varepsilon^2 x) = \sin(\xi) + \mathcal{O}(\varepsilon)$ uniform in $x \in [0, A\varepsilon^{-1}]$ for any positive constant A.
- $\sin(\varepsilon x + \varepsilon) = \varepsilon x + \varepsilon + \mathcal{O}(\varepsilon^3)$ is only uniform on a finite interval and, moreover, does not show any of the inherent periodicity. If we introduce $X = \varepsilon x$, we get the much better $\sin(X+\varepsilon) = \sin(X) + \mathcal{O}(\varepsilon)$, which is even uniform in $\mathbb{R}$.
- $e^{-x/\varepsilon} = 0 + o(\varepsilon^n)$ is a singular expansion on $x > 0$, but if we introduce $\xi = x/\varepsilon$, it becomes the regular expansion $e^{-\xi} = \mathcal{O}(1)$ on $\xi > 0$.
- On $x > 0$ we have $\frac{2}{\pi}\arctan(\frac{x}{\varepsilon}) + \sin(\varepsilon x)/(1+x^2)$ equal to $1 + \varepsilon\big(x/(1+x^2) - 2/\pi x\big) + \mathcal{O}(\varepsilon^2)$ in x, $1 + \varepsilon^2\big(\sin(X)/X^2 - 2/\pi X\big) + \mathcal{O}(\varepsilon^3)$ in $X = \varepsilon x$, and $\frac{2}{\pi}\arctan(x) + \varepsilon^2 x + \mathcal{O}(\varepsilon^3)$ in $\xi = x/\varepsilon$.
- $e^{-\varepsilon x}\sin(x\sqrt{1+\varepsilon}\,) = \sin x + \varepsilon x(\frac{1}{2}\cos x - \sin x) + \mathcal{O}(\varepsilon^2)$ is only uniform on an interval $[0, A]$. This cannot be improved by a single other choice of independent variable. However, if we introduce *two* variables, $x_1 = \varepsilon x$ and $x_2 = (1 + \frac{1}{2}\varepsilon)x$, we get the much better $e^{-x_1}\sin x_2 - \frac{1}{8}\varepsilon^2 x_2\,e^{-x_1}\cos x_2 + \mathcal{O}(\varepsilon^4)$. □

15.3 Regular Perturbation Problems

If a function $\Phi(x;\varepsilon)$ is implicitly given by an equation (usually a differential equation with boundary conditions), say

$$\mathcal{L}[\Phi](x;\varepsilon)=0 \qquad \text{on a domain} \quad \mathcal{D}, \tag{15.4}$$

and both Φ and $\mathcal{L}[\Phi](x;\varepsilon)$ have a regular asymptotic expansion on $\mathcal{D}$ with the same gauge functions, (15.4) is called a *regular perturbation problem* [65]. The shape functions Φ_n are determined as follows. We expand $\mathcal{L}[\Phi]$ as

$$\begin{aligned}\mathcal{L}[\Phi](x;\varepsilon)&=\mu_0(\varepsilon)\mathcal{L}_0[\Phi_0](x)\\&\quad+\ \mu_1(\varepsilon)\mathcal{L}_1[\Phi_1,\Phi_0](x)+\mu_2(\varepsilon)\mathcal{L}_2[\Phi_2,\Phi_1,\Phi_0](x)+\cdots=0.\end{aligned} \tag{15.5}$$

According to Theorem 15.13, each term vanishes, and the sequence (Φ_n) can be determined by induction:

$$\mathcal{L}_0[\Phi_0](x)=0,\quad \mathcal{L}_1[\Phi_1,\Phi_0](x)=0,\quad \mathcal{L}_2[\Phi_2,\Phi_1,\Phi_0](x)=0,\quad \ldots. \tag{15.6}$$

It should be noted that in many interesting cases the problem is only regular after a suitable coordinate transformation. The major task when solving the problem is then to find this scaled or shifted coordinate. Practically important solution methods of this type are the *method of slow variation*, for geometrically stretched or slowly varying configurations, and the *Lindstedt–Poincaré method*, for solutions that are periodic in time with an unknown ε-dependent period.

If (15.4) is not a regular perturbation problem, we call it a singular perturbation problem. Practically important solution methods for singular perturbation problems are the MAE method, where regular expansions exist locally but not in the whole region considered, and the method of multiple scales, where two or more distinct long and short length scales are intertwined.

15.3.1 Method of Slow Variation

Suppose we have a function $\varphi(\boldsymbol{x};\varepsilon)$ of spatial coordinates $\boldsymbol{x}$ and a small parameter ε such that the typical variation in one direction, say x, is of the order of the length scale ε^{-1}. We can express this behaviour most conveniently by writing $\varphi(x,y,z;\varepsilon)=\Phi(\varepsilon x,y,z;\varepsilon)$. Now if we were to expand Φ for small ε, we might, for example, get something like

$$\Phi(\varepsilon x,y,z;\varepsilon)=\Phi(0,y,z;0)+\varepsilon(x\Phi_x(0,y,z;0)+\Phi_\varepsilon(0,y,z;0))+\cdots,$$

which is only uniform in x on an interval $[0,L]$ if $L=\mathcal{O}(1)$, and the inherent slow variation on the longer scale of $x=\mathcal{O}(\varepsilon^{-1})$ would be masked. It is clearly much better to introduce the scaled variable $X=\varepsilon x$, and an (assumed) regular expansion of $\Phi(X,y,z;\varepsilon)$

$$\Phi(X,y,z;\varepsilon)=\mu_0(\varepsilon)\varphi_0(X,y,z)+\cdots \tag{15.7}$$

now retains the slow variation in X in the shape functions of the expansion.

This situation frequently happens when the geometry involved is slender [163]. The theories of one-dimensional gas dynamics, lubrication flow, and sound propagation in horns (Webster's horn equation) are important examples, although they are usually not derived systematically without explicit reference to the slender geometry. We will illustrate the method for heat flow in a varying bar, quasi one-dimensional gas flow, and the shallow-water problem.

Example 15.23 (heat flow in a bar) Consider the stationary problem of the temperature distribution T in a long heat-conducting bar with outward surface normal $\boldsymbol{n}$ and slowly varying cross section $\mathcal{A}$. The bar is kept at a temperature difference such that a given heat flux is maintained, but it is otherwise isolated. As there is no leakage of heat, the flux is constant. With spatial coordinates made dimensionless on a typical bar cross section, we have the following equation and boundary conditions:

$$\nabla^2 T = 0, \qquad \nabla T \cdot \boldsymbol{n} = 0, \qquad \iint_{\mathcal{A}} \frac{\partial T}{\partial x} \mathrm{d}S = Q.$$

After integrating $\nabla^2 T$ over a slice $x_1 \leq x \leq x_2$ and applying Gauss's theorem, we find that the axial flux Q is indeed independent of x. The typical length scale of diameter variation is assumed to be much larger than a diameter. We introduce the ratio between a typical diameter and this length scale as the small parameter ε and write for the bar surface

$$S(X, r, \theta) := r - R(X, \theta) = 0, \qquad X = \varepsilon x,$$

where (x, r, θ) form a cylindrical coordinate system (see Figure 15.3). By writing R as a continuous function of the slow variable X, rather than x, we have made our formal assumption of slow variation explicit in a convenient and simple way, since $R_x = \varepsilon R_X = \mathcal{O}(\varepsilon)$.

The crucial step will now be the assumption that the temperature is *only* affected by the geometric variation induced by R. Any initial or entrance effects are ignored or have disappeared. As a result, the temperature field T is a function of X, rather than x, and its axial gradient scales on ε as $T_x = \mathcal{O}(\varepsilon)$.

We introduce the gradient ∇S and the transverse gradient $\nabla_\perp S$:

$$\nabla S = -\varepsilon R_X \boldsymbol{e}_x + \boldsymbol{e}_r - r^{-1} R_\theta \boldsymbol{e}_\theta, \qquad \nabla_\perp S := S_r \boldsymbol{e}_r + r^{-1} S_\theta \boldsymbol{e}_\theta = \boldsymbol{e}_r - r^{-1} R_\theta \boldsymbol{e}_\theta.$$

At the bar surface $S = 0$ the gradient ∇S is a vector normal to the surface, while the transverse gradient $\nabla_\perp S$, directed in the plane of a cross section $X =$ constant, is normal to the circumference $S(X = c, r, \theta) = 0$. Inside the bar we have the rescaled heat equation

$$\varepsilon^2 T_{XX} + \nabla_\perp^2 T = 0. \qquad (*)$$

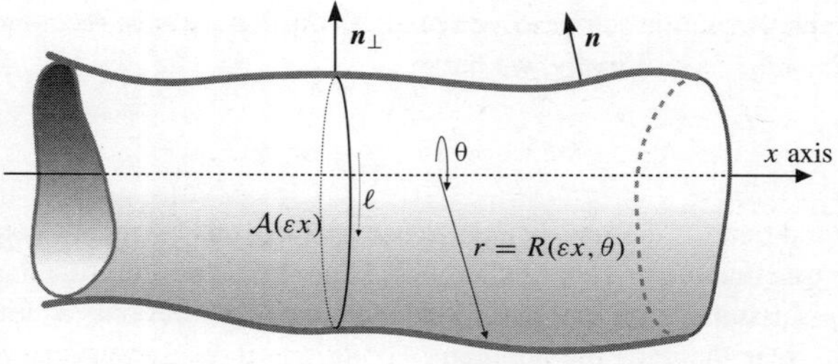

Figure 15.3. *Slowly varying bar.*

At the wall the boundary condition of vanishing heat flux is

$$\nabla T \cdot \nabla S = \varepsilon^2 T_X S_X + \nabla_\perp T \cdot \nabla_\perp S = 0 \quad \text{at} \quad S = 0. \tag{\dagger}$$

The flux condition, for lucidity rewritten with $Q = \varepsilon q$, is given by

$$\iint_{\mathcal{A}} \frac{\partial T}{\partial X}\, \mathrm{d}S = q.$$

This problem is too difficult in general, so we try to utilize the small parameter ε in a systematic manner. Since the perturbation terms are $\mathcal{O}(\varepsilon^2)$, we assume the asymptotic expansion

$$T(X, r, \theta; \varepsilon) = T_0(X, r, \theta) + \varepsilon^2 T_1(X, r, \theta) + \mathcal{O}(\varepsilon^4).$$

After substitution in $(*)$ and boundary condition $(\dagger)$, further expansion in powers of ε^2, and equating like powers of ε, we obtain to leading order a Laplace equation in (r, θ):

$$\nabla_\perp^2 T_0 = 0 \quad \text{with} \quad \nabla_\perp T_0 \cdot \nabla_\perp S = 0 \quad \text{at} \quad S = 0.$$

An obvious solution is $T_0(X, r, \theta) \equiv 0$. Since solutions of Laplace's equation with vanishing normal derivatives at the boundary are unique up to a constant (here a function of X), we have

$$T_0 = T_0(X).$$

We could substitute this directly in the flux condition to find that $AT_{0X} = q$, where $A(X)$ is the area of cross section $\mathcal{A}(X)$. For the present exposition, however, it is of interest to show that this result also emerges from the equations as follows. To obtain an equation for T_0 in X we continue with the $\mathcal{O}(\varepsilon^2)$ equation and corresponding boundary condition

$$\nabla_\perp^2 T_1 + T_{0,XX} = 0, \qquad \nabla_\perp T_1 \cdot \nabla_\perp S = -T_{0,X} S_X. \tag{\ddagger}$$

The boundary condition can be rewritten as

$$\nabla_\perp T_1 \cdot \boldsymbol{n}_\perp = \frac{T_{0,X} R_X}{|\nabla_\perp S|} = \frac{T_{0,X} R R_X}{\sqrt{R^2 + R_\theta^2}},$$

where $\boldsymbol{n}_\perp = \nabla_\perp S / |\nabla_\perp S|$ is the transverse unit normal vector. By integrating $(\ddagger)$ over a cross section $\mathcal{A}$ of area $A(X)$, using Gauss's theorem, and noting that $A = \int_0^{2\pi} \frac{1}{2} R^2\, \mathrm{d}\theta$ and that a circumferential line element is given by $\mathrm{d}\ell = (R^2 + R_\theta^2)^{1/2} \mathrm{d}\theta$, we obtain

$$\iint_{\mathcal{A}} \left(\nabla_\perp^2 T_1 + T_{0,XX}\right) \mathrm{d}S = \oint_{\partial\mathcal{A}} \nabla_\perp T_1 \cdot \boldsymbol{n}_\perp\, \mathrm{d}\ell + AT_{0,XX}$$

$$= T_{0,X} \int_0^{2\pi} R R_X\, \mathrm{d}\theta + AT_{0,XX} = A_X T_{0,X} + AT_{0,XX} = \frac{\mathrm{d}}{\mathrm{d}X}\left(A \frac{\mathrm{d}}{\mathrm{d}X} T_0\right) = 0.$$

The finally obtained equation can be solved easily. Note that we have recovered the conservation law of heat flux $AT_{0X} = q$. Finally, we have

$$T_0(X) = \int^X \frac{q}{A(z)}\, \mathrm{d}z + T_{\text{ref}}.$$

It should be noted that we did not include in our analysis any boundary conditions at the ends of the bar. It is true that the present method fails here. The found solution is uniformly valid on $\mathbb{R}$ (since $R(X)$ is assumed continuous and independent of ε), but only as long as we stay away from any end. Near the ends the boundary conditions induce transverse gradients of $\mathcal{O}(1)$, which make the prevailing length scale x again, rather than X. This region is asymptotically of boundary layer type and should be treated differently (see below). □

Example 15.24 (quasi one-dimensional gas dynamics) Consider a slowly varying duct with irrotational inviscid isentropic flow, described (in dimensionless form) by the velocity potential φ and density ρ satisfying the equation for mass conservation and the compressible form of Bernoulli's equation (see Chapter 7); i.e.,

$$\nabla\cdot(\rho\nabla\varphi) = 0, \qquad \frac{1}{2}|\nabla\varphi|^2 + \frac{\rho^{\gamma-1}}{\gamma-1} = E.$$

The parameter γ is a gas constant (1.4 for air) and E is a constant of the problem. With the same notation as in the previous example, the duct wall is given by $S(\varepsilon x, r, \theta) = 0$, while at the impermeable wall we have $\nabla\varphi\cdot\nabla S = 0$. The mass flux, the same at any cross section A, is given by

$$\iint_A \rho\varphi_x \,\mathrm{d}S = F.$$

We introduce the slow variable $X = \varepsilon x$ and assume φ and ρ to depend essentially on X rather than x. The dimensionless axial flow velocity φ_x, the density ρ, the cross-sectional area A, the flux F, and the thermodynamical constant E are $\mathcal{O}(1)$. So we have to rescale φ and write

$$\varphi(x, y, z; \varepsilon) = \varepsilon^{-1}\Phi(X, y, z; \varepsilon).$$

The equations for Φ and ρ become

$$\varepsilon^2\frac{\partial}{\partial X}\left(\rho\frac{\partial}{\partial X}\Phi\right) + \nabla_\perp\cdot(\rho\nabla_\perp\Phi) = 0, \qquad \frac{1}{2}\Phi_X^2 + \frac{1}{2}\varepsilon^{-2}|\nabla_\perp\Phi|^2 + \frac{\rho^{\gamma-1}}{\gamma-1} = E,$$

with boundary condition

$$\nabla\Phi\cdot\nabla S = \varepsilon^2\Phi_X S_X + \nabla_\perp\Phi\cdot\nabla_\perp S = 0 \qquad \text{at} \quad S = 0.$$

We assume the expansions

$$\Phi(X, y, z; \varepsilon) = \Phi_0(X, y, z) + \mathcal{O}(\varepsilon^2), \qquad \rho(X, y, z; \varepsilon) = \rho_0(X, y, z) + \mathcal{O}(\varepsilon^2).$$

From Bernoulli's equation it follows that $|\nabla_\perp\Phi_0|^2 = 0$, so $\Phi_0 = \Phi_0(X)$, and hence $\rho_0 = \rho_0(X)$. From the mass flux equation we get $\Phi_{0X} = F/\rho_0 A$, while the algebraic equation

$$\frac{F^2}{2\rho_0^2 A^2} + \frac{\rho_0^{\gamma-1}}{\gamma-1} = E$$

finally determines ρ_0 (in general to be solved numerically). □

Example 15.25 (shallow-water equations) The irrotational motion under gravity of a horizontal layer of inviscid, incompressible water is described by the equation for mass conservation and the Bernoulli equation (see (7.11)):

$$\nabla^2\varphi = 0, \qquad \frac{\partial\varphi}{\partial t} + \frac{1}{2}|\nabla\varphi|^2 + \frac{p}{\rho_0} + gz = C(t),$$

where φ is the velocity potential with velocity $\boldsymbol{v} = \nabla\phi$, ρ_0 is the density, p is the pressure, g is the gravitational acceleration, and C is an unimportant function of time. The boundary conditions are provided by the impermeability of the bottom at $z = 0$, the assumption that the free surface $z = h(x, y, t)$ consists of streamlines (particles remain there), and the fact that the pressure is uniformly constant along the free surface (the big difference between the densities of water and air makes the water insensitive to any air motion). As any constant in pressure may be absorbed by C, we may assume that the surface pressure is zero. These conditions result in

$$z=0: \quad \frac{\partial\varphi}{\partial z}=0,$$

$$z=h: \quad \frac{\partial\varphi}{\partial z}=\frac{\partial h}{\partial t}+\frac{\partial\varphi}{\partial x}\frac{\partial h}{\partial x}+\frac{\partial\varphi}{\partial y}\frac{\partial h}{\partial y} \quad \text{and} \quad p=0.$$

We assume that the typical horizontal velocities U are so large and the frequencies f are so low that the corresponding typical length scale $L=U/f$ is large compared to the typical water depth D (e.g., tidal motion). To quantify this slenderness we introduce the small parameter $\varepsilon=D/L$. We are interested in the situation where pressure is coupled to both the inertia of the flow and the effects of gravity. This corresponds to the assumption that p scales on $\rho_0 U^2$ and the inverse-squared Froude number $\gamma=gD/U^2=O(1)$. Suppose we scale and nondimensionalise as follows:

$$x=LX,\, y=LY,\, z=DZ,\, t=LU^{-1}\tau,\, \varphi=UL\psi,\, p=\rho_0 U^2 P,\, h=DH.$$

We also introduce the operator $\nabla_\perp := \boldsymbol{e}_x\frac{\partial}{\partial X}+\boldsymbol{e}_y\frac{\partial}{\partial Y}$, which is ∇ restricted to X and Y. Then we obtain

$$\frac{\partial^2\psi}{\partial Z^2}+\varepsilon^2\nabla_\perp^2\psi=0, \qquad \frac{\partial\psi}{\partial\tau}+\frac{1}{2}|\nabla_\perp\psi|^2+\frac{1}{2}\varepsilon^{-2}\left(\frac{\partial\psi}{\partial Z}\right)^2+P+\gamma Z=\tilde{C}(t),$$

with boundary conditions

$$Z=0: \quad \frac{\partial\psi}{\partial Z}=0,$$

$$Z=H: \quad \frac{\partial\psi}{\partial Z}=\varepsilon^2\left(\frac{\partial H}{\partial\tau}+\nabla_\perp\psi\cdot\nabla_\perp H\right) \quad \text{and} \quad P=0.$$

Assuming no interfering $\mathcal{O}(\varepsilon)$ effects (e.g., from initial or boundary conditions), we expand in powers of ε^2 the only small parameter that occurs:

$$\psi=\Psi_0+\varepsilon^2\Psi_1+\mathcal{O}(\varepsilon^4), \quad H=H_0+\varepsilon^2 H_1+\mathcal{O}(\varepsilon^4), \quad P=P_0+\varepsilon^2 P_1+\mathcal{O}(\varepsilon^4).$$

From this we obtain to leading order $\Psi_{0,ZZ}=0$, which integrates to $\Psi_{0,Z}=B_0(X,Y,\tau)=0$ because of the boundary conditions. So we finally have

$$\Psi_0=A_0(X,Y,\tau).$$

To first order we have $\Psi_{1,ZZ}=-\nabla_\perp^2\Psi_0$, which integrates to

$$\frac{\partial}{\partial Z}\Psi_1=-Z\nabla_\perp^2\Psi_0$$

when we take into account the boundary condition at $Z=0$. Next we expand and substitute these results into Bernoulli's equation (note that $\varepsilon^{-2}(\psi_Z)^2=\mathcal{O}(\varepsilon^2)$) and get

$$\frac{\partial}{\partial t}\Psi_0+\frac{1}{2}\left|\nabla_\perp\Psi_0\right|^2+P_0+\gamma Z=\tilde{C}(t).$$

By the pressure boundary condition at $Z=H_0$ this yields

$$\frac{\partial}{\partial t}\Psi_0+\frac{1}{2}\left|\nabla_\perp\Psi_0\right|^2+\gamma H_0=\tilde{C}(t). \qquad (*)$$

Finally, the streamline condition produces

$$\frac{\partial}{\partial Z}\Psi_1 = -H_0\nabla_\perp^2\Psi_0 = \frac{\partial}{\partial\tau}H_0 + \nabla_\perp\Psi_0\cdot\nabla_\perp H_0$$

or

$$\frac{\partial}{\partial\tau}H_0 + \nabla_\perp\cdot\big(H_0\nabla_\perp\Psi_0\big) = 0. \tag{$\dagger$}$$

Equations ($*$) and ($\dagger$) are known as a form of the *shallow-water equations* (see also Chapter 12). They are not generally solvable, and their behaviour requires extensive analysis.

A family of simple wave solutions (i.e., along a characteristic) may be found as follows (cf. [169]). We look for a plane wave in, say, the $\xi = X\cos\theta + Y\sin\theta$ direction, so we have, after rewriting the equations in terms of ξ, velocity $V_0 = \Psi_{0,\xi}$, and height H_0, that

$$\frac{\partial}{\partial\tau}H_0 + \frac{\partial}{\partial\xi}(V_0H_0) = 0, \qquad \frac{\partial}{\partial\tau}V_0 + \frac{\partial}{\partial\xi}\left(\frac{1}{2}V_0^2 + \gamma H_0\right) = 0.$$

Evidently, both equations may be written in the same characteristic form if

$$\frac{\mathrm{d}}{\mathrm{d}H_0}(V_0H_0) = \frac{\mathrm{d}}{\mathrm{d}V_0}\left(\frac{1}{2}V_0^2 + \gamma H_0\right).$$

This is satisfied if $(\frac{\mathrm{d}}{\mathrm{d}H_0}V_0)^2H_0 = \gamma$, or

$$V_0 = \pm 2\sqrt{\gamma H_0} + c_0,$$

where c_0 is a constant. We can now assume that we have $H_0 = H_0(\eta)$ and the corresponding $V_0 = V_0(\eta)$, where $\eta = \eta(x,t)$ satisfies

$$\frac{\partial}{\partial\tau}\eta + \big(V_0 \pm \sqrt{\gamma H_0}\big)\frac{\partial}{\partial\xi}\eta = 0.$$

Using the result of Exercise 1.5, we obtain solutions implicitly described by

$$\eta = F(\xi - f(\eta)\tau), \qquad \text{where} \quad f(\eta) = V_0(\eta) \pm \sqrt{\gamma H_0(\eta)}$$

and F is any suitable differentiable function. □

15.3.2 Lindstedt–Poincaré Method

When we have a function f depending on a small parameter ε and periodic in t with fundamental frequency $\omega(\varepsilon)$, we can write f as a Fourier series

$$f(t;\varepsilon) = \sum_{n=-\infty}^{\infty} A_n(\varepsilon)\,\mathrm{e}^{\mathrm{i}n\omega(\varepsilon)t}. \tag{15.8}$$

If amplitudes and frequency have an asymptotic expansion for small ε, say

$$A_n(\varepsilon) = A_{n,0} + \varepsilon A_{n,1} + \cdots, \qquad \omega(\varepsilon) = \omega_0 + \varepsilon\omega_1 + \cdots, \tag{15.9}$$

we have a natural asymptotic series expansion for f of the form

$$f(t;\varepsilon) = \sum_{n=-\infty}^{\infty} A_{n,0}\,\mathrm{e}^{\mathrm{i}n\omega_0t} + \varepsilon\sum_{n=-\infty}^{\infty}\Big(A_{n,1} + \mathrm{i}n\omega_1 tA_{n,0}\Big)\,\mathrm{e}^{\mathrm{i}n\omega_0t} + \cdots. \tag{15.10}$$

This expansion, however, is only uniform in t on an interval $[0, T]$, where $T = o(\varepsilon^{-1})$. On a larger interval, e.g., $[0, \varepsilon^{-1}]$, the asymptotic hierarchy in the expansion becomes invalid because $\varepsilon t = \mathcal{O}(1)$. This happens because of the occurrence of algebraically growing oscillatory terms, called "secular terms." *Secular* means "occurring once in a century," and *saeculum* means "generation," referring to its astronomical origins.

Definition 15.26. *The terms proportional to $t^m \sin(n\omega_0 t)$, $t^m \cos(n\omega_0 t)$ are called* secular terms. *More generally, the name refers to any algebraically growing terms that limit the region of validity of an asymptotic expansion.*

It is therefore far better to apply a coordinate transformation $\tau = \omega(\varepsilon)t$, introduce $F(\tau;\varepsilon) = f(t;\varepsilon)$, and expand F, rather than f, asymptotically. We get the uniformly valid approximation

$$F(\tau;\varepsilon) = \sum_{n=-\infty}^{\infty} A_n(\varepsilon)\,\mathrm{e}^{in\tau} = \sum_{n=-\infty}^{\infty} A_{n,0}\,\mathrm{e}^{in\tau} + \varepsilon \sum_{n=-\infty}^{\infty} A_{n,1}\,\mathrm{e}^{in\tau} + \cdots . \tag{15.11}$$

The method is called the *Lindstedt–Poincaré method* or the *method of strained coordinates*. In practical situations the frequency ω is of course unknown and has to be found. Therefore, when constructing the solution, we have to allow for an unknown coordinate transformation. In order to construct the unknown $\omega(\varepsilon)$, we expand τ, e.g., like

$$\tau = (\omega_0 + \varepsilon\omega_1 + \varepsilon^2\omega_2 + \cdots)t, \tag{15.12}$$

but this depends of course on the problem. Note that the purpose of the scaling is to render the asymptotic expansion of F regular, so it is no restriction to assume $\omega_0 = 1$. The other coefficients are determined from the additional condition that the asymptotic hierarchy should be respected as long as possible. In other words, secular terms should not occur. We will illustrate this with the following example.

Example 15.27 **(the pendulum)** Consider the motion of the pendulum described by the initial value problem (see Example 7.2)

$$\ddot{\theta} + K^2 \sin(\theta) = 0, \qquad \text{with} \quad \theta(0) = \varepsilon, \quad \theta'(0) = 0,$$

where $0 < \varepsilon \ll 1$. After the transformation $\tau = \omega t$ and noting that $\theta = \mathcal{O}(\varepsilon)$, we have

$$\omega^2\theta'' + K^2\left(\theta - \frac{1}{6}\theta^3 + \cdots\right) = 0.$$

We expand

$$\omega = 1 + \varepsilon^2\omega_1 + \cdots, \qquad \theta = \varepsilon\theta_0 + \varepsilon^3\theta_1 + \cdots$$

and find, after substitution, the equations for the first two orders:

$$\theta_0'' + K^2\theta_0 = 0, \qquad \theta_0(0) = 1, \quad \theta_0'(0) = 0,$$

$$\theta_1'' + K^2\theta_1 = -2\omega_1\theta_0'' + \frac{1}{6}K\theta_0^3, \qquad \theta_1(0) = 0, \quad \theta_1'(0) = 0.$$

The solution θ_0 is obviously given by

$$\theta_0 = \cos(K\tau),$$

leading to (see the appendix, Section B) the following equation for θ_1:

$$\theta_1'' + K^2\theta_1 = 2K\left(\omega_1 K + \frac{1}{16}\right)\cos(K\tau) + \frac{1}{24}K\cos(3K\tau).$$

At this point it is essential to observe that the right-hand side consists of two forcing terms: one with frequency $3K$ and one with K, the resonance frequency of the left-hand side. This resonance leads to secular terms, as the solutions will behave like $\tau\sin(K\tau)$ and $\tau\cos(K\tau)$. Therefore, in order to suppress the occurrence of secular terms, the amplitude of the resonant forcing term should vanish, which yields the next-order terms ω_1 and θ_1:

$$\omega_1 = -\frac{1}{16K} \qquad \text{leading to} \quad \theta_1 = \frac{1}{192K}\big(\cos(K\tau) - \cos(3K\tau)\big). \qquad \square$$

15.4 Singular Perturbation Problems

If the solution of the problem considered does not allow a regular expansion, the problem is singular and the solution has no uniform Poincaré expansion in the same variable. We will consider two classes of problems. In the first one the singular behaviour is of boundary layer type and the solution can be built up from locally regular expansions. The solution method is called the method of Matched Asymptotic Expansions. In the other one more time or length scales occur together and a solution is constructed by considering these length scales as if they were independent. The solution method is called the method of multiple scales.

15.4.1 Matched Asymptotic Expansions

Very often it happens that a simplifying limit applied to a more comprehensive model gives a correct approximation for the main part of the domain but not everywhere: the limit is *nonuniform*. This nonuniformity may be in space, in time, or in any other variable. For the moment we think of nonuniformity in space, let's say a small region near $x = 0$. If this region of nonuniformity is crucial for the problem, e.g., because it contains a boundary condition, or a source, the primary reduced problem (which does not include the region of nonuniformity) is not sufficient. This, however, does not mean that no use can be made of the inherent small parameter. The local nature of the nonuniformity itself often makes another reduction possible. In such a case we have apparently two physically connected but different problems, called *inner and outer problems*, as far as the dominating mechanism is concerned. Depending on the problem, we now have two simpler problems, serving as boundary conditions to each other via continuity or *matching* conditions.

Nonuniform Asymptotic Approximations Suppose that a given sufficiently smooth function $\Phi(x;\varepsilon)$, with $0 \le x \le 1$, and $0 < \varepsilon \le \varepsilon_1$, does not have a uniform limit $\varepsilon \to 0$, $x \to 0$. Typically, such a function will depend, apart from x, on combinations like $x/\delta(\varepsilon)$, where $\delta = o(1)$.

Assume that this function does not have a regular asymptotic expansion on the whole interval $[0, 1]$ but only on partial intervals $x \in [\eta(\varepsilon), 1]$, where $\eta = o(1)$ and $\delta = o(\eta)$:

$$\Phi(x;\varepsilon) = \sum_{k=0}^{n} \mu_k(\varepsilon)\varphi_k(x) + o(\mu_n) \qquad \varepsilon \to 0, \quad x = \mathcal{O}(1). \tag{15.13}$$

We call this expansion the *outer expansion*, principally valid in the "$x = \mathcal{O}(1)$" outer region, but extendible to $[\mathcal{O}(\eta), 1]$. Now consider the *stretched coordinate*

$$\xi = \frac{x}{\delta(\varepsilon)}. \tag{15.14}$$

Assume that the transformed $\Psi(\xi;\varepsilon) = \Phi(x;\varepsilon)$ has a nontrivial regular asymptotic expansion on partial intervals $\xi \in [0, \zeta(\varepsilon)/\delta(\varepsilon)]$, where $\eta(\varepsilon) < \zeta(\varepsilon)$:

$$\Psi(\xi;\varepsilon) = \sum_{k=0}^{m} \lambda_k(\varepsilon)\psi_k(\xi) + o(\lambda_m) \qquad \varepsilon \to 0, \quad \xi = \mathcal{O}(1). \tag{15.15}$$

We call this expansion the *inner expansion*, principally valid in the "$\xi = \mathcal{O}(1)$" inner region, but extendible to $[0, \mathcal{O}(\zeta)]$.

Example 15.28

$$\begin{aligned}
\Phi(x;\varepsilon) &= \arctan\left(\frac{x}{\varepsilon}\right) + \sin(x+\varepsilon) \\
&= \frac{\pi}{2} + \sin x + \varepsilon\cos x - \frac{\varepsilon}{x} + \mathcal{O}(\varepsilon^{3/2}) \quad \text{on} \quad \frac{1}{2}\varepsilon^{1/2} \le x \le 1; \\
\Psi(\xi;\varepsilon) &= \arctan(\xi) + \sin(\varepsilon\xi + \varepsilon) \\
&= \arctan(\xi) + \varepsilon(\xi+1) + \mathcal{O}(\varepsilon^{3/2}) \quad \text{on} \quad 0 \le \xi \le \frac{2}{\varepsilon^{1/2}}.
\end{aligned}$$ □

The adjective "nontrivial" is essential: the expansion must be *significant*, i.e., different from the outer expansion in φ_n rewritten in the inner variable ξ. This determines the choice of the inner variable $\xi = x/\delta(\varepsilon)$. The scaling $\delta(\varepsilon)$ is the asymptotically *largest* gauge function with this property. We call the expansion for Ψ the inner expansion or *boundary layer* expansion, where the region $\xi = \mathcal{O}(1)$ or $x = \mathcal{O}(\delta)$ is the boundary layer with thickness δ and ξ is the boundary layer variable. Boundary layers may be nested and may occur at internal points of the domain of Φ. Then they are called *internal layers*. The assumption $\eta < \zeta$, i.e., that the inner and outer expansions may be extended to regions that overlap, is called the *overlap hypothesis*.

Suppose $\Phi(x;\varepsilon)$ has an outer expansion

$$\Phi(x;\varepsilon) = \sum_{k=0}^{n} \mu_k(\varepsilon)\varphi_k(x) + o(\mu_n) \tag{15.16}$$

and a boundary layer $x = \mathcal{O}(\delta)$ with inner expansion

$$\Psi(\xi;\varepsilon) = \sum_{k=0}^{m} \lambda_k(\varepsilon)\psi_k(\xi) + o(\lambda_m). \tag{15.17}$$

Suppose that both expansions are complementary; i.e., there is no other boundary layer in between $x = \mathcal{O}(1)$ and $x = \mathcal{O}(\delta)$. Then the overlap hypothesis says that the expansions represent the same function in an intermediate region of overlap. This overlap region may be described by a stretched variable $x = \eta(\varepsilon)\sigma$, asymptotically in between $\mathcal{O}(1)$ and $\mathcal{O}(\delta)$, so

$\delta \ll \eta \ll 1$. In the overlap region both expansions match, which means that the expansions are asymptotically equivalent and reduce to the same expressions.

A widely used procedure, which is simpler to use than matching by means of intermediate variables but not as universal when the expansions involve logarithms, is Van Dyke's matchings rule [162, 38]: the outer expansion, rewritten in the inner variable, has a regular series expansion that is *equal* to the regular asymptotic expansion of the inner expansion rewritten in the outer variable. Suppose that

$$\sum_{k=0}^{n} \mu_k(\varepsilon)\varphi_k(\delta\xi) = \sum_{k=0}^{m} \lambda_k(\varepsilon)\eta_k(\xi) + o(\lambda_m), \tag{15.18a}$$

$$\sum_{k=0}^{n} \lambda_k(\varepsilon)\psi_k(x/\delta) = \sum_{k=0}^{n} \mu_k(\varepsilon)\theta_k(x) + o(\mu_n). \tag{15.18b}$$

Then the expansion of η_k back to x,

$$\sum_{k=0}^{n} \lambda_k(\varepsilon)\eta_k(x/\delta) = \sum_{k=0}^{n} \mu_k(\varepsilon)\zeta_k(x) + o(\mu_n), \tag{15.19}$$

is such that $\zeta_k = \theta_k$ for $k = 0, \ldots, n$.

The idea of matching is very important because it allows one to move smoothly from one regime to the other. The method of constructing local, but matching, expansions is therefore called Matched Asymptotic Expansions (MAE) [81].

Constructing Asymptotic Solutions The most important application of this concept of inner and outer expansions is that approximate solutions of certain differential equations can be constructed for which the limit under a small parameter is apparently nonuniform. The main lines of argument for constructing an MAE solution to a differential equation satisfying some boundary conditions are as follows. Suppose Φ is given by the equation

$$\mathcal{D}[\Phi', \Phi](x; \varepsilon) = 0 + \text{boundary conditions}, \tag{15.20}$$

where $\Phi' = \frac{\mathrm{d}}{\mathrm{d}x}\Phi$. Then we try to construct an outer solution by looking for *nontrivial degenerations* of $\mathcal{D}$ under $\varepsilon \to 0$; i.e., find $\mu_0(\varepsilon)$ and $\nu_0(\varepsilon)$ such that

$$\lim_{\varepsilon \to 0} \nu_0^{-1}(\varepsilon)\mathcal{D}[\mu_0\varphi_0', \mu_0\varphi_0](x; \varepsilon) = \mathcal{D}_0[\varphi_0', \varphi_0](x) = 0 \tag{15.21}$$

has a nontrivial solution φ_0. A series $\varphi = \mu_0\varphi_0 + \mu_1\varphi_1 + \cdots$ is constructed by repeating the process for $\mathcal{D} - \nu_0\mathcal{D}_0$, etc.

Suppose the approximation is nonuniform. For example, not all boundary conditions can be satisfied. Then we start looking for an inner expansion if we have reason to believe that the nonuniformity is of boundary layer type. The presence, location, and size of the boundary layer(s) are now found by the *correspondence principle*, i.e., the (heuristic) idea that if Φ behaves somehow differently in the boundary layer, then the defining equation must also be essentially different. Therefore we search for *significant degenerations* or *distinguished limits* of $\mathcal{D}$. These are degenerations of $\mathcal{D}$ under $\varepsilon \to 0$, with scaled x and Φ, that contain a maximum of information in the sense that they result from at least two effects (i.e., terms) that balance and dominate.

Example 15.29 Under the limit $\varepsilon \to 0$, the equation $\varepsilon y' + y = \sin x$, $y(0) = 1$, reduces to $y = \sin x$ with $y(0) \neq 1$. After the scaling $x = \varepsilon\xi$, the equation reduces to the essentially different $y_\xi + y = 0$. □

The next step is then to select from these distinguished limits the one(s) allowing a solution that matches with the outer solution and satisfies any applicable boundary conditions. Symbolically, we have the following algorithm:

find $x_0, \delta(\varepsilon), \lambda(\varepsilon), \kappa(\varepsilon)$

with $x = x_0 + \delta\xi, \qquad \Phi(x;\varepsilon) = \lambda(\varepsilon)\Psi(\xi;\varepsilon)$

such that $B_0(\psi_0', \psi_0, \xi) = \lim_{\varepsilon\to 0} \kappa^{-1}\mathcal{D}[\delta^{-1}\lambda\Psi', \lambda\Psi](x_0+\delta\xi;\varepsilon)$ is a significant degeneration,

while $B_0(\psi_0', \psi_0, \xi) = 0$ has a solution satisfying the boundary and matching conditions.

Again, an asymptotic expansion may be constructed inductively by repeating the argument. It is of practical importance to note that the order estimate λ of Φ in the boundary layer is often determined a posteriori by boundary or matching conditions. We can illustrate some of the main arguments by considering

$$\mathcal{D}[\varphi', \varphi](x;\varepsilon) = \varepsilon\frac{\mathrm{d}^2\varphi}{\mathrm{d}x^2} + \frac{\mathrm{d}\varphi}{\mathrm{d}x} - 2x = 0, \qquad \varphi(0) = \varphi(1) = 2. \tag{15.22}$$

The leading-order outer equation is evidently (with $\mu_0 = \nu_0 = 1$)

$$\mathcal{D}_0 = \frac{\mathrm{d}\varphi_0}{\mathrm{d}x} - 2x = 0, \tag{15.23}$$

with solution

$$\varphi_0 = x^2 + A. \tag{15.24}$$

The integration constant A can be determined by the boundary condition $\varphi_0(0) = 2$ at $x = 0$ or $\varphi_0(1) = 2$ at $x = 1$, but not both, so we expect a boundary layer at either end. By trial and error we find that no solution can be constructed if we assume a boundary layer at $x = 1$, so, inferring a boundary layer at $x = 0$, we have to use the boundary condition at $x = 1$ and we find that

$$\varphi_0 = x^2 + 1. \tag{15.25}$$

The structure of the equation suggests a correction of $\mathcal{O}(\varepsilon)$, so we try the expansion

$$\varphi = \varphi_0 + \varepsilon\varphi_1 + \varepsilon^2\varphi_2 + \cdots. \tag{15.26}$$

For φ_1 this results in the equation

$$\frac{\mathrm{d}\varphi_1}{\mathrm{d}x} + \frac{\mathrm{d}^2\varphi_0}{\mathrm{d}x^2} = 0, \tag{15.27}$$

with $\varphi_1(1) = 0$ (the $\mathcal{O}(\varepsilon)$ term of the boundary condition), which has the solution

$$\varphi_1 = 2 - 2x. \tag{15.28}$$

Higher orders are straightforward; i.e.,

$$\frac{\mathrm{d}\varphi_n}{\mathrm{d}x} = 0, \qquad \text{with} \quad \varphi_n(1) = 0, \tag{15.29}$$

leading to solutions $\varphi_n \equiv 0$. We find for the outer expansion

$$\varphi = x^2 + 1 + 2\varepsilon(1 - x) + \mathcal{O}(\varepsilon^N). \tag{15.30}$$

We continue with the inner expansion, and find, with $x_0 = 0$, $\varphi = \lambda\psi$, and $x = \delta\xi$, that

$$\frac{\varepsilon\lambda}{\delta^2}\frac{\mathrm{d}^2\psi}{\mathrm{d}\xi^2} + \frac{\lambda}{\delta}\frac{\mathrm{d}\psi}{\mathrm{d}\xi} - 2\delta\xi = 0. \tag{15.31}$$

Both from the matching condition ($\varphi_{\text{outer}} \to 1$ for $x \downarrow 0$) and from the boundary condition ($\varphi(0) = 2$) we have to conclude that $\varphi_{\text{inner}} = \mathcal{O}(1)$ and so $\lambda = 1$. Furthermore, the boundary layer only has a reason for existence if it comprises new effects not described by the outer solution. From the correspondence principle we expect that new effects are only included if $(\mathrm{d}^2\psi/\mathrm{d}\xi^2)$ is included. So $\varepsilon\delta^{-2}$ must be at least as large as δ^{-1}, the largest of δ^{-1} and δ. From the principle that we should look for the equation with the richest structure, it must be exactly as large, implying a boundary layer thickness $\delta = \varepsilon$. Thus we have $\kappa = \varepsilon^{-1}$, and the inner equation is

$$\frac{\mathrm{d}^2\psi}{\mathrm{d}\xi^2} + \frac{\mathrm{d}\psi}{\mathrm{d}\xi} - 2\varepsilon^2\xi = 0. \tag{15.32}$$

From this equation it would *seem* that we have a series expansion without the $\mathcal{O}(\varepsilon)$ term, since the equation for this order would be the same as for the leading order. However, from matching with the outer solution, i.e.,

$$\varphi_{\text{outer}} \to 1 + 2\varepsilon + \varepsilon^2(\xi^2 - 2\xi) + \cdots \qquad (x = \varepsilon\xi, \quad \xi = \mathcal{O}(1)), \tag{15.33}$$

we see that an additional $\mathcal{O}(\varepsilon)$ term is to be included. So we substitute the series expansion

$$\varphi = \psi_0 + \varepsilon\psi_1 + \varepsilon^2\psi_2 + \cdots. \tag{15.34}$$

It is a simple matter to find

$$\frac{\mathrm{d}^2\psi_0}{\mathrm{d}\xi^2} + \frac{\mathrm{d}\psi_0}{\mathrm{d}\xi} = 0, \qquad \psi_0(0) = 2 \Rightarrow \psi_0 = 2 + A_0(\mathrm{e}^{-\xi} - 1), \tag{15.35a}$$

$$\frac{\mathrm{d}^2\psi_1}{\mathrm{d}\xi^2} + \frac{\mathrm{d}\psi_1}{\mathrm{d}\xi} = 0, \qquad \psi_1(0) = 0 \Rightarrow \psi_1 = A_1(\mathrm{e}^{-\xi} - 1), \tag{15.35b}$$

$$\frac{\mathrm{d}^2\psi_2}{\mathrm{d}\xi^2} + \frac{\mathrm{d}\psi_2}{\mathrm{d}\xi} = 2\xi, \qquad \psi_2(0) = 0 \Rightarrow \psi_2 = \xi^2 - 2\xi + A_2(\mathrm{e}^{-\xi} - 1), \tag{15.35c}$$

where the constants $A_0, A_1, A_2, \ldots$ are to be determined from the matching condition that the inner and outer solutions should be asymptotically equivalent in the region of overlap. We can follow the method of intermediate variables and rewrite outer expansion (15.30) and inner expansion (15.34) in terms of an intermediate variable $x = \eta(\varepsilon)\sigma$, where $\varepsilon \ll \eta \ll 1$, and reexpand as follows:

$$x^2 + 1 + 2\varepsilon(1 - x) + \mathcal{O}(\varepsilon^3) \simeq 1 + 2\varepsilon + \eta^2\sigma^2 - 2\varepsilon\eta\sigma + \mathcal{O}(\varepsilon^3), \tag{15.36a}$$

$$2 + A_0(\mathrm{e}^{-\xi} - 1) + \varepsilon A_1(\mathrm{e}^{-\xi} - 1) + \varepsilon^2\big(\xi^2 - 2\xi + A_2(\mathrm{e}^{-\xi} - 1)\big) + \mathcal{O}(\varepsilon^3)$$
$$\simeq 2 - A_0 - \varepsilon A_1 + \eta^2\sigma^2 - 2\varepsilon\eta\sigma - \varepsilon^2 A_2 + \mathcal{O}(\varepsilon^3). \tag{15.36b}$$

Alternatively, we can follow Van Dyke's matching rule and rewrite outer expansion (15.30) in terms of inner variable ξ and inner expansion (15.34) in terms of outer variable x, reexpand, and rewrite the result in x, giving

$$x^2 + 1 + 2\varepsilon(1 - x) + \mathcal{O}(\varepsilon^3) \simeq 1 + 2\varepsilon + x^2 - 2\varepsilon x + \mathcal{O}(\varepsilon^3), \tag{15.36c}$$

$$2 + A_0(\mathrm{e}^{-\xi} - 1) + \varepsilon A_1(\mathrm{e}^{-\xi} - 1) + \varepsilon^2\big(\xi^2 - 2\xi + A_2(\mathrm{e}^{-\xi} - 1)\big) + \mathcal{O}(\varepsilon^3)$$
$$\simeq 2 - A_0 - \varepsilon A_1 + x^2 - 2\varepsilon x - \varepsilon^2 A_2 + \mathcal{O}(\varepsilon^3). \tag{15.36d}$$

In either case the resulting reduced expressions, (15.36a) and (15.36b) (respectively (15.36c) and (15.36d)), must be functionally equivalent. A full matching is thus obtained if we choose $A_0 = 1$, $A_1 = -2$, and $A_2 = 0$.

Example 15.30 (integration across a boundary layer) The same ideas of overlap and intermediate variables are exploited for an integral across a boundary layer. Take, e.g., the above function $\phi(x;\varepsilon)$ on $x \in [0, 1]$. We break up the integration interval at a point in the region of overlap, say at $x = \eta(\varepsilon)$. With matching outer and inner expansions denoted as before we obtain

$$\int_0^1 \phi(x;\varepsilon)\,\mathrm{d}x = \int_0^\eta \phi(x;\varepsilon)\,\mathrm{d}x + \int_\eta^1 \phi(x;\varepsilon)\,\mathrm{d}x = \varepsilon\int_0^{\eta/\varepsilon} \psi(\xi;\varepsilon)\,\mathrm{d}\xi + \int_\eta^1 \phi(x;\varepsilon)\,\mathrm{d}x$$
$$= \varepsilon\int_0^{\eta/\varepsilon} (\psi_0 + \varepsilon\psi_1 + \varepsilon^2\psi_2 + \cdots)\,\mathrm{d}\xi + \int_\eta^1 (\phi_0 + \varepsilon\phi_1 + \varepsilon^2\phi_2 + \cdots)\,\mathrm{d}x.$$

The result will contain terms depending on the auxiliary function η, but these will disappear after reexpanding the result up to $\mathcal{O}(\varepsilon^2)$ while taking η asymptotically about halfway $\mathcal{O}(\varepsilon)$ and $\mathcal{O}(1)$, say $\eta = \sqrt{\varepsilon}$. □

Example 15.31 (Prandtl's boundary layer analysis) The start of modern boundary layer theory is Prandtl's analysis of uniform incompressible low-viscosity flow along a flat plate. Consider the stationary two-dimensional version of (7.3)

$$u_x + v_y = 0, \quad uu_x + vu_y = -p_x + \varepsilon(u_{xx} + u_{yy}), \quad uv_x + vv_y = -p_y + \varepsilon(v_{xx} + v_{yy}),$$

with $\varepsilon = Re^{-1}$ small, subject to boundary conditions $u = v = 0$ at $y = 0, 0 < x < 1$, and outer solution for $y = \mathcal{O}(1)$ to leading order given by $(u, v, p) = (1, 0, 0)$. When we scale $x = X$, $y = \varepsilon^n Y$, $u = U$, $v = \varepsilon^m V$, and $p = \varepsilon^k P$, we find

$$U_X + \varepsilon^{m-n} V_Y = 0, \qquad UU_X + \varepsilon^{m-n} VU_Y = -\varepsilon^k P_X + \varepsilon U_{XX} + \varepsilon^{1-2n} U_{YY},$$
$$\varepsilon^m UV_X + \varepsilon^{2m-n} VV_Y = -\varepsilon^{k-n} P_Y + \varepsilon^{1+m} V_{XX} + \varepsilon^{1+m-2n} V_{YY}.$$

This yields the distinguished limits $m = n = \frac{1}{2}$ and $k = 1$, with the significant degeneration

$$U_X + V_Y = 0, \qquad UU_X + VU_Y = U_{YY}$$

known as *Prandtl's boundary layer equations*. The same equations, but with other boundary conditions, are valid in Goldstein's viscous wake $x > 1$, $y = \mathcal{O}(\varepsilon^{1/2})$. The trailing-edge region around $x = 1$, $y = 0$ is far more complicated. Here the boundary layer structure consists of three layers $y = \mathcal{O}(\varepsilon^{5/8})$, $\mathcal{O}(\varepsilon^{4/8})$, $\mathcal{O}(\varepsilon^{3/8})$ within $x - 1 = \mathcal{O}(\varepsilon^{3/8})$. This is known as Stewartson's triple deck. □

Logarithmic switchback It is not always evident from just the structure of the equation what the necessary expansion will look like. Sometimes it is well concealed and we are only made aware of an invalid initial choice by a matching failure. In fact, it is also the matching process itself that reveals the required sequence of scaling functions. An example of such a back reaction is known as *logarithmic switchback*.

Consider the following problem for $y = y(x; \varepsilon)$ on the unit interval:

$$\varepsilon y'' + x(y' - y) = 0, \qquad 0 < x < 1, \quad y(0; \varepsilon) = 0, \quad y(1; \varepsilon) = \mathrm{e}. \tag{15.37}$$

The outer solution appears to have the expansion

$$y(x; \varepsilon) = y_0(x) + \varepsilon y_1(x) + \varepsilon^2 y_2(x) + \mathcal{O}(\varepsilon^3). \tag{15.38}$$

By trial and error, we find that the boundary layer appears to be located near $x = 0$, so the governing equations and boundary conditions are then

$$y_0' - y_0 = 0, \qquad y_0(1) = \mathrm{e}, \tag{15.39a}$$

$$y_n' - y_n = -x^{-1} y_{n-1}'', \qquad y_n(1) = 0, \tag{15.39b}$$

with general solution

$$y_n(x) = A_n \,\mathrm{e}^x + \int_x^1 z^{-1} \mathrm{e}^{x-z} y_{n-1}''(z)\,\mathrm{d}z \tag{15.40a}$$

such that

$$y_0(x) = \mathrm{e}^x, \tag{15.40b}$$

$$y_1(x) = -\,\mathrm{e}^x \ln(x), \tag{15.40c}$$

$$y_2(x) = \mathrm{e}^x \left(\frac{1}{2} \ln(x)^2 + \frac{3}{2} - 2x^{-1} + \frac{1}{2} x^{-2} \right), \tag{15.40d}$$

etc. The boundary layer thickness is found from the assumed scaling $x = \varepsilon^m t$ and from noting that $y = \mathcal{O}(1)$ because of the matching with the outer solution. This leads to the significant degeneration of $m = \frac{1}{2}$, or $x = \varepsilon^{\frac{1}{2}} t$. The boundary layer equation for $y(x; \varepsilon) = Y(t; \varepsilon)$ is thus

$$Y'' + tY' - \varepsilon^{\frac{1}{2}} tY = 0, \qquad Y(0; \varepsilon) = 0. \tag{15.41}$$

The obvious choice of expansion of Y in powers of $\varepsilon^{\frac{1}{2}}$ is not correct, as the found solution does not match the outer solution. Therefore we consider the outer solution in more detail for small x. When $x = \varepsilon^{\frac{1}{2}}t$, we have for the outer solution

$$y(\varepsilon^{\frac{1}{2}}t;\varepsilon) = 1 + \varepsilon^{\frac{1}{2}}t + \varepsilon\left(-\frac{1}{2}\ln\varepsilon + \frac{1}{2}t^2 - \ln t + \frac{1}{2}t^{-2} + \cdots\right) + \mathcal{O}(\varepsilon^{\frac{3}{2}}\ln\varepsilon). \quad (15.42)$$

(The dots indicate powers of t^{-2} that appear with higher-order y_n.) So we apparently need at least

$$Y(t;\varepsilon) = Y_0(t) + \varepsilon^{\frac{1}{2}}Y_1(t) + \varepsilon\ln(\varepsilon)Y_2(t) + \varepsilon Y_3(t) + o(\varepsilon), \quad (15.43)$$

with equations and boundary conditions

$$Y_0'' + tY_0' = 0, \qquad Y_0(0) = 0, \quad (15.44a)$$
$$Y_1'' + tY_1' = tY_0, \qquad Y_1(0) = 0, \quad (15.44b)$$
$$Y_2'' + tY_2' = 0, \qquad Y_2(0) = 0, \quad (15.44c)$$
$$Y_3'' + tY_3' = tY_1, \qquad Y_3(0) = 0, \quad (15.44d)$$

etc. Hence the inner expansion is given by

$$Y_0(t) = A_0\,\mathrm{erf}\left(\frac{t}{\sqrt{2}}\right), \quad (15.45a)$$

$$Y_1(t) = A_1\,\mathrm{erf}\left(\frac{t}{\sqrt{2}}\right) + A_0\left[t\,\mathrm{erf}\left(\frac{t}{\sqrt{2}}\right) + 2\left(\frac{2}{\pi}\right)^{\frac{1}{2}}(\mathrm{e}^{-\frac{1}{2}t^2} - 1)\right], \quad (15.45b)$$

$$Y_2(t) = A_2\,\mathrm{erf}\left(\frac{t}{\sqrt{2}}\right), \quad (15.45c)$$

$$Y_3(t) = A_3\,\mathrm{erf}\left(\frac{t}{\sqrt{2}}\right) + \int_0^t \mathrm{e}^{-\frac{1}{2}z^2}\int_0^z \mathrm{e}^{\frac{1}{2}\xi^2}\,\xi Y_1(\xi)\,\mathrm{d}\xi\,\mathrm{d}z. \quad (15.45d)$$

Unfortunately, Y_3 cannot be expressed in closed form. However, for demonstration purposes it is sufficient to derive the behaviour of Y_3 for large t. As $\mathrm{erf}(z) \to 1$ exponentially fast for $z \to \infty$, we obtain

$$Y_1(t) = A_0 t + A_1 - 2\left(\frac{2}{\pi}\right)^{\frac{1}{2}} A_0 + \text{exponentially small terms.}$$

If Y_3 behaves algebraically for large t, then $tY_3' \gg Y_3''$, so $Y_3' = Y_1 - t^{-1}Y_3'' \simeq A_0 t$. By successive substitution it follows that

$$Y_3(t) = \frac{1}{2}A_0t^2 + \left(A_1 - 2\left(\frac{2}{\pi}\right)^{\frac{1}{2}} A_0\right)t - A_0\ln(t) + \cdots$$

To match the inner solution we introduce the intermediate variable $\eta = \varepsilon^{-\alpha}x = \varepsilon^{\frac{1}{2}-\alpha}t$, where $0 < \alpha < \frac{1}{2}$, and compare it with expression (15.42). We have

$$A_0 + \varepsilon^{\frac{1}{2}}\left(A_1 - 2\left(\frac{2}{\pi}\right)^{\frac{1}{2}} A_0\right) + \varepsilon^{\alpha} A_0 \eta + \varepsilon \ln(\varepsilon) A_2 + \frac{1}{2}\varepsilon^{2\alpha} A_0 \eta^2$$

$$+ \varepsilon^{\frac{1}{2}+\alpha}\left(A_1 - 2\left(\frac{2}{\pi}\right)^{\frac{1}{2}} A_0\right)\eta - \varepsilon A_0 \ln \eta + \varepsilon\left(\frac{1}{2} - \alpha\right) A_0 \ln \varepsilon$$

$$\equiv 1 + \varepsilon^{\alpha}\eta + \frac{1}{2}\varepsilon^{2\alpha}\eta^2 - \varepsilon \ln \eta - \alpha\varepsilon \ln(\varepsilon) + \frac{1}{2}\varepsilon^{2-2\alpha}\eta^{-2}. \quad (15.46)$$

Noting that $2 - 2\alpha > 1$, we find a full matching with

$$A_0 = 1, \quad A_1 = 2\left(\frac{2}{\pi}\right)^{\frac{1}{2}}, \quad A_2 = -\frac{1}{2}. \quad (15.47)$$

This problem is an example where intermediate matching is preferable.

The Role of Matching It is important to note whether a matching is possible at all! Only part of the terms can be matched by selection of the undetermined constants. Other terms are already equal, without free constants, and there is no way to repair a possibly incomplete matching here. This is an important consistency check on the found solution, at least as long as no real proof is available. If no matching appears to be possible, one of the assumptions made with the construction of the solution almost certainly has to be reconsidered. Particularly notorious are logarithmic singularities of the outer solution, as we saw above. See, for other examples, [86].

In summary, matching of inner and outer expansions plays an important role in the following ways:

(i) It provides information about the sequence of order (gauge) functions $\{\mu_k\}$ and $\{\lambda_k\}$ of the expansions.

(ii) It allows us to determine unknown constants of integration.

(iii) It provides a check on the consistency of the solution, giving us confidence in the correctness.

15.4.2 Multiple Scales

Suppose a function $\varphi(x;\varepsilon)$ depends on more than one length scale acting together, e.g., x, εx, and $\varepsilon^2 x$. Then the function does not have a regular expansion on the full domain of interest, $x \leq \mathcal{O}(\varepsilon^{-2})$ say. It is not possible to bring these different length scales together by a simple coordinate transformation, like in the method of slow variation or the Lindsted–Poincaré method, or to split up our domain in subdomains, like in the MAE method. Therefore we have to find another way to construct asymptotic expansions that is valid in the full domain of interest. The approach that is followed in the *method of multiple scales* is at first sight rather radical: the various length scales are temporarily considered as independent variables—$x_1 = x$, $x_2 = \varepsilon x$, $x_3 = \varepsilon^2 x$—and the original function φ is identified with a more general function $\psi(x_1, x_2, x_3; \varepsilon)$ depending on a higher-dimensional independent variable.

Example 15.32

$$\varphi(x;\varepsilon) = A(\varepsilon)\,\mathrm{e}^{-\varepsilon x}\cos(x+\theta(\varepsilon)) \text{ becomes } \psi(x_1,x_2;\varepsilon) = A(\varepsilon)\,\mathrm{e}^{-x_2}\cos(x_1+\theta(\varepsilon)). \quad \square$$

Since this identification is not unique, we may add constraints such that this auxiliary function ψ does have a Poincaré expansion on the full domain of interest. After having constructed this expansion, we may associate it with the original function along the line $x_1 = x$, $x_2 = \varepsilon x$, $x_3 = \varepsilon^2 x$.

This technique, utilizing this difference between small-scale and large-scale behaviour is the method of multiple scales. As with most approximation methods, this method has grown from practice and works well for certain types of problems. Typically, the multiple scale method is applicable to problems with, on the one hand, a certain global quantity (energy, power), which is conserved or almost conserved, controlling the amplitude, and, on the other hand, two rapidly interacting quantities (kinetic and potential energy) controlling the phase. Usually, this describes slowly varying waves affected by small effects during a long time. Intuitively, it is clear that over a short distance (a few wavelengths) the wave only sees constant conditions and will propagate approximately as in the constant case, but over larger distances it will somehow have to change its shape in accordance with its new environment.

We will illustrate the method by considering a damped harmonic oscillator

$$\frac{\mathrm{d}^2 y}{\mathrm{d}t^2} + 2\varepsilon\frac{\mathrm{d}y}{\mathrm{d}t} + y = 0 \qquad (t \ge 0), \quad y(0) = 0, \quad \frac{\mathrm{d}y(0)}{\mathrm{d}t} = 1, \tag{15.48}$$

with $0 < \varepsilon \ll 1$. The exact solution is readily found to be

$$y(t) = \mathrm{e}^{-\varepsilon t}\,\frac{\sin\left(\sqrt{1-\varepsilon^2}\,t\right)}{\sqrt{1-\varepsilon^2}}. \tag{15.49}$$

A naive approximation of this $y(t)$ for small ε and fixed t is

$$y(t) = \sin t - \varepsilon t \sin t + \mathcal{O}(\varepsilon^2), \tag{15.50}$$

which appears to be useful for $t = \mathcal{O}(1)$ only. For large t the approximation becomes incorrect:

1. If $t \ge \mathcal{O}(\varepsilon^{-1})$, the second term is of equal importance to, or larger than, the first term and nothing is left over of the slow exponential decay.
2. If $t \ge \mathcal{O}(\varepsilon^{-2})$, the phase has an error of $\mathcal{O}(1)$ or larger, giving an approximation of which even the sign may be in error.

We would obtain a far better approximation if we adopted two different time variables, $T = \varepsilon t$ and $\tau = \sqrt{1-\varepsilon^2}\,t$, and changed to $y(t;\varepsilon) = Y(\tau, T;\varepsilon)$, where

$$Y(\tau, T;\varepsilon) = \mathrm{e}^{-T}\,\frac{\sin(\tau)}{\sqrt{1-\varepsilon^2}}.$$

It is easily verified that a Taylor series of Y in ε yields a regular expansion for all t.

If we construct a straightforward approximate solution directly from (15.48), we will get the same approximation as in (15.50), which is too limited for most applications.

However, knowing the character of the error, we may try to avoid it and look for the auxiliary function Y instead of y. As we, in general, do not know the occurring time scales, their determination becomes part of the problem.

Suppose we can expand

$$y(t;\varepsilon) = y_0(t) + \varepsilon y_1(t) + \varepsilon^2 y_2(t) + \cdots . \tag{15.51}$$

Substituting in (15.48) and collecting equal powers of ε gives

$$\mathcal{O}(\varepsilon^0): \quad \frac{\mathrm{d}^2 y_0}{\mathrm{d}t^2} + y_0 = 0 \qquad \text{with} \quad y_0(0) = 0, \qquad \frac{\mathrm{d}y_0(0)}{\mathrm{d}t} = 1,$$

$$\mathcal{O}(\varepsilon^1): \quad \frac{\mathrm{d}^2 y_1}{\mathrm{d}t^2} + y_1 = -2\frac{\mathrm{d}y_0}{\mathrm{d}t} \qquad \text{with} \quad y_1(0) = 0, \qquad \frac{\mathrm{d}y_1(0)}{\mathrm{d}t} = 0.$$

We then find

$$y_0(t) = \sin t, \quad y_1(t) = -t \sin t, \quad \text{etc.,}$$

which indeed reproduces expansion (15.50). The straightforward Poincaré-type expansion (15.51) breaks down for large t when $\varepsilon t \geq \mathcal{O}(1)$. It is important to note that this is caused by the fact that any y_n is excited in its eigenfrequency (by the "source"-terms $-2\mathrm{d}y_{n-1}/\mathrm{d}t$), resulting in resonance. We recognise the algebraically generated growing terms of the type $t^n \sin t$ and $t^n \cos t$ called *secular terms* (Definition 15.26). Apart from being of limited validity, the expansion reveals nothing of the real structure of the solution, and we change our strategy to look for an auxiliary function dependent on different time scales. We start with the hypothesis that, besides a fast time scale t, we have the slow time scale

$$T := \varepsilon t. \tag{15.52}$$

Then we identify the solution y with another suitably chosen function Y that depends on both variables t and T:

$$Y(t, T; \varepsilon) := y(t; \varepsilon). \tag{15.53}$$

There exist infinitely many functions $Y(t, T; \varepsilon)$ that are equal to $y(t, \varepsilon)$ along the line $T = \varepsilon t$ in (t, T) space. So we now have some freedom to prescribe additional conditions. It is natural to think of conditions to be chosen such that no unwelcome secular terms occur when we construct an approximation.

Since the time derivatives of y turn into partial derivatives of Y, i.e.,

$$\frac{\mathrm{d}y}{\mathrm{d}t} = \frac{\partial Y}{\partial t} + \varepsilon \frac{\partial Y}{\partial T}, \tag{15.54}$$

equation (15.48) becomes for Y

$$\frac{\partial^2 Y}{\partial t^2} + Y + 2\varepsilon\left(\frac{\partial Y}{\partial t} + \frac{\partial^2 Y}{\partial t \partial T}\right) + \varepsilon^2\left(\frac{\partial^2 Y}{\partial T^2} + 2\frac{\partial Y}{\partial T}\right) = 0. \tag{15.55}$$

We assume the expansion

$$Y(t, T; \varepsilon) = Y_0(t, T) + \varepsilon Y_1(t, T) + \varepsilon^2 Y_2(t, T) + \cdots \tag{15.56}$$

and substitute this into (15.55) to obtain to leading order

$$\frac{\partial^2 Y_0}{\partial t^2} + Y_0 = 0,$$

$$\frac{\partial^2 Y_1}{\partial t^2} + Y_1 = -2\frac{\partial Y_0}{\partial t} - 2\frac{\partial^2 Y_0}{\partial t \partial T},$$

with initial conditions

$$Y_0(0,0) = 0, \qquad \frac{\partial}{\partial t} Y_0(0,0) = 1,$$

$$Y_1(0,0) = 0, \qquad \frac{\partial}{\partial t} Y_1(0,0) = -\frac{\partial}{\partial T} Y_0(0,0).$$

The solution for Y_0 is easily found to be

$$Y_0(t,T) = A_0(T)\sin(t + \theta_0(T)) \qquad \text{with} \quad A_0(0) = 1, \theta_0(0) = 0. \tag{15.57}$$

This gives a right-hand side for the Y_1 equation of

$$-2\left(A_0 + \frac{\partial A_0}{\partial T}\right)\cos(t+\theta_0) + 2A_0\frac{\partial \theta_0}{\partial T}\sin(t+\theta_0).$$

No secular terms occur (no resonance between Y_1 and Y_0) if these terms vanish:

$$A_0 + \frac{\partial A_0}{\partial T} = 0 \;\text{ yielding }\; A_0 = \mathrm{e}^{-T}, \qquad \frac{\partial \theta_0}{\partial T} = 0 \;\text{ yielding }\; \theta_0 = 0. \tag{15.58}$$

Putting this together, we have indeed constructed an approximation of (15.49) that is valid for $t \leq \mathcal{O}(\varepsilon^{-1})$:

$$y(t;\varepsilon) = \mathrm{e}^{-\varepsilon t}\sin t + \mathcal{O}(\varepsilon). \tag{15.59}$$

Note (this is typical of this approach) that we determined Y_0 only on the level of Y_1 without having to solve Y_1 itself.

The present approach is by and large the multiple scale technique in its simplest form. Variations on this theme are sometimes necessary. For example, we have not completely gotten rid of secular terms. On a longer time scale ($t = \mathcal{O}(\varepsilon^{-2})$) we again have resonance in Y_2 because of the "source" $\mathrm{e}^{-T}\sin t$, yielding terms $\mathcal{O}(\varepsilon^2 t)$. We see that a second time scale $T_2 = \varepsilon^2 t$ is necessary. From the exact solution we may infer that these longer time scales are not really independent and it may be worth trying a fast time of strained coordinate type: $\tau = \omega(\varepsilon)t = (1 + \varepsilon^2\omega_1 + \varepsilon^4\omega_4 + \cdots)t$. In the present example we would recover $\omega(\varepsilon) = \sqrt{1-\varepsilon^2}$.

The method fails when the slow variation is due to external effects, like a slowly varying problem parameter, as is demonstrated by the next example.

Example 15.33 Consider the problem

$$\ddot{x} + \kappa(\varepsilon t)^2 x = 0, \qquad x(0;\varepsilon) = 1, \quad \dot{x}(0;\varepsilon) = 0,$$

where $\kappa = \mathcal{O}(1)$. It seems plausible to assume two time scales: a fast one $\mathcal{O}(\kappa^{-1}) = \mathcal{O}(1)$ and a slow one $\mathcal{O}(\varepsilon^{-1})$. So we introduce besides to t the slow scale $T = \varepsilon t$ and rewrite $x(t;\varepsilon) = X(t,T;\varepsilon)$. We expand $X = X_0 + \varepsilon X_1 + \cdots$, and obtain $X_0 = A_0(T)\cos(\kappa(T)t - \theta_0(T))$. Suppressing secular terms in the equation for X_1 requires $A_0' = \kappa' t - \theta_0' = 0$, which is impossible. □

Here the fast time scale is slowly varying itself and the fast variable is to be strained locally by a suitable strain function as follows:

$$\tau = \int^t \omega(\varepsilon t'; \varepsilon)\, \mathrm{d}t' = \frac{1}{\varepsilon}\int^T \omega(z; \varepsilon)\, \mathrm{d}z, \qquad \text{where } T = \varepsilon t, \tag{15.60}$$

while for $x(t; \varepsilon) = X(\tau, T; \varepsilon)$ we have

$$\dot{x} = \omega X_\tau + \varepsilon X_T \qquad \text{and} \qquad \ddot{x} = \omega^2 X_{\tau\tau} + \varepsilon\omega_T X_\tau + 2\varepsilon\omega X_{\tau T} + \varepsilon^2 X_{TT}. \tag{15.61}$$

Example 15.34 Reconsider 15.33. After expanding $X = X_0 + \varepsilon X_1 + \cdots$ and $\omega = \omega_0 + \varepsilon\omega_1 + \cdots$, we obtain

$$\omega_0^2 X_{0\tau\tau} + \kappa^2 X_0 = 0,$$
$$\omega_0^2 X_{1\tau\tau} + \kappa^2 X_1 = -2\omega_0\omega_1 X_{0\tau\tau} - \omega_{0T} X_{0\tau} - 2\omega_0 X_{0\tau T}. \tag{$*$}$$

The leading-order solution is $X_0 = A_0(T)\cos(\lambda(T)\tau - \theta_0(T))$, where $\lambda = \kappa/\omega_0$. The right-hand side of $(*)$ is then

$$2\omega_0 A_0\lambda(\omega_1\lambda + \lambda_T\tau - \theta_{0T})\cos(\lambda\tau - \theta_0) + (A_0\lambda)^{-1}(\omega_0 A_0^2\lambda^2)_T \sin(\lambda\tau - \theta_0).$$

Suppression of secular terms requires $\lambda_T = 0$. Without loss of generality we can take $\lambda = 1$ or $\omega_0 = \kappa$. Then we need $\omega_1 = \theta_{0T}$, which just yields that $\lambda\tau - \theta_0 = \tau - \theta_0 = \varepsilon^{-1}\int^T \omega(z)\,\mathrm{d}z - \int^T \omega_1(z)\,\mathrm{d}z = \varepsilon^{-1}\int^T \omega_0(z)\,\mathrm{d}z + \mathcal{O}(\varepsilon)$. In other words, we may just as well take $\omega_1 = 0$ and $\theta_0 =$ constant. Finally, we have $\omega_0 A_0^2\lambda^2 = \kappa A_0^2 =$ constant. □

For linear wave-type problems we may anticipate the structure of the solution and assume the WKB hypothesis or *ray approximation*

$$y(t; \varepsilon) = A(T; \varepsilon)\, \mathrm{e}^{\mathrm{i}\varepsilon^{-1}\int_0^T \omega(\tau;\varepsilon)\,\mathrm{d}\tau}. \tag{15.62}$$

The method is again illustrated by the example of the damped oscillator (15.48), but now in complex form, so we consider the real part of (15.62). After substitution and suppressing the exponential factor, we get

$$\begin{aligned}
&(1 - \omega^2)A + \mathrm{i}\varepsilon\left(2\omega\frac{\partial A}{\partial T} + \frac{\partial\omega}{\partial T}A + 2\omega A\right) + \varepsilon^2\left(\frac{\partial^2 A}{\partial T^2} + 2\frac{\partial A}{\partial T}\right) = 0, \\
&\mathrm{Re}(A) = 0, \quad \mathrm{Re}(\mathrm{i}\omega A + \varepsilon A') = 1 \qquad \text{at } T = 0.
\end{aligned} \tag{15.63}$$

Unlike in the multiple scales method the secular terms will not be explicitly suppressed, at least not to leading order. The underlying additional condition here is that the solution of the present type *exists* and that each higher-order correction is no more secular than its predecessor. The solution is expanded as

$$\begin{aligned}
A(T; \varepsilon) &= A_0(T) + \varepsilon A_1(T) + \varepsilon^2 A_2(T) + \cdots, \\
\omega(T; \varepsilon) &= \omega_0(T) + \varepsilon^2\omega_2(T) + \cdots.
\end{aligned} \tag{15.64}$$

Note that ω_1 may be set to zero since the factor $\exp(\mathrm{i}\int_0^T \omega_1(\tau)\,\mathrm{d}\tau)$ may be incorporated in A. By a similar argument, i.e., by reexpanding the exponential for small ε, all other terms $\omega_2, \omega_3, \ldots$ could be absorbed by A (this is often done). This is perfectly acceptable for the time scale $T = \mathcal{O}(1)$, but for larger times we will not be able to suppress higher-order secular terms. So we will find it more convenient to include these terms and use them whenever convenient.

We substitute the expansions and collect equal powers of ε to obtain, to $\mathcal{O}(\varepsilon^0)$, that

$$(1-\omega_0^2)A_0 = 0, \tag{15.65a}$$

with solution $\omega_0 = 1$ (or -1, but that is equivalent for the result). To $\mathcal{O}(\varepsilon^1)$ we then have

$$A_0' + A_0 = 0 \quad \text{with} \quad \mathrm{Re}(A_0) = 0, \qquad \mathrm{Im}(\omega_0 A_0) = -1 \quad \text{at} \quad T = 0, \tag{15.65b}$$

with solution $A_0 = -\mathrm{i}\,\mathrm{e}^{-T}$. To order $\mathcal{O}(\varepsilon^2)$ the equation reduces to

$$A_1' + A_1 = -\mathrm{i}\left(\frac{1}{2} + \omega_2\right)\mathrm{e}^{-T}, \quad \text{with} \quad \mathrm{Re}(A_1) = 0, \quad \mathrm{Im}(\omega_0 A_1) = \mathrm{Re}(A_0') \quad \text{at} \quad T = 0, \tag{15.65c}$$

with solution

$$\omega_2 = -\frac{1}{2}, \qquad A_1 = 0. \tag{15.65d}$$

Note that if we had chosen $\omega_2 = 0$, the solution would have been $A_1 = -\frac{1}{2}T\,\mathrm{e}^{-T}$. Although it is by itself correct for $T = \mathcal{O}(1)$, it renders the asymptotic hierarchy invalid for $T \geq \mathcal{O}(1/\varepsilon)$ and is therefore better avoided. The solution that emerges is indeed consistent with the exact solution.

Example 15.35 (the air-damped resonator) In dimensionless form this is given by

$$\frac{\mathrm{d}^2 y}{\mathrm{d}t^2} + \varepsilon\frac{\mathrm{d}y}{\mathrm{d}t}\left|\frac{\mathrm{d}y}{\mathrm{d}t}\right| + y = 0, \qquad \text{with} \quad y(0) = 1, \quad \frac{\mathrm{d}y(0)}{\mathrm{d}t} = 0. \tag{$*$}$$

By rewriting the equation in the form

$$\frac{\mathrm{d}}{\mathrm{d}t}\left[\frac{1}{2}(y')^2 + \frac{1}{2}y^2\right] = -\varepsilon(y')^2|y'|$$

and assuming that y and $y' = \mathcal{O}(1)$, we may infer that the damping acts on a time scale of $\mathcal{O}(\varepsilon^{-1})$. So we conjecture the presence of the slow time variable $T = \varepsilon t$ and introduce a new dependent variable Y that depends on both t and T. We have

$$T = \varepsilon t, \quad y(t;\varepsilon) = Y(t,T;\varepsilon), \quad \frac{\mathrm{d}y}{\mathrm{d}t} = \frac{\partial Y}{\partial t} + \varepsilon\frac{\partial Y}{\partial T},$$

and we obtain for $(*)$

$$\frac{\partial^2 Y}{\partial t^2} + Y + \varepsilon\left(2\frac{\partial^2 Y}{\partial t \partial T} + \frac{\partial Y}{\partial t}\left|\frac{\partial Y}{\partial t}\right|\right) + \mathcal{O}(\varepsilon^2) = 0,$$

$$Y(0,0;\varepsilon) = 1, \qquad \left(\frac{\partial}{\partial t} + \varepsilon\frac{\partial}{\partial T}\right)Y(0,0;\varepsilon) = 0.$$

The error of $\mathcal{O}(\varepsilon^2)$ results from the approximation $\frac{\partial}{\partial t}Y + \varepsilon\frac{\partial}{\partial T}Y = \frac{\partial}{\partial t}Y + \mathcal{O}(\varepsilon)$ and is of course only valid outside a small neighbourhood of the points where $\frac{\partial}{\partial t}Y = 0$. We expand

$$Y(t,T;\varepsilon) = Y_0(t,T) + \varepsilon Y_1(t,T) + \mathcal{O}(\varepsilon^2)$$

to find for the leading order

$$\frac{\partial^2 Y_0}{\partial t^2} + Y_0 = 0, \qquad \text{with} \quad Y_0(0,0) = 1, \quad \frac{\partial}{\partial t} Y_0(0,0) = 0.$$

The solution is given by

$$Y_0 = A_0(T)\cos(t - \Theta_0(T)), \qquad \text{where} \quad A_0(0) = 1, \quad \Theta_0(0) = 0.$$

For the first order we have the equation

$$\begin{aligned}\frac{\partial^2 Y_1}{\partial t^2} + Y_1 &= -2\frac{\partial^2 Y_0}{\partial t \partial T} - \frac{\partial Y_0}{\partial t}\left|\frac{\partial Y_0}{\partial t}\right| \\ &= 2\frac{\mathrm{d}A_0}{\mathrm{d}T}\sin(t-\Theta_0) - 2A_0\frac{\mathrm{d}\Theta_0}{\mathrm{d}T}\cos(t-\Theta_0) + A_0^2\sin(t-\Theta_0)|\sin(t-\Theta_0)|,\end{aligned}$$

with corresponding initial conditions. The secular terms are suppressed if the first harmonics of the right-hand side cancel. For this we use the Fourier series expansion

$$\sin(t)\,|\sin(t)| = -\frac{8}{\pi}\sum_{n=0}^{\infty}\frac{\sin(2n+1)t}{(2n-1)(2n+1)(2n+3)}.$$

We obtain the equations

$$2\frac{\mathrm{d}A_0}{\mathrm{d}T} + \frac{8}{3\pi}A_0^2 = 0 \quad \text{and} \quad \frac{\mathrm{d}\Theta_0}{\mathrm{d}T} = 0,$$

with solution $\Theta_0(T) = 0$ and

$$A_0(T) = \frac{1}{1 + \frac{4}{3\pi}T}.$$

Altogether we have the approximate solution

$$y(t;\varepsilon) = \frac{\cos(t)}{1 + \frac{4}{3\pi}\varepsilon t} + \mathcal{O}(\varepsilon).$$

This approximation appears to be remarkably accurate. See Figure 15.4, where plots, made for a parameter value of $\varepsilon = 0.1$, of the approximate and a numerically "exact" solution are hardly distinguishable. A maximum difference of 0.03 is found. □

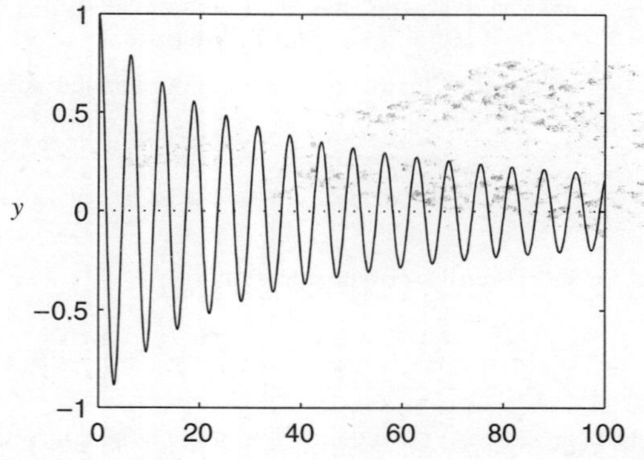

Figure 15.4. *Plots of the approximate solution and a numerically "exact" solution $y(t;\varepsilon)$ of the air-damped resonator problem for $\varepsilon = 0.1$.*

Example 15.36 (sound propagation in a slowly varying duct) Consider a hard-walled circular cylindrical duct with a slowly varying diameter described in polar coordinates (x, r, θ) as $r = R(\varepsilon x)$ with ε a dimensionless small parameter. In this duct we have an acoustic medium with constant mean pressure and sound speed c. Sound waves of circular frequency ω are most conveniently written in complex form, so the physical pressure perturbation is given by $\mathrm{Re}(p(\boldsymbol{x})\,\mathrm{e}^{\mathrm{i}\omega t})$, where p is the complex pressure amplitude described by Helmholtz's equation (see Example 1.8iii):

$$\nabla^2 p + \kappa^2 p = 0. \tag{$*$}$$

$\kappa = \omega/c$ is the free-field wave number, and the boundary condition of a vanishing normal gradient at the wall yields

$$\frac{\partial p}{\partial r} - \varepsilon R'(\varepsilon x)\frac{\partial p}{\partial x} = 0 \quad \text{at} \quad r = R(\varepsilon x).$$

For constant R and constant κ the general solution can be built up from a sum of right- and left-running modes (see Example 7.21) of the type

$$p = \sum_{m=-\infty}^{\infty} \mathrm{e}^{-\mathrm{i}m\theta} \sum_{\mu=1}^{\infty} J_m(\alpha_{m\mu} r)\Big(A_{m\mu}\,\mathrm{e}^{-\mathrm{i}\kappa_{m\mu}x} + B_{m\mu}\,\mathrm{e}^{\mathrm{i}\kappa_{m\mu}x}\Big),$$

$$\alpha_{m\mu} = j'_{m\mu}/R, \quad \kappa^2_{m\mu} = \kappa^2 - \alpha^2_{m\mu}, \quad \mathrm{Re}(\kappa_{m\mu}) \ge 0, \quad \mathrm{Im}(\kappa_{m\mu}) \le 0.$$

J_m denotes the mth order Bessel function [1] and $J'_m(j'_{m\mu}) = 0$. For the present problem we consider only a single mode and we assume, following the previous section, that the solution for the straight duct is locally close to the one for the varying duct. We introduce the slow variable $X = \varepsilon x$ so that $R = R(X)$, and we seek a solution of slowly varying modal type:

$$p = A(X, r; \varepsilon)\,\mathrm{e}^{-\mathrm{i}m\theta}\,\mathrm{e}^{-\mathrm{i}\varepsilon^{-1}\int_0^X \gamma(\xi)\,\mathrm{d}\xi}.$$

For simplicity we assume γ independent of ε. Since

$$\frac{\partial^2 p}{\partial x^2} = \Big(-\gamma^2 A - 2\mathrm{i}\varepsilon\gamma\frac{\partial A}{\partial X} - \mathrm{i}\varepsilon\gamma' A + \varepsilon^2\frac{\partial^2 A}{\partial X^2}\Big)\exp\big(\cdots\big)$$

we have for $(*)$

$$\Big[-\gamma^2 A - 2\mathrm{i}\varepsilon\gamma\frac{\partial A}{\partial X} - \mathrm{i}\varepsilon\gamma' A + \varepsilon^2\frac{\partial^2 A}{\partial X^2} + \frac{\partial^2 A}{\partial r^2} + \frac{1}{r}\frac{\partial A}{\partial r} - \frac{m^2}{r^2}A + \kappa^2 A\Big]\exp\big(\cdots\big) = 0.$$

After we suppress the exponential factor, this expression and the accompanying boundary condition are up to order $\mathcal{O}(\varepsilon)$,

$$\mathcal{L}[A] = \frac{\mathrm{i}\varepsilon}{A}\frac{\partial}{\partial X}(\gamma A^2), \qquad \frac{\partial A}{\partial r} + \mathrm{i}\varepsilon R'\gamma A = 0 \quad \text{at} \quad r = R(X). \tag{$\dagger$}$$

Here we introduced for short the Bessel-type operator

$$\mathcal{L}[A] := \frac{\partial^2 A}{\partial r^2} + \frac{1}{r}\frac{\partial A}{\partial r} + \Big(\kappa^2 - \gamma^2 - \frac{m^2}{r^2}\Big)A$$

and rewrote the right-hand side in a form that will turn out to be convenient later. We expand

$$A(X, r; \varepsilon) = A_0(X, r) + \varepsilon A_1(X, r) + \mathcal{O}(\varepsilon^2),$$

substitute in (†), and collect like powers of ε to obtain

$$\mathcal{O}(1):\ \mathcal{L}[A_0]=0,\qquad \frac{\partial A_0}{\partial r}=0\qquad \text{at } r=R(X), \tag{‡}$$

$$\mathcal{O}(\varepsilon):\ \mathcal{L}[A_1]=\frac{\mathrm{i}}{A_0}\frac{\partial}{\partial X}\gamma A_0^2,\qquad \frac{\partial A_1}{\partial r}=-\mathrm{i}R'\gamma A_0\quad \text{at } r=R(X). \tag{♯}$$

Since the variable X plays no other role in (‡) than that of a parameter, we have for A_0 the "almost mode"

$$A_0(X,r)=P_0(X)J_m(\alpha(X)r),\qquad \alpha(X)=j'_{m\mu}/R(X),$$

$$\gamma^2(X)=\kappa^2-\alpha^2(X),\qquad \mathrm{Re}(\gamma)\geq 0,\quad \mathrm{Im}(\gamma)\leq 0.$$

The amplitude P_0 is still undetermined and follows from A_1. We do not, however, have to solve A_1 if we can find a sufficient solvability condition for A_1. For this we multiply the left- and right-hands side of (♯) by rA_0 and integrate with respect to r from 0 to $R(X)$. For the left-hand side we utilize the self-adjointness of $\mathcal{L}$:

$$\begin{aligned}\int_0^R rA_0\mathcal{L}[A_1]\,\mathrm{d}r &= \int_0^R rA_0\mathcal{L}[A_1]-rA_1\mathcal{L}[A_0]\,\mathrm{d}r\\ &=\left[rA_0\frac{\partial A_1}{\partial r}-rA_1\frac{\partial A_0}{\partial r}\right]_0^R=-\mathrm{i}\gamma RR'A_0^2.\end{aligned}$$

For the right-hand side we apply Leibniz's rule, i.e.,

$$\int_0^R\frac{\partial}{\partial X}(\mathrm{i}\gamma A_0^2)r\,\mathrm{d}r=\frac{\mathrm{d}}{\mathrm{d}X}\int_0^R \mathrm{i}\gamma A_0^2 r\,\mathrm{d}r-\mathrm{i}\gamma RR'A_0^2.$$

Hence

$$\frac{\mathrm{d}}{\mathrm{d}X}\int_0^R r\gamma A_0^2\,\mathrm{d}r=0,$$

and so, using properties of the Bessel function [1], we have

$$\int_0^R r\gamma A_0^2\,\mathrm{d}r=\left[\frac{1}{2}\gamma P_0^2\left(r^2-\frac{m^2}{\alpha^2}\right)J_m(\alpha r)^2\right]_0^R=\frac{1}{2}\gamma P_0^2R^2\left(1-\frac{m^2}{j'^{\,2}_{m\mu}}\right)J_m(j'_{m\mu})^2=C$$

(where C denotes any constant) or

$$P_0(X)=\frac{C}{R(X)\sqrt{\gamma(X)}}.$$

The present problem may be generalised for annular ducts with acoustically lined walls and mean flow or modeling sound propagation in the inlet or exhaust duct of a turbofan aircraft engine [127, 128, 132]. See Chapter 16, Section 16.16. □

In more dimensions the assumed form of (15.62), where an integral occurs in the argument of the exponential, is not practical. In this case it is more convenient to write

$$\varphi(\boldsymbol{x},t;\varepsilon):=A(\boldsymbol{X},T;\varepsilon)\,\mathrm{e}^{\mathrm{i}\varepsilon^{-1}\Omega(\boldsymbol{X},T)}, \tag{15.66}$$

while for clarity of notation we leave Ω independent of ε and introduce the slowly varying frequency and wave vector

$$\omega := \frac{\partial \Omega}{\partial T}, \qquad \boldsymbol{\kappa} := -\overline{\nabla}\Omega,$$

where $\overline{\nabla} := \frac{\partial}{\partial X}\boldsymbol{e}_x + \frac{\partial}{\partial Y}\boldsymbol{e}_y + \frac{\partial}{\partial Z}\boldsymbol{e}_z$. This relation implies the continuity equation of waves

$$\frac{\partial \boldsymbol{\kappa}}{\partial T} + \overline{\nabla}\omega = \boldsymbol{0}. \tag{15.67}$$

Consider the following example of a one-dimensional wave equation with slowly varying coefficients:

$$\frac{\partial}{\partial t}\left(m(X,T)\frac{\partial}{\partial t}\varphi\right) = \frac{\partial}{\partial x}\left(C(X,T)\frac{\partial}{\partial x}\varphi\right) + B(X,T)\varphi, \tag{15.68}$$

where $X = \varepsilon x$ and $T = \varepsilon t$ are slow variables. We assume that the solution φ takes the form given by (15.66). This yields the equation

$$-\omega^2 mA + \frac{\mathrm{i}\varepsilon}{A}\frac{\partial}{\partial T}(\omega m A^2) = -\kappa^2 CA - \frac{\mathrm{i}\varepsilon}{A}\frac{\partial}{\partial X}(\kappa C A^2) + BA + \mathcal{O}(\varepsilon^2). \tag{15.69}$$

As before, we expand

$$A = A_0 + \varepsilon A_1 + \mathcal{O}(\varepsilon^2).$$

After substitution and collecting equal powers of ε, we get from leading order the slowly varying dispersion relation for ω and κ, or the eikonal-type equation for Ω:

$$\omega^2 m = \kappa^2 C - B. \tag{15.70}$$

Equation (15.67) turns into

$$\frac{\partial \kappa}{\partial T} + V(\kappa)\frac{\partial \kappa}{\partial X} = 0,$$

showing that both κ and ω propagate with the group velocity (see (3.38))

$$V = \frac{\mathrm{d}\omega}{\mathrm{d}\kappa} = \frac{\kappa C}{\omega m}. \tag{15.71}$$

The next order yields a conservation-type equation for A_0:

$$\frac{\partial}{\partial T}(\omega m A_0^2) + \frac{\partial}{\partial X}(\kappa C A_0^2) = 0. \tag{15.72}$$

(It should be noted that this result reflects the underlying physics and therefore depends on the original equation. In general the resulting equation is not of conserved type.) The pair $\omega m A_0^2$ and $\kappa C A_0^2$ are called *adiabatic invariants* because they correspond to the density and flux of a quantity that is conserved on the slow time and length scales. This is seen as follows. When we integrate (15.72) between the moving boundaries $X = X_1(T)$ and $X = X_2(T)$, we obtain

$$\int_{X_1}^{X_2} \frac{\partial}{\partial T}(\omega m A_0^2) + \frac{\partial}{\partial X}(\kappa C A_0^2)\,\mathrm{d}X = \frac{\mathrm{d}}{\mathrm{d}T}\int_{X_1}^{X_2} \omega m A_0^2\,\mathrm{d}X$$
$$- V_2[\omega m A_0^2]_{X_2} + V_1[\omega m A_0^2]_{X_1} + [\kappa C A_0^2]_{X_2} - [\kappa C A_0^2]_{X_1} = 0,$$

where $V_1 = \frac{\mathrm{d}}{\mathrm{d}T}X_1$ and $V_2 = \frac{\mathrm{d}}{\mathrm{d}T}X_2$. This reduces to

$$\frac{\mathrm{d}}{\mathrm{d}T}\int_{X_1}^{X_2} \omega m A_0^2\,\mathrm{d}X = 0$$

if the velocity of either endpoint is equal to the group velocity (15.71). In other words, the integral of $\omega m A_0^2$ is conserved between two points moving with the local group velocity.

Suppose, for definiteness, that ϕ denotes position and m denotes mass. Then if ω and κ are real, we can derive from (15.72) with (15.70) the conservation laws for *wave action*

$$\frac{\partial}{\partial T}\left(\frac{1}{2}\omega m|A_0|^2\right) + \frac{\partial}{\partial X}\left(\frac{1}{2}\kappa C|A_0|^2\right) = 0 \qquad (15.73a)$$

($\frac{1}{2}\omega m|A_0|^2$ is the wave action density [8]) and *energy*

$$\frac{\partial}{\partial T}\left(\frac{1}{2}\omega^2 m|A_0|^2\right) + \frac{\partial}{\partial X}\left(\frac{1}{2}\omega\kappa C|A_0|^2\right) = 0 \qquad (15.73b)$$

($\frac{1}{2}\omega^2 m|A_0|^2$ is the wave energy density) by substituting $A_0 = |A_0|\exp(\mathrm{i}\arg A_0)$, dividing out the complex exponent, and taking the real part of what remains.

Example 15.37 (ray acoustics in a temperature gradient) A very important application of waves in a slowly varying medium is the theory of *rays*. When a sound wave propagates in free space through a medium that varies on a much larger scale than the typical wavelength (typically temperature gradients or wind with shear), the same ideas of multiple scales may be applied. In contrast to the duct, where the wave is confined by the duct walls, the waves may now freely refract and follow curved paths. These paths are called rays.

Consider an infinite three-dimensional medium with varying temperature (typical length scale L) but otherwise with a constant mean pressure, so that we have for the acoustic pressure perturbations the equation (7.7), i.e.,

$$\frac{\partial^2}{\partial t^2}p = \nabla\cdot\big(c(\boldsymbol{X})^2\nabla p\big), \qquad (*)$$

where $\boldsymbol{X} = \varepsilon\boldsymbol{x}$ for small ε and the sound speed c is slowly varying when compared to the acoustic length scale. This means that ω, the frequency considered, is high enough for the corresponding wavelength $\lambda \sim 2\pi c/\omega$ to be small compared to L, so $\varepsilon \sim \lambda/L$. Assuming the field to be locally plane, we try an approximate solution (a *ray approximation*)

$$p(\boldsymbol{x}, t) = A\,\mathrm{e}^{\mathrm{i}\omega t - \mathrm{i}\theta/\varepsilon} \qquad (\dagger)$$

with the form of a plane wave but with slowly varying amplitude $A = A(\boldsymbol{X};\varepsilon)$ and phase $\varepsilon^{-1}(\omega T - \theta(\boldsymbol{X}))$, where $T = \varepsilon t$. The surfaces $\theta - \omega T =$ constant describe the propagating wave front, with normal vectors $\nabla\theta$. We define the operator $\overline{\nabla} = (\frac{\partial}{\partial X}, \frac{\partial}{\partial Y}, \frac{\partial}{\partial Z})$ so that $\nabla = \varepsilon\overline{\nabla}$. Then we define the local wave vector $\overline{\nabla}\theta = \boldsymbol{\kappa}$. Since

$$\nabla p = \left(-\mathrm{i}\boldsymbol{\kappa}A + \varepsilon\overline{\nabla}A\right)\mathrm{e}^{\mathrm{i}\omega t - \mathrm{i}\theta/\varepsilon},$$
$$\nabla^2 p = \left(-|\boldsymbol{\kappa}|^2 A - 2\mathrm{i}\varepsilon\boldsymbol{\kappa}\cdot\overline{\nabla}A - \mathrm{i}\varepsilon(\overline{\nabla}\cdot\boldsymbol{\kappa})A + \varepsilon^2\overline{\nabla}^2 A\right)\mathrm{e}^{\mathrm{i}\omega t - \mathrm{i}\theta/\varepsilon},$$

we obtain by substituting (†) in (∗) the equation

$$(\omega^2 - c^2|\boldsymbol{\kappa}|^2)A = \frac{\mathrm{i}\varepsilon}{A}\overline{\nabla}\cdot(c^2\boldsymbol{\kappa}A^2) + \mathcal{O}(\varepsilon^2). \qquad (\ddagger)$$

We expand

$$A = A_0 + \varepsilon A_1 + \mathcal{O}(\varepsilon^2)$$

and collect like powers in (‡). We find to leading orders for $\boldsymbol{\kappa}$ and A_0 that

$$c^2|\boldsymbol{\kappa}|^2 = \omega^2, \qquad (\sharp)$$
$$\overline{\nabla}\cdot(c^2\boldsymbol{\kappa}A_0^2) = 0, \qquad (\sharp\sharp)$$

Written in terms of θ, (♯) is the *eikonal equation*, which determines the wave fronts and the ray paths. Equation (♯♯) is called the *transport equation*. It implies, like in (15.73), the equations for *conservation of wave action* and *conservation of energy*, since $\boldsymbol{\kappa}$ is real. It relates the amplitude variation to diverging or converging rays (see the problem considered in Section 16.9 of Chapter 16). The eikonal equation is a nonlinear first order PDE of hyperbolic type, which can always be reduced to a system of ODEs along characteristics (see Chapter 2). □

15.5 Discussion

- Inherent in any modeling is the hierarchy in importance of the various effects that constitute the model. Therefore certain effects in any modeling will be small. Sometimes they are small but not small enough to be ignored, and sometimes they are small but in a nonuniform way such that they are important locally.

- For an efficient solution, and to obtain qualitative insight, it makes sense to utilize this "smallness." Perturbation methods systematically exploit such intrinsic smallness.

- The basis of a systematic approach is formed by the concept of asymptotic approximations and expansions of Poincaré type. If the approximation is uniform on the domain of interest, it is called regular; otherwise it is called singular. The role of the chosen independent variable in this respect cannot be overemphasized.

- Perturbation methods have a long history. Before the time of numerical methods and the computer, perturbation methods were the only way to increase the applicability of available exact solutions to difficult, otherwise intractable problems. Nowadays, perturbation methods have their use as a natural step in the process of systematic modeling, for the insight they provide in the nature of singularities occurring in the problem and typical parameter dependencies, and sometimes in the speed of practical calculations.

- A multitude of practical problems demonstrate the usefulness of asymptotic methods. We refer to groundwater flow (Section 16.6), growing of monocrystals (Section 16.7), catalytic reactions (Section 16.8), ray acoustics (Section 16.9), thin layer flow (Section 16.10), the formation of a glass parison (Section 16.11), nonlinear oscillations of overhead transmission lines (Section 16.5), sudden compression of a viscoelastic fluid (Section 16.4), and sound propagation in aircraft turbofan engine ducts (Section 16.16).

- The present chapter is just an introduction to the subject, and various methods are left unmentioned, such as *homogenization theory* and *asymptotics beyond all orders*. For further reading on perturbation methods and asymptotics we refer to [17, 20, 38, 62, 65, 70, 71, 81, 99, 103, 106, 107, 162, 164, 138], which is a personal and certainly not exhaustive selection.

Exercises

15.1. Determine asymptotic expansions for $\varepsilon \to 0$ of

(a) $\varepsilon/\tan\varepsilon$,

(b) $\varepsilon/(1-\varepsilon^{\varepsilon})$,

(c) $\log(\sin\varepsilon)$,

(d) $(1-\varepsilon+\varepsilon^2\ln\varepsilon)/(1-\varepsilon\ln\varepsilon-\varepsilon+\varepsilon^2\ln\varepsilon)$.

15.2. Determine, if possible, uniform asymptotic expansions for $\varepsilon \to 0$ and $x \in [0, 1]$ of

(a) $\sin(\varepsilon x)$,

(b) $1/(\varepsilon + x)$,

(c) $x\log(\varepsilon x)$,

(d) $\mathrm{e}^{-\sin(x)\varepsilon}$,

(e) $\mathrm{e}^{-\sin(x)/\varepsilon}$.

15.3. Derive asymptotic solutions (for $\varepsilon \to 0$) of the equation

$$\varepsilon x^3 - x + 2 = 0.$$

15.4. Derive, step by step, by iteratively scaling $x(\varepsilon) = \mu_0(\varepsilon)x_0 + \mu_1(\varepsilon)x_1 + \cdots$ and balancing, that a third order asymptotic solution (for $\varepsilon \to 0$) of the equation

$$\ln(\varepsilon x) + x = a$$

is given by

$$x(\varepsilon) = \ln\varepsilon^{-1} - \ln\big(\ln\varepsilon^{-1}\big) + a + o(1).$$

Find a more efficient expansion based on an alternative asymptotic sequence of gauge functions by combining $\mathrm{e}^{-a}\,\varepsilon$.

15.5. Analyse asymptotically for $\varepsilon \to 0$ the zeros of

$$\mathrm{e}^{-x/\varepsilon^2} + x - \varepsilon.$$

15.6. Derive the *Webster's horn equation* for sound of long wavelength propagating in slowly varying horns, by the method of slender approximation [86]. The reduced wave equation for pressure perturbations p and wave number k is given by

$$\nabla^2 p + k^2 p = 0$$

within a duct given in cylindrical polar coordinates by

$$S = r - R(X,\theta) = 0, \qquad X = \varepsilon x, \quad \varepsilon \text{ is small.}$$

The wave number is $\mathcal{O}(\varepsilon)$, so we scale $k = \varepsilon\kappa$. The duct wall is hard, so we have the boundary condition

$$\nabla p \cdot \nabla S = 0 \text{ at } S = 0.$$

15.7. Consider the incompressible Navier–Stokes equations to describe *lubrication flow* in a two-dimensional narrow and slowly varying channel with prescribed volume flux. (In actual practice this flux is created by a pressure difference.)

(a) Make the channel height and volume flux dimensionless and scale the pressure gradient such that viscous forces are balanced by the pressure gradient, so the Reynolds number $Re \le O(1)$. Verify that we obtain in dimensionless form that

$$Re(\boldsymbol{v} \cdot \nabla \boldsymbol{v}) + \nabla p = \nabla^2 \boldsymbol{v}, \qquad \nabla \cdot \boldsymbol{v} = 0,$$

for the velocity $\boldsymbol{v} = (u, v)^T$ and pressure p in the channel given by $-\infty < x < \infty$ and $g(\varepsilon x) \le y \le h(\varepsilon x)$, where ε is a small parameter. (End conditions in x are not important.) Boundary conditions are as follows: no slip at the walls, i.e., $u = v = 0$ at $y = g(\varepsilon x)$ and $y = h(\varepsilon x)$, and a flux

$$\int_{g(\varepsilon x)}^{h(\varepsilon x)} u(x, y)\, dy = 1.$$

(b) We rewrite $X = \varepsilon x$ and assume that the field varies slowly in X (any end effects are local and irrelevant for the x's considered). Rescale u, v, and p. The order of magnitude of the pressure can be found from the observation that a pressure gradient is necessary to have a flow. The crosswise velocity v will be much smaller than the axial velocity u.

(c) Assume for rescaled u, v, and p an obvious asymptotic expansion in ε and solve up to leading order.

15.8. Consider the function

$$f(x;\varepsilon) = \mathrm{e}^{-x/\varepsilon}(1 + x) + \pi \cos(\pi x + \varepsilon) \qquad \text{for} \quad 0 \le x \le 1.$$

(a) Construct an outer and an inner expansion of f with error $\mathcal{O}(\varepsilon^3)$.

(b) Integrate f from $x = 0$ to $x = 1$ exactly and expand the result up to $\mathcal{O}(\varepsilon^3)$.

(c) Compare this with the integral that is obtained by integrating the inner and outer expansions following the method described in Example 15.30.

15.9. Construct an asymptotic approximation for small ε of

$$\int_0^\infty \frac{1}{x^2 + x^6 + \varepsilon^2}\,\mathrm{d}x.$$

15.10. Reconsider (†) of Example 7.4 to describe a stationary suspended flexible bar of length L.

(a) First we consider a cable with clamped ends at equal height. This is described by adding boundary conditions $x = y = \phi = 0$ at $s = 0$ and $x = D$, and $y = \phi = 0$ at $s = L$, where $1 - D/L$ is positive and not small. Note that for given D the necessary horizontal force H is unknown and to be determined. Make the problem dimensionless by scaling lengths on L and forces on QL. Introduce the parameters $\varepsilon = \sqrt{(EI/QL^3)}$, $h = h(\varepsilon) = H_0/QL$, and $v = v(\varepsilon) = V_0/QL$. If ε is small, the equation describes a suspended cable. Find the solution asymptotically to leading order for small ε.

(b) Do the same for a cable with hinged ends, i.e., with $\phi'(0) = \phi(L)' = 0$.

(c) The same differential equation represents a model for laying submarine gas and oil pipelines from a laybarge. The pipe is freely suspended over an unknown length L, with prescribed curvature R at the liftoff point at height $y = W$ and a prescribed horizontal tension H in order to avoid buckling of the pipe. Both the angle ϕ and the curvature ϕ' vanish at the touchdown point $s = 0$. We have thus $\phi(0) = \phi'(0) = 0$, $\phi'(L) = -R^{-1}$, $y(0) = 0$, and $y(L) = W$, while L is unknown. Make dimensionless and solve the resulting problem asymptotically for small ε.

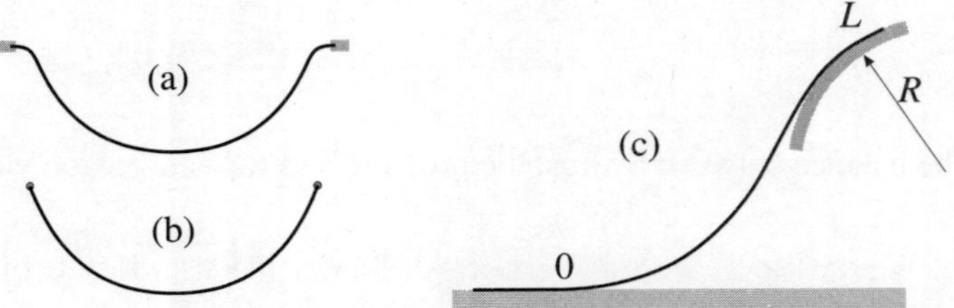

15.11. Determine the asymptotic approximation of the solution $y(x;\varepsilon)$ (first or first and second order terms for positive small parameter $\varepsilon \to 0$) of the following singularly perturbed problems. Let α and β be nonzero constants that are independent of ε. Provide arguments for the determined boundary layer thickness and location, and show how free constants are determined by the matching procedure.

(a) $\varepsilon y'' - y' = 2x$, $\quad y(0;\varepsilon) = \alpha$, $\quad y(1;\varepsilon) = \beta$.

(b) $\varepsilon y' + y^2 = \cos x$, $\quad y(0;\varepsilon) = 0$, $\quad 0 \le x \le 1$.

(c) $\varepsilon y'' + (2x+1)y' + y^2 = 0$, $\quad y(0;\varepsilon) = \alpha$, $\quad y(1;\varepsilon) = \beta$.

15.12. Determine at least two terms of an asymptotic approximation for small $\varepsilon > 0$ of the solution $\psi(x)$ of the equation

$$\varepsilon(\psi'' + \psi\psi') = -2 \quad \text{along} \quad 0 \le x \le 1,$$

with boundary conditions

$$\psi(0) = 1, \qquad \psi(1) = 1.$$

Note that there are boundary layers at both ends. Hint: Airy functions play a role, i.e., solutions of $y'' - xy = 0$. Consult [1] for their asymptotic behaviour.

15.13. When we stir a cup of tea, the surface of the fluid deforms until equilibrium is attained between gravity, centrifugal force, and surface tension. This last force is only important near the wall. (Note: Surface tension is $\sigma = 75$ mN/m for water.) Consider for this problem the following model.

A cylinder (radius a, axis vertical) with fluid (density ρ, surface tension σ) rotates around its axis $\boldsymbol{e}_z$ (angular velocity Ω) in a gravity field $-g\,\boldsymbol{e}_z$. By the gravity and the centrifugal force the surface deforms to something that looks like a paraboloid. Within a small neighbourhood of the cylinder wall the contact angle $\theta = \frac{1}{2}\pi - \alpha$ is felt by means of the surface tension (see (6.50)). Because of symmetry we can describe the surface by a radial tangent angle ψ with the horizon, parametrized by arc length s, such that $s = 0$ corresponds to the axis and $s = L$ corresponds to the wall of the cylinder. L is unknown. Select the origin on the axis at the surface such that the vertical and radial coordinates are given by $Z(s) = \int_0^s \sin\psi(s')\,\mathrm{d}s'$ and $R(s) = \int_0^s \cos\psi(s')\,\mathrm{d}s'$, respectively.

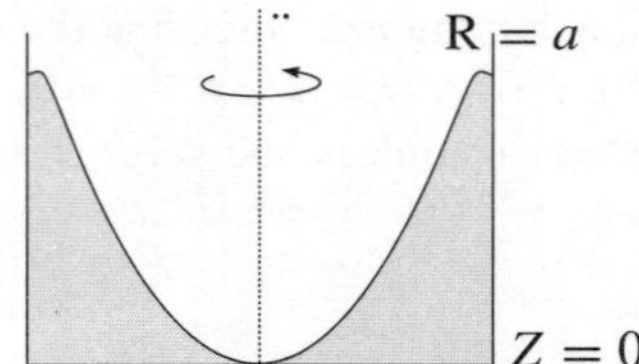

The balance between hydrostatic pressure and surface tension yields the equation

$$p_0 - \rho g Z + \frac{1}{2}\rho\Omega^2 R^2 = -\sigma\left(\frac{\mathrm{d}\psi}{\mathrm{d}s} + \frac{\sin\psi}{R}\right)$$

with unknown p_0. Boundary conditions are $\psi(0) = 0$, $\psi(L) = \alpha$, and $R(L) = a$.

(a) Scale lengths on a, i.e., $s = at$, $R = ar$, $Z = az$, and $L = a\lambda$, and introduce $\beta = p_0/\rho g a$ and the dimensionless parameters $\varepsilon^2 = \sigma/\rho g a^2$ and $\mu = \Omega^2 a/g$. Can you identify the common names of these dimensionless numbers?

(b) Solve the resulting problem asymptotically for $\varepsilon \to 0$, while $\mu = O(1)$.

15.14. Consider the *van der Pol* equation for variable $y = y(t;\varepsilon)$ in t and small parameter ε:

$$y'' + y - \varepsilon(1 - y^2)y' = 0.$$

Construct by using the Lindstedt–Poincaré method an $\mathcal{O}(\varepsilon^2)$ approximation of a periodic solution.

15.15. Consider the system governed by the equation of motion

$$y'' + y + \alpha y^3 = \varepsilon^2 \cos\omega t.$$

Find, asymptotically for $\varepsilon \to 0$ and $\alpha = \mathcal{O}(1)$, to leading order the steady solution
(a) for $\omega^2 = 1 + \mathcal{O}(1)$, i.e., away from resonance;
(b) for $\omega^2 = 1 + \varepsilon\mu$, i.e., near resonance.
Hint: For part (b), introduce the variable $\tau = \omega t$.

15.16. Following Example 15.36, derive a multiple scales solution of sound waves in a slowly varying duct while the sound speed is also a slowly varying function of x. The pertinent equation is therefore $(*)$ of Example 15.37.

15.17. (a) Rewrite the eikonal equation ($\sharp$) of Example 15.37 in characteristic form by using Theorem 12.6.
(b) Prove that in a medium with a linearly varying sound speed the paths of rays are circles.

15.18. Analyse the error $u = y - \tilde{y}$ of (3.43), described by (3.44), but now including the effects of the slowly varying coefficients a, b, and c. Formulate your result in the form of adiabatic invariants.

15.19. Consider the Fisher travelling-wave problem (10.70) with large c

$$U'' + c^2 U' + c^2 U(1 - U) = 0.$$

(a) Derive an approximate solution on $(-\infty, \infty)$ with $U(-\infty) = 1$, $U(\infty) = 0$. It is no restriction to assume that $U(0) = \frac{1}{2}$.
(b) Repeat part (a) on $[0, \infty)$ with $U(0) = 0$, while the previous solution is the outer solution.

15.20. A field ϕ (satisfying the Helmholtz or reduced wave equation; see Example 1.8iii), radiated by a source $q(\boldsymbol{x})$ that is distributed within a finite region Ω of typical diameter L, is given by

$$\phi(\boldsymbol{x}; \kappa) = -\frac{1}{4\pi} \int_\Omega q(\boldsymbol{y}) \frac{\mathrm{e}^{-\mathrm{i}\kappa r}}{r} \, \mathrm{d}V, \qquad \text{where} \quad r = \|\boldsymbol{x} - \boldsymbol{y}\|.$$

With the origin inside Ω, evaluate ϕ asymptotically for small κL for different regions in $\boldsymbol{x}$. Distinguish in particular the near or static zone, $\kappa L \ll \kappa\|\boldsymbol{x}\| \ll 1$; the intermediate or induction zone, $\kappa L \ll \kappa\|\boldsymbol{x}\| = \mathcal{O}(1)$; and the far or radiation zone, $\kappa L \ll 1 \ll \kappa\|\boldsymbol{x}\|$.

15.21. The equation

$$y'' + \frac{1}{r} y' + \left(\alpha^2 - \frac{m^2}{r^2}\right) y = 0$$

has solutions in the form of Bessel functions of order m and argument αr. Find asymptotic solutions of WKB type for $\alpha \to \infty$ and $r = \mathcal{O}(1)$ with $r > m/\alpha$. Consider (i) $m^2 = \mathcal{O}(1)$, (ii) $m^2/\alpha = \mathcal{O}(1)$, and (iii) $m^2/\alpha^2 = \mathcal{O}(1)$.

Chapter 16

Modeling, Analysing, and Simulating Problems from Practice

16.1 Production of Resin-Containing Panels

There exists a variety of synthetic panels for both interior and exterior use. Typical applications are office desktops and facade cladding. One type of panel is made of polymerised resin reinforced with wood fibres or sulphate paper.

In manufacturing these panels we can distinguish two stages. In the first stage sheets of resin-impregnated paper enclosed by two layers of padded paper are pressed together. During this pressing stage the components are compressed to within 10 percent of their final thickness. A schematic three-layer model of a panel is given in Figure 16.1. A panel thus has two bolsters of padded paper and a core consisting of impregnated paper. In the second stage high temperatures are applied at the boundaries of the panel, which causes heating of the panel. This induces a polymerisation reaction in the core. Heat is released during the polymerisation, which again leads to an increase of temperature in the panel. This process is called *resin curing* and continues until the chemicals in the core are depleted. When the polymerisation reaction is finished, the temperature finally tends to a steady state. In this section we introduce a model for the resin-curing process.

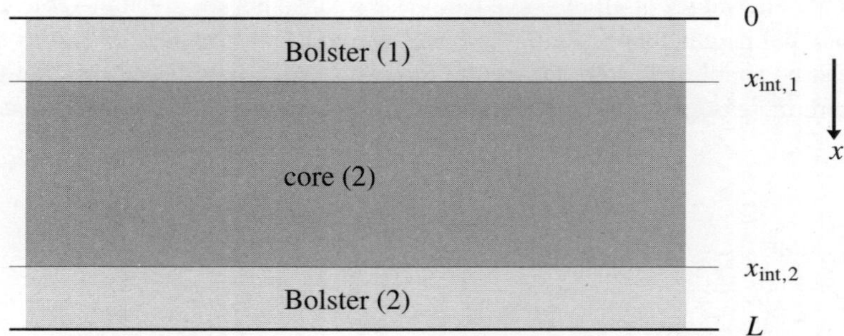

Figure 16.1. *Schematic representation of a three-layer panel.*

16.1.1 Modeling the Curing Process

The curing process can be described by the temperature T in the panel and the concentration C of chemicals involved in the polymerisation reaction in the core. To keep the model feasible we make the following assumptions:

1. The materials are incompressible.
2. Resin curing is a first order reaction.
3. Diffusion of resin is negligible.
4. All variables depend only on the transverse space coordinate x.

The governing equations are the following. In both bolsters the heat equation holds, which in our particular case is given by [104]

$$\rho c_p \frac{\partial T}{\partial t} = \frac{\partial}{\partial x}\left(\lambda \frac{\partial T}{\partial x}\right), \tag{16.1}$$

with ρ, c_p, and λ the density, specific heat, and thermal conductivity, respectively, of the bolster material. These parameters are assumed to be constant. In the core, heat conduction is coupled with the polymerisation reaction, and the governing equations read

$$\rho c_p \frac{\partial T}{\partial t} = \frac{\partial}{\partial x}\left(\lambda \frac{\partial T}{\partial x}\right) + \Delta H k C, \tag{16.2a}$$

$$\frac{\partial C}{\partial t} = -kC, \tag{16.2b}$$

with ΔH and k the enthalpy of polymerisation and the reaction rate, respectively. We assume also that the parameters ρ, c_p, λ, and ΔH in (16.2a) and (16.2b) are constant. The reaction rate is given by the Arrhenius expression

$$k = A\mathrm{e}^{-E_\mathrm{a}/RT}, \tag{16.3}$$

with A, E_a, and R the preexponential factor, the activation energy, and the universal gas constant, respectively. The reaction rate k as a function of the temperature T is shown in Figure 16.2. The values of all physical parameters/constants are given in Table 16.1.

Since the parameters ρ, c_p, λ, and ΔH are constant, the heat equations (16.1) and (16.2a) can be simplified. Introducing the thermal conductivity a and the coefficient b of temperature increase per unit concentration decrease, in short the temperature coefficient, by

$$a := \frac{\lambda}{\rho c_p}, \qquad b := \frac{\Delta H}{\rho c_p}, \tag{16.4}$$

we find

$$\frac{\partial T}{\partial t} = a \frac{\partial^2 T}{\partial x^2} \tag{16.5}$$

for the heat equation in both bolsters and

$$\frac{\partial T}{\partial t} = a \frac{\partial^2 T}{\partial x^2} + bkC \tag{16.6}$$

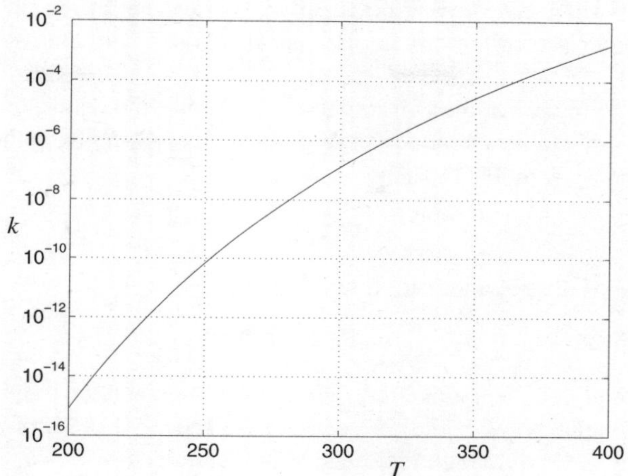

Figure 16.2. *The reaction rate k as a function of the temperature T.*

Table 16.1. *Parameters involved in the curing process.*

Parameter	Value	Units	Parameter	Value	Units
$(\rho c_p)_{1,3}$	1.830×10^6	$\mathrm{J/(m^3 \cdot K)}$	E_a	9.4×10^4	J/mol
$(\rho c_p)_2$	2.123×10^6	$\mathrm{J/(m^3 \cdot K)}$	R	8.314	$\mathrm{J/(mol \cdot K)}$
$\lambda_{1,3}$	1.9032×10^{-1}	$\mathrm{J/(m \cdot s \cdot K)}$	$x_{\mathrm{int},1}$	2.0×10^{-3}	m
λ_2	3.16327×10^{-1}	$\mathrm{J/(m \cdot s \cdot K)}$	$x_{\mathrm{int},2}$	1.6×10^{-2}	m
ΔH	8.0×10^4	J/mol	L	1.8×10^{-2}	m
A	3×10^9	$\mathrm{s^{-1}}$			

in the core. Equation (16.6) is coupled with (16.2b) for the concentration.

The model (16.5), (16.6), and (16.2b) has to be completed with initial and boundary conditions and conditions at the interfaces between core and bolsters. As initial conditions, we choose a constant temperature and concentration:

$$T(x,0) = T_0, \qquad 0 < x < L, \tag{16.7}$$

$$C(x,0) = C_0, \qquad x_{\mathrm{int},1} < x < x_{\mathrm{int},2}. \tag{16.8}$$

At the boundaries we impose a constant temperature:

$$T(0,t) = T_{\mathrm{top}}, \quad T(L,t) = T_{\mathrm{bot}}, \qquad t > 0. \tag{16.9}$$

Finally, at the interfaces we require that both the temperature T and the heat flux $\lambda \frac{\partial T}{\partial x}$ be continuous. This results in the conditions

$$T(x-,t) = T(x+,t), \quad \left(\lambda \frac{\partial T}{\partial x}\right)(x-,t) = \left(\lambda \frac{\partial T}{\partial x}\right)(x+,t), \qquad x = x_{\mathrm{int},1}, x_{\mathrm{int},2}. \tag{16.10}$$

These conditions mean that there is no accumulation of heat at the interfaces. Note that we do not have interface conditions for the concentration C and that we do not need them either.

Before we compute a numerical solution, we estimate the typical time scales involved. The time scale τ_{C} due to heat conduction from the walls follows from the balance between the terms $\frac{\partial}{\partial t}T$ and $\frac{\partial^2}{\partial x^2}T$ and is given by

$$\tau_{\mathrm{C}} = L^2/a. \tag{16.11}$$

The time scale τ_{P} of the polymerisation reaction is

$$\tau_{\mathrm{P}}(T) = k(T)^{-1} \tag{16.12}$$

and depends strongly on the temperature. Heat conduction and the polymerisation reaction are essentially coupled when both time scales are of the same order of magnitude. This is the case when the order of magnitude of the temperature is as follows:

$$T \sim T^* := T_{\mathrm{a}}/\ln\left(AL^2/a\right), \tag{16.13}$$

where the temperature T^* follows from $\tau_{\mathrm{P}}(T^*) = \tau_{\mathrm{C}}$ and where $T_{\mathrm{a}} := E_{\mathrm{a}}/R$ is the *activation temperature*. In this problem we typically have $\tau_{\mathrm{C}} = 2000$ s and $T^* = 383$ K.

16.1.2 Numerical Solution Method

In order to compute a numerical solution of the initial boundary value problem (16.5), (16.6), (16.2b), and (16.7) to (16.10), we introduce the spatial grid

$$\begin{aligned}
&\text{bolster (1):} \quad && x_j &&= (j-1)\Delta x_1 && (j = 1, 2, \dots, N_1),\\
&\text{core (2):} \quad && x_{j+N_1} &&= x_{\mathrm{int},1} + (j-1)\Delta x_2 && (j = 1, 2, \dots, N_2),\\
&\text{bolster (3):} \quad && x_{j+N_1+N_2} &&= x_{\mathrm{int},2} + (j-1)\Delta x_3 && (j = 1, 2, \dots, N_3),
\end{aligned} \tag{16.14}$$

with Δx_r $(r = 1, 2, 3)$ the grid size in region r; see Figure 16.1. We have chosen two grid points at each interface, i.e., $x_{N_1} = x_{N_1+1} = x_{\mathrm{int},1}$ and $x_{N_1+N_2} = x_{N_1+N_2+1} = x_{\mathrm{int},2}$. Moreover, we choose the discrete time levels

$$t^n = n\Delta t \qquad (n = 0, 1, 2, \dots, M), \quad \Delta t = \frac{t_{\mathrm{fin}}}{M}, \tag{16.15}$$

with t_{fin} the final simulation time. To discretise the space derivatives in the governing equations we employ central differences and for time integration we use the ϑ method. On the other hand, to discretise the interface conditions we use one-sided differences. In the following T_j^n and C_j^n denote the numerical approximations of $T(x_j, t^n)$ and $C(x_j, t^n)$, respectively. A subscript r $(r = 1, 2, 3)$ on a parameter refers to the region it relates to.

For the heat equation (16.5) in both bolsters we find the difference equations

$$\begin{aligned}
T_j^{n+1} &- \vartheta d_r\left(T_{j+1}^{n+1} - 2T_j^{n+1} + T_{j-1}^{n+1}\right)\\
&= T_j^n + (1-\vartheta) d_r\left(T_{j+1}^n - 2T_j^n + T_{j-1}^n\right) \qquad (r = 1, 3).
\end{aligned} \tag{16.16}$$

The coefficient d_r in (16.16) is defined by

$$d_r := \frac{a_r \Delta t}{\Delta x_r^2} \tag{16.17}$$

and might be called the conduction number. Likewise, for the equations (16.6) and (16.2b) in the core we have

$$\begin{aligned} T_j^{n+1} - \vartheta\Big(d_2\big(T_{j+1}^{n+1} - 2T_j^{n+1} + T_{j-1}^{n+1}\big) + \Delta t b_2 k(T_j^{n+1}) C_j^{n+1}\Big) \\ = T_j^n + (1-\vartheta)\Big(d_2\big(T_{j+1}^n - 2T_j^n + T_{j-1}^n\big) + \Delta t b_2 k(T_j^n) C_j^n\Big), \end{aligned} \tag{16.18a}$$

$$C_j^{n+1} + \vartheta \Delta t k(T_j^{n+1}) C_j^{n+1} = C_j^n - (1-\vartheta)\Delta t k(T_j^n) C_j^n. \tag{16.18b}$$

The coefficient d_2 in (16.18a) is defined in (16.17). Finally, at the interfaces the following discrete equations hold:

$$T_{N_1}^m = T_{N_1+1}^m \qquad (m = n, n+1), \tag{16.19a}$$

$$\frac{\lambda_1}{\Delta x_1}\big(T_{N_1}^m - T_{N_1-1}^m\big) = \frac{\lambda_2}{\Delta x_2}\big(T_{N_1+2}^m - T_{N_1+1}^m\big), \tag{16.19b}$$

$$T_{N_1+N_2}^m = T_{N_1+N_2+1}^m, \tag{16.19c}$$

$$\frac{\lambda_2}{\Delta x_2}\big(T_{N_1+N_2}^m - T_{N_1+N_2-1}^m\big) = \frac{\lambda_3}{\Delta x_3}\big(T_{N_1+N_2+2}^m - T_{N_1+N_2+1}^m\big). \tag{16.19d}$$

The first two equations apply to the interface $x = x_{\text{int},1}$ and the second two apply to the interface $x = x_{\text{int},2}$. The complete set of equations is (16.16), (16.18), and (16.19) for the unknowns T_j^{n+1} and C_j^{n+1}. In fact, this is a nonlinear system due to the $k(T_j^{n+1})$ terms. In order to avoid solving a nonlinear system, we replace the nonlinear term $k(T_j^{n+1})$ in (16.18a) and (16.18b) by $k(T_j^n)$. This corresponds to one Picard iteration step.

As an example, we have computed a numerical solution for a symmetrically heated panel. The initial temperature and concentration are $T_0 = 293$ K and $C_0 = 1250$ mol/m^3, respectively. Then, at $t = 0$, the temperatures at top and bottom of the panel are set to $T_{\text{top}} = T_{\text{bot}} = 433$ K. The numerical parameters in our numerical simulation are $\vartheta = 1$, $N_1 = N_3 = 11$, $N_2 = 41$, $t_{\text{fin}} = 1500$, and $M = 25$. The results are presented in Figures 16.3 and 16.4. In Figure 16.3 we show the temperature and concentration distributions at the discrete time levels t^n $(n = 0, 1, 2, \ldots, M)$. The time histories of the temperature and concentration at the centre of the panel are shown in Figure 16.4. From these two figures we can see that the temperature in the core initially increases through conduction. The higher temperature initiates a chemical reaction, resulting in heat production and consumption of chemicals. This process continues until the chemicals in the core are depleted. Then the temperature tends to the constant temperature imposed at the boundaries.

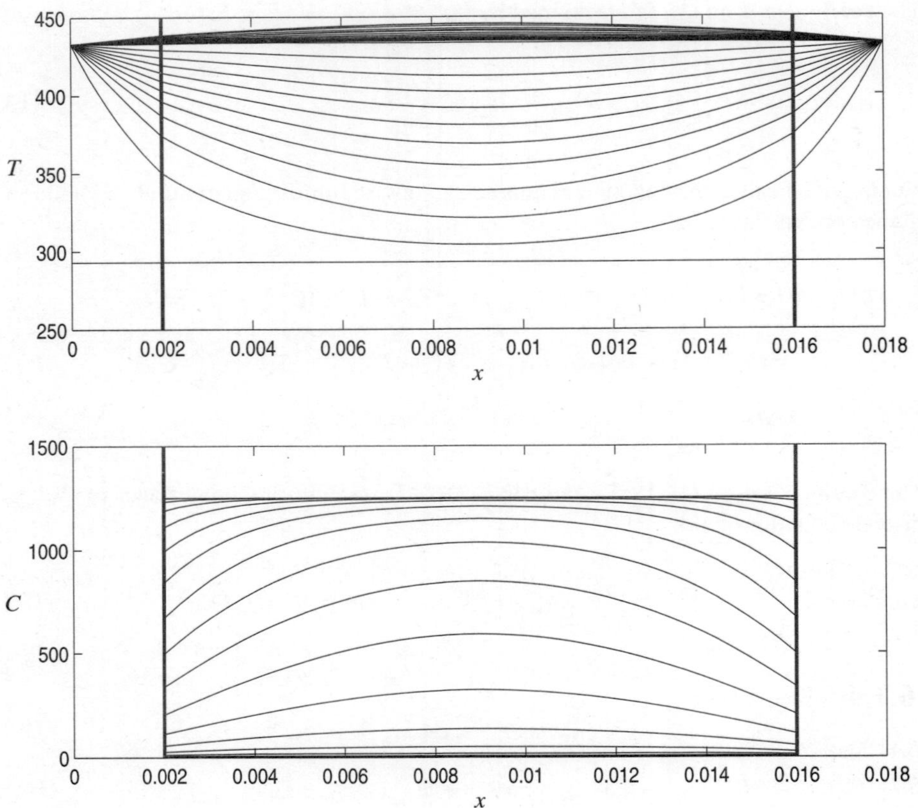

Figure 16.3. *Temperature and concentration distribution in a symmetrically heated panel.*

16.1.3 Discussion and Related Problems

- The proposed model is rather elementary. In the real resin-curing process more complex chemistry, involving more chemical species and reactions, should be taken into account. Also the diffusion of all species should be modeled. Finally, the one-dimensional approximation is most likely not accurate near the ends of the panel. A three-dimensional model should be used instead.

- A more detailed description of resin-containing panels, including deformation through moisture transport, is given in [54, 23].

- The resin-curing process is an example of a reaction-diffusion problem. These problems occur in many fields of science and engineering, e.g., in models for flame propagation, biochemical reactions, and population dynamics. Examples in this text are the thermal explosion in a vessel (Section 16.3) and the catalyzer pellet problem (Section 16.8). Moreover, we discussed Fisher's equation, which is a model problem from population dynamics, in Section 10.6.2.

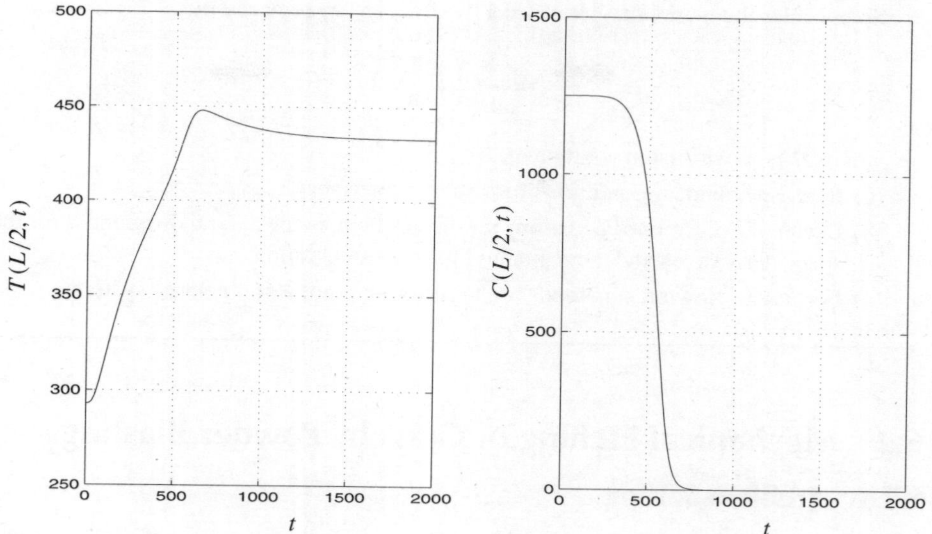

Figure 16.4. *Temperature and concentration at the centre of a symmetrically heated panel.*

16.1.4 Exercises

16.1. Derive expression (16.13) for T^*.

16.2. To make the equations (16.2) dimensionless we can employ the scaling

$$\tilde{x} := x/L, \quad \tilde{t} := t/\tau_c, \quad u := (T - T^*)/\Delta T, \quad \tilde{C} := C/C_0,$$

with $\Delta T := T^* - T_0$. Show that the resulting dimensionless equations (omitting the tilde $\tilde{}$) read

$$\frac{\partial u}{\partial t} = \frac{\partial^2 u}{\partial x^2} + \gamma \exp\left(\frac{\beta u}{u+\alpha}\right) C,$$

$$\frac{\partial C}{\partial t} = -\exp\left(\frac{\beta u}{u+\alpha}\right) C,$$

where the coefficients α, β, and γ are defined by

$$\alpha := T^*/\Delta T, \quad \beta = T_a/T^*, \quad \gamma := bC_0/\Delta T.$$

Thus β is a dimensionless activation energy and γ is a dimensionless heat release parameter.

16.3. Formulate the method of Newton for the solution of the nonlinear algebraic system (16.16), (16.18), and (16.19).

16.4. A more refined model for resin curing would include diffusion of the chemical species.

So consider the model in Section 16.1.1, with (16.2b) replaced by

$$\frac{\partial C}{\partial t} = \frac{\partial}{\partial x}\left(D\frac{\partial C}{\partial x}\right) - kC,$$

where D is the diffusion coefficient.

(a) Propose boundary conditions for C at the interfaces $x_{\text{int},1}$ and $x_{\text{int},2}$.

(b) Derive the discretised equations for this problem using central differences for the space derivatives and the ϑ method for time integration.

(c) Formulate Newton's method for the resulting nonlinear algebraic system.

16.2 Mechanical Etching of Glass by Powder Blasting

16.2.1 Problem Setting

Some modern television displays have a vacuum enclosure that is internally supported by a glass plate. This plate must not hinder the display function. For that reason it has to be accurately patterned with small trenches or holes so that electrons can move from the cathode to the screen. One method of manufacturing such a glass plate is to cover it with an erosion-resistant mask and blast it with an abrasive powder. In this section we present a mathematical model for this *powder-blasting* or *powder-erosion* process and solve it using the method of characteristics.

In powder erosion of brittle materials sharp and abrasive particles in the powder cause deformations and microcracks in the target material. Chipping of target material above the cracks is the dominant erosion mechanism. The rate at which substrate material is removed is determined by the erosion rate E, which is defined as the ratio of the mass loss of the substrate and the mass of the erodent used. For the erosion rate we take the model [140]

$$E = C\big(|\boldsymbol{v}|\cos\theta\big)^k, \tag{16.20}$$

where $\boldsymbol{v}$ is the particle impact velocity, θ, the angle between $\boldsymbol{v}$ and the inward normal $\boldsymbol{n}$ to the surface of the substrate, and C and k are constants. Experimentally, it is observed that $2 \le k \le 4$ [43]. This is in agreement with theoretical models, which predict the value $k = 7/3$; see, e.g., [139].

In Figure 16.5 we show the formation of a pattern in a glass plate by masked erosion. The mask considered has a circular hole and is exposed to a constant particle mass flux; i.e., the particle flux and the impact velocity are constant. Time-lapse photography has exposed a picture at fixed time intervals. Some characteristics of the erosion process can be seen from these photographs. First, the hole is shallow close to the edge of the mask and has a sharp tip in the middle. The reason for this is that, due to the finite particle size, not all particles contribute to the erosion process close to the mask. On the other hand, for a wide, shallow hole we would get a flat bottom in the centre. Second, with increasing depth, the rate of growth of the hole decreases. This can be explained by the dependence of the erosion rate on the angle of impact θ. Finally, with increasing depth, the hole takes a nipple shape.

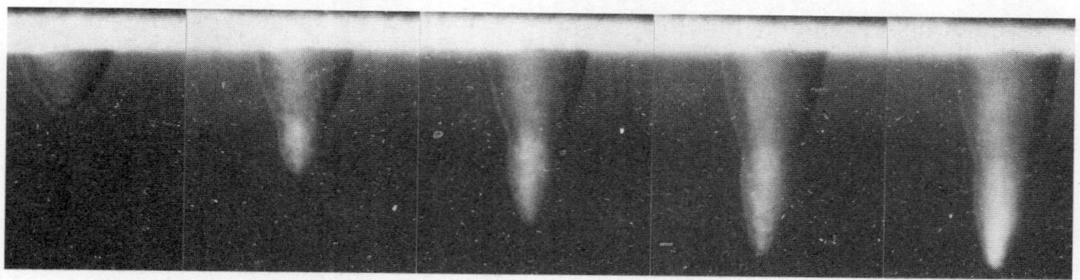

Figure 16.5. *Pattern formation in glass by masked erosion (courtesy P.J. Slikkerveer, Philips Research Laboratories, Eindhoven).*

This is probably caused by rebounding particles from the steep slopes at the sides of the pattern [140]. These particles are focused on the centre of the pattern, where they generate additional erosion. We will not consider this so-called second-strike effect.

16.2.2 Mathematical Model for Powder Erosion

Consider an initially flat substrate of brittle material, covered with a mask. In Figure 16.6 we introduce an (x, y, z) coordinate system, where the (x, y) plane coincides with the initial substrate and where the positive z axis is directed into the material. A flux of particles with velocity $\boldsymbol{v}$ in the positive z direction hits the substrate and removes material. The position of the surface at time t is described by the function $z = \zeta(x, y, t)$. We now derive a kinematic condition for the surface. Let c denote the velocity of the surface in the direction of the unit normal $\boldsymbol{n}$ on the surface, which is directed into the material and is given by

$$\boldsymbol{n} = \frac{1}{\sqrt{1 + \zeta_x^2 + \zeta_y^2}} \begin{pmatrix} -\zeta_x \\ -\zeta_y \\ 1 \end{pmatrix}. \tag{16.21}$$

Consider a point $\boldsymbol{x}(t) = (x(t), y(t), z(t))^T$ on the surface at time t. The displacement of this point satisfies the equation

$$\frac{\mathrm{d}\boldsymbol{x}}{\mathrm{d}t} = c\boldsymbol{n}. \tag{16.22}$$

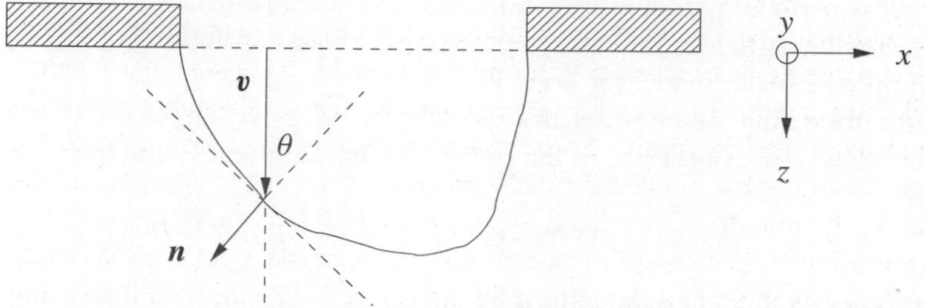

Figure 16.6. *Cross section of a hole in the substrate.*

On the other hand, from the relation $z(t) = \zeta(x(t), y(t), t)$ we readily obtain

$$\frac{\mathrm{d}z}{\mathrm{d}t} = \zeta_x \frac{\mathrm{d}x}{\mathrm{d}t} + \zeta_y \frac{\mathrm{d}y}{\mathrm{d}t} + \zeta_t. \tag{16.23}$$

Inserting (16.22) in (16.23), we find $cn_3 = c(\zeta_x n_1 + \zeta_y n_2) + \zeta_t$, where n_1, n_2, and n_3 denote the components of $\boldsymbol{n}$, and, combining this relation with (16.21), we find the following equation for ζ:

$$\zeta_t - c\sqrt{1 + \zeta_x^2 + \zeta_y^2} = 0. \tag{16.24}$$

We will specify initial and boundary conditions for (16.24) later.

For c we assume the relation [140]

$$c = \frac{1}{\rho_\mathrm{s}} E \boldsymbol{\Phi} \cdot \boldsymbol{n} = \frac{1}{\rho_\mathrm{s}} E \Phi \cos\theta, \tag{16.25}$$

where ρ_s is the mass density of the substrate, E is the erosion rate, $\boldsymbol{\Phi} = \Phi \boldsymbol{e}_z$ is the particle mass flux in the direction of $\boldsymbol{v}$, and θ is the angle between $\boldsymbol{v}$ and $\boldsymbol{n}$. This means that only the particle flux perpendicular to the surface contributes to the velocity c of the surface. Likewise, the erosion rate E depends only on the normal component $\boldsymbol{v} \cdot \boldsymbol{n}$ of the particle velocity $\boldsymbol{v} = v\boldsymbol{e}_z$ and is given by (16.20). Substitution of (16.25) and (16.20) in (16.24) gives

$$\zeta_t - \frac{C}{\rho_s} v^k \Phi \left(1 + \zeta_x^2 + \zeta_y^2\right)^{-k/2} = 0, \tag{16.26}$$

where we have used that $\cos\theta = 1/\sqrt{1 + \zeta_x^2 + \zeta_y^2}$.

Next we make (16.26) dimensionless. Introducing a characteristic length scale L, a characteristic particle mass flux $\tilde{\Phi}$, and the characteristic time $T = L\rho_\mathrm{s}/(C\tilde{\Phi}v^k)$, we obtain in a straightforward way

$$\zeta_t - \Phi(x, y)\left(1 + \zeta_x^2 + \zeta_y^2\right)^{-k/2} = 0. \tag{16.27}$$

The characteristic length L can be the width of a two-dimensional trench or the radius of a rotationally symmetric hole. Note that T is the time needed to propagate the surface over a distance L when the particles hit the surface perpendicularly with mass flux $\tilde{\Phi}$. In the derivation of (16.27) we have assumed that v is constant, which implies that the dimensionless particle mass flux $\Phi(x, y)$ is equal to one. However, later we will see that this choice for the particle mass flux does not give the correct solution of (16.27). Therefore we will propose a better model for $\Phi(x, y)$.

In the following we consider the special case of a two-dimensional trench. Let x denote the transverse coordinate in the trench, as indicated in Figure 16.6. Then (16.27) reduces to

$$\zeta_t + \Phi(x) f(\zeta_x) = 0, \qquad 0 < x < 1, \quad t > 0, \tag{16.28a}$$

where the function $f = f(p)$ is defined by

$$f(p) := -\left(1 + p^2\right)^{-k/2}. \tag{16.28b}$$

Equation (16.28a) is supplemented with the initial and boundary conditions

$$\zeta(x,0) = 0, \qquad 0 < x < 1, \tag{16.28c}$$

$$\zeta(0,t) = \zeta(1,t) = 0, \qquad t > 0. \tag{16.28d}$$

The boundary conditions in (16.28d) mean that the trench cannot grow at the ends $x = 0$ and $x = 1$. We will see that, for a suitable choice of the particle mass flux $\Phi(x)$, boundary conditions for ζ follow directly from the initial condition. For the special case $\zeta(x,0) = 0$ we obtain (16.28d); in this situation these boundary conditions are in fact redundant.

16.2.3 Characteristic Strip Equations

Introducing the variables

$$p := \zeta_x, \qquad q := \zeta_t, \tag{16.29}$$

we can write the PDE (16.28a) in the canonical form

$$F(x,t,\zeta,p,q) := q - \Phi(x)\big(1+p^2\big)^{-k/2} = 0. \tag{16.30}$$

According to Theorem 12.6, the solution of (16.30) can be constructed from the following initial value problem for the characteristic strip equations:

$$\frac{\mathrm{d}x}{\mathrm{d}s} = F_p \qquad = \Phi(x)\frac{kp}{(1+p^2)^{k/2+1}}, \qquad x(0;\sigma) = \sigma, \tag{16.31a}$$

$$\frac{\mathrm{d}t}{\mathrm{d}s} = F_q \qquad = 1, \qquad t(0;\sigma) = 0, \tag{16.31b}$$

$$\frac{\mathrm{d}\zeta}{\mathrm{d}s} = pF_p + qF_q \quad = \Phi(x)\frac{1+(k+1)p^2}{(1+p^2)^{k/2+1}}, \qquad \zeta(0;\sigma) = 0, \tag{16.31c}$$

$$\frac{\mathrm{d}p}{\mathrm{d}s} = -(F_x + pF_\zeta) = \Phi'(x)\frac{1}{(1+p^2)^{k/2}}, \qquad p(0;\sigma) = 0, \tag{16.31d}$$

$$\frac{\mathrm{d}q}{\mathrm{d}s} = -(F_t + qF_\zeta) = 0, \qquad q(0;\sigma) = \Phi(\sigma), \tag{16.31e}$$

where s and σ are the parameters along the characteristics and the initial curve, respectively. The initial condition for q follows from (16.30) and the initial conditions for the other variables. Note that the solution of (16.31b) and (16.31e) is trivial, and we find

$$t(s;\sigma) = s, \quad q(s;\sigma) = \Phi(\sigma). \tag{16.32}$$

The formal solution procedure for the other equations is as follows. We solve (16.31a), (16.31c), and (16.31d), to find $x = x(t;\sigma)$, $\zeta = \zeta(t;\sigma)$, and $p = p(t;\sigma)$. Inverting the first function, we have $\sigma = \sigma(x,t)$, and substitution of the latter expression gives the final solution $\zeta(x,t) := \zeta(t;\sigma(x,t))$ and $p(x,t) := p(t;\sigma(x,t))$. In the following we use the notation $v(t;\sigma)$ or $v(x;\sigma)$ for a generic variable v to indicate that an expression holds only along a characteristic parametrized by t or x. The parameter σ means that the characteristic passes through the point $(\sigma, 0)$. On the other hand, we use the notation $v(x,t)$ if an expression holds in a part of the (x,t) plane.

A first obvious choice for the dimensionless particle mass flux is $\Phi(x) = 1$. The solution of the initial value problem (16.31) in this case is trivial, and we find $\zeta(x,t) = t$ and $p(x,t) = 0$, corresponding to a flat-bottomed hole. This particular solution is in contrast to experimental results, which show slanted sides near the edges of the mask; see Figure 16.5. As observed earlier, this is due to the finite particle size of the eroding powder which causes some particles not to be effective in the erosion process close to the mask. In order to model this phenomenon, we introduce transition regions of thickness δ, where we assume that $\Phi(x)$ increases continuously and monotonically from zero at the boundaries to one at $x = \delta, 1-\delta$. The parameter δ is characteristic of the (dimensionless) particle size and a typical value is $\delta = 0.1$. We adopt the simplest possible choice for $\Phi(x)$; i.e.,

$$\Phi(x) = \begin{cases} x/\delta & \text{if } \; 0 \le x < \delta, \\ 1 & \text{if } \; \delta \le x \le 1-\delta, \\ (1-x)/\delta & \text{if } \; 1-\delta < x \le 1. \end{cases} \tag{16.33}$$

As a result of (16.33), the growth rate of the surface position close to the mask is smaller than in the middle of the hole. Since $\Phi(0) = \Phi(1) = 0$, we obtain from (16.31) the solutions $x(t;0) = \zeta(t;0) = 0$ and $x(t;1) = 1$, $\zeta(t;1) = 0$, implying that the boundary conditions (16.28d) for ζ are automatically satisfied. Moreover, since $\Phi'(0) = 1/\delta$, we can directly compute $p(t;0)$ from (16.31d); likewise we can compute $p(t;1)$. Thus we are not allowed to specify boundary conditions for p. However, the most important consequence of (16.33) is that characteristics originating from the transition regions are directed into the interior domain for increasing t. This implies that the reduced particle mass flux in the transition regions will also influence the solution in the interior domain, which is in agreement with experimental results. We want to emphasize that (16.33) is merely a sensible guess for the particle mass flux that still allows computation of the analytical solution of (16.31). Other choices are possible.

By introducing transition regions, we create intersecting characteristics. Therefore the solution of (16.30) can only be a weak solution and it is anticipated that shocks will emerge from the edges $x = \delta$ and $x = 1-\delta$. Let $x = \xi_{s,1}(t)$ and $x = \xi_{s,2}(t)$ denote the location of the shocks at time t originating at $x = \delta$ and $x = 1-\delta$, respectively. Each point $(\xi_{s,i}(t), t)$ $(i = 1, 2)$ on these shocks is connected to two different characteristics that exist on both sides of the shocks. The speed of these shocks is given by the jump condition

$$\frac{\mathrm{d}\xi_{s,i}}{\mathrm{d}t}[p]_-^+ = -[\Phi(x)(1+p^2)^{-k/2}]_-^+ \qquad (i = 1, 2). \tag{16.34}$$

Thus we can distinguish the following five regions in the (x,t) plane: the left transition region $0 \le x \le \delta$ (region 1), the right transition region $1-\delta \le x \le 1$ (region 2), the interior domain left of the first shock (region 3), the interior domain right of the second shock (region 4), and the region between the two shocks (region 5); see Figure 16.7. Note that the location of the shocks depends on the solution through (16.34) and has to be computed as well.

16.2.4 Solution of the Characteristic Strip Equations

First, consider region 1. If we choose x as the parameter along the characteristics instead of t, we can derive from (16.31) the following differential equations for the slope p and the

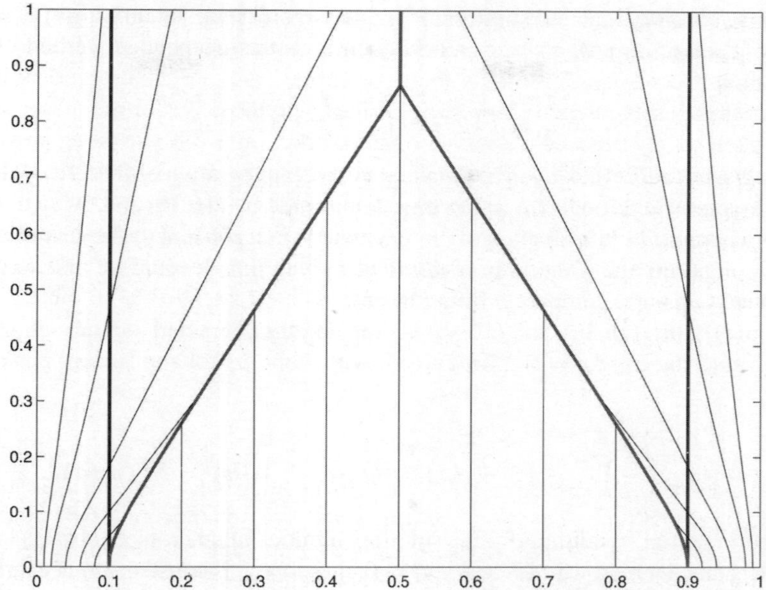

Figure 16.7. *Characteristics and shocks of a two-dimensional trench for $\delta = 0.1$ and $k = 2$.*

surface position ζ:

$$\frac{\mathrm{d}p}{\mathrm{d}x} = \frac{1}{x}\frac{1+p^2}{kp}, \qquad \frac{\mathrm{d}\zeta}{\mathrm{d}x} = \frac{1+(k+1)p^2}{kp}. \tag{16.35}$$

Using the initial conditions $x(0;\sigma) = \sigma$ and $p(0;\sigma) = 0$, we can solve the first equation to find

$$p(x;\sigma) = \sqrt{\left(\frac{x}{\sigma}\right)^{2/k} - 1}. \tag{16.36}$$

Combining both differential equations in (16.35), we obtain the following equation for ζ:

$$\frac{\mathrm{d}\zeta}{\mathrm{d}x} = p + \frac{1+p^2}{kp} = p + x\frac{\mathrm{d}p}{\mathrm{d}x}. \tag{16.37}$$

Together with the initial conditions $\zeta(0;\sigma) = p(0;\sigma) = 0$, (16.37) has the solution

$$\zeta(x;\sigma) = xp(x;\sigma). \tag{16.38}$$

The solutions in (16.36) and (16.38) give p and ζ along the characteristic through the point $(\sigma, 0)$ on the initial curve. The location of this characteristic follows from the differential equation

$$\frac{\mathrm{d}x}{\mathrm{d}t} = \frac{k\sigma}{\delta}\frac{\sqrt{(x/\sigma)^{2/k} - 1}}{(x/\sigma)^{2/k}}, \tag{16.39}$$

which follows readily after substitution of (16.36) in (16.31a). Integrating (16.39) subject to the initial condition $x(0;\sigma)=\sigma$ and choosing p as the independent variable, we arrive at the relation

$$\frac{t}{\delta}=\mathcal{T}(p):=\int_0^p \left(1+r^2\right)^{k/2}\,\mathrm{d}r. \tag{16.40}$$

Alternatively, formula (16.40) can be obtained by direct integration of (16.31d). Points (x,t) on the characteristic through $(\sigma,0)$ are thus determined by (16.36) and (16.40). Note that the slope p in region 1 is a function of t/δ only and is independent of the spatial coordinate x. When computing the solution in region 1 at a given time level t, we first solve (16.40) for p and subsequently compute ζ from (16.38).

From (16.36), (16.38), and (16.40) we see that the characteristic through $(\sigma,0)$ $(0<\sigma<\delta)$ reaches the edge $x=\delta$ at time $t_1(\sigma)$, with slope $p_1(\sigma)$ and surface position $\zeta_1(\sigma)$ given by

$$p_1(\sigma):=\sqrt{\left(\frac{\delta}{\sigma}\right)^{2/k}-1},\quad \zeta_1(\sigma):=\delta p_1(\sigma),\quad t_1(\sigma):=\delta\mathcal{T}(p_1(\sigma)). \tag{16.41}$$

These are the 'initial' conditions for the solution of initial value problem (16.31) in region 3.

Now consider region 3. Since $\Phi'(x)=0$, the slope p is constant along characteristics and, consequently, the differential equations in (16.31) can easily be solved. We find

$$p(x;\sigma)=p_1(\sigma), \tag{16.42a}$$

$$\zeta(x;\sigma)=\zeta_1(\sigma)+\frac{1+(k+1)p_1^2(\sigma)}{kp_1(\sigma)}(x-\delta), \tag{16.42b}$$

$$t(x;\sigma)=t_1(\sigma)+\frac{\left(1+p_1^2(\sigma)\right)^{k/2+1}}{kp_1(\sigma)}(x-\delta). \tag{16.42c}$$

The characteristics in this region are straight lines and are displayed in Figure 16.7. To compute the solution at a given point (x,t) we first have to solve (16.42c) for $p_1(\sigma)$ and subsequently compute $p(x,t)$ and $\zeta(x,t)$ from the other two relations in (16.42).

The solution of (16.31) in region 5 is trivial and is given by

$$x(t;\sigma)=\sigma,\quad p(x,t)=0,\quad \zeta(x,t)=t; \tag{16.43}$$

see Figure 16.7. The characteristics are now vertical lines through $(\sigma,0)$, corresponding to a flat surface.

Since the particle mass flux (16.33) is symmetric around $x=0.5$, it is also clear that the solution of the initial value problem (16.31) is symmetric around $x=0.5$. Consequently, the solution in regions 2 and 4 can easily be obtained from the corresponding solutions in regions 1 and 3, respectively. However, for the sake of brevity, we omit the explicit formulas.

Finally, we have to determine the location of the shocks. We only consider the first shock, emanating from the edge $x=\delta$. The computation of the second shock is completely analogous. The evolution of the first shock is determined by the jump condition (16.34), which in this case leads to the initial value problem

$$\frac{\mathrm{d}\xi_{\mathrm{s},1}}{\mathrm{d}t}=\frac{1}{p_{\mathrm{s},1}}\left(1-\left(1+p_{\mathrm{s},1}^2\right)^{-k/2}\right),\qquad \xi_{\mathrm{s},1}(0)=\delta, \tag{16.44}$$

with $p_{s,1} := p_1(\sigma)$ being the value of the slope just left of the shock on the characteristic through $(\delta, t_1(\sigma))$. Thus the following relation holds between $p_{s,1}$ and $\xi_{s,1}$:

$$\frac{\xi_{s,1} - \delta}{t - t_1(\sigma)} = \frac{k p_{s,1}}{\left(1 + p_{s,1}^2\right)^{k/2+1}}. \tag{16.45}$$

Combining relation (16.45) with the formulas for $t_1(\sigma)$ given in (16.41), differentiating the resulting equation with respect to t, and substituting (16.44), we find the following initial value problem for $p_{s,1}$:

$$\frac{\mathrm{d}p_{s,1}}{\mathrm{d}t} = \frac{p_{s,1}^2}{\left(1 + p_{s,1}^2\right)^{k/2}} \frac{\left(k - \left(1 + p_{s,1}^2\right)\frac{\left(1+p_{s,1}^2\right)^{k/2} - 1}{p_{s,1}^2}\right) \Big/ \left(1 - p_{s,1}^2\right)}{\delta - \frac{(k+1)p_{s,1}^2 - 1}{p_{s,1}^2 - 1}\xi_{s,1}}, \qquad p_{s,1}(0) = 0. \tag{16.46}$$

The initial value for $p_{s,1}$ follows, e.g., from (16.40). The propagation of the shock is determined by the differential equations (16.44) and (16.46), which we have to solve numerically. The result is shown in Figure 16.7.

We have collected the results of this section in Figure 16.8, which gives the solution for ζ and p at time levels $t = 0.0, 0.1, \ldots, 1.0$ for $\delta = 0.1$ and $k = 2.33$. This figure nicely displays the features of the solution: a slanted surface in the transition regions, a flat bottom in the interior domain, and a curved surface in between. Also, the inwardly propagating shocks are clearly visible.

To validate our model we compare in Figure 16.9 the analytical solution with experimental results at four different time levels for the erosion of a trench. All surfaces are dimensionless according to the scaling in Section 16.2.2. We have computed the surfaces with $k = 2.33$ and $\delta = 0.1$. We see a nice qualitative agreement between analytical and experimental results. Both show the shallow, flat-bottomed solution in the middle of the hole at $t = 0.5$. However, the analytically computed hole at $t = 0.5$ is somewhat deeper than the experimental one. The reason for this is probably the shadow effect of the mask, which is not included in the analytical model. For $t \geq 1.6$ the experimental surfaces show

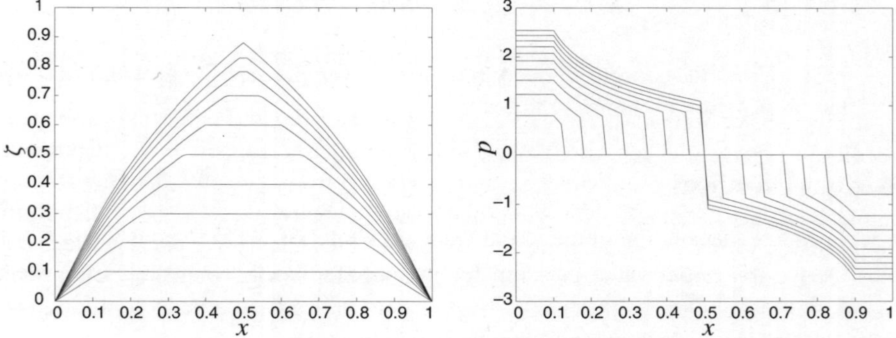

Figure 16.8. *Solution for the surface position (left) and its slope (right) of a one-dimensional trench. Parameter values are $\delta = 0.1$ and $k = 2.33$.*

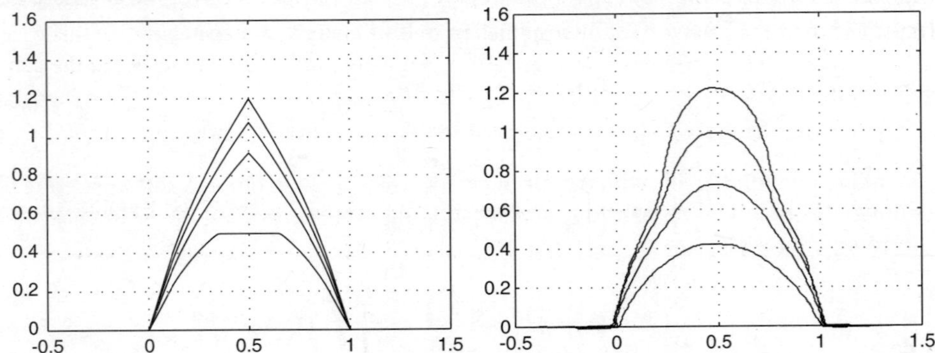

Figure 16.9. *Analytical solution (left) and measurements (right) of the surface position for a two-dimensional trench at* $t = 0.5, 1.1, 1.6,$ *and* 2.1 *(experimental results by P.J. Slikkerveer* [155]*).*

a nipple shape in the middle of the hole, which is probably caused by rebounding particles from the slopes at the sides of the surface. This *second-strike effect* is also not included in our model. Further differences are the round top of the experimental holes due to the finite particle size and the widening of the experimental holes due to mask wear.

16.2.5 Discussion and Related Problems

- Etching is a technique for removing material from a layer by means of either chemical or mechanical processes. It is used to accurately pattern layers and is often applied in the manufacture of semiconductor devices. A mathematical description of etching processes is given in, e.g., [75].

- All etching techniques involve a moving surface. The kinematic condition describing the evolution of such a surface is a first order nonlinear PDE like (16.28a). In this section we solved this equation analytically using the method of characteristics. Alternatively, we could solve it numerically. Another approach is to apply level set methods [137, 108].

- Moving interfaces occur in many problems. Examples are the growth of crystals or the propagation of flames [137].

16.2.6 Exercises

16.5. Compute the solution of the initial value problem (16.31) in regions 2 and 4.

16.6. Derive the initial value problem for the second shock, emanating from the edge $x = 1 - \delta$.

16.7. Show that $p := \zeta_x$ satisfies the quasi-linear equation

$$p_t + \big(\Phi(x) f(p)\big)_x = 0.$$

Use this equation to derive the jump condition (16.34).

16.8. Give an explicit expression for the slope $p(x, t)$ in region 1 for the special case $k = 2$.

16.9. The theory in this section also applies to a rotationally symmetric hole. In this case it is convenient to use polar coordinates (r, ϕ), where the origin is located at the centre of the hole.

(a) Give the PDEs for the surface position $\zeta = \zeta(r, t)$ and its slope $p(r, t) := \zeta_r(r, t)$.

(b) Give the corresponding characteristic strip equations.

(c) Choose a suitable particle mass flux $\Phi = \Phi(r)$ and solve these equations.

16.3 Thermal Explosion in a Vessel

16.3.1 Problem Formulation

In a vessel containing a mixture of reacting species the temperature can increase tremendously, causing great damage to the plant where it is situated. To be more precise, if an exothermic reaction takes place, heat is released, which causes an increase of temperature. Often, the reaction rate increases with increasing temperature, leading to more production of heat, etc. In this way we get a self-accelerating chemical reaction, which continues until the reactant is depleted. Some mixtures can be characterized by the existence of a critical temperature T_{crit}, below which there is virtually no reaction, whereas above this temperature the reaction proceeds extremely fast. Thus at the critical temperature the reaction rate changes very suddenly, practically discontinuously. This process is called a *thermal explosion*. In this section we study a simple model for this type of process.

16.3.2 Mathematical Model

As an example, we consider a spherical vessel containing a reacting mixture. Assuming spherical symmetry and a one-step irreversible reaction and neglecting the flow of the mixture, we have the initial boundary value problem [171]

$$\frac{\partial T}{\partial t} = \frac{a}{r^2}\frac{\partial}{\partial r}\left(r^2\frac{\partial T}{\partial r}\right) + qw(C, T), \qquad 0 < r < L, \quad t > 0, \tag{16.47a}$$

$$\frac{\partial C}{\partial t} = \frac{b}{r^2}\frac{\partial}{\partial r}\left(r^2\frac{\partial C}{\partial r}\right) - w(C, T), \tag{16.47b}$$

$$T(r, 0) = T_0, \quad C(r, 0) = C_0, \qquad 0 < r < L, \tag{16.47c}$$

$$\frac{\partial T}{\partial r}(0, t) = 0, \quad \frac{\partial C}{\partial r}(0, t) = 0, \qquad t > 0, \tag{16.47d}$$

$$T(L, t) = T_{\text{amb}}, \quad \frac{\partial C}{\partial r}(L, t) = 0, \qquad t > 0, \tag{16.47e}$$

where T is the temperature of the mixture and C is the concentration of the reactants. The constants a, b, and q are the thermal diffusivity, the diffusion coefficient, and the heat release

Table 16.2. *Parameters for a thermal explosion.*

Parameter	Value	Units	Parameter	Value	Units
a	1.0	m^2/s	A	3.15×10^{10}	s^{-1}
b	2.0	m^2/s	L	1	m
q	8.1×10^3	$m^3 \cdot K/mol$	T_0	293	K
E_a	7.2×10^4	J/mol	C_0	0.3	mol/m^3
R	8.314	$J/(mol \cdot K)$	T_{amb}	308	K

parameter, respectively. The variable w is the reaction rate and is given by

$$w(C, T) = AC\exp(-E_a/RT), \tag{16.48}$$

with A, E_a, and R being the preexponential factor, the activation energy, and the universal gas constant, respectively.

The values of all physical parameters/constants are given is Table 16.2. The initial boundary value problem (16.47) is a characteristic model for thermal explosions; see Chapter 7. Its solution has the characteristic features of a thermal explosion; however, details of, e.g., the chemistry or flow are not included. These require a much more complicated model.

16.3.3 Estimating the Induction Time

The solution of the initial boundary value problem (16.47) is characterized by an initial period of virtually no reaction called the *induction period*. Then, suddenly, a very intense chemical reaction takes place in which the temperature increases tremendously. In this section we will estimate the *induction time* t_{ind} at which this temperature increase occurs. For this we assume that the vessel is a *perfectly stirred reactor*; i.e., the temperature T and the concentration C are independent of the spatial coordinate r. Then (16.47a) and (16.47b) reduce to

$$\frac{dT}{dt} = qACe^{-T_a/T}, \qquad \frac{dC}{dt} = -ACe^{-T_a/T}, \tag{16.49}$$

where $T_a := E_a/R$ is the *activation temperature*. Taking a suitable linear combination of both equations in (16.49), we obtain the integral

$$T(t) + qC(t) = T_0 + qC_0 = T_{\max}, \tag{16.50}$$

where $T_{\max}$ is the maximum temperature at which the reactant is depleted, i.e., when $C = 0$. Substituting this relation in the ODE for T and integrating the resulting equation, we obtain

$$\int_{T_0}^{T} \frac{1}{T_b - u} e^{T_a/u}\, du = At. \tag{16.51}$$

Note that the integral in (16.51) diverges for $T \to T_b$, implying that the mixture reaches the depleted state $T = T_{\max}$, $C = 0$ for $t \to \infty$. Substituting $v = T_a/u$ and splitting the integrand into partial fractions,we obtain the relation

$$\mathrm{Ei}\left(\frac{T_\mathrm{a}}{T}\right) - \mathrm{Ei}\left(\frac{T_\mathrm{a}}{T_0}\right) - \mathrm{e}^{T_\mathrm{a}/T_\mathrm{b}}\left[\mathrm{Ei}\left(T_\mathrm{a}\left(\frac{1}{T} - \frac{1}{T_\mathrm{b}}\right)\right) - \mathrm{Ei}\left(T_\mathrm{a}\left(\frac{1}{T_0} - \frac{1}{T_\mathrm{b}}\right)\right)\right] = At, \tag{16.52}$$

where $\mathrm{Ei}(x)$ is the exponential integral defined by

$$\mathrm{Ei}(x) := \int_{-\infty}^{x} \frac{\mathrm{e}^t}{t}\,\mathrm{d}t. \tag{16.53}$$

The induction time t_ind is now defined as the time needed to reach the temperature $T = T^* := T_\mathrm{max}(1 - T_\mathrm{max}/T_\mathrm{a})$ is where the reaction rate $w^*(T)$, defined by

$$w^*(T) := w\big((T_\mathrm{max} - T)/q, T\big) = A(T_\mathrm{max} - T)\exp(-T_\mathrm{a}/T),$$

is maximal. Substituting this expression for T^* in (16.52), using the asymptotic approximation (see, e.g., [1] or Example 15.11)

$$\mathrm{Ei}(x) = \frac{\mathrm{e}^x}{x}\left(1 + \mathcal{O}(x^{-1})\right) \qquad \text{for} \quad x \to \infty \tag{16.54}$$

and taking into account that T_a/T^*, T_a/T_0, $T_\mathrm{a}(1/T^* - 1/T_\mathrm{max})$, and $T_\mathrm{a}(1/T_0 - 1/T_\mathrm{max})$ are large, we obtain asymptotically that

$$\left(\frac{T^*}{T_\mathrm{a}} - \frac{T^*T_\mathrm{max}}{T_\mathrm{a}(T_\mathrm{max} - T^*)}\right)\mathrm{e}^{T_\mathrm{a}/T^*} - \left(\frac{T_0}{T_\mathrm{a}} - \frac{T_0T_\mathrm{max}}{T_\mathrm{a}(T_\mathrm{max} - T_0)}\right)\mathrm{e}^{T_\mathrm{a}/T_0} \simeq At_\mathrm{ind}. \tag{16.55}$$

Finally, taking into account that $\mathrm{e}^{T_\mathrm{a}(1/T^* - 1/T_0)}$ is exponentially small, we find the approximation

$$t_\mathrm{ind} \simeq \frac{1}{A}\frac{T_0^2}{T_\mathrm{a}\big(T_\mathrm{max} - T_0\big)}\mathrm{e}^{T_\mathrm{a}/T_0}, \tag{16.56}$$

with T_max given in (16.50). For the parameter values in Table 16.2 we find $t_\mathrm{ind} = 0.8883$ s.

16.3.4 Numerical Solution Method

For the numerical solution of initial boundary value problem (16.47) we employ the finite volume method for space discretisation in combination with the ϑ method for time integration. Thus we cover the interval $[0, L]$ with control volumes V_j as follows:

$$V_j := (r_{j-1/2}, r_{j+1/2}) = ((j-1)\Delta r, j\Delta r), \qquad j = 1, 2, \ldots, N_r, \tag{16.57}$$

with $\Delta r = L/N_r$ the spatial grid size. Each control volume V_j is a spherical shell with $r_{j-1/2} \le r \le r_{j+1/2}$ centred around $r_j = (j - 1/2)\Delta r$. We introduce the discrete time levels

$$t^n := n\Delta t \qquad (n = 0, 1, 2, \ldots, M), \quad \Delta t = \frac{t_\mathrm{fin}}{M}, \tag{16.58}$$

with t_fin the final simulation time. As usual, $T_j(t)$ and T_j^n denote the semi- and fully discrete approximations of $T(r_j, t)$ and $T(r_j, t^n)$, respectively. $C_j(t)$ and C_j^n are defined likewise.

Integrating (16.47a) and (16.47b) over a control volume V_j and applying Gauss's theorem (see the appendix, Section J), we find

$$\frac{\mathrm{d}}{\mathrm{d}t}\int_{V_j} T\,\mathrm{d}V = a\oint_{\partial V_j}\frac{\partial T}{\partial \boldsymbol{n}}\,\mathrm{d}S + q\int_{V_j} w(C,T)\,\mathrm{d}V, \tag{16.59a}$$

$$\frac{\mathrm{d}}{\mathrm{d}t}\int_{V_j} C\,\mathrm{d}V = b\oint_{\partial V_j}\frac{\partial C}{\partial \boldsymbol{n}}\,\mathrm{d}S - \int_{V_j} w(C,T)\,\mathrm{d}V. \tag{16.59b}$$

Next we have to approximate the integrals in (16.59). For the volume integrals we use the midpoint rule. So we have for the first integral in (16.59a)

$$\int_{V_j} T\,\mathrm{d}V \doteq T(r_j,t)|V_j| = \frac{4}{3}\pi\left(r_{j+1/2}^3 - r_{j-1/2}^3\right)T(r_j,t), \tag{16.60}$$

with $|V_j|$ the volume of V_j. The same goes for the other volume integrals. Computing the surface integrals is slightly more complicated. Consider, e.g., the surface integral in (16.59a). Each control volume is bounded by the spheres $r = r_{j+1/2}$ and $r = r_{j-1/2}$ with outward unit normals $\boldsymbol{e}_r$ and $-\boldsymbol{e}_r$, respectively. Applying the midpoint rule once more, we obtain

$$\begin{aligned}\oint_{V_j}\frac{\partial T}{\partial \boldsymbol{n}}\,\mathrm{d}S &= \int_{r=r_{j+1/2}}\frac{\partial T}{\partial r}\,\mathrm{d}S - \int_{r=r_{j-1/2}}\frac{\partial T}{\partial r}\,\mathrm{d}S \\ &\doteq 4\pi\left(r_{j+1/2}^2\frac{\partial T}{\partial r}(r_{j+1/2},t) - r_{j-1/2}^2\frac{\partial T}{\partial r}(r_{j-1/2},t)\right).\end{aligned} \tag{16.61}$$

For the spatial derivatives at the interface of V_j we use standard central differences. This way, we obtain

$$\oint_{V_j}\frac{\partial T}{\partial \boldsymbol{n}}\,\mathrm{d}S \doteq \frac{4\pi}{\Delta r}\Big(r_{j+1/2}^2\big(T(r_{j+1},t) - T(r_j,t)\big) - r_{j-1/2}^2\big(T(r_j,t) - T(r_{j-1},t)\big)\Big). \tag{16.62}$$

A similar expression holds for the surface integral in (16.59b). Inserting the above quadrature rules in (16.59), we obtain the semidiscrete system

$$\frac{\mathrm{d}T_j}{\mathrm{d}t} = a\big(\alpha_{j+1/2}(T_{j+1} - T_j) - \beta_{j-1/2}(T_j - T_{j-1})\big) + qw(C_j,T_j), \tag{16.63a}$$

$$\frac{\mathrm{d}C_j}{\mathrm{d}t} = b\big(\alpha_{j+1/2}(C_{j+1} - C_j) - \beta_{j-1/2}(C_j - C_{j-1})\big) - w(C_j,T_j), \tag{16.63b}$$

where the geometrical factors $\alpha_{j+1/2}$ and $\beta_{j-1/2}$ are defined by

$$\alpha_{j+1/2} := \frac{3}{\Delta r^3}\frac{r_{j+1/2}^3}{r_{j+1/2}^2 + r_{j+1/2}r_{j-1/2} + r_{j-1/2}^2}, \tag{16.64a}$$

$$\beta_{j+1/2} := \frac{3}{\Delta r^3}\frac{r_{j-1/2}^3}{r_{j+1/2}^2 + r_{j+1/2}r_{j-1/2} + r_{j-1/2}^2}. \tag{16.64b}$$

Finally, we apply the ϑ method to (16.63) to obtain

$$T_j^{n+1} - \vartheta \Delta t \left(a \left(\alpha_{j+1/2} \left(T_{j+1}^{n+1} - T_j^{n+1}\right) - \beta_{j-1/2} \left(T_j^{n+1} - T_{j-1}^{n+1}\right)\right) + qw \left(C_j^{n+1}, T_j^{n+1}\right)\right)$$
$$= T_j^n + (1-\vartheta)\Delta t \left(a \left(\alpha_{j+1/2} \left(T_{j+1}^n - T_j^n\right) - \beta_{j-1/2} \left(T_j^n - T_{j-1}^n\right)\right) + qw \left(C_j^n, T_j^n\right)\right), \quad (16.65a)$$

$$C_j^{n+1} - \vartheta \Delta t \left(b \left(\alpha_{j+1/2} \left(C_{j+1}^{n+1} - C_j^{n+1}\right) - \beta_{j-1/2} \left(C_j^{n+1} - C_{j-1}^{n+1}\right)\right) - w \left(C_j^{n+1}, T_j^{n+1}\right)\right)$$
$$= C_j^n + (1-\vartheta)\Delta t \left(b \left(\alpha_{j+1/2} \left(C_{j+1}^n - C_j^n\right) - \beta_{j-1/2} \left(C_j^n - C_{j-1}^n\right)\right) - w \left(C_j^n, T_j^n\right)\right). \quad (16.65b)$$

The equations in (16.65) constitute a nonlinear system coupled through the reaction term $w(C_j^{n+1}, T_j^{n+1})$. Introducing the vector of unknowns $\mathbf{u} := (T_1^{n+1}, C_1^{n+1}, T_2^{n+1}, C_2^{n+1}, \ldots, T_{N_r}^{n+1}, C_{N_r}^{n+1})^{\mathrm{T}}$, we can write (16.65) in the form

$$\mathbf{A}(\mathbf{T}^{n+1})\mathbf{u} = \mathbf{b}, \quad (16.66a)$$

where the coefficient matrix $\mathbf{A}$ depends only on the temperature vector $\mathbf{T}^{n+1} := (T_1^{n+1}, T_2^{n+1}, \ldots, T_{N_r}^{n+1})^{\mathrm{T}}$. This system could be solved using (damped) Newton iteration with the solution at time level t^n as initial guess. However, since the reaction term $w(C, T)$ is extremely sensitive to the temperature T, it is expected that Newton iteration would not converge unless the initial guess was very close to the exact solution, i.e., the time step Δt was extremely small. Alternatively, we solve the linear system

$$\mathbf{A}(\mathbf{T}^n)\mathbf{u} = \mathbf{b}, \quad (16.66b)$$

which we obtain by simply replacing $\mathbf{T}^{n+1}$ by $\mathbf{T}^n$ in the coefficient matrix. This corresponds to one Picard iteration for (16.66a). Moreover, to improve the robustness of the method, we bound the numerical solution according to $T_0 \leq T \leq T_{\max}$ and $0 \leq C \leq C_0$.

As an example, we have computed a solution for $T_0 = 293$ K, $C_0 = 0.3\,\mathrm{mol/m^3}$, and $T_{\mathrm{amb}} = 308$ K. The numerical parameters are $\vartheta = 1$, $N_r = 100$, $t_{\mathrm{fin}} = 2$, and $M = 1000$. The results are presented in Figures 16.10 and 16.11. The temperature profiles in Figure 16.10 are plotted every 10 time steps. The solution displays the following behaviour. Initially, the temperature increases close to the boundary $r = 1$ through conduction. Gradually, the temperature also slightly increases in the interior domain. However, the concentration is still virtually 0.3 everywhere. The increasing temperature induces a chemical reaction, which in turn induces heat production. This process is very slow, until $t \approx 0.80$, which is approximately 10 percent off the induction time computed in Section 16.3.3; see Figure 16.11. Then, all of a sudden, the temperature jumps to its maximum value and the concentration rapidly decreases to zero. Next the temperature decreases slowly through conduction. Finally, the temperature and the concentration tend to their steady values $T = T_{\mathrm{amb}}$ and $C = 0$, respectively.

16.3.5 Discussion and Related Problems

- Although our model for a thermal explosion is a simple one, taking into account only conduction/diffusion and a one-step chemical reaction, it contains the essential characteristics. On the other hand, a more realistic model for thermal explosions should

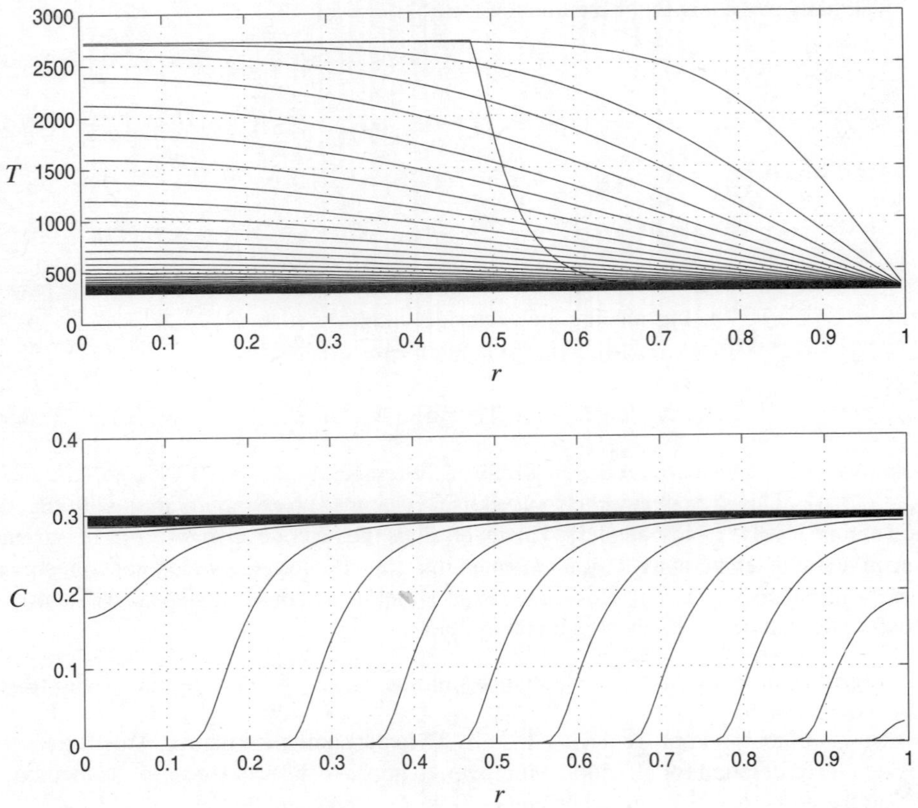

Figure 16.10. *Temperature and concentration distribution in a reactor vessel.*

contain a description of the (turbulent) flow; more complex chemistry, involving many species and reactions; and possibly more complicated boundary conditions.

- The thermal explosion is an example of a reaction-diffusion process. Many of these processes occur in, e.g., chemistry, combustion theory, and population dynamics. In this sense it is related to the resin-curing process (Section 16.1) and the reaction in a catalyzer pellet (Section 16.8).

- The activation energy E_a is usually very large, making the problem even more nonlinear. On the other hand, this makes the problem suitable for *large activation energy asymptotics* [170].

- The system (16.65) is nonlinear in the temperature, putting very high requirements on a nonlinear solver. Newton iteration is most likely not to converge unless the initial guess is extremely close to the exact solution. Therefore Newton iteration has to be combined with *continuation methods* to provide a good initial guess (see, e.g., [3]) and *embedding techniques* to improve the convergence behaviour.

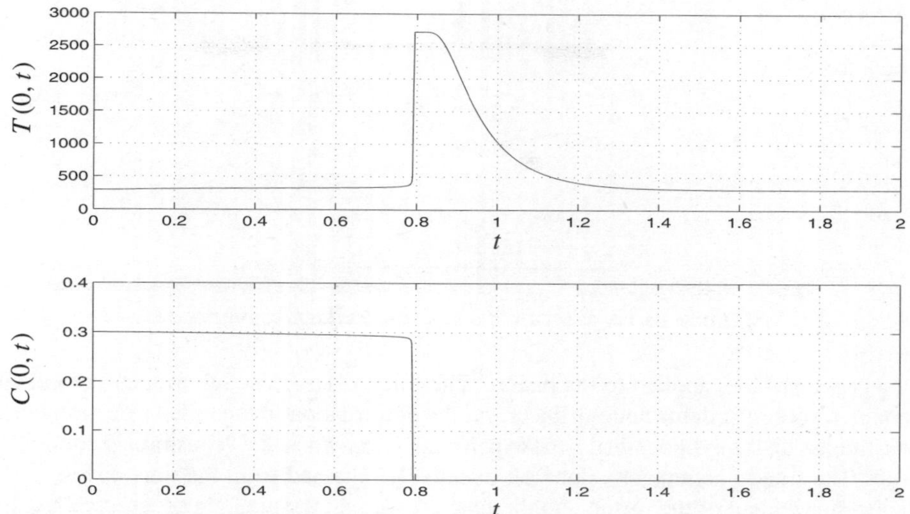

Figure 16.11. *Temperature and concentration at the centre of a reactor vessel.*

- The initial boundary value problem (16.47) can suitably be solved using *moving grid techniques*, since its solution is rapidly varying in a small and moving part of the domain.

16.3.6 Exercises

16.10. The semidiscrete equations (16.63) hold only for control volumes in the interior of $[0, L]$, i.e., for control volumes not adjacent to the boundary. Give the derivation of the semidiscrete equations for the control volumes V_1 and V_{N_r} neighbouring $r = 0$ and $r = L$, respectively.

16.11. Formulate the method of Newton to compute $\mathbf{T}^{n+1}$ and $\mathbf{C}^{n+1}$ from (16.66a). An obvious initial guess would be $\mathbf{T}^n$ and $\mathbf{C}^n$. Could you come up with a better initial guess?

16.4 Determining Viscoelastic Material Parameters

16.4.1 Problem Formulation

In order to measure the stress parameters of a viscoelastic material of Kelvin–Voigt type (see Section 6.8.3), the following experiment is done. A cylinder, filled with the material in question, is closed at one end, $z = 0$, by a solid wall, while at the other end, $z = L$, it is closed by a freely movable piston; see Figure 16.12. Before time $t = 0$, the medium is at rest, undeformed, and free of stress. As of $t = 0$ the medium is suddenly compressed

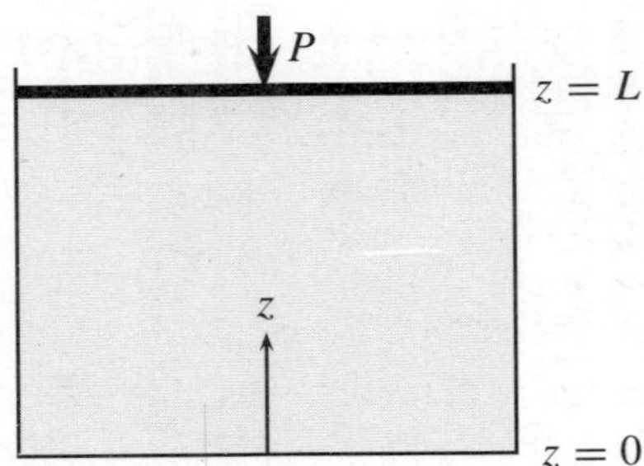

Figure 16.12. *Viscoelastic material suddenly compressed.*

by a constant force applied to the piston. The force is large enough to measure but small enough to keep the deformations linear. As the deformation in the radial direction remains constrained by the cylinder walls, the experiment is known as the "constrained compression test." The sought parameter values are eventually obtained from the time history and the final displacement of the piston and the final pressure at the wall. In order to collect these values accurately, we need a model that relates the parameters to the measured data.

The usual approach is to apply a linear model for the deformations, neglecting any acceleration effects. This is called quasi-static model. Apart from the question of when inertia may be neglected (it depends on the parameters that we are going to measure), this approach is not consistent with an initial value problem because the elastic reaction in the material exists only with deformation. So right at the beginning, when there is no reaction yet, the material deforms with a large acceleration, while at the same time the deformation and stresses predicted by the quasi-static model are completely wrong.

For this reason we are interested in analysing a model that includes the inertia and investigating its role in the initial phase and beyond.

16.4.2 The Model

We introduce $\mathcal{T}$ denoting the stress tensor (see (6.19)), $\mathcal{E}$ the deformation tensor, $\boldsymbol{s}$ the deviatoric stress tensor (see the appendix, Section L.2), $\boldsymbol{\epsilon}$ the deviatoric deformation tensor, $\boldsymbol{u}$ the displacement vector, κ the bulk modulus, η the viscosity, ρ the density, and γ the relaxation rate. The parameters to be determined are thus κ, η, and γ. The constitutive equations of a Kelvin–Voigt medium are given by (6.53a), with $3\lambda\theta_1 + 2\mu\theta_2 = 0$, $\kappa = \lambda + \frac{2}{3}\mu$, $\eta = \mu\theta_2$, and $\gamma = \theta_2^{-1}$. We obtain the equations

$$\mathrm{tr}(\mathcal{T}) = 3\kappa\,\mathrm{tr}(\mathcal{E}) \quad \text{or} \quad \sum_{k=1}^{3} t_{kk} = 3\kappa \sum_{k=1}^{3} e_{kk}, \tag{16.67}$$

$$\frac{\boldsymbol{s}}{2\eta} = \dot{\boldsymbol{\epsilon}} + \gamma\boldsymbol{\epsilon} \quad \text{or} \quad \frac{s_{ij}}{2\eta} = \dot{\epsilon}_{ij} + \gamma\epsilon_{ij} \quad (i, j = 1, 2, 3). \tag{16.68}$$

The linear relations involve $\mathcal{E} = \frac{1}{2}\nabla\boldsymbol{u} + \frac{1}{2}(\nabla\boldsymbol{u})^T$. Initially, we have $\boldsymbol{\epsilon} = \mathbf{0}$. As we will investigate the role of inertia, we use Cauchy's equation of motion (6.25):

$$\nabla\cdot\mathcal{T} = \rho\ddot{\boldsymbol{u}} \quad \text{or} \quad \sum_{j=1}^{3} t_{ij,j} = \rho\ddot{u}_i \quad (i = 1, 2, 3), \tag{16.69}$$

where we have used the compact notation that an index before the comma denotes a vectorial component and an index after the comma denotes a partial derivative. So $u_{i,j} = \frac{\partial}{\partial x_j} u_i$. Furthermore, the dot denotes a time derivative, so $\ddot{u}_i = \frac{\partial^2}{\partial t^2} u_i$. In cylindrical coordinates (r, ϕ, z) the axial, radial, and circumferential displacements are given by (u_z, u_r, u_ϕ). From the symmetry of the problem there is no displacement or variation in ϕ. Furthermore, we assume no displacement or variation in the radial direction due to the rigid container wall. This assumption may be not justified if the fluid pulls at the wall, i.e., if the normal stress in the radial direction is positive. In this case the fluid may separate from the wall.

The radial component of (16.69) is then $(\nabla \cdot \mathcal{T})_r = t_{rr} - t_{\phi\phi} = 0$, and so $t_{\phi\phi} = t_{rr}$. The wall is assumed perfectly smooth, resulting in no shear stresses. From

$$s_{zz} = \frac{2}{3} t_{zz} - \frac{2}{3} t_{rr} = \frac{4}{3} \eta (u_{z,zt} + \gamma u_{z,z}), \qquad t_{zz} + 2t_{rr} = 3\kappa u_{z,z}, \tag{16.70}$$

we obtain

$$t_{zz} = \frac{4}{3} \eta u_{z,zt} + \left(\kappa + \frac{4}{3} \eta \gamma \right) u_{z,z}, \tag{16.71}$$

and so

$$t_{zz,z} = \rho \ddot{u}_z = \frac{4}{3} \eta u_{z,zzt} + \left(\kappa + \frac{4}{3} \eta \gamma \right) u_{z,zz}, \tag{16.72}$$

while the normal stress in the radial direction t_{rr} (which is minus the pressure at the wall) is given by

$$t_{rr} = \left(\kappa - \frac{2}{3} \eta \gamma \right) u_{z,z} - \frac{2}{3} \eta u_{z,zt}. \tag{16.73}$$

At $z = L$ we apply the external pressure p, so we have the boundary condition

$$t_{zz} = \frac{4}{3} \eta u_{z,zt} + \left(\kappa + \frac{4}{3} \eta \gamma \right) u_{z,z} = -p \qquad \text{at} \quad z = L. \tag{16.74}$$

We make expression (16.72) dimensionless as follows:

$$z = Lz^*, \quad t = \frac{\frac{4}{3}\eta}{\kappa + \frac{4}{3}\eta\gamma} t^*, \quad u_z = \frac{pL}{\kappa + \frac{4}{3}\eta\gamma} w \tag{16.75}$$

(asterisks $*$ will be suppressed in the following) and introduce the parameter

$$\varepsilon = \left(\frac{3L}{4\eta} \right)^2 \left(\kappa + \frac{4}{3} \eta \gamma \right) \rho. \tag{16.76}$$

Neglect of inertia is equivalent to $\varepsilon = 0$. So we are interested in the solution for ε small. We obtain in dimensionless form the initial boundary value problem

$$\frac{\partial^3 w}{\partial z^2 \partial t} + \frac{\partial^2 w}{\partial z^2} = \varepsilon \frac{\partial^2 w}{\partial t^2} \qquad \text{for} \quad t > 0, \quad 0 < z < 1, \tag{16.77a}$$

$$w(z, 0) = \frac{\partial w}{\partial t}(z, 0) = 0 \qquad \text{for} \quad 0 < z < 1, \tag{16.77b}$$

$$w(0, t) = 0 \qquad \text{for} \quad t > 0, \tag{16.77c}$$

$$\frac{\partial w}{\partial z}(1, t) + \frac{\partial^2 w}{\partial z \partial t}(1, t) = -1 \qquad \text{for} \quad t > 0, \tag{16.77d}$$

where $0 < \varepsilon \ll 1$. For completeness the dimensional normal stresses in the z and r directions are given by

$$t_{zz} = p\left(\frac{\partial^2 w}{\partial z \partial t} + \frac{\partial w}{\partial z}\right), \tag{16.78a}$$

$$t_{rr} = p\left(\alpha\frac{\partial w}{\partial z} - \frac{1}{2}\frac{\partial^2 w}{\partial z \partial t}\right), \qquad \text{where} \quad \alpha = \frac{\kappa - \frac{2}{3}\eta\gamma}{\kappa + \frac{4}{3}\eta\gamma}. \tag{16.78b}$$

16.4.3 Analysis and Solution

An explicit solution in series or integral form may be constructed (see the exercises). This is, unfortunately, by its form, of limited value if insight and understanding are the goal. Therefore we will further reduce the complexity of the problem by utilizing the small parameter ε and constructing an approximate solution by perturbation methods. The problem will be seen to be of singular perturbation type with boundary layer behaviour in the time coordinate near $t = 0$, and the solution will be a matched asymptotic expansion (MAE) (see Section 15.4.1).

First, we will assume the existence of an asymptotic expansion

$$w(z, t; \varepsilon) = w_0(z, t) + \varepsilon w_1(z, t) + \mathcal{O}(\varepsilon^2), \tag{16.79}$$

which starts with an $\mathcal{O}(1)$ term in view of the boundary condition at $z = 1$. The fact that the correction term is of $\mathcal{O}(\varepsilon)$ is an assumption based on the observation that each correction in the equation is of $\mathcal{O}(\varepsilon)$. After substituting the expansion in (16.77) and collecting like powers of ε, we obtain for w_0 the equation with boundary conditions

$$\frac{\partial^3 w_0}{\partial z^2 \partial t} + \frac{\partial^2 w_0}{\partial z^2} = 0, \tag{16.80}$$

$$w_0(0, t) = 0, \qquad \frac{\partial}{\partial z}w_0(1, t) + \frac{\partial^2}{\partial z \partial t}w_0(1, t) = -1.$$

Since we are alarmed by the fact that the small parameter ε multiplies the largest derivative in time, we have skipped for the moment the initial conditions. By neglecting this second time derivative under the limit of $\varepsilon \to 0$, we will not see the freedom or capacity any more to satisfy the initial conditions. The present approximate solution (the outer solution) is not uniformly valid, and the equation is of singularly perturbed type. A boundary layer is necessary to describe the local behaviour near $t = 0$ and the effects of the initial condition. After integrating (16.80) and applying the boundary conditions, we obtain for w_0

$$w_0(z, t) = R_0(z)\,\mathrm{e}^{-t} - z, \qquad \text{where} \qquad R_0(0) = 0. \tag{16.81}$$

In a similar way the $\mathcal{O}(\varepsilon)$ term is found as

$$\frac{\partial^3 w_1}{\partial z^2 \partial t} + \frac{\partial^2 w_1}{\partial z^2} = \frac{\partial^2 w_0}{\partial t^2}, \tag{16.82}$$

$$w_1(0, t) = 0, \qquad \frac{\partial}{\partial z}w_1(1, t) + \frac{\partial^2}{\partial z \partial t}w_1(1, t) = 0.$$

By integrating the equation and applying the boundary conditions, we find

$$w_1(z,t) = (S_0(z)t + R_1(z))\,\mathrm{e}^{-t}, \tag{16.83}$$
$$\text{where} \quad S_0'' = R_0, \quad S_0(0) = 0, \quad S_0'(1) = 0, \quad R_1(0) = 0.$$

Note that $R_0(z)$ and $R_1(z)$ are still to be determined. In the quasi-static approach R_0 would be determined by the initial conditions, while all other $\mathcal{O}(\varepsilon)$ terms would be neglected. Although it is asymptotically inconsistent, we will see that this result is correct, at least for R_0. For the complete asymptotic problem applying the initial conditions is not possible. We could satisfy one condition, but not both. So we do need a boundary layer near $t = 0$ and we leave *both* conditions for the moment unresolved.

The boundary layer corresponds to a scaling $t = \delta(\varepsilon)\tau$ and $w = \beta(\varepsilon)\Omega$ (where $\tau, \Omega = \mathcal{O}(1)$) such that the resulting equation and boundary conditions degenerate under $\varepsilon \to 0$ to another significant limit than the one for the above outer expansion.

The second order time derivative has to play a role in applying two initial conditions. So the first order time derivative, which is $\sim \delta^{-1}$, has to balance ε times the second order time derivative, which is $\sim \varepsilon\delta^{-2}$. Therefore it is clear that $\delta(\varepsilon) = \varepsilon$. Both the boundary condition at $z = 1$, containing the only inhomogeneous term -1, and the matchings condition for $\tau \to \infty$, $t \to 0$, requiring that $w_t = \mathcal{O}(1)$, yield the choice of $\beta(\varepsilon) = \varepsilon$. So altogether we have the inner problem

$$t = \varepsilon\tau, \qquad w = \varepsilon\Omega,$$
$$\frac{\partial^3\Omega}{\partial z^2\partial\tau} + \varepsilon\frac{\partial^2\Omega}{\partial z^2} = \frac{\partial^2\Omega}{\partial\tau^2}, \tag{16.84}$$
$$\Omega(z,0) = 0, \qquad \frac{\partial}{\partial\tau}\Omega(z,0) = 0,$$
$$\Omega(0,\tau) = 0, \qquad \varepsilon\frac{\partial}{\partial z}\Omega(1,\tau) + \frac{\partial^2}{\partial z\partial\tau}\Omega(1,\tau) = -1.$$

As before, we assume the existence of an asymptotic expansion

$$\Omega(z,\tau;\varepsilon) = \Omega_0(z,\tau) + \varepsilon\Omega_1(z,\tau) + \mathcal{O}(\varepsilon^2), \tag{16.85}$$

which gives, after substituting and collecting like powers of ε,

$$\frac{\partial^3\Omega_0}{\partial z^2\partial\tau} = \frac{\partial^2\Omega_0}{\partial\tau^2} \qquad \text{for } 0 < z < 1, \quad \tau > 0, \tag{16.86}$$
$$\Omega_0(z,0) = 0, \quad \frac{\partial}{\partial\tau}\Omega_0(z,0) = 0 \qquad \text{for } \tau = 0,$$
$$\Omega_0(0,\tau) = 0, \quad \frac{\partial^2}{\partial z\partial\tau}\Omega_0(1,\tau) = -1 \qquad \text{for } 0 < z < 1.$$

Integrating the equation gives

$$\frac{\partial^2\Omega_0}{\partial z^2} = \frac{\partial\Omega_0}{\partial\tau} - C_0(z). \tag{16.87}$$

The function C_0 vanishes identically, as can be deduced from the initial conditions. Since $\Omega_0 = 0$ at $\tau = 0$, we also have $\Omega_{0,zz} = 0$. From the defining equation it thus follows that $\Omega_{0,\tau} = \Omega_{0,zz} + C_0(z) = C_0(z) = 0$. In the same way the condition $\Omega_{0,\tau} = \Omega_{0,zz} = 0$ along $\tau = 0$ can be skipped because it is implied by $\Omega_0 = 0$.

The condition $\Omega_{0,z\tau} = -1$ along $z = 1$ can be integrated to $\Omega_{0,z} = -\tau + a_0$ with an unknown constant a_0. In order to appreciate the meaning of this constant, we consider $\Omega_{0,z}$ in $(z, \tau) = (1, 0)$. In the limit $\tau \downarrow 0$ along $z = 1$ we find $\Omega_{0,z} = a_0$. On the other hand, in the limit $z \to 1$ along $\tau = 0$ we obtain $\Omega_{0,z} = 0$ since Ω_0 is identically zero. So a nonzero a_0 corresponds to $\frac{\partial}{\partial z}\Omega_0$ being discontinuous in $(z, \tau) = (1, 0)$. This is physically unacceptable because it would yield singular stresses (16.78), so apparently this constant is $a_0 = 0$. It is, however, not necessary to add conditions of regularity to our problem. This is already included in the problem and will become available to the inner solution via matching with the outer solution, as we will see below. So we will continue with an unknown a_0 and eventually show via matching that $a_0 = 0$.

We introduce the auxiliary variable $\chi(z, \tau)$ by

$$\Omega_0(z, \tau) = -z\tau - \frac{1}{6}z^3 + \left(\frac{1}{2} + a_0\right) z + \chi(z, \tau). \tag{16.88}$$

So the problem (16.86) reduces to

$$\begin{aligned} &\frac{\partial^2}{\partial z^2}\chi = \frac{\partial}{\partial \tau}\chi && \text{for } 0 < z < 1, \quad \tau > 0, \\ &\chi(z, 0) = \frac{1}{6}z^3 - \left(\frac{1}{2} + a_0\right) z && \text{for } \tau = 0, \\ &\chi(0, \tau) = 0, \quad \frac{\partial}{\partial z}\chi(1, \tau) = 0 && \text{for } \tau > 0. \end{aligned} \tag{16.89}$$

This has a general solution, without considering the initial conditions yet, of

$$\chi(z, \tau) = \sum_{n=0}^{\infty} k_n \, \mathrm{e}^{-\lambda_n^2 \tau} \sin(\lambda_n z), \qquad \lambda_n = \frac{1}{2}\pi + n\pi. \tag{16.90}$$

From Fourier series expansions for z and z^3 we find at $\tau = 0$ that

$$\sum_{n=0}^{\infty} k_n \sin \lambda_n z = \frac{1}{6}z^3 - \left(\frac{1}{2} + a_0\right) z = -2\sum_{n=0}^{\infty}(-1)^n \left(\frac{1}{\lambda_n^4} + \frac{a_0}{\lambda_n^2}\right) \sin \lambda_n z, \tag{16.91}$$

which yields k_n by inspection.

Having found the inner solution, we may determine the unknown constants and unknown functions of the outer solution. After expanding the outer solution, written in the inner variable $\tau = t/\varepsilon$, we have

$$w(z, \varepsilon\tau; \varepsilon) \simeq R_0(z)(1 - \varepsilon\tau) - z + \varepsilon R_1(z) + \mathcal{O}(\varepsilon^2). \tag{16.92}$$

After expanding the inner solution, written in the outer variable $t = \varepsilon\tau$, asymptotically for small ε, we have

$$\varepsilon\Omega(z, t/\varepsilon; \varepsilon) \simeq -zt - \frac{1}{6}\varepsilon z^3 + \left(\frac{1}{2} + a_0\right) \varepsilon z + \mathcal{O}(\varepsilon^2). \tag{16.93}$$

Equality of both expressions requires that

$$R_0(z) = z, \qquad R_1(z) = -S_0(z) = -\frac{1}{6}z^3 + \left(\frac{1}{2} + a_0\right) z. \tag{16.94}$$

The constant a_0 vanishes because of the boundary condition $S_0'(1) = a_0 = 0$ of (16.83).

A composite expansion may be assembled as follows:

$$w(z,t;\varepsilon) = (\mathrm{e}^{-t} - 1)z + \varepsilon\left(\frac{1}{2}z - \frac{1}{6}z^3\right)(1-t)\,\mathrm{e}^{-t} - 2\varepsilon\sum_{n=0}^{\infty}\frac{(-1)^n}{\lambda_n^4}\,\mathrm{e}^{-\lambda_n^2\tau}\sin(\lambda_n z) + \mathcal{O}(\varepsilon^2). \tag{16.95}$$

The corresponding normal stresses are then

$$t_{zz} = p\left(-1 + 2\sum_{n=0}^{\infty}\frac{(-1)^n}{\lambda_n}\,\mathrm{e}^{-\lambda_n^2\tau}\cos(\lambda_n z)\right) + \mathcal{O}(\varepsilon), \tag{16.96}$$

$$t_{rr} = p\left(-\alpha + \left(\alpha + \frac{1}{2}\right)\mathrm{e}^{-t} - \sum_{n=0}^{\infty}\frac{(-1)^n}{\lambda_n}\,\mathrm{e}^{-\lambda_n^2\tau}\cos(\lambda_n z)\right) + \mathcal{O}(\varepsilon). \tag{16.97}$$

16.4.4 Example

An example of results is plotted in Figure 16.13 for $\kappa = 1 \times 10^9$ Pa, $\eta = 1 \times 10^5$ Pa · s, $\gamma = 1 \times 10^2\,\mathrm{s}^{-1}$, $\rho = 1 \times 10^3\,\mathrm{kg/m}^3$, $p = 2 \times 10^5$ Pa, and $L = 0.04$ m. This yields $\varepsilon = 0.0912$ and $\alpha = 0.9803$.

The axial displacement u_z and radial and axial normal stresses t_{rr} and t_{zz} for various z positions are plotted against time. Along the right-hand side the time scale is used that corresponds to the viscoelastic phenomena. This is much longer than the initial time scale in the boundary layer, which is determined by inertia. Therefore the initial part of the plots, rescaled to this boundary layer time, is given along the left-hand side. Note that the boundary layer behaviour is primarily present in the derivative of u_z and is therefore difficult to see in u_z itself.

16.4.5 Discussion and Related Problems

- An interesting (and rather unexpected) result of Figure 16.13 is the sign of the normal radial stress t_{rr}, i.e., minus the pressure at the wall. For small time this is positive, which means that the material pulls at the wall. This may be possible at the lee side of a passing wave in reaction to the compression, but at the wave front it is unlikely. It is probably a failure of the viscoelastic model.

- If the material detaches from the wall, the assumption of a vanishing radial deformation is apparently not justified. A model that would include these deformations is, unfortunately, not one-dimensional any more and probably too complicated for analytical treatment. So we would sacrifice the relatively clear results that are obtainable now.

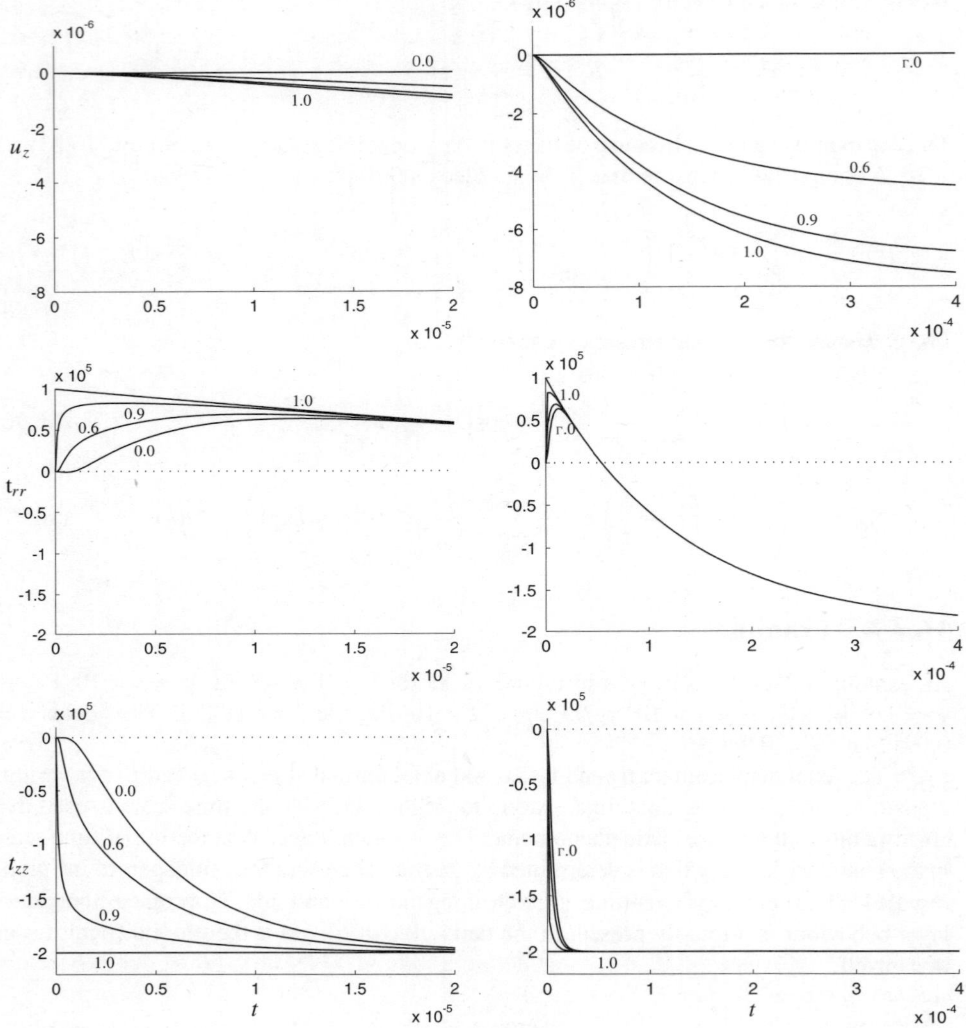

Figure 16.13. *u_z, t_{rr}, and t_{zz} at $z = L$, $z = 0.9L$, $z = 0.6L$, and $z = 0$, plotted as a function of time. At the left side are the results in the boundary layer, while at the right side are the results for larger time.*

16.4.6 Exercises

16.12. Formulate a strategy to determine the problem parameters from the measurements. Is it possible to use only the final status (deformation and stresses), i.e., for $t \to \infty$?

16.13. Derive the Fourier series for z and z^3 used in (16.91).

16.14. Until we match the inner and outer solutions, we are not able to determine the unknown constant a_0 without invoking physical arguments. Physically interpret the solution of (16.86) with arbitrary a_0. Note the appearance of the delta function and its derivative along $\tau = 0$ by the representation

$$\sum_{n=0}^{\infty}(-1)^n \sin \lambda_n z = \sum_{n=-\infty}^{\infty}\Big\{\delta(z+1+4n) - \delta(z-1+4n)\Big\}.$$

16.15. Why is the error in the asymptotic results (16.96) and (16.97) for the normal stresses only $\mathcal{O}(\varepsilon)$?

16.16. Derive the found asymptotic solution directly from the exact solution due to A.A.F. van de Ven (found by Laplace transformation), given by the expression

$$w(z,t;\varepsilon) = -z \\ - 4\varepsilon \sum_{n=0}^{\infty} \frac{(-1)^n \sin(\lambda_n z)}{\lambda_n^2}\left[\frac{\mathrm{e}^{-\lambda_n^2(1-\sigma_n)t/2\varepsilon}}{\lambda_n^2(1-\sigma_n)-4\varepsilon} + \frac{\mathrm{e}^{-\lambda_n^2(1+\sigma_n)t/2\varepsilon}}{\lambda_n^2(1+\sigma_n)-4\varepsilon}\right],$$

where $\sigma_n = (1-4\varepsilon\lambda_n^{-2})^{1/2}$. Note the different regions $t = \mathcal{O}(1)$ and $t = \mathcal{O}(\varepsilon)$.

16.5 Galloping Transmission Lines

16.5.1 Problem Formulation

Overhead transmission lines for transport of high-voltage electricity are cables of aluminum alloy suspended between high towers in the countryside [123]. One cable of several kilometers is carried by several towers in a row, and a section between two towers is called a span. As the cable is connected to the towers by a freely movable suspension string or isolator, the dynamical motions of neighbouring spans are coupled. In wintertime, the cables are vulnerable to large-scale vertical vibrations, induced by crosswind, when the cable is covered by snow or ice. This is an aeroelastic instability known as galloping. For high enough amplitudes neighbouring conductors may get close enough for the air-insulation to break down, causing a short circuit and structural damage to the cables.

It is known from observations that only a small wind force is sufficient to maintain a galloping vibration. The motion of the cable is therefore very close to a free vibration, which is what we will consider here.

16.5.2 The Model

Differential Equations and Boundary Conditions Consider a cable that is fixed at the outer ends and divided into N equal spans by $N-1$ equal supports (Figure 16.14). N is not large (in the asymptotic sense to be explained below). The supports are inextensible suspension strings of length a and negligible weight, suspended from fixed pivots separated

Figure 16.14. *Impression of a suspended cable consisting of three interconnected spans.*

by a distance S, the span size. The cable is linearly elastic, with negligible bending stiffness, of uniform undeformed cross section A, mass per unit length m, and Young's modulus E. It has a length per span L when the cable is free of tension.

We parametrize the position along the cable by the variable $\ell \in [0, NL]$ such that this is just the arc length when the cable is unstretched. The (dimensional) time variable is t. We will here only consider cable motion in a vertical plane, which is provided with a Cartesian coordinate system oriented such that the gravity vector $-g\boldsymbol{e}_y$ points in the negative y direction (in reality the cable moves along a slightly tilted ellipse).

The cable position is given by the position vector $\boldsymbol{X}(\ell, t) = (X(\ell, t), Y(\ell, t))$ with a corresponding tension vector $\boldsymbol{T}(\ell, t) = T(\ell, t)(\cos\psi, \sin\psi)$, where ψ is the angle between the cable tangent and the horizontal. The tension vector is tangent to the cable because of the assumed negligible bending stiffness.

Now consider a small cable element $\mathrm{d}\ell$. Due to gravity, cable tension, and inertial forces, this element is stretched (but the mass remains the same). According to Hooke's law, a cable element is elongated in proportion to the tension (Figure 16.15), so

$$\left(\mathrm{d}X^2 + \mathrm{d}Y^2\right)^{1/2} = \left(1 + \frac{T}{EA}\right)\mathrm{d}\ell. \tag{16.98}$$

According to Newton's law, the internal (tension) and external (gravity) forces are in equilibrium with the inertial forces, so

$$\mathrm{d}\boldsymbol{T} = (g\,\boldsymbol{e}_y + \ddot{\boldsymbol{X}})m\,\mathrm{d}\ell, \tag{16.99}$$

where $\ddot{()}$ denotes a second derivative with respect to time. When we introduce

$$\frac{\partial X}{\partial \ell} = \left(1 + \frac{T}{EA}\right)\cos\psi, \qquad \frac{\partial Y}{\partial \ell} = \left(1 + \frac{T}{EA}\right)\sin\psi, \tag{16.100}$$

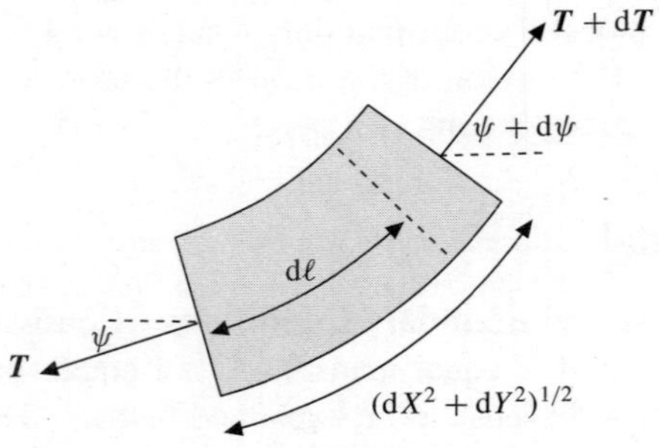

Figure 16.15. *Sketch of cable element stretched under tension.*

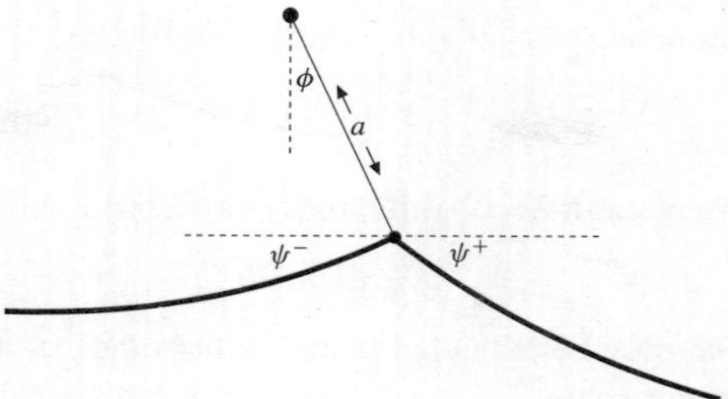

Figure 16.16. *Sketch of cables connected at a suspension string.*

the equations that finally result in the limit $d\ell \to 0$ are given by

$$\frac{\partial}{\partial \ell}\left(\frac{T}{1+T/EA}\frac{\partial X}{\partial \ell}\right) = m\frac{\partial^2 X}{\partial t^2}, \tag{16.101a}$$

$$\frac{\partial}{\partial \ell}\left(\frac{T}{1+T/EA}\frac{\partial Y}{\partial \ell}\right) = m\frac{\partial^2 Y}{\partial t^2} + mg, \tag{16.101b}$$

$$\left(\frac{\partial X}{\partial \ell}\right)^2 + \left(\frac{\partial Y}{\partial \ell}\right)^2 = \left(1+\frac{T}{EA}\right)^2. \tag{16.101c}$$

The boundary and coupling conditions are (see Figure 16.16) as follows:

(i) The supports are fixed at $\ell = 0$ and $\ell = NL$:

$$X = 0, \quad Y = 0 \quad (\ell = 0), \tag{16.102a}$$

$$X = NS, \quad Y = 0 \quad (\ell = NL). \tag{16.102b}$$

(ii) At the suspension strings $\ell = nL$, where $n = 1, \ldots, N-1$, position vector $\boldsymbol{X}$ is continuous, the motion is restricted to a circle, while the force component normal to the string is continuous:

$$[\boldsymbol{X}]_{\ell=nL-}^{\ell=nL+} = 0, \tag{16.103a}$$

$$(X - nS)^2 + (Y - a)^2 = a^2, \tag{16.103b}$$

$$\Big[T\cos(\phi - \psi)\Big]_{\ell=nL-}^{\ell=nL+} = 0. \tag{16.103c}$$

ϕ denotes the angle of the suspension string with the vertical and $\ell = nL\pm$ denotes the limits from either side of the support.

Small Parameters and the Mode of Vibration The type of motion of interest allows further reduction of the model and is specified as follows:

- The ratio of sag D (the maximum stationary vertical displacement) and cable length L is small (typically on the order of $\frac{1}{30}$), so the slenderness

$$\varepsilon := D/L \to 0 \tag{16.104}$$

is a small parameter in the problem. This parameter will be used later to further reduce the problem.

- The total vertical nonstationary displacement is of the order of the sag, so

$$Y/L = \mathcal{O}(\varepsilon). \tag{16.105}$$

- The transversal wavelength λ_T is of the order of L, so

$$\lambda_T/L = \mathcal{O}(1). \tag{16.106}$$

  This is exactly what may be expected from the boundary conditions at the span ends: $\frac{1}{2}$, 1, 2, etc., waves per span.
- Apart from the slenderness, there is still another small parameter in the problem, which is rather more invisible, as it is related to the relative elasticity.

  In general, it takes some time before a change of tension has been distributed as a longitudinal tensional wave through the cable. This time, however, is very short, since the propagation speed of these waves (the sound speed) is high and N is not large. In other words, the longitudinal wavelength λ_L is large compared to L. A second small parameter, however, is usually not very convenient in an asymptotic analysis, so we use the estimate, which is consistent with practice, that

$$L/\lambda_L = \mathcal{O}(\varepsilon). \tag{16.107}$$

- The string length a is of the order of the sag, so

$$a/L = \mathcal{O}(\varepsilon). \tag{16.108}$$

- The relative amplitude δ of the time-dependent perturbation will be taken small but bigger than the orders of ε neglected. We will assume a nearly harmonic vibration with a single dominating frequency ω (which is to be found!). Since we are interested in the intrinsic nonlinear interaction between the harmonics, we will analyse the generated higher harmonics by a Lindstedt–Poincaré series in δ; see Section 15.3.2. To keep the results clear higher harmonics not generated by the first one will be excluded.

16.5.3 Asymptotic Analysis

Reduced Problem The basic small parameter ε will be used to reduce the above general problem to an asymptotic model. This model, assumed independent of ε, will subsequently be analysed asymptotically for small relative amplitude δ. (So ε is *not* the small parameter that we will use later in the dynamical analysis.)

Since the longitudinal wave speed is $c_L = (EA/m)^{1/2}$, while $L/\lambda_L = \omega L/c_L = \mathcal{O}(\varepsilon)$, we can introduce the reference frequency $\omega_{\text{ref}} = \varepsilon(EA/m)^{1/2}/L$. The dimensionless frequency and time variables are then given by

$$\omega = \omega_{\text{ref}}\,\omega^*, \qquad t = t^*/\omega_{\text{ref}}. \tag{16.109}$$

Since $\lambda_T/L = \mathcal{O}(1)$, the spatial coordinate ℓ obviously scales on L. (Note that L is an unknown of the problem!) Since $Y/L = \mathcal{O}(\varepsilon)$, Y scales on εL. The transversal wave velocity is $c_T = (T/m)^{1/2}$, so $\lambda_T/L = c_T/\omega L = \mathcal{O}(1)$, yielding that the tension scales on $T_{\text{ref}} = \varepsilon^2 EA$. Altogether we have for $\ell \in \{\text{the } n\text{th section}\}$ that

$$\ell = (n-1+s)L, \quad Y(\ell,t) = \varepsilon L Y^*(s,t^*;n), \quad T(\ell,t) = T_{\text{ref}}\,T^*(s,t^*;n). \tag{16.110}$$

The suitable scaling for x is more subtle. Of course, the slender geometry suggests immediately that $X \approx \ell$. However, we are only interested in the x displacement, which is very small. By substituting the above estimates in (16.101c), we have $\frac{\partial}{\partial \ell}X = 1 + \mathcal{O}(\varepsilon^2)$, and so for X from the nth section we have

$$X(\ell, t) = (n-1)S + Ls + \varepsilon^2 L X^*(s, t^*; n). \tag{16.111}$$

Finally, we investigate the role of gravity. If we substitute the present estimates into (16.101b), we find the term $mgL/EA\varepsilon^3$ next to terms of $\mathcal{O}(1)$. So this term has to be $\mathcal{O}(1)$ or smaller. Suppose it is smaller. Then the stationary solution would be to leading order $Y \equiv 0$, so $D = 0$, which is contradictory to our scaling assumptions. So the term is $\mathcal{O}(1)$ and we introduce

$$\mu = \frac{mgL}{8EA\varepsilon^3} = \mathcal{O}(1) \tag{16.112}$$

(where the factor 8 is included for convenience later). While we omit the superscript asterisks $*$ from here on, we obtain, under the approximation for small ε, the reduced version of problem (16.101) as follows:

$$\frac{\partial T}{\partial s} = 0, \quad \frac{\partial}{\partial s}\left(T\frac{\partial Y}{\partial s}\right) = 8\mu + \frac{\partial^2 Y}{\partial t^2}, \quad \frac{\partial X}{\partial s} + \frac{1}{2}\left(\frac{\partial Y}{\partial s}\right)^2 = T. \tag{16.113a,b,c}$$

In practice, the sag D and span size S are known, while the cable length L is to be determined. So ε and μ are to be determined from the stationary solution. Moreover, all spans are assumed geometrically the same and so is the stationary solution. It is therefore useful to split up the solution into a stationary and a nonstationary part

$$\begin{gathered} X(s,t;n) = X_0(s) + x(s,t;n), \quad Y(s,t;n) = Y_0(s) + y(s,t;n), \\ T(s,t;n) = T_0(s) + \tau(s,t;n), \end{gathered} \tag{16.114}$$

with boundary conditions for the stationary part of $Y_0(0) = Y_0(1) = 0$ and $X_0(0) = 0$. As we scaled Y on D and, from symmetry, anticipate the location of maximum deflection halfway, we have of course $Y_0(\frac{1}{2}) = -1$. The condition for X_0 at $s = 1$ is not really a boundary condition to the scaled problem but implicitly determines the unknown length L by the relation $S = L(1 + \varepsilon^2 S_0)$ and (up to the ε approximation pursued so far) $X_0(1) = S_0$. The result is easily found (a parabola) and is given by

$$T_0(s) = \mu, \quad Y_0(s) = -4(s - s^2), \quad X_0(s) = \mu s - \frac{4}{3}(1 + (2s-1)^3), \quad S_0 = \mu - \frac{8}{3}. \tag{16.115a,b,c,d}$$

If we substitute (16.115) into (16.113), we obtain our fundamental nonstationary problem

$$\frac{\partial \tau}{\partial s} = 0, \quad (\mu + \tau)\frac{\partial^2 y}{\partial s^2} + 8\tau = \frac{\partial^2 y}{\partial t^2}, \quad \frac{\partial x}{\partial s} + 4(2s-1)\frac{\partial y}{\partial s} + \frac{1}{2}\left(\frac{\partial y}{\partial s}\right)^2 = \tau. \tag{16.116a,b,c}$$

The boundary conditions at the rigid supports follow readily:

$$x(0,t;1) = x(1,t;N) = y(0,t;1) = y(1,t;N) = 0. \tag{16.117}$$

For the conditions at the suspension string ($s = 0$, $n = 2, 3, \ldots, N$, and $s = 1$, $n = 1, 2, \ldots, N-1$) we observe that $a\phi/L = \mathcal{O}(\varepsilon^2)$ as it is of the order of the x variation, so $\phi = \mathcal{O}(\varepsilon)$, and we obtain for small ε the conditions

$$y(0,t;n) = y(1,t;n) = 0, \quad x(1,t;n) = x(0,t;n+1), \quad \tau(1,t;n) = \tau(0,t;n+1), \tag{16.118}$$

with the important consequences that

$$\sum_{n=1}^{N}\big[x(1,t;n) - x(0,t;n)\big] = 0, \qquad \tau = \text{ constant in } s \text{ and independent of } n. \tag{16.119}$$

Weakly Harmonic Motion The Lindstedt–Poincaré method (Section 15.3.2) involves the assumption of a periodic solution with fundamental (as yet unknown!) frequency ω. The amplitude δ will be introduced as a small parameter, and based on this a solution will be constructed as a power series in δ, where ω also depends on δ.

In order to apply the condition of periodicity, it is necessary to introduce the inherent time scale

$$t' := \omega t. \tag{16.120}$$

Now we assume $y = \mathcal{O}(\delta)$ and expand

$$y = \delta y_1 + \delta^2 y_2 + \cdots. \tag{16.121}$$

We can do likewise for x and τ. We consider $y_1, y_2, \ldots$ normalized, e.g., as $y_1(\frac{1}{2}) = y_2(\frac{1}{2}) = \cdots = 1$. We expand ω as

$$\omega = \omega_0 + \delta^2\omega_2 + \cdots, \tag{16.122}$$

where we used the result that $\omega_1 = 0$. This could have been guessed from the intuitive argument that ω should not depend on the sign of the vibrational amplitude.

We introduce the notation $\{\}' = \frac{\partial}{\partial s}\{\}$ and $\{\dot{}\} = \frac{\partial}{\partial t'}\{\}$. We substitute the δ series into (16.116b,c), and collect like powers of δ:

$$\mu y_1'' + 8\tau_1 = \omega_0^2\ddot{y}_1, \qquad x_1' + 4(2s-1)y_1' = \tau_1, \tag{16.123}$$

$$\mu y_2'' + 8\tau_2 + \tau_1 y_1'' = \omega_0^2\ddot{y}_2, \qquad x_2' + 4(2s-1)y_2' + \tfrac{1}{2}(y_1')^2 = \tau_2, \tag{16.124}$$

with $\tau_1, \tau_2, \ldots$ constant in s. The interesting solutions here are based on the harmonic solution of the linearized problem (i.e., for y_1, x_1, τ_1). Since any phase shift in time is unimportant, we may assume y_1, x_1, τ_1 to vary in time like $\sin t'$. The next question is now what time dependence this induces in the second order. In (16.124) the terms $\tau_1 y_1''$ and $(y_1')^2$ act as an inhomogeneous (source) term. Since they vary in time like $\sin(t')^2 = \frac{1}{2} - \frac{1}{2}\cos(2t')$, we expect therefore a constant and a $\cos(2t')$ contribution. So we put

$$y_1(s,t';n) = y_{11}(s;n)\sin t', \qquad y_2(s,t';n) = y_{20}(s;n) + y_{22}(s;n)\cos(2t'), \tag{16.125}$$

etc. We have similar results for x and τ.

By substituting (16.125) into (16.123) and (16.124) and collecting the harmonics, we get

$$\begin{aligned} \mu y_{11}'' + 8\tau_{11} + \omega_0^2 y_{11} &= 0, \\ x_{11}' + 4(2s-1)y_{11}' &= \tau_{11}, \end{aligned} \tag{16.126}$$

$$\begin{aligned} \mu y_{20}'' + 8\tau_{20} + \frac{1}{2}\tau_{11}y_{11}'' &= 0, \\ x_{20}' + 4(2s-1)y_{20}' + \frac{1}{4}(y_{11}')^2 &= \tau_{20}, \end{aligned} \tag{16.127}$$

$$\begin{aligned} \mu y_{22}'' + 8\tau_{22} - \frac{1}{2}\tau_{11}y_{11}'' + 4\omega_0^2 y_{22} &= 0, \\ x_{22}' + 4(2s-1)y_{22}' - \frac{1}{4}(y_{11}')^2 &= \tau_{22}. \end{aligned} \tag{16.128}$$

The boundary conditions are the same as for x, y, and τ ((16.117), (16.118), and (16.119)), but now taken for each harmonic independently.

16.5.4 Solutions

In spite of the many assumptions and reductions made, the possible solutions are still numerous. This is exactly in agreement with reality, where the cables are known to vibrate in many possible modes. This indeterminacy is also a very important problem to be dealt with in any numerical simulation.

For illustration we will pick here a possible family of solutions. Others may be obtained in analogous ways.

A very important class of solutions is one in which the tension's first harmonic τ_{11} vanishes. This is rather unexpected (the visible parts x, y of the cable do vibrate in their first harmonic) and is typical of coupled spans (for a single span it only occurs for higher-order modes). We have then for the nth span ($1 \le n \le N$) the general solution for the jth mode

$$\begin{aligned} &\tau_{11} = 0, \\ &y_{11} = A_n \sin(ks), \qquad \text{where } k := \omega_0/\mu^{1/2} \text{ and } k = k_j := (2j+1)\pi, \\ &x_{11} = B_n - 4(2s-1)A_n \sin(ks) - 8A_n \cos(ks)/k, \\ &\qquad \text{where } B_n = 8(2A_1 + 2A_2 + \cdots + 2A_{n-1} + A_n)/k; \end{aligned} \tag{16.129}$$

$$\begin{aligned} &\tau_{20} = \frac{3}{8}\omega_0^2 \overline{A^2}/(16+3\mu), \\ &y_{20} = 4\tau_{20}(s-s^2)/\mu, \\ &x_{20} = C_n + \tau_{20}s + \frac{8}{3}\tau_{20}(1+(2s-1)^3)/\mu - \frac{1}{8}k^2A_n^2(s+\sin(2ks)/2k), \\ &\qquad \text{where } C_n = (n-1)\tau_{20}(1+16/3\mu) - \frac{1}{8}k^2(A_1^2 + \cdots + A_{n-1}^2) \\ &\qquad \text{and } \overline{A^2} = (A_1^2 + A_2^2 + \cdots + A_N^2)/N; \end{aligned} \tag{16.130}$$

$$\begin{aligned}
\tau_{22} &= \frac{1}{8}k^2\omega_0^2\overline{A^2}/(16-\omega_0^2),\\
y_{22} &= 2\tau_{22}(\cos(2ks)-1)/\omega_0^2,\\
x_{22} &= D_n + \tau_{22}s - 8\tau_{22}(1+(2s-1)\cos(2ks) - \sin(2ks)/k)/\omega_0^2 \qquad (16.131)\\
&\qquad + \frac{1}{8}k^2A_n^2(s+\sin(2ks)/2k),\\
&\text{where } D_n = -(n-1)\tau_{22}(16/\omega_0^2 - 1) + \frac{1}{8}k^2(A_1^2+\cdots+A_{n-1}^2).
\end{aligned}$$

From (16.119) it follows that the amplitudes are further restricted by the condition that

$$A_1 + A_2 + \cdots + A_N = 0. \qquad (16.132)$$

For a two-span configuration ($N = 2$) with equal vibrational energy in each section this becomes $A_1 = -A_2 = 1$, while $\overline{A^2} = 1$.

The frequency is, to leading order, equal to $\omega_0 = (2j+1)\pi\mu^{1/2}$. The higher-order shift ω_2 is not yet determined in the above results.

16.5.5 Examples

Two practically relevant examples are given in Figure 16.17. In both cases we have $m =$ 1 kg/m, $g = 9.8\,\text{m/s}^2$, $EA = 2.156\times10^7$ N, and $S = 320$ m, while $A = \overline{A^2} = 1$ and the lowest mode ($j = 0$) is selected. Shown are the dimensional varying parts of vertical displacement $dY := \varepsilon Ly$ and tension $dT := \varepsilon^2 EA\tau$. In the left column we have a stationary tension $T = 12{,}600$ N and sag $D = 10$ m with $\delta = 0.5$. In this case $L = 320.7$ m, $\mu =$ 0.6011, $\varepsilon = 0.0312$, and the frequency is $\omega = 2.436$ or 0.175 Hz. We observe a mild contribution of the higher harmonics in dY. These are more present in the case displayed in the right column, where we have $T = 16{,}500$ N and $D = 7.61$ m with $\delta = 0.3$. Now $L = 320.3$ m, $\mu = 1.354$, $\varepsilon = 0.0238$, and the frequency is 3.655 or 0.201 Hz.

16.5.6 Discussion and Related Problems

- In this section we considered an example of asymptotic modeling. The modeling process starts with a global model derived from first principles. Then two small parameters are identified, geometrical slenderness and relative elasticity, that allow us to remove irrelevant elements. In the final model all parts contribute significantly. The solution is based on the assumption that the cable vibrates harmonically with weakly nonlinear perturbations.

- The above solution is a gravity mode: $\tau_{11} = 0$. The elasticity does not have much of a role; an inelastic catenary would allow practically the same motion. This is not true for elastogravity modes, where $\tau_{11} \neq 0$ and the elastic tension and the vertical motion have the same frequency. See [123].

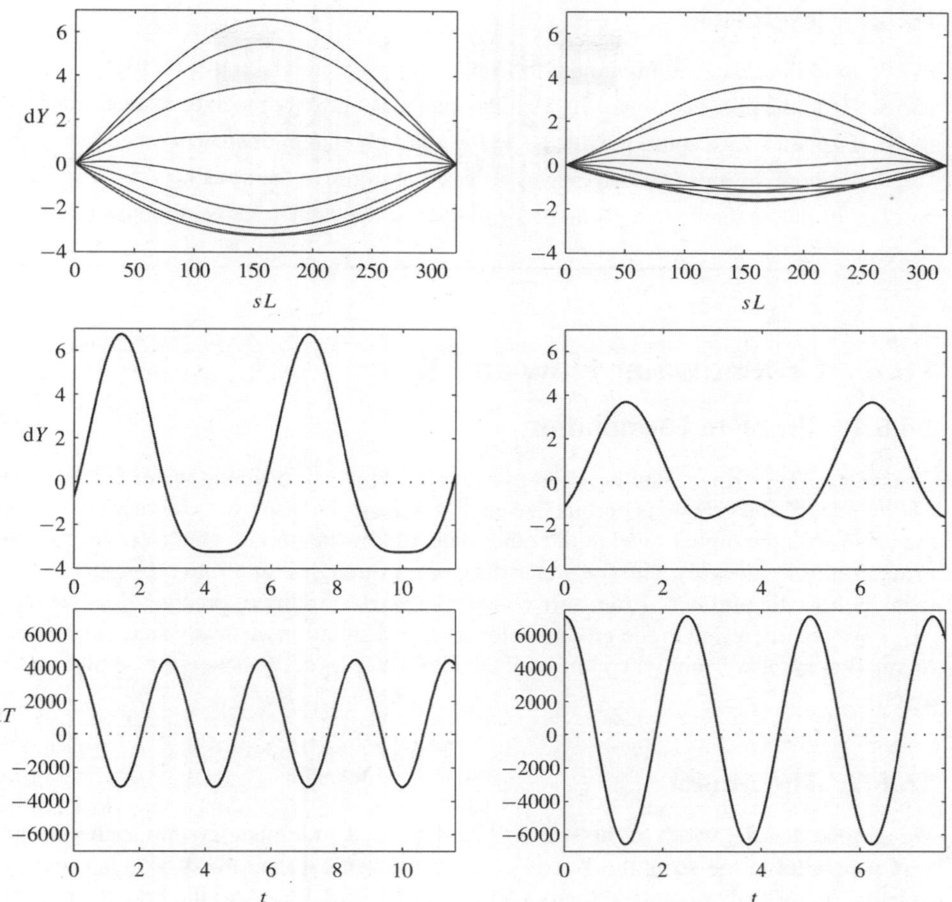

Figure 16.17. *An oscillating cable section (from top to bottom): the vertical displacement* dY *in* sL *for various time steps, the position at the middle* ($s = \frac{1}{2}$) *of the displacement in time, and the varying part of the tension* dT *in time. All variables are dimensional.*

- See the expression for τ_{22} in (16.131). An interesting feature in this solution is the singularity for $\omega_0 = 4$, apparently indicating a breakdown of the present approximation. There is a resonance between the first and second harmonics, where the second order is not small any more compared to the first order; see Figure 16.17.

- A more complete modeling would include the driving force of the wind and the effect of air friction. This is a small effect, but near resonance the resulting amplitudes may be very large.

16.5.7 Exercises

16.17. Find the exact solution to the full stationary problem given by (16.101).

16.18. Redo the above analysis for a single-span elastogravity mode of lowest order.

16.19. Explain why a solution with $\tau_{11} \neq 0$ is unlikely in a multispan geometry.

16.20. Propose various external driving forces that improve the model.

16.21. Formulate the related problem for motion of the cable in three dimensions.

16.6 Groundwater Flow and Rain

16.6.1 Problem Formulation

Consider a long strip of land between two parallel canals of different water levels. Above a certain level, the ground is porous like sand, allowing the water to flow very slowly from the canal with the higher level to the other one. Below the porous top layer, the ground is semipermeable like clay, and the water disappears through it at a rate proportional to the local hydrostatic pressure. From above there is a vertical influx of precipitation, i.e., rain.

We are interested in the groundwater level in a situation of steady state for moderate to small amounts of rain when one of the canal water levels is below the semipermeable layer.

16.6.2 The Model

We assume that the width of the strip is L and select a coordinate system such that the Z axis is parallel to the strip, the Y axis is along the vertical, and the X axis is orthogonal to both. $X = 0$ corresponds to the border of canal 0 and $X = L$ to the border of canal 1; see Figure 16.18. We assume that the groundwater level in the Z direction is constant. The semipermeable layer is situated at $Y = 0$, the groundwater level is at $Y = h(X)$, and the water levels of canals 0 and 1 are at $Y = h_0 < 0$ and $Y = h_1 \geq 0$, respectively. Above $Y = h(x)$, the ground is not saturated with water. Here, any available water slowly creeps downward under the effect of gravity only until it merges with the saturated region $0 \leq Y \leq h(x)$. Water flows through the semipermeable layer at a rate proportional to the local hydrostatic pressure, which is in turn proportional to h, the water height. So the vertical mass flux density of water at $Y = 0$ may be given by

$$\Phi_0 = \alpha h, \tag{16.133}$$

where α is a constant. From the top surface of the strip water comes in due to rainfall. Although, of course, it does not rain continuously, the time scales of the rainfall and the water flow in the unsaturated region $Y > h$ are so much apart that we consider the average vertical mass flux density from the top of

$$\Phi_1 = N \tag{16.134}$$

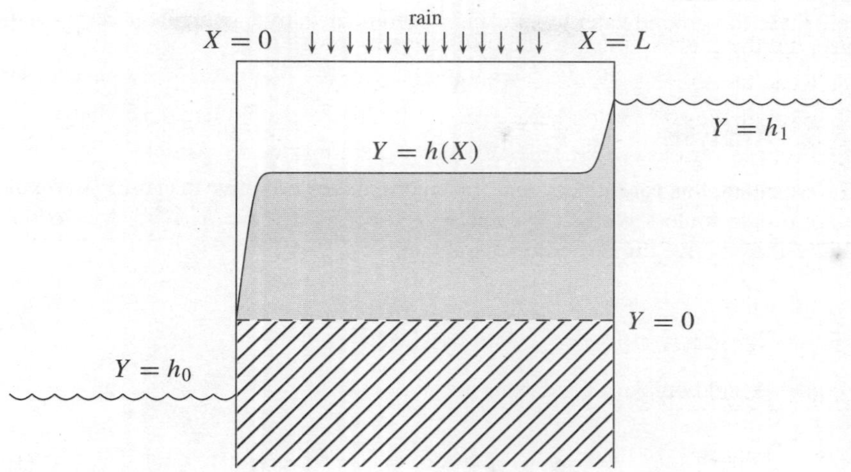

Figure 16.18. *Sketch of groundwater and rain problem.*

to be a constant in the problem. Thus we neglect variations in N due to different top surface coverage (green, buildings, roads).

Consider a slice of ground between X and $X + \mathrm{d}X$. The flux into the slice from the top is equal to $\Phi_1 \mathrm{d}X$ and the flux of mass leaving the slice below is $\Phi_0 \mathrm{d}X$. Between two neighbouring positions X and $X + \mathrm{d}X$ we have a small difference in groundwater level and therefore a small pressure gradient. According to Darcy's law, this produces a slow flow in the X direction with velocity v proportional (and opposite) to the pressure gradient and dependent on the porosity of the ground. In the present model we assume that the porosity is uniformly the same and that the water and the ground are incompressible. Since the pressure difference is the same for all $Y \in [0, h(X)]$, the flow velocity is uniform in Y and we have

$$h(X + \mathrm{d}X) - h(X) \sim p(X + \mathrm{d}X) - p(X) \sim -v(X)\,\mathrm{d}X. \tag{16.135}$$

So the horizontal mass flux density Φ_h, which is proportional to v, is

$$\Phi_h = -D\frac{\mathrm{d}h}{\mathrm{d}X}, \tag{16.136}$$

where D is a positive material constant. The flux balance across all sides of a slice $\mathrm{d}X$ is then given by

$$\big[h\Phi_h\big]_X^{X+\mathrm{d}X} = (\Phi_1 - \Phi_0)\,\mathrm{d}X.$$

After taking the limit $\mathrm{d}X \to 0$ in combination with relations (16.133), (16.134), and (16.136), we obtain

$$\frac{\mathrm{d}}{\mathrm{d}X}\left(Dh\frac{\mathrm{d}h}{\mathrm{d}X}\right) = \alpha h - N. \tag{16.137}$$

The boundary condition at $X = L$ is evidently $h = h_1$. From the presence of the semipermeable layer the groundwater level cannot become negative, so the boundary condition at $X = 0$ is just $h = 0$.

16.6.3 Analysis

As D is constant, this parameter can be taken out of the derivative in (16.137). We make the problem dimensionless with L, h_1, and α—$X = Lx$, $h(X) = h_1\phi(x)$, $N = \alpha h_1 K$—and introduce the positive dimensionless parameter

$$\varepsilon := \frac{Dh_1}{2\alpha L^2}. \tag{16.138}$$

The equation and boundary conditions are

$$\varepsilon \frac{\mathrm{d}^2}{\mathrm{d}x^2}\phi^2 - \phi + K = 0, \qquad 0 \le x \le 1, \tag{16.139a}$$

$$\phi(0) = 0, \qquad \phi(1) = 1. \tag{16.139b}$$

When the strip width L is not very small compared to h_1 and the porosity of the semipermeable layer is much smaller than the porosity of the ground such that the vertical flux $\sim Dh_1^2/L$ is much bigger than the horizontal flux $\sim \alpha h_1 L$, then ε is small. We will exploit this smallness and analyse the problem asymptotically under the limit $\varepsilon \to 0$. As the small parameter multiplies the highest derivative, the problem will appear to be a singular perturbation problem of boundary layer type (Section 15.4.1).

We will distinguish two cases: (i) moderate amounts of rain such that $K = \mathcal{O}(1)$ and (ii) small amounts of rain such that $K = \mathcal{O}(\varepsilon)$.

Moderate Amounts of Rain Assume that $\phi(x;\varepsilon) = \phi_0(x) + \mu_1(\varepsilon)\phi_1(x) + \cdots$. Then

$$\phi_0(x) = K, \quad \phi_1(x) \equiv 0 \qquad \text{for every} \quad \mu_1(\varepsilon). \tag{16.140}$$

As far as we can write $\phi(x;\varepsilon) - K$ as a regular asymptotic approximation, it is equivalent to zero. Since ϕ_0 does not satisfy the boundary conditions at $x = 0$ and $x = 1$, we expect boundary layers there, while ϕ_0 is an outer solution. We start with $x = 0$. Suppose $x = \delta\xi$ and $\phi(x;\varepsilon) = \psi(\xi;\varepsilon)$. Since $\phi_0 = \mathcal{O}(1)$, we expect $\psi = \mathcal{O}(1)$. We find a significant degeneration $\delta = \sqrt{\varepsilon}$.

We substitute $\psi(\xi;\varepsilon) = \psi_0(\xi) + \lambda_1(\varepsilon)\psi_1(\xi) + \cdots$ and find to leading order that

$$\frac{\mathrm{d}^2}{\mathrm{d}\xi^2}\psi_0^2 - \psi_0 + K = 0. \tag{16.141}$$

We multiply by $\frac{\mathrm{d}}{\mathrm{d}\xi}(\psi_0^2)$ and integrate to

$$\frac{1}{2}\left(\frac{\mathrm{d}}{\mathrm{d}\xi}(\psi_0^2)\right)^2 - \frac{2}{3}\psi_0^3 + K\psi_0^2 = C_0. \tag{16.142}$$

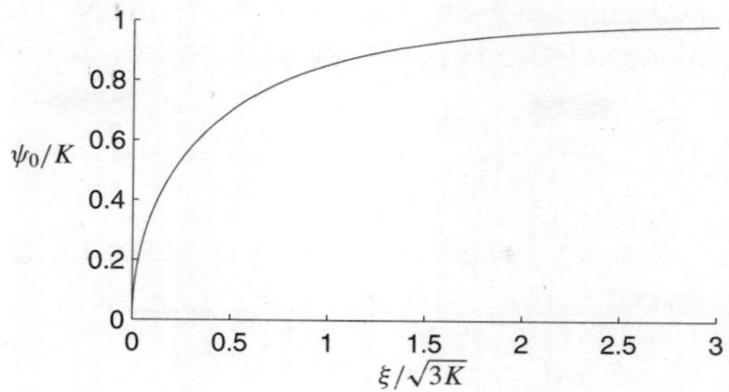

Figure 16.19. *Plot of boundary layer profile $\psi_0(\xi)$ for $K = \mathcal{O}(1)$.*

C_0 follows from the matching condition $\psi_0 \to K$ such that $(\psi_0^2)' \to 0$, and hence

$$C_0 = \frac{1}{3}K^3,$$

$$\frac{\mathrm{d}}{\mathrm{d}\xi}\psi_0^2 = \left(\frac{2}{3}K^3 - 2K\psi_0^2 + \frac{4}{3}\psi_0^3\right)^{1/2} = \frac{2}{\sqrt{3}}(K - \psi_0)\left(\frac{1}{2}K + \psi_0\right)^{1/2}. \tag{16.143}$$

A positive sign prevails because ψ_0 increases from 0 to K. Note that with this choice of C_0 the following integral has a nonintegrable singularity at $\eta = 1$, and so $\xi \to \infty$ when $\psi_0/K \to 1$. We integrate

$$\sqrt{3K}\int_0^{\psi_0/K} \frac{\eta\,\mathrm{d}\eta}{(1-\eta)(\frac{1}{2}+\eta)^{1/2}} = \xi \tag{16.144}$$

to get the implicit solution (see Figure 16.19)

$$\left[-2\sqrt{\tfrac{1}{2}+\eta} + \sqrt{\tfrac{2}{3}}\ln\frac{\sqrt{\frac{3}{2}} + \sqrt{\frac{1}{2}+\eta}}{\sqrt{\frac{3}{2}} - \sqrt{\frac{1}{2}+\eta}}\right]_0^{\psi_0/K} = \frac{\xi}{\sqrt{3K}}. \tag{16.145}$$

The boundary layer near $x = 1$ is analogous. Assume $x = 1 + \sqrt{\varepsilon}\,\zeta$, $\phi(x;\varepsilon) = \chi(\zeta;\varepsilon)$, and $\chi(\zeta;\varepsilon) = \chi_0(\zeta) + \cdots$. Then

$$\frac{\mathrm{d}}{\mathrm{d}\zeta}\chi_0^2 = \pm\frac{2}{\sqrt{3}}(K - \chi_0)\left(\frac{1}{2}K + \chi_0\right)^{1/2}, \tag{16.146}$$

with a sign given by the sign of $1 - K$, the difference between the boundary value at $x = 1$ and the outer solution.

Small Amounts of Rain We write explicitly $K = \varepsilon k$, where k is independent of ε. Suppose $\phi(x;\varepsilon) = \varepsilon\phi_0(x) + \mu_1(\varepsilon)\phi_1(x) + \cdots$. Then

$$\phi_0(x) = k, \quad \phi_1(x) \equiv 0 \qquad \text{for every} \quad \mu_1(\varepsilon) = o(\varepsilon). \tag{16.147}$$

We search for significant degenerations by writing $x = a + \delta(\varepsilon)\xi$, $\phi(x;\varepsilon) = \lambda(\varepsilon)\psi(\xi;\varepsilon)$ (where a is the as yet unknown location of the boundary layer). Then we have

$$\varepsilon\lambda^2\delta^{-2}\frac{\mathrm{d}^2}{\mathrm{d}\xi^2}\psi^2 - \lambda\psi + \varepsilon k = 0 \tag{16.148}$$

such that from the outer solution (εk) and the boundary condition ($\phi(1) = 1$) we can expect two types of boundary layers:

$$\begin{aligned} &\text{(a)} \qquad \lambda = \varepsilon, \qquad \delta = \varepsilon, \\ &\text{(b)} \qquad \lambda = 1, \qquad \delta = \sqrt{\varepsilon}. \end{aligned}$$

Near $x = 0$ there are no problems. Suppose $x = \varepsilon\xi$ and $\phi(x;\varepsilon) = \varepsilon\psi_0(x) + \cdots$. Then

$$\frac{\mathrm{d}^2}{\mathrm{d}\xi^2}\psi_0^2 - \psi_0 + k = 0, \qquad \psi_0(0) = 0, \quad \psi_0 \to k \quad (\xi \to \infty), \tag{16.149}$$

with solution given by

$$\frac{1}{2}\sqrt{3}\ln\left(\frac{\sqrt{3} - \sqrt{1 + 2\psi_0/k}}{\sqrt{3} + \sqrt{1 + 2\psi_0/k}}\frac{\sqrt{3}+1}{\sqrt{3}-1}\right) + \sqrt{1 + 2\psi_0/k} - 1 = -\frac{\xi}{\sqrt{6k}}. \tag{16.150}$$

The boundary layer near $x = 1$ is less obvious. In view of the boundary condition, we start with the assumption that $\phi = \mathcal{O}(1)$:

$$x = 1 + \sqrt{\varepsilon}\zeta, \quad \phi(x;\varepsilon) = \chi_0(\zeta) + \cdots \qquad (-\infty < \zeta \le 0), \tag{16.151}$$

with

$$\frac{\mathrm{d}^2}{\mathrm{d}\zeta^2}\chi_0^2 - \chi_0 = 0, \qquad \chi_0(0) = 1. \tag{16.152}$$

The solution, up to a constant C given by

$$\int_{\chi_0}^{1} \frac{\eta\,\mathrm{d}\eta}{\sqrt{\frac{1}{3}\eta^3 + C}} = -\zeta, \tag{16.153}$$

does *not* satisfy the matching condition $\chi_0 \to 0$ for $\zeta \to -\infty$. Define

$$\zeta_0(C) = -\int_{0}^{1} \frac{\eta\,\mathrm{d}\eta}{\sqrt{\frac{1}{3}\eta^3 + C}}. \tag{16.154}$$

Then ζ_0 is finite for every C, and χ_0 is apparently only defined (with these boundary conditions) along the interval $\zeta_0 \le \zeta \le 0$. So there is something wrong.

We consider several possibilities: there is no solution and therefore no approximation (this is, however, not the case); the $\mathcal{O}(\sqrt{\varepsilon})$ boundary layer is inside another, wider, boundary layer (this is not the case; the only other possibility is $\mathcal{O}(\varepsilon)$, hence smaller); our assumption of an outer solution of $\mathcal{O}(\varepsilon)$ is not right (this is also not the case; the other possibility of $\mathcal{O}(\varepsilon^{-1})$ leads to nothing); we hit a rare example of inner and outer solutions without region of overlap (very hypothetical but not totally impossible).

The solution, however, appears to be an *inner* boundary layer of $\mathcal{O}(\varepsilon)$ around $\zeta = \zeta_0$ where $\phi = \mathcal{O}(\varepsilon)$. This is rather unusual: the outer and inner solutions are connected via a small "interface" interior layer.

The behaviour of χ_0 near $\zeta = \zeta_0$ is given by

$$C \neq 0: \qquad \chi_0 \simeq \left(2\sqrt{C}(\zeta - \zeta_0)\right)^{1/2}, \tag{16.155a}$$

$$C = 0: \qquad \chi_0 = (1 - \zeta/\zeta_0)^2 \quad \text{with} \quad \zeta_0 = -2\sqrt{3} \tag{16.155b}$$

(for $C = 0$, χ_0 is exact). The value of C will be determined from matching. Assume $x = 1 + \sqrt{\varepsilon}\zeta_0 + \varepsilon z$ and $\phi(x; \varepsilon) = f(z; \varepsilon)$. Expand

$$f(z; \varepsilon) = \varepsilon f_0(z) + \cdots . \tag{16.156}$$

Then

$$\frac{d^2}{dz^2} f_0^2 - f_0 + k = 0 \tag{16.157}$$

with matching conditions in both directions:

$$f_0 \simeq \phi_0 = k \qquad (z \to -\infty) \tag{16.158a}$$

$$\varepsilon f_0 \simeq \chi_0 \qquad (z \to \infty). \tag{16.158b}$$

The solution that approaches the constant k in an exponential way for $z \to -\infty$ is

$$\frac{1}{3}\sqrt{3}\ln\left(\frac{\sqrt{1 + 2f_0/k} - \sqrt{3}}{\sqrt{1 + 2f_0/k} + \sqrt{3}}\right) + \sqrt{1 + 2f_0/k} - A = \frac{z}{\sqrt{6k}}. \tag{16.159}$$

If $z \to \infty$, then $f_0 \simeq \frac{1}{2}k(A + z/\sqrt{6k})^2$, which is to be compared with (16.155). A successful matching is apparently obtained with $C = 0$ and $A = 0$.

We summarize the behaviour for small amounts of rain: via a boundary layer of $\mathcal{O}(\varepsilon)$ near $x = 0$ the solution becomes a constant outer solution that continues via an interior boundary layer of $\mathcal{O}(\varepsilon)$ at $x = 1 - 2\sqrt{3\varepsilon}$ into the parabolic behaviour of the boundary layer of $\mathcal{O}(\sqrt{\varepsilon})$ near $x = 1$. This boundary layer has the particular property that it is defined on a *finite* interval.

16.6.4 Discussion and Related Problems

- In order to allow an asymptotic analysis, the proposed model is kept elementary. It is, however, not too difficult to introduce some straightforward generalisations. For example, D and N may be functions of X without any serious change of the analysis.
- Furthermore, the current one-dimensional problem in X may be reformulated to two dimensions in (X, Z) by taking the mass fluxes across small rods $[dX \times dZ \times h(X, Z)]$ rather than slices. Wherever the boundaries are smooth, the boundary layers retain their one-dimensional character, and again the analysis will not be essentially different.

- The present model is rather ad hoc (an example of a constructing model, Section 7.2). For physically more refined models, books on groundwater and porous media flow should be consulted (e.g., [7]). Note that the present model is essentially one of slow variation in X and might well be derived systematically from the fundamental equations of porous media flow by the method of slow variation (Section 15.3.1).

- One of the interesting features that appeared in the analysis is the inner layer for small K, separating the asymptotically constant outer solution from a boundary layer where diffusion from the canal dominates. A very similar transition layer may be found in the catalyzer pellet problem of Section 16.8.

- An interesting detail in the matching problem of (16.143) can be generalised as follows. Consider a boundary layer equation in $Y(\xi)$, $0 \le \xi < \infty$, of the form

$$\frac{\partial^2}{\partial \xi^2} Y + F'(Y) = 0, \qquad \text{which may be integrated to} \quad \frac{1}{2}\left(\frac{\partial}{\partial \xi} Y\right)^2 + F(Y) = E.$$

  If Y should be matched for $\xi \to \infty$ to an outer solution $y(x)$ of $\mathcal{O}(1)$, then the integration constant E may be found by observing that $\frac{\partial}{\partial \xi} Y$ should vanish and $E = F(y(0))$. An important condition for consistency is that the final integral

$$\int_{Y(0)}^{Y} \frac{\mathrm{d}\eta}{\sqrt{E - F(\eta)}} = \pm\sqrt{2}\,\xi$$

  diverge at $\eta = y(0)$ in order to have $\xi \to \infty$. This did *not* occur, for example, in (16.152) and (16.153). The unsuccessful matching was indeed the warning sign for the more complex structure with an internal transition layer.

16.6.5 Exercises

16.22. Derive expression (16.150).

16.23. Solve (16.146).

16.24. Reformulate the asymptotic analysis for $D = D(X)$ and $N = N(X)$.

16.25. Reformulate the problem generalised to two surface dimensions X and Z. Assume that $D = D(X, Z)$ and $N = N(X, Z)$. Try to solve the problem.

16.26. The boundary value problem (16.139) can also be solved numerically.

  (a) Give the central difference scheme for (16.139a). Taking into account the boundary condition (16.139b), write the difference equations as a nonlinear algebraic system $\mathbf{f}(\boldsymbol{\phi}) = \mathbf{0}$.

  (b) Formulate Newton's method for the system in part (a).

  (c) Compute a numerical solution of boundary value problem (16.139) for $\varepsilon = 10^{-1}$ and $K = 1$, for $\varepsilon = 10^{-3}$ and $K = 1$, and for $\varepsilon = 10^{-3}$ and $K = 10^{-3}$. Compare with the asymptotic approximation.

(d) An alternative approach is to reformulate the problem in terms of the variable $u = \sqrt{\phi}$. Do the same as above. What are the (numerical) advantages and (possible) disadvantages of this approach?

16.7 Cooling a Monocrystalline Bar

16.7.1 Problem Formulation

A common way to grow large silicon monocrystals is to pull the crystal very slowly from a melt [73]. The temperatures at which these processes take place are usually rather high. Silicon, for example, solidifies at 1693 K. This is why the crystal will lose most of its heat by radiation. At the same time, the radiation from the surface is small compared to the conduction in the material. In order to investigate the typical temperature variation along the crystalline bar, we consider the following model, originally proposed by [74].

16.7.2 The Model

Consider a long cylindrical bar $z \geq 0, 0 \leq r \leq a, 0 \leq \phi < 2\pi$, where (r, ϕ, z) are cylindrical coordinates. The bar is assumed to be long enough for a semi-infinite approximation. How long is sufficient is not clear yet, but this will follow from the analysis below.

The bar is heated from one end at $z = 0$, where the melting front is situated, while it is cooled by radiation from the other side $r = a$; see Figure 16.20. At these high temperatures cooling by natural convection of the surrounding air is very small compared to radiation. The melting front will be considered planar, although in reality it is slightly curved. The monocrystal is slowly pulled away with constant velocity V in the positive z direction. The heat flow in and out of the bar is in equilibrium, so we have a steady state. Inside the bar $z > 0, r < a$, we have the stationary convective heat equation for temperature T

$$\rho C_p V \frac{\partial T}{\partial z} = \nabla \cdot (\kappa(T) \nabla T), \tag{16.160}$$

where ρ denotes the crystal density and C_p is the specific heat, which are nearly constant for the temperatures considered. κ is, in general, the temperature-dependent thermal conductivity (typically inversely proportional to the temperature [73]). To make progress we

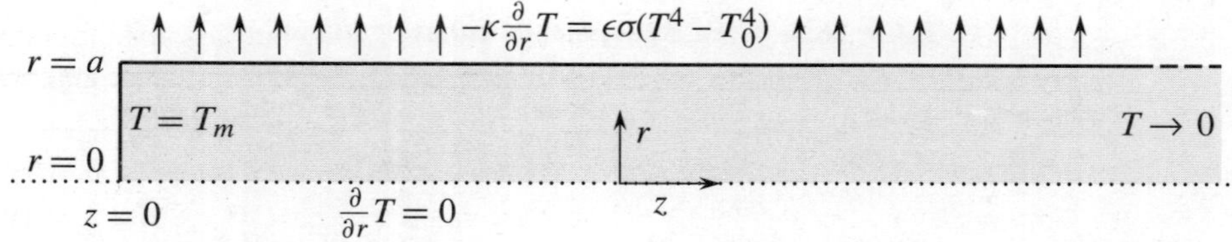

Figure 16.20. *Sketch of a silicon bar.*

assume κ constant [74] and equal to the value at or near the melting temperature. To complete the problem formulation we have the following boundary conditions to quantify the heating and radiation and the symmetry condition at the axis:

$$T = T_m \quad \text{at} \quad z = 0, \tag{16.161a}$$

$$-\kappa \frac{\partial T}{\partial r} = \epsilon\sigma(T^4 - T_0^4) \quad \text{at} \quad r = a, \tag{16.161b}$$

$$T \to T_0 \quad \text{for} \quad z \to \infty, \tag{16.161c}$$

$$\frac{\partial T}{\partial r} = 0 \quad \text{at} \quad r = 0. \tag{16.161d}$$

The temperature T_0 of the environment represents the heating by radiation from the surroundings back into the crystal. This is obviously small in our region of interest as T_0 is much lower than the melting temperature T_m. So we will ignore T_0, although it will imply that for large z the model will become inaccurate. The universal constant σ in the law of radiation is the Stefan–Boltzmann constant. The material constant ϵ is called the emission coefficient or emissivity. It is equal to unity for black bodies and between zero and one otherwise. By symmetry, the problem is independent of ϕ.

We make the problem dimensionless using

$$r = ar^*, \quad z = az^*, \quad T = T_m T^*, \tag{16.162}$$

and we introduce the dimensionless parameters

$$\varepsilon = \frac{\epsilon\sigma T_m^3 a}{\kappa}, \qquad Pe = \frac{\rho C_p V a}{\kappa}. \tag{16.163}$$

The relative importance between heat transfer by radiation and that by conduction is quantified by ε and may be identified with a Biot number. The relative importance of convection and conduction is given by the Péclet number Pe. Using the typical parameter values given in Table 16.3, we find

$$\varepsilon = 9.67 \times 10^{-3}, \qquad Pe = 1.91 \times 10^{-2}.$$

We see that both are small, but it is not yet clear if they are small enough to be neglected. It does not seem to be a good idea to neglect ε, because without it we would have no

Table 16.3. *Typical parameter values of silicon.*

Parameter	Value	Units	Parameter	Value	Units
ρ	2.33×10^3	kg/m^3	T_0	3.0×10^2	K
C_p	9.22×10^2	J/(kg · K)	a	1.15×10^{-2}	m
κ	2.16×10^1	J/(m · s · K)	V	1.67×10^{-5}	m/s
T_m	1.693×10^3	K	σ	5.6704×10^{-8}	J/(s · m^2 · K^4)
ϵ	0.66				

cooling at all. At the same time, Pe is of the same order of magnitude as ε, so it would seem at least consistent to include Pe if we include ε. So we will consider the problem asymptotically for small ε and Pe, but not as two independent parameters. It is almost always inadvisable to retain several independent small parameters. This creates undesirable and useless complexities in the asymptotic analysis. Therefore we introduce

$$Pe = \varepsilon c, \tag{16.164}$$

which is based on the observation that $Pe \simeq \varepsilon$ for the chosen characteristic parameter values. The velocity V we adopted corresponds to 1 mm/min. In [73] higher values (up to 5 mm/min) are considered, leading to $Pe = \mathcal{O}(\varepsilon^{1/2})$. We will not consider this case because it leads to a more difficult problem. Ignoring the asterisks $*$, we have in dimensionless form the problem

$$\frac{1}{r}\frac{\partial}{\partial r}\left(r\frac{\partial}{\partial r}T\right) + \frac{\partial^2}{\partial z^2}T = \varepsilon c\frac{\partial}{\partial z}T, \qquad \text{for} \quad 0 < r < 1, \quad 0 < z < \infty \tag{16.165a}$$

$$\frac{\partial}{\partial r}T = -\varepsilon T^4 \qquad \text{at} \quad r = 1, \tag{16.165b}$$

$$T = 1 \qquad \text{at} \quad z = 0, \tag{16.165c}$$

$$T \to 0 \qquad \text{for} \quad z \to \infty, \tag{16.165d}$$

$$\frac{\partial}{\partial r}T = 0 \qquad \text{at} \quad r = 0. \tag{16.165e}$$

In the following this problem will be analysed asymptotically for $\varepsilon \to 0$.

16.7.3 Asymptotic Analysis

At the outset we can anticipate a region near $z = 0$, where the boundary condition at $z = 0$ dominates the solution. Away from this initial zone the temperature profile is evened out radially and sideways radiation is balanced by axial conduction. In this region the temperature variation scales on another, longer, length scale, say $\delta(\varepsilon)^{-1}$, that is as yet unknown. To write the problem as a singular perturbation problem it is convenient to determine this length scale and rescale the axial variable. Hence we introduce the variables

$$Z = \delta(\varepsilon)z, \qquad T(r, z) = \mathcal{T}(r, Z). \tag{16.166}$$

In (r, Z) coordinates we obtained the following singular perturbation problem (see Chapter 15), with δ to be determined:

$$\frac{1}{r}\frac{\partial}{\partial r}\left(r\frac{\partial}{\partial r}\mathcal{T}\right) + \delta^2\frac{\partial^2}{\partial Z^2}\mathcal{T} = \varepsilon\delta c\frac{\partial}{\partial Z}\mathcal{T} \qquad \text{for} \quad 0 < Z < \infty, \tag{16.167a}$$

$$\frac{\partial}{\partial r}\mathcal{T} = -\varepsilon\mathcal{T}^4 \qquad \text{at} \quad r = 1, \tag{16.167b}$$

$$\mathcal{T} = 1 \qquad \text{at} \quad Z = 0, \tag{16.167c}$$

$$\mathcal{T} \to 0 \qquad \text{for} \quad Z \to \infty, \tag{16.167d}$$

$$\frac{\partial}{\partial r}\mathcal{T} = 0 \qquad \text{at} \quad r = 0. \tag{16.167e}$$

We start by assuming an asymptotic expansion of $\mathcal{T}$. In terms of variable Z this is an outer expansion. Apart from the first one (which has to be $\mathcal{O}(1)$ because the bar's temperature decays only slowly from one at the end), we do not know the order functions in the expansion yet. However, from the equation and boundary conditions it is clear that δ^2 and ε will appear somewhere. So we start with the crude estimate

$$\mathcal{T}(r, Z; \varepsilon) = \mathcal{T}_0(r, Z) + \mathcal{O}(\delta^2, \varepsilon\delta). \tag{16.168}$$

This yields, after substitution in (16.167a), to leading order,

$$\frac{1}{r}\frac{\partial}{\partial r}\left(r\frac{\partial}{\partial r}\mathcal{T}_0\right) = 0, \tag{16.169}$$

with solution $\mathcal{T}_0(r, Z) = \mathcal{Q}_0(Z)$. From the radiation boundary condition (16.167b) we find

$$\frac{\partial}{\partial r}\big[\mathcal{Q}_0 + \mathcal{O}(\delta^2, \varepsilon\delta)\big]_{r=1} = \mathcal{O}(\delta^2, \varepsilon\delta) = \mathcal{O}(\varepsilon), \tag{16.170}$$

which can only be true if

$$\delta = \varepsilon^{1/2}. \tag{16.171}$$

This means that the convection term in (16.167a) is asymptotically smaller than δ^2. We return to the asymptotic expansion and write

$$\mathcal{T}(r, Z; \varepsilon) = \mathcal{T}_0(r, Z) + \varepsilon\mathcal{T}_1(r, Z) + \cdots. \tag{16.172}$$

We have found $\mathcal{T}_0$ already. For $\mathcal{T}_1$ we have

$$\frac{1}{r}\frac{\partial}{\partial r}\left(r\frac{\partial}{\partial r}\mathcal{T}_1\right) = -\mathcal{Q}_0'', \tag{16.173}$$

with solution

$$\mathcal{T}_1(r, Z) = \mathcal{Q}_1(Z) - \frac{1}{4}r^2\mathcal{Q}_0''(Z). \tag{16.174}$$

Since this r-dependent term is the first to contribute to the heat flux at $r = 1$ due to radiation, we arrive at the following equation for $\mathcal{Q}_0$:

$$\frac{\partial \mathcal{T}}{\partial r} = \frac{\partial}{\partial r}\left[\mathcal{Q}_0 + \varepsilon\mathcal{Q}_1 - \frac{1}{4}\varepsilon r^2\mathcal{Q}_0''\right]_{r=1} = -\frac{1}{2}\varepsilon\mathcal{Q}_0'' = -\varepsilon\mathcal{Q}_0^4. \tag{16.175}$$

This equation can be integrated once to

$$\left(\mathcal{Q}_0'\right)^2 = k_1 + \frac{4}{5}\mathcal{Q}_0^5.$$

From the far-field condition (16.167d) it follows that $k_1 = 0$. As $\mathcal{T}$ has to decay from one to zero, we know that $\mathcal{Q}_0' < 0$, and we can integrate another time to obtain the solution

$$\mathcal{T}_0(r, Z) = \mathcal{Q}_0(Z) = \left(\frac{\frac{1}{3}\sqrt{5}}{Z + k_0}\right)^{2/3}, \tag{16.176}$$

where k_0 is still to be determined. Near $z = 0$, T scales on both r and z, so we assume an asymptotic expansion of $T(r, z; \varepsilon)$. In terms of the "compressed" variable Z, z is a stretched variable, and this expansion is an inner expansion. Had we used a more general entrance temperature profile than (16.167c), we would have immediately recognized the need for a boundary layer from the inability of the outer solution to satisfy this condition (see the exercises below).

From the radiation condition it is clear that, in the sequence of order functions, ε will show up somewhere. This appears, however, not to be enough. We will also have to consider the condition of matching with the outer solution $\mathcal{T}_0$. If $Z = \mathcal{O}(\delta)$, we have

$$\mathcal{T}_0 \sim \mathcal{Q}_0(0) + \delta z \mathcal{Q}_0'(0) + \cdots, \tag{16.177}$$

so *before* the $\mathcal{O}(\varepsilon)$ term there is an $\mathcal{O}(\varepsilon^{1/2})$ term. We write, therefore,

$$T(r, z; \varepsilon) = \vartheta_0(r, z) + \varepsilon^{1/2}\vartheta_1(r, z) + \varepsilon\vartheta_2(r, z) + \cdots. \tag{16.178}$$

Upon substitution of this expansion in the equation and boundary conditions, it follows that for $n = 0$ and $n = 1$ we have

$$\frac{1}{r}\frac{\partial}{\partial r}\left(r\frac{\partial}{\partial r}\vartheta_n\right) + \frac{\partial^2}{\partial z^2}\vartheta_n = 0, \qquad \frac{\partial \vartheta_n}{\partial r} = 0 \quad \text{at} \quad r = 1, \qquad \frac{\partial \vartheta_n}{\partial r} = 0 \quad \text{at} \quad r = 0, \tag{16.179}$$

where $\vartheta_0 = 1$ and $\vartheta_1 = 0$ at $z = 0$. This problem has a general solution given by the exponential Bessel function series

$$\vartheta_n(r, z) = C_n z + D_n + \sum_{k=1}^{\infty} J_0(\alpha_k r)\big(A_{nk}\,\mathrm{e}^{-\alpha_k z} + B_{nk}\,\mathrm{e}^{\alpha_k z}\big), \tag{16.180}$$

where J_0 is the ordinary Bessel function of the first kind of order zero; α_k is the kth positive zero of $J_0' = -J_1$; and A_{nk}, B_{nk}, C_n, and D_n are constants to be determined from the boundary conditions.

From the matching condition for $z \to \infty$ to the outer expansion of $\mathcal{T}$ it follows that $B_k^{1/\alpha_k} = o(\mathrm{e}^{-z})$ for $k \to \infty$ and $z \to \infty$. This is only possible if $B_{nk} = 0$. From the orthogonality of the set $\{J_0(\alpha_k r)\}$ on $[0, 1]$, applied to the boundary condition at $z = 0$, it is readily found that we have

$$D_0 = 1, \qquad A_{0k} = \frac{\int_0^1 r J_0(\alpha_k r)\,\mathrm{d}r}{\frac{1}{2}J_0(\alpha_k)^2} = \frac{\alpha_k^{-1}J_1(\alpha_k)}{\frac{1}{2}J_0(\alpha_k)^2} = 0, \tag{16.181a}$$

$$D_1 = 0, \qquad A_{1k} = 0. \tag{16.181b}$$

Furthermore, for $z = \mathcal{O}(\delta^{-1})$ the linear leading-order term $C_0 z$ is too big to match with $\mathcal{T}_0$, which is $\mathcal{O}(1)$, so $C_0 = 0$. Altogether, our first terms of the inner expansion are greatly simplified to

$$\vartheta_0(r, z) = 1, \qquad \vartheta_1(r, z) = C_1 z. \tag{16.182}$$

Under matching with outer solution $\mathcal{T}_0$, this leads to $\lim_{Z\to 0}\mathcal{Q}_0(Z) = 1$, which requires

$$k_0 = \frac{1}{3}\sqrt{5}. \tag{16.183}$$

To determine C_1 of ϑ_1 we note that from matching with the outer solution we have

$$\mathcal{Q}_0(0) + \delta z \mathcal{Q}_0'(0) + \cdots \sim 1 + \varepsilon^{1/2} C_1 z + \cdots . \tag{16.184}$$

The correspondence in z requires

$$C_1 = \mathcal{Q}_0'(0) = -\frac{2}{3} k_0^{-1} = -\frac{2}{5}\sqrt{5}. \tag{16.185}$$

Altogether we have the following description of the sought solution:

$$T(r,z) = \begin{cases} \left(1 + \frac{3}{5}\sqrt{5}Z\right)^{-2/3} + \cdots & \text{if } Z = \mathcal{O}(1), \\ 1 - \frac{2}{5}(5\varepsilon)^{1/2} z + \cdots & \text{if } Z = \mathcal{O}(\varepsilon^{1/2}). \end{cases} \tag{16.186}$$

To leading order there is effectively no boundary layer. Apparently, the boundary condition at $z = 0$ is simple enough to be satisfied by the outer solution up to $\mathcal{O}(\varepsilon)$ and we may as well write

$$T(r,z) = \frac{1}{\left(1 + \frac{3}{5}(5\varepsilon)^{1/2} z\right)^{2/3}} + \mathcal{O}(\varepsilon) \qquad \text{for} \quad 0 \le z < \infty. \tag{16.187}$$

Finally, as a consistency test, it is of interest to check whether the heat flux into the bar at $z = 0$ is indeed equal to the total flux out of the bar along $r = 1$. This is indeed the case:

$$2\pi \int_0^1 \frac{\partial T}{\partial z}(r,0) r \, \mathrm{d}r \simeq -2\pi \int_0^1 \frac{2}{5}(5\varepsilon)^{1/2} r \, \mathrm{d}r = -\frac{2}{5}\pi(5\varepsilon)^{1/2}, \tag{16.188a}$$

$$2\pi \int_0^\infty -\varepsilon T^4(r,z) \, \mathrm{d}z \simeq -2\pi\varepsilon^{1/2} \int_0^\infty \left(1 + \frac{3}{5}\sqrt{5}Z\right)^{-8/3} \mathrm{d}Z = -\frac{2}{5}\pi(5\varepsilon)^{1/2}. \tag{16.188b}$$

An example is plotted in Figure 16.21 for the values listed in Table 16.3, where in particular $\varepsilon = 9.6688 \times 10^{-2}$. We see that for $z = 0.34$ m the temperature drops below the environment temperature of (say) 300 K as a result of our neglecting T_0. Physically it is impossible for an object to release heat by radiation to a temperature lower than its environment.

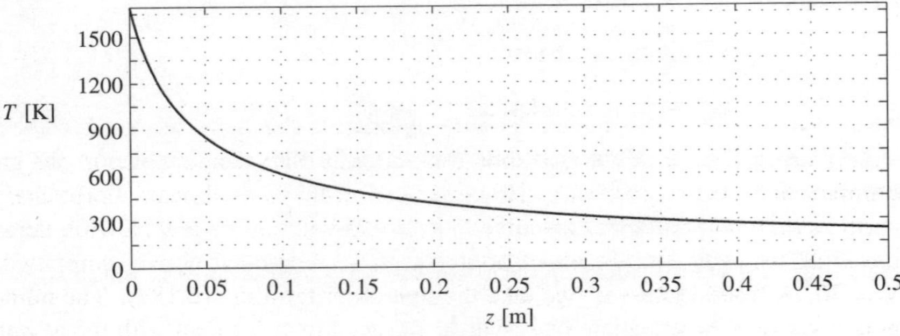

Figure 16.21. *Temperature variation along the monocrystalline bar.*

16.7.4 Numerical Solution Method

An interesting approach to solving the boundary value problem (16.165) numerically is to use alternating direction implicit (ADI); see Section 9.4. Take L large enough and divide the interval $(0, L)$ into $M+1$ intervals of length $\Delta z := L/(M+1)$. Define in addition $\Delta r := 1/(N+1)$, $r_j := j\Delta r$ $(j = 0, 1, \dots, N+1)$, and $z_k := k\Delta z$ $(k = 0, 1, \dots, M+1)$. For the second derivative with respect to r we use the difference approximation

$$\frac{1}{r}\frac{\partial}{\partial r}\left(r\frac{\partial T}{\partial r}\right)(r_j, z_k) \doteq \frac{1}{\Delta r^2}\left(\beta_j T_{j+1,k} - 2T_{j,k} + \alpha_j T_{j-1,k}\right), \tag{16.189a}$$

where the geometrical factors α_j and β_j are defined by

$$\alpha_j := 1 - \frac{\Delta r}{2r_j}, \quad \beta_j := 1 + \frac{\Delta r}{2r_j}, \qquad j = 2, 3, \dots, N. \tag{16.189b}$$

Note that $\frac{\partial}{\partial r}T(0, z) = 0$ and therefore we have, for $j = 1$, that

$$\frac{1}{r}\frac{\partial}{\partial r}\left(r\frac{\partial T}{\partial r}\right)(r_1, z_k) \doteq \frac{4}{\Delta r^2}\left(T_{1,k} - T_{0,k}\right); \tag{16.190}$$

see (5.47a). For the z derivatives we take the standard central difference approximations

$$\frac{\partial^2 T}{\partial z^2}(r_j, z_k) \doteq \frac{1}{\Delta z^2}\left(T_{j,k+1} - 2T_{j,k} + T_{j,k-1}\right), \tag{16.191a}$$

$$\frac{\partial T}{\partial z}(r_j, z_k) \doteq \frac{1}{2\Delta z}\left(T_{j,k+1} - T_{j,k-1}\right). \tag{16.191b}$$

Using the boundary conditions in (16.165), we thus obtain a system of equations of the form

$$\mathbf{A}\mathbf{u} = \boldsymbol{f}(\mathbf{u}), \tag{16.192}$$

where $\mathbf{u}$ is the vector of unknown values $T_{j,k}$, $\mathbf{A}$ is the matrix containing the coefficients defined by the discretisations in (16.189) to (16.191), and $\boldsymbol{f}(\mathbf{u})$ contains the boundary value terms. Note that $\boldsymbol{f}(\mathbf{u})$ is nonlinear due to the boundary condition at $r = 1$. We can split $\mathbf{A}$ into a part containing r derivative approximations and a part containing z derivative approximations, say (cf. (9.66))

$$\mathbf{A} = \mathbf{A}_r + \mathbf{A}_z. \tag{16.193}$$

The ADI scheme for (16.192) then reads

$$\mathbf{u}^{l+1} = \mathbf{u}^l + \tau\left(\mathbf{A}_r\mathbf{u}^l + \mathbf{A}_z\mathbf{u}^{l+1} - \boldsymbol{f}(\mathbf{u}^l)\right), \tag{16.194a}$$

$$\mathbf{u}^{l+2} = \mathbf{u}^{l+1} + \tau\left(\mathbf{A}_r\mathbf{u}^{l+2} + \mathbf{A}_z\mathbf{u}^{l+1} - \boldsymbol{f}(\mathbf{u}^{l+2})\right). \tag{16.194b}$$

As we found out in Section 11.8.1, τ should be chosen $\mathcal{O}(\Delta r)$ and $\mathcal{O}(\Delta z)$. We can directly compute $\mathbf{u}^{l+1}$ from (16.194a). However, (16.194b) is nonlinear. Fortunately, the nonlinearity is quite mild and we can solve this system with only a few Newton iterations. In Figure 16.22 we show the result of applying ADI as outlined above, computed with $M = N = 30$. As initial guess $\mathbf{u}^0$ we take the first order term in (16.187). The numerical solution is practically independent of r, which is indeed in agreement with the asymptotic analysis of the previous section.

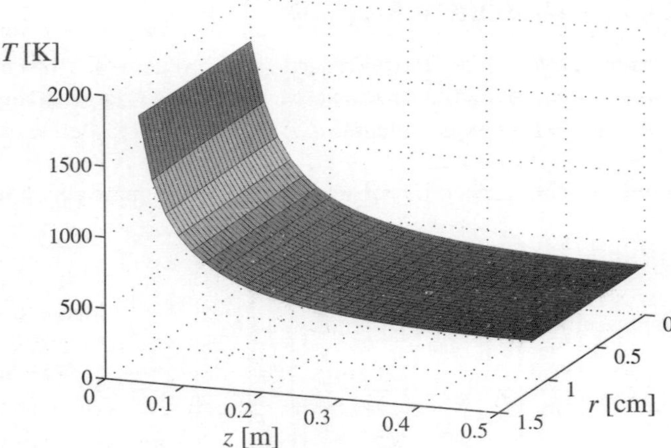

Figure 16.22. *Temperature distribution in the monocrystalline bar.*

16.7.5 Discussion and Related Problems

- An important result is that the role of V (via Pe) is negligible for the adopted velocities of 1 mm/min when $Pe = \mathcal{O}(\varepsilon)$, but it is certainly to be included for larger values when $Pe \geq \mathcal{O}(\varepsilon^{1/2})$.
- The present model is well able to provide insight into the behaviour of the heat flow entering the bar from the reservoir of fluid crystal at $z = 0$ and leaving the bar by radiation. Some simplifications, however, shoud be removed for more accurate results. For example, one should consider a temperature-dependent κ, a finite environment temperature T_0, the effect of slighty larger velocities V, and a finite length of the bar if the bar is relatively short compared to the diameter a or the slow length scale $a/\varepsilon^{1/2}$. Heat release by natural convection of the surrounding air may be considered for the part of the bar with a temperature range near or just above T_0.

16.7.6 Exercises

16.27. Formulate the inner solution for an arbitrary initial (r, ϕ)-dependent temperature profile and verify the consistency check (16.188).

16.28. The present model does not include a varying κ. Formulate the same problem for $\kappa(T) = \kappa_m T_m / T$ and solve the outer solution. Note that condition (16.167d) is inadequate to determine one integration constant. It is always satisfied for any two integration constants. Since the inner solution is only able to determine one condition, we need to invoke an extra condition. In [73] this is solved by taking the bar of finite length and assuming a zero heat flux out of the far end. What is the real cause of this strange behaviour?

16.29. Estimate the order of magnitude of cooling by natural convection (the relevant dimensionless number is a Biot number). Compare this with cooling due to radiation.

16.30. Consider (16.165) with (16.165b) replaced by a linear boundary condition.
(a) Show that ADI as outlined in Section 16.7.4 converges for τ small enough.
(b) Determine the optimal relaxation value of the parameter τ.

16.8 A Catalytic Reaction Problem in Pellets

16.8.1 Problem Formulation

A catalytic reaction is a chemical reaction between reactants, of which one—the catalyst—returns to its original state after the reaction. Its role is entirely to enable the reaction to happen. An example of a catalyst is platinum. The primary reactant is usually a liquid or a gas. As the catalyst and the reactant are immiscible, the reaction occurs at the catalyst surface, which is therefore made as large as possible. A way to achieve this is to apply the catalyst to the pores of porous pellets. The reactant diffuses from the surface to the inside of the pellet. Meanwhile, being in contact with the catalyst, the reactant is converted to the final product.

Assume that reactant A reacts in an equilibrium reaction with catalyst S at the pellet pore surface to the intermediate product AS in a way described by

$$A + S \rightleftarrows AS$$

with concentrations $[A]$ mol/m^3, $[S]$ mol/kg, and $[AS]$ mol/kg satisfying

$$[AS] = K[A][S], \quad \text{where} \quad [S] = [S]_0 - [AS],$$

for equilibrium constant K m^3/mol and initially available catalyst $[S]_0$. It follows that

$$[AS] = \frac{K[A][S]_0}{1 + K[A]}.$$

If AS reacts in a first order reaction to the final product B with reaction rate constant $\tilde{k}$, we have the reaction equation

$$\frac{\partial}{\partial t}[AS] = -\tilde{k}[AS] = -\tilde{k}[S]_0 \frac{K[A]}{1 + K[A]} = -k\frac{K[A]}{1 + K[A]}.$$

The term $K[A]/(1 + K[A])$ is called a Langmuir isotherm. This reaction acts as a source term for species B or, equivalently, as a sink term for A. We define, for ease of notation, the concentration of A inside a pellet as $C := [A]$. Under the additional assumption of a well-stirred fluid in order to maintain a constant concentration $C = C_R$ and $[B] = 0$ at the outer surface of spherical pellets (see Figure 16.23), we obtain the initial boundary value problem

$$\frac{\partial C}{\partial t} - \nabla\cdot(D\nabla C) = -k\frac{KC}{1 + KC}, \qquad 0 < \tilde{r} < R, \quad t > 0, \tag{16.195a}$$

$$C(r, 0) = 0, \qquad 0 < \tilde{r} < R, \tag{16.195b}$$

$$C(R, t) = C_R, \quad \frac{\partial}{\partial \tilde{r}}C(0, t) = 0, \qquad t > 0, \tag{16.195c}$$

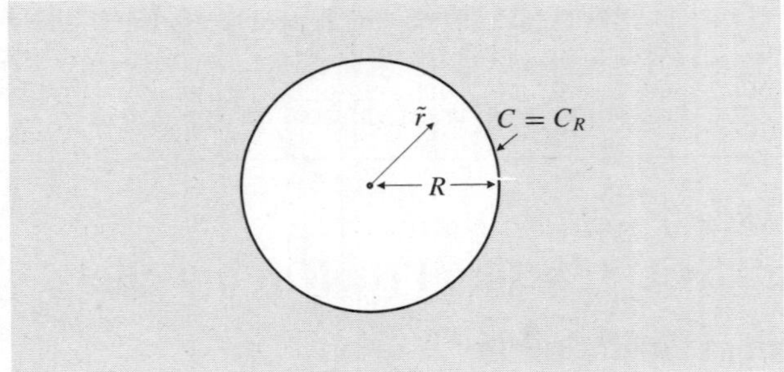

Figure 16.23. *A spherical catalyzer pellet.*

where D is the diffusion coefficient of C inside the pellet. After a time $t \gg \max(R^2/D, (1 + KC_R)/kK)$, the concentration C is in stationary equilibrium. Assuming spherical symmetry and a constant diffusion coefficient D, C satisfies the boundary value problem

$$D\frac{1}{\tilde{r}^2}\frac{\mathrm{d}}{\mathrm{d}\tilde{r}}\left(\tilde{r}^2\frac{\mathrm{d}C}{\mathrm{d}\tilde{r}}\right) = k\frac{KC}{1+KC}, \qquad 0 < \tilde{r} < R, \tag{16.196a}$$

$$C(R) = C_R, \qquad \frac{\mathrm{d}}{\mathrm{d}\tilde{r}}C(0) = 0. \tag{16.196b}$$

A point of concern mathematically is the singularity at $C = -1/K$. Physically, of course, this is just an irrelevant anomaly of the model because C is only meant to be positive, but it is not clear beforehand if the model per se precludes negative C. We therefore assume whenever necessary that the right-hand side of (16.196a) is zero for $C < 0$.

We make the problem dimensionless as follows:

$$c := \frac{C}{C_R}, \quad r := \frac{\tilde{r}}{R}, \quad \lambda = \frac{kR^2}{DC_R}, \quad \alpha = \frac{1}{KC_R} \tag{16.197}$$

such that

$$\frac{1}{r^2}\frac{\mathrm{d}}{\mathrm{d}r}\left(r^2\frac{\mathrm{d}c}{\mathrm{d}r}\right) = \lambda\frac{cH(c)}{\alpha + c}, \qquad 0 < r < 1, \tag{16.198a}$$

$$c(1) = 1, \qquad c'(0) = 0, \tag{16.198b}$$

where the prime ($'$) denotes differentiation with respect to r and H denotes the Heaviside function (4.23). The net mass flux into the pellet, an important final result, is given by $-4\pi R^2 D\frac{\mathrm{d}}{\mathrm{d}r}C(R)$ (Fick's law; see (6.40)). In dimensionless quantities this is thus equal to $-4\pi RDC_R c'(1)$.

We will consider in detail the behaviour of the solution for the particular limit of $\lambda = \mathcal{O}(1)$, $\alpha \to 0$. This is an interesting limit because, as we will see, a critical value of λ exists above which a (steady) front exists.

In addition, the approximate analytical results will be compared with probably more accurate numerically generated data. It should be remembered, however, that the analytical results take the form of a functional relation among c, r, λ, and α. This relationship is therefore different from just the numerical value at a given point. The information it provides includes trends, behaviour at infinity or near singular points, etc., and has therefore to be interpreted in a much wider sense than just at the level of "numerical outcome."

16.8.2 Numerical Solution Method

To discretise (16.198a) we employ the finite volume method; see Chapter 5. In this section we restrict ourselves to a short outline. Integrating (16.198a) over a control volume $V_j := (r_{j-\frac{1}{2}}, r_{j+\frac{1}{2}}) = ((j-\frac{1}{2})\Delta r, (j+\frac{1}{2})\Delta r)$, centred around grid point $r_j = j\Delta r$ with Δr the grid size, and applying Gauss's theorem, we obtain the integral balance

$$\oint_{\partial V_j} \frac{\partial c}{\partial \boldsymbol{n}}\,\mathrm{d}S = \int_{V_j} f(c)\,\mathrm{d}V, \qquad f(c) := \lambda \frac{cH(c)}{\alpha + c}, \tag{16.199}$$

where $\boldsymbol{n}$ is the outward unit normal on the boundary ∂V_j of the control volume. Next we have to approximate both integrals in (16.199). For the surface integral we have

$$\begin{aligned} \oint_{\partial V_j} \frac{\partial c}{\partial \boldsymbol{n}}\,\mathrm{d}S &= \int_{r=r_{j+\frac{1}{2}}} \frac{\partial c}{\partial r}\,\mathrm{d}S - \int_{r=r_{j-\frac{1}{2}}} \frac{\partial c}{\partial r}\,\mathrm{d}S \\ &\doteq \frac{\partial c}{\partial r}(r_{j+\frac{1}{2}}) 4\pi r_{j+\frac{1}{2}}^2 - \frac{\partial c}{\partial r}(r_{j-\frac{1}{2}}) 4\pi r_{j-\frac{1}{2}}^2 \\ &\doteq \frac{4\pi}{\Delta r}\Big(r_{j+\frac{1}{2}}^2 (c_{j+1} - c_j) - r_{j-\frac{1}{2}}^2 (c_j - c_{j-1}) \Big), \end{aligned} \tag{16.200}$$

where c_j denotes the numerical approximation of $c(r_j)$. To derive (16.200) we have used that $\boldsymbol{n} = \boldsymbol{e}_r$ at the sphere $r = r_{j+\frac{1}{2}}$ and $\boldsymbol{n} = -\boldsymbol{e}_r$ at the sphere $r = r_{j-\frac{1}{2}}$. Furthermore, we have approximated the integrals and derivatives involved by the midpoint rule and the central difference scheme, respectively. To approximate the volume integral we use the midpoint rule to obtain

$$\int_{V_j} f(c)\,\mathrm{d}V \doteq f(c_j)|V_j| = \frac{4}{3}\pi f(c_j)\left(r_{j+\frac{1}{2}}^3 - r_{j-\frac{1}{2}}^3 \right). \tag{16.201}$$

Inserting (16.200) and (16.201) into (16.199) and taking into account the relation $x^3 - y^3 = (x-y)(x^2 + xy + y^2)$, we obtain the discrete conservation law

$$\frac{1}{\Delta r^2}\Big(\beta_{j+\frac{1}{2}}(c_{j+1} - c_j) - \alpha_{j-\frac{1}{2}}(c_j - c_{j-1}) \Big) = f(c_j), \tag{16.202}$$

where the coefficients $\beta_{j+\frac{1}{2}}$ and $\alpha_{j-\frac{1}{2}}$ are defined by

$$\beta_{j+\frac{1}{2}} := \frac{3r_{j+\frac{1}{2}}^2}{r_{j+\frac{1}{2}}^2 + r_{j+\frac{1}{2}} r_{j-\frac{1}{2}} + r_{j-\frac{1}{2}}^2}, \qquad \alpha_{j-\frac{1}{2}} := \frac{3r_{j-\frac{1}{2}}^2}{r_{j+\frac{1}{2}}^2 + r_{j+\frac{1}{2}} r_{j-\frac{1}{2}} + r_{j-\frac{1}{2}}^2}. \tag{16.203}$$

Note that the diffusion term, the left-hand side of (16.198a), is not singular in $r = 0$ because $c(r) = a + br^2 + \mathcal{O}(r^4)$. Then we have $r^{-2}(r^2 c')'' = c'' + 2c'/r = 6b + \mathcal{O}(r^{*2})$, and we have to adjust the numerical scheme (16.202) accordingly. Finally, we use Newton's method to solve the nonlinear system (16.202); see the exercises.

16.8.3 Asymptotic Analysis

The limit we will consider is one of a strong nonlinear reaction and corresponds to $\lambda = \mathcal{O}(1)$, $\alpha \to 0$. We note the following:

(i) As long as $c \neq 0$, the small parameter α is not multiplied by the highest derivative. From the physics, we do know, however, that $c > 0$, although c may be small. So we have to be aware of singular behaviour if $c \leq \mathcal{O}(\alpha)$ somewhere.

(ii) The reduced problem ($\alpha = 0$) is corrected by terms of $\mathcal{O}(\alpha)$.

(iii) From the boundary condition $c(1) = 1$ we can, at least in the neighbourhood of $r = 1$, estimate the order of magnitude of c as $c = \mathcal{O}(1)$.

We start with the assumption (see Chapter 15) that we can expand c as follows:

$$c(r;\alpha) = c_0(r) + \alpha c_1(r) + \mathcal{O}(\alpha^2). \tag{16.204}$$

Since $c(1) = 1$, this will probably be valid in the neighbourhood of $r = 1$. Equation (16.198a) becomes

$$\frac{1}{r^2}(r^2 c_0')' + \alpha \frac{1}{r^2}(r^2 c_1')' + \mathcal{O}(\alpha^2) = \lambda \frac{c}{\alpha + c} = \lambda - \lambda \frac{\alpha}{c_0} + \mathcal{O}(\alpha^2). \tag{16.205}$$

Note, in addition to point (i) above, that the geometrical series expansion used is not valid any more if $c \leq \mathcal{O}(\alpha)$.

To leading order we get

$$\frac{1}{r^2}(r^2 c_0')' = \lambda \tag{16.206}$$

with a first integral

$$c_0' = \frac{1}{3}\lambda r + \frac{A_0}{r^2}. \tag{16.207}$$

If c_0 is a valid approximation everywhere, we can apply the boundary condition in $r = 0$ such that $A_0 = 0$. Integrating (16.207) and applying the boundary condition at $r = 1$, we find

$$c_0(r) = 1 - \frac{1}{6}\lambda(1 - r^2). \tag{16.208}$$

We see now that there is an important distinction to be made between a solution for $\lambda < 6$ and one for $\lambda > 6$:

$\lambda < 6$: $c_0 \neq 0$ everywhere, so our assumption is indeed satisfied.

$\lambda > 6$: c_0 has a zero in $r_0 = \sqrt{1 - 6/\lambda}$, and the solution is certainly not valid near and below r_0.

As the solution found is (to leading order) complete for $\lambda < 6$, we continue with the assumption $\lambda > 6$. The question that immediately arises is whether this solution, invalid near r_0, is at least valid for r well above r_0. The stunning answer is that this solution is not valid anywhere!

As soon as the interval is cut into two pieces by the zero r_0, we cannot apply to c_0 the boundary condition at $r = 0$. One might say that c_0 has no contact any more with the other end. So the constant of integration A_0 (16.207) remains undetermined yet, and so is any zero $r = r_1$ of this c_0. So we have for $r > r_1$ that

$$c_0(r) = 1 - \frac{1}{6}\lambda(1 - r^2) + A_0\left(1 - \frac{1}{r}\right). \tag{16.209}$$

For the interval $0 \le r < r_1$ we have to consider the fact that physically c represents a positive quantity (a concentration) and therefore cannot be negative. (Strictly speaking, this is not a relevant argument mathematically since we are dealing with a mathematical model that may just be invalid for $\lambda > 6$. A mathematical proof of the solution being positive is better; see the exercises.) So we assume that $0 \le c \le \mathcal{O}(\alpha)$. If we scale c on a gauge function $\alpha\gamma(\alpha)$ to be determined, so $c = \alpha\gamma\tilde{c}$, then we have

$$\alpha\frac{1}{r^2}\frac{\mathrm{d}}{\mathrm{d}r}\left(r^2\frac{\mathrm{d}\tilde{c}}{\mathrm{d}r}\right) = \lambda\frac{\tilde{c}}{1+\gamma\tilde{c}}, \tag{16.210}$$

and we see that for any choice of γ every asymptotic expansion $\tilde{c}(r;\alpha) = \tilde{c}_0(r) + \cdots$ leads to $c(r) \equiv 0$. Apparently, the solution here is asymptotically equivalent to zero.

What remain to be determined are the position and neighbourhood of the transition point r_1, where both parts of the solution come together. As noted before, this point is still unknown since we do not know the constant A_0 yet.

The only way that the solution, asymptotically equivalent to zero, can be set free is by a quick transition in a boundary layer (Section 15.4.1), because only then are the derivatives of c big enough to outweigh the small factor α. Therefore we will search for a boundary layer

$$c = \alpha^n\psi, \qquad r = r_1 + \alpha^m\xi \tag{16.211}$$

that yields a reduced equation rich enough to allow this transition. We substitute

$$\frac{\alpha^{n-2m}}{(r_1+\alpha^m\xi)^2}\frac{\mathrm{d}}{\mathrm{d}\xi}\left((r_1+\alpha^m\xi)^2\frac{\mathrm{d}\psi}{\mathrm{d}\xi}\right) = \lambda\frac{\alpha^n\psi}{\alpha+\alpha^n\psi}, \tag{16.212}$$

which becomes, under $\alpha \to 0$,

$$\alpha^{-2m}\frac{\mathrm{d}^2\psi}{\mathrm{d}\xi^2} = \lambda\frac{\psi}{\alpha+\alpha^n\psi}. \tag{16.213}$$

Apparently, the richest structure (distinguished limit) is found for $m = \frac{1}{2}$ and $n = 1$; thus $c = \alpha\psi$ and $r = r_1 + \sqrt{\alpha}\xi$ and

$$\frac{\mathrm{d}^2\psi}{\mathrm{d}\xi^2} = \lambda\frac{\psi}{1+\psi}. \tag{16.214}$$

Of remarkable importance now are the matching conditions. First, ψ has to connect to the zero solution to the left of r_1, so $\psi \to 0$ if $\xi \to -\infty$. At the right-hand side of r_1, however, ψ has to connect to $c_0(r)$, the behaviour of which is described near r_1 by

$$\begin{aligned} c_0(r) &= c_0(r_1) + (r-r_1)c_0'(r_1) + \frac{1}{2}(r-r_1)^2c_0''(r_1) + \cdots \qquad (r \downarrow r_1) \\ &= \sqrt{\alpha}\,\xi c_0'(r_1) + \frac{1}{2}\alpha\xi^2c_0''(r_1) + \cdots \\ &\simeq \alpha\,\psi(\xi) \qquad (\xi \to \infty) \end{aligned} \tag{16.215}$$

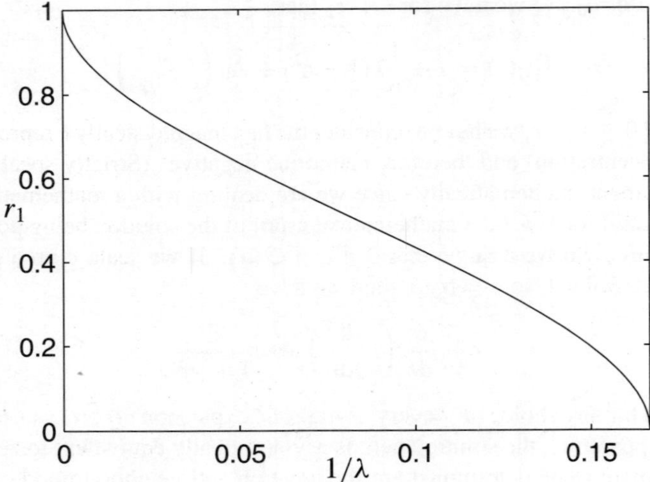

Figure 16.24. r_1 *as a function of* $1/\lambda$.

This provides us with the following important information. The $\mathcal{O}(\sqrt{\alpha})$ term has to be equal to zero, so $c_0'(r_1) = \frac{1}{3}\lambda r_1 + A_0/r_1^2 = 0$, implying that $A_0 = -\frac{1}{3}\lambda r_1^3$. Physically, this makes sense: if there is no source at the border of the region where the concentration vanishes, then the flux there ($\sim c'$) cannot be anything else but zero. Next we can determine r_1 from $c_0(r_1) = 0$, which is now $\frac{1}{3}r_1^3 - \frac{1}{2}r_1^2 + \frac{1}{6} - \lambda^{-1} = 0$ and has the solution [1]

$$r_1 = \frac{1}{2} + \cos\left(\frac{1}{3}\arccos(12\lambda^{-1} - 1) - \frac{2}{3}\pi\right). \tag{16.216}$$

In Figure 16.24 r_1 is plotted as a function of $1/\lambda$ (note that $r_1 = 0$ if $\lambda = 6$).

We are now ready as far as the most important information is concerned. The outer solution c_0 is described by

$$c_0(r) = 1 - \frac{1}{6}\lambda(1 - r^2), \qquad \lambda \le 6, \tag{16.217a}$$

$$c_0(r) = \left\{\begin{array}{ll} 0 & \text{if } 0 \le r < r_1 \\ 1 - \frac{1}{6}\lambda(1 - r^2) - \frac{1}{3}\lambda r_1^3\left(1 - \frac{1}{r}\right) & \text{if } r_1 < r \le 1 \end{array}\right\}, \qquad \lambda \ge 6. \tag{16.217b}$$

The mass flux into the pellet is found from $c_0'(1) = \frac{1}{3}\lambda$ if $\lambda \le 6$ and $c_0'(1) = \frac{1}{3}\lambda(1 - r_1^3)$ if $\lambda \ge 6$.

The boundary layer, or inner solution, near $r = r_1$ is rather complicated so we leave its determining as an exercise. Interestingly enough, the behaviour of the boundary layer

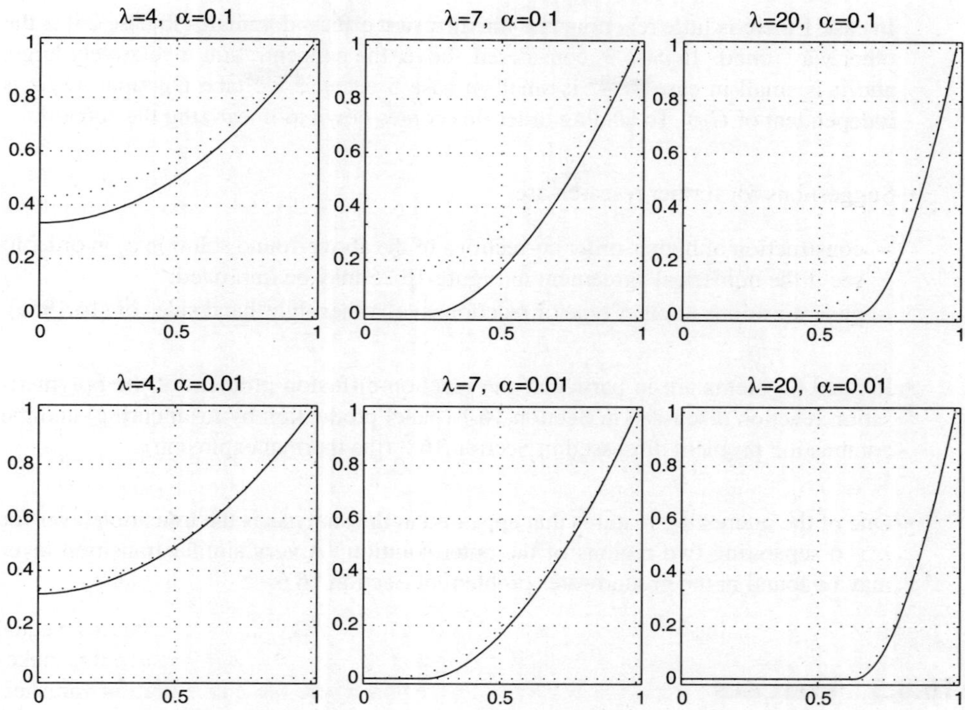

Figure 16.25. *Leading-order approximation for $\alpha \to 0$ and $\lambda = \mathcal{O}(1)$.*

was essential for fixing the outer solution, although the inner problem itself need not be solved!

Some examples of the outer solution (solid line), with a numerical "exact" solution (dotted line) as a reference, are given in Figure 16.25. We see that for $\alpha = 0.1$ the approximation isn't very good, although it is indeed just the expected error of $\mathcal{O}(\alpha)$. For $\alpha = 0.01$ the analytical prediction of a transition layer is completely confirmed by the numerical solution.

16.8.4 Discussion and Related Problems

- We have considered one particular asymptotic combination of λ and α because it contained the interesting (steady) reaction-diffusion front if $\lambda > 6$. Other combinations may be considered too. We may distinguish the following five cases:

  1. $\lambda \to 0,\quad \alpha = \mathcal{O}(1)$ (regular),
  2. $\lambda \to \infty,\quad \alpha = \mathcal{O}(1)$ (singular),
  3. $\lambda = \mathcal{O}(1),\quad \alpha \to 0$ (singular and regular),
  4. $\lambda = \mathcal{O}(1),\quad \alpha \to \infty$ (regular).
  5. $\lambda \to \infty,\quad \alpha \to \infty\quad \lambda/\alpha = \mathcal{O}(1)$ (linearization, regular).

In case 1 there is little reaction, and the diffusion effects dominate. In case 2 it is the other way round. In case 3, considered above, the concentration is relatively large, and it is small in case 4. C is small in case 5 because we take C_R small (λ/α is independent of C_R). To leading order this comes down to linearizing the equation.

- Suggestions for further research are
  - construction of higher-order corrections of the above-found solution c_0 in order to see if the numerical agreement in Figure 16.25 may be improved;
  - generalisations to other type of reactions, i.e., other right-hand sides of (16.195a).

- Related problems are in particular the reaction-diffusion problem of the polymerisation reaction discussed in Section 16.1 (panel production by resin curing) and the combustion problem discussed in Section 16.3 (the thermal explosion).

- One of the interesting features that appeared in the analysis is the transition layer for $\lambda > 6$ separating two regions of the outer solution. A very similar transition layer may be found in the groundwater problem of Section 16.6.

16.8.5 Exercises

16.31. Analyse asymptotically the regular perturbation problems of cases 1, 4, and 5.

16.32. Analyse asymptotically the singular (boundary layer type) perturbation problem of case 2.

16.33. Give the discretisation of (16.202) for $j = 0$.

16.34. Formulate Newton's method for the nonlinear system (16.202).

16.35. Determine the inner solution (implicitly, in the form of an integral) given by the problem (16.214). Find the integration constants by matching arguments.

16.36. We may prove uniqueness and positivity of the solution of the problem

$$\nabla^2 c = f(c) \quad \text{for} \quad \boldsymbol{x} \in \Omega, \qquad c = b \quad \text{for} \quad \boldsymbol{x} \in \partial\Omega,$$

where Ω is the unit sphere, b is a nonnegative constant, and f is a sufficiently smooth function with $f(0) = 0$ if $c \leq 0$ and $f'(c) > 0$ for $c > 0$.

(a) Assume there exist two solutions c_1 and c_2 satisfying the same boundary condition. Let $u = c_2 - c_1$. Apply Green's first identity to $u\nabla^2 u$ with an integral over Ω. Apply a monotonicity argument to $f(c_2) - f(c_1)$ and show that $c_2 = c_1$.

(b) Observe that $c \equiv 0$ is the unique solution for $b = 0$. Assume a solution c with positive boundary value. If $c(r_1) = 0$ for some $r = r_1 < 1$, argue that c remains zero for $0 \leq r < r_1$. In other words, $c \geq 0$.

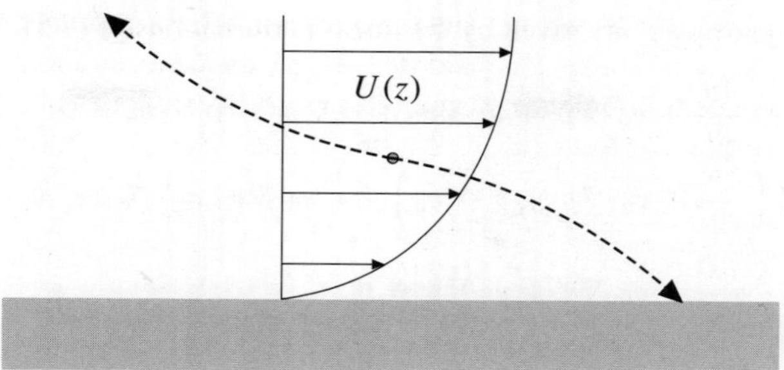

Figure 16.26. *Acoustic refraction in shear flow.*

16.9 Outdoor Noise Enhancement by Atmospheric Conditions

16.9.1 Problem Formulation

The propagation of sound waves in the atmosphere is greatly affected by wind (see Figure 16.26). For example, the communication between two people, one downstream and one upstream, is not symmetric. The one upstream is easier to understand for the one downstream than the other way around. This is *not* because the wind "carries the waves faster," but it is due to *refraction* by the wind gradient (the atmospheric boundary layer). It may be compared to the refraction by a sound speed or temperature gradient described in Example 15.37, but the refraction is now directionally dependent. We can study this problem by the following ray model, a generalisation of Example 15.37.

16.9.2 The Model

Consider an inviscid, perfect, non-heat-conducting, subsonic, and steady mean flow with density ρ_0, velocity $\boldsymbol{v}_0$, pressure p_0, and entropy s_0, satisfying the steady version of (7.5), i.e., the equations

$$\nabla\cdot(\rho_0\boldsymbol{v}_0) = 0, \tag{16.218a}$$

$$\rho_0\boldsymbol{v}_0\cdot\nabla\boldsymbol{v}_0 + \nabla p_0 = 0, \tag{16.218b}$$

$$\boldsymbol{v}_0\cdot\nabla s_0 = 0, \tag{16.218c}$$

$$s_0 = C_V \ln p_0 - C_P \ln \rho_0. \tag{16.218d}$$

We introduce the sound speed c_0 given by

$$c_0^2 = \frac{C_P}{C_V}\frac{p_0}{\rho_0} = \gamma\frac{p_0}{\rho_0}, \tag{16.218e}$$

where the gas constants C_P and C_V are the specific heats at constant pressure and constant volume, respectively, and $\gamma = C_P/C_V$. Acoustic perturbations in pressure p', density ρ', velocity $\boldsymbol{v}'$, and entropy s' are given by the linear perturbations of (7.5), i.e., the equations

$$\frac{\partial \rho'}{\partial t} + \boldsymbol{v}_0 \cdot \nabla \rho' + \boldsymbol{v}' \cdot \nabla \rho_0 + \rho_0 \nabla \cdot \boldsymbol{v}' + \rho' \nabla \cdot \boldsymbol{v}_0 = 0, \tag{16.219a}$$

$$\rho_0 \left(\frac{\partial \boldsymbol{v}'}{\partial t} + \boldsymbol{v}_0 \cdot \nabla \boldsymbol{v}' + \boldsymbol{v}' \cdot \nabla \boldsymbol{v}_0 \right) + \rho' \boldsymbol{v}_0 \cdot \nabla \boldsymbol{v}_0 = -\nabla p', \tag{16.219b}$$

$$\frac{\partial s'}{\partial t} + \boldsymbol{v}_0 \cdot \nabla s' + \boldsymbol{v}' \cdot \nabla s_0 = 0, \tag{16.219c}$$

$$s' = \frac{C_V}{p_0} p' - \frac{C_P}{\rho_0} \rho'. \tag{16.219d}$$

If ω_{ref} is a typical frequency of the perturbations and c_{ref} and ρ_{ref} are typical values of mean sound speed and density, we can make (16.218) and (16.219) dimensionless by scaling velocities on c_{ref}, densities on ρ_{ref}, pressures on $\rho_{\text{ref}} c_{\text{ref}}^2$, time on ω_{ref}^{-1}, length on $c_{\text{ref}}/\omega_{\text{ref}}$, and entropy on C_P (but that is less important). The equations are now of exactly the same form as before, so we will not repeat them here. Thus in the following we will regard system (16.219) as the equations for the dimensionless perturbations.

We assume that the typical wavelength $c_{\text{ref}}/\omega_{\text{ref}}$ is short compared to the typical length scale L of the mean flow variations. In order to quantify this, we introduce the small parameter $\varepsilon := c_{\text{ref}}/\omega_{\text{ref}} L \ll 1$. We introduce the slow variables $\boldsymbol{X} := \varepsilon \boldsymbol{x}$ and $T := \varepsilon t$. The mean flow variables depend on $\boldsymbol{X}$ only.

We assume that the perturbations are harmonic in time with frequency ω, equal to or of the order of ω_{ref}. Next we introduce the *ray approximation* (see Section 15.4.2)

$$p', \rho', \boldsymbol{v}', s' = P(\boldsymbol{X}; \varepsilon), R(\boldsymbol{X}; \varepsilon), \boldsymbol{V}(\boldsymbol{X}; \varepsilon), S(\boldsymbol{X}; \varepsilon) \times \mathrm{e}^{\mathrm{i}(\omega T - \vartheta(\boldsymbol{X}; \varepsilon))/\varepsilon}. \tag{16.220}$$

16.9.3 Asymptotic Analysis

We define the operator $\overline{\nabla} := (\frac{\partial}{\partial X}, \frac{\partial}{\partial Y}, \frac{\partial}{\partial Z})$ so that $\nabla = \varepsilon \overline{\nabla}$. We introduce the *local wave vector* $\boldsymbol{\kappa} := \overline{\nabla}\vartheta$, $\kappa := |\boldsymbol{\kappa}|$, and the *phase plane normal field* $\boldsymbol{n} := \overline{\nabla}\vartheta / |\overline{\nabla}\vartheta| = \boldsymbol{\kappa}/\kappa$. The phase planes $\vartheta - \omega T =$ constant describe the propagating wave front. We define the shifted frequency $\Omega := \omega - \boldsymbol{v}_0 \cdot \boldsymbol{\kappa}$. Since

$$\nabla p' = \left(-\mathrm{i} \boldsymbol{\kappa} P + \varepsilon \overline{\nabla} P \right) \mathrm{e}^{\mathrm{i}(\omega T - \vartheta)/\varepsilon},$$

etc., we obtain, by substituting (16.220) in (16.219) and dropping the exponentials, the equations

$$\mathrm{i}\Omega R - \mathrm{i}\rho_0 \boldsymbol{V} \cdot \boldsymbol{\kappa} + \varepsilon \big(\boldsymbol{v}_0 \cdot \overline{\nabla} R + \boldsymbol{V} \cdot \overline{\nabla} \rho_0 + \rho_0 \overline{\nabla} \cdot \boldsymbol{V} + R \overline{\nabla} \cdot \boldsymbol{v}_0 \big) = 0, \tag{16.221a}$$

$$\mathrm{i}\rho_0 \Omega \boldsymbol{V} - \mathrm{i} P \boldsymbol{\kappa} + \varepsilon \big(\rho_0 \boldsymbol{v}_0 \cdot \overline{\nabla} \boldsymbol{V} + \rho_0 \boldsymbol{V} \cdot \overline{\nabla} \boldsymbol{v}_0 + R \boldsymbol{v}_0 \cdot \overline{\nabla} \boldsymbol{v}_0 + \overline{\nabla} P \big) = \boldsymbol{0}, \tag{16.221b}$$

$$\mathrm{i}\Omega S + \varepsilon \big(\boldsymbol{v}_0 \cdot \overline{\nabla} S + \boldsymbol{V} \cdot \overline{\nabla} s_0 \big) = 0, \tag{16.221c}$$

$$S = \frac{C_V}{p_0} P - \frac{C_P}{\rho_0} R. \tag{16.221d}$$

We expand in the usual way (cf. (15.64)) to obtain

$$P(\boldsymbol{X}; \varepsilon) = P(\boldsymbol{X}) + \mathcal{O}(\varepsilon), \qquad \vartheta(\boldsymbol{X}; \varepsilon) = \vartheta(\boldsymbol{X}) + \mathcal{O}(\varepsilon), \tag{16.222}$$

etc., where we adopt the convention of using the same notation for both the variable and its leading-order approximation. Then we get for the perturbations to $\mathcal{O}(1)$

$$\Omega R - \rho_0 \boldsymbol{V} \cdot \boldsymbol{\kappa} = 0, \tag{16.223a}$$

$$\rho_0 \Omega \boldsymbol{V} - P \boldsymbol{\kappa} = \boldsymbol{0}, \tag{16.223b}$$

$$\Omega S = 0. \tag{16.223c}$$

Using (16.221d), this leads to the relations

$$S = 0, \quad P = c_0^2 R, \quad \boldsymbol{V} = \frac{P\kappa}{\rho_0 \Omega} \boldsymbol{n} \tag{16.224}$$

together with the eikonal equation

$$c_0^2 \kappa^2 = \Omega^2. \tag{16.225a}$$

Written as a differential equation for ϑ this is

$$c_0^2 |\overline{\nabla} \vartheta|^2 = (\omega - \boldsymbol{v}_0 \cdot \overline{\nabla} \vartheta)^2. \tag{16.225b}$$

In a subsonic flow with $|\boldsymbol{v}_0| < c_0$ and positive ω and κ we can only have the positive branch of (16.225a), so we may assume that $c_0 \kappa = \Omega$ and thus $\boldsymbol{V} = (P/\rho_0 c_0) \boldsymbol{n}$.

If we interpret (16.225a) as a dispersion relation $\omega = \omega(\boldsymbol{\kappa})$, the group velocity is given by

$$\boldsymbol{v}_{\mathrm{g}} = \frac{\partial \omega}{\partial \boldsymbol{\kappa}} = c_0 \boldsymbol{n} + \boldsymbol{v}_0. \tag{16.226}$$

To rewrite equation (16.225b) in characteristic form, i.e., as a system of ODEs along rays, we apply Theorem 12.6. It is enlightening to make sure that we obtain a parametrization of the rays that corresponds as much as possible to physical time. It is easily verified that (16.225b), rewritten in the form (see Theorem 12.6)

$$F(\boldsymbol{X}, \vartheta, \boldsymbol{\kappa}) := \frac{\frac{1}{2} c_0^2 \kappa^2}{\omega - \boldsymbol{v}_0 \cdot \boldsymbol{\kappa}} - \frac{1}{2} (\omega - \boldsymbol{v}_0 \cdot \boldsymbol{\kappa}) = 0,$$

yields for characteristic variable s the equation

$$\frac{\mathrm{d}\vartheta}{\mathrm{d}s} = \boldsymbol{\kappa} \cdot \frac{\partial F}{\partial \boldsymbol{\kappa}} = \omega;$$

in other words, $\vartheta = \omega s + \text{constant}$. Since we know that $\vartheta = \omega T + \text{constant}$ along the phase surfaces, it follows that s may be identified with the slow time scale T. The ray equations are then

$$\vartheta(\boldsymbol{X}_{\mathrm{r}}) = \omega T + \text{constant} \tag{16.227a}$$

along the ray $\boldsymbol{X} = \boldsymbol{X}_{\mathrm{r}}(T)$ given by

$$\frac{\mathrm{d}\boldsymbol{X}_{\mathrm{r}}}{\mathrm{d}T} = \frac{\partial F}{\partial \boldsymbol{\kappa}} = c_0 \boldsymbol{n} + \boldsymbol{v}_0, \tag{16.227b}$$

$$\frac{\mathrm{d}\boldsymbol{\kappa}}{\mathrm{d}T} = -\boldsymbol{\kappa} \frac{\partial F}{\partial \vartheta} - \overline{\nabla} F = -\kappa \overline{\nabla} c_0 - \boldsymbol{\kappa} \cdot \frac{\partial \boldsymbol{v}_0}{\partial \boldsymbol{X}}, \tag{16.227c}$$

where $\boldsymbol{\kappa}\cdot\frac{\partial}{\partial X}\boldsymbol{v}_0 := (\boldsymbol{\kappa}\cdot\frac{\partial}{\partial X}\boldsymbol{v}_0, \boldsymbol{\kappa}\cdot\frac{\partial}{\partial Y}\boldsymbol{v}_0, \boldsymbol{\kappa}\cdot\frac{\partial}{\partial Z}\boldsymbol{v}_0)^T$. Note that the propagation speed of information, $c_0\boldsymbol{n} + \boldsymbol{v}_0$, is indeed just the group velocity (16.226).

We continue with the perturbations to $\mathcal{O}(\varepsilon)$ in order to get equations for the amplitudes. We have

$$\overline{\nabla}\cdot(\boldsymbol{v}_0 R + \rho_0 \boldsymbol{V}) = 0, \tag{16.228a}$$

$$\rho_0(\boldsymbol{v}_0\cdot\overline{\nabla}\boldsymbol{V} + \boldsymbol{V}\cdot\overline{\nabla}\boldsymbol{v}_0) + R\boldsymbol{v}_0\cdot\overline{\nabla}\boldsymbol{v}_0 + \overline{\nabla}P = 0, \tag{16.228b}$$

$$\boldsymbol{V}\cdot\overline{\nabla}s_0 = 0. \tag{16.228c}$$

Equation (16.228a), expressed in P and $\boldsymbol{n}$, becomes a conservation law for P equivalent to mass conservation; i.e.,

$$\overline{\nabla}\cdot\left(\frac{P(\boldsymbol{v}_0 + c_0\boldsymbol{n})}{c_0^2}\right) = 0. \tag{16.229}$$

This does not, however, include the effect of inertia, i.e., the conservation of momentum. For this we need some rather subtle, albeit technical, manipulations. First, we have from (16.228a) and (16.228b) that

$$\overline{\nabla}\cdot\left(\frac{\boldsymbol{v}_0}{c_0^2}P + \rho_0\boldsymbol{V}\right) = 0, \tag{16.230a}$$

$$\rho_0\big(\boldsymbol{v}_0\cdot\overline{\nabla}\boldsymbol{V} + \boldsymbol{V}\cdot\overline{\nabla}\boldsymbol{v}_0\big) + \overline{\nabla}P - \frac{P}{\rho_0 c_0^2}\overline{\nabla}p_0 = \boldsymbol{0}. \tag{16.230b}$$

We multiply (16.230a) by P/ρ_0 and dot-multiply (16.230b) by $\boldsymbol{V}$ and add. Using the identity $\boldsymbol{V}\cdot(\boldsymbol{v}_0\cdot\overline{\nabla}\boldsymbol{V}) = \frac{1}{2}\boldsymbol{v}_0\cdot\overline{\nabla}V^2$, where $V := |\boldsymbol{V}|$, and the relation $\boldsymbol{V}\cdot\overline{\nabla}p_0 = c_0^2\boldsymbol{V}\cdot\overline{\nabla}\rho_0$, we get

$$\frac{1}{2}\rho_0\boldsymbol{v}_0\cdot\overline{\nabla}V^2 + \rho_0\boldsymbol{V}\cdot(\boldsymbol{V}\cdot\overline{\nabla}\boldsymbol{v}_0) + \boldsymbol{V}\cdot\overline{\nabla}P - \frac{P}{\rho_0}\boldsymbol{V}\cdot\overline{\nabla}\rho_0 + \frac{P}{\rho_0}\overline{\nabla}\cdot\left(\frac{\boldsymbol{v}_0}{c_0^2}P + \rho_0\boldsymbol{V}\right) = 0.$$

Next we introduce the (energy-like) auxiliary variable

$$E := \frac{\rho_0 V^2}{2} + \frac{P^2}{2\rho_0 c_0^2}. \tag{16.231}$$

Since

$$\overline{\nabla}E = \frac{1}{2}\rho_0\overline{\nabla}V^2 + \frac{1}{2}V^2\overline{\nabla}\rho_0 + \frac{1}{2}\frac{1}{\rho_0 c_0^2}\overline{\nabla}P^2 + \frac{1}{2}P^2\overline{\nabla}\left(\frac{1}{\rho_0 c_0^2}\right),$$

we have

$$\boldsymbol{v}_0\cdot\overline{\nabla}E - \frac{1}{2}V^2\boldsymbol{v}_0\cdot\overline{\nabla}\rho_0 - \frac{1}{2}P^2\boldsymbol{v}_0\cdot\overline{\nabla}\left(\frac{1}{\rho_0 c_0^2}\right) + \rho_0\boldsymbol{V}\cdot\big(\boldsymbol{V}\cdot\overline{\nabla}\boldsymbol{v}_0\big) + \overline{\nabla}\cdot(P\boldsymbol{V})$$
$$+\frac{P^2}{\rho_0}\overline{\nabla}\cdot\left(\frac{\boldsymbol{v}_0}{c_0^2}\right) = 0.$$

Rewriting P and $\boldsymbol{V}$ in E and $\boldsymbol{n}$, i.e., $P^2 = \rho_0 c_0^2 E$, $V^2 = E/\rho_0$, $P\boldsymbol{V} = c_0 E\boldsymbol{n}$, we obtain

$$\boldsymbol{v}_0\cdot\overline{\nabla}E - \frac{1}{2}E\boldsymbol{v}_0\cdot\left[\frac{1}{\rho_0}\overline{\nabla}\rho_0 + \rho_0 c_0^2\overline{\nabla}\left(\frac{1}{\rho_0 c_0^2}\right)\right] + E\boldsymbol{n}\cdot\big(\boldsymbol{n}\cdot\overline{\nabla}\boldsymbol{v}_0\big) + \overline{\nabla}\cdot(c_0 E\boldsymbol{n})$$
$$+c_0^2 E\overline{\nabla}\cdot\left(\frac{\boldsymbol{v}_0}{c_0^2}\right) = 0.$$

Then we use the relation

$$-\frac{1}{2}\left[\frac{1}{\rho_0}\overline{\nabla}\rho_0 + \rho_0 c_0^2 \overline{\nabla}\left(\frac{1}{\rho_0 c_0^2}\right)\right] = -\frac{1}{2}\overline{\nabla}(\ln \rho_0 - \ln \rho_0 c_0^2) = \overline{\nabla} \ln c_0 = \frac{1}{c_0}\overline{\nabla} c_0$$

to get

$$\boldsymbol{v}_0 \cdot \overline{\nabla} E + \frac{E}{c_0}\boldsymbol{v}_0 \cdot \overline{\nabla} c_0 + E\boldsymbol{n}\cdot\left(\boldsymbol{n}\cdot\overline{\nabla}\boldsymbol{v}_0\right) + \overline{\nabla}\cdot(c_0 E\boldsymbol{n}) + c_0^2 E\overline{\nabla}\cdot\left(\frac{\boldsymbol{v}_0}{c_0^2}\right) = 0.$$

Next we eliminate the term $\boldsymbol{n}\cdot(\boldsymbol{n}\cdot\overline{\nabla}\boldsymbol{v}_0)$. First, we rewrite, by using the vectorial identities (J.5) and (J.6) and the fact that $\overline{\nabla}\times\boldsymbol{\kappa} = \overline{\nabla}\times\overline{\nabla}\vartheta = \mathbf{0}$, the following expression in $\boldsymbol{\kappa}$:

$$\begin{aligned}
\boldsymbol{\kappa}\cdot(\boldsymbol{\kappa}\cdot\overline{\nabla}\boldsymbol{v}_0) &= \boldsymbol{\kappa}\cdot\overline{\nabla}(\boldsymbol{\kappa}\cdot\boldsymbol{v}_0) - \frac{1}{2}\boldsymbol{v}_0\cdot\overline{\nabla}\kappa^2 - \boldsymbol{v}_0\cdot\left((\overline{\nabla}\times\boldsymbol{\kappa})\times\boldsymbol{\kappa}\right) \\
&= \boldsymbol{\kappa}\cdot\overline{\nabla}(\boldsymbol{\kappa}\cdot\boldsymbol{v}_0) - \frac{1}{2}\boldsymbol{v}_0\cdot\overline{\nabla}\kappa^2 \\
&= \boldsymbol{\kappa}\cdot\overline{\nabla}(\omega - \Omega) - \kappa\boldsymbol{v}_0\cdot\overline{\nabla}\kappa \\
&= -\boldsymbol{\kappa}\cdot\overline{\nabla}\Omega - (\Omega/c_0)\boldsymbol{v}_0\cdot\overline{\nabla}(\Omega/c_0).
\end{aligned}$$

From this we have

$$\boldsymbol{n}\cdot(\boldsymbol{n}\cdot\overline{\nabla}\boldsymbol{v}_0) = -\frac{c_0}{\Omega}\boldsymbol{n}\cdot\overline{\nabla}\Omega - \frac{c_0}{\Omega}\boldsymbol{v}_0\cdot\overline{\nabla}\left(\frac{\Omega}{c_0}\right) = c_0\Omega\boldsymbol{n}\cdot\overline{\nabla}\left(\frac{1}{\Omega}\right) + \frac{\Omega}{c_0}\boldsymbol{v}_0\cdot\overline{\nabla}\left(\frac{c_0}{\Omega}\right),$$

which yields

$$\begin{aligned}
&\boldsymbol{v}_0\cdot\overline{\nabla}E + \frac{E}{c_0}\boldsymbol{v}_0\cdot\overline{\nabla}c_0 + Ec_0\Omega\boldsymbol{n}\cdot\overline{\nabla}\left(\frac{1}{\Omega}\right) + E\frac{\Omega}{c_0}\boldsymbol{v}_0\cdot\overline{\nabla}\left(\frac{c_0}{\Omega}\right) + \overline{\nabla}\cdot(c_0 E\boldsymbol{n}) \\
&\quad + c_0^2 E\overline{\nabla}\cdot\left(\frac{\boldsymbol{v}_0}{c_0^2}\right) = 0.
\end{aligned}$$

This may finally be simplified to

$$\Omega\overline{\nabla}\cdot\left(\frac{E}{\Omega}\boldsymbol{v}_0\right) + \Omega\overline{\nabla}\cdot\left(\frac{E}{\Omega}c_0\boldsymbol{n}\right) = \Omega\overline{\nabla}\cdot\left(\frac{E}{\Omega}(\boldsymbol{v}_0 + c_0\boldsymbol{n})\right) = \Omega\overline{\nabla}\cdot\left(\frac{E}{\Omega}\boldsymbol{v}_{\mathrm{g}}\right) = 0$$

or

$$\overline{\nabla}\cdot\left(\frac{P^2(c_0\boldsymbol{n} + \boldsymbol{v}_0)}{\rho_0 c_0^2\Omega}\right) = 0. \tag{16.232}$$

This is a classic result in ray acoustics known as the *Blokhintzev invariant*. It allows us to connect the variation of the amplitude with the rate of divergence of the rays.

At time $T = T_1$ the initial phase surface is given by $\vartheta = C_0$. At this surface we consider a small patch A_1 (see Figure 16.27). We trace the rays that start from A_1 until a time $T = T_2$, where they form the surface A_2, the projection of A_1 on the phase surface $\vartheta = \omega(T_2 - T_1) + C_0$. When A_1 is small enough to approximate the ray's amplitude P as constant over a cross section, this bundle of rays forms a *ray tube*. When we integrate

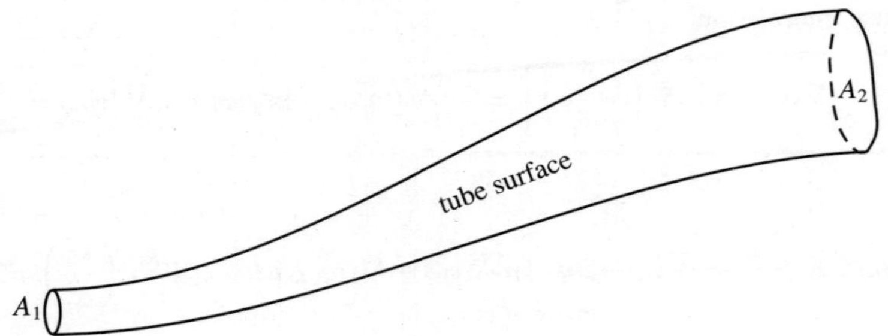

Figure 16.27. *A ray tube.*

(16.232) over the ray tube, closed by A_1 and A_2, and apply Gauss's theorem, we obtain

$$\iint_{A_1 \cup A_2 \cup \text{tube wall}} \frac{P^2}{\rho_0 c_0^2 \Omega} (c_0 \boldsymbol{n} + \boldsymbol{v}_0) \cdot \mathrm{d}\boldsymbol{S} = 0.$$

Since $c_0 \boldsymbol{n} + \boldsymbol{v}_0$ is tangential to the ray tube surface, there is no contribution from the tube wall. Since $\boldsymbol{n}$ is normal to any cross section A, while $(c_0 \boldsymbol{n} + \boldsymbol{v}_0) \cdot \boldsymbol{n} = c_0 + (\boldsymbol{v}_0 \cdot \boldsymbol{\kappa})/\kappa = \omega/\kappa$, we thus have

$$\frac{P^2}{\rho_0 c_0^3} \frac{\omega A}{\kappa^2} = \text{constant} \tag{16.233}$$

along the ray tube. A mathematically more satisfactory form [19] may be obtained by taking the limit $|A| \to 0$. Assume a smooth parametrisation $\boldsymbol{X} = \Xi(T, \xi, \eta)$ such that any curve $(\xi, \eta) = $ constant corresponds to a ray. Then the change of surface elements between surfaces given by $T = $ constant is equal to the Jacobian $|\frac{\partial}{\partial \xi} \Xi \times \frac{\partial}{\partial \eta} \Xi|$. As a result we have along a ray that

$$\frac{P^2}{\rho_0 c_0^3} \frac{\omega}{\kappa^2} \left| \frac{\partial \Xi}{\partial \xi} \times \frac{\partial \Xi}{\partial \eta} \right| = \text{constant}. \tag{16.234}$$

16.9.4 Solutions

As an example we compute the solution of the ODE system (16.227b) and (16.227c) for an initially vertical wave front located in the YZ plane. We will consider waves of 108.2 Hz ($\omega_{\text{ref}} = 680$ Hz) propagating along an atmospheric boundary layer, modeled as uniform shear flow and dimensionally given by $17 \tanh(z/L)$ m/s, where the typical mean flow length scale is $L = 50$ m. The mean sound speed is uniform and equal to $c_{\text{ref}} = 340$ m/s. Hence $\varepsilon = 10^{-2}$. The typical wavelength is then $\lambda = 2\pi c_{\text{ref}}/\omega_{\text{ref}} = 3.14$ m, which is indeed much smaller than L. Nondimensionally, we have the mean flow velocity given by

$$\boldsymbol{v}_0(\boldsymbol{X}) = (U_0(Z), 0, 0)^T, \qquad U_0(Z) = U_\infty \tanh(Z), \tag{16.235}$$

where $U_\infty = 5 \times 10^{-2}$ and we have a constant sound speed $c_0 = 1$. More specifically, we compute the rays starting from the circle (in terms of the slow variables) of radius ε and

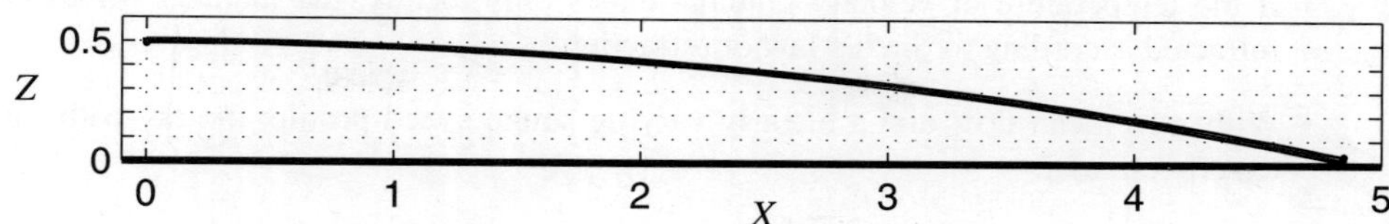

Figure 16.28. *A ray tube starting from a circle of radius ε in the YZ plane, refracting downward in right-running shear flow.*

centre (0, 0, 0.5). In summary, we have the initial value problem

$$\frac{dX_r}{dT} = \frac{c_0\kappa_X}{\kappa} + U_0(Z_r), \quad \frac{dY_r}{dT} = \frac{c_0\kappa_Y}{\kappa}, \quad \frac{dZ_r}{dT} = \frac{c_0\kappa_Z}{\kappa}, \tag{16.236a}$$

$$\frac{d\kappa_X}{dT} = \frac{d\kappa_Y}{dT} = 0, \qquad \frac{d\kappa_Z}{dT} = -U_0'(Z_r)\kappa_X, \tag{16.236b}$$

$$X_r(0) = 0, \quad Y_r(0) = \varepsilon\cos\alpha, \quad Z_r(0) = 0.5 + \varepsilon\sin\alpha, \tag{16.236c}$$

$$\kappa_X(0) = 1, \quad \kappa_Y(0) = 0, \quad \kappa_Z(0) = 0, \tag{16.236d}$$

where $\kappa = (\kappa_X^2 + \kappa_Y^2 + \kappa_Z^2)^{1/2}$ and $\alpha = \alpha_n = n\pi/4$ for $n = 0, 1, 2, \ldots, 7$. The equations for κ_X, κ_Y, and Y_r are easily solved and we find $\kappa_X(T) = 1$, $\kappa_Y(T) = 0$, and $Y_r(T) = \varepsilon\cos\alpha$. The ODE system for the other variables reduces to

$$\frac{dX_r}{dT} = U_0(Z_r) + \frac{c_0}{\sqrt{1+\kappa_Z^2}}, \quad \frac{dZ_r}{dT} = c_0\frac{\kappa_Z}{\sqrt{1+\kappa_Z^2}}, \quad \frac{d\kappa_Z}{dT} = -U_0'(Z_r). \tag{16.237}$$

We computed numerically a solution of (16.237) for $0 \le T \le 4.8$ ($\equiv 4.8/\varepsilon\omega_{\text{ref}} = 0.71$ s dimensionally) and the result is presented in Figure 16.28. We can see that the phase plane normal $\boldsymbol{n}$ is initially directed horizontally, $\boldsymbol{n}_0 = (1, 0, 0)$, and at the end it is bent down to $\boldsymbol{n}_1 = (0.9785, 0, -0.2061)$.

Clearly, since we have a positive wind shear, i.e., $U_0'(Z) > 0$, the Z component $\kappa_Z = \frac{\partial}{\partial Z}\vartheta$ decreases. In other words, the rays will bend toward the low wind speed regions. Propagating with the wind (as in the example), the waves bend down and remain near the ground (by repeated reflections, not included here); against the wind they bend up and disappear into free space.

It happens that the tube cross-sectional surface (i.e., the surface enclosed by the bundle of eight rays) increased in size by a factor of 1.4167, while the modulus of the tube wave vector κ (i.e., averaged over the rays) increased by a factor of 1.0219. From (16.233) we obtain that the amplitude P decreased by a factor of 0.8586, which is about 1.3 dB.

16.9.5 Discussion and Related Problems

- The above analysis refers to sound propagating in a medium with smooth mean flow and temperature gradients. Scattering at edges is to be included by other ways (see the literature on rays, e.g., [69, 89, 37, 114, 22]). Also, rays that creep along smoothly curved surfaces into the shadow zone are to be treated differently.

- If the temperature or velocity gradient exists only locally, the incident waves are refracted according to Snell's law of refraction.
- With zero mean flow and a linearly varying sound speed profile, the ray paths are exactly circular.
- Salinity together with temperature and pressure gradients in the ocean may create layers of minimum sound speed where sound rays remain trapped.
- The solution of any problem that can be described by a first order scalar PDE, and can therefore be solved along characteristics, is ray like. In this sense the etching problem of Section 16.2 and the thin-layer flow of Section 16.10 are related problems.

16.9.6 Exercises

16.37. Prove that for subsonic flow the negative branch $\Omega = -c_0\kappa$ as solution of (16.225b) cannot hold.

16.38. Show the existence of circular ray paths for linearly varying sound speed profiles.

16.39. Derive Snell's law for a temperature change across a thin layer.

16.40. Repeat Exercise 16.39 for a velocity shear layer.

16.10 Thin-Layer Flow Along a Curved Surface

An example of a regular perturbation problem is the flow of a thin layer of water along a curved surface, e.g., in a sink or a toilet [125].

The flow of a moderately thin layer of water of sufficiently high speed along a curved surface, like a sink or a toilet, is presumably dominated by a balance between inertia and gravity, while viscosity and surface tension are less important. Of course, this is not entirely correct. The Reynolds number of water flow (the ratio between inertia and viscous forces), based on a density of 10^3 kg/m^3, a thickness of 1 mm, and a velocity of 1 m/s, is 1000, which is indeed large but may not be large enough to ignore accumulated friction effects. The Weber number (the ratio between inertia and surface tension), based on similar estimates, is around 15, which is again large but may not be large enough in all cases, e.g., to avoid drop formation.

Assuming irrotational steady flow allowing a velocity potential, and assuming a small layer thickness compared to other length scales, we may describe the flow may be described by a regular perturbation in terms of the dimensionless typical layer thickness ε. The resulting leading-order equations are similar to the eikonal and energy equations in ray theory. These hyperbolic equations can be integrated along streamlines, with explicit results for various simple geometries.

In these equations the smoothing back reaction of the pressure is decoupled from the problem, leading to the possibility of singular lines being the envelope of crossing streamlines. This crossing of streamlines seems unphysical. However, in reality crossing streamlines do appear to exist: the streamlines cross at different levels.

16.10.1 The Model

Consider along a surface $\tilde{S}(\tilde{x}) = 0$ in a gravity field $-g\boldsymbol{e}_z$ a thin layer of incompressible, inviscid, irrotational, and stationary flow (water). At the free surface, given by $\tilde{F}(\tilde{x}) = 0$ (we write dimensional variables with a tilde), the pressure is equal to the constant atmospheric pressure p_0.

Inside the layer we have a potential $\tilde{\Phi}$, a pressure $\tilde{p}$, and a constant density ρ_0 satisfying the equation of mass conservation and Bernoulli's equation, respectively:

$$\nabla^2\tilde{\Phi} = 0, \tag{16.238a}$$

$$\tilde{p} + \frac{1}{2}\rho_0|\nabla\tilde{\Phi}|^2 + \rho_0 g\tilde{z} = \text{constant along streamlines}, \tag{16.238b}$$

with boundary conditions

$$\nabla\tilde{\Phi}\cdot\nabla\tilde{S} = 0 \qquad \text{at} \quad \tilde{S} = 0, \tag{16.238c}$$

$$\nabla\tilde{\Phi}\cdot\nabla\tilde{F} = 0 \qquad \text{at} \quad \tilde{F} = 0, \tag{16.238d}$$

$$\tilde{p} = p_0 \qquad \text{at} \quad \tilde{F} = 0. \tag{16.238e}$$

When we make lengths dimensionless on a typical length L related to the size of the curved surface, make velocities dimensionless on a typical velocity U, and write the pressure as $\tilde{p} := p_0 + \rho_0 U^2 p$, we obtain

$$\nabla^2\Phi(x, y, z) = 0, \tag{16.239a}$$

$$p + \frac{1}{2}|\nabla\Phi|^2 + \gamma z = \text{constant along streamlines}, \tag{16.239b}$$

with boundary conditions

$$\nabla\Phi\cdot\nabla S = 0 \qquad \text{at} \quad S = 0, \tag{16.239c}$$

$$\nabla\Phi\cdot\nabla F = 0 \qquad \text{at} \quad F = 0, \tag{16.239d}$$

$$p = 0 \qquad \text{at} \quad F = 0, \tag{16.239e}$$

where the inverse of the squared Froude number

$$\gamma := \frac{gL}{U^2} \tag{16.240}$$

is supposed to be finite, i.e., $\mathcal{O}(1)$ in terms of the asymptotic expansion to follow.

16.10.2 Curvilinear Coordinates

We assume that we can define a curvilinear orthogonal coordinate system (σ, τ, ν) given by

$$x = x(\sigma, \tau, \nu), \quad y = y(\sigma, \tau, \nu), \quad z = z(\sigma, \tau, \nu) \tag{16.241}$$

and attached to the surface such that $\nu = 0$ corresponds to $S = 0$ and $\nu > 0$ is the wet side. The free surface $F = 0$ corresponds to $\nu = \varepsilon H(\sigma, \tau)$ and ε is small. For simplicity we will write

$$\Phi(x, y, z) = \Phi(\sigma, \tau, \nu). \tag{16.242}$$

To rewrite the problem in the new coordinates we utilize the calculus of vectors and tensors in curvilinear orthogonal coordinates [15]:

$$\text{scale factors:}\quad h_\alpha^2 = \sum_\xi \left(\frac{\partial \xi}{\partial \alpha}\right)^2, \qquad \alpha = \sigma, \tau, \nu, \quad \xi = x, y, z;$$

$$\text{unit vectors:}\quad \boldsymbol{e}_\alpha = \frac{1}{h_\alpha}\sum_\xi \frac{\partial \xi}{\partial \alpha}\boldsymbol{e}_\xi, \qquad \frac{\boldsymbol{e}_\beta}{h_\beta}\cdot\frac{\partial \boldsymbol{e}_\alpha}{\partial \beta} = \sum_\xi \left(\frac{1}{h_\beta^2}\frac{\partial \xi}{\partial \beta}\frac{\partial}{\partial \beta}\left(\frac{1}{h_\alpha}\frac{\partial \xi}{\partial \alpha}\right)\right);$$

$$\text{gradient:}\qquad \nabla = \sum_\alpha \frac{\boldsymbol{e}_\alpha}{h_\alpha}\frac{\partial}{\partial \alpha}.$$

This yields for the Laplace equation (16.239a)

$$\nabla^2\Phi = \sum_\alpha \frac{1}{h_\alpha}\frac{\partial}{\partial \alpha}\left(\frac{1}{h_\alpha}\frac{\partial \Phi}{\partial \alpha}\right) + \sum_\alpha \left(\frac{1}{h_\alpha}\frac{\partial \Phi}{\partial \alpha}\sum_\beta \left(\frac{\boldsymbol{e}_\beta}{h_\beta}\cdot\frac{\partial \boldsymbol{e}_\alpha}{\partial \beta}\right)\right) = 0. \tag{16.243}$$

The normal vectors of the surfaces are given by

$$\nabla S \propto \nabla \nu \qquad = \frac{\boldsymbol{e}_\nu}{h_\nu} \qquad\qquad \text{at}\quad \nu = 0,$$

$$\nabla F \propto \nabla(\nu - \varepsilon H) = \frac{\boldsymbol{e}_\nu}{h_\nu} - \varepsilon\frac{\boldsymbol{e}_\sigma}{h_\sigma}\frac{\partial H}{\partial \sigma} - \varepsilon\frac{\boldsymbol{e}_\tau}{h_\tau}\frac{\partial H}{\partial \tau} \qquad \text{at}\quad \nu = \varepsilon H.$$

As the surface $F = 0$ consists of streamlines, we may use Bernoulli's equation, and we obtain from (16.239c) to (16.239e) that

$$\frac{1}{h_\nu^2}\frac{\partial \Phi}{\partial \nu} = 0 \qquad \text{at}\quad \nu = 0, \tag{16.244a}$$

$$\frac{1}{h_\nu^2}\frac{\partial \Phi}{\partial \nu} - \varepsilon\frac{1}{h_\sigma^2}\frac{\partial H}{\partial \sigma}\frac{\partial \Phi}{\partial \sigma} - \varepsilon\frac{1}{h_\tau^2}\frac{\partial H}{\partial \tau}\frac{\partial \Phi}{\partial \tau} = 0 \qquad \text{at}\quad \nu = \varepsilon H, \tag{16.244b}$$

$$\tfrac{1}{2}\sum_\alpha \frac{1}{h_\alpha^2}\left(\frac{\partial \Phi}{\partial \alpha}\right)^2 + \gamma z = \text{constant} \qquad \text{at}\quad \nu = \varepsilon H. \tag{16.244c}$$

16.10.3 Perturbation Analysis

For small ε we scale the normal coordinate

$$\nu = \varepsilon\eta \tag{16.245}$$

and substitute this in (16.243) and (16.244), resulting in

$$\nabla^2\Phi = \frac{1}{h_\sigma}\frac{\partial}{\partial\sigma}\left(\frac{1}{h_\sigma}\frac{\partial\Phi}{\partial\sigma}\right) + \frac{1}{h_\sigma}\frac{\partial\Phi}{\partial\sigma}\sum_\alpha\left(\frac{\boldsymbol{e}_\alpha}{h_\alpha}\cdot\frac{\partial\boldsymbol{e}_\sigma}{\partial\alpha}\right) + \frac{1}{h_\tau}\frac{\partial}{\partial\tau}\left(\frac{1}{h_\tau}\frac{\partial\Phi}{\partial\tau}\right)$$
$$+\frac{1}{h_\tau}\frac{\partial\Phi}{\partial\tau}\sum_\alpha\left(\frac{\boldsymbol{e}_\alpha}{h_\alpha}\cdot\frac{\partial\boldsymbol{e}_\tau}{\partial\alpha}\right) + \frac{1}{\varepsilon^2}\frac{1}{h_\nu^2}\frac{\partial^2\Phi}{\partial\eta^2} + \frac{1}{\varepsilon}\frac{1}{h_\nu}\frac{\partial}{\partial\nu}\left(\frac{1}{h_\nu}\right)\frac{\partial\Phi}{\partial\eta}$$
$$+\frac{1}{\varepsilon}\frac{1}{h_\nu}\frac{\partial\Phi}{\partial\eta}\sum_\alpha\left(\frac{\boldsymbol{e}_\alpha}{h_\alpha}\cdot\frac{\partial\boldsymbol{e}_\nu}{\partial\alpha}\right) = 0$$

and the boundary conditions

$$\frac{\partial\Phi}{\partial\eta} = 0 \qquad \text{at} \quad \eta = 0,$$
$$\frac{1}{h_\nu^2}\frac{\partial\Phi}{\partial\eta} - \varepsilon^2\left(\frac{1}{h_\sigma^2}\frac{\partial H}{\partial\sigma}\frac{\partial\Phi}{\partial\sigma} + \frac{1}{h_\tau^2}\frac{\partial H}{\partial\tau}\frac{\partial\Phi}{\partial\tau}\right) = 0 \qquad \text{at} \quad \eta = H(\sigma,\tau),$$
$$\frac{1}{2}\left(\frac{1}{h_\sigma^2}\left(\frac{\partial\Phi}{\partial\sigma}\right)^2 + \frac{1}{h_\tau^2}\left(\frac{\partial\Phi}{\partial\tau}\right)^2 + \frac{1}{\varepsilon^2}\frac{1}{h_\nu^2}\left(\frac{\partial\Phi}{\partial\eta}\right)^2\right) + \gamma z = \text{constant} \qquad \text{at} \quad \eta = H(\sigma,\tau).$$

It seems reasonable to assume that our rescaling of ν yields a regular expansion (see Definition 15.19) for Φ, i.e.,

$$\Phi(\sigma,\tau,\eta;\varepsilon) = \Phi_0(\sigma,\tau,\eta) + \varepsilon\Phi_1(\sigma,\tau,\eta) + \varepsilon^2\Phi_2(\sigma,\tau,\eta) + \cdots \tag{16.246}$$

if the surface is sufficiently smooth, such that

$$h_\alpha = \mathcal{O}(1), \qquad \frac{\partial}{\partial\alpha}\left(\frac{\boldsymbol{e}_\alpha}{h_\alpha}\right) = \mathcal{O}(1). \tag{16.247}$$

We expand

$$h_\alpha(\sigma,\tau,\varepsilon\eta) = h_\alpha(\sigma,\tau,0) + \mathcal{O}(\varepsilon), \qquad \boldsymbol{e}_\alpha(\sigma,\tau,\varepsilon\eta) = \boldsymbol{e}_\alpha(\sigma,\tau,0) + \mathcal{O}(\varepsilon). \tag{16.248}$$

Substituting these expansions in the equations and boundary conditions and collecting like powers of ε, we obtain that Φ_0 and Φ_1 are functions independent of η, as follows:

$$\varepsilon^0\!: \quad \frac{\partial^2\Phi_0}{\partial\eta^2} = 0 \quad \text{while} \quad \frac{\partial\Phi_0}{\partial\eta} = 0 \quad \text{at} \quad \eta = 0 \Longrightarrow \Phi_0 = \phi(\sigma,\tau),$$
$$\varepsilon^1\!: \quad \frac{\partial^2\Phi_1}{\partial\eta^2} = 0 \quad \text{while} \quad \frac{\partial\Phi_1}{\partial\eta} = 0 \quad \text{at} \quad \eta = 0 \Longrightarrow \Phi_1 = \phi_1(\sigma,\tau).$$

The condition of constant pressure is, to leading order,

$$\frac{1}{2}\left(\frac{1}{h_\sigma^2}\left(\frac{\partial\phi}{\partial\sigma}\right)^2 + \frac{1}{h_\tau^2}\left(\frac{\partial\phi}{\partial\tau}\right)^2\right) + \gamma z(\sigma,\tau) = \text{constant}. \tag{16.249}$$

To fully determine the zeroth order problem we go on to the $\mathcal{O}(\varepsilon^2)$ term. First we have the equation

$$\frac{1}{h_\nu^2}\frac{\partial^2\Phi_2}{\partial\eta^2} = -\left[\frac{1}{h_\sigma}\frac{\partial}{\partial\sigma}\left(\frac{1}{h_\sigma}\frac{\partial\Phi_0}{\partial\sigma}\right) + \frac{1}{h_\sigma}\frac{\partial\Phi_0}{\partial\sigma}\sum_\alpha\left(\frac{\boldsymbol{e}_\alpha}{h_\alpha}\cdot\frac{\partial\boldsymbol{e}_\sigma}{\partial\alpha}\right)\right.$$
$$\left. + \frac{1}{h_\tau}\frac{\partial}{\partial\tau}\left(\frac{1}{h_\tau}\frac{\partial\Phi_0}{\partial\tau}\right) + \frac{1}{h_\tau}\frac{\partial\Phi_0}{\partial\tau}\sum_\alpha\left(\frac{\boldsymbol{e}_\alpha}{h_\alpha}\cdot\frac{\partial\boldsymbol{e}_\tau}{\partial\alpha}\right)\right], \tag{16.250}$$

which can be integrated immediately because Φ_0, h_α, and $\boldsymbol{e}_\alpha$ (evaluated at $\nu = 0$) are independent of η. With the boundary condition $\frac{\partial}{\partial\eta}\Phi_2 = 0$ at $\eta = 0$ we obtain

$$\frac{1}{h_\nu^2}\frac{\partial\Phi_2}{\partial\eta} = \eta\Big(\text{right-hand side of (16.250)}\Big).$$

We combine this result with the boundary condition for Φ_2 at the free surface $\eta = H$, i.e.,

$$\frac{1}{h_\nu^2}\frac{\partial\Phi_2}{\partial\eta} = \frac{1}{h_\sigma^2}\frac{\partial H}{\partial\sigma}\frac{\partial\Phi_0}{\partial\sigma} + \frac{1}{h_\tau^2}\frac{\partial H}{\partial\tau}\frac{\partial\Phi_0}{\partial\tau} \quad \text{at} \quad \eta = H, \tag{16.251}$$

to eliminate $\frac{\partial}{\partial\eta}\Phi_2$ and arrive at the following equation for H and ϕ:

$$\frac{1}{h_\sigma}\frac{\partial}{\partial\sigma}\left(\frac{H}{h_\sigma}\frac{\partial\phi}{\partial\sigma}\right) + \left(\frac{H}{h_\sigma}\frac{\partial\phi}{\partial\sigma}\right)\sum_\alpha\left(\frac{\boldsymbol{e}_\alpha}{h_\alpha}\cdot\frac{\partial\boldsymbol{e}_\sigma}{\partial\alpha}\right)$$
$$+ \frac{1}{h_\tau}\frac{\partial}{\partial\tau}\left(\frac{H}{h_\tau}\frac{\partial\phi}{\partial\tau}\right) + \left(\frac{H}{h_\tau}\frac{\partial\phi}{\partial\tau}\right)\sum_\alpha\left(\frac{\boldsymbol{e}_\alpha}{h_\alpha}\cdot\frac{\partial\boldsymbol{e}_\tau}{\partial\alpha}\right) = 0. \tag{16.252}$$

Note that this is just the expected

$$\nabla_S\cdot(H\nabla_S\phi) = 0, \tag{16.253}$$

where subscript S means restricted to the surface $S = 0$. Equations (16.249) and (16.253) are very similar to the *eikonal* and *transport* equations of ray theory (Example 15.37, (♯), (♯♯)). Equation (16.249) is of first order hyperbolic type and may be solved by integration along characteristics or (what they denote here) streamlines. We apply Theorem 12.6 to (16.249). Define

$$q_\sigma := \frac{\partial\phi}{\partial\sigma}, \qquad q_\tau := \frac{\partial\phi}{\partial\tau}. \tag{16.254}$$

Consider the curve $(\sigma(t), \tau(t))$ parametrized by t (indeed it will appear to be equivalent to time). We obtain the system of ODEs

$$\frac{\mathrm{d}\sigma}{\mathrm{d}t} = \frac{q_\sigma}{h_\sigma^2}, \tag{16.255a}$$

$$\frac{\mathrm{d}\tau}{\mathrm{d}t} = \frac{q_\tau}{h_\tau^2}, \tag{16.255b}$$

$$\frac{\mathrm{d}\phi}{\mathrm{d}t} = \frac{q_\sigma^2}{h_\sigma^2} + \frac{q_\tau^2}{h_\tau^2}, \tag{16.255c}$$

$$\frac{\mathrm{d}q_\sigma}{\mathrm{d}t} = \frac{1}{h_\sigma^3}\left(\frac{\partial h_\sigma}{\partial \sigma}\right) q_\sigma^2 + \frac{1}{h_\tau^3}\left(\frac{\partial h_\tau}{\partial \sigma}\right) q_\tau^2 - \gamma \frac{\partial z}{\partial \sigma}, \tag{16.255d}$$

$$\frac{\mathrm{d}q_\tau}{\mathrm{d}t} = \frac{1}{h_\sigma^3}\left(\frac{\partial h_\sigma}{\partial \tau}\right) q_\sigma^2 + \frac{1}{h_\tau^3}\left(\frac{\partial h_\tau}{\partial \tau}\right) q_\tau^2 - \gamma \frac{\partial z}{\partial \tau}, \tag{16.255e}$$

with suitable initial (point source) conditions at $t = 0$

$$\sigma(0) = \sigma^0, \quad \tau(0) = \tau^0, \quad \frac{\partial \phi}{\partial \sigma} = \phi_\sigma^0, \quad \frac{\partial \phi}{\partial \tau} = \phi_\tau^0, \quad \phi = \phi^0. \tag{16.255f}$$

Note that the curve is indeed a streamline: the local velocity of a point following the curve

$$\frac{\mathrm{d}\boldsymbol{x}}{\mathrm{d}t} = \frac{\mathrm{d}\sigma}{\mathrm{d}t}\frac{\partial \boldsymbol{x}}{\partial \sigma} + \frac{\mathrm{d}\tau}{\mathrm{d}t}\frac{\partial \boldsymbol{x}}{\partial \tau} = \frac{\mathrm{d}\sigma}{\mathrm{d}t} h_\sigma \boldsymbol{e}_\sigma + \frac{\mathrm{d}\tau}{\mathrm{d}t} h_\tau \boldsymbol{e}_\tau = \frac{\boldsymbol{e}_\sigma}{h_\sigma}\frac{\partial \phi}{\partial \sigma} + \frac{\boldsymbol{e}_\tau}{h_\tau}\frac{\partial \phi}{\partial \tau} = \nabla \phi \tag{16.256}$$

is just the flow velocity.

The height H is found from Gauss's theorem applied to (16.253). Consider a small line segment of length ℓ_1, where $\phi = c_1$ (constant), and connect this via streamlines ($//\nabla\phi$) to another line segment of length ℓ_2, where $\phi = c_2$ (Figure 16.29). We have now (with $\boldsymbol{n}$ denoting the in-plane normal vector to ℓ_1, ℓ_2 and the streamlines, respectively)

$$\begin{aligned} \text{at } \ell_1, \ell_2: &\quad \nabla\phi \cdot \boldsymbol{n} = |\nabla\phi|, \\ \text{along streamlines:} &\quad \nabla\phi \cdot \boldsymbol{n} = 0, \end{aligned}$$

so that H is given by

$$H|\nabla_S \phi|\ell = \text{constant}. \tag{16.257}$$

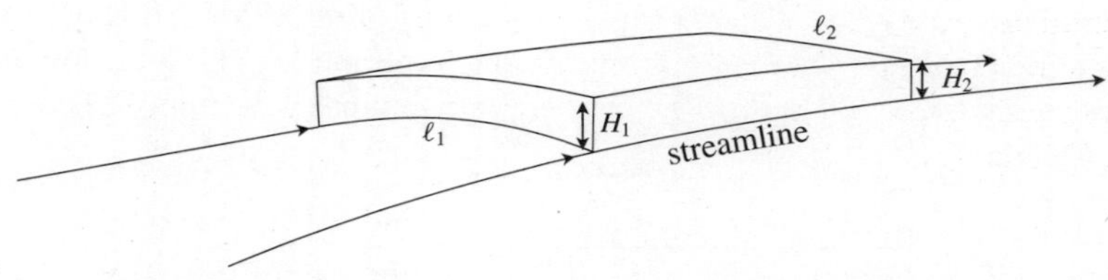

Figure 16.29. *Mass conservation between streamlines.*

In the following we will consider examples of a flow field that issues from a point source, so the streamlines may be parametrized by the azimuthal angle θ and formally given by $\Sigma(\sigma, t, \theta) = 0$ and $T(\tau, t, \theta) = 0$.

As noted before, there is no mechanism in the model that precludes streamlines from crossing. The envelope of crossing streamlines (caustics) is formed by the locus of the point of intersection of two arbitrarily closely spaced streamlines [169]. This implies that Σ is stationary in σ for small changes in t and θ. In other words, $\Sigma(\sigma, t + \mathrm{d}t, \theta + \mathrm{d}\theta) \simeq \mathrm{d}t\, \Sigma_t + \mathrm{d}\theta\, \Sigma_\theta = 0$. The same goes for T. This leads to the condition

$$\Sigma_\theta / \Sigma_t = T_\theta / T_t. \tag{16.258}$$

16.10.4 The Inclined Plane Surface

Probably the simplest example, with a completely analytical solution, is that of a plane surface inclined under an angle κ (Figure 16.30). The local coordinates

$$x = \sigma \cos \kappa - \nu \sin \kappa, \quad y = \tau, \quad z = \sigma \sin \kappa + \nu \cos \kappa \tag{16.259}$$

have unit scale factors, i.e., $h_\sigma = h_\tau = h_\nu = 1$, so we have the equations

$$\frac{\partial}{\partial \sigma}\left(H \frac{\partial \phi}{\partial \sigma}\right) + \frac{\partial}{\partial \tau}\left(H \frac{\partial \phi}{\partial \tau}\right) = 0, \qquad \frac{1}{2}\left(\left(\frac{\partial \phi}{\partial \sigma}\right)^2 + \left(\frac{\partial \phi}{\partial \tau}\right)^2\right) + \gamma \sigma \sin \kappa = \text{constant}, \tag{16.260}$$

which are in characteristic form described by

$$\frac{\mathrm{d}\sigma}{\mathrm{d}t} = q_\sigma, \quad \frac{\mathrm{d}q_\sigma}{\mathrm{d}t} = -\gamma \sin \kappa, \quad \frac{\mathrm{d}\tau}{\mathrm{d}t} = q_\tau, \quad \frac{\mathrm{d}q_\tau}{\mathrm{d}t} = 0. \tag{16.261}$$

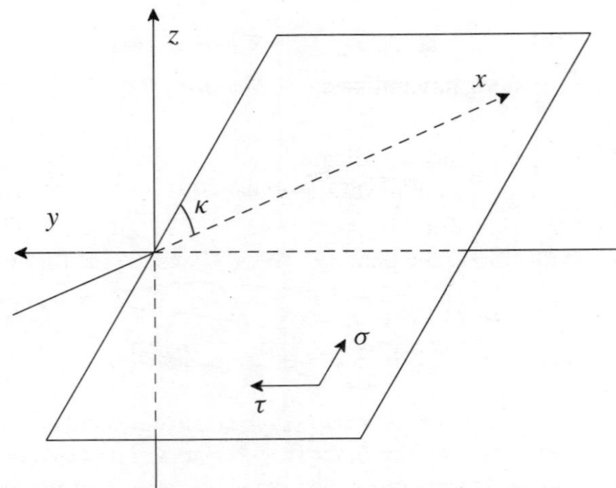

Figure 16.30. *Inclined plane surface.*

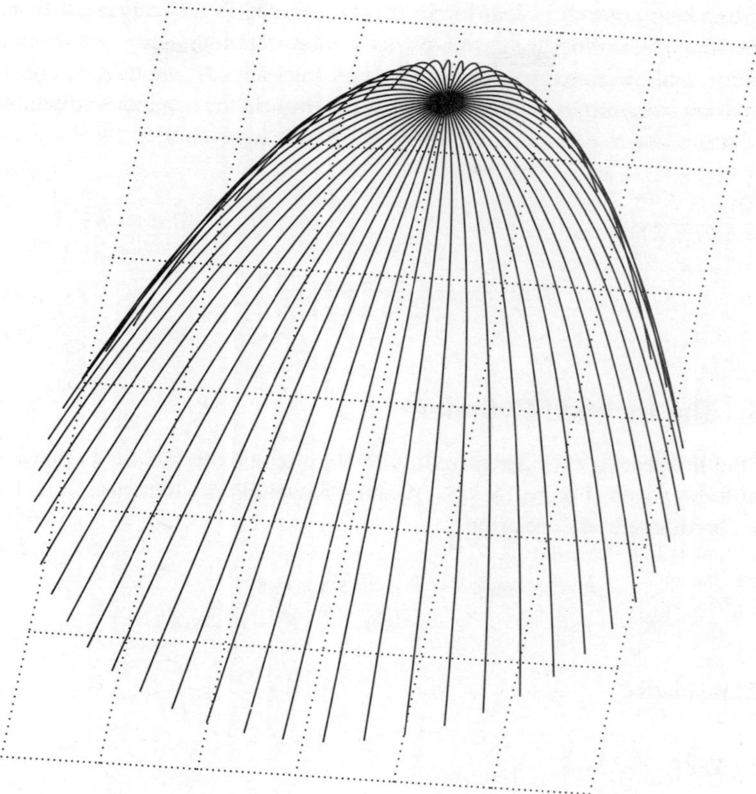

Figure 16.31. *Streamlines along a flat plate.*

They have the solution

$$q_\sigma = \phi_\sigma^0 - \gamma t \sin\kappa, \quad q_\tau = \phi_\tau^0, \quad \sigma = \sigma^0 + \phi_\sigma^0 t - \frac{1}{2}\gamma t^2 \sin\kappa, \quad \tau = \tau^0 + \phi_\tau^0 t. \tag{16.262}$$

The quadratic relation between σ and t indicates that a streamline may cross a given point (σ, τ) twice, making the solution double valued which may not be realistic (Figure 16.31). For a point source at the origin, i.e., $\sigma^0 = \tau^0 = 0$, of strength $\phi_\sigma^0 = Q\sin\theta$ and $\phi_\tau^0 = Q\cos\theta$, this crossing occurs along the parabola-shaped caustic (see Figure 16.31)

$$\sigma = \frac{\gamma \sin\kappa}{2Q^2}\left(\frac{Q^4}{\gamma^2 \sin^2\kappa} - \tau^2\right). \tag{16.263}$$

In general it is safe to say that as soon as two or more streamlines cross, the present theory, based on the assumption that the pressure in the flow is completely governed by the outside atmospheric pressure, is not valid any more, and some interaction of the confluent flows is to be taken into account. It seems possible, however, for interaction to be avoided and the flows to cross each other at two different levels, generating a two-layered flow, in which case the crossing streamlines do have a physical interpretation.

In order to allow an expression for the layer thickness H, we have to consider a flow without crossing streamlines. For example, parallel flow in the negative σ direction, starting at the coordinate line $\sigma = 0$ as $\phi_\sigma^0 = -U_0$ and $\phi_\tau^0 = 0$, is given by

$$\phi = \frac{1}{3\gamma \sin\kappa}(U_0^2 - 2\gamma\sigma \sin\kappa)^{3/2}, \qquad \frac{\partial \phi}{\partial \sigma} = -(U_0^2 - 2\gamma\sigma \sin\kappa)^{1/2},$$
$$H = \frac{H_0 U_0}{(U_0^2 - 2\gamma\sigma \sin\kappa)^{1/2}}. \tag{16.264}$$

16.10.5 The Inclined Cylinder

Consider the inside of a cylinder of radius R, with its axis in the (x, z) plane making an angle κ with the x axis (Figure 16.32). (Assume for simplicity that $\cos\kappa > 0$.)

The coordinate transformation

$$x = \sigma \cos\kappa + (R - \nu)\sin\kappa \cos\tau,$$
$$y = (R - \nu)\sin\tau, \qquad z = \sigma \sin\kappa - (R - \nu)\cos\kappa \cos\tau \tag{16.265}$$

has the Jacobi matrix

$$\begin{pmatrix} \partial x/\partial\sigma & \partial y/\partial\sigma & \partial z/\partial\sigma \\ \partial x/\partial\tau & \partial y/\partial\tau & \partial z/\partial\tau \\ \partial x/\partial\nu & \partial y/\partial\nu & \partial z/\partial\nu \end{pmatrix} = \begin{pmatrix} \cos\kappa & 0 & \sin\kappa \\ -(R-\nu)\sin\kappa \sin\tau & (R-\nu)\cos\tau & (R-\nu)\cos\kappa \sin\tau \\ -\sin\kappa \cos\tau & -\sin\tau & \cos\kappa \cos\tau \end{pmatrix}.$$

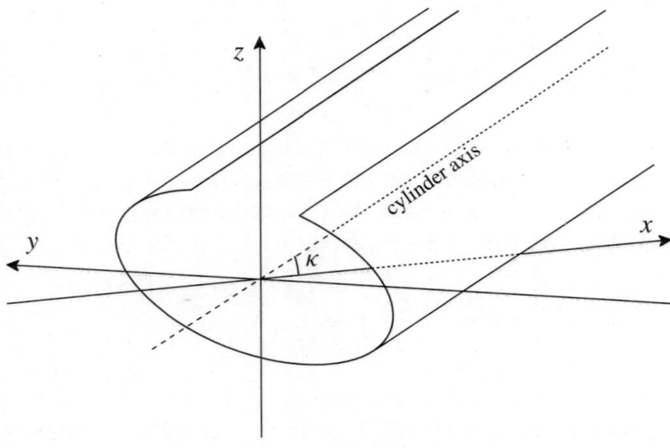

Figure 16.32. *Inclined cylinder.*

This yields

$$h_\sigma = 1, \quad h_\tau = R - \nu, \quad h_\nu = 1,$$
$$\boldsymbol{e}_\sigma = \boldsymbol{e}_x \cos\kappa + \boldsymbol{e}_z \sin\kappa,$$
$$\boldsymbol{e}_\tau = -\boldsymbol{e}_x \sin\kappa \sin\tau + \boldsymbol{e}_y \cos\tau + \boldsymbol{e}_z \cos\kappa \sin\tau,$$
$$\boldsymbol{e}_\nu = -\boldsymbol{e}_x \sin\kappa \cos\tau - \boldsymbol{e}_y \sin\tau + \boldsymbol{e}_z \cos\kappa \cos\tau,$$
$$\sum_\alpha \left(\frac{\boldsymbol{e}_\alpha}{h_\alpha}\cdot\frac{\partial \boldsymbol{e}_\sigma}{\partial\alpha}\right) = 0, \quad \sum_\alpha \left(\frac{\boldsymbol{e}_\alpha}{h_\alpha}\cdot\frac{\partial \boldsymbol{e}_\tau}{\partial\alpha}\right) = 0, \quad \sum_\alpha \left(\frac{\boldsymbol{e}_\alpha}{h_\alpha}\cdot\frac{\partial \boldsymbol{e}_\nu}{\partial\alpha}\right) = \frac{-1}{R-\nu}$$

such that

$$\frac{1}{2}\left(\left(\frac{\partial\phi}{\partial\sigma}\right)^2 + \frac{1}{R^2}\left(\frac{\partial\phi}{\partial\tau}\right)^2\right) + \gamma(\sigma\sin\kappa - R\cos\kappa\cos\tau) = \text{constant}, \qquad (16.266)$$

$$\frac{\partial}{\partial\sigma}\left(H\frac{\partial\phi}{\partial\sigma}\right) + \frac{1}{R^2}\frac{\partial}{\partial\tau}\left(H\frac{\partial\phi}{\partial\tau}\right) = 0. \qquad (16.267)$$

This becomes in characteristic form

$$\frac{\mathrm{d}\sigma}{\mathrm{d}t} = q_\sigma, \quad \frac{\mathrm{d}q_\sigma}{\mathrm{d}t} = -\gamma\sin\kappa, \quad \frac{\mathrm{d}\tau}{\mathrm{d}t} = \frac{q_\tau}{R^2}, \quad \frac{\mathrm{d}q_\tau}{\mathrm{d}t} = -\gamma R\cos\kappa\sin\tau. \qquad (16.268)$$

The σ direction has solution

$$q_\sigma = q_\sigma^0 - \gamma t\sin\kappa, \qquad \sigma = \sigma^0 + q_\sigma^0 t - \frac{1}{2}\gamma t^2 \sin\kappa. \qquad (16.269)$$

The τ direction may be combined into the oscillator equation

$$\frac{\mathrm{d}^2\tau}{\mathrm{d}t^2} + \Omega^2\sin\tau = 0 \qquad \text{with frequency} \quad \Omega = \left(\frac{\gamma}{R}\cos\kappa\right)^{1/2}. \qquad (16.270)$$

This equation has the energy integral

$$\frac{1}{2}\left(\frac{\mathrm{d}\tau}{\mathrm{d}t}\right)^2 - \Omega^2\cos\tau = E_\tau = \frac{1}{2}\left(\frac{q_\tau^0}{R^2}\right)^2 - \Omega^2\cos\tau^0. \qquad (16.271)$$

We have a left- or right-spiraling flow if the total energy is high enough, i.e., $E_\tau > \Omega^2$:

$$\Omega t = \pm\frac{\sqrt{2}}{\sqrt{1+E_\tau/\Omega^2}}\, F\left(\frac{1}{2}\tau, \frac{\sqrt{2}}{\sqrt{1+E_\tau/\Omega^2}}\right) + \text{constant}. \qquad (16.272)$$

We have an oscillatory flow if $-\Omega^2 < E_\tau < \Omega^2$:

$$\Omega t = \pm F\left(\arcsin\sqrt{\frac{2\sin^2(\frac{1}{2}\tau)}{1+E_\tau/\Omega^2}}, \frac{\sqrt{1+E_\tau/\Omega^2}}{\sqrt{2}}\right) + \text{constant}. \qquad (16.273)$$

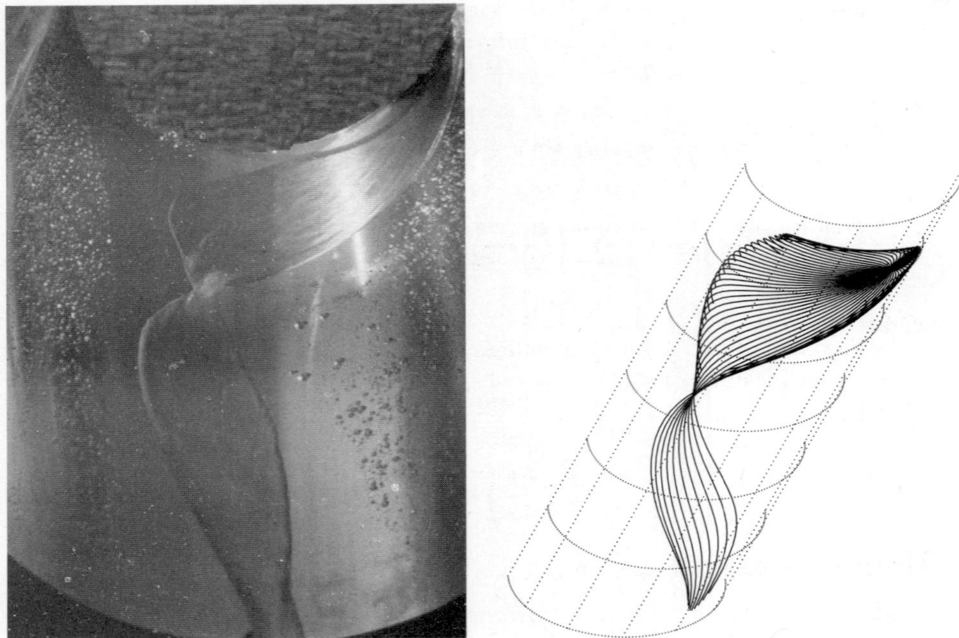

Figure 16.33. *Streamlines along an inclined cylinder issuing from a point source in the wall. Right: analytical solution; left: experiments by one of the authors (SWR). Note the realistic but (strictly speaking) not modeled crossing streamlines!*

$F(\varphi, k)$ denotes the elliptic integral of the first kind [1]. In the case of the oscillatory solution the sign of t changes each time τ reaches its maximum or minimum, and the solution continues on the other branch. This is clearly illustrated by the linearized version of the problem (small τ), for which the explicit solution is given by

$$\tau = \tau^0 \cos \Omega t + \frac{q_\tau^0}{R^2 \Omega} \sin \Omega t. \tag{16.274}$$

Again, the situation of crossing streamlines requires special consideration.

An illustration of this oscillatory motion is given in Figure 16.33 (right), to be compared with experimental results (left) from one of the authors (SWR). The qualitative agreement is remarkable.

16.10.6 Discussion and Related Problems

- If inertial effects due to the geometry of the basin become negligible, like the flow in a canal, the Froude number $U/\sqrt{gL}$ should be based on the water depth rather than L. In the present limit this would result in a high Froude number flow, which is known as "supercritical." This Froude number turns out to be the ratio between flow speed and the speed of the fastest surface waves on the fluid. This means that supercritical flow is faster than the waves: information about downstream disturbances cannot propagate upstream.

- The related problem of thin-layer flow along a horizontal plane is described by the shallow-water equation [150] (see Example 15.25 or Section 12.12.5).
- Wall friction is left out of the present (asymptotic) model. It is possible, however, to include some form of friction (in an ad hoc way) along the streamlines (characteristics).
- The issue of the crossing streamlines is far from resolved. Various mechanisms and competing effects may be proposed.
- Other problems described by similar nonlinear first order hyperbolic equations are ray propagation (Section 16.9) and etching of glass (Section 16.2).

16.10.7 Exercises

16.41. Do a similar analysis as above for the sphere given by

$$x = (R - \nu)\cos\tau\cos\sigma, \quad y = (R - \nu)\sin\tau\cos\sigma, \quad z = (R - \nu)\sin\sigma.$$

16.42. The present formulation is based on orthogonal coordinates along the surface. This is convenient but not necessary. Try to rewrite the present formulas in a slightly more general way with nonorthogonal curvilinear coordinates (σ, τ) along the surface [143].

16.43. Propose wall friction laws that may be included in the model.

16.44. Derive (16.258).

16.45. Derive expression (16.263).

16.46. Derive the linearised result (16.274).

16.47. Rewrite (16.257) by means of a Jacobian in a form similar to (16.234).

16.11 Forming Container Glass

16.11.1 Problem Formulation

For many years, glass technology based on empirical knowledge and the expertise of craftspeople was sufficient to ensure competitiveness. However, over the last 20 years mathematical modeling of various aspects of glass production has become increasingly important. Two thirds of the glass produced is used for packaging in the form of jars and bottles, one quarter of the production is in the form of float glass used for panes, and the rest is for specialist products such as cathode ray tubes (CRTs) and fibers.

Since production costs are dominated by the cost of melting raw material, ever more stringent rules regarding the weight and thickness of container glass are being imposed for its production. As a result, modeling has become an essential tool to meet these demands. In the glass production process, sand grains and additives, such as soda, are heated in a tank, which can be several tens of meters long and a few meters high and wide. Gas burners or electric heaters are used to raise the temperature of the material to about 1200°C. The liquid glass emerges at one end of the tank and is fed to pressing or blowing machines, or onto a bed of liquid tin, where it spreads out to become float glass. In this section we will consider the problem of pressing liquid glass in a mould. Often the forming is done in two stages. After the preform (*parison*) has been pressed, it is blown or further pressed into its final shape (bottle, jar, etc.). At each stage the (pre)form is cooled through the walls of the mould.

16.11.2 Governing Equations

The motion of glass at temperatures above 600°C can be described by the Navier–Stokes equations for incompressible Newtonian fluids [77], which are given by

$$\rho\left(\frac{\partial \boldsymbol{v}}{\partial t}+\boldsymbol{v}\cdot\nabla\boldsymbol{v}\right)=-\nabla p+\mu\nabla^2\boldsymbol{v},\qquad \nabla\cdot\boldsymbol{v}=0, \tag{16.275}$$

where ρ, $\boldsymbol{v}$, and p denote density, velocity, and pressure; see (7.3). This expression is valid for a constant viscosity μ, which is to be verified first. Since the viscosity depends significantly on the temperature [157, 10], we have to make sure that the temperature is constant in the problem. This is not immediately evident. The walls differ in temperature from the glass and may supply or absorb heat by conduction, while the high viscous forces may generate heat by friction. Also, heat transport by radiation may play a role.

To make progress we will exclude any significant effect of radiation by assuming that the walls are highly reflective. Of course, this may vary in other cases [80]. As for the other effects, a suitable model, describing heat conduction according to Fourier's law, heat convection by the incompressible glass flow, and heat generation by viscous friction, is given by

$$\rho C_p\left(\frac{\partial T}{\partial t}+\boldsymbol{v}\cdot\nabla T\right)=\kappa\nabla^2 T+2\mu(\boldsymbol{\mathcal{D}}:\boldsymbol{\mathcal{D}}) \tag{16.276}$$

(see (6.45h) with relations (6.46b) and (6.46c)). The equations are valid in the region between the fixed mould and the axially moving plunger that contains the glass (see Figure 16.34). In view of the geometry of plunger and mould, we choose cylindrical coordinates (r, ϕ, z), while v, w, u will denote the r, ϕ, z components of the velocity $\boldsymbol{v}$. We assume the problem to be axisymmetric, so that both w and all ϕ derivatives vanish and the problem reduces to a two-dimensional one in the (r, z) plane and $\boldsymbol{v} = v\boldsymbol{e}_r + u\boldsymbol{e}_z$.

$r = R_\mathrm{m}(z)$ is the mould surface, $z = z_\mathrm{p}(t)$ is the varying position of the plunger top, $r = R_\mathrm{p}(z - z_\mathrm{p})$ is the plunger surface, and $z = b(t)$ is the position of the glass surface. As the plunger moves down, the decreasing volume between mould and plunger presses the glass upward, so the glass level b increases in time. If we define $R_\mathrm{p}(z) = 0$ if $z \le 0$, the region of definition of our problem is $0 \le z \le b(t)$ and $R_\mathrm{p} \le r \le R_\mathrm{m}$.

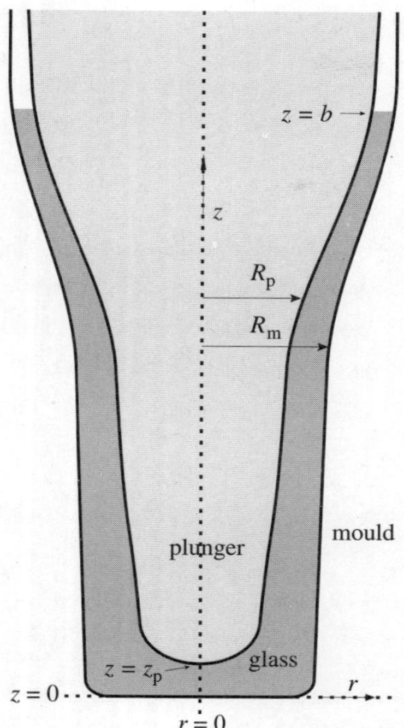

Figure 16.34. *Sketch of configuration.* $R_\mathrm{p} = R_\mathrm{p}(z - z_\mathrm{p})$ *is the plunger surface,* $R_\mathrm{m} = R_\mathrm{m}(z)$ *is the mould surface,* $z_\mathrm{p}(t)$ *is the position of the plunger top, and* $b(t)$ *is the position of the glass surface.*

16.11.3 Slender Geometry Approximation

We will concentrate our analysis on the flow in the narrow annular duct between plunger and mould ($z > z_\mathrm{p}$; see Figure 16.34). This region is very slender and therefore amenable to asymptotic analysis (the method of slow variation; see Section 15.3.1), while at the same time the flow in the region between mould bottom and plunger top is practically stagnant and therefore less important.

We will make (16.275) dimensionless by a suitable scaling. From the geometry of plunger and mould we have two relevant length scales: the characteristic thickness of the parison (D) and the length of the plunger (L), with $D \ll L$. Except right near the plunger top, any variation in z scales on L and any variation in r scales on D. Therefore we scale z with L and r with D and we introduce the small parameter

$$\varepsilon := \frac{D}{L}. \tag{16.277}$$

The axial velocity u scales on a typical velocity V, while from the equation of mass conservation it follows that the radial velocity v has to scale on εV. As density ρ and viscosity μ are constant, they are parameters of the problem. Pressure p is to be scaled on $\mu V L / D^2$

Table 16.4. *Typical values of the parison problem parameters.*

D	typical parison length scale	10^{-2} m
ρ	glass density	2500 kg/m^3
V	typical plunger velocity	10^{-1} m/s
μ_g	glass dynamic viscosity	10^4 kg/(m · s)
L	typical plunger length scale	10^{-1} m
T_g	glass temperature	800°C,
C_p	glass heat capacity	1100 J/(kg · °C)
T_m	mould temperature	500°C,
κ	glass thermal conductivity	1.7 W/(m · °C)

(rather than ρV^2) because the glass flow is highly viscous, as we will see below. We have then

$$\begin{aligned} z &:= Dz^*, \quad Z := \varepsilon z^*, \quad r := Dr^*, \quad u := Vu^*, \\ v &= \varepsilon V v^*, \quad p = \frac{\mu V L}{D^2} p^*, \quad t = \frac{L}{V} t^*. \end{aligned} \tag{16.278}$$

Note that in order to maintain a transparent bookkeeping of the equations, it is advisable not to scale z on L directly, but via a slow variable $Z = \varepsilon z^*$. Now we substitute the above scalings in (16.275) and henceforth ignore the asterisks * to obtain the dimensionless equations

$$\varepsilon Re \left(\frac{\partial u}{\partial t} + u \frac{\partial u}{\partial Z} + v \frac{\partial u}{\partial r} \right) = -\frac{\partial p}{\partial Z} + \varepsilon^2 \frac{\partial^2 u}{\partial Z^2} + \frac{1}{r} \frac{\partial}{\partial r} \left(r \frac{\partial u}{\partial r} \right), \tag{16.279a}$$

$$\varepsilon^3 Re \left(\frac{\partial v}{\partial t} + u \frac{\partial v}{\partial Z} + v \frac{\partial v}{\partial r} \right) = -\frac{\partial p}{\partial r} + \varepsilon^4 \frac{\partial^2 v}{\partial Z^2} + \varepsilon^2 \frac{\partial}{\partial r} \left(\frac{1}{r} \frac{\partial}{\partial r} (rv) \right), \tag{16.279b}$$

$$\frac{\partial u}{\partial Z} + \frac{1}{r} \frac{\partial}{\partial r} (rv) = 0, \tag{16.279c}$$

where $Re = \rho V D / \mu$ is the Reynolds number. Typical values of the parameters (Table 16.4) yield $\varepsilon = 0.1$, $Re = 2.5 \times 10^{-4}$, $\varepsilon Re = 2.5 \times 10^{-5}$, and $\varepsilon^3 Re = 2.5 \times 10^{-7}$, and we can ignore the inertia and radial friction terms to obtain

$$\frac{\partial p}{\partial Z} = \frac{1}{r} \frac{\partial}{\partial r} \left(r \frac{\partial u}{\partial r} \right), \quad \frac{\partial p}{\partial r} = 0, \quad \frac{\partial u}{\partial Z} + \frac{1}{r} \frac{\partial}{\partial r} (rv) = 0, \tag{16.280}$$

with an assumed error of $\mathcal{O}(\varepsilon^2)$ because $Re < \varepsilon$. This set of equations may be recognized as Reynolds's lubrication flow equations in cylindrical coordinates [47, p. 83].

16.11.4 Constant Temperature

To investigate the assumption of a constant viscosity we will analyse the energy equation and estimate the order of magnitude of the various terms. In addition to (16.278) we introduce scaled temperature and viscosity

$$T = T_{\mathrm{m}} + \Delta T\, T^*, \qquad \mu = \mu_{\mathrm{g}}\mu^*, \tag{16.281}$$

where $\Delta T = T_{\mathrm{g}} - T_{\mathrm{m}}$ and μ_{g} denotes the glass viscosity at the bulk temperature. (Note that in the rest of this section we will use a constant viscosity $\mu = \mu_{\mathrm{g}}$.) We find (ignoring the asterisks $*$) that

$$\varepsilon\left[\frac{\partial T}{\partial t} + \boldsymbol{v}\cdot\nabla T\right] = \frac{1}{Pe}\left[\frac{1}{r}\frac{\partial}{\partial r}\left(r\frac{\partial T}{\partial r}\right) + \varepsilon^2\frac{\partial^2 T}{\partial Z^2}\right] + \frac{Ec}{Re}\mu\left[\left(\frac{\partial u}{\partial r} + \varepsilon^2\frac{\partial v}{\partial Z}\right)^2 + 2\varepsilon^2\left(\left(\frac{\partial v}{\partial r}\right)^2 + \left(\frac{v}{r}\right)^2 + \left(\frac{\partial u}{\partial Z}\right)^2\right)\right], \tag{16.282}$$

where the Reynolds number is $Re = \rho V D/\mu_{\mathrm{g}}$, the Eckert number is $Ec = V^2/C_p\Delta T$, and the Péclet number is $Pe = \rho V D C_p/\kappa$. Note that Pe and Re are related through the Prandtl number as $Pe = Re\,Pr$, while $Pe\,Ec/Re = Br$ is called the Brinkman number. When we substitute the values given in Table 16.4, typical of glass at 800°C, we get the corresponding dimensionless numbers

$$\varepsilon = 10^{-1}, \qquad \frac{1}{Pe} = 6.2\times 10^{-4}, \qquad \frac{Ec}{Re} = 1.2\times 10^{-4}. \tag{16.283}$$

Both $1/\varepsilon Pe$ and $Ec/\varepsilon Re$ are very small numbers, so we can ignore heat conduction and thermal production terms, the second and third terms in (16.282), in the bulk of the flow against the convection (first) term. Hence the energy equation simplifies to

$$\frac{\partial T}{\partial t} + \boldsymbol{v}\cdot\nabla T = \frac{\mathrm{d}T}{\mathrm{d}t} = 0, \tag{16.284}$$

indicating that the temperature is preserved following the flow. So if we start with a uniform temperature field, it will remain uniform everywhere, and it follows that the viscosity also remains constant. Only very close to the wall is this not true. Here the temperature varies from the value at the wall to the bulk temperature, and the conduction term is comparable with the convection term. We will consider this thermal boundary layer small enough to be ignored.

16.11.5 Boundary Conditions

At the walls of the plunger and the mould we have to apply boundary conditions. First, we introduce suitably scaled plunger and mould surface functions. In order to make their slowly varying behaviour in the z direction explicit, we write

$$R_{\mathrm{p}}(z) = DR_{\mathrm{p}}^*(z/L) = DR_{\mathrm{p}}^*(\varepsilon z^*) = DR_{\mathrm{p}}^*(Z); \tag{16.285}$$

R_{m} is written likewise. (The asterisks $*$ will be ignored from now on.) Subscript p and m will denote the values at the plunger and mould, respectively. We recall that the moving plunger surface is defined by

$$r = R_{\mathrm{p}}(Z - Z_{\mathrm{p}}(t)),$$

where $Z = Z_p(t) = \varepsilon z_p(t)$ denotes the position of the top of the plunger at time t. (Unless indicated otherwise, we will write here $R_p = R_p(Z - Z_p)$.) The velocity of a point moving with the plunger is then $\boldsymbol{v}_p = \dot{z}_p \boldsymbol{e}_z$.

We now return to the physics of the flowing glass. Depending on problem parameters like wall temperature, fluid pressure, surface tension, or the presence of a lubricant like graphite powder [34, 160, 161, 40], the glass flow will slip completely, partly, or not at all along the wall. This means that the tangential component of the glass velocity $\boldsymbol{v}$ at the wall may differ from the wall velocity $\boldsymbol{v}_w$, where the difference is called the slip velocity. For a complete model this should be included [130], but for now we will simply assume no slip. In other words, $\boldsymbol{v}$ is just equal to the wall velocity and we have

$$\begin{aligned} u &= u_p, & v &= 0 & &\text{at} \quad r = R_p(Z - Z_p), \\ u &= 0, & v &= 0 & &\text{at} \quad r = R_m(Z). \end{aligned} \tag{16.286}$$

As the blob of glass does not initially fill the mould completely (see Figure 16.34), there is a free surface of glass moving into the annular duct between mould and plunger. At the free boundary the normal stress must be equal to the external pressure p_0, which is assumed to be constant. In the present incompressible model it is no restriction to take $p_0 = 0$. In our slender geometry approximation the exact shape of the free surface cannot be determined. In its neighbourhood the flow scales in both the r and Z directions on the thickness D of the annular channel. In other words, the flow is not slowly varying in Z any more, and the present approximation is not valid.

Within the present approximation (and assumption of circular symmetry), we will deal with the average level b of the free surface as follows:

$$p = 0 \quad \text{at} \quad Z = b(t), \tag{16.287}$$

where b (scaled on L) is a function of the time-dependent geometry, implicitly defined in such a way that the (incompressible) glass volume between $Z = 0$ and $Z = b(t)$ is constant for all t.

16.11.6 Solution

We start by deriving an expression for the axial flux. Consider the flux at some level z through cross section S_1 (see Figure 16.35). The value of this flux depends on z, since the plunger goes down and causes the glass to move upward through a varying cross section. We will use this value later to find the pressure gradient.

Since glass is an incompressible fluid, we have with Gauss's divergence theorem that

$$0 = \int_\Omega \nabla \cdot \boldsymbol{v} \, dV = \oint_{\partial\Omega} \boldsymbol{v} \cdot \boldsymbol{n} \, dS = \int_{S_1} \boldsymbol{v} \cdot \boldsymbol{n} \, dS + \int_{S_2} \boldsymbol{v} \cdot \boldsymbol{n} \, dS + \int_{S_3} \boldsymbol{v} \cdot \boldsymbol{n} \, dS, \tag{16.288}$$

where $\boldsymbol{v} = u\boldsymbol{e}_z + \varepsilon v \boldsymbol{e}_r$. Since the mould is stationary and impermeable, we have a zero flux through S_2. For the flux through S_3 we note that it would make no difference for the amount of glass displaced by the plunger if the plunger were filled with glass, because glass is just as incompressible as the solid plunger. Therefore, instead of control surface S_3, we can just

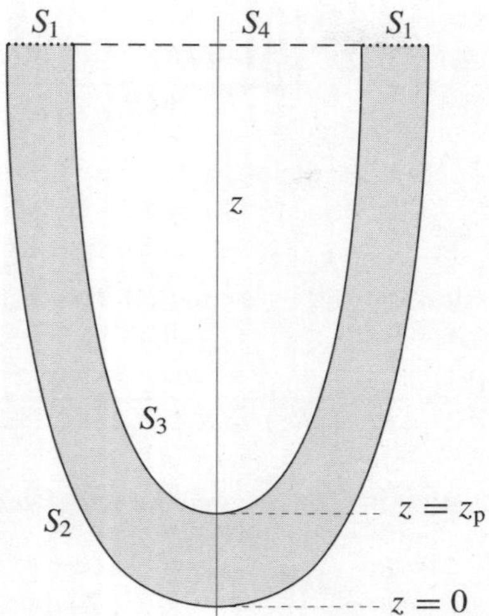

Figure 16.35. *Sketch of the control surfaces to calculate the axial flux.*

as well use surface S_4, which yields more easily the result

$$\int_{S_3} \boldsymbol{v} \cdot \boldsymbol{n} \, \mathrm{d}S = \int_{S_4} \boldsymbol{v} \cdot \boldsymbol{n} \, \mathrm{d}S = \pi u_{\mathrm{p}} R_{\mathrm{p}}^2 \, .$$

It follows that the axial flux, i.e., through S_1, is given by

$$\int_{S_1} \boldsymbol{v} \cdot \boldsymbol{n} \, \mathrm{d}S = 2\pi \int_{R_{\mathrm{p}}}^{R_{\mathrm{m}}} r u(r, z) \, \mathrm{d}r = -\pi u_{\mathrm{p}} R_{\mathrm{p}}^2 \, . \tag{16.289}$$

Now we are ready to solve (16.280) with boundary conditions (16.286). At first, we note that p is a function of z only. Then the equation

$$\frac{\mathrm{d}p}{\mathrm{d}Z} = \frac{1}{r} \frac{\partial}{\partial r} \left(r \frac{\partial u}{\partial r} \right)$$

can be integrated to

$$u = \frac{1}{4} r^2 \frac{\mathrm{d}p}{\mathrm{d}Z} + A(Z) \ln(r) + B(Z).$$

Using boundary conditions (16.286), we obtain

$$u = \frac{1}{4} \frac{\mathrm{d}p}{\mathrm{d}Z} \left[r^2 - \frac{R_{\mathrm{m}}^2 \ln(r/R_{\mathrm{p}}) + R_{\mathrm{p}}^2 \ln(R_{\mathrm{m}}/r)}{\ln(R_{\mathrm{m}}/R_{\mathrm{p}})} \right] + u_{\mathrm{p}} \frac{\ln(R_{\mathrm{m}}/r)}{\ln(R_{\mathrm{m}}/R_{\mathrm{p}})}. \tag{16.290}$$

We have the following integral for ru

$$\int ru(r,Z)\,\mathrm{d}r = \frac{1}{8}r^2\frac{\mathrm{d}p}{\mathrm{d}Z}\left(\frac{1}{2}r^2 - \frac{R_\mathrm{m}^2\ln(r/R_\mathrm{p}) + R_\mathrm{p}^2\ln(R_\mathrm{m}/r)}{\ln(R_\mathrm{m}/R_\mathrm{p})} + \frac{R_\mathrm{m}^2 - R_\mathrm{p}^2}{2\ln(R_\mathrm{m}/R_\mathrm{p})}\right) + \frac{1}{2}r^2u_\mathrm{p}\frac{\ln(R_\mathrm{m}/r)+\frac{1}{2}}{\ln(R_\mathrm{m}/R_\mathrm{p})}. \quad (16.291)$$

With (16.291) and expression (16.289), we can find the pressure gradient

$$\frac{\mathrm{d}p}{\mathrm{d}Z} = \frac{4u_\mathrm{p}}{(R_\mathrm{m}^2 + R_\mathrm{p}^2)\ln(R_\mathrm{m}/R_\mathrm{p}) - (R_\mathrm{m}^2 - R_\mathrm{p}^2)}. \quad (16.292)$$

Finally, the continuity equation in (16.280) may be integrated to yield the expression

$$v = -\frac{1}{r}\frac{\mathrm{d}}{\mathrm{d}Z}\int_r^{R_\mathrm{m}} ru(r,Z)\,\mathrm{d}r, \quad (16.293)$$

which includes, with expression (16.291), an explicit, albeit complicated, expression for v.

16.11.7 Example

An example of the pressure gradient (16.292) is plotted in Figure 16.36 for mould and plunger defined by

$$R_\mathrm{m}(Z) = \sqrt{\frac{2\sqrt{Z} + 2.56Z^4}{1 + 4Z - 6.4Z^2 + 2.56Z^4}}, \qquad R_\mathrm{p}(Z) = 1.2\sqrt{\frac{Z + Z^4}{1 + 4Z - 6.4Z^2 + 2.56Z^4}},$$

while $\varepsilon = 0.2$. The plunger enters the mould from right to left with (dimensionless) velocity $u_\mathrm{p} = -1$. The scaled plunger top position Z_p runs through the values $0.9, \ldots, 0.1$ with steps of -0.1.

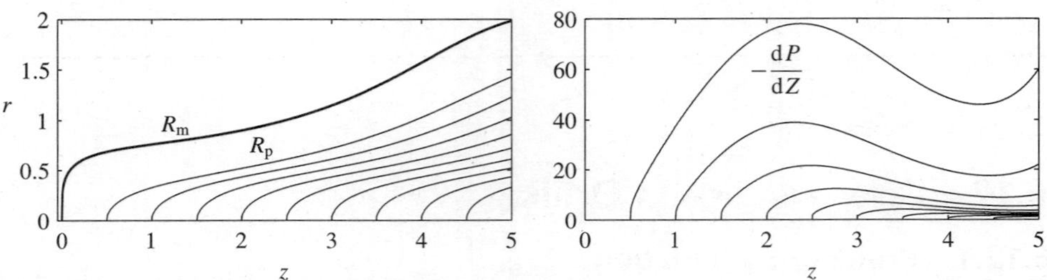

Figure 16.36. *Dimensionless position (left) and minus pressure gradient (right) corresponding to a plunger entering a mould fully filled with fluid glass.*

16.11.8 Discussion and Related Problems

- A major simplification in the present analysis is the adopted no-slip boundary condition. This has, however, no bearing on the approximation applied, and it is indeed possible to include slip conditions (see [130]).
- Details of the flow near the free surface $z = b(t)$ are known to play a role in certain production faults (local cracks or fractures). The present approach is not able to describe these details and a numerical full three-dimensional simulation is typically our next tool of choice [78, 79].
- In practice the plunger is pushed by a piston. This means that we may assume that it is moving by virtue of a constant force or a constant power (force $\times$ velocity). As this force is counteracted by the viscous forces in the glass, we need to couple this motion to the glass flow.
- The cooling of the parison becomes important during the so-called dwell. This is the time period after the pressing has come to a stop, while the mould is still closed. In general the heat exchange is a very complicated matter, as the heat flux also comes from radiation. For some further reading about this see, e.g., [80]

16.11.9 Exercises

16.48. Evaluate expression (16.293).

16.49. Explain why the pressure gradient in Figure 16.36 grows so quickly when the distance between the mould and plunger wall becomes small. (Hint: Write $R_{\mathrm{m}} = R_{\mathrm{p}} + \delta$ and expand (16.292) for small δ.)

16.50. A practically important quantity is the total force $F_{\mathrm{tot}}(t)$ on the plunger (see Section 16.11.8 above). Write this as a single integral, consistent with the present approximation. Show that its time integral

$$\int_{t_0}^{t_1} F_{\mathrm{tot}}(t)\,\mathrm{d}t$$

only depends on the end positions $z_{\mathrm{p}}(t_0)$ and $z_{\mathrm{p}}(t_1)$ and is independent of the way z_{p} actually varies in time.

16.12 Laser Percussion Drilling

16.12.1 Problem Formulation

Lasers are often used to machine materials when conventional methods fail. Laser percussion drilling is used, e.g., to drill cooling holes in fans, that are part of a gas turbine;

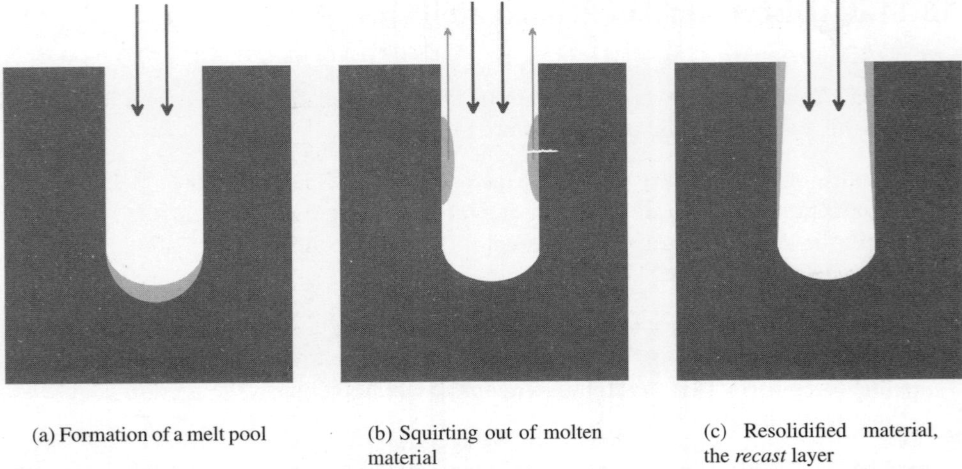

(a) Formation of a melt pool (b) Squirting out of molten material (c) Resolidified material, the *recast* layer

Figure 16.37. *The three events in a laser percussion drilling process where a melting model is needed:* (a) *melting,* (b) *splashing, and* (c) *resolidification.*

such fans are typically made of super alloys. The term "percussion" refers to the repeated operation of the laser in short pulses (10^{-3} s), separated by longer time periods (10^{-2} s). The energy supplied by the laser is bounded, and pulsewise behaviour allows for large bursts of energy. These energy bursts cause local melting and splashing of material and sometimes (unwanted) local recasting. A laser percussion drilling process may in fact be split into three stages. First, a thin region of molten material is formed by absorption of laser energy at the target surface. After some time, the surface of this melt pool reaches vaporisation temperature. The second stage is then the squirting out of the molten material, caused by the sudden expansion of the vapor evaporating from the surface. On its way out some part of this molten material may resolidify at the walls, which is the third stage. For each of these three stages a melting model is needed. These events are depicted in Figure 16.37. In this section we will only consider the first stage, studying the free boundary evolution between the solid and the liquid material.

16.12.2 The Model

The lasers used in practice to drill holes typically produce a Gaussian intensity distribution, which is, ideally, axisymmetric. So let us take an axisymmetric coordinate system, where $z = 0$ denotes the surface of the irradiated material. The density ρ, the specific heat c, and the thermal conductivity κ of the material are known and assumed to be constant. The temperature $T = T(r, z, t)$ in both the solid and the liquid material is governed by the heat equation. In cylindrical coordinates, and assuming axisymmetry, it is given by

$$\rho c \frac{\partial T}{\partial t} = \frac{\kappa}{r} \frac{\partial}{\partial r} \left(r \frac{\partial T}{\partial r} \right) + \kappa \frac{\partial^2 T}{\partial z^2}. \tag{16.294}$$

Table 16.5. *Parameters for the laser drilling process applied to aluminium using an Nd:YAG-laser.*

Parameter	Value	Units	Parameter	Value	Units
ρ	2.7×10^3	kg/m^3	L_f	3.6×10^5	J/kg
c	9.0×10^2	J/(kg · K)	w_0	1.0×10^{-3}	m
κ	2.3×10^2	J/(m · s · K)	I_{ref}	1.5×10^{10}	J/(m^2· s)
T_m	9.3×10^2	K	ε	6.4×10^{-4}	
T_v	2.5×10^3	K	λ_f	0.25	

The intensity distribution of the laser beam is denoted as $I = I(r, t)$. The laser energy is supplied at the surface $z = 0$, yielding the boundary condition

$$\kappa \frac{\partial T}{\partial z}(r, 0, t) = -I(r, t). \tag{16.295}$$

The model has to be completed with conditions at the solid-liquid interface. This will be discussed in the next section.

To make the model dimensionless we introduce some scaling parameters. Let us consider the melting problem for aluminium, which is good enough to test our numerical method. Typical parameters are given in Table 16.5. For the temperature we need the *vaporisation temperature* T_v and the *melting temperature* T_m. For the radial coordinate we use the *waist*, denoted by w_0, of the (Gaussian) laser beam to define a length scale. Furthermore, let I_{ref} be a typical intensity, say its maximum value. From this we can define the following dimensionless variables (indicated by an asterisk $*$):

$$z =: Lz^*, \quad r =: w_0 r^*, \quad t =: \tau t^*, \quad T =: T_m + (T_v - T_m)T^*, \tag{16.296a}$$

where L is a typical length scale in the z direction and τ is the diffusive time scale, defined by

$$L := \kappa \frac{T_v - T_m}{I_{\text{ref}}}, \qquad \tau := \frac{\rho c}{\kappa} L^2, \tag{16.296b}$$

respectively. The length scale L follows from balancing the two terms in the boundary condition (16.295). Applying this scaling to (16.294) and (16.295), we obtain (omitting the asterisk $*$ for ease of notation)

$$\frac{\partial T}{\partial t} = \frac{\varepsilon}{r} \frac{\partial}{\partial r}\left(r \frac{\partial T}{\partial r} \right) + \frac{\partial^2 T}{\partial z^2}, \tag{16.297}$$

$$\frac{\partial T}{\partial z}(r, 0, t) = -J, \tag{16.298}$$

where the parameter ε and the function $J = J(r, t)$ are defined by

$$\varepsilon := \left(\frac{L}{w_0} \right)^2 = \left(\frac{k(T_v - T_m)}{w_0 I_{\text{ref}}} \right)^2, \qquad J := \frac{I}{I_{\text{ref}}}, \tag{16.299}$$

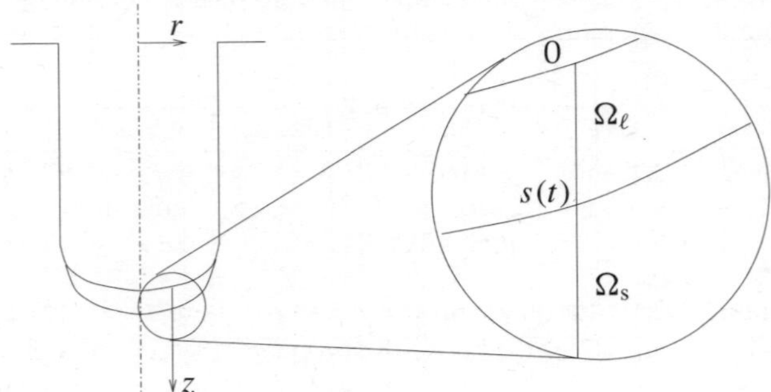

Figure 16.38. *The geometry of the laser-induced melting problem.*

respectively. For a typical laser percussion drilling process, the parameter ε is much smaller than one; see Table 16.5. Thus radial diffusion can be neglected and the initial stage of the laser percussion drilling process can in fact be modeled as a one-dimensional problem.

16.12.3 The Stefan Problem

Clearly, the problem at hand is a *Stefan problem*; see Section 10.5. Let $s(t)$ denote the position of the solid-liquid interface, as sketched in Figure 16.38. Let Ω_ℓ denote the liquid region, $0 \le z < s(t)$, and Ω_s the solid region, $s(t) < z < \infty$. The temperature in the liquid and solid regions is governed by the (dimensionless) one-dimensional heat equation

$$\frac{\partial T_p}{\partial t} = \frac{\partial^2 T_p}{\partial z^2}, \qquad z \in \Omega_p \quad (p = \ell, s). \tag{16.300}$$

Here the subscripts ℓ and s refer to the liquid and solid regions, respectively. At the boundary $z = 0$ where the laser hits the surface we have the boundary condition

$$\frac{\partial T_\ell}{\partial z}(0, t) = -J. \tag{16.301a}$$

The extra heat to melt the material, in addition to the amount needed to raise the temperature, is called the *latent heat of fusion* L_f. In dimensionless form it is defined as

$$\lambda_f := \frac{L_f}{c(T_v - T_m)}. \tag{16.301b}$$

A typical value of λ_f for aluminium can be found in Table 16.5. At the liquid-solid interface we need an additional condition that expresses the absorption of heat needed for phase change. So the *Stefan condition* at $z = s(t)$ is given by (cf. (11.151g))

$$\frac{\partial T_s}{\partial z}(s(t)+, t) - \frac{\partial T_\ell}{\partial z}(s(t)-, t) = \lambda_f \frac{ds}{dt}. \tag{16.301c}$$

Moreover, since the temperature is assumed to be continuous across the interface, we have

$$T_s(s(t)+, t) = T_\ell(s(t)-, t) = 0. \tag{16.301d}$$

We assume the domain to be large enough so that the solid material far away from the interface has the same temperature as its surrounding, T_{amb} say. Hence for some "cutoff" point z_{b} we have

$$T_{\mathrm{s}}(z_{\mathrm{b}}, t) = T_{\mathrm{amb}}. \tag{16.301e}$$

To complete our model we prescribe a known initial temperature distribution

$$T(z, 0) = T_0(z). \tag{16.301f}$$

To obtain a numerical solution we proceed as in Section 11.10. However, contrary to the approach in that section, we now choose the number of grid points on each of the two regions fixed, say, on the domains $0 \leq z \leq s(t)$ and $s(t) \leq z \leq z_{\mathrm{b}}$, so we have N_ℓ and N_{s} equispaced points, respectively. So in the liquid domain we define

$$z_{\ell,j} := j\Delta z_\ell, \qquad j = 0, 1, \ldots, N_\ell \quad \text{with} \quad \Delta z_\ell = \Delta z_\ell(t) := \frac{s(t)}{N_\ell}, \tag{16.302a}$$

and likewise for the solid we have

$$z_{\mathrm{s},j} := s(t) + j\Delta z_{\mathrm{s}}, \qquad j = 0, 1, \ldots, N_{\mathrm{s}} \quad \text{with} \quad \Delta z_{\mathrm{s}} = \Delta z_{\mathrm{s}}(t) = \frac{z_{\mathrm{b}} - s(t)}{N_{\mathrm{s}}}. \tag{16.302b}$$

Applying the implicit Euler scheme for time integration and central differences for space discretisation, we have

$$T^{n+1}_{\mathrm{p},j} - \frac{\Delta t}{\Delta z^2_{\mathrm{p}}}\left(T^{n+1}_{\mathrm{p},j+1} - 2T^{n+1}_{\mathrm{p},j} + T^{n+1}_{\mathrm{p},j-1}\right) = T^n_{\mathrm{p},j}, \qquad \mathrm{p} = \ell, \mathrm{s}. \tag{16.303}$$

We provide these equations with the boundary conditions

$$T^n_{\ell,N_\ell} = T^n_{\mathrm{s},0} = 0, \qquad T^n_{\mathrm{s},N_{\mathrm{s}}} = T_{\mathrm{amb}},$$

and a discretised form of (16.301a); see Section 11.10. So we find an approximate temperature at time t^{n+1}. Next we have to update the interface position and recompute the grids in both domains. Let s^n be the numerical approximation of $s(t^n)$. Then, using the explicit Euler scheme and one-sided differences for the Stefan condition in (16.301c) and applying the boundary conditions above, we obtain

$$s^{n+1} = s^n + \frac{\Delta t}{\lambda_{\mathrm{f}}}\left(\frac{1}{\Delta z^n_\ell}T^n_{\ell,N_\ell-1} + \frac{1}{\Delta z^n_{\mathrm{s}}}T^n_{\mathrm{s},1}\right), \tag{16.304}$$

where $\Delta z^n_{\mathrm{p}} = \Delta z_{\mathrm{p}}(t^n)$ ($\mathrm{p} = \ell, \mathrm{s}$). One should note that the time step restriction $\Delta t = \mathcal{O}(\Delta z^n_{\mathrm{p}})$ ($\mathrm{p} = \ell, \mathrm{s}$) implies that the grid size at the next time level t^{n+1} is only $\mathcal{O}((\Delta z^n_{\mathrm{p}})^2)$ away from the grid size at t^n; see (16.302). Hence the approximations at the new grids are still $\mathcal{O}((\Delta z^n_{\mathrm{p}})^2)$ correct and thus the free boundary is found with the same accuracy.

In a practical implementation we cannot have Δt being constant (unless it is unduly small) because of the stability restriction for explicit Euler. Hence we should use the implicit Euler scheme.

16.12.4 Finding Suitable Initial Conditions

We still have to provide suitable initial conditions. In order to find these, let us look at the premelting problem. In the heating-up stage, the temperature T in the solid material satisfies the heat equation (16.300), the boundary condition $T(z,t) \to T_{\text{amb}}$ for $z \to \infty$, and

$$\frac{\partial T}{\partial z}(0,t) = -J, \tag{16.305}$$

where J is the (dimensionless) energy I/I_{ref} supplied at the surface. Moreover, for the initial condition we take $T(z,0) = T_{\text{amb}}$. We can solve this initial boundary value problem using the Laplace transform [70] and find

$$T(z,t) = J\left(2\sqrt{\frac{t}{\pi}}\exp\left(-\frac{z^2}{4t}\right) - z\,\text{erfc}\left(\frac{z}{2\sqrt{t}}\right)\right) + T_{\text{amb}}. \tag{16.306}$$

Therefore, when latent heat is neglected, the position s of the liquid-solid interface is found from

$$J\left(2\sqrt{\frac{t}{\pi}}\exp\left(-\frac{s^2}{4t}\right) - s\,\text{erfc}\left(\frac{s}{2\sqrt{t}}\right)\right) + T_{\text{amb}} = 0 \tag{16.307}$$

for all t. In Figure 16.39 this position is sketched for aluminium. This yields the correct initial value of $\frac{ds}{dt}$. From this we may compute $s(\Delta t)$ by an explicit Euler step. Furthermore, substituting $s = 0$ in (16.306), we conclude that the surface starts to melt at time t_{m} given by

$$t_{\text{m}} = \frac{1}{4}\pi\left(T_{\text{amb}}/J\right)^2. \tag{16.308}$$

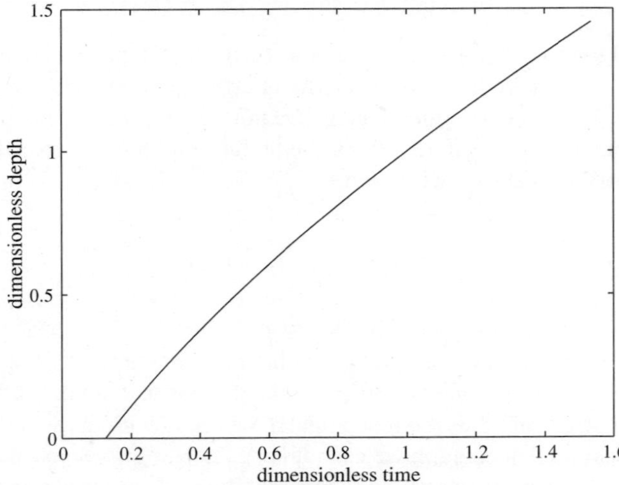

Figure 16.39. *The position of the solid-liquid interface for the case of aluminium. The absorption of latent heat is neglected.*

In order to compute the initial condition for the temperature distribution, we once more employ the Stefan condition (16.301c), i.e.,

$$\frac{\partial T_{\mathrm{s}}}{\partial z}(s(t)+, t) = \lambda_{\mathrm{f}} \frac{\mathrm{d}s}{\mathrm{d}t} + \frac{\partial T_{\ell}}{\partial z}(s(t)-, t) \approx \lambda_{\mathrm{f}} \frac{\mathrm{d}s}{\mathrm{d}t} - J =: -J_* \tag{16.309}$$

for s small. Using (16.309) as boundary condition, we obtain, similarly to (16.306), an expression for the temperature. In particular, we find

$$T^1_{\mathrm{s},j} = T_{\mathrm{amb}} \exp\left(-\frac{(z_{\mathrm{s},j} - s)^2 J_*^2}{T_{\mathrm{amb}}^2 \pi}\right) - J_*(z_{\mathrm{s},j} - s)\,\mathrm{erfc}\left(\frac{(z_{\mathrm{s},j} - s) J_*}{T_{\mathrm{amb}}\sqrt{\pi}}\right) + T_{\mathrm{amb}}, \quad j = 1, 2, \ldots, N_{\mathrm{s}} - 1, \tag{16.310}$$

as the initial temperature in the solid at time level is $t^1 = \Delta t$. Because $s(\Delta t)$ is small, we take the temperature distribution in the liquid at $t^1 = \Delta t$ to be linear; i.e.,

$$T^1_{\ell,j} = J\big(s(t^1) - z^1_{\ell,j}\big), \qquad j = 0, 1, \ldots, N_{\mathrm{liq}} - 1. \tag{16.311}$$

Clearly this satisfies the boundary conditions at $z = 0$ and $z = s(t)$. Note that this approach cannot easily be extended to cover problems with no constant energy supply.

16.12.5 The Enthalpy Formulation

The *enthalpy* H is defined as the sum of the sensible and the latent heat in a substance. If a material is liquid, it contains latent heat of fusion per unit mass L_{f} in addition to the sensible heat (per unit mass) cT. Figure 16.40 shows the relation between the temperature and the enthalpy for two different types of materials. Figure 16.40(a) shows this relation for pure, i.e., nonalloy, substances with a single melting-point temperature, whereas Figure 16.40(b) illustrates this relation for a material where the phase change takes place over an extended temperature range from the solid temperature T_{sol} to the liquid temperature T_{liq}. The region

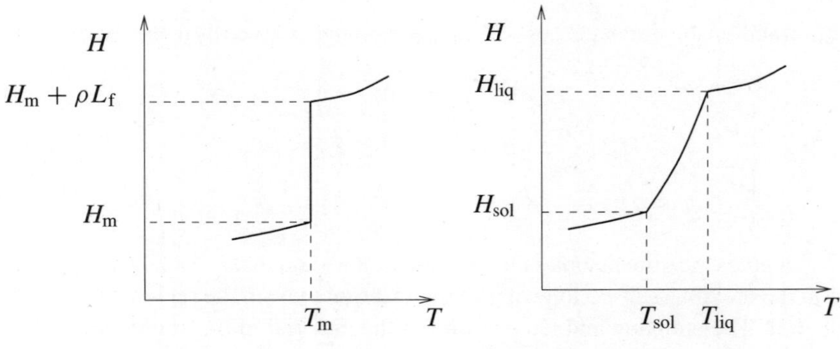

(a) Pure crystalline substances (b) Glassy substances and alloys

Figure 16.40. *Relation between enthalpy and temperature for pure crystalline substances and for glassy substances and alloys.*

with temperature between the solid and the liquid temperatures is referred to as the *mushy region*. In order to nondimensionalise the enthalpy, we introduce two typical parameters. The first is the *enthalpy at vaporisation temperature* H_v, which is given by

$$H_v := \rho c T_v + \rho L_f. \tag{16.312}$$

For materials with a discrete melting point the second number is the *enthalpy at melting temperature* H_m, given by

$$H_m := \rho c T_m, \tag{16.313a}$$

whereas for glassy substances and alloys this second number is the *enthalpy at solid temperature*

$$H_{sol} := \rho c T_{sol}. \tag{16.313b}$$

Hence the dimensionless enthalpy H^* is defined for pure materials by

$$H =: H_m + (H_v - H_m)H^* \tag{16.314a}$$

and for materials with a melting range by

$$H =: H_{sol} + (H_v - H_{sol})H^*. \tag{16.314b}$$

For later use we introduce the constants D_1 and D_2:

$$D_1 := \frac{\rho c(T_v - T_m)}{H_v - H_m}, \qquad D_2 := \frac{\rho c(T_v - T_{sol})}{H_v - H_{sol}}. \tag{16.315}$$

We will drop the asterisk $*$ on the dimensionless enthalpy from now on.

The relationship between the dimensionless enthalpy and the dimensionless temperature for pure materials is now given by (see Figure 16.40(a)),

$$H(T) = \begin{cases} D_1 T & \text{if } T < 0, \\ [0, D_1\lambda_f] & \text{if } T = 0, \\ D_1 T + D_1\lambda_f & \text{if } T > 0. \end{cases} \tag{16.316a}$$

Likewise, this relationship for materials with a melting range is given by (see Figure 16.40(b))

$$H(T) = \begin{cases} D_2 T & \text{if } T \leq 0, \\ D_2 T + D_2\lambda_f \dfrac{T}{T_{liq}} & \text{if } 0 \leq T \leq T_{liq}, \\ D_2 T + D_2\lambda_f & \text{if } T_{liq} \leq T. \end{cases} \tag{16.316b}$$

Note that T_{liq} denotes the dimensionless liquid temperature and that $\lambda_f = L_f/(c(T_v - T_{sol}))$.

Like in the previous subsection we take the material to be in the region $z \geq 0$ with the surface at $z = 0$. The enthalpy and temperature of the material in this region are governed by the single energy equation in enthalpy form, which, in dimensionless form, is given by

$$\frac{\partial H}{\partial t} = D_1 \frac{\partial^2 T}{\partial z^2}, \qquad z > 0, \tag{16.317}$$

for materials with a discrete melting point. For materials with a melting range D_2 takes the role of D_1.

Our problem to solve is (16.317), with the enthalpy given in (16.316a) or (16.316b), plus the boundary values derived before. Let us discretise (16.317) with the implicit Euler method on a fixed grid with $\Delta z := z_b/N$ for some N; i.e.,

$$H_j^{n+1} - \frac{\Delta t}{\Delta z^2}\left(T_{j+1}^{n+1} - 2T_j^{n+1} + T_{j-1}^{n+1}\right) = H_j^n. \tag{16.318}$$

From (16.316a) or (16.316b) we can express the temperature as a function of the enthalpy. Thus we have a relation of the form

$$T_j^n := T(H_j^n). \tag{16.319}$$

We use the boundary conditions (16.301a) and (16.301e) to obtain H_N^n as well as H_0^n. Writing the approximate vector of enthalpies at time t^n as $\mathbf{v}^n$ and of temperatures as $\mathbf{u}^n$, we have a system of equations like

$$\mathbf{v}^{n+1} - \mathbf{A}\mathbf{u}^{n+1} = \mathbf{v}^n.$$

Again using (16.319) we can rewrite this as the nonlinear system

$$\mathbf{v}^{n+1} - \mathbf{F}(\mathbf{v}^{n+1}) - \mathbf{v}^n = \mathbf{0}. \tag{16.320}$$

This can be solved by Newton's method. If no phase change occurs, the numerical method derived from the enthalpy problem is satisfactory. For phase change problems it is correct on average, since heat balances are fulfilled throughout. However, the calculated positions of the solid-liquid interface, the temperature, and the enthalpy oscillate with a period corresponding to the time the interface needs to travel through a certain grid interval. This phenomenon can be seen in Figure 16.41. The plateau generated propagates to adjacent intervals and smoothes out only after the solid-liquid interface has travelled a sufficient distance from the point under consideration. These problems can be overcome by special approaches.

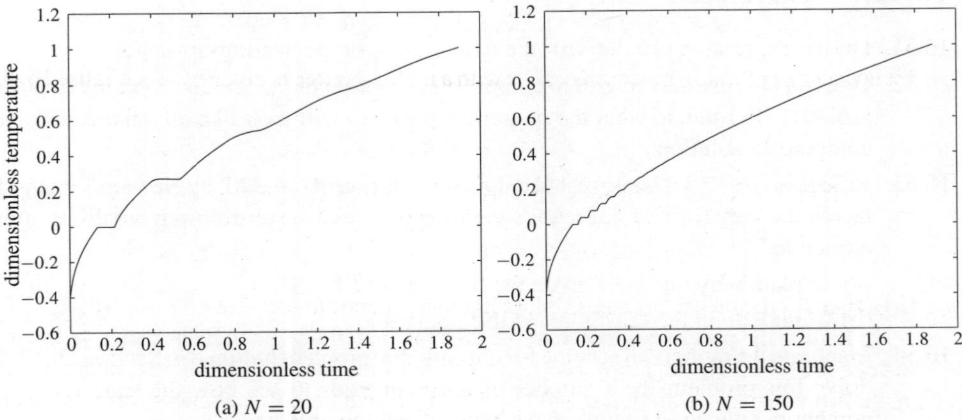

Figure 16.41. *Time history plots for temperature at the interface in the enthalpy problem.*

16.12.6 Discussion and Related Problems

- The radiation coming from the laser can have a variety of appearances. The form of the bundle is responsible for the shape of the melt pool. Two typical forms are a top hat and a Gaussian. If the melt pool, or even worse, the drilled hole, is not flat, there will be reflections by the walls that may result in secondary strikes. The radiation bundle is most likely divergent; i.e., not all rays are parallel; this may also cause secondary strikes.
- Most materials do not have a single melting point but a *mushy* region. (For an alloy this can be understood from the fact that the various constituting components have a different melting point.) This requires a special approach. In fact, one needs to resort to the enthalpy formulation entirely.
- Splashing takes place if enough vapor has been formed. Note that this is in fact a three-phase problem. In order to predict the moment of splashing, one needs to calculate the pressure buildup by vaporization.
- One of the problems (in particular for deeper holes) is that the material will resolidify along the walls on its way up. This is a very interesting problem to model. Indeed, there are two free boundaries involved: one being the top of the moving fluid and the other being the top of the recast layer. In fact, shocks may occur.
- Free-boundary problems occur in a large variety of problems. A large class can be found in books on porous media. Here one often encounters two kinds of fluids, like water and oil, or salty and fresh water. Then convection is usually the dominant phenomenon.
- Another important class of Stefan problems is where we have two different chemical substances rather than phase changes. Typical examples are etching processes, where a chemical (like an acid) reacts with a material.

16.12.7 Exercises

16.51. Derive expression (16.306) for the temperature in the heating-up stage.

16.52. One can also use a fixed grid to compute the solution of the Stefan problem (16.300), (16.301). Formulate what the numerical problem will look like on a fixed grid and compute its solution.

16.53. In Section 16.12.5 it is suggested to find the solution of (16.320) by Newton's method. Given the very form of it, Picard's iteration (successive substitution) would be quite attractive.

(a) Explain why this is so given the form of $H(T)$.

(b) Comment on the feasibility of this approach.

16.54. Elaborate the numerical scheme for the enthalpy problem outlined in Section 16.12.5. Solve this problem for a number of different grids to see how the staircase phenomenon, noticed in Figure 16.41, depends on the grid size.

16.13 Determining Chemical Composition by Electrophoresis

16.13.1 Introduction

The purpose of analytical chemistry is to separate, quantify, and identify chemical components in complicated mixtures. Important areas of application include the monitoring of environmental pollution and the quality control of products ranging from food and beverages to pharmaceuticals and virtually all materials made by modern industry. For diagnosis of diseases, and biomedical research in general, analytical instrumentation is indispensable. A wide range of instrumentation is available, the choice of which is among other things determined by the properties of the chemical components to be determined. For volatile components gas chromatography is the most important and widespread separation technique, the separation making use of differences in evaporation upon heating of the mixture. Many components in biochemistry, such as proteins and nucleic acids (DNA, RNA), however, are sensitive to heat treatment, but they are not volatile. On the other hand, these molecules, as well as many low molecular weight components, carry one or more charges when dissolved in water. The differences in their charge, size, and shape make it possible to separate them in an electric field using a family of techniques called *electrophoresis* [102]. This separation technique was originally developed by Arne Tiselius, who received the Nobel Prize for Chemistry for this in 1948. It can be carried out in glass tubes, in capillaries, in gels, and on paper impregnated with an aqueous buffer solution, the pH of which can be used to selectively alter the effective charge of different components and thus enhance separation.

The component property relevant for separation is the *mobility* μ $[\mathrm{m}^2/(\mathrm{V}\cdot\mathrm{s})]$, which is defined as the ratio between the linear velocity v [m/s] of the molecule in the solution and the electric field E [V/m], i.e.,

$$v = \mu E. \tag{16.321}$$

Depending on the chemical composition of the buffer solution(s) in the separation system, there are a few variations of the technique. One is called *isotachophoresis* [39]. Here one uses a leading electrolyte, with a higher mobility than any of the ions in the sample, as well as a terminating electrolyte, with a lower mobility than any of those ions. The sample is placed in between. A buffering counterion is present in the whole system. In the first stage of isotachophoresis the various components will migrate with different velocities and thus separate, and eventually zones with only one component (in addition to the common counterion) will appear. The length (or volume) of the zones is proportional to their respective amounts. This is called a steady state as all components will travel with the same speed (hence its name: $\iota\sigma o$ means *equal* and $\tau\alpha\chi o\varsigma$ means *speed*). In this steady state the electric field increases, therefore going from the leading electrolyte to the terminating one, and is apparently inversely proportional to the mobility of the respective component. Strictly speaking, isotachophoresis is not a separation technique, but rather the end result of one. However, the intermediate stages leading to the steady state are very interesting for gaining valuable insight into the separation mechanism, not only of isotachophoresis but of electrophoresis in general.

16.13.2 Mathematical Model of Capillary Electrophoresis

The most important issue in electrophoresis is to determine the concentration of various components in a mixture. In this section we consider a simple one-dimensional model of capillary electrophoresis involving n components. The concentrations $c_1, c_2, \ldots, c_n$ [mol/m^3] of these components depend on the spatial coordinate x along the capillary and the time t and satisfy the following set of advection-diffusion equations:

$$\frac{\partial c_i}{\partial t} + \frac{\partial}{\partial x}(v_i\, c_i) = D_i \frac{\partial^2 c_i}{\partial x^2} \qquad (i = 1, 2, \ldots, n). \tag{16.322}$$

In (16.322) v_i is the advection velocity of the ith component and D_i [m^2/s] is the diffusion coefficient, assumed to be constant for all components. As noted before, the velocities v_i are related to the mobility and electric field. In particular, we have the relation (cf. (16.321))

$$v_i = (\mu_e + \mu_i)E, \tag{16.323}$$

with μ_e the (constant) electro-osmotic mobility, μ_i the effective mobility of the ith component, and E the electric field. For this problem we can take $\mu_e = 0$. The equations in (16.322) are coupled through the advection terms, since the mobilities μ_i and the electric field E depend in a complicated way on the concentrations c_i, as specified below.

Let component i have n_i subspecies; the acidity is responsible for the degree to which they are dissociated. Let us denote the concentration of (ionised) hydrogen by c_H. Then the degree of dissociation $\alpha_{i,j}$ [−] of subspecies j in component i is given by

$$\alpha_{i,j} = \frac{\prod_{p=1}^{j} \frac{K_{i,p}}{c_H}}{1 + \sum_{q=1}^{j} \prod_{r=1}^{q} \frac{K_{i,r}}{c_H}} \qquad (j = 1, 2, \ldots, n_i - 1). \tag{16.324a}$$

Here $K_{i,j}$ [mol/m^3] is the acidic constant for the latter subspecies. The dissociation degrees then give the mobilities from the relation

$$\mu_i = \sum_{j=1}^{n_i} \alpha_{i,j}(\mu_{i,j}^0 - \mu_{i,j-1}^0), \tag{16.324b}$$

where $\mu_{i,j}^0$ is the mobility of the subspecies at infinite dilution and full subspecies dissociation. These mobilities and the concentrations are needed to compute the conductivity κ [S/m] from the relation

$$\kappa = F\left(\mu_H c_H - \frac{\mu_{OH} K_w}{c_H} + \sum_{i=1}^{n} c_i \sum_{j=1}^{n_i} \alpha_{i,j} Z_{i,j} \mu_{i,j}^0\right), \tag{16.325a}$$

where F is Faraday's constant; μ_H and μ_{OH} are the mobilities of hydrogen and hydroxide ions, respectively; K_w is the equilibrium constant of water; and $Z_{i,j}$ [−] is the charge number of subspecies j in component i. These constants have the values $F = 9.648456 \times 10^4$ C/mol, $\mu_H = 3.5 \times 10^{-7}$ m^2/(V · s), $\mu_{OH} = 2 \times 10^{-7}$m^2/(V · s), and $K_w = 10^{-8}\,\text{mol}^2/\text{m}^6$. Given a (constant) current density J [A/m^2], the electric field follows from the modified Ohm's law; i.e.,

$$E = \frac{J}{\kappa}. \tag{16.325b}$$

The expressions in (16.325) determine E as a function of the concentrations c_i and c_{H}.

Finally, to compute c_{H}, and thus pH, we use Gauss's law from electromagnetism, describing conservation of charge. Assuming electroneutrality, it reduces in our application to

$$0 = c_{\mathrm{H}} - \frac{K_w}{c_{\mathrm{H}}} + \sum_{i=1}^{n}\sum_{j=1}^{n_i} Z_{i,j}\alpha_{i,j}c_i. \tag{16.326}$$

This relation specifies c_{H} as a function of the concentrations c_i.

16.13.3 Numerical Solution Method

In this section we outline a numerical solution method for the electrophoresis problem from Section 16.13.2. We focus on the advection-diffusion equation (16.322). In order to choose the proper discretisation scheme for (16.322), we note the following. First, the diffusion term should be taken implicitly, since an explicit scheme would lead to a severe time step restriction; e.g., the explicit Euler scheme in combination with central difference space discretisation would require an unduly small time step Δt given by $\Delta t \leq \frac{1}{2} D \Delta x^2$, with Δx the grid size; cf. Section 13.3.1. Second, since all equations are coupled through the advection term, we have to take it explicitly. Finally, a scheme discretising the advection term by central differences is prone to spurious oscillations when the Péclet number Pe, defined by

$$Pe := \max_{i,j} |v_{i,j}| \Delta x / D, \tag{16.327}$$

is large; cf. Section 13.3.1. In (16.327) $v_{i,j}$ denotes the numerical approximation of v_i at grid point x_j. Based on these observations we choose a mixed time integration scheme (explicit advection and implicit diffusion) (see Section 11.8.2), a central difference approximation for the diffusion term, and upwinding for the advection term. If $v_i > 0$, we have

$$c_{i,j}^{n+1} - \frac{D\Delta t}{\Delta x^2}\left(c_{i,j+1}^{n+1} - 2c_{i,j}^{n+1} + c_{i,j-1}^{n+1}\right) = c_{i,j}^{n} - \frac{\Delta t}{\Delta x}\left(v_{i,j}^{n} c_{i,j}^{n} - v_{i,j-1}^{n} c_{i,j-1}^{n}\right), \tag{16.328}$$

where $c_{i,j}^n$ denotes the numerical approximation of $c_i(x_j, t^n)$, etc. A similar scheme holds for $v_i < 0$. The stability condition for this scheme reads

$$\frac{\Delta t}{\Delta x} \max_{i,j} |v_{i,j}^n| \leq 1, \tag{16.329}$$

where the maximum should be taken over all components i and all grid points x_j. Furthermore, (16.328) leads to a symmetric positive definite tridiagonal system, which can be solved very efficiently using LU-decomposition; see Section 11.3.1.

Once we have computed the concentrations at the new time level $t^{n+1} = (n+1)\Delta t$, we have to update the velocities as follows. First, we compute c_H from (16.326). Next we compute the mobilities μ_i from (16.324) and the electric field E from (16.324a) and (16.325). Substituting these in (16.323), we find the advection velocities v_i.

16.13.4 A Practical Example

The Swedish Department of Occupational Safety and Health in Stockholm has been monitoring brief exposure to styrene and xylene of factory workers in a construction plant that manufactures plastic recreational boats, where organic solvents were in common use [144]. This monitoring was carried out by simply analyzing the urine collected during 24 h. In the human body styrene and xylene are metabolized to mandelic acid and methyl-hippuric acid, respectively. These metabolytes are subsequently excreted in the urine. Their concentration is therefore a good quantitative indication of styrene/xylene exposure.

The samples were analysed using isotachophoresis. An example of experimental conditions was as follows:

- The leading electrolyte was $10\,\text{mol/m}^3$ hydrochloric acid, buffered to $\text{pH} = 3.10$ using $12.5\,\text{mol/m}^3$ β-alanine as counterion.
- The terminator electrolyte was $10\,\text{mol/m}^3$ propionic acid, buffered to $\text{pH} = 4$ with $5\,\text{mol/m}^3$ β-alanine as counterion.

We have computed numerical solutions of a 30 cm capillary with internal diameter of 75 μm and with a constant applied current of 10 μA. The detection system was a DC conductivity detector recording field strength. A mixture of mandelic acid and methyl-hippuric acid was injected at the interface between leading and terminating electrolyte, and the current was switched on. A solution of $5\,\text{mol/m}^3$ of each of the two components was injected over a length of 1 mm, as a perfectly cylindrical plug. Component data of this experiment are given in Table 16.6, where $pK_{i,j} := -\,{}^{10}\log(K_{i,j})$. Furthermore, the charge number is $Z_{i,j} = 0$ if $\mu_{i,j} = 0$; otherwise we have $Z_{i,j} = 1$.

One can see from the tabulated data that separation at $\text{pH} > 8$ would be easiest. The methyl-hippuric acid molecule would carry two full charges, whereas mandelic acid would carry a single one. Consequently, methyl-hippuric acid would be twice as fast as mandelic acid (49.7 vs. 24.5). But in practice this does not work, because there is too much

Table 16.6. *Component data in the simulation. The mobilities are in* $10^{-9}\text{m}^2/(\text{V}\cdot\text{s})$ *and the diffusion coefficients are in* $10^{-9}\text{m}^2/\text{s}$.

	component	$\mu^0_{i,0}$	$pK_{i,1}$	$\mu^0_{i,1}$	$pK_{i,2}$	$\mu^0_{i,2}$	D_i
1	hydrochloric acid	0	−2	−79.0			2.03
2	propionic acid	0	4.78	−36.9			0.95
3	β-alanine	36.7	3.55	0	10.24	−30.8	0.94
4	methyl-hippuric acid	0	2.71	−25.0	7.25	−49.7	0.64
5	mandelic acid	0	3.37	−24.5			0.64

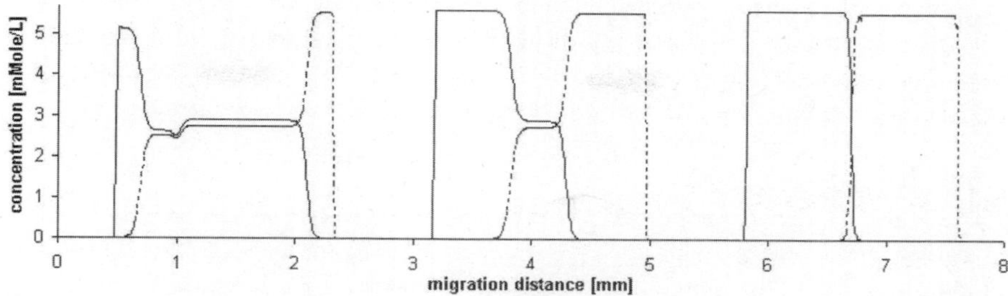

Figure 16.42. *Time development of concentration of methyl-hippuric and mandelic acid after (from left to right)* 1, 3, *and* 5 s *of migration.*

interference from the other components present. For the sake of illustration we choose a separation at pH = 3.10 because it takes longer for the separation to complete, and the intermediate stages can be more easily seen.

16.13.5 Numerical Simulations

The time development of the locations in the capillary of the two sample components is illustrated in Figure 16.42. In the leftmost part one can observe that the components only partially leave the injection compartment (boundary at 1 mm) and that adjustment of concentration is taking place. A short, pure zone of methyl-hippuric acid can already be seen, migrating in front (the fastest component), followed by a mixed zone. In the middle part, after 3 s, the isotachophoretic train is fully adjusted (it has left the injection compartment). The steep front and rear boundaries move at equal velocities, whereas the length of the mixed zone is slowly decreasing with time. On the right-hand side it can be seen that there is already a steady state after 5 s; the lengths of both zones, along the horizontal distance axis, will remain the same and are directly proportional to the amount originally injected.

The rectangular concentration profiles of both components are clearly seen. Also, the front and rear boundaries are much steeper than the boundary in between the components. This is easily explained by plotting the simulated electric field; see Figure 16.43. We see that the larger the electric field gradient at the boundary, the steeper the corresponding concentration gradient, because the self-correcting properties of the field are much stronger there. One can say that the electric field gradient is really the driving force that counteracts the diffusion of components into each other's zone. One should also notice the stepwise increase of the local pH going from the leading to the terminating zone.

The method turns out to be very robust. In this particular case, with five different components (two sample components, leading, terminating, and counterion), the *Pe* number during the separation ranges from 5 to 70 initially, and from 5 (in the leading) to 50 (in the terminating) as soon as the steady state is reached, without oscillations or mass-balance violations.

In contrast to most other chemical separation methods, the antidiffusive power of electromigration in electrophoresis is striking. This is illustrated in the next example, where in the separation system listed in Table 16.6, a steady state 1 mm zone of 5 mol/m^3

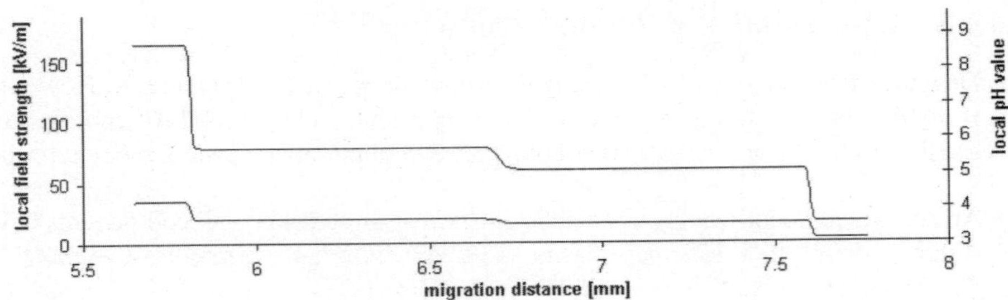

Figure 16.43. *Distribution of local pH (lower trace) and local electric field (upper trace) once the steady state is reached.*

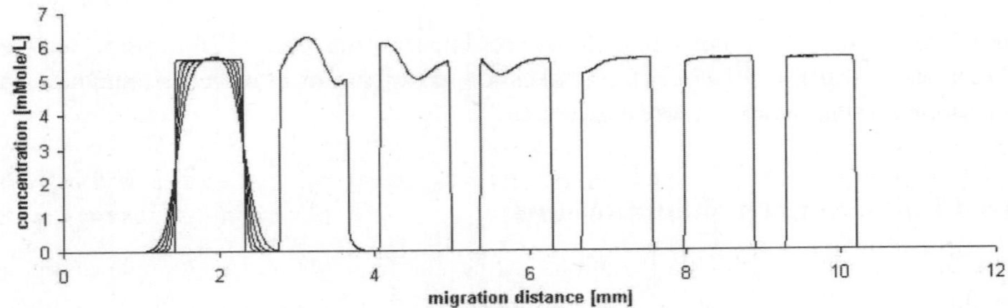

Figure 16.44. *Migration of mandelic acid from steady state with* 2, 5, 10, *and* 20 s *of diffusion (overlapping profiles on the left) and after restoration of the driving current in six* 200 ms *intervals.*

mandelic acid is depicted in Figure 16.44, a perfect rectangular concentration profile located at 2 mm. Next the driving current is reduced from 10 μA to 0.01 μA. As a result, the zone virtually comes to a standstill and the diffusion will have full play. In the figure the subsequent concentration profiles after 2, 5, 10, and 20 s are shown. The outward diffusion flux at the zone boundaries leads to a rapid smoothing. In the middle of the zone the concentration seems to increase beyond the original value. Being physically impossible, this must be a numerical artifact, probably due to the sudden decrease in driving current, and thus the electric field. When the original current of 10 μA is restored, several phenomena are observed, as illustrated by the sequence depicted at intervals of 200 ms in Figure 16.44. The zone is again moving to the right. The steep concentration profiles on both sides of the mandelic acid zone are quickly restored within 100 ms. A pronounced sinusoidal disturbance occurs, originating from the front (right) of the zone, which slowly dies out to the left. The original concentration profile of the steady state is restored after 1.2 s in the profile on the far right, at 10 mm.

These examples show that numerical simulation is a powerful tool to study isotachophoresis. In particular, it enables one to investigate the intermediate stages, which cannot be measured.

16.13.6 Discussion and Related Problems

- Deviation from local electroneutrality is often negligible in electrophoresis. However, if we do not assume electroneutrality, the right-hand side of (16.326) contains the gradient of the electric field, thus coupling the computation of $c_{\rm H}$ and E; see exercises.

- An additional, locally different, electric field is generated since all components have different diffusion coefficients. This additional field has a strength of at most 10 percent of the external electric field.

- In some applications the temperature in the capillary has a radial gradient due to heat loss at the wall, thus introducing two-dimensional effects. In this case the two-dimensional cylindrical symmetric advection-diffusion equation holds. Also, the heat equation should be included.

- To discretise the advection term we could use higher-order schemes, such as, e.g., slope limiter methods; see Chapter 13.

- Equation (16.322), with v_i given by (16.323), is also referred to as a *drift-diffusion equation*. These occur often in physics and electrical engineering. An example is the drift-diffusion model describing semiconductor devices [136].

16.13.7 Exercises

16.55. Give the discretisation scheme of (16.322) when the upwind discretisation of the advection equation is replaced by the slope limiter method from Section 13.8.2.

16.56. Consider the advection-diffusion equation

$$\frac{\partial u}{\partial t} + b\frac{\partial u}{\partial x} = D\frac{\partial^2 u}{\partial x^2},$$

where the advection velocity b is assumed constant.

(a) Give the mixed explicit advection/implicit diffusion time integration scheme. Use central differences for the space derivatives.

(b) Investigate the stability of this scheme.

(c) Investigate when spurious oscillations occur.

16.57. Suppose electroneutrality does not hold. Then (16.326) should be replaced by

$$\frac{\epsilon}{F}\frac{\partial E}{\partial x} = c_{\rm H} - \frac{K_{\rm w}}{c_{\rm H}} + \sum_{i=1}^{n}\sum_{j=1}^{n_i} Z_{i,j}\alpha_{i,j}c_i,$$

with ϵ the dielectric constant. Modify the numerical solution procedure in Section 16.13.3.

16.14 Pulse Tube Refrigerators

16.14.1 Problem Formulation

Many apparatuses produce so much heat that they need cooling to function properly. The market for stand-alone small cooling devices is stiii increasing, in particular for communication equipment. There exist all sorts of cooling machinery. A relatively new small type of so-called cryocooler is the pulse tube, based on a principle originally designed by Stirling. This device typically reaches temperatures of about 120 K, but temperatures below 2 K have been reached as well. The essential elements of a pulse tube refrigerator are shown in Figure 16.45. The pulse tube employs a cyclic compression and expansion of a gas, say helium. Due to heat exchange between gas, regenerator, tube walls, and the two heat exchangers, a temperature difference arises along the tube. The pressure oscillations in the system are generated by a piston compressor or by switching valves. The aftercooler removes the heat of compression so that the regenerator can work more efficiently. The regenerator acts as a buffer: it absorbs heat from the gas during the compression part of the pressure cycle and it returns heat to the gas during the expansion part. The cold heat exchanger (CHX) is the coldest point of the system. Here the heat is extracted from the load to be cooled. In the tube the compressible gas oscillates. If there is a suitable phase relationship between the pressure and the gas flow, heat will be transported from the cold end to the warm end. The hot heat exchanger (HHX) removes the heat carried through the tube. It is maintained at ambient temperature. In the orifice design (see Figure 16.45b) the basic pulse tube is modified by adding a reservoir and an orifice. The reservoir is large compared to the pulse tube volume. Gas flows through the orifice due to a pressure difference. More gas now contributes to the cooling power, which improves the efficiency of the cooler.

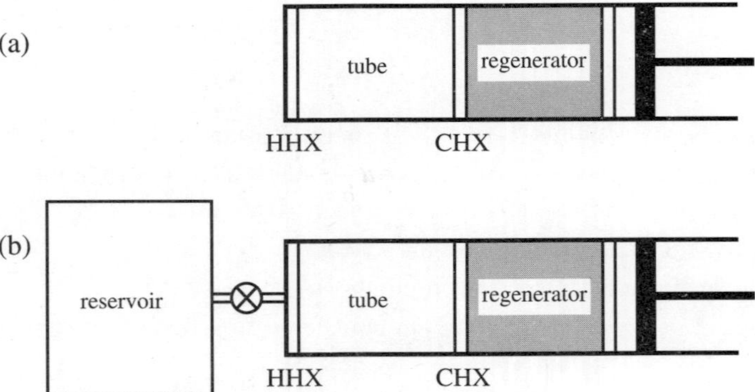

Figure 16.45. *A schematic picture of the Stirling-type pulse tube refrigerator:* (a) *the basic pulse tube and* (b) *the orifice pulse tube. From right to left the basic system consists of a piston, a regenerator, a cold heat exchanger (CHX), a tube, and a hot heat exchanger (HHX). In* (b) *the system is extended with an orifice and a reservoir (buffer).*

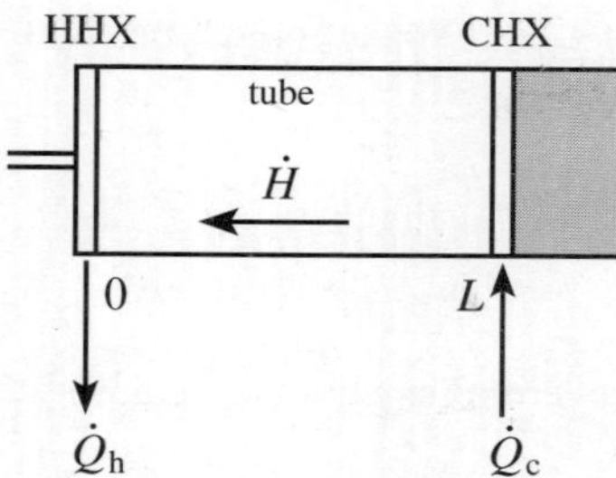

Figure 16.46. *Net heat and enthalpy flows in the tube.*

16.14.2 The Model

We will model the tube section of the pulse tube refrigerator to study the energy transfer from the cold to the hot end; see, e.g., [100, 165, 166, 66, 5, 6, 141]. Figure 16.46 shows the energy flow in the pulse tube. At the HHX heat $\dot{Q}_{\text{h}}$ is extracted from the system. At the CHX heat $\dot{Q}_{\text{c}}$ is put into the system. There is a net enthalpy flow from the CHX to the HHX. According to the first law of thermodynamics, the average cooling power is equal to the enthalpy flow in the pulse tube. To simplify the analysis we will consider a one-dimensional model. Let L be the length of the tube section of the cooler. Then the domain is the interval $0 < x < L$. We assume that the fluid is Newtonian, the gas is ideal, and the flow is laminar. Let us furthermore denote the density by $\rho(x,t)$, the velocity by $u(x,t)$, the temperature by $T(x,t)$, the pressure by $p(x,t)$, the dynamic viscosity by μ, the specific heat of the gas at constant pressure by C_p, the thermal conductivity of the gas by κ_{g}, and the specific ideal gas constant by $\mathcal{R}$. The equations for mass, momentum, and energy conservation and the equation of state are then (cf. Chapter 6)

$$\frac{\partial \rho}{\partial t} + \frac{\partial}{\partial x}(\rho u) = 0, \tag{16.330a}$$

$$\rho\left(\frac{\partial u}{\partial t} + u\frac{\partial u}{\partial x}\right) = -\frac{\partial p}{\partial x} + \frac{4}{3}\frac{\partial}{\partial x}\left(\mu\frac{\partial u}{\partial x}\right), \tag{16.330b}$$

$$\rho C_p\left(\frac{\partial T}{\partial t} + u\frac{\partial T}{\partial x}\right) = \frac{\partial p}{\partial t} + u\frac{\partial p}{\partial x} + \frac{\partial}{\partial x}\left(\kappa_{\text{g}}\frac{\partial T}{\partial x}\right), \tag{16.330c}$$

$$p = \rho \mathcal{R} T. \tag{16.330d}$$

16.14.3 Nondimensionalisation

The system of equations (16.330) will first be made dimensionless. The scaling parameters are chosen as follows: The shortest physical time scale of importance is $1/\omega$, where ω is the angular frequency of the piston movement. We introduce U to be a representative value for the velocity and therefore U/ω to be a typical length scale. Let P be the amplitude of the pressure oscillations, p_0 the average pressure, T_{amb} an ambient temperature, ρ_{ref} a typical density, μ_{ref} a typical viscosity, and κ_{ref} a typical thermal conductivity of the gas. We take the pressure variation caused by the piston or valves to be $p_{\text{p}}(t) = p_0 + Pp^*(t)$, where

$p^*(t)$ is a known function of time. We introduce the following dimensionless variables (indicated by an asterisk $*$):

$$
\begin{aligned}
&\rho := \rho_{\text{ref}}\rho^*, \quad T := T_{\text{amb}}T^*, \quad p := p_0 + Pp^*, \quad u := Uu^*, \\
&x := (U/\omega)x^*, \quad t := t^*/\omega, \quad \mu := \mu_{\text{ref}}\mu^*, \quad \kappa_{\text{g}} := \kappa_{\text{ref}}\kappa_{\text{g}}^*.
\end{aligned}
\tag{16.331}
$$

Using this, we find that the governing equations (16.330) become (leaving out the asterisk notation for simplicity's sake)

$$\frac{\partial \rho}{\partial t} + \frac{\partial}{\partial x}(\rho u) = 0, \tag{16.332a}$$

$$\rho\left(\frac{\partial u}{\partial t} + u\frac{\partial u}{\partial x}\right) = -\frac{1}{Ma^2}\frac{\partial p}{\partial x} + \frac{4}{3Re}\frac{\partial}{\partial x}\left(\mu\frac{\partial u}{\partial x}\right), \tag{16.332b}$$

$$\rho\left(\frac{\partial T}{\partial t} + u\frac{\partial T}{\partial x}\right) = \frac{Ec}{Ma^2}\left(\frac{\partial p}{\partial t} + u\frac{\partial p}{\partial x}\right) + \frac{1}{Pe}\frac{\partial}{\partial x}\left(\kappa_{\text{g}}\frac{\partial T}{\partial x}\right), \tag{16.332c}$$

$$(\mathcal{A} + p)\mathcal{B} = \rho T. \tag{16.332d}$$

Let us, more specifically, consider a pulse tube operating at 20 Hz, with a typical velocity of 1.5 m/s. The average distance that gas particles travel in the tube (displacement length) is 0.012 m. The length of the tube is 0.2 m. From this we can easily derive the following dimensionless numbers with their typical values:

$$
\begin{aligned}
Ma &:= \frac{U}{\sqrt{P/\rho_{\text{ref}}}} \sim 4.6 \times 10^{-3}, & Re &:= \frac{\rho_{\text{ref}}U^2}{\mu_{\text{ref}}\omega} \sim 4 \times 10^3, \\
Ec &:= \frac{U^2}{c_{\text{p}}T_{\text{amb}}} \sim 4.5 \times 10^{-6}, & Pe &:= \frac{\rho_{\text{ref}}U^2C_p}{\omega\kappa_{\text{ref}}} \sim 2.6 \times 10^3, \\
\mathcal{A} &:= \frac{p_0}{P} \sim 6, & \mathcal{B} &:= \frac{P}{\rho_{\text{ref}}\mathcal{R}T_{\text{amb}}} \sim 0.17.
\end{aligned}
$$

The Mach number Ma characterizes the compressibility of the flow, the Reynolds number Re indicates how strong the inertia is with respect to the viscous effects, the Péclet number Pe determines the ratio of advection of heat to conduction, and, finally, the Eckert number Ec gives the kinetic energy against thermal energy.

16.14.4 Derivation of Velocity and Temperature Equations

The momentum equation (16.332b) can be rewritten as

$$\frac{\partial p}{\partial x} = \frac{4Ma^2}{3Re}\frac{\partial}{\partial x}\left(\mu\frac{\partial u}{\partial x}\right) - Ma^2\rho\left(\frac{\partial u}{\partial t} + u\frac{\partial u}{\partial x}\right). \tag{16.333}$$

The first and second terms on the right-hand side represent viscous and inertial forces. Dimensional analysis reveals that the constants Ma^2/Re and Ma^2 are of the order of 10^{-9} and 10^{-5}, respectively. This means that the inertial and viscous forces are too small to produce a significant pressure gradient; so the right-hand side of (16.333) is approximately zero. Therefore the pressure is uniform in space and the momentum equation can be neglected. The system (16.332) now simplifies to

$$\frac{\partial \rho}{\partial t} + \frac{\partial}{\partial x}(\rho u) = 0, \tag{16.334a}$$

$$\frac{\partial p}{\partial x} = 0, \tag{16.334b}$$

$$\rho\left(\frac{\partial T}{\partial t} + u\frac{\partial T}{\partial x}\right) = \frac{Ec}{Ma^2}\frac{\mathrm{d}p}{\mathrm{d}t} + \frac{1}{Pe}\frac{\partial}{\partial x}\left(\kappa_{\mathrm{g}}\frac{\partial T}{\partial x}\right), \tag{16.334c}$$

$$(\mathcal{A} + p)\mathcal{B} = \rho T. \tag{16.334d}$$

Adding (16.334a) and (16.334c) and subsequently substituting (16.334d), we obtain the following equation for the velocity:

$$(\mathcal{A} + p)\mathcal{B}\frac{\partial u}{\partial x} = \left(\frac{Ec}{Ma^2} - \mathcal{B}\right)\frac{\mathrm{d}p}{\mathrm{d}t} + \frac{1}{Pe}\frac{\partial}{\partial x}\left(\kappa_{\mathrm{g}}\frac{\partial T}{\partial x}\right). \tag{16.335}$$

In order to obtain the required equation for the temperature, we eliminate the density in (16.334c) using (16.334d). Finally, assuming that the gas thermal conductivity κ_{g} is constant, we have the equations for the velocity and temperature:

$$\frac{\partial u}{\partial x} = \epsilon(t)\frac{\partial^2 T}{\partial x^2} + s_1(t), \tag{16.336a}$$

$$\frac{\partial T}{\partial t} + u\frac{\partial T}{\partial x} = \epsilon(t)T\frac{\partial^2 T}{\partial x^2} + s_2(t)T, \tag{16.336b}$$

where $s_1(t)$, $s_2(t)$, and $\epsilon(t)$ are defined by

$$s_1(t) := \frac{-1}{\gamma}\frac{1}{\mathcal{A} + p(t)}\frac{\mathrm{d}p}{\mathrm{d}t}, \tag{16.337a}$$

$$s_2(t) := \frac{\gamma - 1}{\gamma}\frac{1}{\mathcal{A} + p(t)}\frac{\mathrm{d}p}{\mathrm{d}t}, \tag{16.337b}$$

$$\epsilon(t) := \frac{1}{(\mathcal{A} + p(t))\mathcal{B}}\frac{1}{Pe} \ll 1, \tag{16.337c}$$

with $\gamma = c_p/c_v$ the specific heat ratio. The temperature equation (16.336b) is a nonlinear advection-diffusion equation where the variable velocity $u(x,t)$ represents the presence of advection and the diffusion coefficient $\epsilon(t)T$ represents the presence of diffusion. Note that the diffusion coefficient is small, as follows from our dimensional analysis. The temperature equation is then mostly of an advective nature and is therefore close to hyperbolic.

To complete the system of equations we introduce boundary and initial conditions. We only need one boundary condition for (16.336a), i.e., the velocity at the hot end $u_{\mathrm{h}}(t)$. To find the velocity at the hot end we consider the volume flow through the orifice. From [166] the volume flow in a linear approximation is given by

$$\dot{V}_{\mathrm{h}} = -C_{\mathrm{or}}(p - p_{\mathrm{b}}), \tag{16.338}$$

where p is the *tube pressure*, p_b is the *buffer pressure*, and C_{or} is the *flow conductance* of the orifice. The velocity is then given by

$$u(0,t) = u_h(t) = -\frac{C_{or}}{A}\big(p(t) - p_b(t)\big), \tag{16.339}$$

where A is the cross-sectional area of the tube. Special care has to be taken in determining the buffer pressure such that the net mass flow over the hot end is zero. We nondimensionalise (16.339) by taking $p = p_0 + Pp^*$ and $p_b = p_0 + Pp_b^*$, just like in Section 16.14.3. In dimensionless form (asterisks $*$ are omitted) the boundary condition for the velocity is

$$u_h(t) = -\mathcal{C}\big(p(t) - p_b(t)\big), \qquad \mathcal{C} := \frac{C_{or}P}{AU}. \tag{16.340}$$

Furthermore, for the temperature boundary conditions we take

$$\begin{cases} T(0,t) = T_h & \text{if } u(0,t) \geq 0, \\ \dfrac{\partial T}{\partial x}(0,t) = 0 & \text{if } u(0,t) < 0, \end{cases} \qquad \begin{cases} \dfrac{\partial T}{\partial x}(L,t) = 0 & \text{if } u(L,t) > 0, \\ T(L,t) = T_c & \text{if } u(L,t) \leq 0. \end{cases} \tag{16.341}$$

We only have to specify an initial condition for the temperature, which we assume to be linear:

$$T(x,0) = T_h - (T_h - T_c)\frac{x}{L}. \tag{16.342}$$

16.14.5 Numerical Solution

To solve the problem numerically we have to discretise the governing equations (16.336) in space and time. Thus we introduce a spatial grid $\{x_j = j\Delta x,\ j = 0, 1, \ldots, N_x\}$ with grid size $\Delta x := L/N_x$ and time levels $t^n = n\Delta t$ $(n = 0, 1, 2, \ldots)$ for some time step $\Delta t > 0$. We denote by u_j^n the numerical approximation of the velocity and by T_j^n that of the temperature at the grid point (x_j, t^n). We use the following scheme to compute the velocity:

$$u_j^n = u_{j-1}^n + \frac{\epsilon^n}{\Delta x}\big(T_{j+1}^n - 2T_j^n + T_{j-1}^n\big) + \Delta x\, s_1(t^n), \tag{16.343}$$

with $\epsilon^n = \epsilon(t^n)$. Before we choose the proper discretisation scheme for the advection-dominated temperature equation, we note the following. First, the diffusion term should be taken implicitly, since an explicit scheme requires a stability condition of the type $\Delta t = \mathcal{O}(\Delta x^2)$. Second, in order to get a sharp approximation of steep gradients, we use a flux limiter scheme similar to the one from Section 13.7. So if, e.g., $u_j^n > 0$, we have for the temperature equation

$$\begin{aligned} T_j^{n+1} = T_j^n &- c_j^n\left(1 + \frac{1}{2}(1 - c_j^n)\left(\frac{\Phi_{j+\frac{1}{2}}^n}{r_{j+\frac{1}{2}}^n} - \Phi_{j-\frac{1}{2}}^n\right)\right)\big(T_j^n - T_{j-1}^n\big) \\ &+ \epsilon^n \frac{\Delta t}{\Delta x^2} T_j^n \big(T_{j+1}^{n+1} - 2T_j^{n+1} + T_{j-1}^{n+1}\big) + \Delta t s_2(t^{n+1}) T_j^{n+1}, \end{aligned} \tag{16.344}$$

where the local Courant number c_j^n is defined by $c_j^n := \Delta t u_j^n / \Delta x$. Furthermore, the limiter $\Phi_{j+\frac{1}{2}}$ is given by $\Phi^n_{j+\frac{1}{2}} = \Phi(r^n_{j+\frac{1}{2}})$ with $\Phi(r)$ a suitable limiter function and with the variable $r^n_{j+\frac{1}{2}}$ defined by

$$r^n_{j+\frac{1}{2}} := \begin{cases} \dfrac{T_j^n - T_{j-1}^n}{T_{j+1}^n - T_j^n} & \text{if} \quad u_j^n > 0, \\ \dfrac{T_{j+2}^n - T_{j+1}^n}{T_{j+1}^n - T_j^n} & \text{if} \quad u_j^n < 0; \end{cases} \tag{16.345}$$

cf. Section 13.7. A similar scheme holds if $u_j^n < 0$.

In a practical implementation it is worthwhile to choose a variable time step. To do so we compute $\max_j |u_j^n|$ at each time level t^n and determine Δt such that the condition

$$\max_j |u_j^n| \Delta t \leq \Delta x \tag{16.346}$$

is satisfied. We then compute the temperature T_j^{n+1} from (16.344) and subsequently the velocity u_j^{n+1} from (16.343).

The time integration proceeds until the system reaches the periodic steady state. We require that the absolute difference in temperature between two consecutive full cycles at the grid points be less than a predefined tolerance. The parameter values for the typical single-inlet pulse tube we used in our simulation are given in Table 16.7. In our numerical experiments we took a sinusoidal pressure variation, as shown in Figure 16.47, and a constant buffer pressure. Our model now allows us to study the temperature dynamics in the tube. In Figure 16.48 temperature profiles are displayed at times $t = 4\pi$, 4.5π, 5π, and 5.5π, which correspond to the various parts of the pressure cycle: 4π for pressure increases, 4.5π for maximum pressure, 5π for pressure decreases, and 5.5π for minimum pressure. As for determining the periodic steady state for $\Delta x = 0.015$ with a tolerance tol $= 10^{-4}$, this was achieved after nine cycles.

Table 16.7. *Physical data for a typical single-inlet pulse tube (values at* 300 K*).*

Parameter	Value	Units	Parameter	Value	Units
ρ_{ref}	4.7	kg/m^3	c_p	5.2×10^3	J/(kg · K)
T_{amb}	300	K	$\mathcal{R}$	2.1×10^3	J/(kg · K)
p_0	3×10^6	Pa	L	0.2	m
P	5×10^5	Pa	C_{or}	10^{-8}	m^3/(Pa · s)
U	1.5	m/s	A	2×10^{-3}	m^2
ω	1.2566×10^2	s^{-1}	p_{b}	3.025×10^6	Pa
μ_{ref}	2.0×10^{-5}	Pa · s	T_{h}	300	K
κ_{ref}	1.58×10^{-1}	W/(m · K)	T_{c}	70	K
γ	1.4			70	

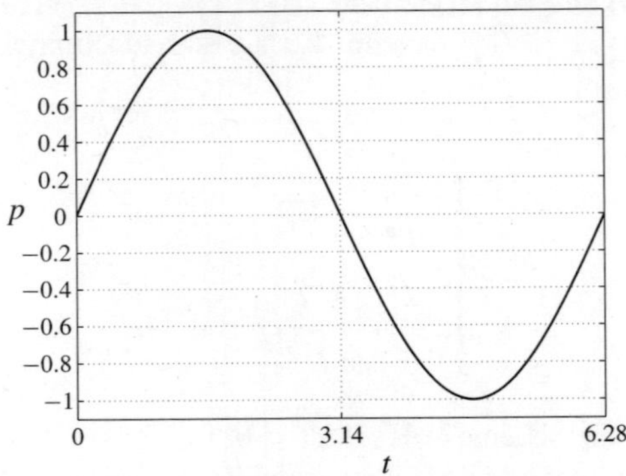

Figure 16.47. *Dimensionless pressure variation used in the simulation.*

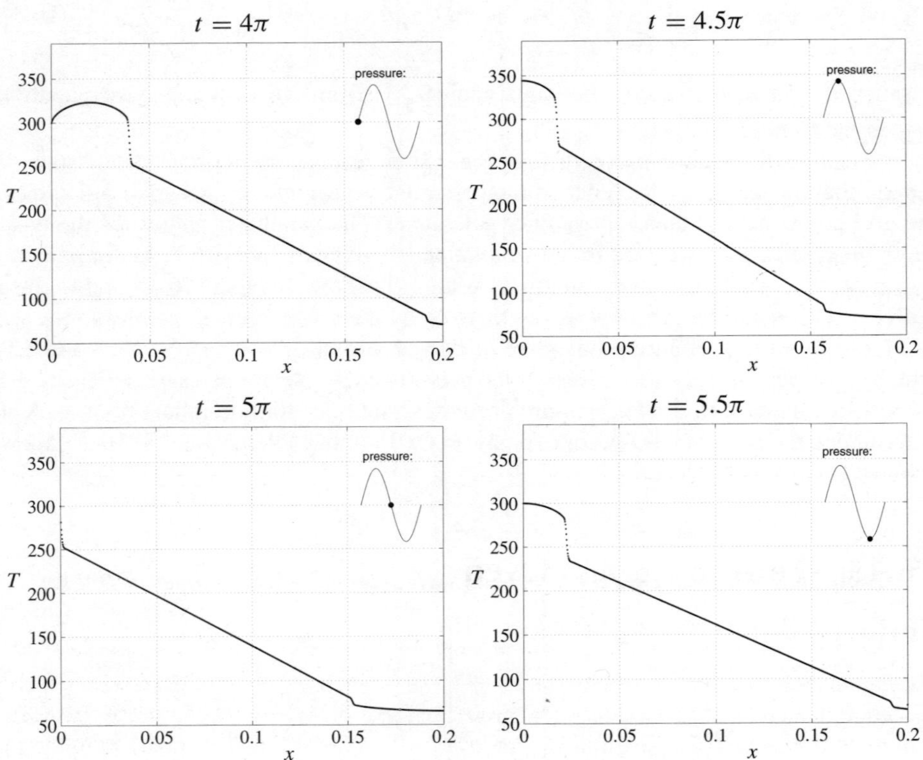

Figure 16.48. *Temperature profiles for the different parts of the pressure cycle.*

16.14.6 Discussion and Related Problems

- There are more parts of the pulse tube that need further analysis, e.g., the regenerator, which plays a crucial role in the entire process. Actually, the flow as such is much more complicated than modeled here. One should at least include the valve and the reservoir. Cryocoolers often have more reservoirs that may be included as well.

- An important question for any manufacturer is how to design an optimal product. Pulse tubes are no exception. This is a subject outside the scope of this book as it involves optimisation techniques. *Shape optimisation* requires a way to solve a series of problems, say by Newton's method. An option is to use a simpler analytical approximating solution to the problem, such as a harmonic balance approach based on Fourier series [165].
- In practice one has to resort to higher-dimensional situations. In particular, the flow is locally turbulent and as such requires a three-dimensional approach. This falls outside the scope of this book. It is still an active area of research as the full heat transfer problem may crucially hinge on the mixing effect of this turbulence.

16.14.7 Exercises

16.58. One can also derive a time-dependent buffer pressure by considering expansion and compression as adiabatic. To see this use Poisson's law to derive the volume flow through the orifice. The resulting equation for the pressure is of Riccati type (and can therefore be solved exactly).

16.59. Repeat the calculations, now with a square wave as the (dimensionless) pressure, i.e., a piecewise constant function being either $+1$ or -1, which results from instantaneous valve operation.

16.60. Rather than one-sided differences, one may try to use central differences for the velocity in (16.343). Give this scheme. Carry out numerical experiments and compare the results.

16.61. Perform numerical tests to see whether the number of iterations to reach the periodic steady state depends on the accuracy, i.e., on the chosen grid, both in time and space.

16.15 Flow in a Glass Oven

16.15.1 Problem Formulation

Glass is manufactured in *tank* furnaces in which raw materials like sand and soda are melted in a complicated physical and chemical process. The typical length of furnaces varies from 6 m to 30 m, the width from 3 m to 15 m, and the height (of the glass melt) from 0.6 m to 1.5 m. Raw materials, the *batch*, are put into the furnace via a so-called doghouse. These materials, being lighter than the glass melt, initially stay on top and gradually melt under the influence of heat coming from burners and already molten glass. Melting gradually, the glass changes its viscosity from around 1000 [Pa · s] to 10 [Pa · s], which triggers the flow. The changes in density due to the heating process are less dramatic. In fact, the density changes by about 10 percent of its value, say from 2600 [kg/m^3] to 2350 [kg/m^3]. However, these density gradients are essential for establishing a (mixing) flow.

Since the density of the molten glass is higher than that of the batch, the glass flow is first directed downward, then it flows forward along the bottom, and somewhere around the middle of the furnace it flows up and toward the batch (backflow). This particular pattern is caused by buoyancy due to the temperature and density differences. The heating process is organized in such a way that the hottest spot on the surface of the melt stays in the middle of the furnace. This particular arrangement ensures the backflow of the glass, which is crucial for the proper melting of the batch. The glass needs to stay in the furnace for a certain period of time (up to 24 h) to melt and mix properly. Due to the viscous nature of the glass melt, its typical velocity is relatively small, say in the range of 0.01 [m/s] to 0.04 [m/s]. It is very important to obtain proper insight into the glass flow as this determines the mixing and thus the quality of the glass.

16.15.2 The Model

In order to model the flow in a glass oven, we employ the conservation equations of mass and momentum. The inflow of raw material and the outflow of the melt can be assumed constant, implying a stationary flow. Therefore these equations read

$$\nabla\cdot(\rho\boldsymbol{v}) = 0, \tag{16.347a}$$

$$\nabla\cdot(\rho\boldsymbol{v}^T\boldsymbol{v}) = \nabla\cdot\mathcal{T} - \rho g\boldsymbol{e}_z, \tag{16.347b}$$

where ρ is the density, $\boldsymbol{v}$ is the velocity, $\mathcal{T}$ is the stress tensor, and g is the gravitational acceleration, assumed to be directed in the negative z direction. To solve the momentum equation (16.347b) we need to relate the stress tensor $\mathcal{T}$ to the velocity $\boldsymbol{v}$ and the pressure p. We assume the fluid to be Newtonian, so

$$\mathcal{T} = -p\boldsymbol{I} + 2\mu\mathcal{D}, \qquad \mathcal{D} = \frac{1}{2}\left(\nabla\boldsymbol{v} + \nabla\boldsymbol{v}^T\right), \tag{16.348}$$

where $\boldsymbol{I}$ and $\mathcal{D}$ are the identity tensor and the rate of deformation tensor, respectively, and where μ is the dynamic viscosity. Substituting (16.348) in (16.347b), we find for the momentum equation

$$\nabla\cdot(\rho\boldsymbol{v}^T\boldsymbol{v}) - \nabla\cdot(\mu\nabla\boldsymbol{v}) + \nabla p = -\rho g\boldsymbol{e}_z. \tag{16.349}$$

The scaling parameters to make the equations dimensionless are a length scale L representative of the furnace, a characteristic velocity V, a characteristic density ρ_{ref}, and a characteristic viscosity μ_{ref}. We introduce the following dimensionless variables (indicated with an asterisk $*$):

$$\boldsymbol{x} =: L\boldsymbol{x}^*, \quad \boldsymbol{v} =: V\boldsymbol{v}^*, \quad \rho =: \rho_{\text{ref}}\rho^*, \quad \mu =: \mu_{\text{ref}}\mu^*, \quad p =: \rho_{\text{ref}}V^2p^*. \tag{16.350}$$

Applying this scaling, we find that the governing equations (16.347a) and (16.349) become (omitting the asterisk $*$ for simplicity's sake)

$$\nabla\cdot(\rho\boldsymbol{v}) = 0, \tag{16.351a}$$

$$\nabla\cdot(\rho\boldsymbol{v}^T\boldsymbol{v}) - \frac{1}{Re}\nabla\cdot(\mu\nabla\boldsymbol{v}) + \nabla p = -\frac{1}{Fr^2}\rho\boldsymbol{e}_z, \tag{16.351b}$$

where *Re* and *Fr* are the Reynolds number and Froude number, respectively, given by

$$Re := \frac{\rho_{\text{ref}} V L}{\mu_{\text{ref}}}, \qquad Fr := \frac{V}{\sqrt{gL}}. \tag{16.352}$$

In order to assess (16.351b), we use the following typical values of the scaling parameters: $L = 5\,[\text{m}]$, $V = 0.05\,[\text{m/s}]$, $\rho_{\text{ref}} = 2300\,[\text{kg/m}^3]$, and $\mu_{\text{ref}} = 10\,[\text{Pa}\cdot\text{s}]$. Hence the dimensionless numbers are $Re = 57.5$ and $Fr = 7.14 \times 10^{-3}$. The convection part dominates over diffusion and obviously cannot be neglected; i.e., the Stokes approximation (cf. Section 16.11) cannot be used.

We note that the density and the viscosity are temperature dependent. The temperature field is not known a priori and is to be computed from the energy equation. However, this equation makes the model far more complicated and results, among other things, in a nonlinear coupling of all equations. We therefore simplify the model by using simulated (given) data for the temperature instead. The density and the viscosity are then computed using this given temperature field. Having fixed the temperature field as $T = T(\boldsymbol{x})$, we treat the density and the viscosity as functions depending on space:

$$\rho = \rho(\boldsymbol{x}) = \rho(T(\boldsymbol{x})), \qquad \mu = \mu(\boldsymbol{x}) = \mu(T(\boldsymbol{x})).$$

The density of glass ρ is a linear function of the temperature given by

$$\rho(\boldsymbol{x}) := \rho_0\big(1 - \beta(T(\boldsymbol{x}) - T_0)\big). \tag{16.353}$$

Here T_0 is the reference temperature, ρ_0 is the density of the glass at T_0, and β is the thermal expansion coefficient. The value of the density typically ranges from $2300\,[\text{kg/m}^3]$ to $2500\,[\text{kg/m}^3]$. Although the density variations are not large, they cause buoyancy effects and thus are the driving force for the glass flow. The gravitational force in the momentum equation (16.351b) changes proportionally with the density. The viscosity of the glass decreases exponentially with increasing temperature according to the Vogel–Fulcher–Tamman relation [157].

16.15.3 Defining the Domain

Since the oven has a rectangular structure we can model it as a union of, say, *S* nonintersecting blocks; i.e., the domain Ω can be written as

$$\Omega := \bigcup_{s=1}^{S} \Omega^{(s)}, \qquad \Omega^{(s)} := \big(x_0^{(s)}, x_1^{(s)}\big) \times \big(y_0^{(s)}, y_1^{(s)}\big) \times \big(z_0^{(s)}, z_1^{(s)}\big). \tag{16.354}$$

The boundary of Ω, which we denote by Γ, can be split into several parts. We distinguish the *walls* of the furnace, the *top layer*, the *inflow* area, and the *outflow* area:

$$\Gamma = \Gamma_{\text{wall}} \cup \Gamma_{\text{top}} \cup \Gamma_{\text{in}} \cup \Gamma_{\text{out}}. \tag{16.355}$$

We now specify the boundary conditions. At any point of the boundary we can write the velocity vector $\boldsymbol{v}$ in a coordinate system constituted by the outer unit normal vector $\boldsymbol{n}$

and two unit vectors $\boldsymbol{t}_1$ and $\boldsymbol{t}_2$ in the boundary plane; i.e., $\boldsymbol{v} = v_n \boldsymbol{n} + v_{t_1} \boldsymbol{t}_1 + v_{t_2} \boldsymbol{t}_2$ for all $\boldsymbol{x} \in \Gamma$. We thus have the following situations.

- **Walls.** The walls are impermeable; i.e., $v_n = 0$. Moreover, we assume the no-slip condition $v_{t_1} = v_{t_2} = 0$. As a result, we have

$$\boldsymbol{v} = \boldsymbol{0}, \qquad \boldsymbol{x} \in \Gamma_{\text{wall}}. \tag{16.356a}$$

- **Top layer.** We model the top layer as a symmetry plane, i.e., $v_n = 0$, and the normal derivative of the tangential velocity components is zero as well. The boundary condition is thus given by

$$v_n = 0, \qquad \frac{\partial v_{t_1}}{\partial \boldsymbol{n}} = \frac{\partial v_{t_2}}{\partial \boldsymbol{n}} = 0, \qquad \boldsymbol{x} \in \Gamma_{\text{top}}. \tag{16.356b}$$

- **Inflow.** The inflow area is characterised by a perpetual supply of glass flowing into the furnace with a constant speed. We assume that the inflow is directed along the inner unit normal, which means that the tangential components of the inflow are equal to zero; i.e.,

$$v_n = v_0 < 0, \quad v_{t_1} = v_{t_2} = 0, \qquad \boldsymbol{x} \in \Gamma_{\text{in}}. \tag{16.356c}$$

- **Outflow.** The boundary conditions for the outflow are similar to the ones for the inflow. We assume that the glass melt floats out of the furnace along the outer unit normal with a constant speed. These boundary conditions thus read

$$v_n = v_1 > 0, \quad v_{t_1} = v_{t_2} = 0, \qquad \boldsymbol{x} \in \Gamma_{\text{out}}. \tag{16.356d}$$

  The outflow velocity v_1 cannot be chosen arbitrarily; otherwise we would encounter mass loss or mass gain effects in the furnace. Applying Gauss's law to (16.351a) and taking into account the boundary conditions at the walls and the top layer, we find the following relation between inflow velocity v_0 and outflow velocity v_1:

$$v_0 \int_{\Gamma_{\text{in}}} \rho \, \mathrm{d}S + v_1 \int_{\Gamma_{\text{out}}} \rho \, \mathrm{d}S = 0. \tag{16.356e}$$

The domain Ω can be enclosed by a single block. Indeed, the corner points of this block are given by

$$\begin{aligned} x_\ell &:= \min_s x_0^{(s)}, & y_\ell &:= \min_s y_0^{(s)}, & z_\ell &:= \min_s z_0^{(s)}, \\ x_\mathrm{r} &:= \max_s x_0^{(s)}, & y_\mathrm{r} &:= \max_s y_0^{(s)}, & z_\mathrm{r} &:= \max_s z_0^{(s)}. \end{aligned} \tag{16.357}$$

By construction, the closure of Ω is a subset of the closure of the block defined by (16.357); i.e.,

$$\bar{\Omega} \subset \Omega', \qquad \Omega' := [x_\ell, x_\mathrm{r}] \times [y_\ell, y_\mathrm{r}] \times [z_\ell, z_\mathrm{r}]. \tag{16.358}$$

Note that Ω' is the smallest block that encloses Ω.

Let us define a grid Ω_Δ as follows:

$$\Omega_\Delta := \{(x, y, z) | \, x = x_\ell + (i-1)\Delta x,\ y = y_\ell + (j-1)\Delta y,\ z = z_\ell + (k-1)\Delta z, \\ i = 1, \ldots, N_x, \quad j = 1, \ldots, N_y, \quad k = 1, \ldots N_z\}, \quad (16.359)$$

where x_ℓ, y_ℓ, and z_ℓ are defined in (16.357) and N_x, N_y, and N_z are the numbers of grid points in the x, y and z directions, respectively. The grid sizes Δx, Δy, and Δz are thus defined by

$$\Delta x := \frac{x_\mathrm{r} - x_\ell}{N_x - 1}, \qquad \Delta y := \frac{y_\mathrm{r} - y_\ell}{N_y - 1}, \qquad \Delta z := \frac{z_\mathrm{r} - z_\ell}{N_z - 1}.$$

In order to facilitate our discussion, we introduce the *material property function* $m(\boldsymbol{x})$ for $\boldsymbol{x} \in \Omega_\Delta$ as follows:

$$m(\boldsymbol{x}) = \begin{cases} 0 & \text{if} \quad \boldsymbol{x} \notin \bar{\Omega}, \\ \text{wall} & \text{if} \quad \boldsymbol{x} \in \Gamma_{\text{wall}}, \\ \text{inflow} & \text{if} \quad \boldsymbol{x} \in \Gamma_{\text{in}}, \\ \text{top} & \text{if} \quad \boldsymbol{x} \in \Gamma_{\text{top}}, \\ \text{outflow} & \text{if} \quad \boldsymbol{x} \in \Gamma_{\text{out}}, \\ \text{glass} & \text{otherwise.} \end{cases} \qquad (16.360)$$

We have to construct the discrete equations based on the information provided by (16.360). Thus if, e.g., $m(\boldsymbol{x}) = \text{glass}$, we interpret this as $\boldsymbol{x} \in \Omega \backslash \Gamma$ and use the full set of equations. If, however, $m(\boldsymbol{x}) = \text{inflow}$ or $m(\boldsymbol{x}) = \text{outflow}$, the prescribed boundary values of the velocity should be used at the corresponding point.

16.15.4 Discretising the Convection-Diffusion Operator

Rather than writing down the complete discrete momentum equations, we derive the discretisation for the x component of the convection-diffusion operator only. It reads

$$\mathcal{M}(\boldsymbol{v})[v_x] := \nabla \cdot (\rho \boldsymbol{v} v_x) - \frac{1}{Re} \nabla \cdot (\mu \nabla v_x). \qquad (16.361)$$

We will treat the diffusion and convection parts of (16.361) separately. At a point $\boldsymbol{x} \in \Omega_\Delta$ for which $m(\boldsymbol{x}) = \text{glass}$ we approximate the diffusion part of (16.361), i.e., $-\nabla \cdot (\mu \nabla v_x)/Re$, by central differences, i.e.

$$
\begin{aligned}
&-\frac{1}{Re}\nabla\cdot(\mu\nabla v_x)(\boldsymbol{x})\\
&\quad\doteq \frac{1}{Re\Delta x^2}\left(\mu\left(\boldsymbol{x}-\frac{1}{2}\boldsymbol{h}_x\right)(v_x(\boldsymbol{x})-v_x(\boldsymbol{x}-\boldsymbol{h}_x))\right.\\
&\quad\left.+\,\mu\left(\boldsymbol{x}+\frac{1}{2}\boldsymbol{h}_x\right)(v_x(\boldsymbol{x})-v_x(\boldsymbol{x}+\boldsymbol{h}_x))\right)\\
&\quad+\frac{1}{Re\Delta y^2}\left(\mu\left(\boldsymbol{x}-\frac{1}{2}\boldsymbol{h}_y\right)(v_x(\boldsymbol{x})-v_x(\boldsymbol{x}-\boldsymbol{h}_y))\right.\\
&\quad\left.+\,\mu\left(\boldsymbol{x}+\frac{1}{2}\boldsymbol{h}_y\right)(v_x(\boldsymbol{x})-v_x(\boldsymbol{x}+\boldsymbol{h}_y))\right)\\
&\quad+\frac{1}{Re\Delta z^2}\left(\mu\left(\boldsymbol{x}-\frac{1}{2}\boldsymbol{h}_z\right)(v_x(\boldsymbol{x})-v_x(\boldsymbol{x}-\boldsymbol{h}_z))\right.\\
&\quad\left.+\,\mu\left(\boldsymbol{x}+\frac{1}{2}\boldsymbol{h}_z\right)(v_x(\boldsymbol{x})-v_x(\boldsymbol{x}+\boldsymbol{h}_z))\right),
\end{aligned}
\tag{16.362}
$$

where $\boldsymbol{h}_x := (\Delta x, 0, 0)^T$, etc. For a collocated grid any variable computed at the middle between two neighbouring grid points is approximated by linear interpolation; i.e.,

$$
g\left(\boldsymbol{x}+\frac{1}{2}\boldsymbol{h}\right) := \frac{1}{2}(g(\boldsymbol{x})+g(\boldsymbol{x}+\boldsymbol{h})),
\tag{16.363}
$$

where $\boldsymbol{h}$ is equal to $\boldsymbol{h}_x$, $\boldsymbol{h}_y$, or $\boldsymbol{h}_z$.

The discrete diffusion operator on a collocated grid in (16.362) is identical to the discrete diffusion operator on a staggered grid (cf. Figure 9.9). The important difference lies, however, in computing the viscosity μ at the midpoints. When a staggered grid is used, no interpolation is required.

For the convection part of (16.361), i.e., $\nabla\cdot(\rho\boldsymbol{v}v_x)$, we use central differences as well. Thus we have

$$
\begin{aligned}
&\nabla\cdot(\rho\boldsymbol{v}v_x)(\mathbf{x})\\
&\quad\doteq \frac{1}{\Delta x}\left((\rho v_x)\left(\boldsymbol{x}+\frac{1}{2}\boldsymbol{h}_x\right)\frac{1}{2}(v_x(\boldsymbol{x})+v_x(\boldsymbol{x}+\boldsymbol{h}_x))\right.\\
&\quad\left.-\,(\rho v_x)\left(\boldsymbol{x}-\frac{1}{2}\boldsymbol{h}_x\right)\frac{1}{2}(v_x(\boldsymbol{x})+v_x(\boldsymbol{x}-\boldsymbol{h}_x))\right)\\
&\quad+\frac{1}{\Delta y}\left((\rho v_y)\left(\boldsymbol{x}+\frac{1}{2}\boldsymbol{h}_y\right)\frac{1}{2}(v_x(\boldsymbol{x})+v_x(\boldsymbol{x}+\boldsymbol{h}_y))\right.\\
&\quad\left.-\,(\rho v_y)\left(\boldsymbol{x}-\frac{1}{2}\boldsymbol{h}_y\right)\frac{1}{2}(v_x(\boldsymbol{x})+v_x(\boldsymbol{x}-\boldsymbol{h}_y))\right)\\
&\quad+\frac{1}{\Delta z}\left((\rho v_z)\left(\boldsymbol{x}+\frac{1}{2}\boldsymbol{h}_z\right)\frac{1}{2}(v_x(\boldsymbol{x})+v_x(\boldsymbol{x}+\boldsymbol{h}_z))\right.\\
&\quad\left.-\,(\rho v_z)\left(\boldsymbol{x}-\frac{1}{2}\boldsymbol{h}_z\right)\frac{1}{2}(v_x(\boldsymbol{x})+v_x(\boldsymbol{x}-\boldsymbol{h}_z))\right).
\end{aligned}
\tag{16.364}
$$

For this term the mass fluxes ρv_x, ρv_y, and ρv_z have to be approximated by linear interpolation.

16.15.5 Discretising the Pressure Gradient and the Continuity Equation

Discretising the pressure gradient on a collocated grid is rather straightforward. The components of the discrete gradient have to be computed at all points of Ω_Δ where the corresponding velocity component is not prescribed. This means that at an arbitrary point $\boldsymbol{x} \in \Omega_\Delta$ where $m(\boldsymbol{x}) = \text{glass}$ we need to discretise three components of the gradient. At a point $\boldsymbol{x}$ on the top layer Γ_{top}, assumed to be horizontal, only two components (x and y) of the gradient need to be discretised; the third (z) will not be used in the solution method because $v_z(\boldsymbol{x}) = 0$ when $m(\boldsymbol{x}) = \text{top}$. If $\boldsymbol{x}$ is a *glass* point, the discrete gradient there is defined by

$$\nabla p(\boldsymbol{x}) \doteq \left(\frac{1}{2\Delta x}(p(\boldsymbol{x} + \boldsymbol{h}_x) - p(\boldsymbol{x} - \boldsymbol{h}_x)), \right.$$
$$\left. \frac{1}{2\Delta y}(p(\boldsymbol{x} + \boldsymbol{h}_y) - p(\boldsymbol{x} - \boldsymbol{h}_y)), \frac{1}{2\Delta z}(p(\boldsymbol{x} + \boldsymbol{h}_z) - p(\boldsymbol{x} - \boldsymbol{h}_z)) \right). \quad (16.365)$$

Discretising the continuity equation is somewhat more complicated. Ideally, we would like to use central differences for each of the three terms. A problem with central differences, however, appears when the grid point at which we discretise the continuity equation is a boundary point; this means that one of the grid points involved in the central difference approximation lies outside the computational domain. Hence a one-sided difference should be used instead. If $\boldsymbol{x}$ is a point in the interior, i.e.,

$$m(\boldsymbol{x} - \boldsymbol{h}_x) \neq 0 \quad \text{and} \quad m(\boldsymbol{x} + \boldsymbol{h}_x) \neq 0,$$

the first term in the continuity equation, $\partial(\rho v_x)/\partial x$, is discretised as

$$\frac{1}{2\Delta x}((\rho v_x)(\boldsymbol{x} + \boldsymbol{h}_x) - (\rho v_x)(\boldsymbol{x} - \boldsymbol{h}_x)).$$

If, however, $\boldsymbol{x}$ belongs to the left boundary, i.e.,

$$m(\boldsymbol{x} - \boldsymbol{h}_x) = 0 \quad \text{and} \quad m(\boldsymbol{x}) \neq 0,$$

no velocity is defined at $\boldsymbol{x} - \boldsymbol{h}_x$; therefore we use the forward difference

$$\frac{1}{\Delta x}((\rho v_x)(\boldsymbol{x} + \boldsymbol{h}_x) - (\rho v_x)(\boldsymbol{x})). \quad (16.366)$$

Similarly, we employ the backward difference approximation at the right boundary. Discretising the two other components of the continuity equation, $\partial(\rho v_y)/\partial y$ and $\partial(\rho v_z)/\partial z$, and adding up the result, we end up with the discrete continuity equation. For instance, at an arbitrary glass point $\boldsymbol{x}$ the discrete divergence reads

$$\frac{1}{2\Delta x}((\rho v_x)(\boldsymbol{x} + \boldsymbol{h}_x) - (\rho v_x)(\boldsymbol{x} - \boldsymbol{h}_x)) + \frac{1}{2\Delta y}((\rho v_y)(\boldsymbol{x} + \boldsymbol{h}_y) - (\rho v_y)(\boldsymbol{x} - \boldsymbol{h}_y))$$
$$+ \frac{1}{2\Delta z}((\rho v_z)(\boldsymbol{x} + \boldsymbol{h}_z) - (\rho v_z)(\boldsymbol{x} - \boldsymbol{h}_z)) = 0. \quad (16.367)$$

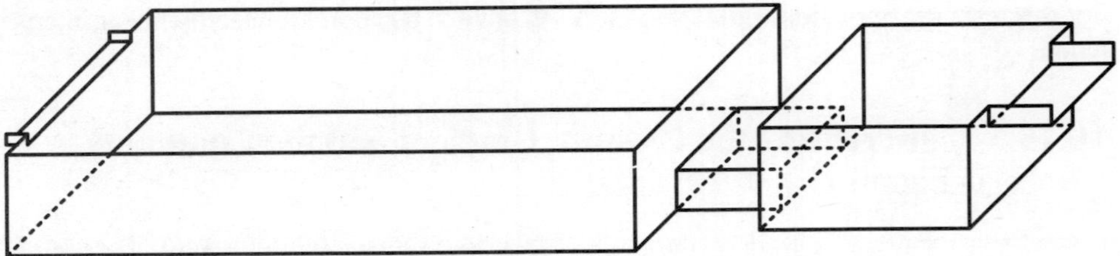

Figure 16.49. *Sketch of a model furnace.*

16.15.6 Simulations

To illustrate the potential of the discretisation technique we consider a model furnace, see Figure 16.49. It measures $10.5 \times 4 \times 1$ and consists of two tanks connected by a small channel: a larger melting tank and a smaller postprocessing tank. The melting tank measures $8 \times 4 \times 1$, while the postprocessing tank measures $2 \times 2 \times 1$.

The inflow area Γ_{in} is located at the left end of the furnace and is of rectangular shape. The outflow area Γ_{out}, in turn, is located at the right end of the furnace. These areas measure 3.8×0.2 and 3.4×0.3, respectively. The inflow velocity is directed normal to the boundary and is equal in absolute value to 0.01. We can find the velocity for the outflow area from (16.356e).

In order to have a reasonable approximation for the temperature field, the lowest temperature in the furnace occurs near the inflow area and furthermore the temperature is lower near the walls and the hottest spot occurs somewhere in the middle of the top layer. For a unit cube a suitable approximation is then given by the function

$$T = \frac{1}{2}\left(1 + \sin\left(\frac{1}{2}\pi(3x-1)\right)\right) + 4\left(x - \frac{1}{5}\right)(1-x)\left(\frac{6}{5} + \sin\left(\frac{1}{2}\pi(2z-1)\right)\right).$$

For the realistic problem of an actual furnace we need to scale this relation first.

The computational domain is covered with a grid Ω_Δ the sizes of which, Δx, Δy, and Δz, are equal to

$$\Delta x = \Delta y =: h, \quad \Delta z = \frac{1}{2}h, \qquad h = 0.2.$$

We have applied the pressure correction method [45] to solve the discretised continuity and momentum equations; see also Section 9.7, where we applied the method to the Stokes equations. The computed velocity is shown in Figure 16.50. As can be seen from this picture, the velocity of the flow near the inflow area is directed downward, which is explained by the fact that the cold glass flow is heavier than the warm one. The molten glass flows along the bottom and then up to the top, approximately in the middle of the melting tank. Then the flow splits: part of the melt flows toward the input area, thus forming the backflow, and part flows after a convection loop toward the postprocessing tank. These results reflect the actual behaviour of glass melt in furnaces.

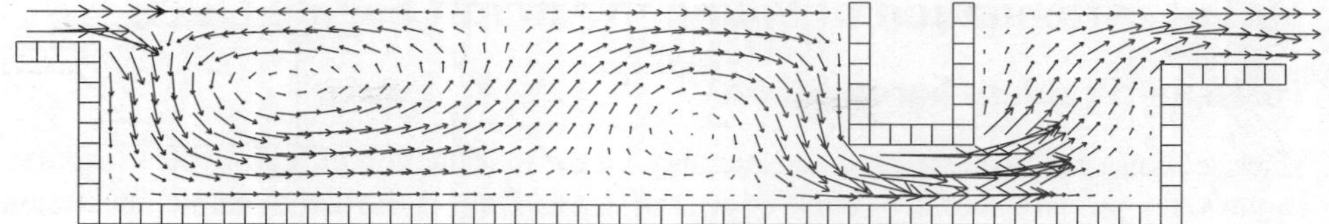

Figure 16.50. *Velocity field as a function of* (x, z) *for* $y = 2.0$.

16.15.7 Discussion and Related Problems

- The temperature that was left out in the model above, for simplicity's sake, is, of course, a crucial quantity for obtaining proper insight into the problem itself. Indeed, the differences in density driving the flow are caused by temperature differences. Besides heating the glass from above, one also often employs *boosters*, i.e., electric heaters in the tank to enforce this flow locally. For a more complete model one should also take into account that the material at the inlet is not a fluid yet (not even glass, as the chemical process takes place only gradually). This batch floats on top of the bath to some extent. Besides this shielding by the batch, a foam blanket is often formed that hinders the heat transfer of the burners above.

- The mixing in the tank is enhanced in several ways. Besides using boosters, one may also stir the glass mechanically. Both for this and to decrease the gas content in the melt one uses a bubbling technique in which a regular flow of air bubbles is injected at some spots. The global flow is furthermore directed by a special shape in the tank, such as a threshold somewhere on the bottom.

- As with many problems where confined material flows, there are difficult questions about what is happening at the contacts. In particular, there is always some friction and so we often have to impose some combined stick/slip condition (Robin type). Given the strong dependence of the glass on the temperature it is to be expected that the warmer regions have more slip than the cooler ones.

16.15.8 Exercises

16.62. The continuity and momentum equations can also be discretised on a staggered grid, as outlined in Section 9.7.

(a) Give the discretisation of (16.351b) on a staggered grid.

(b) Formulate the pressure correction method for the resulting algebraic systems.

(c) Simplify the setting by ignoring the y dependency so that the problem becomes two-dimensional. Compute the flow in the oven specified in Section 16.15.6.

16.63. Change the geometry of the furnace by putting a block shaped threshold somewhere on the bottom of the tank. Compute the flow in this tank.

16.16 Attenuation of Sound in Aircraft Engine Ducts

16.16.1 Problem Formulation

Since jet noise has been reduced considerably by the introduction of high-bypass turbofan engines, the primary source of noise of aircraft aeroengines is the aerodynamic interaction of fan and bypass stator (Figure 16.51). This noise radiates upstream via the inlet duct and downstream via the bypass duct, away into the far field. It causes considerable nuisance to the community during approach and takeoff. Aircraft noise is therefore subject to very strict regulations, which already limit the capacity of a number of major airports around the world. One of the many measures taken to reduce the noise is the application of acoustic lining in the inlet and bypass ducts. This lining usually consists of an array of small resonators designed to attenuate the sound waves that pass along the wall. In order to optimize the damping properties, we need to model the sound propagation through the inlet and bypass ducts. This will be done here by the well-established model of inviscid mean flow and acoustic perturbations, with assumed absent vorticity and uniform entropy. The inherent slow variation of the geometry of a typical flow duct provides the basis of an asymptotic analysis favoring analytic solutions. As far as the asymptotic reduction is part of the modeling process, it is an example of asymptotic modeling (Section 7.2).

16.16.2 The Model

Especially in the inlet, both mean flow and acoustic perturbations are quite well modeled by an inviscid, homentropic, irrotational flow (7.5), while the mean flow velocity is nearly uniform. So we can introduce a velocity potential and integrate the momentum equation to obtain Bernoulli's equation (7.12). We write velocity, density, and pressure (made dimensionless as shown below) as the sum of a steady component and a harmonic perturbation of frequency ω (dimensionless; also known as the Helmholtz number); i.e.,

$$[V, D, P] + \mathrm{Re}\big([\nabla\phi, \rho, p]\,\mathrm{e}^{\mathrm{i}\omega t}\big). \tag{16.368}$$

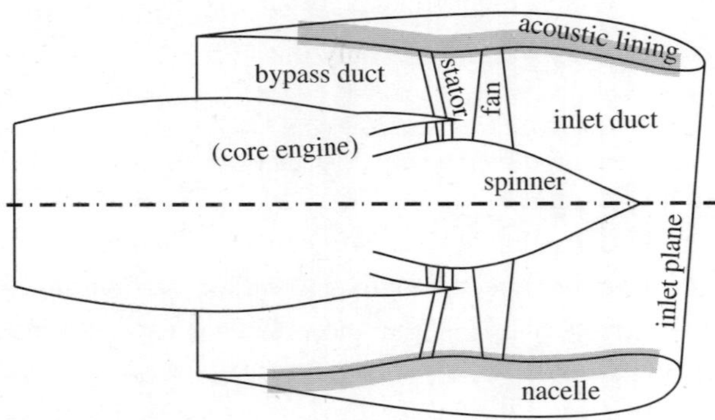

Figure 16.51. *Sketch of an aircraft turbofan engine.*

(There is no need to introduce explicitly a mean flow potential.) After linearisation we find for the mean flow the (dimensionless) equations

$$\frac{1}{2}V^2 + \frac{C^2}{\gamma - 1} = E, \quad \nabla\cdot(DV) = 0, \quad \frac{P}{D^\gamma} = \text{constant}, \quad C^2 = \gamma\frac{P}{D}, \tag{16.369}$$

where γ is the specific heat ratio, C denotes the mean sound speed, and E is the integration constant in Bernoulli's equation. For the acoustic perturbations we have

$$\big(\mathrm{i}\omega + \boldsymbol{V}\cdot\nabla\big)\rho + \rho\nabla\cdot\boldsymbol{V} + \nabla\cdot\big(D\nabla\phi\big) = 0,\ D\big(\mathrm{i}\omega + \boldsymbol{V}\cdot\nabla\big)\phi + p = 0,\ p = C^2\rho. \tag{16.370}$$

By means of the second and third equations we eliminate p and ρ from the first. By using $\nabla\cdot(D\boldsymbol{V}) = 0$, we obtain

$$D^{-1}\nabla\cdot\big(D\nabla\phi\big) - \big(\mathrm{i}\omega + \boldsymbol{V}\cdot\nabla\big)\Big[C^{-2}\big(\mathrm{i}\omega + \boldsymbol{V}\cdot\nabla\big)\phi\Big] = 0. \tag{16.371}$$

The model is valid inside the slowly varying hollow or annular duct given in cylindrical coordinates (x, r, θ) by

$$0 \le R_1(X) \le r \le R_2(X), \qquad X = \varepsilon x, \tag{16.372}$$

where ε is small and R_1, R_2 do not depend on ε other than via X. (Note that technically ε is only a bookkeeping parameter. Its size is of the order of the derivative of R_i with respect to x.) For definiteness we will define the position of the source at $x = 0$. Lengths are made dimensionless on a typical outer diameter, usually at the source plane. We may have, e.g., $R_2(0) = 1$. Velocities and densities are made dimensionless on a typical value of sound speed and density, e.g., at the source plane.

The duct wall is impermeable for the mean flow, while it is lined by acoustic material of impedance Z_1 and Z_2. The acoustic boundary condition is, however, not the regular impedance condition—the ratio between pressure and velocity—but Myers's impedance condition [98] to include the effect of refraction through the vanishing mean flow shear layer along the wall. This yields the following boundary conditions at $r = R_i$, $i = 1, 2$:

$$(\boldsymbol{V}\cdot\boldsymbol{n}_i) = 0, \qquad \mathrm{i}\omega Z_i(\nabla\phi\cdot\boldsymbol{n}_i) = [\mathrm{i}\omega + \boldsymbol{V}\cdot\nabla - \boldsymbol{n}_i\cdot(\boldsymbol{n}_i\cdot\nabla\boldsymbol{V})]p. \tag{16.373}$$

The mean flow will be defined by a given mass flux F and the assumption that it starts far upstream as a uniform flow and varies only through X (like in Example 15.24). The acoustic field that will be considered is a slowly varying duct mode, i.e., a mode that would be a strict solution of the equations in a straight duct (like in Example 15.36).

16.16.3 Asymptotic Solution

The Mean Flow The mean flow is found by using the method of slow variation (Section 15.3.1). Its derivation is given in detail in Example 15.24. For now we only need to know that

$$\begin{aligned} &\boldsymbol{V} = U_0(X)\boldsymbol{e}_x + \varepsilon V_1(X, r)\boldsymbol{e}_r + \mathcal{O}(\varepsilon^2), \qquad U_0(X) = F/D_0(X)A(X),\\ &\frac{\partial}{\partial X}(rD_0U_0) + \frac{\partial}{\partial r}(rD_0V_1) = 0, \qquad \text{and} \qquad V_1 = U_0R_i' \quad \text{along} \quad r = R_i, \end{aligned} \tag{16.374}$$

where R_i' denotes the derivative of R_i with respect to X, A is the surface of a duct cross section, $C = C_0(X) + \mathcal{O}(\varepsilon^2)$, $D = D_0(X) + \mathcal{O}(\varepsilon^2)$, and D_0 is the solution of the algebraic equation

$$\frac{F}{2D_0^2A^2} + \frac{D_0^{\gamma-1}}{\gamma - 1} = E. \tag{16.375}$$

The Acoustic Field A straight-duct modal wave form is a function of r multiplied by a complex exponential in θ and x. The mode-like wave we are looking for here is obtained by assuming the amplitude and axial and radial wave numbers to be slowly varying, i.e., depending on X. So we introduce the so-called WKB ansatz (15.62)

$$\phi(x, r, \theta; \varepsilon) = \Phi(X, r; \varepsilon) \exp\left(-im\theta - \mathrm{i}\varepsilon^{-1}\int^X \kappa(\xi)\,\mathrm{d}\xi\right), \tag{16.376}$$

where m and κ denote, respectively, the circumferential periodicity and the slowly varying axial wave number of the mode of interest. Substituting in (16.371), suppressing the exponential, we obtain, and collecting like powers of ε up to order ε^2,

$$D_0\mathcal{L}[\Phi] = \frac{\mathrm{i}\varepsilon}{\Phi}\left\{\frac{\partial}{\partial X}\left[\left(\frac{U_0\Omega}{C_0^2} + \kappa\right) D_0\Phi^2\right] + \frac{1}{r}\frac{\partial}{\partial r}\left[r\frac{V_1\Omega}{C_0^2}D_0\Phi^2\right]\right\}, \tag{16.377}$$

where

$$\Omega = \omega - \kappa U_0 \tag{16.378}$$

and the operator $\mathcal{L}$ is defined by

$$\mathcal{L} = \frac{\partial^2}{\partial r^2} + \frac{1}{r}\frac{\partial}{\partial r} + \frac{\Omega^2}{C_0^2} - \kappa^2 - \frac{m^2}{r^2}. \tag{16.379}$$

With $\boldsymbol{n}_i \cdot (\boldsymbol{n}_i \cdot \nabla \boldsymbol{V}) = \varepsilon\frac{\partial}{\partial r}V_1 + \mathcal{O}(\varepsilon^2)$, the boundary conditions (16.373) are, up to $\mathcal{O}(\varepsilon^2)$,

$$\mathrm{i}\omega\frac{\partial\Phi}{\partial r} \pm \frac{\Omega^2D_0\Phi}{Z_i} = \varepsilon\omega\kappa R_i'\Phi \pm \frac{\mathrm{i}\varepsilon}{\Phi}\left[U_0\frac{\partial}{\partial X} + V_1\frac{\partial}{\partial r} - \frac{\partial V_1}{\partial r}\right]\left(\frac{\Omega D_0\Phi^2}{Z_i}\right), \tag{16.380}$$

where the plus sign relates to $r = R_1$ and the minus sign to $r = R_2$. We now assume

$$\Phi(X, r; \varepsilon) = \Phi_0(X, r) + \varepsilon\Phi_1(X, r) + \cdots. \tag{16.381}$$

Substitution into (16.377) yields for the $\mathcal{O}(1)$ and $\mathcal{O}(\varepsilon)$ terms, respectively,

$$\mathcal{L}[\Phi_0] = 0, \tag{16.382a}$$

$$D_0\mathcal{L}[\Phi_1] = \frac{\mathrm{i}}{\Phi_0}\left\{\frac{\partial}{\partial X}\left[\left(\frac{U_0\Omega}{C_0^2} + \kappa\right) D_0\Phi_0^2\right] + \frac{1}{r}\frac{\partial}{\partial r}\left[r\frac{V_1\Omega}{C_0^2}D_0\Phi_0^2\right]\right\}, \tag{16.382b}$$

subject to the boundary conditions

$$\mathrm{i}\omega\frac{\partial\Phi_0}{\partial r} \pm \frac{\Omega^2 D_0 \Phi_0}{Z_i} = 0, \tag{16.383a}$$

$$\mathrm{i}\omega\frac{\partial\Phi_1}{\partial r} \pm \frac{\Omega^2 D_0 \Phi_1}{Z_i} = \omega\kappa R_i' \Phi_0 \pm \frac{\mathrm{i}}{\Phi_0}\left[U_0\frac{\partial}{\partial X} + V_1\frac{\partial}{\partial r} - \frac{\partial V_1}{\partial r}\right]\left(\frac{\Omega D_0 \Phi_0^2}{Z_i}\right). \tag{16.383b}$$

The leading-order equation (16.382a) is, up to a radial coordinate stretching, Bessel's equation in r, with X acting only as a parameter. The mode-like solution we are looking for is then

$$\Phi_0(X, r) = N(X)\psi(X, r), \tag{16.384}$$

where ψ is given by

$$\psi(X, r) = a(X) J_m(\alpha(X)r) + b(X) Y_m(\alpha(X)r) \tag{16.385}$$

and J_m and Y_m are the mth order Bessel function of the first and second kind, respectively [1]. α is defined by the dispersion relation

$$\alpha^2 = \Omega^2/C_0^2 - \kappa^2. \tag{16.386}$$

Boundary conditions (16.383a) produce the following equations for the slowly varying eigenvalue α considered and corresponding amplitudes a and b:

$$\frac{\alpha R_2 J_m'(\alpha R_2) - \zeta_2 J_m(\alpha R_2)}{\alpha R_2 Y_m'(\alpha R_2) - \zeta_2 Y_m(\alpha R_2)} = \frac{\alpha R_1 J_m'(\alpha R_1) + \zeta_1 J_m(\alpha R_1)}{\alpha R_1 Y_m'(\alpha R_1) + \zeta_1 Y_m(\alpha R_1)} = -\frac{b(X)}{a(X)}, \tag{16.387}$$

where $\zeta_i = \Omega^2 D_0 R_i/\mathrm{i}\omega Z_i$. ψ is scaled such that for any X

$$\int_{R_1}^{R_2} \psi^2(X, r) r \,\mathrm{d}r = 1. \tag{16.388}$$

This can be done explicitly (see [1]) by using the relation

$$\int \psi^2 r \,\mathrm{d}r = \frac{1}{2}(r^2 - (m/\alpha)^2)\psi^2 + \frac{1}{2}(r/\alpha)^2\left(\frac{\partial}{\partial r}\psi\right)^2, \tag{16.389}$$

where, of course, $\frac{\partial}{\partial r}\psi$ can be eliminated by substituting the boundary condition (16.383a). For notational convenience we introduce the reduced axial wave number

$$\sigma = \sqrt{1 - (C_0^2 - U_0^2)\frac{\alpha^2}{\omega^2}}, \tag{16.390}$$

which is roughly equal to κ for a comoving observer and scaled by ω, such that

$$\kappa = \omega\frac{C_0\sigma - U_0}{C_0^2 - U_0^2}, \quad \frac{U_0\Omega}{C_0^2} + \kappa = \frac{\omega\sigma}{C_0}, \quad \Omega = \omega C_0\frac{C_0 - U_0\sigma}{C_0^2 - U_0^2}, \quad (C_0^2 - U_0^2)\frac{\alpha^2}{\omega^2} + \sigma^2 = 1.$$

Finally, we have to determine $N = N(X)$ by the usual solvability argument applied to Φ_1. Since the operator $r\mathcal{L}$ is self-adjoint in r, we have

$$\int_{R_1}^{R_2} \Phi_0 \mathcal{L}[\Phi_1] r \,\mathrm{d}r = R_2 \left[\Phi_0 \frac{\partial \Phi_1}{\partial r} - \Phi_1 \frac{\partial \Phi_0}{\partial r} \right]_{r=R_2} - R_1 \left[\Phi_0 \frac{\partial \Phi_1}{\partial r} - \Phi_1 \frac{\partial \Phi_0}{\partial r} \right]_{r=R_1}.$$

Further evaluation of this expression (using (16.383b) and the corresponding right-hand side of (16.382b)) gives, finally, after a considerable amount of algebra, the adiabatic invariant (see (15.72))

$$\frac{\mathrm{d}}{\mathrm{d}X}\left[D_0 \frac{\omega\sigma}{C_0} \int_{R_1}^{R_2} \Phi_0^2(X,r) r \,\mathrm{d}r + \frac{D_0 U_0}{\Omega} \left(\zeta_2 \Phi_0^2(X,R_2) + \zeta_1 \Phi_0^2(X,R_1) \right) \right] = 0. \quad (16.391)$$

We have made use of the defining equations (16.374) of U_0 and V_1, Leibniz's rule, and the identity $U_0 \frac{\partial}{\partial X} + V_1 \frac{\partial}{\partial r} = U_0 \frac{\mathrm{d}}{\mathrm{d}X}$ along $r = R_i$.

Equation (16.391) can be integrated immediately with a constant of integration Q^2. This constant is determined by the initial amplitude of the mode entering the duct. Since the solution is linear, it is irrelevant here. The remaining integral over Φ_0^2 is absorbed by the normalisation of ψ. We eventually obtain for N the remarkably simple expression

$$N(X) = \frac{Q}{\sqrt{\dfrac{\omega\sigma D_0}{C_0} + \dfrac{D_0^2 \Omega U_0}{\mathrm{i}\omega} \left(\dfrac{R_2}{Z_2} \psi^2(X,R_2) + \dfrac{R_1}{Z_1} \psi^2(X,R_1) \right)}}. \quad (16.392)$$

The chosen branch of the square root is not important (Q is to be chosen anyway), but we have to make sure to remain at the same branch along the duct.

It should be noted that expression (16.392) is quite general. It is equally valid for no flow ($U_0 = 0$), a hard-walled duct ($Z_i \to \infty$), and a hollow duct ($R_1 \equiv 0$).

16.16.4 Numerical Evaluation of a Typical Example

The found analytic expression is not entirely explicit, since it requires solving two algebraic equations. Numerical evaluation involves the following steps:

(i) We define the duct in terms of the inner (R_1) and outer (R_2) walls, from which we obtain the cross section $A = \pi(R_2^2 - R_1^2)$.

(ii) We define the mean flow parameters E and F and solve (16.375) for D_0 at any X. This defines the axial mean flow U_0. In aircraft applications at takeoff or approach we will be interested in a Mach number between 0.2 and 0.5.

(iii) We select the Helmholtz number ω. This is determined by the shaft rotational speed being such that the fan blade tip speed is just about transonic and depends on the diameter of the duct and the harmonic order of interest (one, two, or sometimes three times the blade passing frequency). Usual values for high-bypass engines vary between 20 and 70.

(iv) We select the prevailing circumferential mode number m. This number is related to the number of fan blades and stator vanes and also depends on the frequency of the fan we are interested in. Typical values vary between 10 and 30.

(v) We select a class of candidate wall impedances. Hard walls correspond to $|Z| = \infty$ and any passive wall has $\mathrm{Re}(Z) \geq 0$, while a purely imaginary wall is not practically feasible. A typical impedance is found in a rectangle of about $0 < \mathrm{Re}(Z) < 6$, $-6 < \mathrm{Im}(Z) < 6$. Furthermore, if we consider several frequencies, we should take account of Z's inherent dependence on ω. This requires knowledge of the way the lining is actually constructed.

(vi) Finally, we determine (or guess) the source, i.e., the amplitudes of the radial modes. For each relevant mode we have to solve the eigenvalue equation (16.387) and trace α as a function of X. This gives us the axial wave number κ, which has to be integrated with respect to X for the modal phase (16.376).

We computed numerically a plot (Figure 16.52) of a pressure field (at a given time) for a typical inlet duct given by

$$\begin{aligned} R_2(x) &= 1 - 0.18453\,\xi^2 + 0.10158\,(\mathrm{e}^{-11(1-\xi)} - \mathrm{e}^{-11})/(1 - \mathrm{e}^{-11}), \\ R_1(x) &= \max\bigl[0, 0.64212 - (0.04777 + 0.98234\,\xi^2)^{1/2}\bigr], \end{aligned} \tag{16.393}$$

where $\xi = x/L$ and $L = 1.86393$ is the length of the duct. (Note that R_i is written as a function of x rather than X.) The corresponding value of ε is about 0.1, since R_2' varies between -0.1 and 0.1 over $0 \leq x \leq 1.73$, the larger part of the interval. Although $|R_1'|$ attains higher values near the spinner top (almost 0.5), this is not as relevant because the sound field practically vanishes near the duct axis due to the usual high values of m, while the mode behaves like $\sim r^m$ for $r \to 0$.

By selecting $F = -0.2462$ and $E = 2.545$, we obtain a Mach number $M = -0.3$ at $x = 0$ (in the picture: flow is from right to left). The outer wall impedance is $Z_2 = 2 - \mathrm{i}$ while the inner wall (the spinner) is hard. The acoustic field is given by $m = 10$

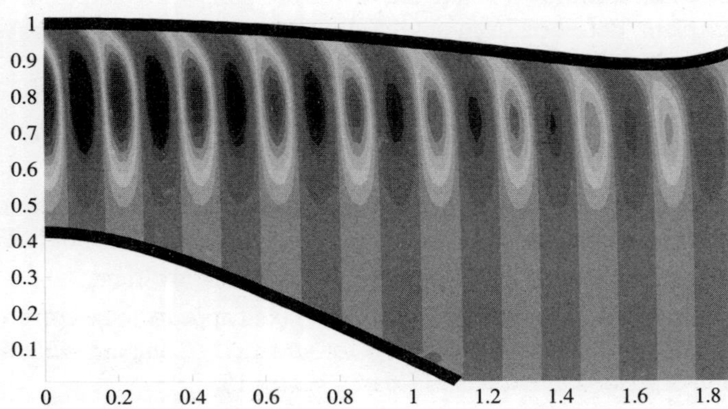

Figure 16.52. *Acoustic pressure field in an aircraft engine inlet duct in the (x, r) plane.*

and $\omega = 25$, while a right-running radial mode is initially selected with the eigenvalue $\alpha = 14.834 + 0.645\mathrm{i}$ and axial mode number $\kappa = 30.905 - 0.464\mathrm{i}$. The total attenuation between source and inlet plane in this example is 9.48 dB.

16.16.5 Discussion and Related Problems

- The present example is a shortened version of [127], while the simplification using a normalised ψ is taken from [132]. The main result (16.392) is to be compared with equation (49) in [132], including the correction of a misprinted minus sign.
- The present solution may be generalised to very challenging problems of nonpotential flows with mean swirl [28], noncircular ducts [132, 113] and nonhomentropic flow [27].
- The present solution breaks down where the mode passes a point where the denominator of (16.392) vanishes. This typically happens for hard-walled ducts ($Z_i = \infty$) at a point of "duct resonance" (this is only a real resonance in the case of a straight duct), where $\sigma = 0$. As the mode cannot continue and has to reflect, this point is called a turning point. A local analysis is required where incident and reflected mode couple [129, 109, 110].
- The acoustic design of aircraft engines is usually done by fully numerical modeling, allowing a more flexible variation of geometry and mean flow properties. Nevertheless, the present solution has been shown to be a very interesting alternative [128] and served as an independent check for a number of numerical solutions. It also provides a tool to produce a matching procedure to connect computational fluid dynamics type source data to the acoustic field [111].
- For high frequencies that occur in aircraft applications the eigenvalues defined by (16.387) may be classified in two groups [131]. One consists of typical duct modes, with the imaginary parts of α not large and κ remaining near a corresponding hard-wall value. The other consists of surface waves (for each m, two without mean flow and four with mean flow), where the imaginary part of α is large and κ strays in the complex plain, detached from any hard-wall mode. The surface waves are usually more difficult to find.

16.16.6 Exercises

16.64. Give an explicit, full expression of ψ, i.e., including the normalization, and compare the resulting expression of N with equations (4.11) and (4.14) of [127].

16.65. Give a detailed derivation of (16.391), the adiabatic invariant.

16.66. The present solution is derived for $\omega = \mathcal{O}(1)$ but appears to be correct for relatively large ω. What analysis would result for small ω? Compare with Webster's horn equation described in Exercise 15.6 of Chapter 15.

Appendices

Useful Definitions and Properties

A Asymptotic Order Symbols

A function $f(\varepsilon)$ may be expressed asymptotically for small ε by another (usually simpler) function $\varphi(\varepsilon)$ as follows.

Definition A.1.

1. *$f(\varepsilon) = \mathcal{O}(\varphi(\varepsilon))$ as $\varepsilon \to 0$ if there are positive constants K and ε_1 (both independent of ε) such that*

$$|f(\varepsilon)| \leq K|\varphi(\varepsilon)| \quad \text{for} \quad 0 < \varepsilon < \varepsilon_1.$$

We say, "f is big-O of φ as ε tends to zero."

2. *$f(\varepsilon) = o(\varphi(\varepsilon))$ as $\varepsilon \to 0$ if for every $\delta > 0$ there is an ε_1 (independent of ε) such that*

$$|f(\varepsilon)| \leq \delta|\varphi(\varepsilon)| \quad \text{for} \quad 0 < \varepsilon < \varepsilon_1.$$

We say, "f is little-o of φ as ε tends to zero."

3. *$f(\varepsilon) = \mathcal{O}_s(\varphi(\varepsilon))$ as $\varepsilon \to 0$ if $f(\varepsilon) = \mathcal{O}(\varphi(\varepsilon))$ and $f(\varepsilon) \neq o(\varphi(\varepsilon))$. Note that quite often $\mathcal{O}$ is used where $\mathcal{O}_s$ is actually meant. Further, there is no uniform terminology. We could say, "f is big sharp of φ as ε tends to zero."*

Theorem A.2.

1. *If $f(\varepsilon) = o(\varphi(\varepsilon))$ as $\varepsilon \to 0$, then also $f(\varepsilon) = \mathcal{O}(\varphi(\varepsilon))$.*
2. *If the limit $\lim_{\varepsilon\to 0} \frac{f(\varepsilon)}{\varphi(\varepsilon)}$ exists as a finite number, then $f(\varepsilon) = \mathcal{O}(\varphi(\varepsilon))$ as $\varepsilon \to 0$.*
3. *If the limit $\lim_{\varepsilon\to 0} \frac{f(\varepsilon)}{\varphi(\varepsilon)}$ exists as a finite nonzero number, then $f(\varepsilon) = \mathcal{O}_s(\varphi(\varepsilon))$ as $\varepsilon \to 0$.*
4. *If $\lim_{\varepsilon\to 0} \frac{f(\varepsilon)}{\varphi(\varepsilon)} = 0$, then $f(\varepsilon) = o(\varphi(\varepsilon))$ as $\varepsilon \to 0$.*

Proof. The proof is trivial. □

Example A.3

$$\begin{aligned}
\varepsilon \sin(\varepsilon) &= \mathcal{O}_s(\varepsilon^2), && \varepsilon \to 0, \\
\varepsilon \cos(\varepsilon) &= \mathcal{O}(1), && \varepsilon \to 0, \\
\varepsilon^n &= o(1), && \varepsilon \to 0, \quad \text{for any positive } n, \\
\mathrm{e}^{-1/\varepsilon} &= o(\varepsilon^n), && \varepsilon \to 0, \quad \text{for any positive } n.
\end{aligned}$$

From this last example, $\mathrm{e}^{-1/\varepsilon}$ is called a transcendentally small term or exponentially small term and can be ignored asymptotically against any power of ε. □

B Trigonometric Relations

The real or imaginary parts of the binomial series $(\mathrm{e}^{\mathrm{i}x} \pm \mathrm{e}^{-\mathrm{i}x})^n = \sum_{k=0}^{n} (\pm 1)^k \binom{n}{k} \mathrm{e}^{\mathrm{i}(n-2k)x}$ easily yield trigonometric relations that are useful for recognising resonance terms:

$$\sin^2 x = \frac{1}{2}(1 - \cos 2x), \qquad \sin^2 x \cos^2 x = \frac{1}{8}(1 - \cos 4x),$$

$$\sin x \cos x = \frac{1}{2} \sin 2x, \qquad \sin x \cos^3 x = \frac{1}{8}(2 \sin 2x + \sin 4x).$$

$$\cos^2 x = \frac{1}{2}(1 + \cos 2x), \qquad \cos^4 x = \frac{1}{8}(3 + 4 \cos 2x + \cos 4x),$$

$$\sin^3 x = \frac{1}{4}(3 \sin x - \sin 3x), \qquad \sin^5 x = \frac{1}{16}(10 \sin x - 5 \sin 3x + \sin 5x),$$

$$\sin^2 x \cos x = \frac{1}{4}(\cos x - \cos 3x), \qquad \sin^4 x \cos x = \frac{1}{16}(2 \cos x - 3 \cos 3x + \cos 5x),$$

$$\sin x \cos^2 x = \frac{1}{4}(\sin x + \sin 3x), \qquad \sin^3 x \cos^2 x = \frac{1}{16}(2 \sin x + \sin 3x - \sin 5x),$$

$$\cos^3 x = \frac{1}{4}(3 \cos x + \cos 3x), \qquad \sin^2 x \cos^3 x = \frac{1}{16}(2 \cos x - \cos 3x - \cos 5x),$$

$$\sin^4 x = \frac{1}{8}(3 - 4 \cos 2x + \cos 4x), \qquad \sin x \cos^4 x = \frac{1}{16}(2 \sin x + 3 \sin 3x + \sin 5x),$$

$$\sin^3 x \cos x = \frac{1}{8}(2 \sin 2x - \sin 4x), \qquad \cos^5 x = \frac{1}{16}(10 \cos x + 5 \cos 3x + \cos 5x).$$

C Convergence of Series

The series

$$S(x) := \sum_{n=0}^{\infty} c_n f_n(x), \qquad x \in \Omega, \tag{C.1}$$

is said to converge *pointwise* in $x \in \Omega$ if we can find for any given $\varepsilon > 0$ a sufficiently large number $N \in \mathbb{N}$ such that the remaining part of the series is smaller than ε; i.e.,

$$\left\| S(x) - \sum_{n=0}^{N} c_n f_n(x) \right\| < \varepsilon.$$

In general, N depends on ε and x. If N may be chosen independent of any $x \in \Omega$, the series is said to converge *uniformly* in Ω. A sufficient condition for uniform convergence is Weierstrass's M-test. If

$$\| c_n f_n(x) \| \leq M_n \quad \text{for all } x \in \Omega$$

and

$$\sum_{n=0}^{\infty} M_n$$

converges, then (C.1) is uniformly convergent.

The concept of uniform convergence is important because of the following properties. Suppose (C.1) is uniformly convergent.

(i) If all $f_n(x)$ are continuous, then $S(x)$ is continuous and

$$\lim_{x \to a} S(x) = \sum_{n=0}^{\infty} c_n \lim_{x \to a} f_n(x).$$

(ii) If all $f_n(x)$ are continuous, then $S(x)$ is integrable and

$$\int S(x)\, \mathrm{d}x = \sum_{n=0}^{\infty} c_n \int f_n(x)\, \mathrm{d}x.$$

(iii) If all $f_n(x)$ are differentiable, then $S(x)$ is differentiable and

$$\frac{\mathrm{d}}{\mathrm{d}x} S(x) = \sum_{n=0}^{\infty} c_n \frac{\mathrm{d}}{\mathrm{d}x} f_n(x).$$

D Multistep Formulas

If the grid points ξ_i are equispaced, a simple relation exists between interpolating polynomials and *backward difference* operators, which in turn give explicit expressions for the coefficients of multistep methods.

The backward difference operators ∇^k for $k = 0, 1, \dots$ are defined for a set $f_1, \dots, f_m$ by

$$\nabla^0 f_j := f_j, \; \nabla^l f_j := \nabla^{l-1} f_j - \nabla^{l-1} f_{j-1}, \qquad l = 1, 2, \dots . \tag{D.1}$$

So we have, e.g.,

$$\nabla^1 f_j = f_j - f_{j-1}, \; \nabla^2 f_j = (f_j - f_{j-1}) - (f_{j-1} - f_{j-2}) = f_j - 2 f_{j-1} + f_{j-2}.$$

From this we find

$$f_{m-l} = \sum_{j=0}^{l} (-1)^j \binom{l}{j} \nabla^j f_m. \tag{D.2}$$

The backward difference notation is the same as the nabla operator. However, no confusion should occur as we only use backward difference operators in this Appendix D.

Let $\xi_1, \ldots, \xi_m$ be equispaced grid points with grid size h and let f_j be a function value at the point ξ_j. Then we can rewrite the interpolation polynomial in terms of the $\nabla^l f_m$ as

$$p(\xi) = \sum_{j=0}^{m-1} (-1)^j \binom{-s}{j} \nabla^j f_m, \tag{D.3}$$

where $s := \dfrac{\xi - \xi_m}{h}$.

Now consider the ODE

$$\frac{\mathrm{d}x}{\mathrm{d}t} = f(x, t).$$

We want to find an approximation of the solution $x(t_j)$ on a set of equispaced grid points $t_0, t_1, \ldots, t_i, t_{i+1}, \ldots$. At the point t_{i+1} we can easily find an approximation of $\frac{\mathrm{d}x}{\mathrm{d}t}(t_{i+1})$ from (D.3) by differentiation. As a result, we find, identifying the points ξ_j with $t_{i-m+1+j}$ and denoting the numerical approximation of $x(t_{i+1})$ by x_{i+1}, that

$$\sum_{j=0}^{m-1} \nabla^j x_{i+1} = hf(x_{i+1}, t_{i+1}). \tag{D.4a}$$

The coefficients in (D.4a) have a nice form. One can check that

$$\gamma_0 = 0,$$
$$\gamma_j = \frac{1}{j}, \qquad j \geq 0. \tag{D.4b}$$

The formulas found in (D.4) are the *backward difference formulas (BDF)*. In standard form (cf. (5.75)) they are given by

$$\sum_{j=0}^{k} x_{i-j+1} \doteq hf(t_{i+1}, x_{i+1}). \tag{D.5}$$

For $k = 1, 2, 3, 4$ the coefficients α_j are given in Table D.1. Note that for $k = 1$ we have the Euler backward method. Also, for other multistep methods, the formulation in terms of backward differences enables one to find the coefficients in a simple way:

$$x(t_{i+1}) - x(t_i) = \int_{t_i}^{t_{i+1}} f(x(\tau), \tau)\, \mathrm{d}\tau. \tag{D.6}$$

Table D.1. *Coefficients for the BDF.*

k	a_0	a_1	a_2	a_3	a_4
1	1	-1			
2	$\frac{3}{2}$	-2	$\frac{1}{2}$		
3	$\frac{11}{6}$	-3	$\frac{3}{2}$	$-\frac{1}{3}$	
4	$\frac{25}{12}$	-4	3	$-\frac{4}{3}$	$\frac{1}{4}$

If we apply (D.3) on the interval (t_{i-k+1}, t_i), i.e., approximate $f(x(t), t)$ by a polynomial p of degree $k-1$ there, we obtain

$$x(t_{i+1}) - x(t_i) \doteq h \sum_{j=0}^{k-1} \hat{\gamma}_j \nabla^j f_i, \tag{D.7a}$$

where

$$\hat{\gamma}_j := (-1)^j \int_0^1 \binom{-\tau}{j} \mathrm{d}\tau. \tag{D.7b}$$

This is the k-step *Adams–Bashforth* formula. The $\hat{\gamma}_j$ are simply calculated by recursion:

$$\begin{aligned} \hat{\gamma}_0 &= 1, \\ \hat{\gamma}_j &= 1 - \frac{1}{j+1}\hat{\gamma}_0 - \frac{1}{j}\hat{\gamma}_1 - \cdots - \frac{1}{2}\hat{\gamma}_{j-1}, \qquad j \geq 1. \end{aligned} \tag{D.7c}$$

In standard form we have

$$x(t_{i+1}) - x(t_i) \doteq h \sum_{j=1}^{k} b_j f_{i-j+1}. \tag{D.8}$$

For $k = 1, 2, 3, 4$ the coefficients are given in Table D.2. If we apply (D.3) on the interval (t_{i-k+1}, t_{i+1}), i.e., approximate p by a polynomial of degree k there, we obtain

$$x(t_{i+1}) - x(t_i) \doteq h \sum_{j=0}^{k} \bar{\gamma}_j \nabla^j f_{i+1}, \tag{D.9a}$$

where

$$\bar{\gamma}_j = (-1)^j \int_{-1}^{0} \binom{-\tau}{j} \mathrm{d}\tau\ . \tag{D.9b}$$

This is the k-step *Adams–Moulton* formula.

Table D.2. *Coefficients for the Adams–Bashforth formula.*

k	b_1	b_2	b_3	b_4
1	1			
2	$\frac{3}{2}$	$-\frac{1}{2}$		
3	$\frac{23}{12}$	$-\frac{16}{12}$	$\frac{5}{12}$	
4	$\frac{55}{24}$	$-\frac{59}{24}$	$\frac{37}{24}$	$-\frac{9}{24}$

From this we find

$$\bar{\gamma}_0 = 1,$$

$$\bar{\gamma}_j = -\frac{1}{j+1}\,\bar{\gamma}_0 - \frac{1}{j}\,\bar{\gamma}_1 - \cdots - \frac{1}{2}\,\bar{\gamma}_{j-1}, \qquad j \geq 1. \tag{D.9c}$$

The coefficients $\hat{\gamma}_j$ and the $\bar{\gamma}_j$ are independent of k. They are related by

$$\sum_{j=0}^{k} \bar{\gamma}_j = \hat{\gamma}_k. \tag{D.10}$$

We can write the Adams–Moulton formulas in the standard formulation as

$$x_{i+1}^h - x_i^h = h \sum_{j=0}^{k} b_j f_{i-j+1}. \tag{D.11}$$

For $k = 0, 1, 2, 3$ the coefficients b_j are given in Table D.3. Note that for $k = 0$ we obtain the Euler backward method and for $k = 1$ we find the trapezoidal formula (both one step!).

Table D.3. *Coefficients for the Adams–Moulton formula.*

k	b_0	b_1	b_2	b_3
0	1			
1	$\frac{1}{2}$	$\frac{1}{2}$		
2	$\frac{5}{12}$	$\frac{8}{12}$	$-\frac{1}{12}$	
3	$\frac{0}{14}$	$\frac{19}{24}$	$-\frac{5}{12}$	$\frac{1}{24}$

E Solution of Recursions

Consider the recursion

$$\mathbf{x}_{i+1} = \mathbf{A}_i \mathbf{x}_i + \mathbf{b}_i, \tag{E.1}$$

where $\{\mathbf{A}_i\}$ is a set of square matrices (in particular it may be a set of scalars) and $\{\mathbf{b}_i\}$ is a set of vectors. Then the solution of (E.1) is given by

$$\mathbf{x}_i = \left(\prod_{j=0}^{i-1} \mathbf{A}_j\right) \mathbf{x}_0 + \sum_{m=0}^{i-1} \left\{ \left(\prod_{j=m+1}^{i-1} \mathbf{A}_j\right) \mathbf{b}_j \right\}. \tag{E.2}$$

In (E.2) one has to interpret $\prod_{m+1}^{i-1} \mathbf{A}_j$ as $\mathbf{A}_{i-1}\mathbf{A}_{i-2}\cdots\mathbf{A}_{m+1}$ and as $\mathbf{I}$ if $i-1 < m+1$. Now consider the second order scalar recursion

$$x_{i+1} = a_i x_i + b_i x_{i-1} + c_i. \tag{E.3}$$

If $c_i \equiv 0$ and a_i and b_i are constant, then the solution is given by

$$x_i = \alpha\lambda_1^i + \beta\lambda_2^i, \qquad \alpha, \beta \in \mathbb{R}, \tag{E.4}$$

if the *characteristic equation*

$$\lambda^2 - a\lambda - b = 0 \qquad (\text{N.B. } \forall_i a_i = a, \forall_i b_i = b) \tag{E.5}$$

has two different roots λ_1, λ_2. If it has a double root, the solution can be written as

$$x_i = (\alpha + \beta i)\lambda_1^i. \tag{E.6}$$

The constants in (E.4) and (E.6) are to be found from initial or boundary values.

In general the homogeneous part of (E.3) has two independent basis solutions $\{f_i\}$ and $\{g_i\}$. In terms of these we can give a formal solution of (E.3). We find

$$x_i = \sum_{j=1}^{i} \left\{ \frac{f_{j-1}g_i - g_{j-1}f_i}{f_{j-1}g_j - g_{j-1}f_j} c_j \right\} + \alpha f_i + \beta g_i, \tag{E.7}$$

where α and β are to be determined from initial or boundary conditions. In particular, for constant coefficients and two different roots λ_1, λ_2 we obtain

$$x_i = \sum_{j=1}^{i} c_j \frac{\lambda_2^{i-j} - \lambda_1^{i-j}}{\lambda_2 - \lambda_1} + \alpha\lambda_1^i + \beta\lambda_2^i. \tag{E.8}$$

F Eigenvalues and Eigenvectors of a Tridiagonal Matrix

Consider the matrix $\mathbf{A} \in \mathbb{R}^{n\times n}$:

$$\mathbf{A} = \begin{pmatrix} b & c & & & \\ a & b & c & & \\ & \ddots & \ddots & \ddots & \\ & & a & b & c \\ & & & a & b \end{pmatrix}. \tag{F.1}$$

Let μ be an eigenvalue of $\mathbf{A}$ and $\mathbf{x} = (x_1, \ldots, x_N)^T$ be a corresponding eigenvector. Then we find from $\mathbf{Ax} = \mu\mathbf{x}$ that

$$(b-\mu)x_1 + cx_2 = 0, \tag{F.2a}$$

$$ax_{j-1} + (b-\mu)x_j + cx_{j+1} = 0, \qquad j = 2, \ldots, N-1, \tag{F.2b}$$

$$ax_{N-1} + (b-\mu)x_N = 0. \tag{F.2c}$$

Let us take $x_0 = x_{N+1} = 0$. Then we find from (F.2) that $\{x_j\}_0^{N+1}$ is a solution of the three-term recurrence equation

$$ax_{j-1} + (b-\mu)x_j + cx_{j+1} = 0, \qquad j = 1, \ldots, N, \tag{F.3}$$

with boundary values

$$x_0 = x_{N+1} = 0. \tag{F.4}$$

The recursion (F.3) has the characteristic polynomial

$$\lambda^2 + (b-\mu)\lambda + c = 0 \tag{F.5}$$

with roots λ_1 and λ_2 say, so that a solution is formally given by

$$x_j = \alpha\lambda_1^j + \beta\lambda_2^j, \tag{F.6}$$

where α and β can be found from the boundary values (F.4). This gives $\beta = -\alpha$ and eventually the relation

$$\left(\frac{\lambda_1}{\lambda_2}\right)^{N+1} = 1; \tag{F.7}$$

i.e., $\frac{\lambda_1}{\lambda_2}$ is an $(N+1)$th power unit root. Hence

$$\frac{\lambda_1}{\lambda_2} = \mathrm{e}^{\frac{2\pi i l}{N+1}}, \qquad l = 1, \ldots, N. \tag{F.8}$$

Since the product of the roots equals $\dfrac{c}{a}$, we obtain

$$\lambda_1 = \left(\frac{c}{a}\right)^{1/2} \mathrm{e}^{\frac{\pi i l}{N+1}}, \tag{F.9a}$$

$$\lambda_2 = \left(\frac{c}{a}\right)^{1/2} \mathrm{e}^{-\frac{\pi i l}{N+1}}. \tag{F.9b}$$

We now use the fact that $\lambda_1 + \lambda_2 = -(b-\mu)/a$, so that (F.9) gives for the eigenvalues μ_l

$$\mu_l = b + \sqrt{ac}\left(\mathrm{e}^{\frac{\pi i l}{N+1}} + \mathrm{e}^{-\frac{\pi i l}{N+1}}\right) = b + 2\sqrt{ac}\cos\left(\frac{l\pi}{N+1}\right), \qquad l = 1, \ldots, N. \tag{F.10}$$

For the jth component x_l^j of the corresponding eigenvector $\mathbf{x}_l$ we apparently have

$$x_l^j = \alpha(\lambda_1^j - \lambda_2^j) = \alpha\left(\frac{c}{a}\right)^{j/2}\left(\mathrm{e}^{\frac{\pi i j l}{N+1}} + \mathrm{e}^{-\frac{\pi i j l}{N+1}}\right) = 2\mathrm{i}\alpha\left(\frac{c}{a}\right)^{j/2}\sin\left(\frac{\pi j l}{N+1}\right). \tag{F.11}$$

Note that α can be chosen arbitrarily, so, in particular, if $\operatorname{sign}(a) = \operatorname{sign}(b)$, we may assume x_j to be real for all l and j. If, moreover, $a = c$, we can choose α such that $\sum x_j^2 = 1$, resulting in an orthogonal set of eigenvectors with

$$x_l^j = \frac{2}{\sqrt{N}}\sin\left(\frac{\pi j l}{N+1}\right). \tag{F.12}$$

Remark 1. We still have to check whether the representation (F.6) is correct. Indeed, if $\lambda_1 = \lambda_2$ (so we have a double root), the foregoing is not correct. This situation cannot occur, however. Suppose we have $\lambda_1 = \lambda_2$. Then we would have, instead of (F.6),

$$x_j = (\alpha + \beta j)\lambda_1^j. \tag{F.13}$$

From (F.4) we then immediately see that $\alpha = \beta = 0$, which is not interesting, of course.

Remark 2. We can find out more generally whether a tridiagonal matrix has (geometrically) multiple eigenvalues and real eigenvectors (so it is diagonisable) or not. Indeed consider instead of (F.1) the matrix

$$\mathbf{A} = \begin{pmatrix} b_1 & c_1 & & & \\ a_2 & b_2 & c_2 & & \\ & \ddots & \ddots & \ddots & \\ & & a_{N-1} & b_{N-1} & c_{N-1} \\ & & & a_N & b_N \end{pmatrix}. \tag{F.14}$$

Let $\mathbf{D}$ be a diagonal matrix $\mathbf{D} = \text{diag}(d_1, \ldots, d_N)$, with $d_1, \ldots, d_N$ still to be chosen. We define

$$\tilde{\mathbf{A}} = \mathbf{D}\mathbf{A}\mathbf{D}^{-1}. \tag{F.15}$$

We now require $\mathbf{D}$ to be such that $\tilde{\mathbf{A}}$ is a symmetric matrix. This is so if

$$\left(\frac{d_{j+1}}{d_j}\right)^2 = \frac{c_j}{a_{j+1}}, \qquad 1 \le j \le N-1. \tag{F.16}$$

Hence if, e.g., $\text{sign}(c_j) = \text{sign}(a_{j+1})$, such a matrix $\mathbf{D}$ certainly exists. Apparently, $\tilde{\mathbf{A}}$ has simple eigenvalues. Due to $\mathbf{A}$ and $\tilde{\mathbf{A}}$ being similar, this property carries over to $\mathbf{A}$.

G Norms

Let V be a linear vector space.

Definition G.1. *A norm on V, denoted by $\|\cdot\|$, satisfies the following four conditions:*

(i) $\|\mathbf{x}\| \in \mathbb{R}$ *and* $\|\mathbf{x}\| \ge 0$.

(ii) $\|\mathbf{x}\| = 0$ *if and only if* $\mathbf{x} = \mathbf{0}$.

(iii) $\|\gamma\mathbf{x}\| = |\gamma|\|\mathbf{x}\|$ *for any* $\gamma \in \mathbb{R}$.

(iv) $\|\mathbf{x} + \mathbf{y}\| \le \|\mathbf{x}\| + \|\mathbf{y}\|$ *for any* $\mathbf{x}, \mathbf{y} \in V$ *(triangle inequality).*

If $V = \mathbb{R}^n$ and $\mathbf{x} \in \mathbb{R}^N$ denotes a vector with coordinates $x_1, \ldots, x_N$, say, then we often use *Hölder* norms:

$$\|\mathbf{x}\|_1 = \sum_{j=1}^{N} |x_j|, \tag{G.1a}$$

$$\|\mathbf{x}\|_2 = \left(\sum_{j=1}^{N} |x_j|^2\right)^{1/2}, \tag{G.1b}$$

$$\|\mathbf{x}\|_\infty = \max_{1\le j\le N} |x_j|. \tag{G.1c}$$

The norms $\|\cdot\|_\alpha$ and $\|\cdot\|_\beta$ for some α and β are called *equivalent* if there exist $c_1, c_2 \in \mathbb{R}_+$ such that for all $\mathbf{x}$ $c_1\|\mathbf{x}\|_\alpha \le \|\mathbf{x}\|_\beta \le c_2\|\mathbf{x}\|_\alpha$. For $N < \infty$ all norms are equivalent. In particular, we have

$$\begin{aligned} \|\mathbf{x}\|_2 &\le \|\mathbf{x}\|_1 \le \sqrt{N}\|\mathbf{x}\|_2, \\ \|\mathbf{x}\|_\infty &\le \|\mathbf{x}\|_2 \le \sqrt{N}\|\mathbf{x}\|_\infty. \end{aligned} \tag{G.2}$$

The bounds above are attainable. Hence there is no longer an equivalence if $N \to \infty$.

A consequence of equivalence is that a series that converges in one norm also converges in an equivalent norm. If $V = \mathbb{R}^n$, this implies that convergence considerations are norm independent.

A useful property is the *Cauchy–Schwarz inequality*: For all $\mathbf{x}, \mathbf{y} \in \mathbb{R}^N$ we have

$$|(\mathbf{x}, \mathbf{y})| \le \|\mathbf{x}\|_2\|\mathbf{y}\|_2 \qquad (\,(\cdot,\cdot) \text{ is the Euclidean inner product}). \tag{G.3}$$

Next we consider matrix norms. Let V be the linear space consisting of matrices. A norm on V satisfies conditions similar to those in Definition G.1(i), ..., (iv). A vector norm induces an *associated matrix norm* in a natural way as follows:

$$\|\mathbf{A}\| := \max_{\mathbf{x}\neq 0} \frac{\|\mathbf{A}\mathbf{x}\|}{\|\mathbf{x}\|}. \tag{G.4}$$

As one can easily verify, we have

$$\|\mathbf{A}\| = \max_{\|\mathbf{x}\|=1} \|\mathbf{A}\mathbf{x}\|.$$

On top of (i), ..., (iv), such an associated norm apparently also has a *multiplicity* property

(v) $\|\mathbf{AB}\| \le \|\mathbf{A}\|\|\mathbf{B}\|$.

The most often used associated matrix norms are

$$\|\mathbf{A}\|_1 = \max_j \sum_{i=1}^{n} |a_{ij}|, \tag{G.5a}$$

$$\|\mathbf{A}\|_\infty = \max_i \sum_{j=1}^{n} |a_{ij}|, \tag{G.5b}$$

$$\|\mathbf{A}\|_2 = \left(\rho(\mathbf{A}^T\mathbf{A})\right)^{\frac{1}{2}}, \tag{G.5c}$$

where $\rho(\mathbf{B})$ is the absolutely largest eigenvalue of $\mathbf{B}$; see Definition H.3.

The 2-norm is especially interesting because it is *orthogonally invariant*; i.e., if $\mathbf{Q}_1$ and $\mathbf{Q}_2$ are orthogonal matrices, then the following holds:

$$\|\mathbf{Q}_1\mathbf{A}\mathbf{Q}_2\|_2 = \|\mathbf{A}\|_2.$$

Also for matrices one can show equivalence of norms, at least for finite dimension. In particular we have

$$\begin{aligned}\frac{1}{\sqrt{N}}\|\mathbf{A}\|_1 &\le \|\mathbf{A}\|_2 \le \sqrt{N}\|\mathbf{A}\|_1,\\ \frac{1}{\sqrt{N}}\|\mathbf{A}\|_\infty &\le \|\mathbf{A}\|_2 \le \sqrt{N}\|\mathbf{A}\|_\infty.\end{aligned} \tag{G.6}$$

If V is the linear space of scalar functions $x(t)$, defined on an interval $[\alpha, \beta]$ say, we can introduce analogues of the Hölder norm for the continuous case:

$$\|x\|_p := \left(\int_\alpha^\beta |x(t)|^p \,\mathrm{d}t\right)^{\frac{1}{p}}. \tag{G.7}$$

Clearly, if V is a space of vector functions, one has to replace the modulus by a suitable vector norm in (G.7). For $p = \infty$ we have

$$\|x\|_\infty = \sup_{t\in[a,b]} |x(t)|. \tag{G.8}$$

Each norm on a linear space V generates a *metric* d on V by $d(\mathbf{x}, \mathbf{y}) := \|\mathbf{x} - \mathbf{y}\|$, $\mathbf{x}, \mathbf{y} \in V$. This metric has the property of *translation invariance*: $d(\mathbf{x} + \mathbf{z}, \mathbf{y} + \mathbf{z}) = d(\mathbf{x}, \mathbf{y})$.

H Similarity

Let $\mathbf{A}$ be a matrix. If

$$\mathbf{A}\mathbf{x} = \lambda\mathbf{x} \tag{H.1}$$

for some vector $\mathbf{x}$ and scalar λ, then λ is called an *eigenvalue* and $\mathbf{x}$ an *eigenvector* (belonging to λ) of $\mathbf{A}$.

Property H.1.

(i) *An eigenvalue λ is a zero of the* characteristic polynomial $\det(\mathbf{A} - \lambda\mathbf{I})$.

(ii) *The product of the eigenvalues of* $\mathbf{A}$ *is equal to* $\det(\mathbf{A})$.

(iii) *The sum of the eigenvalues of* $\mathbf{A}$ *is equal to* $\sum_{j=1}^n a_{jj}$, *the* trace *of* $\mathbf{A}$.

For every matrix $\mathbf{A}$ a nonsingular matrix $\mathbf{T}$ exists such that

$$\mathbf{T}^{-1}\mathbf{A}\mathbf{T} = \mathbf{J}, \tag{H.2}$$

where $\mathbf{J}$ is a bidiagonal matrix consisting of blocks containing the same eigenvalues whose dimensions correspond to the (algebraic) multiplicity of those eigenvalues; i.e.,

$$\mathbf{J} := \begin{pmatrix} \lambda_1 & 1 & & & & & & \emptyset \\ & \ddots & 1 & & & & & \\ & & \lambda_1 & & & & & \\ & & & \lambda_2 & \ddots & & & \\ & & & & \ddots & & & \\ & & & & & \lambda_j & \ddots & \\ & & & & & & \ddots & \\ & & \emptyset & & & & & \lambda_p \end{pmatrix}. \tag{H.3}$$

The matrices $\mathbf{T}$ in (H.2) are often chosen such that the "dots" in (H.3) are all equal to one. From a numerical point of view this may not be so meaningful (as it may cause $\mathbf{T}$ to be very skew). The form (H.2) is called the *Jordan normal form.* The *geometric multiplicity* of an eigenvalue is the dimension of the space of independent eigenvectors. If, in particular, the algebraic and geometric multiplicity are the same for all eigenvalues, then $\mathbf{J}$ is a diagonal matrix and each column of $\mathbf{T}$ is then an eigenvector.

If $\mathbf{A}$ is *symmetric*, i.e., $\mathbf{A} = \mathbf{A}^T$, or *skew symmetric*, i.e., $\mathbf{A} = -\mathbf{A}^T$, the transformation matrix $\mathbf{T}$ is orthogonal.

The transformation (H.2) is a special instance of a *matrix similarity transformation.* If $\mathbf{S}$ is nonsingular, the matrix $\mathbf{S}^{-1}\mathbf{A}\mathbf{S} = \mathbf{B}$ is called *similar* to $\mathbf{A}$. Of course it corresponds to viewing a mapping on a different basis.

Property H.2. *Similar matrices have the same eigenvalues.*

Definition H.3. *The absolute value of the absolutely largest eigenvalue of a matrix is called the spectral radius of* $\mathbf{A}$ *and is denoted as* $\rho(\mathbf{A})$.

Property H.4. *If* $\mathbf{A}$ *is symmetric, then* $\|\mathbf{A}\|_2 = \rho(\mathbf{A})$. *If* $\mathbf{A}$ *is not symmetric, then* $\|\mathbf{A}\|_2 = (\rho(\mathbf{A}^T\mathbf{A}))^{\frac{1}{2}}$.

I Estimates of Eigenvalues and Consequences

The spectral radius of a matrix is a useful means of deciding a number of properties. In particular, it is sometimes crucial to know whether this radius is at least smaller than one. This is why we have the following theorems.

Theorem I.1. $\rho(\mathbf{A}) < 1$ *if and only if* $\lim_{i\to\infty} \mathbf{A}^i = 0$.

Proof. First we prove the "if" direction:

$$\left(\rho(\mathbf{A})\right)^i = \rho\left(\mathbf{A}^i\right) \le \|\mathbf{A}^i\|^2.$$

Now we deal with the "only if" direction: Consider the Jordan normal form. Then $\mathbf{A}^i = \mathbf{T}\mathbf{J}^i\mathbf{T}^{-1}$. By studying a single Jordan block and splitting it into diagonal and codiagonal parts, we can easily complete the proof. □

Theorem I.2 (Neumann series). $\lim_{n\to\infty}\sum_{i=0}^{n}\mathbf{A}^i$ *exists if and only if* $\rho(\mathbf{A}) < 1$. *The following holds:* $\sum_{i=0}^{\infty}\mathbf{A}^i = (\mathbf{I}-\mathbf{A})^{-1}$.

Proof. If $\{\sum_0^n \mathbf{A}^i\}$ converges, then $\mathbf{A}^i \to 0$, which implies that $\rho(\mathbf{A}) < 1$. If, on the other hand, $\rho(\mathbf{A}) < 1$, then $\det(\mathbf{I}-\mathbf{A}) \neq 0$, so $(\mathbf{I}-\mathbf{A})^{-1}$ exists. For each n we have $(\mathbf{I}+\mathbf{A}+\cdots+\mathbf{A}^n)(\mathbf{I}-\mathbf{A}) = \mathbf{I}-\mathbf{A}^{n+1}$. □

Corollary I.3. *The matrix* $\mathbf{I}-\mathbf{A}$ *is nonsingular if for any norm* $\|\mathbf{A}\| < 1$.

Definition I.4. *An eigenvalue λ is said to have algebraic multiplicity p if the characteristic polynomial of the matrix has precisely p roots λ. The eigenvalue λ is said to have geometric multiplicity q if there exists a q-dimensional subspace of eigenvectors belonging to λ. A matrix is called defect if the geometric multiplicity is smaller than the algebraic multiplicity.*

Clearly a matrix is defect if and only if it has Jordan blocks of size larger than one.

Definition I.5. *A matrix is called stable if* $\rho(\mathbf{A}) \leq 1$ *and the geometric multiplicity of the eigenvalues with modulus one is equal to the algebraic multiplicity.*

Note that a stable matrix may have more than one eigenvalue equal to one. The identity matrix **I** is the best-known example.

Property I.6. *If* **A** *is stable, then there exists a constant* κ, *say, such that for all* $\mathbf{x} \in \mathbb{R}^N$ $\|\mathbf{A}^i\mathbf{x}\| \leq \kappa\|\mathbf{x}\|$ *(this* κ *depends on N and the norm chosen).*

Finally we give an important theorem to estimate the eigenvalues.

Theorem I.7 (Gershgorin). *An eigenvalue of* **A** *lies in at least one of the closed discs with centre point* a_{ii} *and radius* $\sum_{j\neq i}|a_{ij}|$. *(Note that the discs are in* $\mathbb{C}$.*)*

Proof. Let λ be an eigenvalue and $\mathbf{x} = (x_1, \ldots, x_N)^T$ be a corresponding eigenvector. Then

$$\mathbf{A}\mathbf{x} = \lambda\mathbf{x} \;\Rightarrow\; (\lambda - a_{rr})x_r = \sum_{j\neq r} a_{rj}x_j.$$

Hence $|\lambda - a_{rr}||x_r| \leq \sum_{j\neq r}|a_{rj}||x_j|$ for all r. If m is such that $|x_m| = \max_r |x_r|$ (note that $x_m \neq 0$), then

$$|\lambda - a_{mm}| \leq \sum_{j\neq m}|a_{mj}|\frac{|x_j|}{|x_m|} \leq \sum_{j\neq m}|a_{mj}|.$$

Note that if the matrix is symmetric, the discs are in fact segments of the real axis. □

J Theorems from Vector Calculus

Let $\boldsymbol{a}$, $\boldsymbol{b}$, and $\boldsymbol{c}$ be three-dimensional vectors, let $\boldsymbol{v}$ and $\boldsymbol{w}$ be well-behaved vector functions from $\mathbb{R}^3 \to \mathbb{R}^3$, and let φ and ψ be well-behaved scalar functions mapping $\mathbb{R}^3 \to \mathbb{R}$. Let Ω be a three-dimensional volume with volume element $\mathrm{d}V$, and let $\partial\Omega$ be a closed two-dimensional surface bounding Ω with area element $\mathrm{d}S$ and associated unit outward vector $\boldsymbol{n}$. Let $\mathcal{S}$ denote an open surface with the oriented contour $\mathcal{C}$, with line element $\mathrm{d}\boldsymbol{\ell}$ bounding it. The normal $\boldsymbol{n}$ to $\mathcal{S}$ is defined according to the right-hand screw rule applied to $\mathcal{C}$. Then we have the following vector and integral relations:

$$\boldsymbol{a}\cdot(\boldsymbol{b}\times\boldsymbol{c}) = \boldsymbol{b}\cdot(\boldsymbol{c}\times\boldsymbol{a}) = \boldsymbol{c}\cdot(\boldsymbol{a}\times\boldsymbol{b}), \tag{J.1}$$

$$\boldsymbol{a}\times(\boldsymbol{b}\times\boldsymbol{c}) = \boldsymbol{b}(\boldsymbol{a}\cdot\boldsymbol{c}) - \boldsymbol{c}(\boldsymbol{a}\cdot\boldsymbol{b}), \tag{J.2}$$

$$(\boldsymbol{a}\times\boldsymbol{b})\cdot(\boldsymbol{c}\times\boldsymbol{d}) = (\boldsymbol{a}\cdot\boldsymbol{c})(\boldsymbol{b}\cdot\boldsymbol{d}) - (\boldsymbol{a}\cdot\boldsymbol{d})(\boldsymbol{b}\cdot\boldsymbol{c}), \tag{J.3}$$

$$\nabla(\boldsymbol{v}\cdot\boldsymbol{w}) = \boldsymbol{v}\cdot\nabla\boldsymbol{w} + \boldsymbol{w}\cdot\nabla\boldsymbol{v} + \boldsymbol{v}\times(\nabla\times\boldsymbol{w}) + \boldsymbol{w}\times(\nabla\times\boldsymbol{v}), \tag{J.4}$$

$$\boldsymbol{v}\cdot\nabla\boldsymbol{v} = \frac{1}{2}\nabla(\boldsymbol{v}\cdot\boldsymbol{v}) + (\nabla\times\boldsymbol{v})\times\boldsymbol{v}, \tag{J.5}$$

$$\boldsymbol{a}\cdot\nabla(\boldsymbol{v}\cdot\boldsymbol{w}) = \boldsymbol{v}\cdot(\boldsymbol{a}\cdot\nabla\boldsymbol{w}) + \boldsymbol{w}\cdot(\boldsymbol{a}\cdot\nabla\boldsymbol{v}), \tag{J.6}$$

$$\nabla\cdot(\varphi\boldsymbol{v}) = \varphi\nabla\cdot\boldsymbol{v} + \boldsymbol{v}\cdot\nabla\varphi, \tag{J.7}$$

$$\nabla\cdot(\boldsymbol{v}\times\boldsymbol{w}) = \boldsymbol{w}\cdot(\nabla\times\boldsymbol{v}) - \boldsymbol{v}\cdot(\nabla\times\boldsymbol{w}), \tag{J.8}$$

$$\nabla\times(\boldsymbol{v}\times\boldsymbol{w}) = \boldsymbol{v}(\nabla\cdot\boldsymbol{w}) - \boldsymbol{w}(\nabla\cdot\boldsymbol{v}) - \boldsymbol{v}\cdot\nabla\boldsymbol{w} + \boldsymbol{w}\cdot\nabla\boldsymbol{v}, \tag{J.9}$$

$$\nabla\cdot(\nabla\times\boldsymbol{v}) = 0, \tag{J.10}$$

$$\nabla\times(\nabla\varphi) = 0, \tag{J.11}$$

$$\nabla\times(\nabla\times\boldsymbol{v}) = \nabla(\nabla\cdot\boldsymbol{v}) - \nabla^2\boldsymbol{v}. \tag{J.12}$$

We also have the following results:

Gauss's or divergence theorem:

$$\int_\Omega \nabla\cdot\boldsymbol{v}\,\mathrm{d}V = \oint_{\partial\Omega} \boldsymbol{v}\cdot\boldsymbol{n}\,\mathrm{d}S, \tag{J.13}$$

$$\int_\Omega \nabla\varphi\,\mathrm{d}V = \oint_{\partial\Omega} \varphi\boldsymbol{n}\,\mathrm{d}S, \tag{J.14}$$

$$\int_\Omega \nabla\times\boldsymbol{v}\,\mathrm{d}V = \oint_{\partial\Omega} \boldsymbol{n}\times\boldsymbol{v}\,\mathrm{d}S; \tag{J.15}$$

Green's first identity:

$$\int_\Omega \left(\varphi\nabla^2\psi + \nabla\varphi\cdot\nabla\psi\right)\mathrm{d}V = \oint_{\partial\Omega} \varphi\nabla\psi\cdot\boldsymbol{n}\,\mathrm{d}S; \tag{J.16}$$

Green's second identity:

$$\int_\Omega (\varphi\nabla^2\psi - \psi\nabla^2\varphi)\,\mathrm{d}V = \oint_{\partial\Omega} (\varphi\nabla\psi - \psi\nabla\varphi)\cdot\boldsymbol{n}\,\mathrm{d}S; \tag{J.17}$$

Stokes's theorem:

$$\int_{\mathcal{S}} (\nabla\times\boldsymbol{v})\cdot\boldsymbol{n}\,\mathrm{d}S = \oint_{\mathcal{C}} \boldsymbol{v}\cdot\mathrm{d}\boldsymbol{\ell}, \tag{J.18}$$

$$\int_{\mathcal{S}} \boldsymbol{n}\times\nabla\varphi\,\mathrm{d}S = \oint_{\mathcal{C}} \varphi\mathrm{d}\boldsymbol{\ell}. \tag{J.19}$$

Let $q(\boldsymbol{x},t)$ be a quantity per unit volume of a fluid. Consider a material volume $\Omega(t)$ moving with the flow. Then we have the following:

$$\text{Transport theorem:}\quad \frac{\mathrm{d}}{\mathrm{d}t}\int_{\Omega(t)} q(\boldsymbol{x},t)\,\mathrm{d}V = \int_{\Omega(t)} \left(\frac{\partial}{\partial t}q(\boldsymbol{x},t) + \nabla\cdot(q\boldsymbol{v})(\boldsymbol{x},t)\right)\mathrm{d}V. \quad \text{(J.20)}$$

K Cartesian, Cylindrical, and Spherical Coordinates

Cartesian. Let $\boldsymbol{e}_x$, $\boldsymbol{e}_y$, and $\boldsymbol{e}_z$ be the orthogonal unit vectors associated with the Cartesian x, y, and z coordinates, and let $\boldsymbol{E} = \boldsymbol{e}_x E_x + \boldsymbol{e}_y E_y + \boldsymbol{e}_z E_z$ and φ be smooth functions of (x,y,z). Then

$$(\mathrm{d}\ell)^2 = (\mathrm{d}x)^2 + (\mathrm{d}y)^2 + (\mathrm{d}z)^2, \quad \text{(K.1a)}$$

$$\nabla\varphi = \boldsymbol{e}_x\frac{\partial\varphi}{\partial x} + \boldsymbol{e}_y\frac{\partial\varphi}{\partial y} + \boldsymbol{e}_z\frac{\partial\varphi}{\partial z}, \quad \text{(K.1b)}$$

$$\nabla^2\varphi = \frac{\partial^2\varphi}{\partial x^2} + \frac{\partial^2\varphi}{\partial y^2} + \frac{\partial^2\varphi}{\partial z^2}, \quad \text{(K.1c)}$$

$$\nabla\cdot\boldsymbol{E} = \frac{\partial E_x}{\partial x} + \frac{\partial E_y}{\partial y} + \frac{\partial E_z}{\partial z}, \quad \text{(K.1d)}$$

$$\nabla\times\boldsymbol{E} = \begin{vmatrix} \boldsymbol{e}_x & \boldsymbol{e}_y & \boldsymbol{e}_z \\ \dfrac{\partial}{\partial x} & \dfrac{\partial}{\partial y} & \dfrac{\partial}{\partial z} \\ E_x & E_y & E_z \end{vmatrix}. \quad \text{(K.1e)}$$

Cylindrical. Let $\boldsymbol{e}_r$, $\boldsymbol{e}_\phi$, and $\boldsymbol{e}_z$ be the orthogonal unit vectors associated with the cylindrical r, ϕ, and z coordinates, and let $\boldsymbol{E} = \boldsymbol{e}_r E_r + \boldsymbol{e}_\phi E_\phi + \boldsymbol{e}_z E_z$ and φ be smooth functions of (r,ϕ,z). Then

$$x = r\cos\phi, \quad y = r\sin\phi, \quad z = z, \quad \text{(K.2a)}$$

$$(\mathrm{d}\ell)^2 = (\mathrm{d}r)^2 + (r\,\mathrm{d}\phi)^2 + (\mathrm{d}z)^2, \quad \text{(K.2b)}$$

$$\nabla\varphi = \boldsymbol{e}_r\frac{\partial\varphi}{\partial r} + \boldsymbol{e}_\phi\frac{1}{r}\frac{\partial\varphi}{\partial\phi} + \boldsymbol{e}_z\frac{\partial\varphi}{\partial z}, \quad \text{(K.2c)}$$

$$\nabla^2\varphi = \frac{1}{r}\frac{\partial}{\partial r}\left(r\frac{\partial\varphi}{\partial r}\right) + \frac{1}{r^2}\frac{\partial^2\varphi}{\partial\phi^2} + \frac{\partial^2\varphi}{\partial z^2}, \quad \text{(K.2d)}$$

$$\nabla\cdot\boldsymbol{E} = \frac{1}{r}\frac{\partial}{\partial r}(rE_r) + \frac{1}{r}\frac{\partial E_\phi}{\partial\phi} + \frac{\partial E_z}{\partial z}, \quad \text{(K.2e)}$$

$$\nabla\times\boldsymbol{E} = \frac{1}{r}\begin{vmatrix} \boldsymbol{e}_r & \boldsymbol{e}_\phi r & \boldsymbol{e}_z \\ \dfrac{\partial}{\partial r} & \dfrac{\partial}{\partial\phi} & \dfrac{\partial}{\partial z} \\ E_r & E_\phi r & E_z \end{vmatrix}. \quad \text{(K.2f)}$$

Spherical. Let $\boldsymbol{e}_r$, $\boldsymbol{e}_\theta$, and $\boldsymbol{e}_\phi$ be the orthogonal unit vectors associated with the spherical r, θ, and ϕ coordinates, and let $\boldsymbol{E} = \boldsymbol{e}_r E_r + \boldsymbol{e}_\theta E_\theta + \boldsymbol{e}_\phi E_\phi$ and φ be smooth functions of (r, θ, ϕ). Then

$$x = r\sin\theta\cos\phi, \quad y = r\sin\theta\sin\phi, \quad z = r\cos\theta, \tag{K.3a}$$

$$(\mathrm{d}\ell)^2 = (\mathrm{d}r)^2 + (r\,\mathrm{d}\theta)^2 + (r\sin\theta\,\mathrm{d}\phi)^2, \tag{K.3b}$$

$$\nabla\varphi = \boldsymbol{e}_r\frac{\partial\varphi}{\partial r} + \boldsymbol{e}_\theta\frac{1}{r}\frac{\partial\varphi}{\partial\theta} + \boldsymbol{e}_\phi\frac{1}{r\sin\theta}\frac{\partial\varphi}{\partial\phi}, \tag{K.3c}$$

$$\nabla^2\varphi = \frac{1}{r^2}\frac{\partial}{\partial r}\left(r^2\frac{\partial\varphi}{\partial r}\right) + \frac{1}{r^2\sin\theta}\frac{\partial}{\partial\theta}\left(\sin\theta\frac{\partial\varphi}{\partial\theta}\right) + \frac{1}{r^2\sin^2\theta}\frac{\partial^2\varphi}{\partial\phi^2}, \tag{K.3d}$$

$$\nabla\cdot\boldsymbol{E} = \frac{1}{r^2}\frac{\partial}{\partial r}(r^2E_r) + \frac{1}{r\sin\theta}\frac{\partial}{\partial\theta}(\sin\theta\,E_\theta) + \frac{1}{r\sin\theta}\frac{\partial E_\phi}{\partial\phi}, \tag{K.3e}$$

$$\nabla\times\boldsymbol{E} = \frac{1}{r^2\sin\theta}\begin{vmatrix} \boldsymbol{e}_r & \boldsymbol{e}_\theta r & \boldsymbol{e}_\phi r\sin\theta \\ \dfrac{\partial}{\partial r} & \dfrac{\partial}{\partial\theta} & \dfrac{\partial}{\partial\phi} \\ E_r & E_\theta r & E_\phi r\sin\theta \end{vmatrix}. \tag{K.3f}$$

L Tensors

A three-dimensional vector space over the field of real numbers, equipped with an inner, dot, or scalar product $\boldsymbol{a}\cdot\boldsymbol{b}$ and an outer, cross, or vector product $\boldsymbol{a}\times\boldsymbol{b}$ is called a *Euclidean vector space*.

A *tensor*—strictly speaking of order two—is a linear transformation of a Euclidean vector space into itself. The identity tensor is denoted by $\mathcal{I}$.

If the tensor $\mathcal{A}$ is written as a 3×3 matrix (a_{ij}) on the standard basis $[\boldsymbol{e}_1, \boldsymbol{e}_2, \boldsymbol{e}_3]$ of $\mathbb{R}^3$, we have the *contraction* or *trace* of $\mathcal{A}$, given by $\mathrm{tr}(\mathcal{A}) = a_{11} + a_{22} + a_{33}$, and the *determinant* of $\mathcal{A}$, given by $\det(\mathcal{A}) = \mathcal{A}\boldsymbol{e}_1\cdot(\mathcal{A}\boldsymbol{e}_2\times\mathcal{A}\boldsymbol{e}_3)$. Both are invariants of $\mathcal{A}$ under rotation of axes and may therefore be expressed in terms of the eigenvalues $\lambda_1, \lambda_2, \lambda_3$ of matrix (a_{ij}). Altogether, $\mathcal{A}$ has three invariants.

A (second order) tensor $\mathcal{A}$ is completely determined by its *invariants*

$$\mathrm{tr}(\mathcal{A}) = \lambda_1 + \lambda_2 + \lambda_3, \tag{L.1a}$$

$$\frac{1}{2}[\mathrm{tr}(\mathcal{A})^2 - \mathrm{tr}(\mathcal{A}^2)] = \lambda_1\lambda_2 + \lambda_2\lambda_3 + \lambda_3\lambda_1, \tag{L.1b}$$

$$\det(\mathcal{A}) = \lambda_1\lambda_2\lambda_3. \tag{L.1c}$$

With each $\mathcal{A}$ is associated a traceless tensor (the *deviator* or deviatoric part of $\mathcal{A}$)

$$\mathcal{A}' := \mathcal{A} - \frac{1}{3}\,\mathrm{tr}(\mathcal{A})\mathcal{I}, \tag{L.2}$$

with the same invariants except the first one, which is zero.

The *divergence* of a tensor is a vector whose ith component, in orthogonal coordinates (x_1, x_2, x_3), is given by

$$(\nabla\cdot\mathcal{A})_i = \sum_{j=1}^{3} \frac{\partial}{\partial x_j}\mathcal{A}_{ij} \qquad (i = 1, 2, 3). \tag{L.3}$$

The *inner product* of two tensors $\mathcal{A}$ and $\mathcal{B}$ produces a tensor $\mathcal{A}\cdot\mathcal{B}$ whose components are given by

$$(\mathcal{A}\cdot\mathcal{B})_{ij} := \sum_{k=1}^{3} \mathcal{A}_{ik}\mathcal{B}_{kj}. \tag{L.4}$$

The *double inner product* of two tensors $\mathcal{A}$ and $\mathcal{B}$ produces a scalar $\mathcal{A}:\mathcal{B}$, which can be evaluated as the sum of the nine products of the tensor components:

$$\mathcal{A}:\mathcal{B} = \sum_{i=1}^{3}\sum_{j=1}^{3} \mathcal{A}_{ij}\mathcal{B}_{ij}. \tag{L.5}$$

The *dyadic product* of two vectors $\boldsymbol{a}$ and $\boldsymbol{b}$ produces a tensor $\boldsymbol{a}\boldsymbol{b}^T$ given by

$$(\boldsymbol{a}\boldsymbol{b}^T)_{ij} = a_i b_j \quad \text{or} \quad \boldsymbol{a}\boldsymbol{b}^T = \begin{pmatrix} a_1b_1 & a_1b_2 & a_1b_3 \\ a_2b_1 & a_2b_2 & a_2b_3 \\ a_3b_1 & a_3b_2 & a_3b_3 \end{pmatrix}. \tag{L.6}$$

M Dimensionless Numbers

Archimedes	Ar	$g\Delta\rho L^3/\rho\nu^2$	particles, drops, or bubbles
Arrhenius	Arr	$E/\mathcal{R}T$	chemical reactions
Biot	Bi	hL/κ	heat transfer at surface of body
Biot	Bi	$h_D L/D$	mass transfer
Bodenstein	Bo	VL/D_{ax}	mass transfer with axial dispersion
Bond	Bo	$\rho g L^2/\sigma$	gravity against surface tension
Capillary	Ca	$\mu V/\sigma$	viscous forces against surface tension
Dean	De	$(VL/\nu)(L/2r)^{1/2}$	flow in curved channels
Eckert	Ec	$V^2/C_P\Delta T$	kinetic energy against enthalpy difference
Euler	Eu	$\Delta p/\rho V^2$	pressure resistance
Fourier	Fo	$\alpha t/L^2$	heat conduction
Fourier	Fo	Dt/L^2	diffusion
Froude	Fr	$V/(gL)^{1/2}$	gravity waves
Galileo	Ga	$gL^3\rho^2/\mu^2$	gravity against viscous forces
Grashof	Gr	$\beta\Delta T g L^3/\nu^3$	natural convection
Helmholtz	He	$\omega L/c = kL$	acoustic wave number
Kapitza	Ka	$g\mu^4/\rho\sigma^3$	film flow
Knudsen	Kn	λ/L	low-density flow
Lewis	Le	α/D	combined heat and mass transfer
Mach	Ma	V/c	compressible flow
Nusselt	Nu	hL/κ	convective heat transfer
Ohnesorge	Oh	$\mu/(\rho L\sigma)^{1/2}$	viscous forces, inertia, and surface tension
Péclet	Pe	VL/α	forced-convection heat transfer
Péclet	Pe	VL/D	forced-convection mass transfer
Prandtl	Pr	$\nu/\alpha = C_P\mu/\kappa$	convective heat transfer
Rayleigh	Ra	$\beta\Delta T g L^3/\alpha\nu$	natural convection heat transfer
Reynolds	Re	$\rho VL/\mu$	viscous forces against inertia
Schmidt	Sc	ν/D	convective mass transfer
Sherwood	Sh	$h_D L/D$	convective mass transfer
Stanton	St	$h/\rho C_P V$	forced-convection heat transfer
Stanton	St	h_D/V	forced-convection mass transfer
Stokes	S	ν/fL^2	viscous damping in unsteady flow
Strouhal	Sr	fL/V	hydrodynamic wave number
Weber	We	$\rho V^2 L/\sigma$	film flow, bubble formation, or droplet breakup

In this table we used the following symbols.

c	sound speed	m/s
C_P	specific heat	J/(kg · K)
D	diffusion coefficient	m^2/s
D_{ax}	axial dispersion coefficient	m^2/s
E	activation energy	J/mol
f	frequency	1/s
g	gravitational acceleration	m/s^2
h	heat transfer coefficient	$W/(m^2 \cdot K)$
h_D	mass transfer coefficient	m/s
k	wave number = ω/c	1/m
L	length	m
$p, \Delta p$	pressure	Pa
$\mathcal{R}$	universal gas constant	J/(mol · K)
r	radius of curvature	m
$T, \Delta T$	temperature	K
t	time	s
V	velocity	m/s
$\alpha = \kappa/\rho C_P$	thermal diffusivity	m^2/s
β	coefficient of thermal expansion	K^{-1}
κ	thermal conductivity	W/(m · K)
λ	molecular mean free path	m
μ	dynamic viscosity	Pa · s
$\nu = \mu/\rho$	kinematic viscosity	m^2/s
$\rho, \Delta\rho$	density	kg/m^3
σ	surface tension	N/m
ω	circular frequency = $2\pi f$	1/s

Bibliography

[1] M. Abramowitz and I.A. Stegun, *Handbook of Mathematical Functions*, National Bureau of Standards, Dover Publications, New York, 1964.

[2] R. Aris, *Vectors, Tensors and the Basic Equations of Fluid Dynamics*, Dover Publications, New York, 1962.

[3] U.M. Ascher, R.M.M. Mattheij, and R.D. Russell, *Numerical Solution of Boundary Value Problems for Ordinary Differential Equations*, Classics in Appl. Math. 13, SIAM, Philadelphia, 1995.

[4] H. Ashley and M. Landahl, *Aerodynamics of Wings and Bodies*, Dover Publications, New York, 1985.

[5] S.H. Baek, E.S. Jeong, and S. Jeong, *Two-dimensional model for tapered pulse tubes. Part 1: Theoretical modeling and net enthalpy flow*, Cryogenics, 40 (2000), pp. 379–385.

[6] S.H. Baek, E.S. Jeong, and S. Jeong, *Two-dimensional model for tapered pulse tubes. Part 2: Mass streaming and streaming-driven enthalpy flow loss*, Cryogenics, 40 (2000), pp. 387–392.

[7] J. Bear, *Dynamics of Fluids in Porous Media*, American Elsevier, New York, 1972.

[8] F.P. Bretherton and C.J.R. Garrett, Wavetrains in inhomogeneous moving media, *Proceedings of the Royal Society* A, 302 (1969), pp. 529–554.

[9] O. Axelsson and V.A. Barker, *Finite Element Solution of Boundary Value Problems: Theory and Computation*, Classics in App. Math. 35, SIAM, Philadelphia, 2001.

[10] C.L. Babcock, *Viscosity and electrical conductivity of molten glasses*, Journal of the American Ceramic Society, 17 (1934), pp. 329–342.

[11] I. Babuska, J.E. Flaherty, and W.D. Henshaw, *Modeling, Mesh Generation, and Adaptive Numerical Methods for Partial Differential Equations*, Springer, New York, 1995.

[12] G.I. Barenblatt, *Dimensional Analysis*, Gordon and Breach Science Publishers, New York, 1987.

[13] G.I. Barenblatt, *Scaling, Self-similarity, and Intermediate Asymptotics*, Cambridge University Press, Cambridge, 1996.

[14] G.I. Barenblatt, *Scaling*, Cambridge University Press, Cambridge, 2003.

[15] G.K. Batchelor, *An Introduction to Fluid Mechanics*, Cambridge University Press, Cambridge, 1967.

[16] A.A. Becker, *The Boundary Element Method in Engineering*, McGraw–Hill, London, 1992.

[17] C.M. Bender and S.A. Orszag, *Advanced Mathematical Methods for Scientists and Engineers*, McGraw–Hill, New York, 1978.

[18] C.J. Bouwkamp, *Diffraction theory*, Reports on Progress in Physics, 17 (1954), p. 35.

[19] L.M. Brekhovskikh and O.A. Godin, *Acoustics of Layered Media* II, Vol. 10 of Springer Series on Wave Phenomena, Springer-Verlag, Berlin, 1999.

[20] N.G. de Bruijn, *Asymptotic Methods in Analysis*, North–Holland, Amsterdam, 1958; Dover Publications, New York, 1981.

[21] J.C. Butcher, *The Numerical Analysis of Ordinary Differential Equations. Runge-Kutta and General Linear Methods*, John Wiley & Sons, Chichester, 1987.

[22] S.M. Candel, *Numerical solution of conservation equations arising in linear wave theory: Application to aeroacoustics*, Journal of Fluid Mechanics, 83 (1977), pp. 465–493.

[23] D. Chandra, H.J.J. Gramberg, T. Ivashkova, W.R. Smith, A. Suryanto, J.H.M. ten Thije Boonkkamp, T. Ulicevic, and J.C.J. Verhoeven, *Modelling of moisture induced warp in panels containing wood fibres*, in Proceedings of the Thirty-sixth European Study Group with Industry (ESGI 36), J. Molenaar (ed.), Eindhoven University of Technology, EUT-Report-00-WSK-01, pp. 25–32, 2000.

[24] P. Chadwick, *Continuum Mechanics: Concise Theory and Problems*, Dover Publications, New York, 1999.

[25] D.C. Champeney, *A Handbook of Fourier Theorems*, Cambridge University Press, Cambridge, 1987.

[26] C.J. Chapman, *High Speed Flow*, Cambridge University Press, Cambridge, 2000.

[27] A.J. Cooper, *Effect of mean entropy on unsteady disturbance propagation in a slowly varying duct with mean swirling flow*, Journal of Sound and Vibration, to appear.

[28] A.J. Cooper and N. Peake, *Propagation of unsteady disturbances in a slowly varying duct with mean swirling flow*, Journal of Fluid Mechanics, 445 (2001), pp. 207–234.

[29] R. Courant and D. Hilbert, *Methods of Mathematical Physics, Volume* II*: Partial Differential Equations*, Wiley, New York, 1989.

[30] J. Crank, *The Mathematics of Diffusion*, Clarendon Press, Oxford, 1956.

[31] J. Crank, *Free and Moving Boundary Problems*, Clarendon Press, Oxford, 1984.

[32] G.D. Crapper, *Introduction to Water Waves*, John Wiley & Sons, New York, 1984.

[33] E. Cumberbatch and A.D. Fitt (eds.), *Mathematical Modeling: Case Studies from Industry*, Cambridge University Press, Cambridge, 2001.

[34] W.C. Dowling, H.V. Fairbanks, and W.A. Koehler, *A study of the effect of lubricants on molten glass to heated metals*, Journal of the American Ceramic Society, 33 (1950), pp. 269–273.

[35] P.G. Drazin, *Solitons*, Cambridge University Press, Cambridge, 1983.

[36] G.F.D. Duff and D. Naylor, *Differential Equations of Applied Mathematics*, John Wiley, New York, 1966.

[37] P.A. Durbin, *High-frequency Green-function for aerodynamic noise in moving-media. 1. General theory*, Journal of Sound and Vibration, 91 (1983), pp. 519–525.

[38] W. Eckhaus, *Asymptotic Analysis of Singular Perturbations*, North–Holland, Amsterdam, 1979.

[39] F.M. Everaerts, J.L. Beckers, and Th.P.E.M. Verheggen, *Isotachophoresis, theory, instrumentation and applications*, Journal of Chromatography Library, Vol. 6, Elsevier, Amsterdam, 1976.

[40] M. Falipou, F. Sicloroff, and C. Donnet, *New method for measuring the friction between hot viscous glass and metals*, Glastechnische Berichte, 72 (1999), pp. 59–66.

[41] M. Feistauer, *Mathematical Methods in Fluid Dynamics*, Longman Scientific & Technical, Harlow, 1993.

[42] M. Feistauer, J. Felcman, and I. Straškraba, *Mathematical and Computational Methods for Compressible Flow*, Clarendon Press, Oxford, 2003.

[43] Z. Feng and A. Ball, *The erosion of four materials using seven erodents—Towards an understanding*, Wear 233–235 (1999), pp. 674–684.

[44] R.T. Fenner, *Engineering Elasticity—Application of Numerical and Analytical Techniques*, Ellis Norwood, Chichester, 1986.

[45] J.H. Ferziger and M. Perić, *Computational Methods for Fluid Dynamics*, Springer, Berlin, 1997.

[46] R. Finn, *Equilibrium Capillary Surfaces*, Springer-Verlag, New York, 1986.

[47] A.C. Fowler, *Mathematical Models in the Applied Sciences*, Cambridge University Press, Cambridge, 1998.

[48] R. Frisch-Fay, *Flexible Bars*, Butterworths, London, 1962.

[49] F.R. Gantmacher, *Theory of Matrices, Volume* II, Chelsea, New York, 1974.

[50] E. Godlewski and P.-A. Raviart, *Numerical Approximation of Hyperbolic Systems of Conservation Laws*, Springer, New York, 1996.

[51] M.A. Goldberg and C.S. Chen, *Discrete Projection Methods for Integral Equations*, Computational Mechanics Publications, Southampton, 1997.

[52] G.H. Golub and C.F. van Loan, *Matrix Computations*, 3rd ed., The Johns Hopkins University Press, London, 1996.

[53] J.B. Goodman and R.J. LeVeque, *A geometric approach to high resolution TVD schemes*, SIAM Journal on Numerical Analysis, 25 (1988), pp. 268–284.

[54] W.R. Smith and H.J.J. Gramberg, *Mathematical modelling of moisture-induced panel deformation*, Journal of Engineering Mathematics, 43 (2002), pp. 347–366.

[55] P.M. Gresho, R.L. Sani, and M.S. Engelman, *Incompressible Flow and the Finite Element Method Advection-Diffusion and Isothermal Laminar Flow*, John Wiley & Sons, Chichester, 1998.

[56] T. Hagstrom, *Radiation boundary conditions for the numerical simulation of waves*, Acta Numerica, 8 (1999), pp. 47–106.

[57] E. Hairer, S.P. Nœrsett, and G. Wanner, *Solving Ordinary Differential Equations* I, Springer, Berlin, 1993.

[58] E. Hairer and G. Wanner, *Solving Ordinary Differential Equations* II, Springer-Verlag, Berlin, 1991.

[59] A. Harten, *High resolution schemes for hyperbolic conservation laws*, Journal of Computational Physics, 49 (1983), pp. 357–393.

[60] A. Harten and J.M. Hyman, *Self adjusting grid methods for one-dimensional hyperbolic conservation laws*, Journal of Computational Physics, 50 (1983), pp. 235–269.

[61] R. Helmig, *Multiphase Flow and Transport Processes in the Subsurface. A Contribution to the Modeling of Hydrosystems*, Springer, Berlin, 1997.

[62] E.J. Hinch, *Perturbation Methods*, Cambridge University Press, 1991.

[63] C. Hirsch, *Numerical Computation of Internal and External Flows, Volume 1: Fundamentals of Numerical Discretization*, Wiley, Chichester, 1988.

[64] C. Hirsch, *Numerical Computation of Internal and External Flows, Volume 2: Computational Methods for Inviscid and Viscous Flows*, Wiley, Chichester, 1990.

[65] M.H. Holmes, *Introduction to Perturbation Methods*, Springer, New York, 1995.

[66] Y. Hozumi and M. Murakami, *Numerical study of gas dynamics inside of a pulse tube refrigerator*, Advances in Cryogenic Engineering, 45 (2000), pp. 167–174.

[67] J. Jackson, *Classical Electrodynamics*, 2nd ed., J. Wiley, New York, 1975.

[68] L.V.K. Kantorovich and G.P. Akilov, *Functional Analysis*, Pergamon Press, Oxford, 1982.

[69] J.B. Keller, *Geometrical theory of diffraction*, Journal of the Optical Society of America, 52 (1962), pp. 116–130.

[70] J. Kevorkian, *Partial Differential Equations. Analytical Solution Techniques*, Wadsworth & Brooks/Cole, Pacific Grove, CA, 1990.

[71] J. Kevorkian and J.D. Cole, *Multiple Scale and Singular Perturbation Methods*, Springer, New York, 1996.

[72] H.O. Kreiss and J. Lorenz, *Initial Boundary Value Problems and the Navier-Stokes Equations*, Academic Press, San Diego, 1989.

[73] H.K. Kuiken and P.J. Roksnoer, *Analysis of the temperature distribution in FZ silicon crystals*, Journal of Crystal Growth, 47 (1979), pp. 29–42.

[74] H.K. Kuiken, *The cooling of low-heat-resistance cylinders by radiation*, Journal of Engineering Mathematics, 13 (1979), pp. 97–106.

[75] H.K. Kuiken, *Mathematical modelling of etching processes*, in Free Boundary Problems: Theory and Applications, Volume 1, K.H. Hoffmann and J. Sprekels, (eds.), Pitman Research Notes in Mathematical Series 185, Longman Scientific & Technical, Harlow, 1990, pp. 89–109.

[76] H.K. Kuiken, *Practical asymptotics*, Journal of Engineering Mathematics, 39 (2001), pp. 1–2.

[77] K. Laevsky and R.M.M. Mattheij, *Mathematical modelling of some glass problems*, in Complex Flows in Industrial Processes, A. Fasano, (ed.), Birkhäuser, Boston, 2000, pp. 191–214.

[78] K. Laevsky and R.M.M. Mattheij, *Numerical simulation of the plunger velocity*, in Progress in Industrial Mathematics at ECMI 2000, A. Anile et al. (eds.), Springer, Berlin, 2002, pp. 230–235.

[79] K. Laevsky and R.M.M. Mattheij, *Determining the velocity as a kinematic boundary condition in a glass pressing problem*, Journal of Computational Methods in Science and Engineering, 2 (2003), pp. 285–298.

[80] B.J. van der Linden and R.M.M. Mattheij, *A new method for solving radiative heat problems in glass*, International Journal of Forming Processes, 2 (1999), pp. 41–61.

[81] P.A. Lagerstrom, *Matched Asymptotic Expansions: Ideas and Techniques*, Springer-Verlag, New York, 1988.

[82] L. Lapidus and G. F. Pinder, *Numerical Solution of Partial Differential Equations in Science and Engineering*, John Wiley & Sons, 1999.

[83] L.D. Landau and E.M. Lifshitz, *Fluid Mechanics*, 2nd ed., Pergamon Press, Oxford, 1987.

[84] L.D. Landau and E.M. Lifshitz, *Theory of Elasticity*, 2nd ed., Pergamon Press, Oxford, 1970.

[85] B. van Leer, *Towards the ultimate conservative difference scheme* II. *Monotonicity and conservation combined in a second order accurate scheme*, Journal of Computational Physics, 14 (1974), pp. 361–370.

[86] M.B. Lesser and D.G. Crighton, *Physical acoustics and the method of matched asymptotic expansions*, in Physical Acoustics, Vol. XI, W.P. Mason and R.N. Thurston, eds., Academic Press, New York, 1975.

[87] R.J. LeVeque, *Numerical Methods for Conservation Laws*, Birkhäuser, Basel, 1990.

[88] R.J. LeVeque, *Finite Volume Methods for Hyperbolic Problems*, Cambridge University Press, Cambridge, 2002.

[89] M.J. Lighthill, *The Fourth Annual Fairey Lecture: The propagation of sound through moving fluids*, Journal of Sound and Vibration, 24 (1972), pp. 471–492.

[90] M.J. Lighthill, *Waves in Fluids*, Cambridge University Press, Cambridge, 1978.

[91] C.C. Lin and L.A. Segel, *Mathematics Applied to Deterministic Problems in the Natural Sciences*, Macmillan, New York, 1974.

[92] M.M. Lipschutz, *Schaum's Outline of Theory and Problems of Differential Geometry*, McGraw–Hill, New York, 1969

[93] H.A. Lorentz, *Eene algemeene stelling omtrent de beweging eener vloeistof met wrijving en eenige daaruit afgeleide gevolgen*. Zittingsverslag van de Koninklijke Akademie van Wetenschappen te Amsterdam, 5 (1896), pp. 168–175. Translated into English by H.K. Kuiken, *A general theorem on the motion of a fluid with friction and a few results derived from it*, Journal of Engineering Mathematics, 30 (1996), pp. 19–24.

[94] P. Lorrain, D.R. Corson, and F. Lorrain, *Electromagnetic Fields and Waves: Including Electric Circuits*, 3rd ed., Freeman, New York, 1988.

[95] A. Majda, *Compressible fluid flow and systems of conservation laws in several space dimensions*, Applied Mathematical Science, 53, Springer-Verlag, Berlin, 1984.

[96] G.I. Marchuk, *Splitting and alternating direction methods*, in Handbook of Numerical Analysis, Volume 1, P.G. Ciarlet and J.L. Lions, (eds.), North–Holland, Amsterdam, 1990, pp. 197–462.

[97] R.M.M. Mattheij and J. Molenaar, *Ordinary Differential Equations in Theory and Practice*, Classics in Appl. Math. 43, SIAM, Philadelphia, 2002.

[98] M.K. Myers, *On the acoustic boundary condition in the presence of flow*, Journal of Sound and Vibration, 71(3) (1980), pp. 429–434.

[99] R.E. Mickens, *An Introduction to Nonlinear Oscillations*, Cambridge University Press, Cambridge, 1981.

[100] E.I. Mikulin, A.A. Tarasov, and M.P. Shkrebyonock, *Low temperature expansion pulse tubes*, Advances in Cryogenic Engineering, 29 (1984), pp. 629–637.

[101] P.M. Morse and K.U. Ingard, *Theoretical Acoustics*, McGraw–Hill, New York, 1968; reprinted by Princeton University Press, Princeton, NJ, 1986.

[102] R.A. Mosher, D.A. Saville, and W. Thormann, *The Dynamics of Electrophoresis Library*, B.J. Radola (ed.), VCH, Weinheim, Germany (1992).

[103] A.H. Nayfeh, *Perturbation Methods*, John Wiley & Sons, New York, 1973.

[104] M. Necati Özişik, *Heat Conduction*, John Wiley, New York, 1993.

[105] O. Oleinik, *Discontinuous solutions of nonlinear differential equations*, American Mathematical Society Translations Series 2, 26 (1957), pp. 95–172.

[106] R.E. O'Malley, *Singular Perturbation Methods for Ordinary Differential Equations*, Springer-Verlag, New York, 1991.

[107] F.W.J. Olver, *Asymptotics and Special Functions*, Academic Press, London, 1974.

[108] S. Osher and R. Fedkiw, *Level Set Methods and Dynamic Implicit Surfaces*, Springer, New York, 2003.

[109] N.C. Ovenden, *Near cut-on/cut-off transitions in lined ducts with flow*, paper AIAA 2002-2445 of the 8th AIAA/CEAS Aeroacoustics Conference, Breckenridge, CO, 17–19 June 2002.

[110] N.C. Ovenden, W. Eversman, and S.W. Rienstra, *Cut-on cut-off transition in flow ducts: Comparing multiple-scales and finite-element solutions*, paper AIAA 2004-2945 of the 10th AIAA/CEAS Aeroacoustics Conference, Manchester, U.K., 10–12 May 2004.

[111] N.C. Ovenden and S.W. Rienstra, *Mode-matching strategies in slowly varying engine ducts*, AIAA Journal, 42(9) (2004), pp. 1832–1840.

[112] A. Papoulis, *The Fourier Integral and Its Applications*, McGraw–Hill, New York, 1962.

[113] N. Peake and A.J. Cooper, *Acoustic propagation in ducts with slowly varying elliptic cross-section*, Journal of Sound and Vibration, 243(3) (2001), pp. 381–401.

[114] A.D. Pierce, *Acoustics: An Introduction to Its Physical Principles and Applications*, McGraw–Hill, New York, 1981. Also available from the Acoustical Society of America.

[115] M.H. Protter and H.F. Weinberger, *Maximum Principles in Differential Equations*, Prentice–Hall, Englewood Cliffs, NJ, 1967.

[116] M. Rahman, *Water Waves: Relating Modern Theory to Advanced Engineering Applications*, IMA monograph series, Clarendon Press, Oxford, 1995.

[117] A. Ralston, *A First Course in Numerical Analysis*, McGraw–Hill, New York, 1965.

[118] R.W. Ramirez, *The FFT, Fundamentals and Concepts*, Prentice–Hall, Englewood Cliffs, NJ, 1985.

[119] I. Richards and H. Youn, *Theory of Distributions: A Non-technical Introduction*, Cambridge University Press, Cambridge, 1990.

[120] R.D. Richtmyer and K.W. Morton, *Difference Methods for Initial Value Problems*, 2nd ed., Wiley-Interscience, New York, 1967.

[121] S.W. Rienstra, *Sound diffraction at a trailing edge*, Journal of Fluid Mechanics, 108 (1981), pp. 443–460.

[122] S.W. Rienstra, *Acoustic radiation from a semi-infinite annular duct in a uniform subsonic mean flow*, Journal of Sound and Vibration 94(2) (1984), pp. 267–288.

[123] S.W. Rienstra, *Non-linear free vibrations of coupled spans of suspended cables*, Proceedings of the Third European Conference on Mathematics in Industry, 27–31 August 1988, Glasgow, J. Manley et al. (eds.), Kluwer Academic Publishers and B.G. Teubner, Stuttgart, pp. 133–144.

[124] S.W. Rienstra, *The shape of a sessile drop for small and large surface tension*, Journal of Engineering Mathematics, 24 (1990), pp. 193–202.

[125] S.W. Rienstra, *Thin layer flow along arbitrary curved surfaces*, Applied Mathematics and Mechanics, Vol. 76, Supplement 5, 1996, pp. 423–424.

[126] S.W. Rienstra, *Geometrical effects in a Joule heating problem from miniature soldering*, Journal of Engineering Mathematics, 31 (1997), pp. 59–80.

[127] S.W. Rienstra, *Sound transmission in slowly varying circular and annular ducts with flow*, Journal of Fluid Mechanics, 380 (1999), pp. 279–296.

[128] S.W. Rienstra and W. Eversman, *A numerical comparison between multiple-scales and finite-element solution for sound propagation in lined flow ducts*, Journal of Fluid Mechanics, 437 (2001), pp. 367–384.

[129] S.W. Rienstra, *Cut-on, cut-off transition of sound in slowly varying flow ducts*, Aerotecnica Missili E Spazio (Journal of the Associazione Italiana di Aeronautica e Astronautica AIDAA), 79(3–4) (2000), pp. 93–96.

[130] S.W. Rienstra and T.D. Chandra, *Analytical approximations to the viscous glass flow problem in the mould-plunger pressing process, including an investigation of boundary conditions*, Journal of Engineering Mathematics, 39 (2001), pp. 241–259.

[131] S.W. Rienstra, *A classification of duct modes based on surface waves*, Wave Motion, 37 (2003), pp. 119–135.

[132] S.W. Rienstra, *Sound propagation in slowly varying lined flow ducts of arbitrary cross section*, Journal of Fluid Mechanics, 495 (2003), pp. 157–173.

[133] P.L. Roe, *Some contributions to the modeling of discontinuous flows*, Lecture Notes Applied Mathematics, 22 (1985), pp. 163–193.

[134] P.L. Roe, *Approximate Riemann solvers, parameter vectors and difference schemes*, Journal of Computational Physics, 43 (1981), pp. 357–372.

[135] Y. Saad, *Iterative Methods for Sparse Linear Systems*, PWS Publishing Company, Boston, 1996. Second edition published by SIAM, Philadelphia, 2003.

[136] S.F.Y. Li, *Capillary Electrophoresis: Principles, Practice and Applications*, Journal of Chromatography Library, Elsevier, Amsterdam, 1992.

[137] J.A. Sethian, *Level Set Methods and Fast Marching Methods*, Cambridge University Press, Cambridge, 2001.

[138] J.G. Simmonds and J.E. Mann, Jr., *A First Look at Perturbation Theory*, 2nd ed., Dover Publications, New York, 1998.

[139] P.J. Slikkerveer, P.C.P. Bouten, F.H. in 't Veld, and H. Scholten, *Erosion and damage by sharp particles*, Wear, 217 (1998), pp. 237–250.

[140] P.J. Slikkerveer and F.H. in 't Veld, *Model for patterned erosion*, Wear, 233–235 (1999), pp. 377–386.

[141] W.R. Smith, *One-dimensional models for heat and mass transfer in pulse tube refrigerators*, Cryogenics, 41 (2001), pp. 573–582.

[142] J. Smoller, *Shock Waves and Reaction-Diffusion Equations*, Springer, New York, 1982.

[143] I.S. Sokolnikoff, *Tensor Analysis*, John Wiley & Sons, New York, 1951.

[144] J. Sollenberg, *Biological Monitoring of Occupational Exposure to Styrene and Xylene*, Arbetslivsinstitutet, Stockholm, 1995.

[145] A. Sommerfeld, *Mathematische Theory der Diffraktion*, Mathematische Annalen, 47 (1896), p. 317.

[146] B.K. Soni, J.F Thompson, and N.P. Weatherhill, *Handbook of Grid Generation*, CRC Press, Boca Raton, FL, 1999.

[147] I. Stakgold, *Green's Functions and Boundary Value Problems*, 2nd ed., John Wiley & Sons, New York, 1998.

[148] H. Stetter, *Analysis of Discretisation Methods for Ordinary Differential Equations*, Springer-Verlag, Berlin, 1973.

[149] J. Stoer and R. Bulirsch, *Introduction to Numerical Analysis*, Springer-Verlag, New York, 1980.

[150] J.J. Stoker, *Water Waves: The Mathematical Theory with Applications*, Interscience Publishers, New York, 1957.

[151] D.M. Stump and G.H.M. van der Heijden, *Matched asymptotic expansions for bent and twisted rods: Application for cable and pipeline laying*, Journal of Engineering Mathematics, 38 (2000), pp. 13–31.

[152] A.B. Tayler, *Mathematical Models in Applied Mechanics*, Clarendon Press, Oxford, 1986.

[153] B.N. Taylor, *Guide for the Use of the International System of Units (SI)*, National Institute of Standards and Technology, NIST Special Publication 811, 1995 Edition (CODEN: NSPUE2). Downloadable from http://physics.nist.gov/Pubs/contents.html.

[154] M. Taylor, *Pseudodifferential Operators*, Princeton University Press, Princeton, 1981.

[155] J.H.M. ten Thije Boonkkamp, and P.J. Slikkerveer, *Mathematical modelling of erosion by powder blasting*, Surveys on Mathematics Industry, 10 (2002), pp. 89–105.

[156] J.W. Thomas, *Numerical Partial Differential Equations: Conservation Laws and Elliptic Equations*, Springer, New York, 1999.

[157] F.V. Tooley, *The Handbook of Glass Manufacture, Vol.* II, Books For Industry and Glass Industry Magazine, New York, 1974.

[158] E.F. Toro, *Riemann Solvers and Numerical Methods for Fluid Dynamics: A Practical Introduction*, Springer, Berlin, 1997.

[159] E.F. Toro, *Shock-Capturing Methods for Free-Surface Shallow Flows*, Wiley, Chichester, 2001.

[160] W. Trier and F. Hassoun, *Mechanik des Gleitens Heißen, Zähflüssigen Glases auf Metalloberflächen*, Glastechnische Berichte, 45 (1972), pp. 271–276.

[161] W. Trier, *Gleitverhalten von Heißem, Zähflüssigem Glas auf Metalloberflächen*, Glastechnische Berichte, 51 (1978), pp. 240–243.

[162] M. Van Dyke, *Perturbation Methods in Fluid Mechanics*, Parabolic Press, Stanford, CA, 1975 (annotated edition from original, Academic Press, 1964).

[163] M. Van Dyke, *Slow variations in continuum mechanics*, Advances in Applied Mechanics, Vol. 25, Academic Press, Orlando, 1987, pp. 1–45.

[164] F. Verhulst, *Methods and applications of singular perturbations*, Texts in Applied Mathematics, Vol. 50, Springer, Berlin, 2005.

[165] A.T.A.M. de Waele, P.P. Steijaert, and J. Gijzen, *Thermodynamical aspects of pulse tubes*, Cryogenics, 37 (1997), pp. 313–324.

[166] A.T.A.M. de Waele, P.P. Steijaert, and J. Gijzen, *Thermodynamical aspects of pulse tubes* II, Cryogenics, 38 (1998), pp. 329–335.

[167] P. Wesseling, *An Introduction to Multigrid Methods*, Wiley, Chichester, 1992.

[168] P. Wesseling, *Principles of Computational Fluid Dynamics*, Springer, Berlin, 2001.

[169] G.B. Whitham, *Linear and Nonlinear Waves*, John Wiley & Sons, New York, 1974.

[170] F.A. Williams, *Combustion Theory*, Addison–Wesley, Redwood City, CA, 1985.

[171] Ya. B. Zeldovich, G.I. Barenblatt, V.B. Librovich, and G.M. Makhviladze, *The Mathematical Theory of Combustion and Explosions*, Consultants Bureau, New York, 1985.

[172] M. Zerroukat and C.R. Chatwin, *Computational Moving Boundary Problems*, John Wiley & Sons, Chichester, 1994.

[173] D. Zwillinger, *Handbook of Differential Equations*, Academic Press, San Diego, 1998.

Index